W9-BUE-982

Electricity and Electronics Technology

Seventh Edition

Peter Buban
Marshall L. Schmitt
Charles G. Carter, Jr.

New York, New York   Columbus, Ohio   Woodland Hills, California   Peoria, Illinois

ABOUT THE AUTHORS

Peter Buban, late instructor of electricity and electronics at the Quincy (Illinois) Area Vocational Technical Center. He received his M.A. degree from Northeast Missouri State University. His many years of practical military and industrial experience in the electrical and electronics fields included such jobs as electronics technician, chief radioman, and electrician.

Marshall L. Schmitt is a former Senior Program Specialist with the U.S. Department of Education. He received his doctorate from Pennsylvania State University. He has taught courses in electricity and electronics for U.S. Navy personnel and at the high school level in New York state. He also taught electricity and electronics in the Industrial Arts Department at North Carolina State University, Raleigh, North Carolina. Now retired, he is a volunteer teacher in a public school. An active participant in multimedia and Internet activities in a computer users group, he is also an amateur radio operator.

Charles G. Carter, Jr. teaches Electricity and Electronics at Wayne Valley High School, Wayne, New Jersey. He received his M.A. in Industrial Arts Education from Montclair State University in Montclair, New Jersey. Mr. Carter also teaches microcomputer repair at the secondary and adult levels. He received the New Jersey Governor's Teacher Recognition Award for 1986-1987 at Wayne Valley High School. He was a NASA NEWMAST Fellow in 1990.

Glencoe/McGraw-Hill

A Division of The ***McGraw-Hill*** *Companies*

Send all inquiries to:
Glencoe/McGraw-Hill
3008 W. Willow Knolls Drive
Peoria, IL 61614-1083

ISBN 0-02-683427-8 (Student text)
ISBN 0-02-683429-4 (Student Workbook)
ISBN 0-02-683428-6 (Teacher's Resource Guide)

Printed in the United States of America

1 2 3 4 5 6 7 8 9 10 003/043 02 01 00 99 98

Preface

Electricity and Electronics Technology is intended for a basic course in electricity and electronics. There is sufficient material for a full-year course. This book has also been designed for easy adaptation to courses of varying lengths.

Two supplements have been correlated with the text to provide a complete basic electricity and electronics learning program. The supplements are the *Student Workbook* and the *Teacher's Resource Guide*. The *Student Workbook* contains procedures for performing 74 experiments.

The experiments help students understand the important relationships between the way components work and how they are used. Questions throughout the experiments require students to think critically about the procedures they are performing.

The *Teacher's Resource Guide* provides a program overview, program development strategies, 32 chapter teaching guides, two visual masters per chapter, 32 chapter tests, and answer keys.

Early in this text the student is introduced to the fundamental theories on which later discussions of practical applications are based. This logical explanation of the subject is particularly effective in relating the technological advances in electricity and electronics to basic concepts in physics, mechanics, and chemistry to which the student is introduced. This approach is strengthened by a judicious use of basic mathematics.

Classroom experience with the previous editions and many useful suggestions from users of the text have helped shape the content of this edition. Career and avocational opportunities in electricity and electronics are explored in detail.

The text presents 32 chapters of instruction and suggestions for the construction of several projects. End-of-chapter materials include a chapter summary, a review of main ideas, apply your knowledge exercises, and connections to communication skills, mathematics, science, social studies, technology, and workplace skills.

The section that reviews main ideas provides a way for students to assess their own progress in each chapter in a challenging and independent fashion. The apply your knowledge section may be used by the teacher to stimulate classroom discussions of specific topics, to highlight particular points, and to develop topics more fully. The connections to other subjects help students appreciate the links between electricity and electronics and other curriculum areas. Use of these connections can be an effective method of extending the student's learning experiences beyond the classroom.

The section titled "Hands-On Experiences" presents suggestions for projects that will provide practical experiences in the design and construction of useful electrical and electronic devices. The projects are grouped by level of difficulty.

The Appendixes present metric conversions and formulas.

As in the previous edition, metric units are included with customary units.

The authors gratefully acknowledge the assistance of the following companies and organizations: Admiral Home Appliances; American National Standards, Inc.; The American Radio Relay League; Ametek; Analog Devices; AT&T; Ball Corp.; Beckman Industries, Inc.; Bell Atlantic Mobile Systems; Black Box Corp.; Burr-Brown Corp.; Electronic Industries Association; Exide Corp.; Exxon Corp.; Fayette Manufacturing Corp.; Federal Communications Commission; The Foxboro Co.; General Electric Co.; Georator Corp.; Global Specialties; GTE Laboratories, Inc.; Hayes Microcomputer Products, Inc.; Hechinger Store #12; Hunter Associates; ICOM AMERICA, Inc.; ILC Data Device Corp.; Infolink Corp.; Institute of Electrical and Electronics Engineers; Interactive Image Technologies Ltd.; Ivex, Design International; Jay Klein Production, Inc.; Lead Industries Association, Inc.; Magnetic Shield Division, Perfection Mica Co.; Mid-Atlantic Finishing, Inc.; Modern School Supply, Inc.; Mostek Corp.; National Instruments; NEC America, Inc.; Ortel Corporation; Pace, Inc.; Parallax, Inc.; Pitney Bowes; Potomac Electric Power Co.; H. C. Protek; Radiant Communications Corp.; RCA Corp.; Reference Technology, Inc.; Sencore; Siemens Opto; Sony Corporation of America; Sunbeam Appliance Co.; Tandy Corp./Radio Shack; Teletype Corp.; 3M, Telecom Systems Division; Toroid Corp. of Maryland; U.S. Department of Energy; Varta Batteries, Inc.; Veeco Instruments, Inc.; Virginia Concrete Co.; Virginia Electric and Power Co.; Wayne Electronics, Inc.; and Zenith Electronics Corp.

The authors wish to express special appreciation to the following individuals: Marshall E. Canaday, Pace, Inc.; Scott Corbett, E-D-Datacom; and high school teachers Dan Wilkins, Robert Kussey, Peter Tucker, Arthur Wilson, Jamie Yoss, Joe Wesolowski, and Clark Payne.

The authors wish to express their sincere appreciation to Doris L. Schmitt and Elizabeth G. Carter, who have provided encouragement and assistance. A special acknowledgment is made to the late Peter Buban. He will be remembered for his dedication to his profession, his concern for students, and his many contributions to the previous editions of this text.

Comments and suggestions for the improvement of this edition are most welcome.

Marshall L. Schmitt
Charles G. Carter, Jr.

Reviewers

Herbert J. Abrams
Instructor, Consumer Electronics Repair
Miami Jackson Senior High School
Miami, FL

James A. Bailey
North Parkway Middle School
Jackson, TN

Arnall Lynn Cox
Technology Studies Instructor
Venice High School
Sarasota County, FL

Larry Davis
Technology Teacher
Northwest Middle School
Winston Salem, NC

Ken Franson
Kingswood Regional High School
Wolfeboro, NH

James D. Genovese
Technology Teacher
Valparaiso High School
Valparaiso, IN

John E. Gray
Technology Education Teacher
Klondike Middle School
West Lafayette, IN

Patrick J. Gunter
Technology Instructor
R. P. Dawkins Middle School
Moore, SC
Adjunct Professor Clemson University

Tina E. Hayden
Technology Teacher
Owen Valley High School
Spencer, IN

Charles Horsken
Technology Education Chairman
Kingswood Regional High School
Wolfeboro, NH

John Klock
Critical Thinking & Multimedia Development
T. A. Dugger Jr. High School
Elizabethton, TN

Ronald F. Logan
Drafting/Design Technology Instructor
Vocational Department Co-Chair
Austin-East Magnet High School
Knoxville, TN

Henry L. Webb
Oak Grove High School
North Little Rock, AR

Acknowledgments

Don H. Grout, Ed.D.
Champlain College
Burlington, VT

Valerie Legare
Morton, IL

Deborah Paul
Worthington, OH

Contents in Brief

Table of Contents

Section 1

Introduction to Electricity and Electronics

Chapter 

Careers

OBJECTIVES

After completing this chapter, you will be able to:

- Describe a variety of careers available within the field of electricity and electronics.
- Discuss common education and training requirements for careers in the field of electricity and electronics.
- Explain why it is a good idea to start planning now for your future career.
- Describe methods of researching careers in electricity and electronics technology.
- Find career information from various sources.

Terms to Study

apprenticeship program

career

electrical engineer

electrician

Federal Communications Commission license

job

occupation

preventive maintenance

technician

Can you imagine yourself designing or programming a robot? How about working with lasers or repairing a computer? The field of electricity and electronics technology includes these jobs and many more. Since most of your adult life will be spent working, it only makes sense to spend time learning more about your potential career opportunities. This chapter can help you plan which courses you need to take to prepare for a career in one of the many areas of electricity and electronics technology. In fact, you may discover some career ideas that you have not considered.

CAREER OPPORTUNITIES

Before you continue, you need to understand a few basic terms. Do you understand the difference between a career and a job? A **job** is merely the work you have been hired to do at a given time in your life. A **career** is a string of related jobs in which you train and pursue a life's work. An **occupation** is your lifetime profession.

The job market has changed rapidly over the past few years. This trend will continue because people and companies are constantly making new technological advancements. The new systems, products, and processes that result from these advancements have a large impact on the workplace. If you want to work with newly developed technology, the field of electricity and electronics technology offers many career opportunities.

As society becomes more dependent on the products of technology, the field of electronics expands. Who knows what new, exciting careers in this field may develop—even within the next ten years? The careers described in this chapter are just a few of the opportunities now available.

Electrical and Electronic Engineering

An *engineer* is someone trained in the planning or building of a specific structure or system. For example, an engineer might be concerned with the planning or building of roads, buildings, or bridges. There are many types of engineers. An **electrical engineer** is trained in the planning and building of electrical systems.

Electrical and electronic engineers plan, design, and direct the production and maintenance of electrical and electronic equipment (Fig. 1-1). Their work may include:

- trying to find ways to make better use of electrical energy
- improving the performance of systems and products already in use
- designing and developing new circuits or other devices used in equipment such as computers, telephones, and stereos

Fig. 1-1. The work of an electrical engineer involves a great deal of planning.

Think of all the products and services you use every day that depend in some way on electricity or electronics. What types of jobs—and careers—might be required to produce those products and services?

Typical Job Characteristics

Some of the many fields open to electrical and electronic engineers are medical electronics, computers, power distribution, and missile guidance. Engineers generally work in offices or in research laboratories. However, some engineers direct operations or conduct inspections in factories or at construction sites.

Education and Training

Engineers must have knowledge not only of engineering and design, but of advanced mathematics, physical science, and computer science as well. They often use computers to simulate and test machines, structures, and systems. Engineers must have good analytical skills, combined with an above-average ability to visualize objects from pictures and descriptions. They must also have good written and verbal communication skills.

A bachelor's degree in engineering from an accredited engineering program is usually required for entry-level engineering positions. To be accepted into an accredited school, you must have a strong high school background in mathematics and science.

Engineers who want to advance beyond entry-level positions usually continue their education in specialized areas of study. The type of program they enter depends on their specialty.

Electrical and Electronics Engineering Technicians

In general, a **technician** is someone who has technical knowledge about a specific field. *Engineering technicians* collect, record, and coordinate technical information to solve engineering problems. They typically work in areas such as research and development, manufacturing, sales, construction, and customer service. Some of the tasks they perform are:

- collecting data
- conducting surveys
- inspecting products or projects
- preparing detailed working drawings of mechanical devices
- researching records and writing technical reports.

Typical Job Characteristics

Many engineering technicians assist engineers and scientists, especially in the area of research and development. Engineering technicians usually start out doing routine work under the supervision of an experienced engineer. As they gain experience, they are given more difficult assignments and work with less supervision. Most engineering technicians work in laboratories, offices, industrial plants, or on construction sites.

Education and Training

To become an engineering technician, you must have a sound knowledge of mathematics, science, and computer science. One of the most important skills you can have is the ability to perform detailed work with a high degree of accuracy. Engineering technicians are often part of a team of engineers and other technicians. They must therefore have good communication skills and be able to work well with others.

Most entry-level positions for engineering technicians require education and training

FASCINATING FACTS

Most electric and power units of measurement in use are named after people who were pioneers in the field. Allesandro Volta (volt), James Watt (watt), Andre Marie Ampere (ampere), James Joule (joule), Michael Faraday (farad), Charles A. de Coulomb (coulomb), and Georg Simon Ohm (ohm) have units named after them.

beyond high school. A foundation built on mathematics and science courses and an associate degree program in engineering technologies will prepare you to be an electricity or electronics engineering technician.

Electronics Technicians

Electronics technicians help design, install, maintain, and repair a wide range of electronic equipment (Fig. 1-2). This equipment includes:

- consumer products such as computers, televisions, stereos, and photocopy machines.
- industrial and military equipment such as lasers and radar systems.

Typical Job Characteristics

Electronics technicians may work alone or with an engineer. They may be involved in a broad number of activities in research, development, quality control, production, and sales. Many of these tasks require precise and detailed work.

Tasks may include:

- using information from blueprints and drawings to test, adjust, and inspect products.
- evaluating equipment.
- ensuring that company standards and product specifications have been met.
- assisting in the development or improvement of electronic products.
- troubleshooting, repairing, and conducting safety checks on defective equipment.

Education and Training

Electronics technicians need a thorough knowledge of mathematics and science. They must have good analytical skills and the ability to solve problems, as well as a solid foundation in electronics and electricity. The ability to understand and work from technical drawings is also very important.

Most electronics technicians complete a two-year applied associate degree in electronics technology to qualify for entry-level positions. A strong high school background in mathematics and science provides a good background for the associate degree.

Laser Technicians

Laser technicians assist with the research and development of lasers. Laser technology

Fig. 1-2. Electronic service technicians generally work for equipment manufacturers or for companies that specialize in repairing electronic equipment.

involves mechanical, electronic, and optical concepts and components. Tasks that laser technicians perform include:

- cleaning and aligning optical parts.
- experimenting with new uses for laser power.
- servicing and repairing laser equipment.

Typical Job Characteristics

Laser technicians are responsible for building, testing, and repairing laser equipment and systems. Lasers are used to cut, weld, and drill many materials. Fiber optics and related technologies are replacing wire cables in communication lines and in many electronic products.

Most laser technicians work in laboratories and assembly areas. However, some work in "clean rooms" in which temperature, humidity, and dust are highly regulated.

Education and Training

Laser technicians need the ability to perform detailed, precise work. They should also be able to visualize objects from drawings or descriptions. Most employers require a two-year degree in electronics technology for entry-level positions. Helpful high school courses include mathematics, physics, computer programming and applications, electronics, and blueprint reading.

Robotics Technicians

Robotics technicians perform a wide variety of tasks relating to the design, production, and maintenance of robots and robotics devices. Typical tasks include:

- reviewing diagrams and blueprints.
- drafting drawings of equipment parts.
- conducting tests on robotics equipment.
- analyzing and recording test results.
- installing robots and programming them to perform specific tasks.
- monitoring robotics activities, identifying problems, and making necessary repairs.

Typical Job Characteristics

Robotics technicians must be able to visualize objects from pictures and drawings. They should be capable of noticing minor changes and differences among technical documents and drawings. Their work must be done precisely and skillfully, and they often use precision hand tools and gauges.

Education and Training

Robotics technicians should have knowledge of hydraulics, electronics, and programming. Students who are considering becoming robotics technicians should include advanced mathematics, physics, and industrial technology in their high school courses. Most employers require entry-level robotics technicians to have one to two years of training in addition to a high school degree. Even experienced technicians often attend training sessions to update their skills because of rapid technological changes.

Broadcast Technicians

Broadcast technicians install, operate, and maintain electronic equipment used to record or transmit radio and television programs. They work with television cameras, microphones, audio- and videotape recorders, light and sound effects, transmitters, antennas, and other equipment.

Typical Job Characteristics

Broadcast technicians who are employed at small stations perform a variety of tasks. At larger stations and networks, their tasks are more specialized. Promotions to top-level technical positions, such as chief engineer, require a very good understanding of all facets of the broadcasting field, an ability to work well with others, and creativity in designing and modifying broadcasting equipment.

Education and Training

Broadcast technicians should have a broad knowledge of electronic principles and a good understanding of broadcast equipment. They must understand computer programming and repair. Broadcast technicians also need the hand coordination necessary to operate technical equipment (Fig. 1-3).

Federal law requires that anyone who operates broadcast transmitters in radio and television stations must have at least a restricted radio telephone operator permit. Prospective technicians should take high school courses in mathematics, physics, and electronics. Most employers prefer to hire broadcast technicians that have some electronics training after high school in addition to a general class **FCC license**. The **Federal Communications Commission** (FCC) regulates radio, television, wire, and cable communications. Television stations generally require two years of training in broadcast technology. Prospective technicians should take high school courses in mathematics, physics, and electronics.

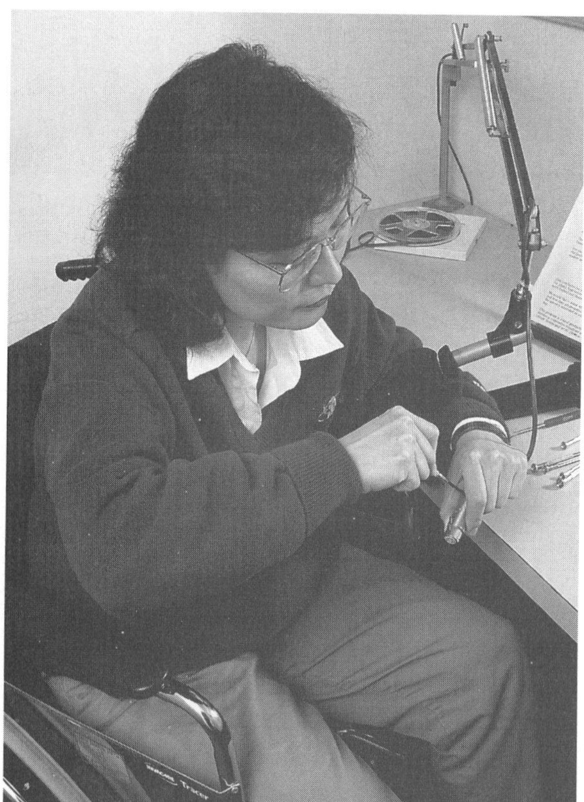

Fig. 1-3. This broadcast technician is repairing the microphones in a sound studio.

Electricians

An **electrician** installs and maintains electrical systems for a variety of purposes, including climate control, security, and communications. Most electricians specialize in either construction or maintenance, although a growing number of them do both. Electricians' work must conform to state and local building codes (Fig. 1-4).

Typical Job Characteristics

Construction electricians follow blueprints to install electrical systems in factories, office buildings, homes, and other structures. They also install coaxial or fiber optic cable for computers and other telecommunication equipment.

Fig. 1-4. Electricians install, inspect, maintain, and repair wiring in residential and commercial buildings.

Maintenance electricians spend much of their time performing **preventive maintenance**, inspecting equipment and locating problems before breakdowns occur. In addition, they perform a variety of other tasks, depending on where they are employed. Some maintenance technicians specialize in residential work, such as rewiring homes. Others may work in factories repairing motors and generators on machine tools.

Education and Training

The best way to learn the electrical trade is to complete a four- or five-year **apprenticeship program**, which may include classroom instruction and on-the-job training. Another way to prepare is to enroll in an applied associate degree program. Licensing is required for all electricians, regardless of how they receive their training.

If you are interested in becoming an electrician, consider taking high school courses in mathematics, electricity, electronics, mechanical drafting, and science. These courses will provide a good background for either an apprenticeship or an associate degree. If your interest lies mostly in maintenance, a background in electronics is becoming increasingly important due to the growing use of complex electronic controls on manufacturing equipment.

Computer Support Technicians

Computer support technicians install, modify, and repair computer hardware and software systems. As home computers become more popular, increasing numbers of computer support technicians are being hired to provide telephone support for software products.

LINKS

For more information on testing, troubleshooting, and repair, see Chapter 23.

Social Studies

Apprenticeships

The type and availability of an apprenticeship may vary greatly, dependent on the country in which an individual trains. In Western countries, governments may promote, prescribe standards, and even offer financial incentives for apprenticeships. However, the employer or the union decides whether to take on an apprentice. The number of occupations with apprenticeships is also rather limited. In contrast, in Germany, Austria, and Switzerland, the governments pay businesses who train employees. In these countries approximately 5% of workers apprentice. Apprenticeships also expand beyond the standard crafts and trades to include entry-level jobs in offices, banks, travel services, hotels, restaurants, and medicine.

In the United States, there are about 90 skilled trade apprenticeship programs encompassing over 350 occupations. An apprenticeship instruction, especially one that combines on-the-job training with classroom instruction, can be an attractive alternative to college. The learner is paid, and these programs are often a first step toward owning a business or gaining a management or leadership position. With increasing worker participation in management, skilled workers may also take responsibility for training others.

Common tasks include:

- providing technical assistance by locating problems and performing minor repairs based on their knowledge of system operations.
- providing instruction to clients on the use of the equipment, software, and manuals.

Typical Job Characteristics

Most computer support technicians either work in an office or travel to various customer sites to work on computer systems. Some technicians specialize in installing new systems.

Others may limit their work to repairing computer systems or modifying or expanding existing systems.

Education and Training

Computer support technicians should have a background in computer applications, operating systems, and networks. They should understand the fundamentals of electronics and especially microelectronics. Technicians must work well with people and be able to communicate technical information to people who do not have a technical background (Fig. 1-5).

Employers prefer to hire computer support technicians who have at least two-year applied associate degrees. Helpful high school courses include mathematics and science, in addition to electronics and any available courses in computer programming and applications.

Fig. 1-5. Computer support technicians often travel to the customer's site to work on equipment.

Line Installers and Line Repairers

Line installers and *line repairers* set up and repair the vast network of electrical, telephone, and cable television lines used by business and residential customers. Their jobs typically include:

- installing transmission cables
- installing fiber optic cables
- repairing and replacing broken or defective lines

Typical Job Characteristics

Line installers and repairers must often climb poles or work underground to install or repair cables. Their jobs, therefore, require that they be in good physical condition (Fig. 1-6).

SAFETY TIP

To work safely and efficiently, line installers and line repairers must be able to work as part of a team. Since they often work with live electrical wiring, they must understand and follow electrical safety guidelines at all times.

Education and Training

Many line installers and line repairers learn their skills through on-the-job training. Construction and electric companies require job applicants to complete an apprenticeship program to be a line installer. Employers generally prefer high school graduates who have a background in mathematics and are in good physical condition. Knowledge of electricity and the ability to read blueprints are also helpful.

Fig. 1-6. This worker is installing cable to supply a cable hookup for a new residential area.

Electronic Service Technicians

Electronic service technicians install, maintain, and repair electronic equipment used in offices, factories, homes, hospitals, and other places. Their actual job titles often refer to the kind of equipment they service or the services they provide. For example, they may be "electronic equipment repairers" or "field service representatives."

Typical Job Characteristics

Electronic service technicians ensure that the equipment operates properly. They keep detailed records that provide a history of tests, performance problems, and repairs for individual machines.

Future employment opportunities will be best for computer and office machine repairers because the number of computers in service is increasing. Improvement in product reliability and low equipment maintenance prices may result in fewer openings for some other service technicians. Electronic home entertainment equipment repairers, telephone installers and repairers, and home appliance service workers are among those who may be affected.

Education and Training

Employers prefer to hire technicians who have completed one or two years of training in electronics after high school. If you are interested in becoming an electronic service technician, you should take as many courses as possible in mathematics, physics, electricity, and electronics. In addition, you should choose courses that prepare you for the specialty area in which you are interested.

Heating, Ventilation, Air-Conditioning, and Refrigeration (HVAC/R) Technicians

Heating, ventilation, air-conditioning, and refrigeration (HVAC/R) technicians install, maintain, and repair systems that control the temperature, humidity, and total air quality in residential, commercial, and industrial buildings. They also service refrigeration systems used to store and transport food, medicine, and other perishable items. These systems consist of many mechanical, electrical, and electronic components (Fig. 1-7).

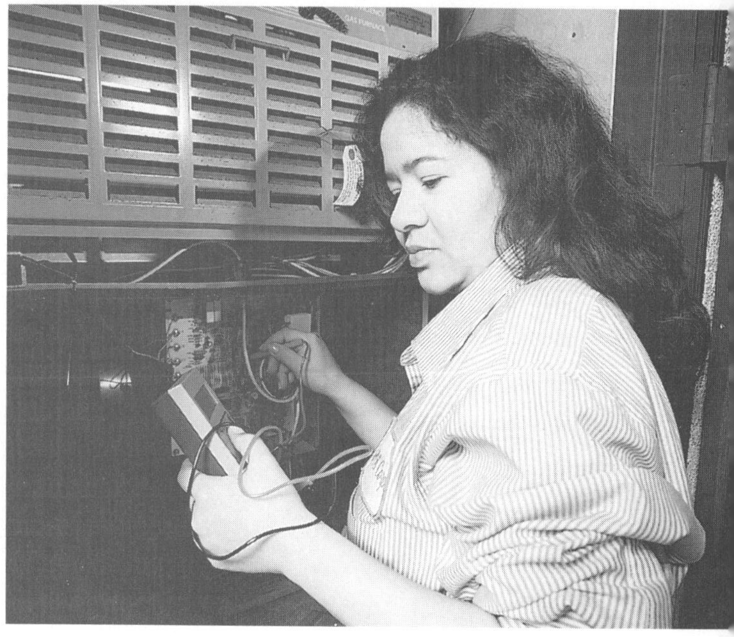

Fig. 1-7. This HVAC/R technician is checking a residential air-conditioning system for proper operation. Preventive maintenance is a common and necessary task, particularly in areas that get very hot in the summer months.

Typical Job Characteristics

HVAC/R technicians generally specialize in either installation or maintenance and repair. However, they are trained to do both. HVAC/R technicians generally work at customer locations. However, some of the larger HVAC/R employers maintain large repair shops that contain state-of-the-art, computerized diagnostic equipment. Technicians who work for these companies may bring parts or units to the shop to be diagnosed and repaired.

Education and Training

Heating, ventilation, air-conditioning, and refrigeration systems are becoming increasingly sophisticated. Employers, therefore, hire technicians who have completed an apprenticeship or an appropriate two-year program at a technical school. To prepare for either educational program, you should take courses in mathematics, applied physics, chemistry, electronics, and computer applications.

BRAIN BOOSTER

Assume that you are fairly sure you will pursue a career in electricity and electronics technology. However, you are still unsure which specific career to choose. What courses can you take that will help you no matter which career you choose?

PLANNING FOR YOUR FUTURE

This chapter has provided you with brief introductions to some of the occupations related to electricity and electronics technology. For the occupations that interest you, take some time to learn more details about them. For example, you may want to know more about the:

- daily work routine.
- setting in which you would be working.
- education and skills required, particularly if you are interested in a specialty within any of these occupations.
- salary you can expect to receive.

Also, keep in mind that to find a career that will satisfy you, it is important to consider your personal interests and abilities. All of this information will help you set your future career and educational goals.

General Educational Requirements

As you may have noticed, most careers in electricity and electronics technology have similar basic requirements. Communication skills, both verbal and written, are necessary in every field. To prepare for further education and training, you will also need a good foundation in mathematics and science. You must also be able to get along with both clients and coworkers.

Many of the careers mentioned in this chapter require applied associate degrees. Although the individual programs vary, most of them include certain basic courses or elements. Among these are courses in:

- electronics-related physics.
- technical and applied mathematics.
- basic and advanced electronic devices, including transistors and microchip technology.
- communication.
- circuitry analysis, including AC/DC circuits, electronic amplifiers, and control circuits and systems.
- electronic instruments and measurements.
- digital electronics.
- electronic design and fabrication,
- technical reporting.

Where to Go from Here

Most jobs require some form of qualifying training. This training provides the basic skills and knowledge needed to begin working in the field of electricity and electronics technology. High school courses in electricity and electronics are obviously important for anyone planning to build a career in this field. Mathematics and science courses are also important in gaining some of the basic knowledge you will need.

Other valuable courses may not be so obvious. Courses in English and speech help you develop your written and verbal communication skills. These skills are important no matter which career you may choose. Most electricity and electronics careers involve at least some amount of communicating with coworkers and customers.

In addition, consider joining electronics, science, or mathematics clubs. Not only can these clubs help you understand more about subjects that interest you, but they also help strengthen your resume.

There are many different ways to obtain education and training beyond the high school level. For example, you can enroll in a junior or community college or technical school for an associate degree. For advanced education in your field, look into four-year colleges and universities. You can also seek apprenticeships in areas that interest you. The armed forces offer training in exchange for a certain number of years of service, usually in the occupation for which you train. Depending on the occupation, training programs may last a few months or several years.

To enter, compete, and advance in your chosen career, you will need the skills and understanding to help you stay on top of current and emerging technologies. Technological advances make it necessary for you to refresh your training in one of various ways. Some employers offer workshops to keep their employees up-to-date. Certification programs provide "continuing education" in the many occupations.

Sources of Information

The following list offers resources that may help as you research various occupations.

- *School and public libraries* provide books and other printed materials that describe careers. They may also offer videotapes and computer software regarding careers and career choices. In addition, the library may have access to the Internet—a good tool for people who are researching various career opportunities. Search on key words related to the career you are researching. Ask the librarian if you need help using the Internet.
- *Government publications* include *The Dictionary of Occupational Titles*, *The Occupational Outlook Handbook*, and *The Guide for Occupational Exploration*. These sources, published by the U.S. Department of

Labor, are available in most school and public libraries.

- *Personal contacts* such as family members, teachers, neighbors, and friends are good sources of information. Interview someone who is employed in the occupation you are considering.
- *Business, industry, government, professional, and service organizations* often provide free brochures on specific occupations. A few professional organizations are listed below.

American Electronics Association
IEEE Consumer Electronics Society
International Society of Certified Electronic Technicians (ISCET)
The Electronics Technician Association International (ETA-I)
Electronic Industries Association
Junior Engineering Technical Society

You will be able to obtain their addresses from your local library. Ask your teacher or counselor for the names and addresses of other groups that may interest you.

Think about what you know about careers in electricity and electronics technology. Apply this knowledge to today's world—read the newspapers, watch news programs. What can you predict about future career opportunities in electricity and electronics technology?

- *Work experience* is probably the best way to find out whether you like a specific occupation. Part-time employment and volunteer work provide first-hand experience. This, too, will help when you apply for a position in the area of your choice.

EMPLOYMENT PROJECTIONS

As you consider various careers and occupations, it pays to look to the future. What occupations in electricity and electronics will be

Table 1-A Projected Employment Opportunities

Occupation	Projected Number Employed[1]
Electrical and Electronic Engineering Technicians	746,000
Electricians	553,000
Electrical and Electronics Engineers	418,000
Electronic Service Technicians	389,000
Line Installers and Repairers	303,700
Heating, Ventilation, Air-Conditioning, and Refrigeration Technicians	299,000
Computer and Office Machine Repairers	166,000
Broadcast Technicians	40,100

[1]The numbers in this table are for selected electrical and electronics technology occupations in the year 2005.

Source: U. S. Department of Labor, *The Occupational Outlook Quarterly*, Spring 1996.

Table 1-B Highest-Paying Occupations in Electricity and Electronics

Highest → Lowest	Occupation
Highest	Electrical and Electronics Engineers
	Electrical and Electronic Engineering Technicians
	Electronics Technicians
	Electricians
	Line Installers and Repairers
	Laser Technicians
Lowest	Broadcast Technicians

expanding? Will the need for people in any of these occupations decrease? If so, why? Are new technologies replacing them? Are you interested in the new technologies?

Where Will the Jobs Be?

Table 1-A shows projected employment opportunities in the year 2005. Study these figures carefully. Where do most of the opportunities seem to lie? How might this affect your career plans?

How Much Will They Pay?

One very practical consideration is the salary level you can expect in various occupations. Table 1-B shows the seven highest-paying occupations in electricity and electronics. Use this table as a general guideline to the top-paying jobs, but remember this represents current salaries. Demand for these occupations may change by the time you are ready to enter your career.

You should be very careful not to let an occupation's pay scale override other equally important factors. However, you should make sure that the career you choose will pay enough to allow you to live comfortably. You may need to do some research into the cost of living. How much will you need to earn per month or per year? Note that if you have "expensive taste," you will need to plan for more education and training. As in many other occupations, the pay scales in electricity and electronics are based on knowledge and skill levels, and sometimes on degrees earned at various learning institutions.

Jobs Related to Electricity and Electronics Technology

As you are researching various career opportunities, you may come across other, related careers that also interest you. Examples of careers related to electricity and electronics technology are:

- computer laboratory technician.
- instrumentation technician.
- drafter.
- electronics communication technician.
- fiber technologist.
- nuclear reactor technician.
- optomechanical technician.
- semiconductor development technician.
- systems testing laboratory technician.
- quality assurance technician.

These careers offer further opportunities. If you become interested in these or other careers, you can use the techniques discussed in this chapter to find out more about them.

Chapter *Review*

Summary

- The field of electricity and electronics technology provides an incredible variety of careers.
- The time to start deciding what career you want to pursue is now.
- As you research various career opportunities, keep practical considerations in mind.
- Career information is available from many sources. Among the best resources, if you have an idea of which career you want to pursue, are the professional organizations.
- Many professional organizations have student chapters that allow people who are still in training to be members of the organization.

Review Main Ideas

Review this chapter's main ideas by writing, on a separate sheet of paper, the word or words that most correctly complete the following statements.

1. A string of related ______ make up a person's ______.
2. People who have technical knowledge and skill in a specific field or area are commonly called ______.
3. Electricians ______ and ______ electrical systems for a variety of purposes.
4. ______ set up the vast network of electrical, telephone, and cable television lines.
5. The term HVAC/R stands for ______ and ______.
6. Five common sources of career information are ______, ______, ______, ______, and ______.
7. Because most fields require at least some interaction with people, most careers require good ______ and ______ communication skills.
8. To prepare for further education and training, you will also need a good foundation in ______ and ______.
9. In planning your career, learn details such as the ______, the setting in which you would be working, ______, and the salary you could expect to receive.

10. Most electricity and electronics careers share courses in physics, mathematics, electronic devices, communication, circuitry analysis, electronic instruments and measurements, ______, electronic design and fabrication, and ______.

Apply Your Knowledge

1. Suppose that you have decided on a career as a computer support technician. Research education and training opportunities in your area. Begin by identifying specific courses that are offered in your high school. Continue by identifying courses and degrees offered by post-secondary schools in your area. Also, identify any apprenticeships or summer/after-school jobs that could help you get the education/training you would need to become a computer support technician. Describe your findings in a short essay.
2. Suppose that, during your career research, you have become interested in becoming an elevator installer or repairer. Identify sources you could use to find out more about this career; then use the resources you have identified to write a short paragraph about the education and training needed to become an elevator installer or repairer.
3. Select at least five electricity/electronics careers discussed in this chapter. Try to choose those you are most interested in. Design a table that defines each career and describes the education necessary. Make sure your table highlights the differences and similarities of each of the five careers.

Make Connections

1. **Communication Skills.** Select ten occupations in electricity and electronics. Using resources from your guidance department, identify the requirements for employment in each occupation. Make a wall chart that presents this information to your class.
2. **Mathematics.** Obtain the latest statistics on the average income in your area. List the types of occupations that provide an income that meets or exceeds the average income. Report your findings to the class.
3. **Science.** Select an occupation in which you have an interest. Research the role of science in this occupation. Report your findings to the class.
4. **Workplace Skills.** Summarize the various technical activities and experiences you have had. List the specific skills that you have acquired that can be applied to an occupation in technology.
5. **Social Studies.** Most careers in electricity and electronics have similar basic requirements. Research the training of electricians and electronics technicians over the last fifty years. Prepare a written report, commenting on any changes.

Chapter 

Safety

Terms to Study

adapter

circuit breaker

electric shock

fuse

ground

ground-fault circuit interrupter

open circuit

OSHA

short circuit

OBJECTIVES

After completing this chapter, you will be able to:

- Explain the dangers of electrical shock.
- List seven general safety rules and practices.
- Explain the differences between open and short circuits.
- Identify the following fuses: Plug, tamperproof, dual-element, and cartridge.
- Compare the operation of circuit breakers and ground-fault interrupters.

Safety is everyone's business. Developing safe work habits depends on the right attitude—a feeling that you want to work safely.

Failing to follow electrical safety rules and practices can injure you or others and destroy property. Shock and burns can result when the body conducts electricity. Property may be damaged by electrical fires started by overheated wires or sparks. These dangers must be prevented if we are to use electric energy safely and efficiently.

HAZARDS

Hazards are generally considered unavoidable dangers or risks. Although unavoidable, many hazards can be minimized so that the risk is not as great. Shock is an electrical hazard. The potential for an electrical shock is always present because that is the nature of electricity; however, if proper precautions are taken, the danger or risk of shock can be minimized.

Note the words "proper precautions." This is how safety is achieved. Safety implies "proper precautions," and safety is the subject of this chapter.

Identification

How many types of safety hazards can you identify? Since this text is about electricity, the first hazard you might think of is electrical. There are, however, other hazards we encounter on a daily basis that may not come to mind immediately. A few of the more obvious ones are mechanical, chemical, and physical hazards. A bit more remote are hazards such as fire, radiation, temperature extremes, noise, and ergonomics.

Examples of most of these hazards can easily be brought to mind. Electrical shocking, tripping over misplaced equipment, soldering burns, splattering flux in the eyes, and microwave and x-ray radiation threaten the electronic workplace. Ergonomic hazards may not be quite so obvious. Ergonomics is the study of human engineering. Therefore, ergonomic hazards are the hazards associated with human engineering, such as; workbench height and seat/chair design, lighting, heating and ventilating, and the movement and position of your body. Even the design of floor materials and computer keyboards fall in the range of ergonomic hazards.

Be on the alert for hazards of all types. Identify and correct those that you can, and report the others. Hazards can have serious consequences for both the individual and the team. This chapter focuses on two critical electrical hazards which are shocks and burns.

Electric Shock

Electric shock is a physical reaction of the nerves to electric current. In minor cases, it is only a harmless spasm of the affected muscles. In severe cases, the breathing and heart muscles become paralyzed. If the muscles are permanently damaged, the result is often electrocution. If the muscles are not permanently damaged, they often can be restored to normal operation by cardiopulmonary resuscitation, or CPR. CPR is an emergency procedure for reviving the heart and lung functions. It involves special training and physical techniques. In the event of severe shock, call a physician or the emergency medical service at once.

The amount of current that can cause serious damage to the muscles of a person's body depends on his or her physical condition. Records show that people have been electrocuted by very low values of current. Serious muscular damage is likely to result if an excessive current passes through the chest area. This happens when the conducting path goes from hand to hand or from one hand to the feet (Fig. 2-1).

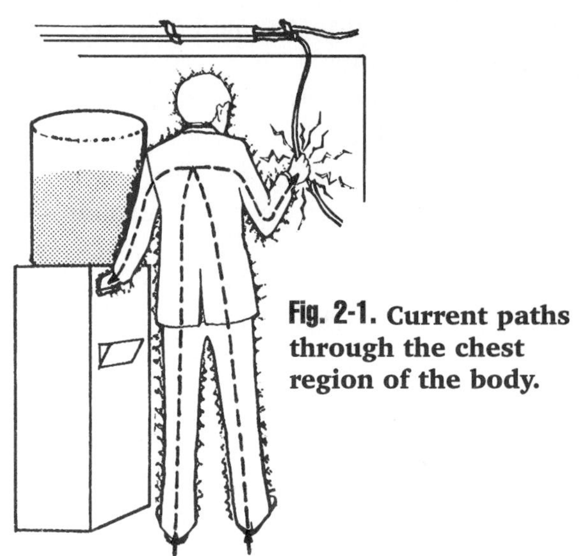

Fig. 2-1. Current paths through the chest region of the body.

The danger of a severe shock increases when the skin is wet. Under normal conditions, the epidermis, or outer layer of skin, presents a high resistance to current. If the epidermis becomes wet from sweat or other fluid, its resistance is greatly lowered. Under this condition, a voltage lower than 120 volts (V) can cause a dangerous amount of current to pass through the body.

Burns

In addition to shock, excessive current passing through the body can cause severe burns. Such burns result from the heat produced by friction between the electrons and body tissues. Electrical burns often occur inside the body along the path followed by current. Burns of this kind can be very painful and difficult to heal.

FIRST AID

First aid falls into two general categories: (1) what the average person can be expected to do, and (2) what the trained person can be expected to do. Most of us have little if any formal training in first aid. Common sense, however, can go a long way. Common sense says that if a live electrical wire is lying on a person you should not touch the wire *or* the person. However, any persons who work with dangerous voltages or current should be trained and certified in Red Cross CPR.

Knowing what to do and knowing how to do it often separate the average person from the trained person. We all know that if a person has received a severe shock, CPR can often save his or her life. But, how many of us can perform CPR? If you have first aid training, you know what to do when the need arises. If you do not have first aid training, the most important thing is to recognize the need for help. Call the nearest emergency medical service.

1. List ten workplace hazards and their remedies.
2. Contact your local Red Cross and find out how to become CPR certified.
3. Locate a copy of the National Electrical Code at your library. Find and copy the regulations for grounding cabinets in an environment where three-phase 220 Vac is commonly used.

RULES AND REGULATIONS

Many safety procedures, as well as standard operating procedures, have been developed by federal and state governments and by national organizations. The Federal government has established the Occupational Safety and Health Act to protect employees in the workplace. Many states have their own versions of this national act. The National Fire Protection Association developed the National Electrical Code. Material Safety Data Sheets (MSDS) are available, and most companies have standard operating procedures dealing with safety including accident reporting, handling of toxic waste, correcting hazards, etc. Every material should be labeled in such a way that identifies its toxicity, flammability, and health hazards.

Occupational Safety and Health Act (OSHA)

As a future worker, you will become familiar with the *Occupational Safety and Health Act*, commonly called **OSHA**. The U.S. Congress passed this act in 1970. Its basic purpose is to ensure that every worker in the nation has safe and healthful working conditions.

This act does several things. First, it encourages employers and employees to reduce workplace hazards. Second, it establishes responsibilities and rights for employers and employees. Third, it develops and enforces required job safety and health standards.

Both the worker and the employer have responsibilities under this act. The worker must read the OSHA posters at the job sites. The worker must obey all OSHA standards that apply. The worker must follow all the employer's safety and health rules and regulations. The employer must tell all employees about OSHA. The employer must make sure employees have and use safe tools and equipment. It is not possible to describe OSHA in detail in this textbook, but much information about OSHA is available from the U.S. Department of Labor, which administers the act.

Safety Color Codes

The Occupational Safety and Health Administration has established the following colors to designate certain cautions and dangers:

Red designates:

- Fire protection equipment and apparatus.
- Portable containers of flammable liquids.
- Emergency stop buttons and switches.

Yellow designates:

- Caution and to mark physical hazards.
- Waste containers and storage cabinets for explosive or combustible materials.
- Caution against starting, using, or moving equipment under repair.
- Identification of the starting point or power source of machinery.

Orange designates:

- Dangerous parts of machines.
- Safety starter buttons.
- The exposed parts of pulleys, gears, rollers, cutting devices, and power jaws.

FASCINATING FACTS

The National Fire Protection Association has been in existence since 1896. The Association develops and distributes over 275 national fire codes. Most of these are adopted at the local or state levels.

Purple designates:

- Radiation hazards.

Green designates:

- Safety
- Locations of first-aid equipment (other than fire fighting equipment).

National Fire Protection Association

The National Fire Protection Association (NFPA) and other organizations provide leadership in developing electrical safety standards. The National Electrical Code, a publication sponsored by NFPA, provides many safety standards. One standard, for example, relates to the installation of electrical conductors, including optical fiber cable and equipment within public and private building and structures. While the National Electrical Code is purely advisory, various federal, state, and local authorities reference it in laws, ordinances, regulations, and administration orders.

LINKS

Chapter 28, Residential Wiring, provides specific applications of residential standards.

SAFETY DEVICES AND CONTROLS

Grounding for Personal Safety

The **ground**, or earth, is a good conducting point for electrons when a path exists between it and a live (charged) wire. Because of this, putting any part of your body between the ground and such a wire presents the danger of severe electrical shock.

If an uninsulated wire of an electrical appliance contacts the bare metal of its cabinet, or frame, the cabinet becomes an extension of the wire. When the cabinet is not connected to the ground with a wire or other device, the line voltage appears between the cabinet and ground. In most homes, the line voltage is about 115 Vac. Thus, there is a danger of severe shock if a person makes contact between the cabinet and any surface that is in electrical contact with the ground.

Grounding Plugs, Outlets, and Wires

Metal cabinets, cases, and frames of appliances, tools, and machines are usually grounded by a separate grounding wire. This grounding wire is part of their power cords. The grounding wire is connected to the cabinet, case, or frame of the device and to the round prong of the cord's plug. It is not connected to any part of the device's electric circuit. When the plug is put into a grounded outlet, the grounding wire of the cord is automatically connected to the building ground. This connection is made through a grounding wire in the cord leading to the outlet (Fig. 2-2A). Since the cabinet, case, or frame is then at the same voltage as the ground, a shock hazard cannot exist between it and any grounded surface.

Three-pronged grounded plugs are called *polarized plugs*. Their prongs will fit into an outlet only when they are properly aligned.

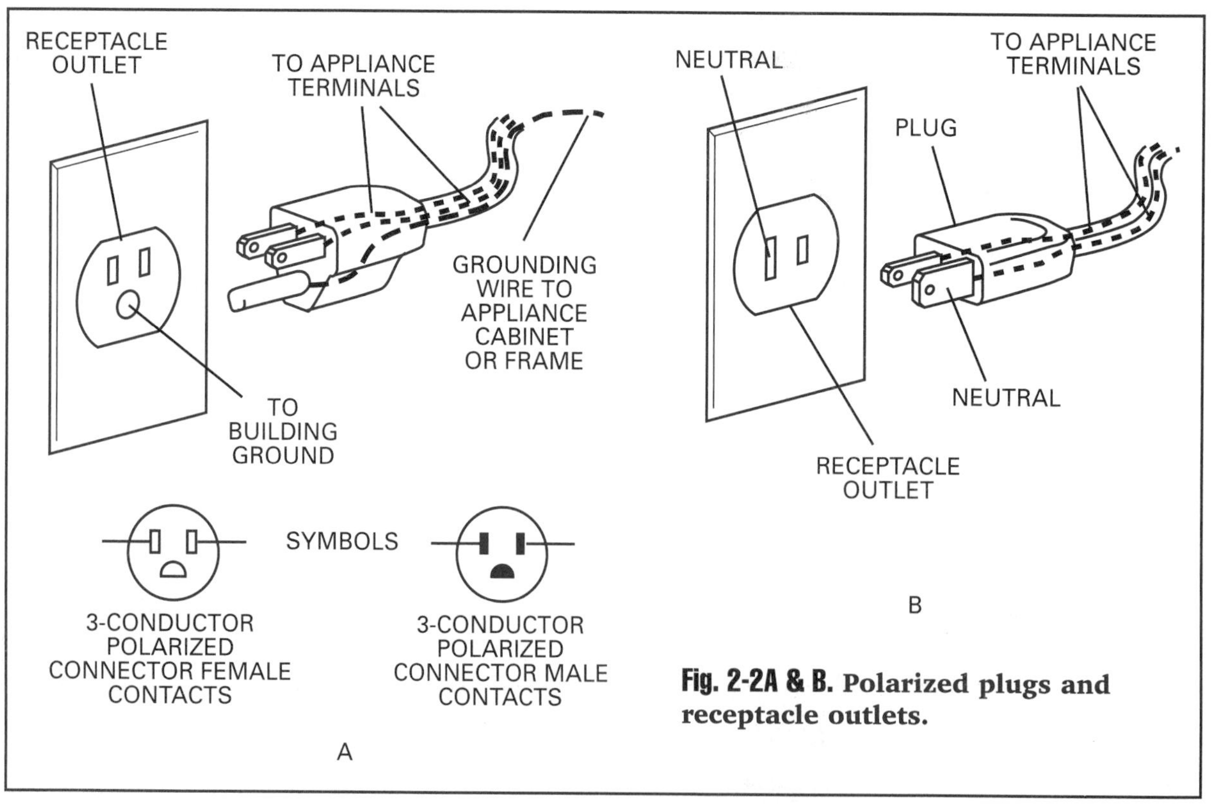

Fig. 2-2A & B. Polarized plugs and receptacle outlets.

LINKS

See Chapter 28, Residential Wiring, for a description of the hot and neutral wires.

The round prong connected to the equipment ground wire can fit only into the round hole in the outlet. When this happens, the other two prongs automatically line up with their slots in the outlet. One is connected to the "hot" wire. The other is connected to the "neutral" wire.

Another type of polarized plug has only two prongs (Fig. 2-2B). One prong is taller than the other prong. Since the two slots of a wall socket are different heights, the plug will only fit into the wall socket one way. This insures that the neutral side of the appliance is connected to the neutral side of the line voltage. Older homes, appliances, and some equipment might not have polarized, or even grounded, plugs or wall sockets.

Frames of machines and metal cabinets of appliances do not always use grounding plugs and outlets. When grounding plugs are not used, the devices are grounded by connecting them directly to ground rods (Fig. 2-3). It is no longer considered safe to use water pipes as grounding conductors. This is because an electrical condition could occur that might electrocute a person who is bathing, showering, or washing dishes.

Two-to-three-wire Adapter

A two-to-three-wire **adapter** makes it possible to use a grounding plug in a two-socket outlet. Such an adapter does not automatically ground the device. It must be connected, using the grounding wire, to the ground point. The grounding wire is usually a green wire and it

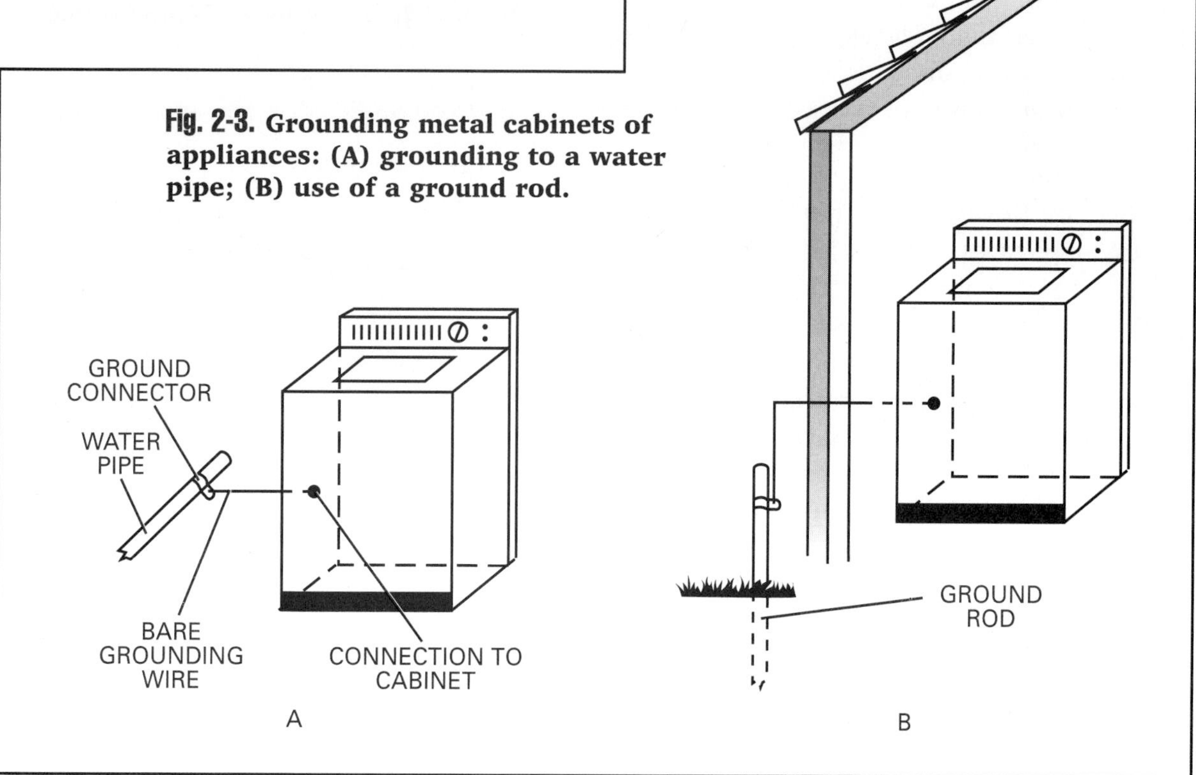

Fig. 2-3. Grounding metal cabinets of appliances: (A) grounding to a water pipe; (B) use of a ground rod.

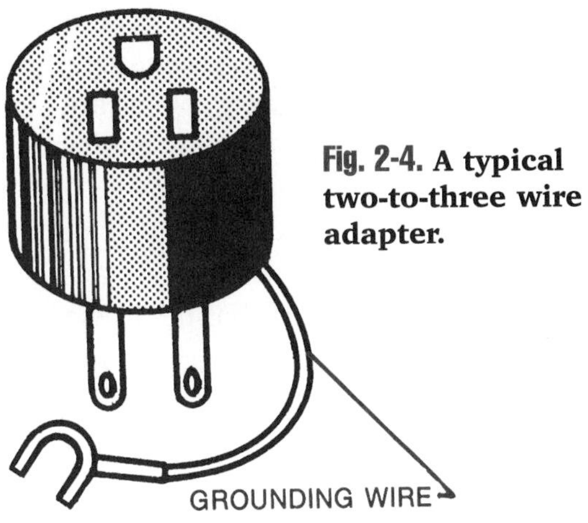

Fig. 2-4. A typical two-to-three wire adapter.

should be connected directly to the building ground (Fig. 2-4). Some appliances also use the green wire as a part of their power cord rather than using a three-pronged grounding plug. If this is the case, be sure the green wire is properly connected to ground and provides adequate protection against electric shock. To test a metal cabinet for voltage, use a voltmeter connected from the cabinet to ground. *Never* touch a cabinet that you suspect is "live." It is best, therefore, to disconnect power entirely and use an ohmmeter to check for a short to chassis.

Double, or Reinforced, Insulation

Many appliances, portable power tools, and test equipment protect against shock by means of double, or reinforced, insulation. In these devices, the wiring is kept from grounding by special kinds of insulation between the wiring and the equipment cabinet or case. To further protect against shock, certain gears and screws that hold the assembly together are often made of nylon or some other insulating material. Very often, the entire case is made of strong plastic material. This kind of construction significantly lessens the danger of severe electric shock.

LINKS

Chapter 22 will teach you how to use a meter.

Open and Short Circuits

An **open circuit** occurs when a wire is broken, or is disconnected at some point, and the current flow stops (Fig. 2-5A). An open circuit can be dangerous. The voltage of the circuit appears across the point at which the circuit is broken. Whenever possible, disconnect the power to equipment with broken connections before handling it.

A **short circuit** occurs if two uninsulated live wires contact each other in such a way as to bypass the load, causing an excessive amount of current to flow in the circuit (Fig. 2-5B). Short circuits are very dangerous when the excess current is high enough to damage the insulation of the wire and start a fire or cause personal injury. In most circuits, circuit breakers and fuses are used to automatically open the circuit if there is excessive current flow.

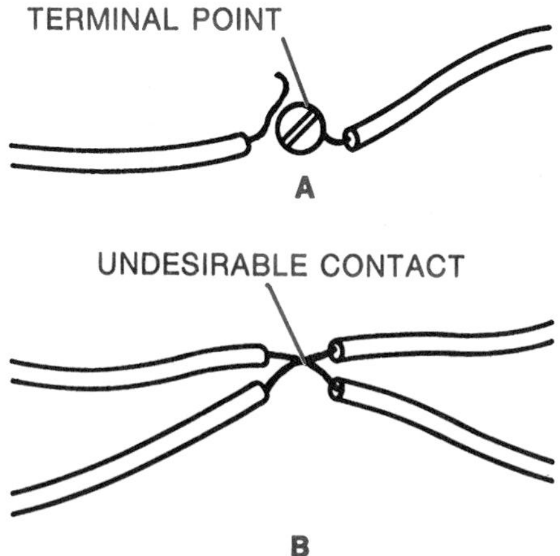

Fig. 2-5A & B. Examples of (A) open circuit; (B) short circuit.

Fuses

A **fuse** is a safety device that works as a switch to turn a circuit off when the current goes over its specified value. Too much current may be the result of a short circuit or a condition known as overloading. Overloading happens when a load demands more current than the conductors can safely carry. Overloading causes the circuit to become overheated and may result in a fire. Overloading generally happens when a load has jammed or too much equipment is connected to one circuit.

Plug Fuses

Plug fuses are commonly used in older household circuits. Such a fuse has a strip of metal called a fuse link, or element (Fig. 2-6). The link, usually made of zinc, is designed to melt or burn apart when more than the rated value level of current passes through it. The electrical size of a fuse is equal to its rated value of current and voltage. Current is given in a unit called an ampere, abbreviated A. Voltage is abbreviated V, and is followed by an ac (alternating current) or dc (direct current). The values of amperage and voltage are usually printed on the body of a fuse, for example, 15 A or 15 Amp and 115 Vac. This means that the fuse will open if the current goes over 15 amperes and that it will withstand voltages up to 115 Vac. Plug fuses are put into a circuit by screwing them into special sockets.

Some plug fuses have a metal screw base like the base of an ordinary light bulb. This kind of plug fuse is known as an Edison, or medium, base fuse. Using medium based plugs is no longer advisable and has been discontinued. Most electrical building codes now prohibit their use; however, they can still be found in older buildings.

Tamperproof Fuse

Tamperproof plug fuses are designed so they cannot be replaced with a fuse of a different electrical size. They are also referred to as non-interchangeable plug fuses. A socket is used that accepts only certain sizes of fuses. For example, 25 A or 30 A fuses will not fit into sockets for 15 A and 20 A fuses. To use these fuses in a medium-base socket, an adapter must be installed (Fig. 2-7). The adapter is first screwed into the fuse socket, then the fuse is screwed into the adapter. Once in place, the adapter cannot be removed easily. Tamperproof fuses are also available as *cartridge fuses*.

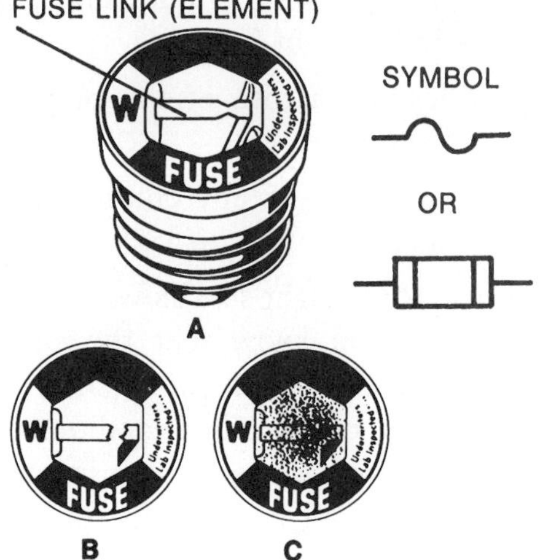

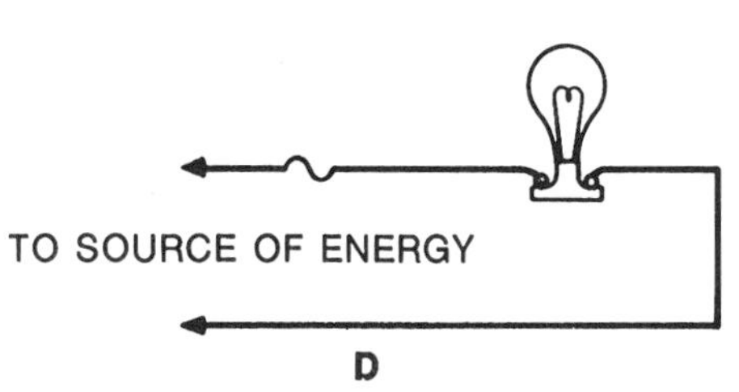

Fig. 2-6. Medium-base plug fuse: (A) fuse in good condition; (B) blown or broken fuse link; (C) blown fuse link indicated by darkened "window"; (D) method of connecting a fuse to a circuit.

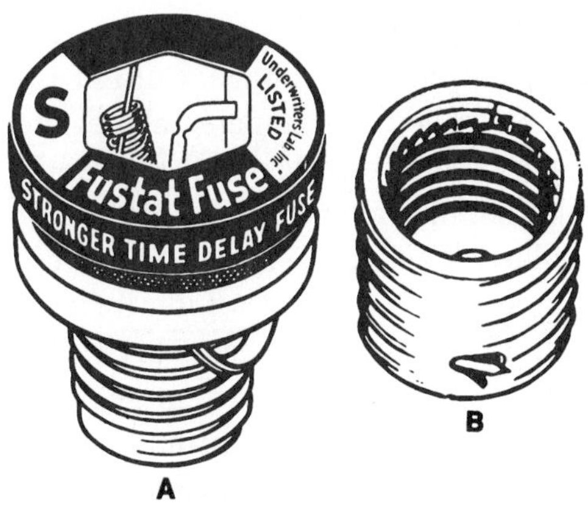

Fig. 2-7. Tamperproof fuse: (A) plug fuse; (B) fuse adapter.

Dual-Element Plug Fuse

A dual-element, or time-delay, plug fuse is designed to withstand a current larger than its rated value for a short time. If the excessive current continues after this time, the fuse blows just like an ordinary plug fuse. It must then be replaced.

The construction of a dual-element plug fuse is shown in Fig. 2-8. When too much current passes through the thermal cutout of the fuse, the solder holding the fuse line to the thermal cutout becomes hot. If this amount of current continues, the solder softens. This allows the spring to pull the fuse link out of the thermal cutout and break its connection.

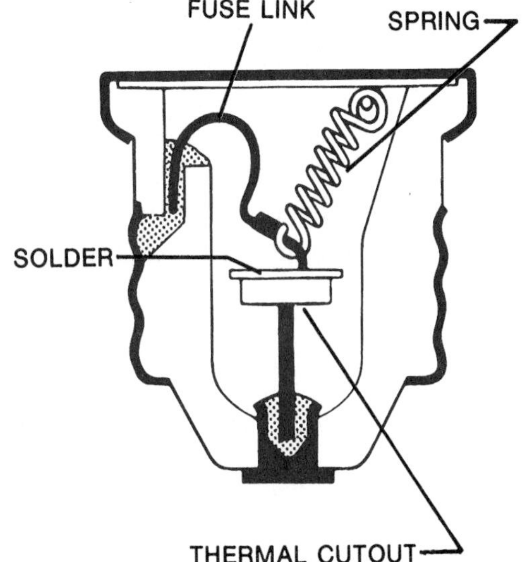

Fig. 2-8. Cutaway view showing construction of a dual-element plug fuse.

Dual-element plug fuses are most often used in motor circuits. A motor draws much more current on starting than when it does after reaching full speed. A dual-element plug fuse will withstand this short period of high speed. It then continues to protect the motor circuit after the current has decreased to its running value.

Cartridge Fuses

Like plug fuses, cartridge fuses are available in either ordinary or dual-element versions (Fig. 2-9A). Cartridge fuses are mounted in clip holders (Fig. 2-9B). Small-size glass cartridge fuses can be used in clip holders or in spring-loaded fuse holders (Fig. 2-9C).

Fuse Cautions

Fuses will protect circuits only if they are used correctly. Observe the following cautions for the safe handling of fuses:

1. Never replace a fuse of the proper size with a larger one.
2. Never put a coin or any other object in a fuse socket.
3. If a fuse of the right size blows out repeatedly, then you know that the circuit is overloaded or that there is a short somewhere in the circuit. Such conditions must be corrected before the circuit is put back into use.
4. Never try to bypass a fuse by connecting a wire between the terminals of its socket or holder. Always keep a supply of spare fuses of the right sizes on hand in case of emergency.
5. Always use fuse pullers when you remove a fuse.

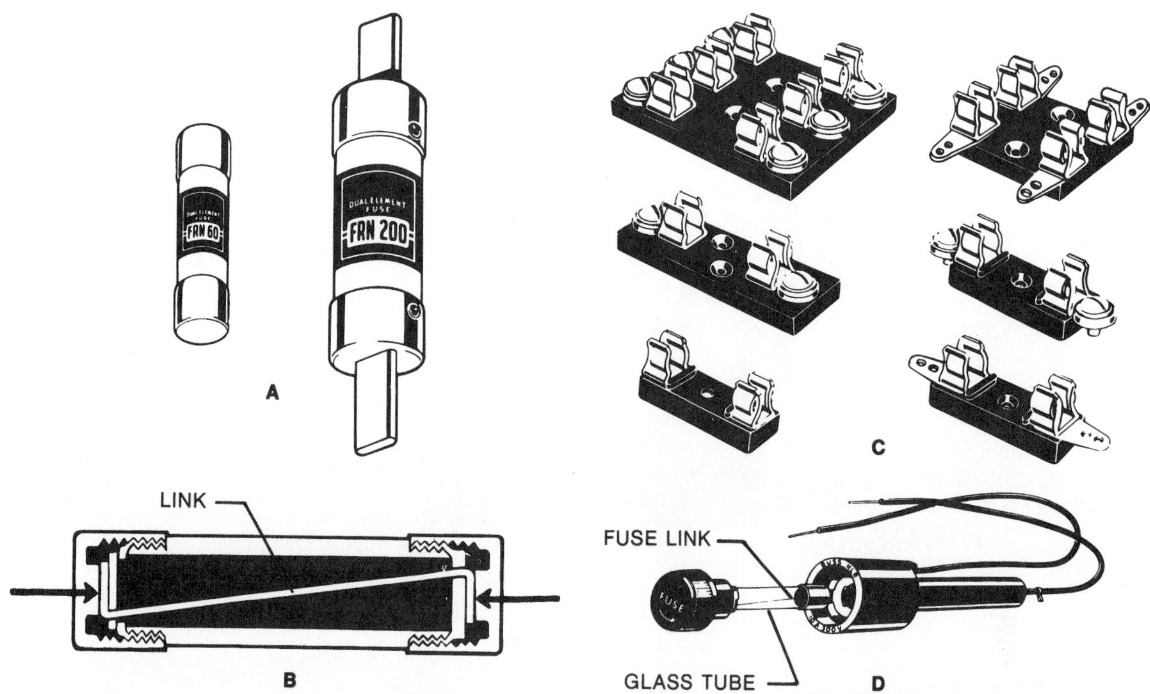

Fig. 2-9A, B, C, D. Cartridge fuses and holders: (A) typical fuse cases; (B) cutaway view of renewable-link fuse; (C) clip-style fuse holders; (D) spring-loaded fuse holder.

Circuit Breakers

A circuit breaker is a mechanical device that performs the same protective function as a fuse. Some circuit breakers have a switch mechanism that works by thermal, or heat, control. Others use a magnetic switch mechanism. Larger circuit breakers have switch-like mechanisms that are both thermally and magnetically controlled (Fig. 2-10).

LINKS

Thermal control involves the use of a bimetallic element, discussed in Chapter 30. Magnetic control involves the use of a solenoid, discussed in Chapter 11.

Like fuses, circuit breakers are rated according to the amount of current in amperes that can pass through them before they are tripped, or opened. However, unlike a fuse, a circuit breaker can be reset after being tripped. Some circuit breakers have adjustable trip sizes. Others have only a single trip value.

When a circuit breaker of the kind shown in Fig. 2-10 trips, its lever moves to or near the off position. To be reset, the lever must be moved fully "off" and then back to "on." Never design the use of a breaker as an on/off switch. Their job is to remove power and to protect against over-current circumstances.

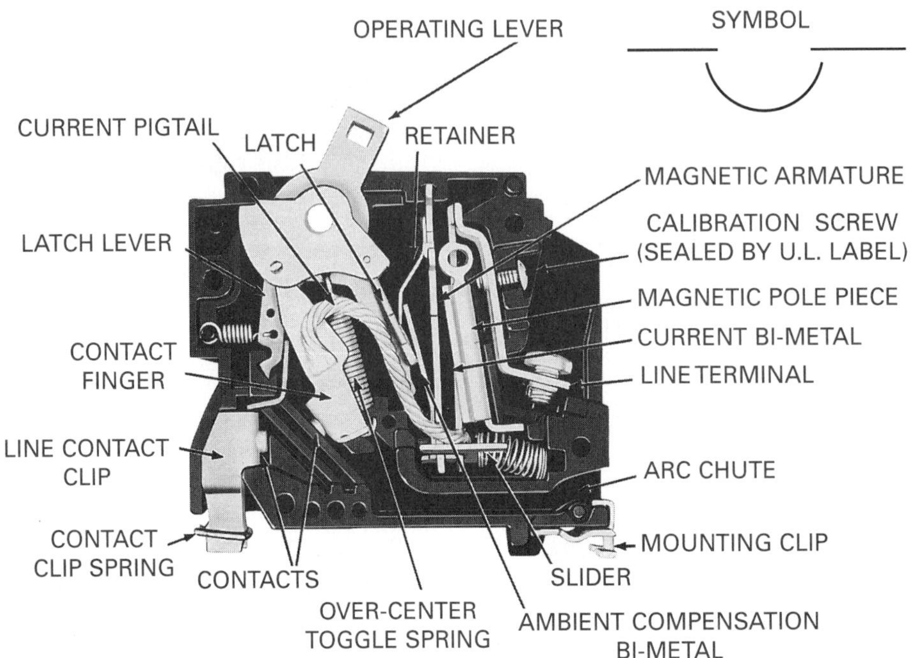

Fig. 2-10. Construction of a typical circuit breaker.

Ground-Fault Circuit Interrupter

The **ground-fault circuit interrupter** (GFCI) is a device that gives extra personal protection against electric shock. Devices such as fuses, equipment grounds, and double insulation are used for personal and equipment protection, but they have shortcomings. An electric current of about 60 milliamperes (mA) can cause even adult hearts to beat irregularly or to stop altogether. A 15 A fuse or circuit breaker disconnects the circuit only when the current exceeds 15 A. This amount is much greater than 60 mA or .060 A.

Figure 2-11 shows a portable ground-fault circuit interrupter. They are also available as built-in circuit breakers and duplex outlets. The portable GFCI can be plugged into a common duplex receptacle; however, it must be properly grounded to the receptacle. Before connecting an electrical cord of an appliance to a GFCI, test the GFCI to see if it is working properly. Do this by following the specific instructions for the GFCI being used.

Fig. 2-11. A portable ground-fault circuit interrupter (GFCI).

Ground-fault circuit interrupters work on the principle that the current flowing in the hot wire to the device should equal the current flowing in the neutral wire from the same device. Any difference between the currents in these wires indicates that the current is taking another path to complete the circuit. Ground-fault circuit interrupters sense the difference between the currents in the hot and neutral wires. If there is a difference of more than 5 mA, a safety switch trips in the GFCI and interrupts the circuit. This protects the person who is operating the electrical equipment from a serious electrical shock. The GFCI does not eliminate the feeling of an electric shock; but it opens the circuit quickly enough to prevent injury to a healthy person. A GFCI provides protection against dangerous currents that do not overload 15 A or 20 A fuses or circuit breakers. A GFCI should not be used in place of a separate grounding wire, and a GFCI does not protect a person who accidentally touches bare hot and neutral wires at the same time.

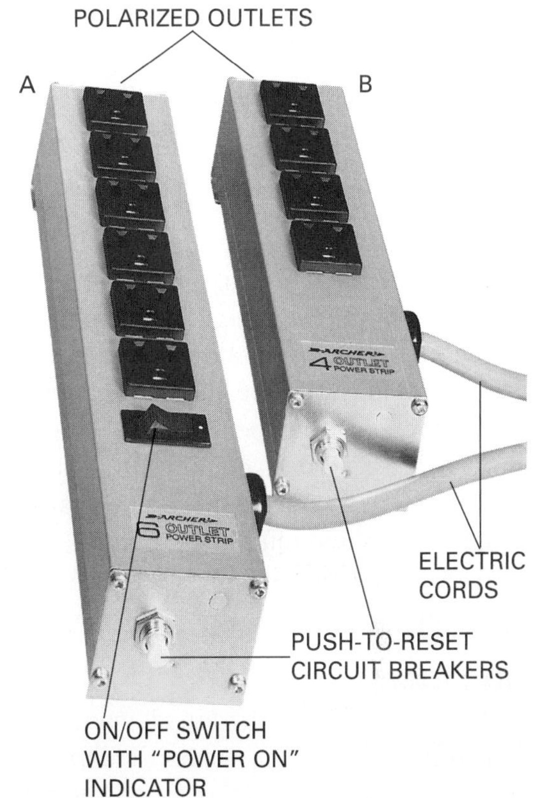

Fig. 2-12A & B. Electrically grounded plug-in strips: (A) six-outlet; (B) four-outlet.

Power Strips

Electrical devices, such as microcomputers, printers, and monitors, function as a complete electrical system. Each device, however, requires an individual electrical outlet. The electrical wiring systems of most homes and business offices do not have a sufficient number of electrical outlets located in one place to meet this new need. The electrically grounded plug-in power strips help meet this new need (Fig. 2-12).

The power strip shown in Fig. 2-12A can accommodate six electrical devices. It has an on/off switch with a power on indicator and a push-to-reset circuit breaker. Power strips used in homes are generally rated at 15 A, 125 Vac.

Some power strips provide surge protection for electrical equipment. Power surges, or excessive current or voltage, are often caused by switching on and off inductive loads, such as air conditioners and furnace fans. Lightning from thunderstorms causes most damage due to the sudden high-voltage discharge of atmospheric electricity. Entire computer systems can be damaged by lightning or other electrical surges unless protection is provided.

Monitors, Alarms, and Sensors

Circuit breakers and fuses provide protection from excess current. Current, however, is not the only hazard people working with electricity may encounter. Radiation is a hazard that is easy to monitor. Employees in radiation areas such as x-ray labs normally wear radiation-sensitive badges which monitor the radiation received. Smoke and carbon monoxide detectors are common examples of alarms that warn of fire and inordinately high levels of carbon monoxide, respectively. Smoke detectors, shown in Fig. 2-13, are required by building codes in most states.

PERSONAL PROTECTIVE EQUIPMENT

The essence of safety is protection. Personal protection requires equipment designed for individual safety. When wearing personal protective equipment, make sure that the equipment is in good condition, that it fits, and that it is approved for the specific task being performed. Since electrical equipment is used in nearly every environment, make sure that your personal protective equipment is appropriate for the environment, as well as for the specific electrical application.

Safety Glasses

Safety glasses are required in all manufacturing facilities and most school laboratories. Safety glasses should be worn when soldering, cutting wires, handling chemicals and television picture tubes, and using machines such as drill presses, portable electric drills, and grinders.

BRAIN BOOSTERS

1. **Locate the circuit over-current protection in your home. List the types and sizes of fuses that you use.**
2. **List the three criteria for personal protective equipment.**
3. **Find information about Material Safety Data Sheets (MSDS) and how to use them.**

Prescription glasses can double as safety glasses *only* if they have been so warranted by the lens manufacturer. Glass lenses and some plastic lenses do not qualify as safety glasses since they can shatter and cause serious eye injury. Goggles should be worn over safety glasses when there is a danger of fluids, dust, or rust falling or splashing into the eyes.

Clothing

Protective clothing is not as essential in the electrical industry as it is in many other occu-

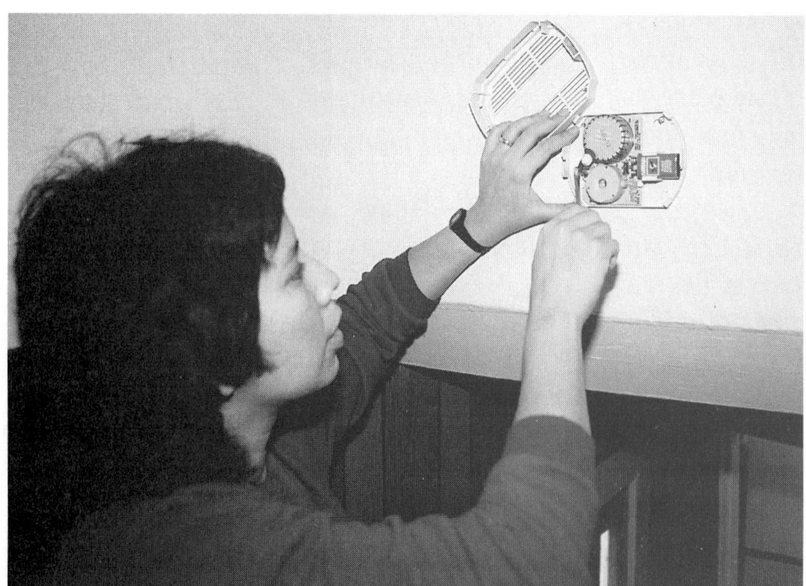

Fig. 2-13. A smoke detector.

pations. Aprons, smocks, and coveralls are appropriate for the protection of clothing. They can cover the body to protect it from solder burns, battery acid, and general external contamination. In some instances, rubber or latex gloves are necessary to protect the skin.

Masks and Respirators

Safety in electronics is not limited to the lab, office, or classroom. Since electronic equipment is found in every environment, some conditions require masks and respirators. A medical electronic technician working on equipment in an active operation room must wear a surgical mask. Respirators must be worn by engineers, technicians, and scientists working in contaminated areas. Space stations, modules, and capsules also require appropriate breathing devices.

WORKPLACE SAFETY

Workplace safety protects people at work. Personal behavior, personal grooming, housekeeping, and rules and procedures generally are the elements of workplace safety.

Personal behavior is a major factor in workplace safety. When dealing with personal behavior one of the first considerations is commonly referred to as "horse play," that is, rough or boisterous play, which is not appropriate in the workplace. Other forms of improper personal behavior include pranks, practical jokes, and all types of harassment. This doesn't mean that the workplace can't be an interesting, exciting, and fun place to be; however, it should not be this way at the expense of others. Observe common courtesies at all times.

Personal grooming from the safety standpoint has little to do with personal grooming from the aesthetic standpoint. Flowing hair, heavy jewelry, and loose clothing may look nice, but that long hair can easily get caught in machinery or obscure your vision. Heavy jewelry can get in the way and conducts electricity. Loose clothing can interfere with movement, restrict activity, and get caught in machinery.

Housekeeping usually requires everyone's help, but your own work area is your responsibility. Keep it neat and orderly for safety's sake.

One final caution for the workplace; follow the rules and observe proper procedures. One of the first assignments you may have when you start a new job is to become familiar with the standard operating procedures for your company. These usually include a section on safety. Read it carefully, and make sure you understand what it says. If you don't, ask.

GENERAL ELECTRICAL SAFETY RULES AND PRACTICES

Safety rules and practices can not cover all cases. The following promote electrical safety:

1. Always treat electrical wires in a circuit as being potentially dangerous.
2. Always disconnect the power from a piece of equipment before removing the case or cabinet in which the wiring is contained.

Safety Engineering

Safety engineering is one of the specialized careers created to help prevent worker injury and reduce employer costs. Safety engineers study the causes of accidental deaths and injuries and devise ways to prevent them. Current safety engineering trends include an increased emphasis on identifying potential hazards and guaranteeing product liability and consumer protection. Other trends include developing legislation in product and consumer safety and the work environment.

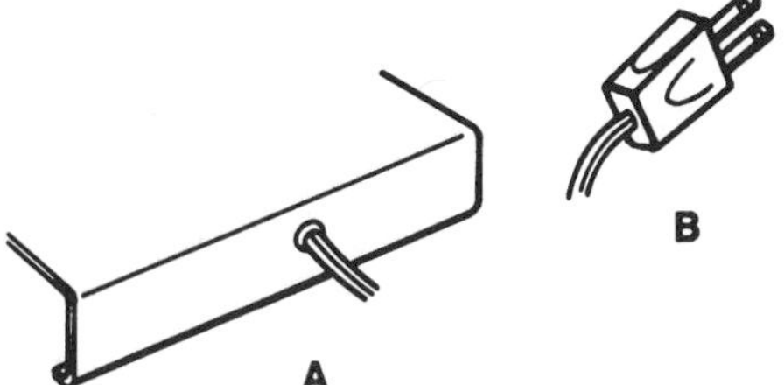

Fig. 2-14. Points at which electrical cord defects most often occur; (A) where cords enter appliances; (B) where cords enter attachment plugs.

3. Do not touch a water or gas pipe, a sink, or bathtub, or any wet surface while handling the metal parts of electrical equipment that is plugged in.
4. Never place any plugged-in equipment where it can fall into a sink or a bathtub containing water.
5. Be sure to read and follow all instructions supplied with electrical or electronic equipment before using it.
6. Make a habit of periodically inspecting all the electrical appliances and tools you use. Periodic inspections will find potentially dangerous conditions such as loose wires, undesirable bare wires, and frayed cords (Fig. 2-14).
7. Always unplug or disconnect equipment that does not seem to be working normally.
8. Do not work on any equipment unless you know the right procedures to follow. It is always a good practice to call in an expert when you are unsure about the safe and proper way to handle the equipment.
9. If you must work with the exposed parts of a circuit that is plugged in, handle the parts with one hand only. Keep the other hand behind your back or in a pocket. If you accidentally touch a live wire, this will keep the current from passing through your chest. Do not wear rings, watches, bracelets, or other jewelry when touching live electric circuits. Do not ever work on circuits higher than 120 V without disconnecting the power.

Fig. 2-15. The Underwriters' Laboratories label.

10. It is always best to use only equipment that has Underwriters' Laboratories labels (Fig. 2-15). This shows that the item has been designed to meet safety standards.
11. Find and learn to work the main power switch in your home and in other places where you work with electricity. This will let you quickly turn off all the circuits in an emergency.

12. During an electric shock, the muscles will often cause the fingers to hold the point of contact tightly. Turn off the power immediately. The possibility of serious muscular damage and burns increases the longer current passes through your body.
13. Extreme care must be taken when removing a person from the point of electrical contact. The path of current can pass through the rescuer, causing her or him to suffer severe shock also.
14. Although an electric shock may not always be dangerous, it can be startling. This may cause a fall or a bump against hard surfaces, resulting in serious injury.

LINKS

Chapter 28 discusses these problems and provides suggestions for reducing them.

General safety rules and practices apply to computers and peripheral equipment. There are, however, two health problems associated with the use of computers. One problem is eye strain. The other problem is repetitive stress injury to the fingers, hands, and wrists.

What are the three most important rules for safety working with electricity?

Chapter Review

Summary

- Hazards are unavoidable risks. Two electrical hazards are shocks and burns. Different amounts of current can cause differing amounts of bodily damage depending on a person's age, gender, size, fitness, and their environmental conditions.
- The Occupational Safety and Health Act supports the protection of employees in the workplace.
- The *National Electrical Code*, developed by the National Fire Protection Association, is referenced in laws, ordinances, regulations, and administrative orders.
- Proper grounding procedures are essential for electrical safety.
- Electrical circuits must have correctly sized and designed over-current protection as well.
- In addition to using safe wiring practices, it is necessary to wear personal protective equipment in the workplace.

Review Main Ideas

Review this chapter's main ideas by writing, on a separate sheet of paper, the word or words that most correctly complete the following statements:

1. An electric shock is a physical sensation caused by the reaction of the nerves to ______.
2. The danger of severe shock is greatly increased if the skin becomes ______.
3. In addition to shock, an excessive current through any part of the body can cause severe ______.
4. Electrical wires, wherever they may be found in a circuit, should always be treated as being potentially ______.
5. It is always safe practice to read and thoroughly understand all available operating ______ supplied with electrical and electronic products.
6. You should make it a habit to periodically ______ all electrical and electronic equipment you use.
7. A plug-in device should be immediately ______ if it does not seem to be working normally.
8. An ______ Laboratories label on a product shows that it has been made according to strict ______ standards.
9. Putting any part of your body between the ______ and a live wire presents the danger of severe shock.
10. A grounding wire is used to prevent a shock hazard from existing between a metal cabinet and ______.
11. Grounding plugs and outlets automatically connect power cords to the building ______.
12. The color of the grounding wire in the power cords of tools and appliances is ______.

13. The use of ______, or ______ insulation in products significantly lessens the danger of severe electric shock.

14. A fuse is a safety device that works as a ______ to turn a circuit ______ when there is too much current.

15. When a circuit is overloaded, it becomes ______.

16. Current in an ordinary plug fuse is given in ______.

17. Dual-element fuses are usually needed in ______ circuits.

18. Never replace a fuse of the proper size with a ______ one.

19. A ______ is a mechanical device that, like a fuse, protects a circuit from excessive current.

20. Ground-fault circuit interrupters sense the difference between the currents in the ______ and ______ wires.

21. OSHA stands for ______.

Apply Your Knowledge

1. Choose the correct type of fuse for each of the following:
   a. Drill press.
   b. A fuse in an older home with medium-based, screw-in fuses.
2. Name the first step in checking for a shorted conductor to chassis, or ground, with an ohmmeter.
3. Describe how a common circuit breaker differs in operation from a ground-fault circuit interrupter.
4. List three types of personal protective equipment.
5. Discuss in depth, any of the fourteen General Electric Safety Rules and Practices. Explain the technical reasoning behind each rule.

Make Connections

1. **Communication Skills.** Write a short report on how safe practices are developed and finally recommended and published in the *National Electrical Code* (NEC).
2. **Mathematics.** Examine the electrical appliances in your home that are used for food preparation. You might examine a toaster, an electric skillet, or an electric can opener. Calculate the total wattage and current for each of these devices. Identify those appliances that can be operated safely all at one time.
3. **Science.** Explain to the class why moisture on the skin greatly increases the danger of severe electric shock.
4. **Technology Skills.** Bring in an electric cord from your home that you feel is damaged. Repair the cord. Test it for grounds and shorts.
5. **Social Studies.** Homes built about forty years ago used electric outlets with only two prongs. Organize a debate (pro and con) to discuss the question: Should owners of older homes be required to install newer grounded outlets to meet ordinances that require grounded outlets in newly built homes?

Chapter 3

Atoms, Electrons, and Electric Charges

Terms to Study

atom

compound

electron

element

ion

matter

molecule

proton

valence electrons

OBJECTIVES

After completing this chapter, you will be able to:

- Differentiate between a compound and an element.
- Define the terms *atom*, *electron*, and *proton*.
- Sketch a hydrogen atom and label its parts.
- Specify the types of electric charges on electrons, protons, and neutrons.
- Explain the purpose of a particle accelerator.
- Identify industrial applications that use static charges.

Have you ever wondered what is inside an atom? Or wondered what energy is? Wood, metal, and glass are common materials. On the surface they look very different, but are there some aspects about these materials that are similar? What about lightning? You can see lightening, but what causes it? This chapter provides some answers to these questions.

MATTER AND ENERGY

Before discussing the basics of electricity and electronics, you need to know what matter and energy are. **Matter** is anything that occupies space and has mass, or weight. Wire, rubber, and glass are examples of matter. Scientists study the properties of matter in order to know how it works.

Matter is made up of tiny particles called **molecules**. Molecules are made up of even smaller particles called *atoms* (Fig. 3-1). In studying electricity and electronics, it is important for you to understand the atom because the *electron* is one of its parts. In this chapter, the structure of the atom and its electrical properties are discussed.

When all the atoms in a substance are alike, the substance is called an **element**. Copper, iron, and carbon are among the more than 100 different elements known to exist. Different elements can combine to form a substance called a **compound**. Water, salt, and plastic materials are examples of compounds.

Scientists are also concerned with the study of energy. Energy is the potential to do work. Energy can make changes in materials, such as putting them together, taking them apart, or simply moving them from one place to another. Energy has many different forms, such as heat, light, and electricity. Electric energy results from the motion of electrons.

Can you name examples of materials you have seen changed by various forms of energy such as heat or light?

ATOMS

The **atom** is the basic unit of matter. An atom is made of tiny particles called electrons, protons, and neutrons. **Electrons** are negatively charged particles that revolve around the nucleus of an atom much like planets around the sun. **Protons**, positively charged particles, and neutrons are held tightly together in the *nucleus*, or center, of the atom. These particles are important because they determine the charge of an atom. The charge of an atom creates the energy used as electricity.

Structure of the Atom

Electrons move around the center of an atom in paths called *shells* (Fig. 3-2). An atom can have several shells around its nucleus. Each of these shells can have only a certain number of electrons. This number is called the *quota* of a shell. When every shell of an atom contains its quota of electrons, the atom is said to be in a stable condition.

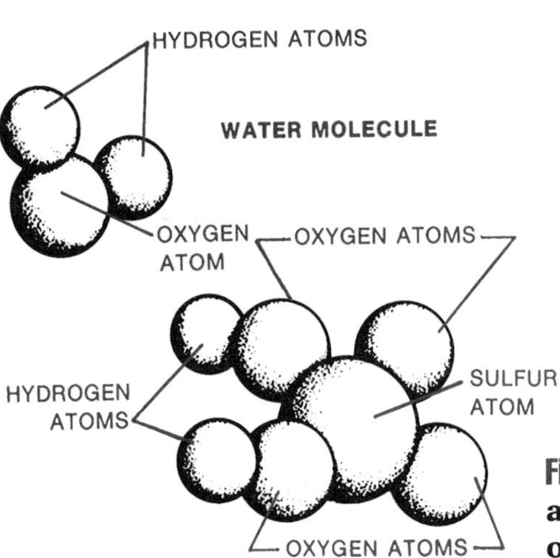

Fig. 3-1. Molecules of water and sulfuric acid. A molecule is made up of a number of atoms.

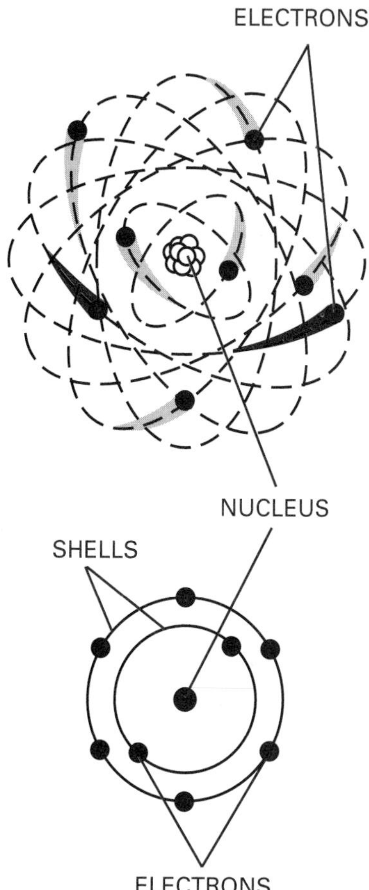

Fig. 3-2. Artist's view of the inside of an atom, containing eighty electrons that move around the nucleus in two shells.

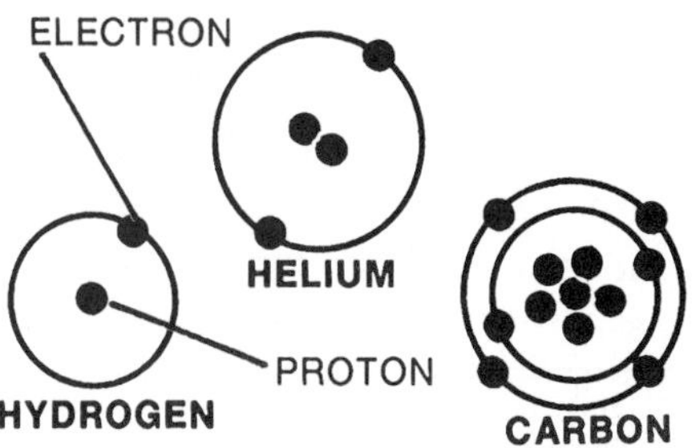

Fig. 3-3. Protons and electrons in hydrogen, helium, and carbon.

All electrons are alike, and all protons are alike. Thus, atoms differ from one another only in the number of electrons and protons they contain (Fig. 3-3). The number of protons in the nucleus is the atomic number of that atom. Neutrons weigh about the same as protons. The term atomic weight refers to the total number of particles (both protons and neutrons) in the nucleus of an atom.

While scientists have provided much information about the atom, the search for more information continues. Experiments conducted in *particle accelerators* provide new insights about the nucleus of an atom. A particle accelerator breaks atoms into smaller pieces. In a particle accelerator facility, scientists send streams of protons in opposite directions at nearly the speed of light. Whenever the protons collide, they shatter. This creates many smaller, more complex, subatomic particles. The term *quark* identifies one class of these subatomic particles, but discussion of subatomic particles is beyond the scope of this text. These particles, however, are being studied intensively by scientists in the field of *high-energy physics* (Fig. 3-4).

Valence Electrons

Valence electrons are those electrons in the outermost shell of an atom. In the study of electricity and electronics, we are concerned mostly with the behavior of valence electrons. They can, under certain conditions, leave their "parent" atoms. The number of valence electrons in atoms also determines important electrical and chemical characteristics of a substance.

FASCINATING FACTS

The theory now termed the shell model of the nucleus was discovered in 1948 by Dr. Maria Goeppert Mayer, a German-American mathematician and physicist. When she received the Nobel prize in 1963 for this theory, Dr. Mayer was the only living woman with a Nobel prize in science. She was also the first woman ever to win it for theoretical physics.

Fig. 3-4. Aerial view of a facility for research in high-energy physics.

Energy Levels and Free Electrons

The electrons in any shell of an atom are said to be located at certain energy levels. These levels are related to the distance of the electron's shell from the nucleus of the atom. When outside energy such as heat, light, or electricity is applied to certain materials, the electrons within the atoms of these materials gain some of that energy. This may cause the electrons to move to a higher energy level. Thus, they move farther from the nuclei (plural of nucleus) of their atoms (Fig. 3-5A).

When an electron has moved to the highest possible energy level (or the outermost shell of its atom), it is least attracted by the charges of the protons within the nucleus of the atom. If enough energy is then applied to the atom, some of the outermost shell, or valence electrons, will leave the atom. Such electrons are called free electrons (Fig. 3-5B).

ELECTRIC CHARGES

Electrons and protons have tiny amounts of energy known as electric charges. Electrons have negative (-) charges. Protons have positive (+) charges. Neutrons have no electrical

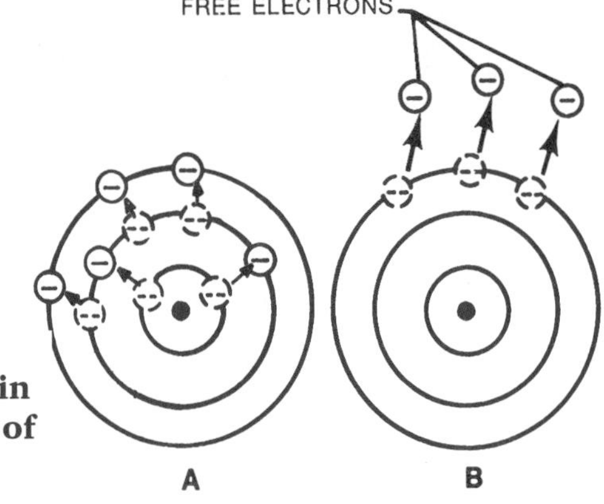

Fig. 3-5. Some electrons within an atom may move to higher energy levels within the atom or leave the atom as a result of the absorption of energy.

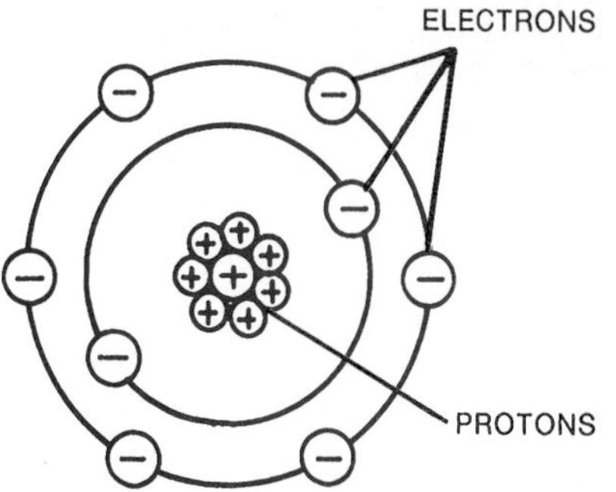

Fig. 3-6. Electronically neutral atoms have equal numbers of protons and electrons.

charge. Thus, they are neutral. The amount of the negative charge of each electron is equal to the amount of the positive charge of each proton. These opposite charges attract each other to hold the atom together.

Under normal conditions, the negative and the positive charges in an atom are equal. This is because the atom has an equal number of electrons and protons. An atom in this condition is said to be electrically neutral (Fig. 3-6).

Ions

If the negative and positive changes are not equal, the atom becomes an ion. In other words, an **ion** is a charged atom. If a neutral atom gains electrons, there are then more electrons than protons in the atom. Thus, the atom becomes a negatively charged ion (Fig. 3-7A). If a neutral atom loses electrons, protons then outnumber the remaining electrons. Thus, the atom becomes a positively charged ion (Fig. 3-7B). Ions with unlike charges attract one another. Ions with like charges repel one another. This process by which atoms either gain or lose electrons is called *ionization*.

FASCINATING FACTS

In the 1700s, the French physicist Charles-Augustin de Coulomb showed the relationship between the amount of electric force two charged objects can exert on each other and the distance between them.

The Coulomb

The size of the electrical charge an object has depends on the number of electrons within the object as compared to the number of protons. The unit of measurement for this comparison is the *coulomb*. A coulomb represents 6,240,000,000,000,000,000 electrons. Thus, if an object contains 6,240 quadrillion more electrons than protons, its charge is one negative coulomb. Also, if an object contains 6,240 quadrillion more protons than electrons, its charge is equal to one positive coulomb. This unit is named after Charles A. Coulomb, a French physicist (1736-1806).

Electric Charges in Action

Flashes of lightning in the sky are a common sight, especially during the stormy summer months. Lightning is a natural electrical discharge in the atmosphere. Following the flashes, you usually hear the roar of thunder. The flashes are the result of strong electric charges

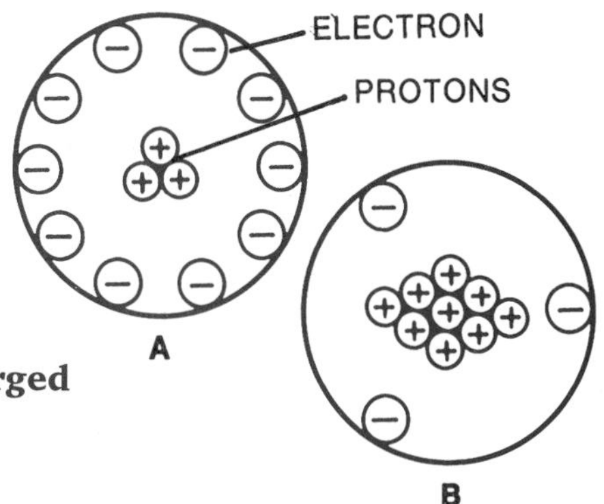

Fig. 3-7. Ions: (A) negatively charged ion; (B) positively charged ion.

built up in the clouds. Generally, negative charges concentrate in the lower part of a cloud and positive charges concentrate in the higher region. When the charges are high enough, they discharge between the clouds and the ground. The thunder you hear is the result of a shock wave generated by the heat of lightning.

A simple way to generate an electric charge is with friction. For example, if you rub a rubber balloon briskly with a wool cloth, electrons will move from the cloth to the balloon, causing the balloon to become negatively charged. If you then put the balloon against a wall, the balloon's negative charge will repel electrons from the surface of the wall (Fig. 3-8). This will, in turn, cause the surface of the wall to become positively charged. Then the attraction between the opposite charges of the balloon and a small surface area of the wall is strong enough to hold the balloon in place. Other substances such as glass, when rubbed briskly with a silk cloth, lose their electrons and therefore become positively charged.

STATIC ELECTRICITY

In the above example, the attraction between the charged balloon and the wall represents work done by *electrostatic energy*, which is often called static electricity. An *electrostatic field* is the energy that surrounds every charged object (Fig. 3-9). In this kind of electricity, there is no movement of electrons between the balloon and the wall. Thus, the electricity is said to be static, or at rest. Although static electricity is sometimes thought of as a nuisance, devices such as capacitors, air cleaners, and industrial processes such as the manufacture of abrasive paper use it.

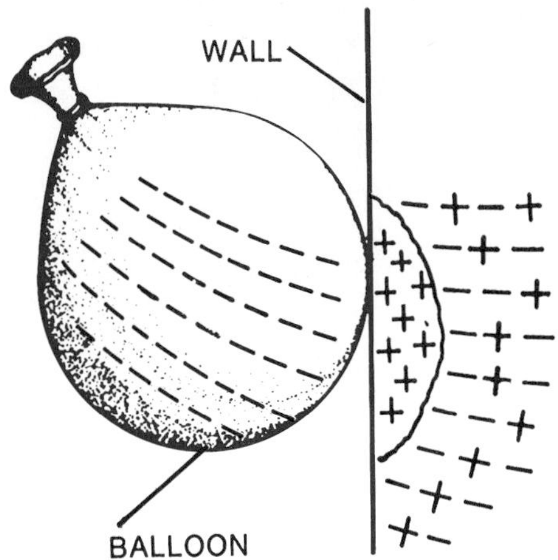

Fig. 3-8. **Opposite electric charges attract each other and thus cause a rubber balloon to "stick" to a wall.**

Lightning

Benjamin Franklin, the American statesman and printer from Philadelphia, Pennsylvania, sent a kite high into the air during a thunderstorm. He did this to determine if lightning performed like electricity. He found that it did and that clouds had positive or negative charges of electricity similar to the charged bodies on earth. Many individuals have since been injured trying to duplicate this experiment.

Franklin used the results of his experiments to make a lightning rod to protect buildings from lightning strikes. His ideas on lightning protection are still used today.

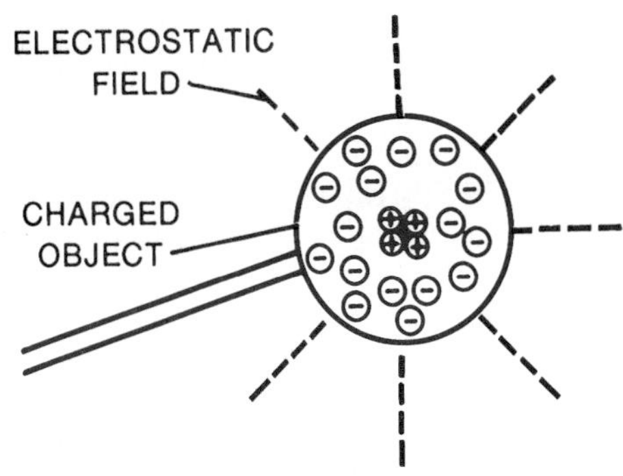

Fig. 3-9. **An invisible electrostatic field is present around every charged body.**

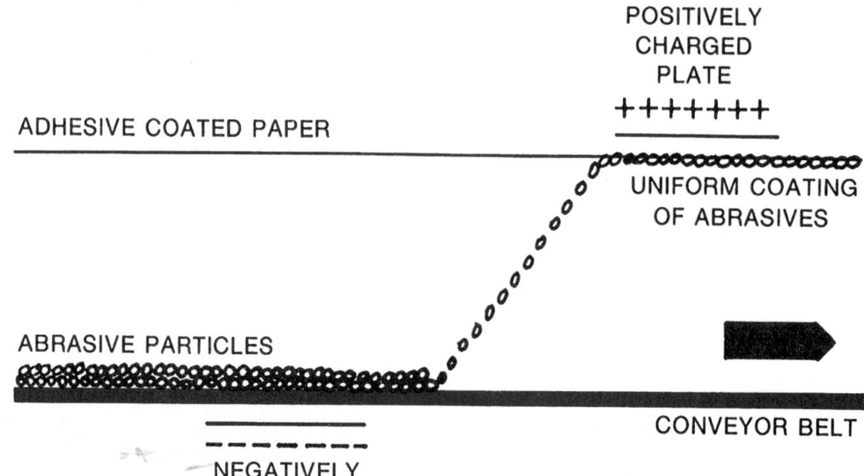

Fig. 3-10. Electrostatic charges used in making abrasive paper.

Manufacturing Processes

Figure 3-10 shows how static charges are used to make sandpaper, or abrasive paper. Such paper is used to smooth the surfaces of wood or metal. A conveyor belt carries the abrasive particles over a negatively charged plate. The abrasive particles are negatively charged by contact. At the same time, the paper that is tacky from an adhesive coating moves under a positively charged plate. This gives the paper a positive charge. Since opposite charges attract, the abrasive particles are attracted to the paper. These particles form a very uniform and dense abrasive surface on the paper.

Environmental Protection

Static electricity can also be used to protect the environment with an air cleaner (Fig. 3-11). Manufacturers use this device to reduce air pollution. It can be used in home heating systems to clean the air as it circulates through the furnace. *Electrostatic air filters* are much more efficient than simple cloth or paper fil-

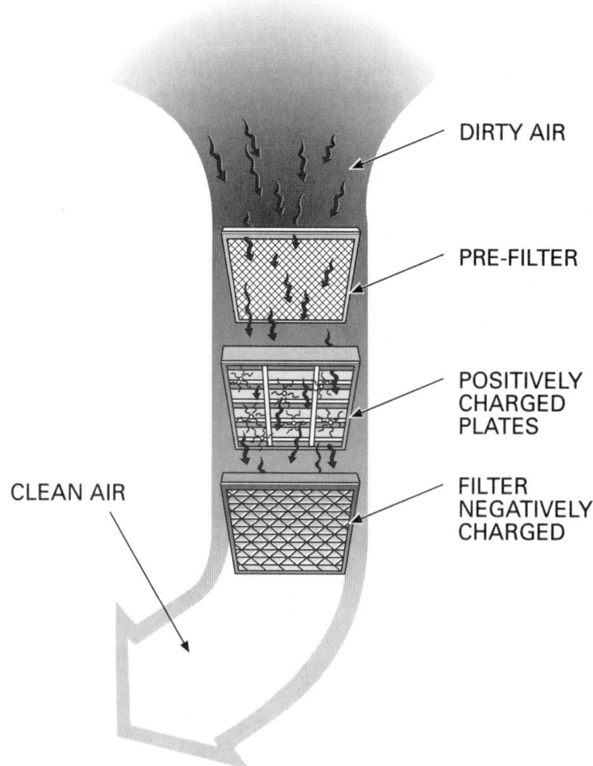

Fig. 3-11. An electrostatic air cleaner.

ters. Cloth and paper remove only large particles from the air. An electrostatic filter can remove tiny particles. As shown in Fig. 3-11, the dirty air passes through a paper prefilter that removes large pieces of dirt and lint. The air then moves through a series of plates that give a positive charge to the dirt and dust particles. Finally the air moves through a negatively charged filter. Microscopic particles become trapped there. When the filters get dirty they must be cleaned.

Another environmental use of static electricity is shown in Fig. 3-12. Spray painting is a common way of producing beautifully finished surfaces. However, it is wasteful. Much of the paint never sticks to the surface. It stays in the air as *spray dust*. Spray dust is not only wasteful but unhealthy. Most of the spray dust can be eliminated by positively charging the object to be painted and negatively charging the paint. Moreover, since the charge decreases as the paint is applied, a uniform thickness of paint is produced.

Although static electricity can be very useful, it cannot operate loads such as lamps, heaters, motors, and other devices. This takes dynamic, or active, electricity that is discussed in Chapter 4.

What types of static electricity have you experienced?

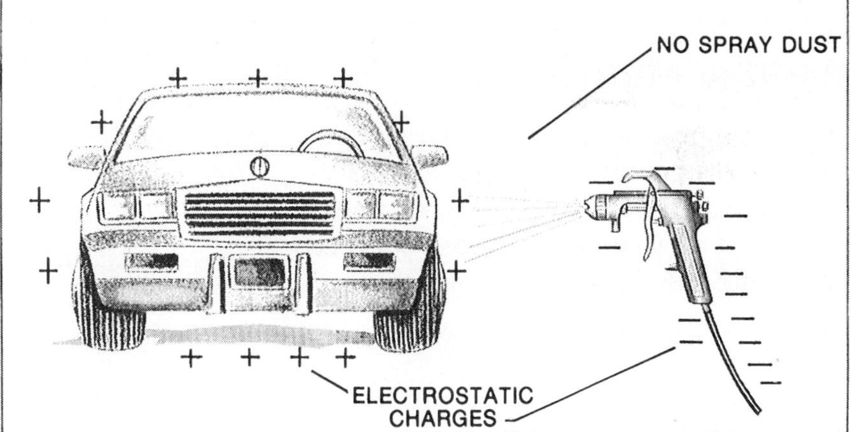

Fig. 3-12. **Electrostatic spray painting.**

Chapter Review

Summary

- Matter is anything that occupies space and has mass.
- The basic unit of matter is the atom, which consists of tiny particles called electrons, protons, and neutrons.
- Electrons and protons have tiny amounts of energy known as charges. Electrons have a negative charge, and protons have a positive charge. Neutrons are neutral.
- Energy, or the potential of matter to do work, results from the motion of electrons within the atom.

Review Main Ideas

Review this chapter's main ideas by writing, on a separate sheet of paper, the word or words that most correctly complete the following statements:

1. ______ is anything that occupies space and has mass.
2. The potential to do work is called ______.
3. Electrons move around the nucleus of an atom in paths that are usually called ______.
4. The nucleus of an atom is made up of particles called ______ and ______.
5. Atoms differ from one another only in the number of ______ and ______ they contain.
6. The number of protons in the nucleus of an atom is known as the atomic ______ of that atom.
7. When all the atoms within a substance are alike, the substance is called an ______.
8. Common examples of elements are ______, ______, and ______.
9. Different elements can combine to form a substance called a ______.
10. The electrons in the outermost shell of an atom are called ______ electrons.
11. Electrons have ______ charges. Protons have ______ charges.
12. An electrically neutral atom is one that has the same number of ______ and ______.
13. The energy ______ of an electron is related to the distance of the electron's shell from the nucleus of an atom.
14. If a neutral atom gains electrons, it becomes a ______ ion.
15. If a neutral atom loses electrons, it becomes a ______ ion.
16. Ions with unlike electric charges ______ each other. Ions with like electric charges ______ each other.

17. The process by which atoms either gain or lose electrons is called ______

18. A simple way of generating an electric charge is by ______.

19. In ______ electricity, there is no movement of electrons between two charged objects.

20. Every charged object is surrounded by an ______ field.

Apply Your Knowledge

1. Name the parts of an atom.
2. Describe how an element is related to a compound.
3. Define a molecule.
4. Explain how free electrons move from atom to atom.
5. State the laws of electrical attraction and repulsion.

Make Connections

1. **Communication Skills.** In 100 words, explain to a young adult how lightning and thunder are produced.
2. **Mathematics.** Write the following numbers using a shorter method of expressing the number: 1000; 1,000,000; and 6,240 quadrillion.
3. **Science.** While watching a storm cloud, a student saw lightning strike the ground. Four seconds later she heard the roar of thunder. How far away from the student was the lightning strike?
4. **Technology Skills.** Obtain two electrostatic air filters that have been used in home furnaces for at least three months. Show the differences between a filter that traps microscopic particles and one that doesn't. Discuss which one would provide the best cost benefit to the people in the home.
5. **Social Studies.** Name one industry that has been identified as a major polluter of air or water. Discuss with the class appropriate action that should be taken by government authorities.

Section 2

Electric Circuits and Devices

Chapter

Circuits, Voltage, and Current

OBJECTIVES

After completing this chapter, you will be able to:

- Sketch a basic circuit and identify its four parts.
- Explain differences between series, parallel, and series-parallel circuits.
- Define voltage and current.
- Differentiate between direct current (dc) and alternating current (ac).
- Explain what sine waves represent.

Terms to Study

alternating current
ampere
conductor
current
cycle
direct current
electric circuit
load
parallel circuit
series circuit
series-parallel circuit
voltage

You may know that a *circuit* is a course or route, but what's an *electric* circuit? Is it like the moon's circuit around the earth? Is it like a person going around performing his duties, such as a circuit court judge? Like the moon or the judge, electricity can travel in a circuit. You will learn about electric circuits in this chapter.

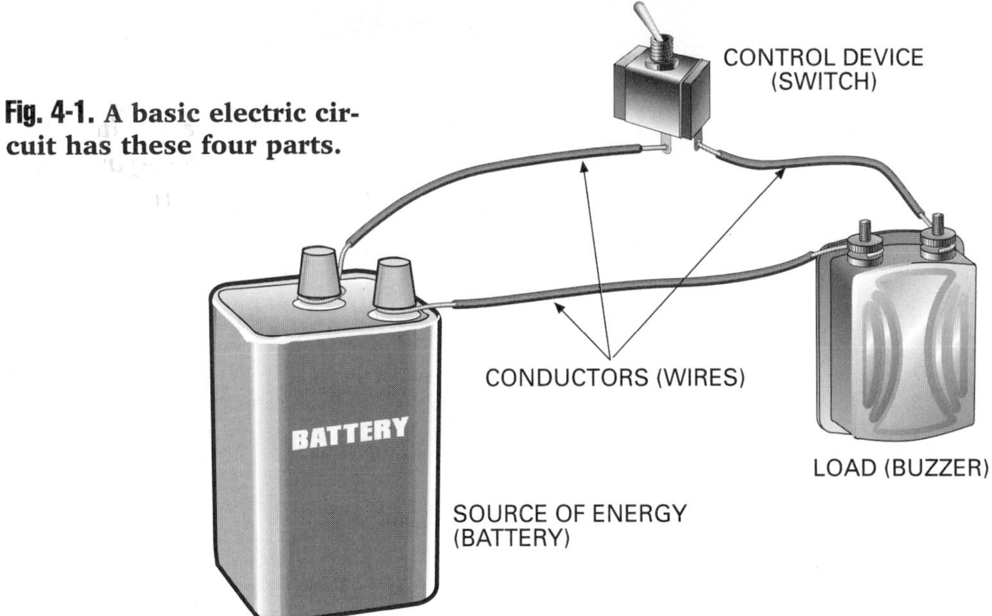

Fig. 4-1. **A basic electric circuit has these four parts.**

PARTS OF A CIRCUIT

An **electric circuit** is a combination of parts connected to form a complete path through which electrons can move. The purpose of a circuit is to make use of the energy of moving electrons. Therefore, a circuit is also a system of parts, or components, by which electric energy can be changed into other forms of energy, such as heat, light, or magnetism. A basic complete circuit has four parts: (1) the *energy source*, (2) the *conductors*, (3) the *load*, and (4) the *control device* (Fig. 4-1).

Energy Source

The energy source in a circuit produces the force that causes electrons to move. It is like a pump that forces water through a pipe. In electricity, this force is called **voltage**, or *electromotive force*. The most common energy sources used in electric circuits are chemical cells and electromechanical generators. These devices do the work needed to move electrons through the parts of the circuit.

Because electrons are negatively charged, they are attracted by positive charges and repelled by negative charges. If two charged objects are connected by a conducting material such as wire, electrons will flow from the negative object to the positive object. The flow of electrons is called **current**. To produce a continuous electric current in the wire, energy must be supplied continuously (Fig. 4-2). Electrons are not used up as they move through a circuit. Therefore, the number of electrons that return to the positive (+) terminal of an energy source equals the number of electrons that leave the negative (-) terminal of that energy source (Fig. 4-3).

LINKS

Chapter 14 explains more about chemical cells. Chapter 15 contains information about generators.

NOTE: This textbook uses the electron theory of matter that indicates current as the flow of electrons from the negative terminal to the positive terminal. There is an older theory called the *conventional theory.* The conventional theo-

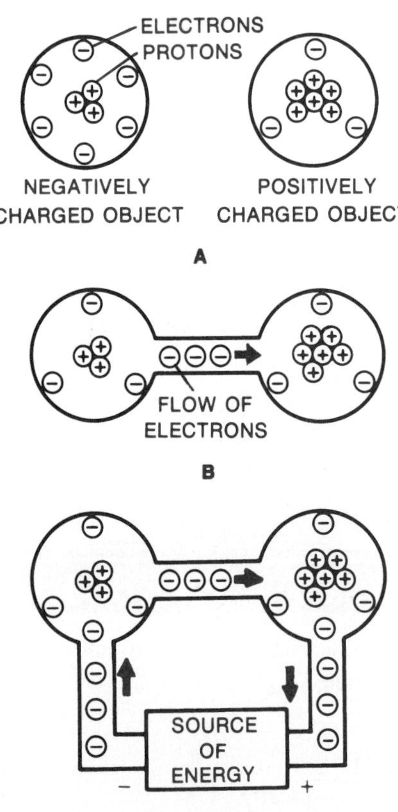

Fig. 4-2. **(A) Two charged objects, (B) Electrons are attracted to the positively charged object. (C) A source of energy provides a continuous supply of electrons.**

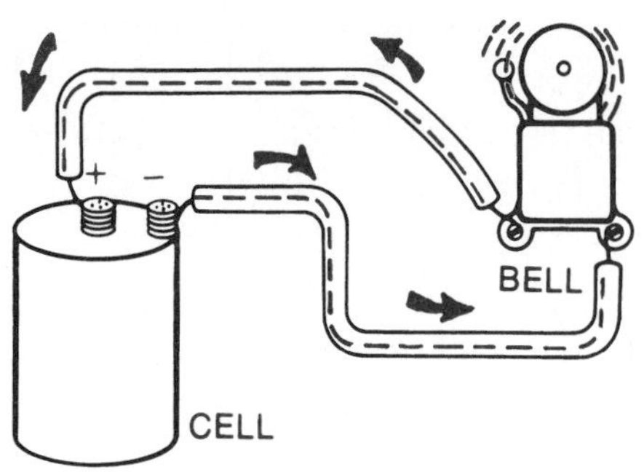

Fig. 4-3. **Electrons are not consumed as they move through a circuit.**

FASCINATING FACTS

Did you know some crystals can be a source of voltage? The voltage is produced by applying varying pressure to the crystal's surface. The process is called the piezoelectric effect. One of the first practical applications of this effect came during World War I, when piezoelectric crystals were used to produce sound waves in seawater to detect submarines. The technique was called sonar (for sound navigation and ranging).

ry indicates current as the flow of positive charges in the opposite direction of the electron flow. Both the conventional current flow theory and the electron flow theory are acceptable and are used for different applications. However, in this textbook, all arrows indicate the direction of current from negative to positive.

Conductors

The **conductors** in a circuit provide an easy path through which electrons can move through the circuit. Copper is the most commonly used conductor material. It is formed into wires, bars, or channels. Copper wire may be bare or covered with some kind of insulating material. The insulation provides a method to prevent the conductors (wires) from touching each other or some other conducting surface. Thus, the insulation prevents a short. (Fig. 4-4).

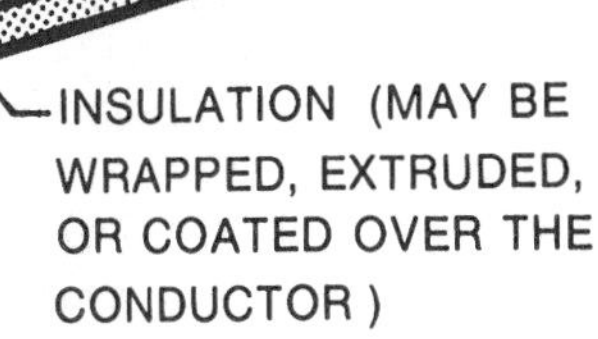

Fig. 4-4. **Insulated wire. Some common insulating materials are neoprene, rubber, nylon, polyethylene, Teflon™, vinyl, cotton, paper, and enamel.**

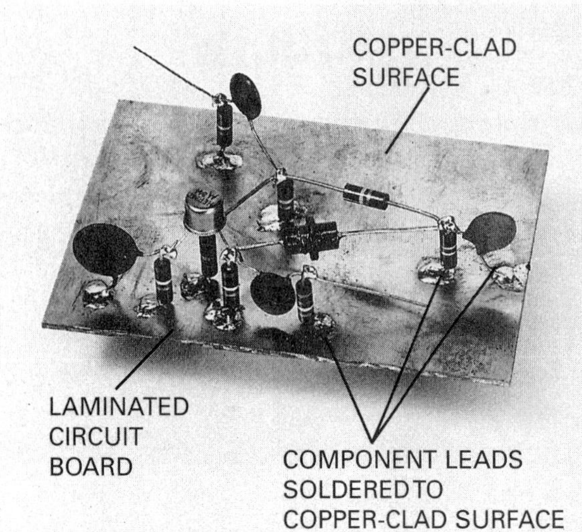

Fig. 4-5. The copper-clad surface of a circuit board provides a common conductor for this circuit.

In some circuits, metal objects other than copper conductors form the conducting paths. In an automobile, for example, the entire car frame serves as a conductor. It completes a number of circuits that connect the voltage source (the car battery) to various electrical and electronic devices. The metal chassis, or frame, that supports the components of electrical devices is also used as a common conductor for the electric circuit. Figure 4-5 shows an experimental circuit that uses a novel method to achieve a common conducting path. Many of the component leads of this circuit are soldered directly to the copper-clad surface of the laminated circuit board. The copper-clad surface forms a common conducting path for these components.

Load

The **load** is the part of a circuit that changes the energy of moving electrons into some other useful form of energy. A light bulb is a very common circuit load. As electrons move through the filament of the lamp, the energy of the electrons in motion is changed into heat energy and light energy.

Loads can be connected into a circuit in *series*, in *parallel*, or in *series-parallel* combinations. A **series circuit** provides only one path, or one loop, through which electrons can move from one terminal of the energy source to the other (Fig. 4-6A). In a **parallel circuit**, there may be two or more different paths, or loops, through which electrons can flow (Fig. 4-6B). A **series-parallel circuit** is shown in Fig. 4-6C. Note that series and parallel circuits are combined to form one circuit with several paths, or loops, that allow electrons to flow through more than one load.

SAFETY TIP

Always treat electrical wires in a circuit as being potentially dangerous.

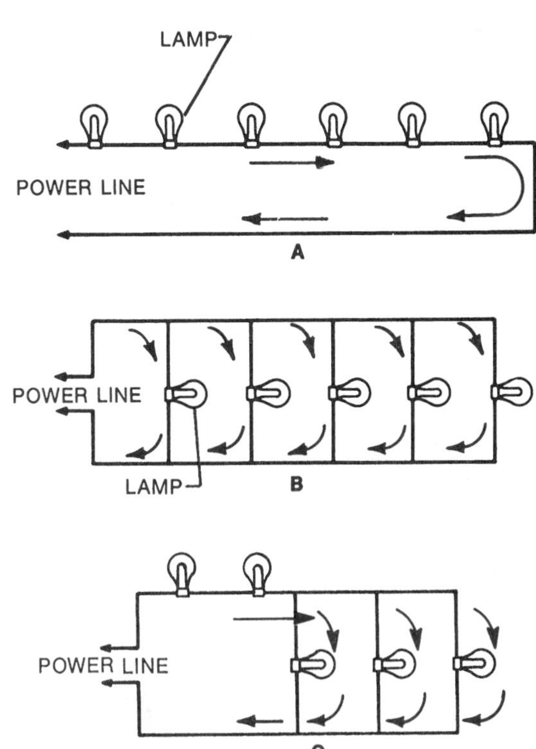

Fig. 4-6. Circuit connections: (A) series circuit; (B) parallel circuit; (C) series-parallel circuit. Arrows indicate a closed current loop.

Fig. 4-7. **An on-off switch is one example of a control device.**

Control Device

The mechanical wall switch is an example of a simple circuit control device. It opens and closes a circuit (Fig. 4-7). When the switch is "on," it acts as a conductor to keep electrons moving through the circuit. The circuit is said to be closed. In this condition, the circuit has a closed loop. When the switch is "off," the circuit path is interrupted. Electrons can no longer move through the circuit. The circuit is said to be open; that is, it has an open loop.

In addition to the familiar on-off switches, other devices can provide a switching action and control the flow of electrons in a circuit. Some of these are electromagnetic relays, diodes, transistors, and logic circuits. Switching devices have many uses. For example, furnaces, ovens, and air conditioning units use switches to automatically maintain temperature.

VOLTAGE

As stated earlier, the energy that forces electrons through a circuit is called the electromotive force. The electromotive force (emf) is measured in units called volts. Because of this, it is often referred to as voltage. When you hear or read the term *voltage*, you will know that it is the force that moves electrons through a circuit.

LINKS

You'll find information about these devices in Chapter 28.

BRAIN BOOSTERS

1. **Name an object you see or use every day that uses only the four parts described in this section for a basic, complete circuit.**
2. **The letter symbol for voltage is E, but its abbreviation is V. Why do you think this is so?**

Most of the electrical and electronic equipment found in homes operates at about 120 volts. Industrial applications need 120 volts, but they also use 208 volt, 480 volt, 4160 volt and even higher levels of electricity. A common flashlight dry cell produces 1.5 volts. A modern automobile battery produces 12 volts. The voltage used to operate a television picture tube may be as high as 30,000 volts.

CURRENT

As you learned earlier, the movement of electrons is called current. It is measured in units called **amperes**, or amps.

One ampere of current equals the movement of one coulomb (6,240 quadrillion) of electrons past any point in a circuit during one second of time. A 100-watt light bulb requires about 0.8 amps of current to operate. A one horse-power electric motor needs about 6 amps of current to operate. The cranking motor of an automobile may use over 200 amps when the starter switch is turned on.

Voltage and Current Requirements

If an electrical device is to work properly, the source of energy must be able to do two things. First, it must supply the voltage. Second, it must deliver the current for which the device was designed. For example, you can connect eight flashlight dry cells together in such a way as to form a battery that produces 12 volts.

PLUG IN TO *History*

Volta

The unit of electromotive force, the volt, was named in honor of Count Alessandro Volta, an Italian physicist who lived from 1745 to 1827. Volta was one of the first scientists to produce electricity by chemical means. He took pairs of unlike metal discs (one copper, one zinc) and stacked them into piles. Between each pair of metal discs was a cardboard disc moistened with a salt solution. When Volta connected the ends of the piles with wire, current flowed. This device, which came to be known as the voltaic pile, was an early battery.

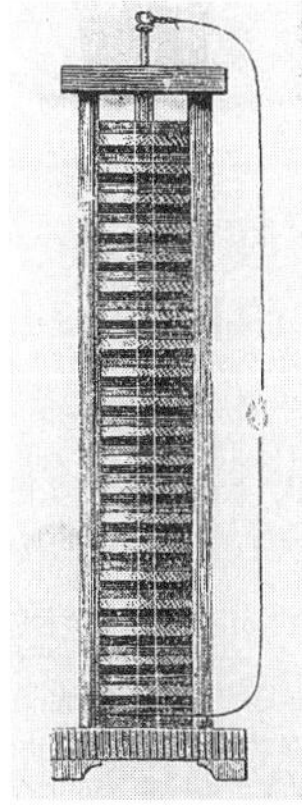

Volta's voltaic pile is shown at the right.

However, if you tried to start an automobile engine with this battery, it would not work. This is because the battery does not have the ability to deliver the large amount of current needed to operate the automobile's electric cranking motor. To do that, a larger battery, also producing 12 volts but having a much larger current-delivering capacity, must be used. It is important, therefore, to know both the voltage and current requirements of electrical equipment and tools. These requirements are often given on the nameplates attached to these products. They can also be found in the product's technical literature.

DIRECT CURRENT (DC)

Direct current (dc) is produced in a circuit by a steady voltage source. That is, the positive and negative terminals, or poles, of the voltage source do not change their charges over time. These terminals are said to have fixed polarity. Therefore, the direction of the current does not change over time. Such a voltage is provided by electric cells, batteries, and dc generators. Direct current may be constant, or steady, in value (Fig. 4-8A). The current also may be varying or pulsating (Fig. 4-8B and C). The applied voltage and the nature of the load determine the kind of direct current supplied.

ALTERNATING CURRENT (AC)

Alternating current (ac) is produced by a voltage source that changes polarity, or alternates, with time. This causes the current in the circuit to move first in one direction and then in the other (Fig. 4-9). The most common source of alternating voltage is the alternating-current generator, or alternator.

1. Find the total voltage required to operate a device by multiplying the number of batteries, connected in series, times the voltage rating of the batteries.

2. What items in your home might use direct current?

Fig. 4-8. Direct current: (A) steady direct current; (B) varying direct current; (C) pulsating direct current.

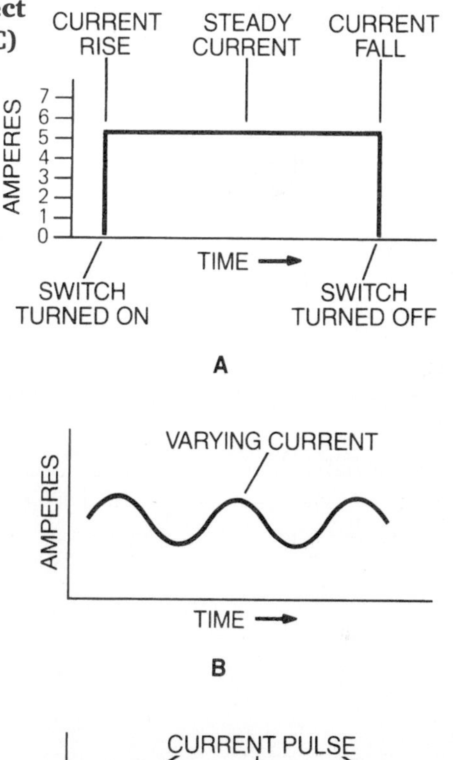

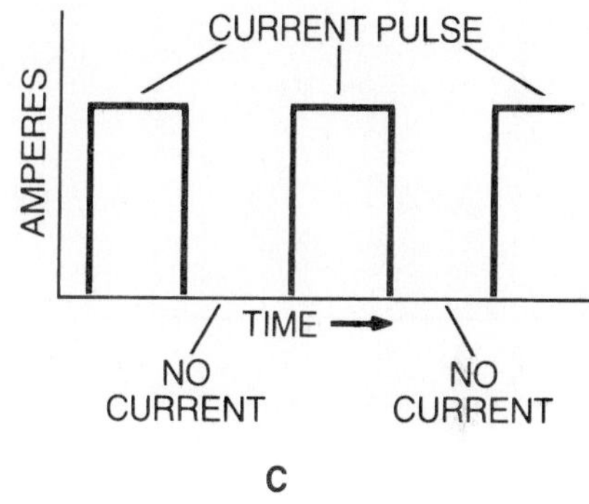

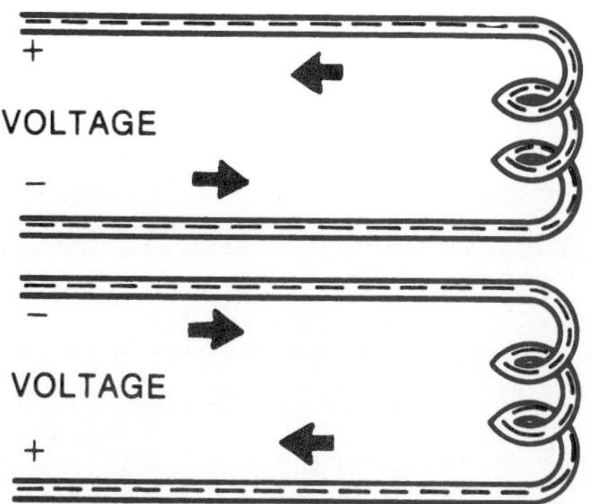

Fig. 4-9. The polarity of the voltage in an AC circuit alternates, or changes, at regular intervals.

Sinusoidal Alternating Current

In addition to changing direction, most kinds of alternating current changes in value with time. For example, the variation of current with time may follow the form of a sine wave. This is called *sinusoidal alternating current*. Most electric utilities in this country supply sinusoidal alternating current and voltage to their customers.

Sine Wave

A graph called a *sine wave* shows the direction and the value of the current that passes through a given point in a circuit during a certain period of time (Fig. 4-10). One complete wave is a **cycle** of alternating current. The time it takes to complete one cycle is the period of the wave.

In Fig. 4-10, points on the vertical line AB represent current values. Base line CD is the time line. The heavy curved line, the sine wave, shows how the current changes in value during

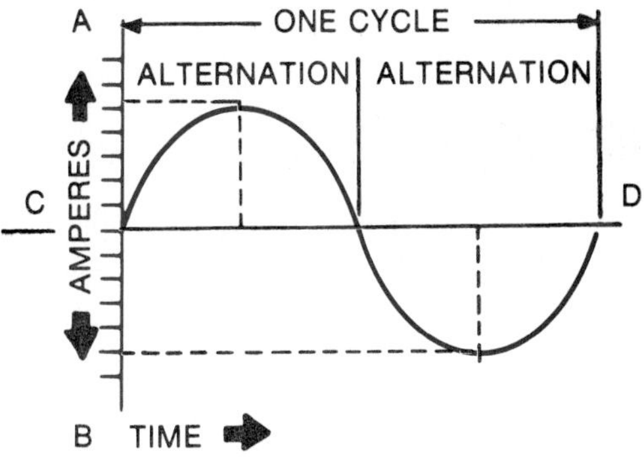

Fig. 4-10. A sine wave describes the variation in the value of current during a period of time.

Unit Prefixes

Very small and very large values of voltage and current are usually expressed in a shorthand way, using a system of decimal prefixes. The most common of these prefixes and the numerical values they represent are given in Table 4-A. As shown in this table, the prefix milli- (m) means one-thousandth (0.001). A current of 0.001 ampere can thus be expressed as a current of 1 milliampere, or more simply as 1 mA. Likewise, a voltage of 1000 volts can be expressed as 1 kilovolt, or 1 kV.

Table 4-A Common Numerical Prefixes Used with Metric and Other Units

Prefix and Symbol	Numerical Equivalent	Powers of 10
Giga (G)	1 000 000 000	= 10^9 (thousand-millions)
Mega (M)	1 000 000	= 10^6 (millions)
Kilo (k)	1 000	= 10^3 (thousands)
Milli (m)	0.001	= 10^{-3}(thousandths)
Micro (μ)	0.000 001	= 10^{-6} (millionths)
Nano (n)	0.000 000 001	= 10^{-9} (billionths or thousand millionths)
Pico (p)	0.000 000 000 001	= 10^{-12} (trillionths or million-millionths)

one cycle. The part of the sine wave above the base line represents the movement of current in one direction. The part of the sine wave below the base line represents the movement of current in the other direction.

During one alternation, or one-half, of the sine-wave cycle, the current moving in one direction increases from zero to a maximum value and then returns to zero. At that time, the current begins to increase again but in the opposite direction. It again increases to a maximum value and then decreases to zero. This completes one cycle.

LINKS

Look ahead to Chapter 15 and see details on how one cycle of current is developed in an ac generator.

Nonsinusoidal Alternating Current

Alternating voltages and currents may have waveforms that are not like sine waves. These are *nonsinusoidal wave forms* (Fig. 4-11). Television sets, radios, and other devices may have circuits in which these voltage waveforms are present.

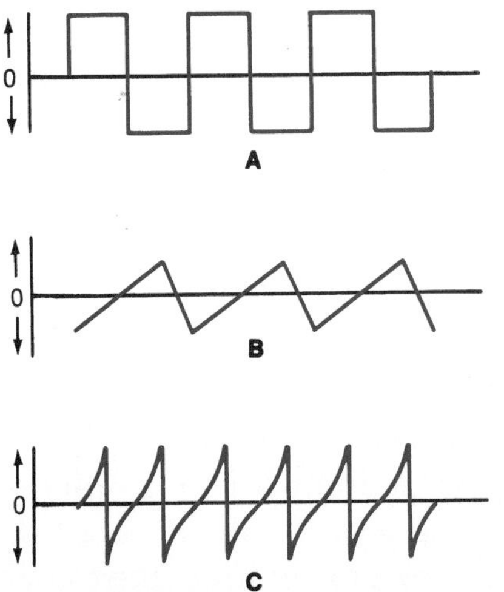

Fig. 4-11. **Nonsinusoidal AC wave-forms: (A) square waveform; (B) sawtooth waveform; (C) peaked waveform.**

Power companies can transmit alternating current more easily and efficiently than direct current. If your power company switched to direct current, how might your utility bill be affected?

USING DIRECT OR ALTERNATING CURRENT

Electric energy is most often supplied in the form of alternating current. Therefore, most electrical and electronic products are designed to operate with it. However, many of these products will work with either alternating or direct current. This is true, for example, of common lamps and almost all equipment with heating elements, such as ovens and irons. The right voltage however, must be applied to these devices.

Direct current is needed to operate electronic devices such as transistors and integrated circuits. Radios that are not battery-operated must be provided with a source of direct current. A special circuit in the radio changes the power-line alternating current into direct current. This special circuit is called a *rectifier circuit*. Direct current must also be used for certain electrochemical processes such as charging batteries and electroplating metals.

Direct-current power supplies are designed to provide direct—voltage outputs that can usually be varied from zero to a certain maximum voltage. These power supplies are used wherever a controlled variable voltage is needed, such as in experiments and circuit testing.

Alternating current must be used to operate the *induction motors* used in manufacturing, such as those on conveyors and hoists. Other motors, called *universal motors*, can be operated with either alternating or direct current. *Transformers* must be operated with alternating current or with varying or pulsating direct current. *Inverters* are used to change direct current to alternating current.

Electric Power and Energy

Power is the rate of doing work. In an electric circuit, power may also be defined in two other ways. First, it is the rate at which electric energy is delivered to a circuit. Second, it is the rate at which an electric circuit does the work of converting the energy of moving electrons into some other form of energy. The basic unit of power is the watt (W). Since electric power is the rate at which energy is delivered to the circuit, it is easy to calculate. You simply multiply the power by some time unit. This will give you the total energy delivered to the circuit. The time unit usually used is the hour. The power unit is usually given in terms of kilowatts (1,000 watts). Therefore, the energy unit on which most electric companies base their bills is the kilowatt-hour (kWh).

Chapter *Review*

Summary

- A basic circuit has four parts: the source, the conductors, the load, and the control device.
- The flow of electrons through a circuit is called current. The force driving that current is called electromotive force, or voltage. The voltage and current of a circuit are supplied by its energy source.
- Conductors provide a path through which electrons can move through the circuit from the source to the load.
- The load is that part of a circuit that changes the energy of the moving electrons into some other useful form of energy.
- Control devices open and close the path for electrons in a circuit.
- A circuit's parts can be connected into series, parallel, or series-parallel circuits.
- Direct and alternating currents are the two main types of electricity. Direct current has a fixed direction, whereas alternating current changes direction with time.

Review Main Ideas

Review this chapter's main ideas by writing, on a separate sheet of paper, the word or words that most correctly complete the following statements:

1. An electric circuit makes it possible to use the energy of moving _______ .
2. In a circuit, electric energy is changed into other forms of energy, such as _______ , _______ , or _______ .
3. The four basic parts of a complete circuit are the _______ , the _______ , the _______ , and the _______ .
4. Electromotive force causes _______ to move through a circuit.
5. Common sources of energy used in electric circuits are _______ and _______ .
6. The flow of electrons in a circuit is called _______ .
7. Electrons flow through a circuit from the _______ terminal of the energy source to the _______ terminal of the energy source.
8. Conductors provide a _______ through which electrons can move in a circuit.
9. The most common conductor material is _______ .
10. That part of a circuit that changes electric energy into another form of energy is the _______ .
11. The series circuit provides _______ path(s) through which electrons can move.

12. In a _______ circuit, there may be two or more electron paths.
13. In a _______ circuit, _______ and _______ circuits are combined.
14. A switch is an example of a circuit _______ device.
15. Electromotive force is measured in units called _______.
16. Current is measured in units called _______ .
17. For an electrical device to operate, an energy source must supply the _______ and the _______ that the device requires.
18. Direct current moves through a circuit in one _______ only.
19. The action of a sinusoidal alternating current is best described by what is called a _______ wave.
20. Examples of devices or products that can be operated with either direct or alternating current are _______ and _______ .
21. Direct current must be used for the operation of electronic devices such as _______ and _______ .
22. A _______ circuit is one that changes alternating current into direct current.
23. Electric companies base their bills on a unit called the _______.

Apply Your Knowledge

1. If you have a circuit with a battery, a light bulb, and a switch, do you have a complete circuit? Why or why not?
2. According to charges, in what direction does current flow in a circuit? See Fig. 4-2. Explain why. (See Chapter 3.)
3. What is the purpose of insulation that surrounds bare conductors (wires)?
4. What role does the load play in a circuit? The control device?
5. An induction motor on a conveyor will not run. The input devices are all in the correct position to allow current flow, but the source has recently been replaced. What could be wrong?

Make Connections

1. **Communication Skills.** Prepare a written report describing series, parallel, and series-parallel circuits. Include sketches and list examples.
2. **Mathematics.** How many electrons will flow past a given point in one second if there are three amperes of current in that circuit?
3. **Science.** Look in a science book to find out how electric energy is changed into light. What happens to the electrons?
4. **Workplace Skills.** Examine the owner's manual for an appliance, such as a coffee maker. If you were writing such a manual, what information would you include about voltage and current? Why?
5. **Social Studies.** The volt and ampere were named for scientists. What do the following units measure and for whom were they named: ohm, henry, watt, tesla? Report your findings to the class.

Chapter 5

Conductance and Resistance

OBJECTIVES

After completing this chapter, you will be able to:

- Define the terms conductor and insulator.
- Sketch an illustration of electrons flowing through a wire.
- Describe the following types of resistors: carbon-composition, film, wire-wound, and precision.
- Determine the resistance value of carbon-composition resistors by color code.

Terms to Study

conductor

dielectrics

insulator

ohm

potentiometers

resistance

resistor

rheostat

superconductivity

tolerance

How do electrons travel or move from one place to another? They aren't visible. So, what is the force that causes this to happen? This chapter discusses how electrons are restricted, or slowed down, in materials called insulators.

Because moving electrons have energy, they can do work. What electrons do and how they can be controlled in a circuit depends mostly on the kinds of materials they flow through. There are also certain materials through which electrons cannot flow easily. Knowing about these materials will help you to better understand electricity.

Name three materials that are poor conductors. What do you think would happen if you applied a large amount of voltage to a poor conductor?

CONDUCTORS

A **conductor** is a material through which electrons flow easily. In such materials, voltage easily removes valence electrons from their parent atoms. To put it another way, a conductor is a material having free electrons.

Three good electrical conductors are silver, copper, and aluminum. In fact, metals are generally good conductors. Certain gases also can be used as conductors under special conditions. For example, neon gas, argon gas, mercury vapor, and sodium vapor are all used in lamps.

FASCINATING FACTS

Henry Cavendish (1731-1810), an English physicist and chemist, showed that iron wire conducted electricity 400 million times better than pure water. His work led to the current use of copper wire, which is one of the best conductors.

Electron Flow

The basic way electrons flow through a wire is shown in Fig. 5-1. The action begins when the positive terminal of the cell attracts a valence electron from atom 3. Atom 3 is now positively charged, having lost part of its negative charge. Thus, it attracts an electron from atom 2. Atom 2, in turn, attracts an electron from atom 1. As atom 1 loses its electron, an electron leaves the negative terminal of the cell and is attracted to atom 1.

In a circuit, this action continues among very large numbers of atoms within the conductor. This produces a steady flow of free electrons through the conductor. The direction of this electron flow is always from the negative to the positive terminal of the energy source.

Electrical Impulse

Individual electrons flow through a conductor relatively slowly, usually at less than 1 inch (25.4 mm) per second. After voltage is applied to a circuit, however, it produces a flow of free electrons at the speed of light through all points in that circuit.

If it is hard for you to imagine an electrical impulse that travels at the speed of light, maybe this example will help. Imagine that you have built a circuit 186,000 miles (300,000 km) long. This is more than seven times the distance around Earth at the equator. One second after you close the switch on the circuit, the electrical impulse takes place throughout it.

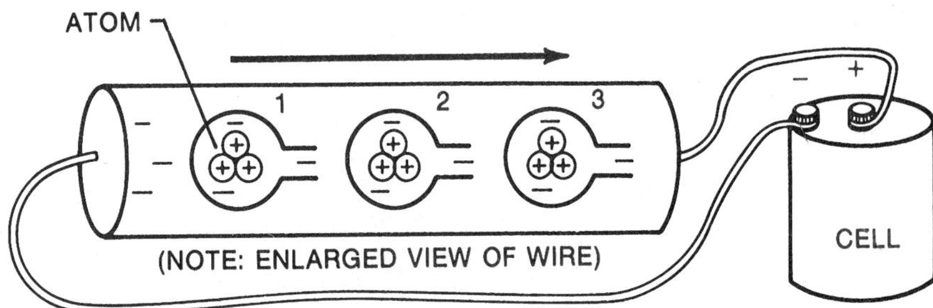

Fig. 5-1. In a conductor material, voltage causes some electrons to be removed from their atoms. These "free" electrons then move through the conductor.

All materials, even insulators, will conduct electricity when the applied voltage is high enough. Name a material not listed in this section that conducts electricity well.

RESISTANCE

Resistance is an opposing force between one thing and another. In an electric circuit the property of a conductor (wire) to oppose the passage of current is called resistance.

As electrons flow through a conductor, they collide with other electrons and with atomic particles. These collisions reduce the number of electrons that flow through the conductor. This is similar to the difficulty that water has in passing through a hose or pipe that is partly filled with dirt or sand.

The basic unit of resistance is the **ohm**. It is named after Georg Simon Ohm. He was a German physicist who lived from 1787 to 1854. The letter symbol for resistance is *R*. One ohm of resistance is that amount of resistance that will limit the current in a circuit to 1 A when 1 V is applied to the circuit (Fig. 5-2). Larger values of resistance are often expressed by using the prefixes *kilo-* and *mega-*. For example, a resistance of 2,000 ohms can be written as 2 kilohms, and a resistance of 2,000,000 can be written as 2 megohms. The Greek capital letter omega (Ω) is used as an abbreviation for ohms behind a numerical value, for example, 25 Ω. A resistance of 2 kilohms can also be written as 2 kΩ A resistance of 2 megohms can be written as 2 MΩ.

Resistance of Metal Conductors

The resistance of a metal conductor depends on four things: (1) the kind of metal from which it is made, (2) its temperature, (3) its length, and (4) its cross-sectional area.

Different metals have different properties of electrical resistance. The resistances of several common metals compared to copper are given in Table 5-A.

Superconductors and Cryogenics

The study of the effects of extremely low temperatures on gases and metals is termed cryogenics. An early contributor to this field was Dutch physicist, Heike Kamerlingh Onnes (1853-1926). In 1894 he founded the Cryogenic Laboratory, which is now named after him.

Onnes studied the effects of extreme cold on a number of gases and metals. He was the first to discover superconductivity. He found that certain metals' electrical resistance practically disappeared at temperatures close to absolute zero, therefore, their electrical conductivity increased. In 1911, he observed that mercury became superconductive when cooled to about 4 degrees above absolute zero.

In 1986 and 1987 it was discovered that certain rare, Earth-based oxide ceramics were superconductive at temperatures near room temperature.

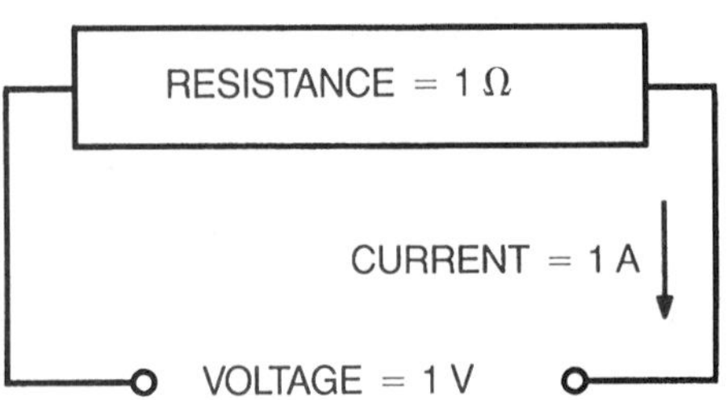

Fig. 5-2. One ohm is the amount of resistance that allows one ampere of current to pass through a circuit when one volt is applied.

In general, as a metal gets hotter, its resistance increases. A hot wire has more resistance than the same wire when it is cold. For example, the resistance of the tungsten wire filament in a 100-W light bulb is about 10 Ω when the bulb is off. When the bulb is lit, the resistance of the white-hot filament increases to about 100 Ω. As in any conductor, heat causes atoms within the wire to move about much more rapidly than usual. This increases the number of collisions between free electrons and other atomic particles within the wire. As a result, the resistance to the flow of electrons through the wire also increases.

Just as resistance increases when temperature increases, a decrease in temperature decreases resistance. If the temperature is reduced to absolute zero (-273.16^{o} C, or -459.69^{o} F), the resistance of the conductor will be zero. This is because all molecular activity in the material stops at absolute zero. There is then nothing to resist the flow of electrons. This condition is called **superconductivity**. Temperatures very close to absolute zero, that have been produced in the laboratory, make it possible to operate circuits having nearly zero resistance.

The resistance of a wire increases as its length increases. If two wires of different lengths have the same cross-sectional area, the longer wire will have higher resistance. There is a good reason for this. The electrons must move a greater distance in the longer wire, thus they will collide with more particles in the longer wire.

If two wires are the same length, the wire with the larger cross-sectional area will have less resistance. This is because the thicker wire has more free electrons. It also has more space through which these electrons can move.

Table 5-A Relative Resistance

Metals	**Relative Resistance Compared with Annealed Copper***
aluminum (pure)	1.70
brass	3.57
copper (hard-drawn)	1.12
copper (annealed)	1.00
iron (pure)	5.65
silver	0.94
tin	7.70
nickel	6.25-8.33

*For example, a silver wire has only 0.94 times, or 94%, as much resistance as the same-size copper wire, where as an aluminum wire has 1.70 times, or 170%, the resistance of copper.

Resistance of Insulators

An electrical **insulator** is a material that does not easily conduct an electric current. Such materials contain valence electrons that are tightly bound to the nuclei of their atoms. As a result, it takes an unusually high voltage to produce significant numbers of free electrons in them. Such materials are also called *nonconductors* and **dielectrics**.

Common insulators are glass, porcelain, mica, rubber, plastics, paper, and wood. These materials are used to separate conductors electrically so the current will flow in the correct paths (Fig. 5-3).

Semiconductor is a term given to materials that are often classified between conductors and insulators. Semiconductors are neither good conductors nor good insulators.

LINKS

You can learn about these important semiconductor materials in Chapter 17.

There is no sharp line dividing conductors and insulators. All insulating materials will conduct electric current if a high enough voltage is applied to them. For example, air is usually thought of as being a fairly good insulator. However, during a thunderstorm, the huge voltage generated between the clouds and the earth cause air to conduct current in the form of lightning.

The ability of a material to insulate is known as its *dielectric strength*, and is actually a form of very high resistance. The dielectric strengths of several common insulators are given in Table 5-B.

Name three types of resistance other than electrical.

Fig. 5-3. **Common insulating materials and their uses.**

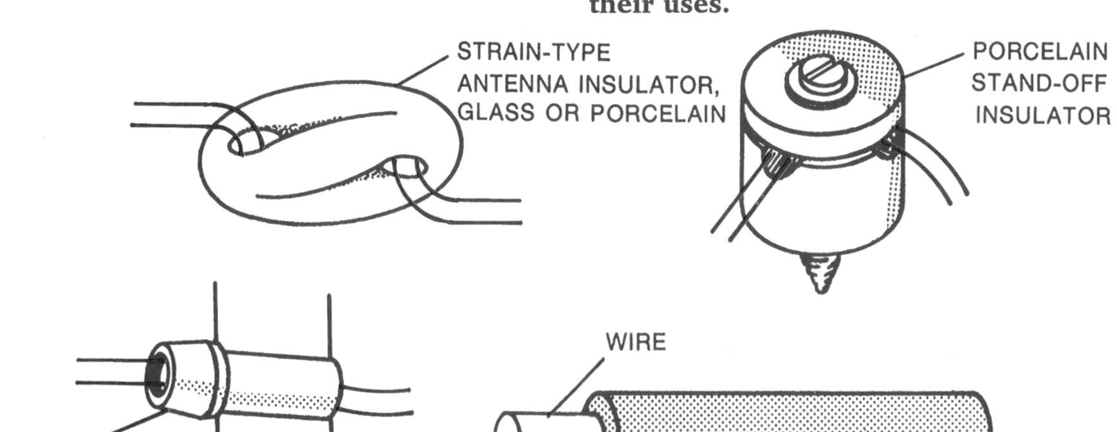

Table 5-B Dielectric Strengths of Common Insulating Materials

Material	Dielectric Strength Breakdown Voltage in Volts per 0.0001 in. (0.0254mm)
Bakelite	300
Formica	450
glass (window)	200-250
mica	3,500-5,000
polystyrene	500-700
porcelain	50-100
Teflon	1,000-2,000
air	75
kraft paper (glazed)	150
wood	125-750

RESISTORS

Resistances are deliberately designed to meet the needs of the circuit. When this is done, the resistance is the circuit load. In an electric range, for example, the heating elements are made of a special wire called *resistance wire*. As current passes through this wire, the energy of moving electrons is changed into useful heat. Resistance is also put into many different circuits and is provided by components called resistors.

However, when too much current passes through a conductor, the conductor may become hot. This can cause a fire or the conductor to burn in two.

Types of Resistors

A **resistor** is a device with a known value of resistance. Resistors are very common parts of many electric and electronic circuits (Fig. 5-4). They are

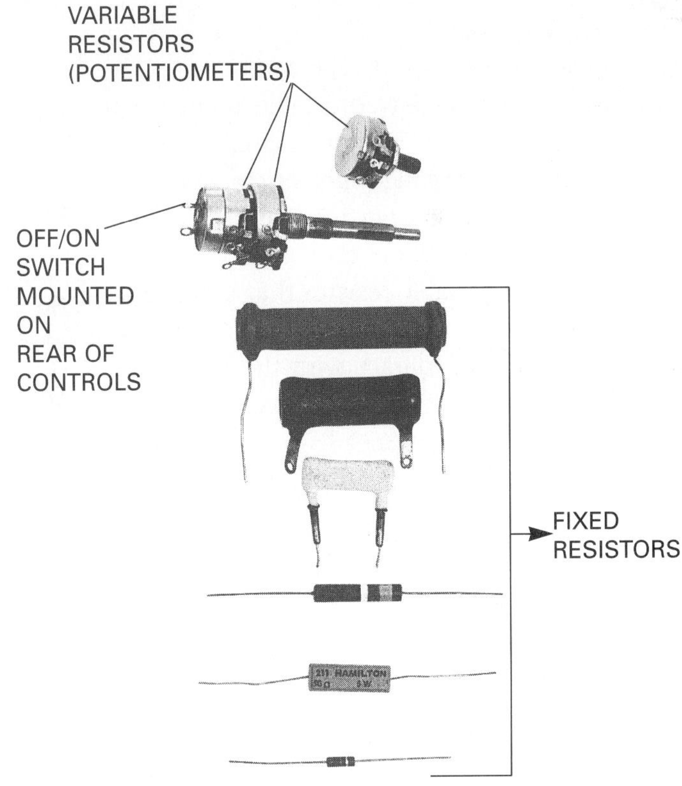

Fig. 5-4. Resistors are made in many sizes and shapes.

FASCINATING FACTS

Thin film coatings that conduct electricity are applied to glass rods or tubes to make resistors for electronic circuits. These same coatings are often applied to aircraft windshields to keep frost from forming.

used to control voltage and current and operate as either *fixed* or *variable* values. They come in many sizes and shapes, and are made from several different materials.

Fixed Resistors

A *fixed resistor* has a single value of resistance which remains the same under normal conditions. Three common kinds of fixed resistors are carbon-composition resistors, wire-wound resistors, and film resistors.

Variable Resistors

Variable resistors can also be carbon, film, or wire, but they have an adjustable amount of resistance in a circuit. The most common variable resistors are called *potentiometers* and *rheostats* (Fig. 5-5). **Potentiometers** generally have carbon-composition resistance elements. A **rheostat** is a variable resistor that is generally made of resistance wire. In both devices, a sliding arm makes contact with the resistance element. In most variable resistors, the arm is attached to a shaft that can be turned in almost a full circle. Precision potentiometers can turn ten times. As the shaft is turned, the point of contact of the sliding arm on the resistance element changes. This changes the resistance between the sliding-arm terminal and the terminals of the element (Fig. 5-6).

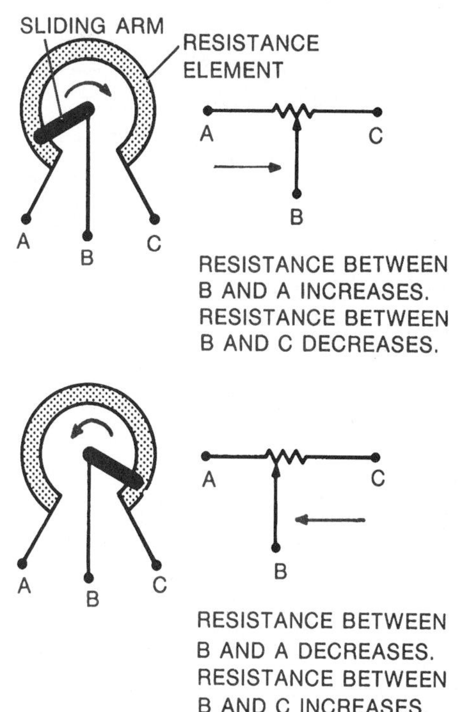

Fig. 5-6. When the sliding arm of a variable resistor is moved, the resistance between the center terminal and the end terminals changes.

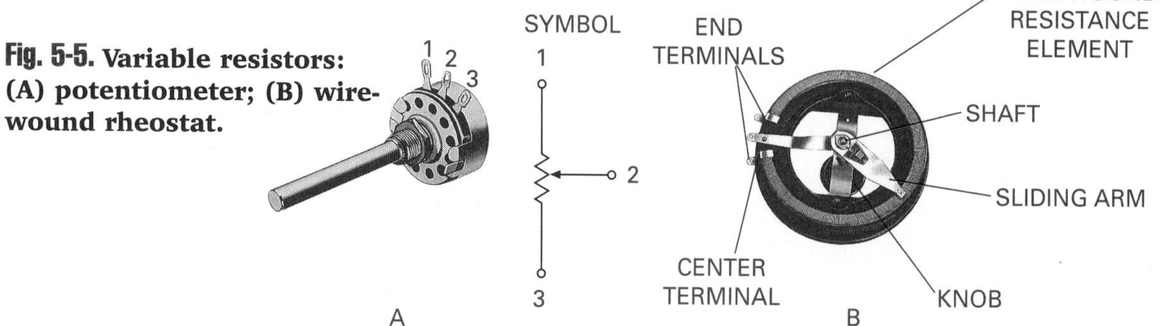

Fig. 5-5. Variable resistors: (A) potentiometer; (B) wire-wound rheostat.

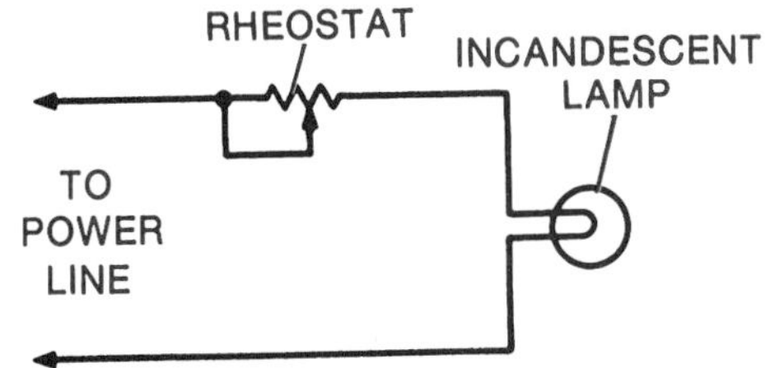

Fig. 5-7. Rheostat used to control current in a lamp circuit.

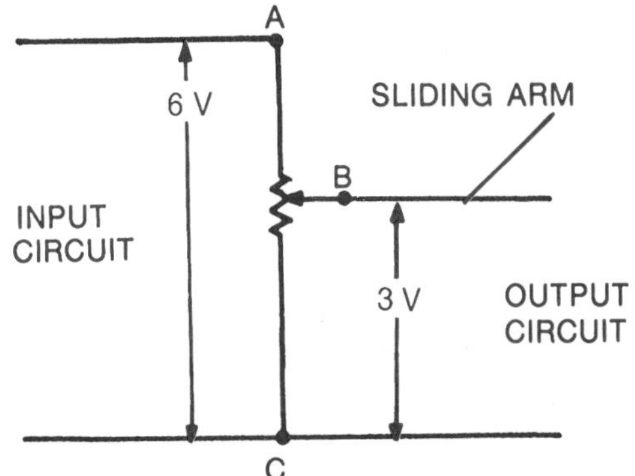

Fig. 5-8. Basic potentiometer action.

Rheostats

Rheostats are commonly used to control higher currents, such as those in motor and lamp circuits. An example of how a rheostat is connected into a lamp circuit is shown in Fig. 5-7. Although similar to a potentiometer, a rheostat is usually larger because its resistance element carries greater currents and gives off greater amounts of heat.

Potentiometers

A potentiometer is used to vary the value of voltage applied to a circuit, as shown in Fig. 5-8. In this circuit, the input voltage is applied across the terminals A–C of the resistance element. When the position of the sliding arm (terminal B) is changed, the voltage across terminals B–C will change. As the sliding arm moves closer to terminal A, the output voltage of the circuit increases. As the sliding arm moves closer to terminal C, the output voltage of the circuit decreases.

Potentiometers are commonly used as control devices in amplifiers, radios, television sets, and different kinds of meters. Typical uses include volume, balance, brightness, and zeroing adjustments.

Materials

The resistance element in a *carbon-composition* resistor is mainly graphite or some other form of solid carbon. The carbon material is measured carefully to provide the right resistance (Fig. 5-9). These resistors generally have resistance values from 0.1 Ω to 22 MΩ.

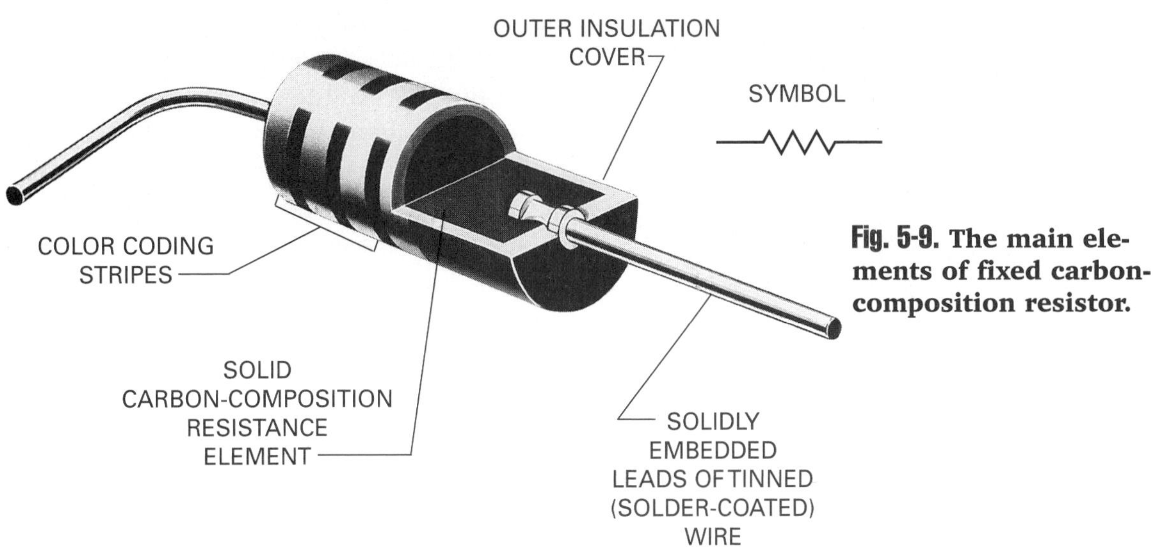

Fig. 5-9. The main elements of fixed carbon-composition resistor.

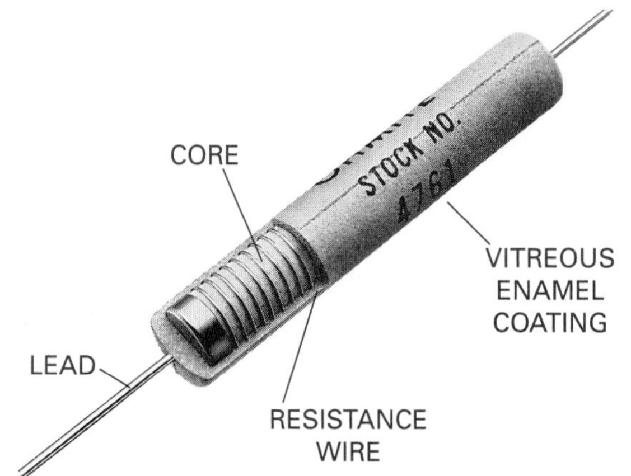

Fig. 5-10. Wire-wound resistor.

The resistance element of a *wire-wound* resistor is usually nickel-chromium wire. This wire is wound around a ceramic core. The whole assembly is usually coated with a ceramic material or a special enamel (Fig. 5-10). Wire-wound resistors generally have resistance values from 1 Ω to 100 kΩ.

Film resistors have a ceramic core called the *substrate*. A film of resistant material such as carbon or metal is deposited on the substrate serves as the resistance element. The film may be carbon or metal. It may also be a mixture of metal and glass called *metal glaze* (Fig. 5-11).

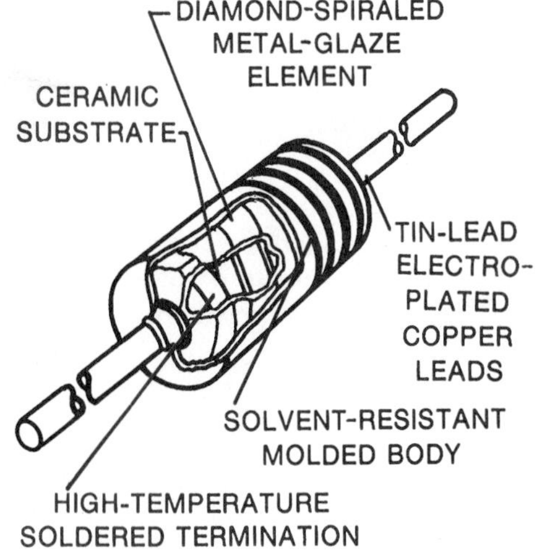

Fig. 5-11. Metal-glaze resistor.

Name three variable resistors in your home, office, or factory. (volume controls, lighting rheostats, speed controls, tone controls).

RATINGS

Resistors have two other ratings in addition to resistance: tolerance and power. These ratings determine the material, compositions, and size of a resistor and are very important to their use. Tolerance expresses how much leeway there is in a resistor's value. *Power* determines the amount of current a resistor can safely conduct.

Tolerance

The actual resistance of a resistor may be greater or less than its rated value, a variation called **tolerance**. Carbon-composition, film, and wire-wound resistors each have a given tolerance. For example, common tolerances of carbon-composition resistors are ±5 percent, ±10 percent, and ±20 percent. This means that a resistor with a stated resistance of 100 Ω and a tolerance of ±5 percent can actually have a resistance of any value between 95 Ω and 105 Ω. General purpose, wire-wound resistors usually have a tolerance of +10 percent.

Resistors having tolerances as high as ±20 percent are used in many electric and electronic circuits. The advantage in using high-tolerance resistors is that they are less expensive to make than low-tolerance resistors. However, they can be used only in circuits where variation is not important.

Precision Resistors

Some wire-wound and film resistors have actual values that are nearly equal to their rated values. These are called *precision resistors*. They are used in special circuits such as in

LINKS

For more information about power, refer to Chapter 9.

test instruments and critical communication devices. Precision resistors are much more expensive than lower rated resistors. Therefore, designers of consumer electronic products will often create designs using less accurate and less expensive resistors. Precision resistors usually have a tolerance between 1 and 5 percent.

Power Rating

The power rating of a resistor indicates how much heat a resistor can handle before it burns in two. Since it is the current that produces heat, the power rating also gives some indication of the maximum current a resistor can safely carry. The power rating of a resistor is given in watts.

Carbon-composition resistors have power ratings from 1/16 W to 2 W. Wire-wound resistors have power ratings from 3 W to hundreds of watts.

The physical size of a resistor has nothing to do with its resistance. A very small resistor can have a very low or a very high resistance. The physical size of a resistor does, however, suggest its wattage rating. For a given value of resistance, the physical size of a resistor increases as the wattage rating increases (Fig. 5-12). With experience, you can soon learn to tell the wattage ratings of resistors by their physical sizes.

Resistor Color Code

The resistance and wattage values of resistors are usually printed directly on them. The resistance values of fixed carbon-composition resistors and some film resistors are shown by a *color code*. The colors of the code and the numbers they stand for are given in Table 5-C. Three stripes are used for resistance values. A fourth stripe is often used to show the tolerance. A fifth stripe, when used, tells the *failure rate*. That is the amount the resistance will change over a period of time, such as 1,000 hours. The color code is read as follows:

1. The first, or end, stripe indicates the first number of the resistance value.
2. The second stripe indicates the *second number* of the resistance value.

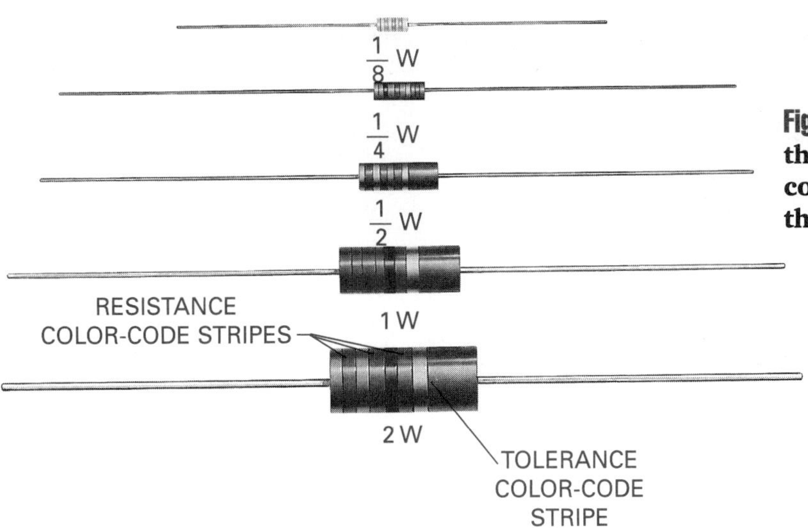

Fig. 5-12. Relationship between the physical size of carbon-composition resistors and their wattage ratings.

Table 5-C Resistor Color Code

Resistance Values, First Three Stripes	Tolerance Values, Fourth Stripe
black = 0	gold = ±5%
brown = 1	silver = ±10%
red = 2	none = ±20%
orange = 3	Failure Rate, Fifth Stripe
yellow = 4	brown = 1%
green = 5	red = 0.1%
blue = 6	orange = 0.01%
violet = 7	yellow = 0.001%
gray = 8	
white = 9	
gold = divide by 10	
silver = divide by 100	

3. The third stripe indicates the *number of zeros* that follow the first two numbers of the resistance value. If the third stripe is black, no zeros are added after the first two numbers since black represents zero.
4. If the third stripe is gold, the number given by the first two stripes is divided by ten.
5. If the third stripe is silver, the number given by the first two stripes is divided by 100.

Several examples of resistance values given by the color code are shown in Fig. 5-13.

1. How do wattage ratings relate to current?
2. What can happen in a circuit if a resistor becomes overheated and bulges?

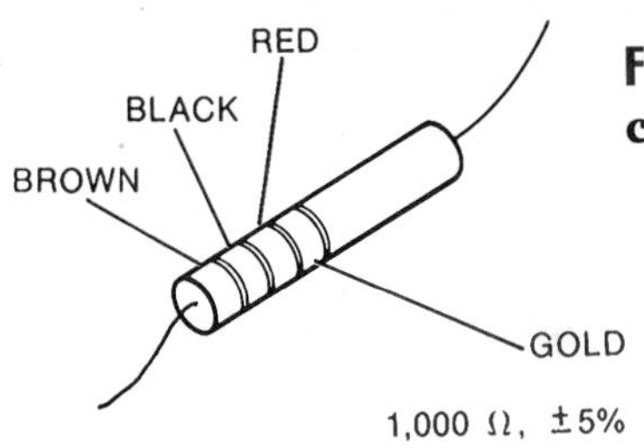

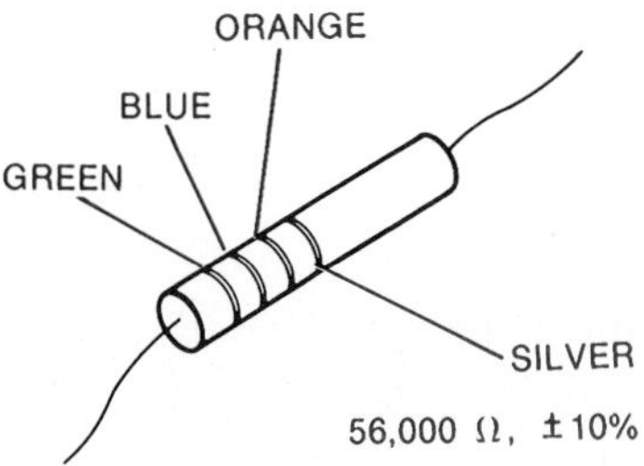

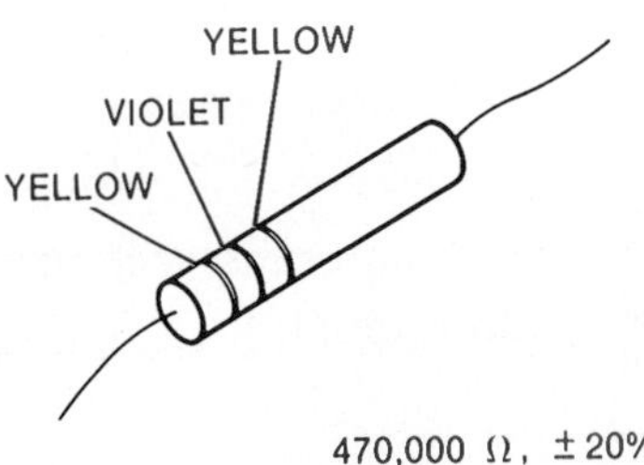

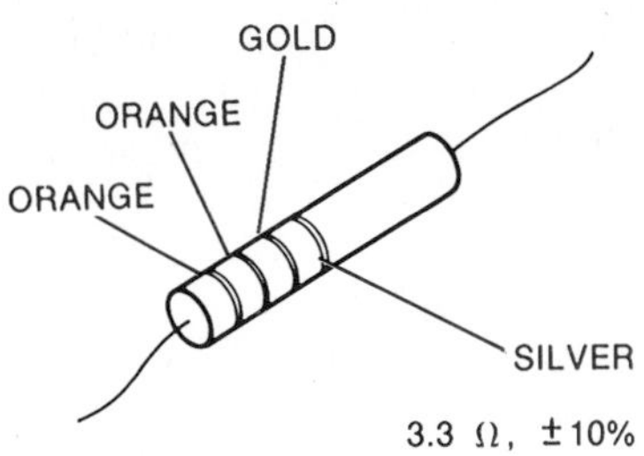

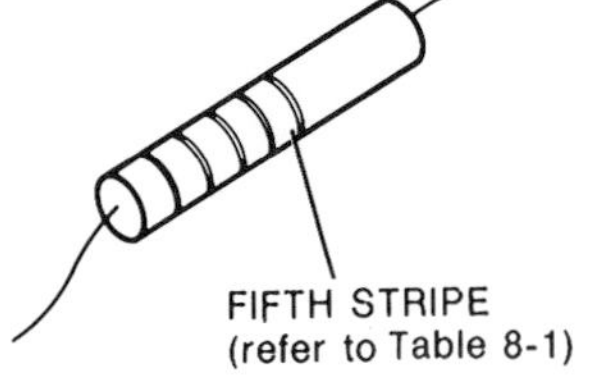

Fig. 5-13. Reading the resistor color code.

RESISTOR DEFECTS

Resistors are rugged devices. They seldom become defective unless too large a current passes through them. This may happen when there is a short circuit somewhere in the circuit.

A carbon-composition or a film resistor that becomes overheated will often be burned completely apart. In other cases, overheating will cause such a resistor to be scorched, to crack, or to bulge out. The value of the resistance may then increase to many times its normal value.

When a wire-wound resistor is overheated, the resistance wire will often burn apart at one point. This makes the resistor open. A carbon-composition, film, or wire-wound resistor can also become defective if one of its wires breaks loose inside the resistor body.

Chapter *Review*

Summary

- A conductor is a material, often metal, that has many free electrons. Electrical current flow is the result of those free electrons being passed from atom to atom.
- Resistance is the opposition to this flow of electrons. If the resistance is high enough, the material is an insulator, or nonconductor.
- Devices called resistors are purposely put into circuits to control current and voltage levels. Each resistor has a different application in a circuit and is selected by designers according to size, value, accuracy, and cost.

Review Main Ideas

Review this chapter's main ideas by writing, on a separate sheet of paper, the word or words that most correctly complete the following statements:

1. A conductor is a material through which electrons can flow ______.
2. In addition to metals, certain ______ are also used as conductors under special conditions.
3. Electrons move through a circuit conductor from the ______ to ______ terminal of the energy source.
4. Electronic collisions reduce the number of electrons that flow through the ______.
5. The basic unit of resistance is the ______.
6. The Greek capital letter ______ is used as an abbreviation for ohms after a numerical value. This letter is written as ______.
7. In almost all metal conductors, the resistance ______ as the metal's temperature increases.
8. With two wires of the same length, the wire with the larger cross-sectional area has ______ resistance.
9. All insulating materials will conduct current if a high enough ______ is applied to them.
10. Resistors are used to control ______ and ______.
11. A fixed resistor has a ______ value of resistance.
12. The two most common kinds of variable resistors are called ______ and ______.
13. The resistance of a potentiometer or a rheostat changes between the ______ terminal and the ______ of the element.
14. The sliding arm of a potentiometer or a rheostat is attached to the ______ of the device.
15. A potentiometer is most often used to vary the value of the ______ applied to a circuit.
16. The resistance element in a carbon composition resistor is mainly ______ or ______.

17. The resistance element of a wire-wound resistor is usually ______ wire.

18. In a film resistor, a film of ______ is deposited on the substance.

19. The actual resistance of a resistor may be greater or less than its rated value, a variation known as the ______ of the resistor.

20. A resistor that has an actual value that is very nearly equal to its rated value is called a ______ resistor.

21. The ______ rating of a resistor indicates how much heat the resistor can handle before ______.

22. The physical size of a resistor has nothing to do with its ______.

23. A larger resistor of a given value has a greater ______ rating than a smaller resistor of the same value.

24. The first three stripes of the resistor color code indicate the ______ value of a resistor. The fourth stripe shows the ______ of the resistor.

25. Resistors seldom become defective unless too large a ______ passes through them.

26. The overheating of a carbon-composition or film resistor often causes the resistance of the resistor to ______.

Apply Your Knowledge

1. Draw a sketch of electrons moving through a conductor.
2. Name three materials that make good conductors.
3. Describe the main difference between an insulator and a conductor.
4. Using Table 5-B, choose the highest dielectric material to use as an insulator.
5. Explain why resistance is important in a circuit.

Make Connections

1. **Communication Skills.** Using your own words, write a definition for each of these terms: insulator, conductor, dielectric, rheostat, and tolerance. The definitions should apply to the field of electricity and electronics.
2. **Mathematics.** Calculate the range of the resistances for each resistor in Fig. 5-13 using the tolerance specified.
3. **Science.** Define the word *force* as it is used in physics.
4. **Workplace Skills.** Design a test to determine the conductance of various sizedwire. Use, for example, AWG gage Nos. 28, 22, 18, and 12. Cut each wire to a length of 1 foot. Measure the current as it is increased for each wire until it breaks.
5. **Social Studies.** List the reasons you believe a government-funded research effort in the area of superconductivity is important for the nation.

Chapter 

Ohm's Law and Power Formulas

OBJECTIVES

After completing this chapter, you will be able to:

- Use the letter symbols for voltage, current, and resistance to show the relationship between them.
- Calculate the current, voltage, and resistance in a circuit when any two are known quantities.
- Determine the wattage of an electric circuit when the current and voltage are known.
- Find the cost of operating an electrical product for a period of time if the cost per kilowatt-hour is known.

Terms to Study

kilowatt-hour

Ohm's law

power

power formula

resistance

watt

watt-hour meter

You have previously studied the electrical effects of voltage, current, and resistance. Do you know that mathematical equations can describe the relationships of those electrical properties studied in previous chapters? Do you know that engineers rely on mathematics to understand and describe how circuits work? This chapter explains the relationship of these electrical properties and how to use mathematical expressions to calculate unknown values.

OHM'S LAW

The opposition to the flow of electrons in a circuit is **resistance**. The law that presents the relationship between *voltage*, *current*, and *resistance* is known as **Ohm's law**. George Simon Ohm discovered this relationship in 1827. According to Ohm's law:

1. The voltage needed to force a given amount of current through a circuit is equal to the product of the current and the resistance of the circuit.
2. The amount of current in a circuit is equal to the voltage applied to the circuit divided by the resistance of the circuit.
3. The resistance of a circuit is equal to the voltage applied to the circuit divided by the amount of current in the circuit.

Ohm's Law Formulas

By using the letter symbols for voltage (E), current (I), and resistance (R), Ohm's law can be expressed by the formulas given here.

To solve for E (voltage), use the formula:

$$E = I \times R$$

To solve for I (current), this same formula can be written as:

$$I = \frac{E}{R}$$

FASCINATING FACTS

How would you explain the properties of electricity? George Simon Ohm, a German mathematician and physicist, compared electricity to water in a hose. He found that just as water flowing from a hose decreases as you close the nozzle, so the flow of electricity decreases as resistance to it increases. Ohm's discovery became the basis for all study of electrical properties.

To solve for R (resistance), arrange it as

$$R = \frac{E}{I}$$

The arrangement of terms in a formula allows us to state certain general rules about the relationship of current, voltage, and resistance. For example, in the Ohm's law formula:

$$I = \frac{E}{R}$$

the E in the numerator tells us that I (current) will change in step with E (voltage), if the denominator R (resistance) stays the same.

Thus, if the voltage in a circuit doubles, the resulting current will be double its original value. For example, if E = 20 V and R = 2 Ω, I = ?

$$I = \frac{E}{R} = \frac{20}{2} = 10 \text{ A}$$

If E = 40 V and R = 2 Ω, I = ?

$$I = \frac{E}{R} = \frac{40}{2} = 20 \text{ A}$$

If the voltage in a circuit is reduced to one-half its original value, the current will adjust itself to one-half of its original value. For example, if E = 120 V and R = 20 Ω, then I = ?

$$I = \frac{E}{R} = \frac{120}{20} = 6 \text{ A}$$

If E = 60 V and R = 20 Ω, then I = ?

$$I = \frac{E}{R} = \frac{60}{20} = 3 \text{ A}$$

Thus, Ohm's law shows us that current is directly proportional to voltage. Provided that the resistance does not change, the current will increase if the voltage increases, and the current will decreases if the voltage decreases.

However, look closely at resistance R in the following formula:

$$I = \frac{E}{R}$$

The R in the denominator states that if E (voltage) remains the same, then I (current) is inversely proportional to R (resistance).

In other words, as resistance increases, current decreases proportionally. If the resistance in a circuit is increased to twice its original value, then the current will decrease to one-half its original value. For example if $E = 16$ V and $R = 4\ \Omega$, then $I = ?$

$$I = \frac{E}{R} = \frac{16}{4} = 4 \text{ A}$$

If $E = 16$ V and $R = 8\ \Omega$, then $I = ?$

$$I = \frac{E}{R} = \frac{16}{8} = 2 \text{ A}$$

Thus, when the voltage remains the same, the current decreases as the resistance increases.

Similarly, a decrease in resistance results in an increase in current. If resistance is decreased to one-third its original value, the current will increase to three times its original value. For example, if $E = 60$ V and $R = 30\ \Omega$, then $I = ?$

$$I = \frac{E}{R} = \frac{60}{30} = 2 \text{ A}$$

If $E = 60$ V and $R = 10\ \Omega$, then $I = ?$

$$I = \frac{E}{R} = \frac{60}{10} = 6 \text{ A}$$

The preceding example shows that current is inversely proportional to resistance. In this case, current increases as resistance decreases when voltage remains the same.

I = ?

Sample Problem 1.

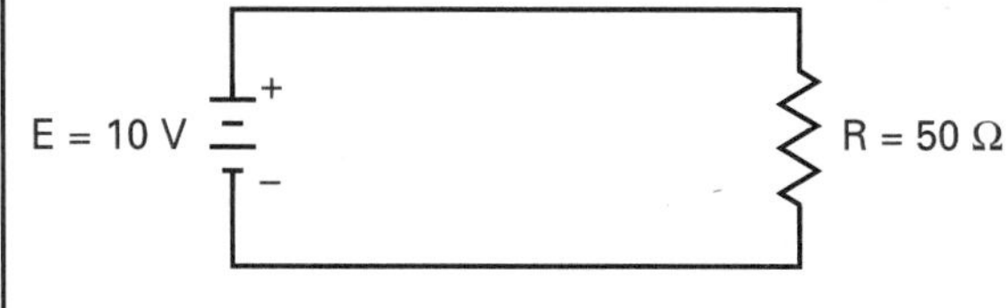

I = ? A
I = ? mA

Sample Problem 2.

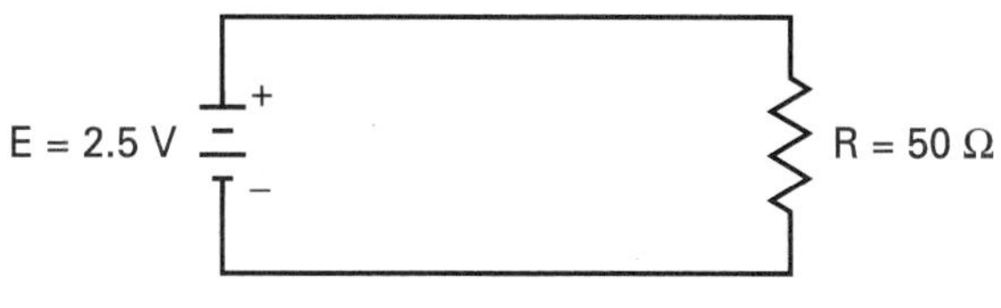

I = ? A
I = ? mA

Sample Problem 3.

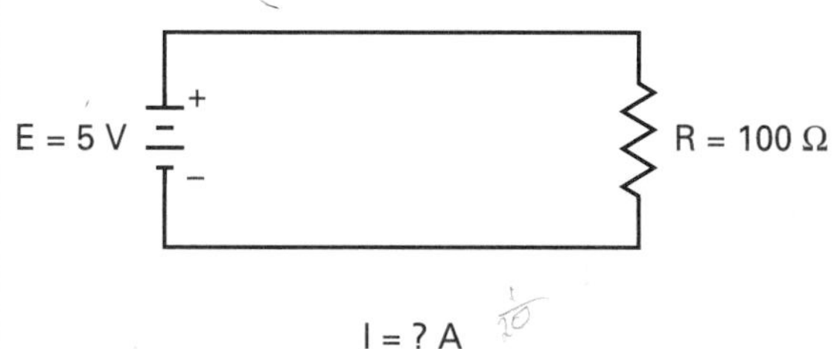

I = ? A
I = ? mA

Sample Problem 4.

Ohm's Law in Use

The Ohm's law formulas can be learned easily by using a divided circle (Fig. 6-1A). To use this circle, cover any one of the quantities E, I, or R. The relationship between the other two quantities in the circle shows how to find the covered quantity (Fig. 6-1B).

E

I R

A

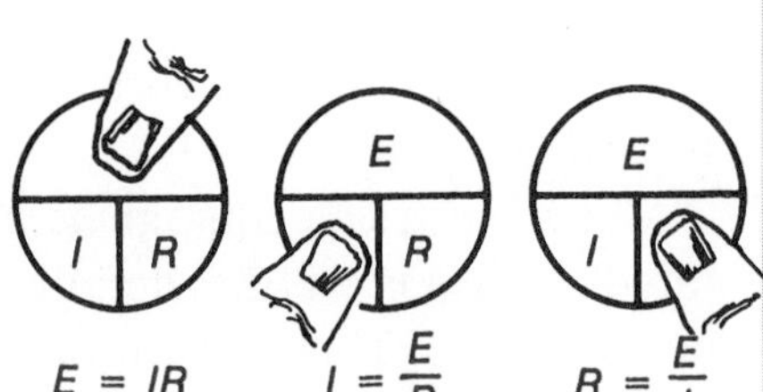

Fig. 6-1. The Ohm's law circle.

B

Ohm's law is important in understanding the behavior of circuits. It is also important because it can be used to find the value of any one unknown of the three basic circuit quantities (voltage, current, or resistance). Thus, circuits and their parts can be designed mathematically. This saves time and prevents equipment damage during the design process. How to use Ohm's law to solve practical circuit problems is shown in the following examples:

Problem 1: An electric light bulb uses 0.5 A of current in a 120 V circuit. What is the resistance of the bulb?

Solution: The first step in solving a circuit problem is to sketch a schematic diagram of the circuit. The second step is to label each of the parts and show the known values (Fig. 6-2).

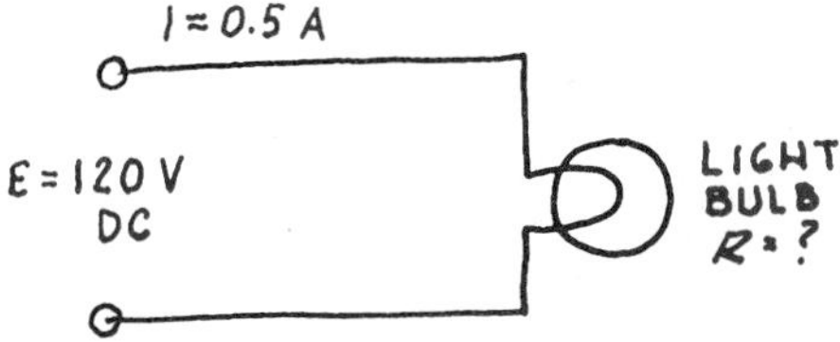

Fig. 6-2. Diagram for Problem 1.

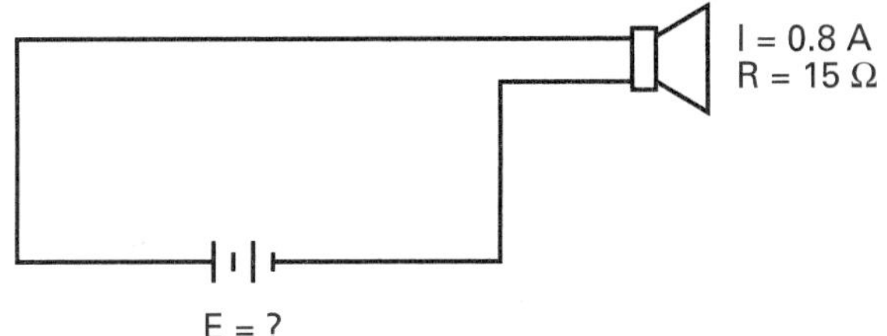

Fig. 6-3. Diagram for Problem 2.

In this problem, the values for I and E are known. To solve for R, we use the formula:

$$R = \frac{E}{I}$$

$$R = \frac{120}{.5}$$

$$= 240\ \Omega$$

Problem 2: An alarm horn has a current rating of 8 A printed on it. The resistance of the horn coil is known to be 15 Ω. Compute the voltage that must be applied to the horn circuit if it is to operate correctly (Fig. 6-3).

Solution: Since voltage is the unknown quantity here, use the formula:

$$E = I \times R$$

$$= 8 \times 15$$

$$= 120\ \text{V}$$

Problem 3: To use the right size fuse in an automobile circuit, it is necessary to find the current needed by a certain device. The device is to be connected to the 12-V battery and has a resistance of 4.35 Ω (Fig. 6-4). Find the current.

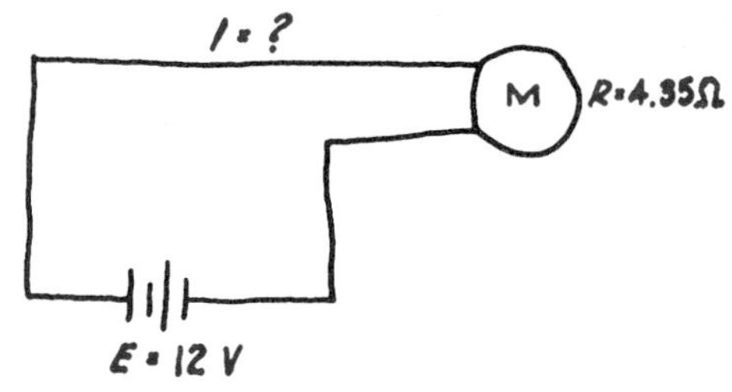

Fig. 6-4. Diagram for Problem 3.

Solution: Since current is the unknown quantity, use the formula:

$$I = \frac{E}{R}$$
$$= \frac{12}{4.35}$$
$$= 2.76 \text{ A}$$

Think about a light bulb. If it is a 100-Ω bulb using 120 V, how much current will it use?

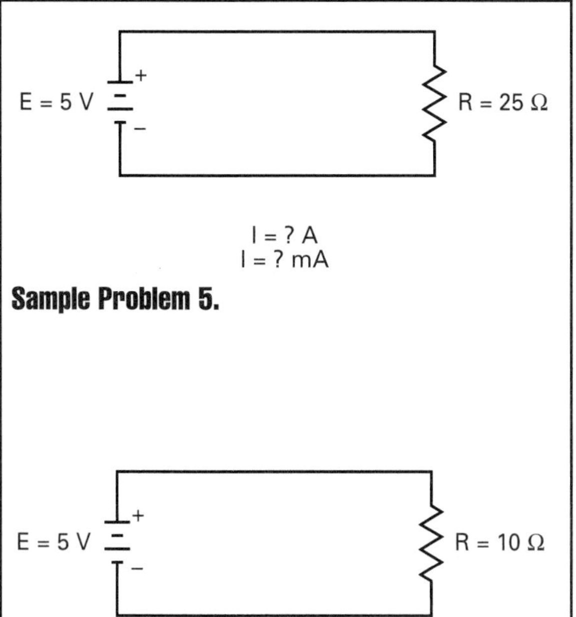

ELECTRIC POWER AND ENERGY

Power is the time rate of doing work. In an electric circuit, power may also be defined in two other ways. First, it is the rate at which electric energy is delivered to a circuit. Second, it is the rate at which an electric circuit does the work of converting the energy of moving electrons into some other form of energy. The basic unit of power is the watt (W). It is named after James Watt. He was a Scottish inventor who lived from 1736 to 1819. The watt is based on the unit of work energy, *joules* per second.

Much electrical equipment that we use is rated in terms of the amount of power, or watts, it uses. Thus, a light bulb is rated as 60 watt, 100 watt, 200 watt, and so on. Motors, ovens, heating elements, and power tools also have certain wattage ratings. Since electric power is the rate at which energy is delivered to the circuit, it is easy to calculate. You simply multiply the power by some time unit. This will give you the total energy delivered to the circuit. The time unit usually used is the hour. The power unit is usually given in terms of kilowatts (1,000 watts). Therefore, the energy unit on which most electric companies base their bills is the kilowatt-hour (kWh).

Power Formulas

The **power formula** shows the relationship between electric power (*P*), voltage (*E*), and current (*I*) in a dc circuit. The basic power formula is:

$$P = E \times I$$

From this formula, it is possible to get two other commonly used power formulas. For

LINKS

Look in the glossary for a fuller definition of *joules*.

PLUG IN TO *History*

Watt and the Steam Engine

James Watt effectively utilized power with his significant improvements on the steam engine.

Watt developed his lifelong interest in steam engines while serving as instrument maker at the University of Glasgow, Scotland. When the university's model engine needed repair in 1764, Watt had the opportunity to refine the steam engine.

To make a faster engine, Watt added valves that admitted steam to each side of the piston. With each admission, the valves released the steam to a separate vessel where it was condensed. This method kept the cylinder from cooling down at each stroke. The condensation created a vacuum that made the new steam more effective.

To change the piston's back-and-forth motion into rotary motion, the piston in Watt's engine drove a connecting rod, and a crank turned an axle. These changes made the engine useful for turning wheels in factories.

Watt also developed the steam governor, which consisted of two heavy balls. They were mounted on swinging arms that were connected to regulate the steam valve. The whole assembly rotated with the engine's motion and maintained a desired engine speed. If the speed increased, centrifugal force drove the balls outward in wider circles, moving the arms. The arms then choked the steam valve, reducing speed. If the engine lagged, the balls lowered and admitted more steam.

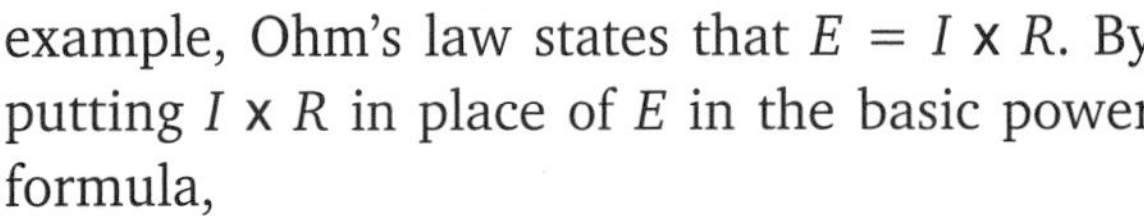

example, Ohm's law states that $E = I \times R$. By putting $I \times R$ in place of E in the basic power formula,

$$P = I \times R \times I$$
$$= I^2R$$

Also, from Ohm's law we know that $I = E/R$. By putting E/R in place of I in the basic power formula,

$$P = E \times \frac{E}{R}$$
$$= \frac{E^2}{R}$$

Power Formulas in Use

As mentioned, power is the time rate of doing work. The basic unit of power is the **watt**. Power can be defined in two ways: the rate at which electric energy is delivered to a circuit and the rate at which an electric circuit uses electric energy, or how much work it can do.

The power formulas can be used to find the wattage ratings of circuit parts. It can also be used to find the value of current in a circuit and the cost of operating electrical and electronic products. How to solve practical circuit problems with these formulas is shown in the following examples.

Problem 1: The current through a 100-Ω resistor that is to be used in a circuit is 0.15 A. What should be the wattage rating of the resistor?

Solution: Since the two known quantities in this problem are current and resistance, use the formula $P = I^2R$:

$$P = I^2R$$
$$= .15^2 \times 100$$
$$= 2.25 \text{ W}$$

To keep a resistor from overheating, its wattage rating should be about twice the wattage computed from a power formula. Thus, the resistor used in this circuit should have a wattage rating of about 5 W.

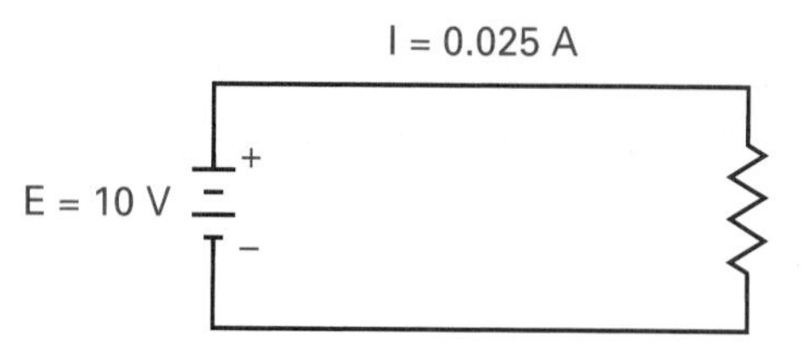

Sample Problem 7.

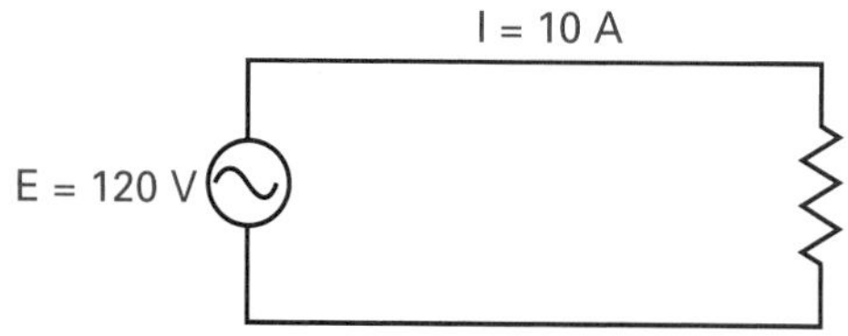

Sample Problem 8.

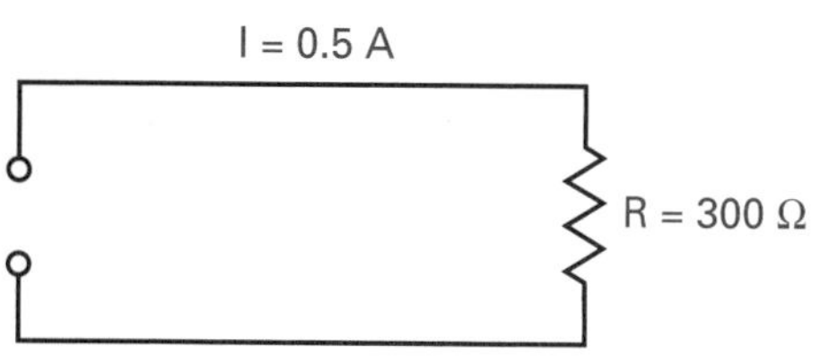

Sample Problem 9.

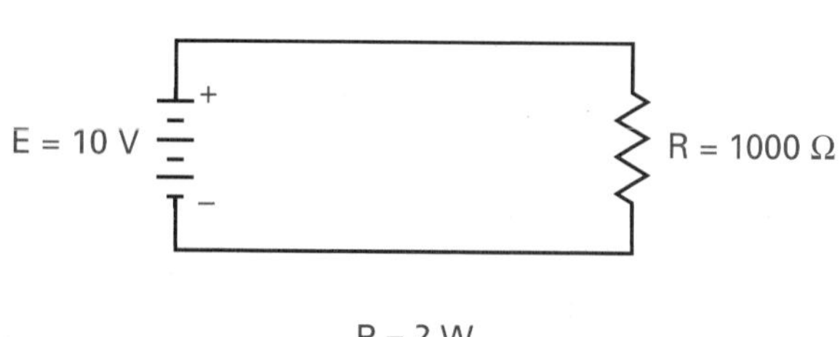

Sample Problem 10.

Problem 2: Find the current that flows through a 60-W incandescent lamp operating at 120 V. Also find the current through a 200-W lamp and a 300-W lamp operating at 120 V.

Solution: In this problem, the power and voltage are known. We wish to find the current. The simplest formula to use, therefore, is $P = E$ x I, from which we solve for I:

$$I = \frac{P}{E}$$

For the 60-W, 120-V lamp:

$$I = \frac{60}{120}$$
$$= 0.5 \text{ A}$$

For the 200-W, 120-V lamp:

$$I = \frac{200}{120}$$
$$= 1.67 \text{ A}$$

For the 300-W, 120-V lamp:

$$I = \frac{300}{120}$$
$$= 2.5 \text{ A}$$

In this problem, the voltage applied to each of the lamps is the same (120 V). As the lamp wattage increases, the current in the circuit also increases. This means that there is a direct relationship between the power dissipated in the load and the current flowing through it.

Problem 3: The current in a house wiring system increases to twice the original value, from 2 A to 4 A. What effect does this have on the temperature of the wires in the circuit? Assume a resistance of 25 Ω.

Solution: According to the formula $P = I^2R$, if the resistance does not change, power is directly proportional to the square of the current. Thus if the current is doubled, the power increases four times.

$$P = I^2R = 2^2 \times 25$$
$$= 4 \times 25 = 100 \text{ W}$$

If the current is doubled from 2 A to 4 A,

$$P = 4^2 \times 25$$
$$= 16 \times 25 = 400 \text{ W}$$

In this problem, doubling the current increases the power four times from 100 to 400 W. Power ratings can be thought of as the rate at which the energy of moving electrons is changed into heat in the wires. If the current is doubled, the wires will become heated to four times the original temperature.

This relationship between power and current is important to know. It means that even a small increase in current will cause a large increase in wire temperature. Overheated wires are a major cause of electrical fires.

Kilowatt-hour Formula

The **kilowatt-hour** (kWh) is the unit of electric energy on which electric companies base their bills. Kilowatts multiplied by the number of hours used is equal to kilowatt-hours. For example, if an oil heater, rated at 1,000 W, is operated for 30 minutes (0.5 hour), the energy used is 1 kW x 0.5 hour = 0.5 kWh.

The amount of energy in kilowatt-hours used by equipment or some other product can be computed by the following formula:

$$\text{kWh} = \frac{\text{wattage rating} \times \text{no. of hours used}}{1000}$$

$$\text{kWh} = \frac{P \times h}{1000}$$

where P = power in watts
h = time in hours

Watt-hour meters measure the electric energy in kilowatt-hours supplied by electric power companies to users. Such a meter is usually mounted on the side of a building. It is monitored, or read, at certain times. This reading is used to calculate the electric bill for a period of time, usually one month.

Using the Kilowatt-hour Formula

The kilowatt-hour formula can be used to find the cost of operating a product for any period of time if the cost per kilowatt-hour is known. This is shown in the problems below.

Problem 1: The price of electric energy in a town is 6 cents per kWh. Find the cost of operating a 250 W television set for 1.5 hours.

Solution:

$$\text{kWh} = \frac{P \times h}{1000}$$
$$= \frac{250 \times 1.5}{1000}$$
$$= 0.375$$

Since the energy costs 6 cents per kWh, the cost of operating the television set for 1.5 hours is 0.375 x 6 = 2.25 cents.

Problem 2: At 10 cents per kWh, what is the cost of operating a 1,200-W steamer for 2 hours?

Solution:

$$\text{kWh} = \frac{P \times h}{1000}$$
$$= \frac{1200 \times 2}{1000}$$
$$= 2.4$$

Total cost = 2.4 x 10, or 24 cents

How much voltage does a 1,000-W heater element need if it operates at 15 A?

Chapter *Review*

Summary

- In a circuit, the relationship between voltage, current, and resistance is expressed by Ohm's law.
- Power is the time rate of doing work, and its basic unit is the watt. The power formula shows the relationship between power, voltage, and current in a dc circuit.
- The kilowatt-hour (kWh) is the unit of electric energy on which electric companies base their bills. Kilowatts multiplied by the number of hours used is equal to kilowatt-hours.

Review Main Ideas

Review this chapter's main ideas by writing, on a separate sheet of paper, the word or words that most correctly complete the following statements:

1. Ohm's law states the relationships among ______, ______, and ______.
2. The voltage applied to a circuit is equal to the product of the ______ and the ______ of the circuit.
3. The current in a circuit is equal to the applied ______ divided by the ______.
4. The resistance of a circuit is equal to the applied ______ divided by the ______.
5. If the voltage applied to a circuit is doubled and the resistance remains constant, the current in the circuit will increase to ______ the original value.
6. If the voltage applied to a circuit remains constant the resistance is doubled, the current will decrease to ______ the original value.
7. The three most commonly used power formulas are ______, ______, and ______.
8. If the current through a conductor is doubled and the resistance remains unchanged, the power dissipated by the conductor will increase to ______ times the original amount.
9. The unit of electric energy on which electric companies base their bills is the ______.
10. The kilowatt-hours of energy used by an appliance can be calculated by multiplying the ______ of the appliance by the ______ it is used and dividing by 1,000.

Chapter Review

Apply Your Knowledge

1. Design a 100-V circuit that operates all day, every day, and uses 1400 kilowatt-hours of energy per week.
2. Differentiate between directly and indirectly proportional.
3. Write Ohm's law.
4. Using Ohm's law, describe how an increase in resistance will cause a decrease in current.
5. Calculate how many watts a heater element should be rated for if it uses 7.7 amperes at 240 Vac.

Make Connections

1. **Communication Skills.** Write a brief report on what it means when the Ohm's Law equation is stated as "current is directly proportional to voltage and inversely proportional to resistance."
2. **Mathematics.** Identify the electrical appliances you use in your home. Determine the wattage rating for each. Estimate the time you use the appliance during one month. Using the information available on the bill for your home, calculate that portion of the bill that reflects the use of the appliance.
3. **Science.** Design an experiment to measure the forms of energy produced by an incandescent light bulb. Compare the total output to the total input. How would you account for any differences?
4. **Technology Skills.** Using common materials, construct a carbon resistor. Determine its resistance value and then change it to have a specific designed value.
5. **Social Studies.** Select the historical person associated with voltage, amperes, ohms, and watts. Research the person and his work.

Answers to Sample Problems

1. .4 A
2. .2 A (200 mA)
3. .05 A (50 mA)
4. .05 A (50 mA)
5. .2 A (200 mA)
6. .5 A (500 mA)
7. .25 W
8. 1200 W
9. 75 W
10. .1 W

Chapter 7

Series Circuits

Terms to Study

aiding voltage sources

closed loop

junction

Kirchhoff's voltage law

node

open loop

opposing voltage sources

polarity

series circuit

voltage divider

voltage drop

OBJECTIVES

After completing this chapter, you will be able to:

- Draw a series circuit, label each part, and identify each part's polarity.
- Determine voltage drops in a series circuit.
- Calculate the voltage across a voltage divider circuit.
- Determine the total source voltage of aiding direct current cells in series.
- Determine the total source voltage of two opposing batteries in series.

Have you ever wondered why, in some strings of lights, when one lamp burns out or is removed all the other lights go out? Have you ever tried to find out how a switch controls a light? This chapter discusses the series circuit in detail and these and other questions. Topics such as open and closed loops, direction of current flow, polarity, nodes, and fundamental laws governing series circuits are explained in this chapter.

THE SERIES CIRCUIT

In previous chapters, you learned that a complete circuit contains four basic parts: (1) the energy source, (2) the control device, (3) the conductor, and (4) the load (Fig. 7-1A). You also learned that loads can be connected in series, in parallel, or in series-parallel combinations.

LINKS

See Chapter 8 for parallel circuits. Series-parallel circuits are discussed in Chapter 9.

FASCINATING FACTS

Since each added resistor in a series circuit causes the current in the entire circuit to drop, this type of wiring is most appropriate for devices which only need low amounts of power to operate. The filaments in some types of Christmas tree lights illustrate loads that are connected in series.

The **series circuit** is an electrical circuit that has only one path through which electrons can flow. If a switch in the circuit is closed, the energy source forces electrons to flow through the closed switch, the conductor, the load, and then back to the energy source. In a direct current circuit, the electron flow (current) is unidirectional, or flows in only one direction from negative to positive.

Open and Closed Loops

Figure 7-1B represents a simplified drawing of 7-1A. It assumes that the switch is closed, a voltage is present, and the circuit is complete. The arrow labeled *current loop* indicates the direction of the current. A **closed loop** is a circuit through which current can flow. The path, or circuit, must be complete. As you learned in Chapter 3, the current flows from the negative (-) terminal of the energy source to the positive (+) terminal of the energy source.

An open circuit, or an **open loop**, means that the circuit is not complete. A broken conductor or an open switch causes a break in the circuit. If a circuit is broken or open at any point, the whole circuit is turned off. In this case the circuit is an open loop. The entire voltage of the energy source can be measured using a voltmeter across the open switch or a broken conductor. This is because the resistance at the open loop is very high compared to any other resistance of the circuit. This feature is used to control and protect electrical systems. Devices such as switches, fuses, and circuit breakers are connected in series with a load to intentionally create open loops as either safety or control devices.

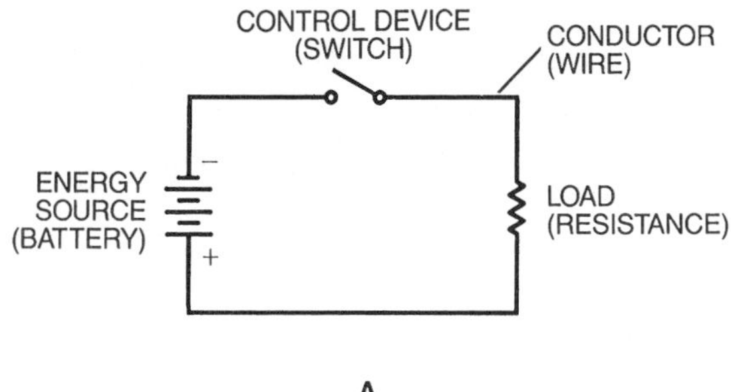

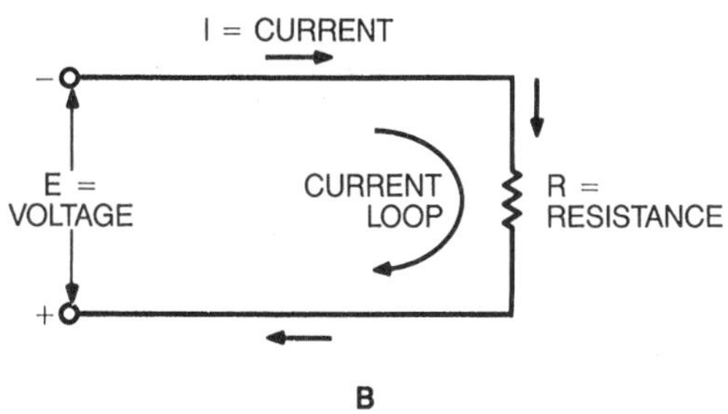

Fig. 7-1. The series circuit: (A) the basic parts of a series circuit; (B) simplified drawing of a series circuit.

Examples of Series Circuits

Figure 7-2 shows several sketches of series circuits. When more than one resistance or load is shown, the loads are connected to the conductors one after the other. Thus, there is only one path through which current can pass as it moves between the negative and positive terminals of the energy source. Notice that different symbols or letters represent the energy sources. Figure 7-2A shows details about voltage and resistance. This sketch is more complete than the others. However, all the sketches are meaningful and represent ways that technicians and engineers convey information.

Polarity

The **polarity**, or electrical charge of an energy source, is indicated on diagrams by common electrical symbols of positive (+) and negative (-) signs. However, polarity charges exist in other parts of the circuit, too. Figure 7-2B shows the polarity (charges) at the resistors, or loads, in a circuit. The polarity is negative (-) at the point where current enters the resistor. The polarity is positive (+) where the current leaves the resistor. It is necessary to know polarity of direct current circuits in order to properly connect instruments and components to the circuit.

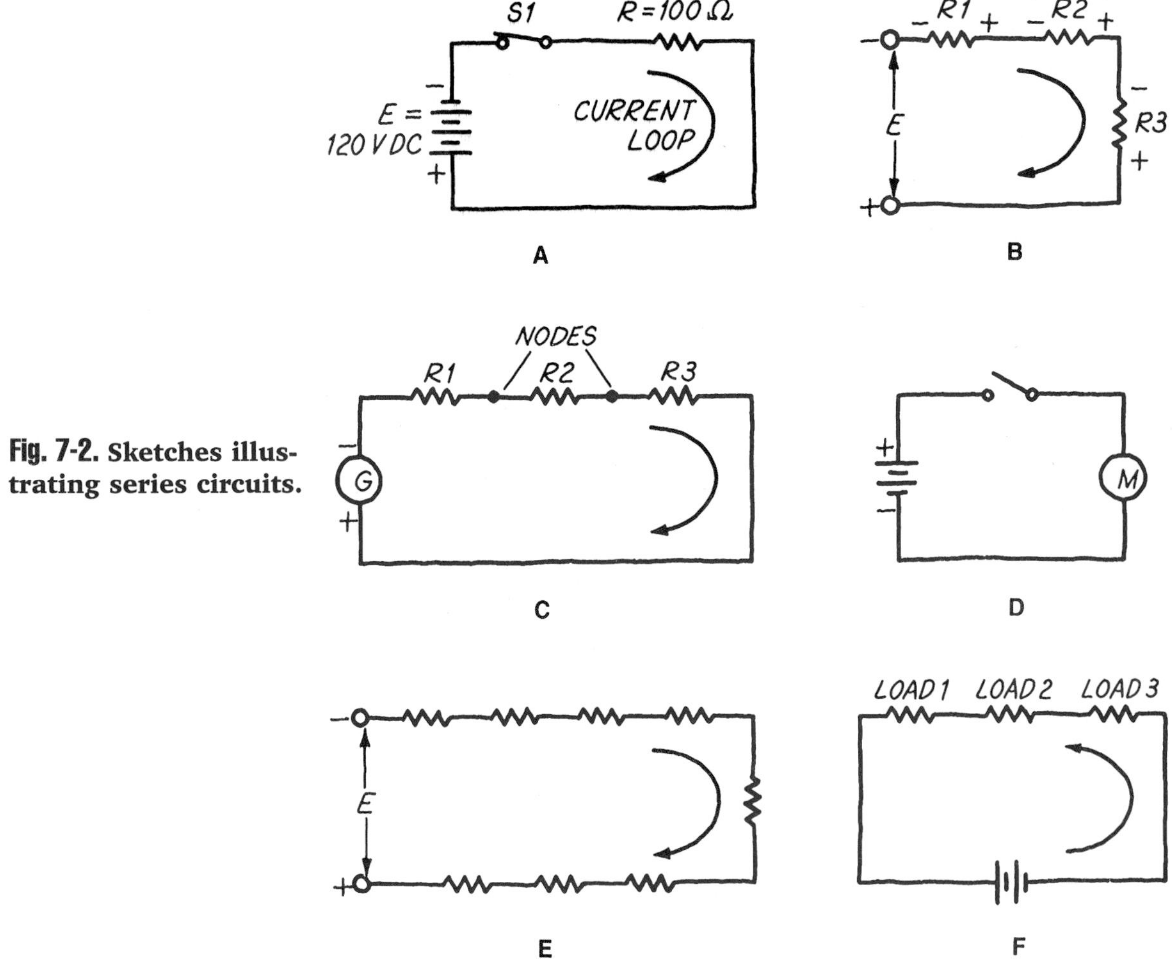

Fig. 7-2. Sketches illustrating series circuits.

Nodes and Junctions

A **node** is a terminal, or connection point, between two or more parts of a circuit. Figure 7-2C illustrates where two nodes are located in the circuit. They are identified by large dots between the resistors R_1, R_2 and R_2, R_3. When three or more circuit elements such as resistors are connected to one point, a **junction** is formed. Nodes and junctions become test points for measuring voltage and current.

LINKS

Junctions are discussed in Chapter 8 on parallel circuits.

Series Resistance

Figure 7-3 shows three resistors in series. Since the circuit has only one path for current to flow, the current must pass through all the resistors in the circuit. Therefore, the total resistance is the sum of all the resistors in the series circuit. The following formula is used to calculate the total resistance (R_t) in a series circuit:

$$R_t = R_1 + R_2 + R_3 + \cdots + R_n$$

NOTE: the small '*n*' in R_n stands for any number of resistors.

Problem 1: Find the total resistance of the three resistors connected in series (Fig. 7-3).

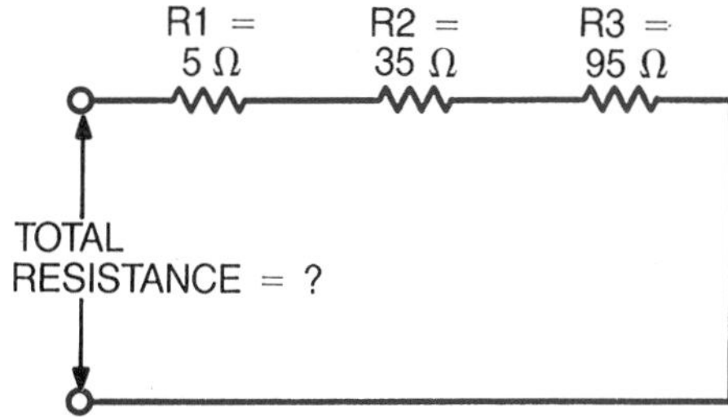

Fig. 7-3. Diagram for Problem 1.

Solution:

$$\begin{aligned} R_t &= R_1 + R_2 + R_3 \\ &= 5\ \Omega + 35\ \Omega + 95\ \Omega \\ &= 135\ \Omega \end{aligned}$$

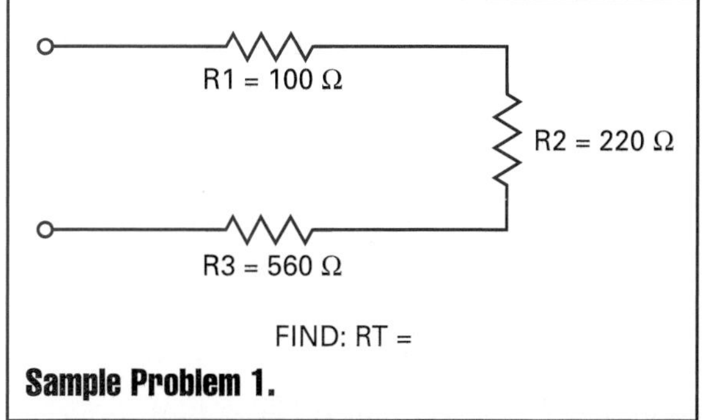

Sample Problem 1.

Problem 2: A series circuit contains six resistors: $R_1 = 6\ \Omega$, $R_2 = 14\ \Omega$, $R_3 = 7\ \Omega$, $R_4 = 10\ \Omega$, $R_5 = 150\ \Omega$, $R_6 = 2\ \Omega$. Calculate the total resistance.

Solution:

$$\begin{aligned} R_t &= R_1 + R_2 + R_3 + R_4 + R_5 + R_6 \\ &= 6\ \Omega + 14\ \Omega + 7\ \Omega + 10\ \Omega + 150\ \Omega + 2\ \Omega \\ &= 189\ \Omega \end{aligned}$$

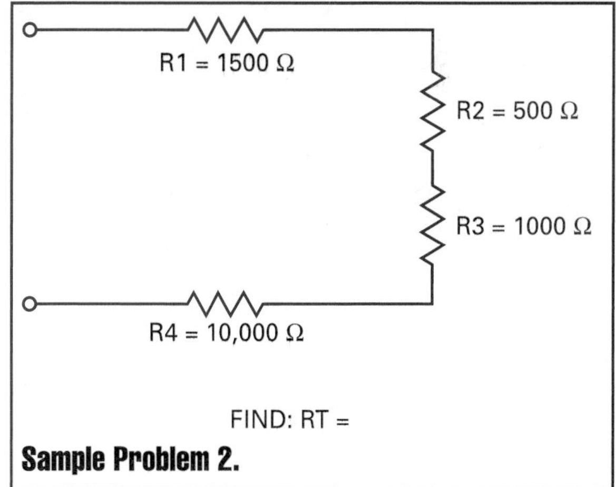

Sample Problem 2.

Problem 3: A series circuit contains five resistors. The total resistance of the circuit is 186 Ω. The values of the four resistors are: R_1 = 20 Ω, R_2 = 14 Ω, R_3 = 10 Ω, R_4 = 100 Ω. What is the value of the resistor R_5? Manipulate the formula for resistance in a series circuit and solve for R_5.

Solution:

$$R_t = R_1 + R_2 + R_3 + R_4 + R_5$$

Rewrite the formula to put R_5 on the left side of the equal sign:

$$R_1 + R_2 + R_3 + R_4 + R_5 = R_t$$

Subtract R_1, R_2, R_3, and R_4 from both sides of the formula:

or $R_5 = R_t - R_1 - R_2 - R_3 - R_4$

$$\begin{aligned} R_5 &= R_t - (R_1 + R_2 + R3 + R4) \\ &= 186\ \Omega - (20\ \Omega + 14\ \Omega + 10\ \Omega + 100\ \Omega) \\ &= 186\ \Omega - 144\ \Omega \\ &= 42\ \Omega \end{aligned}$$

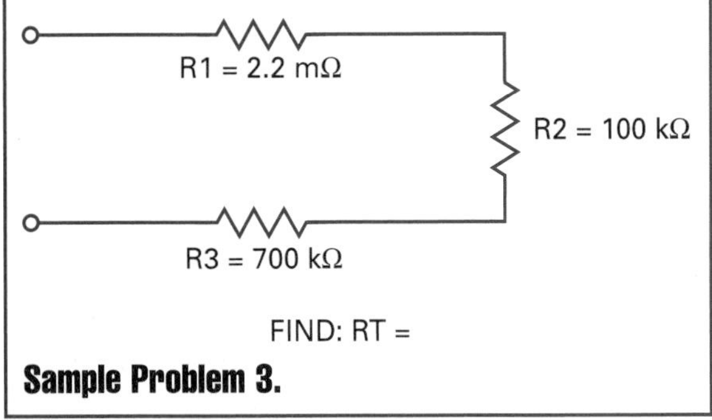

Sample Problem 3.

Series Circuit

Since a series circuit has only one current path, the same number of electrons flow *from* any point of the circuit as flow *to* that point. This means that there is the same amount of current in all parts of a series circuit at any time.

Problem 4: A series circuit has a 100-VDC source and three resistors: R_1 = 6 Ω, R_2 = 12 Ω, R_3 = 7 Ω (Fig. 7-4). Find the total current in the circuit.

Solution:

1. Determine the total resistance of the circuit:

$$\begin{aligned} R_t &= R_1 + R_2 + R_3 \\ &= 6\ \Omega + 12\ \Omega + 7\ \Omega \\ &= 25\ \Omega \end{aligned}$$

2. Since the total resistance and total voltage are known, use Ohm's law to calculate the total current.

$$I_t = \frac{E_t}{R_t} = \frac{100\ \text{V}}{25\ \Omega} = 4\ \text{A}$$

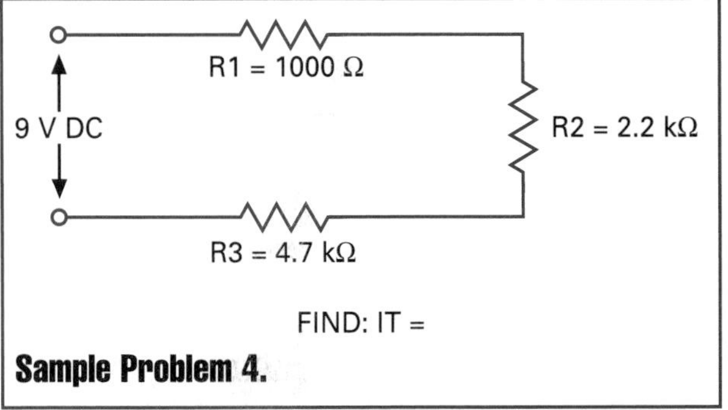

Sample Problem 4.

The total current calculated in Problem 4 is 4 A. If the current is measured at the nodes (Fig. 7-4), you will find it to be the same at each node.

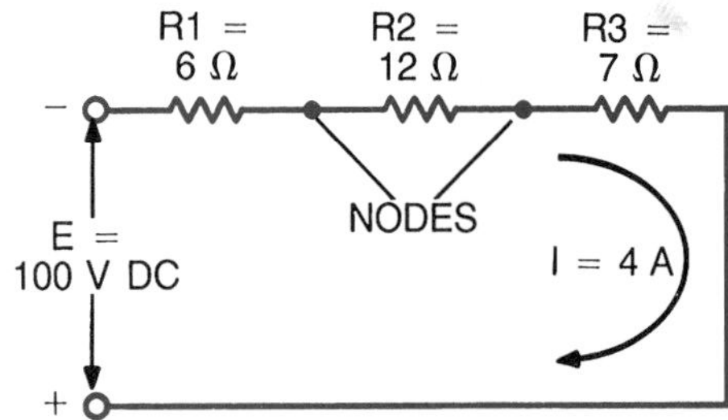

Fig. 7-4. Current in a series circuit is the same in all parts of the circuit.

Draw a series circuit with two aiding and two opposing voltage sources that have an output of 1.75 Vdc.

VOLTAGE

The total voltage applied to a series circuit is automatically divided among the loads and devices in the circuit. The voltage across any load is the amount needed to force the circuit current through the resistance of that load. This is also called a **voltage drop**. The voltage drop across any load in the circuit is equal to the product of the current through the load and the resistance of the load:

E	=	I	x	R
Voltage drop across load		Current through load		Resistance of load in ohms

Kirchhoff's voltage law states that in a series circuit, the sum of the voltage drops across each load is equal to the total voltage applied to the circuit.

Problem 5: Three resistances are connected in series (Fig. 7-5A). Calculate the voltage drop across each resistor. This problem involves several steps.

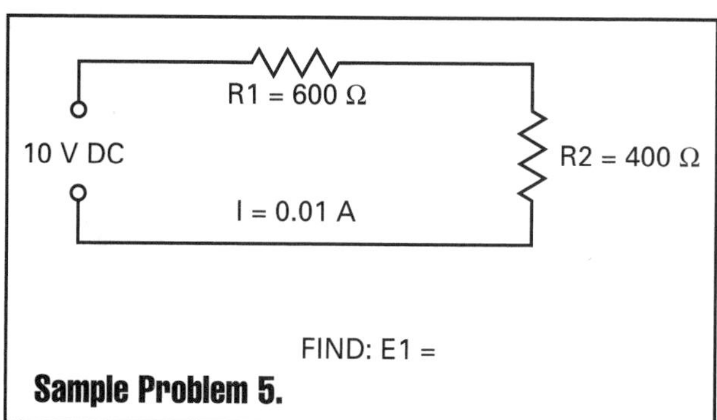

Sample Problem 5.

Kirchhoff and Bunsen

After detailing his laws of currents, voltages, and resistances in 1845, Gustav Kirchhoff (1824-1887) and Robert Bunsen (1811-1899) went on to make other significant discoveries in science. They used a spectroscope (an optical instrument) to study the spectra (or bands of light) of certain substances vaporized in a Bunsen burner flame.

Only minute quantities of an element are needed to make its spectral lines appear. This discovery allows identification of elements in unknown substances and bodies such as the sun and stars. In 1859 Kirchhoff published these findings in his laws of radiation and absorption.

1. Find the total resistance.

$R_t = R_1 + R_2 + R_3 = 1\ \Omega + 2\ \Omega + 3\ \Omega = 6\ \Omega$

2. Since the total voltage and total resistance are known, calculate the total current.

$$I_t = \frac{E_t}{R_t} = \frac{12}{6} = 2\text{A}$$

3. Now that the total current is known, you can figure the voltage drops across each resistor by using Ohm's law for that part of the circuit.

Voltage drop across R_1: $E = IR = 2 \times 1 = 2$ V
Voltage drop across R_2: $E = IR = 2 \times 2 = 4$ V
Voltage drop across R_3: $E = IR = 2 \times 3 = \underline{6\text{ V}}$
Total voltage drops = 12 V

Notice that the sum of the voltage drops across the resistors equals the applied voltage (Fig. 7-5A). If any one of the resistors is changed in such a circuit, all the voltage drops

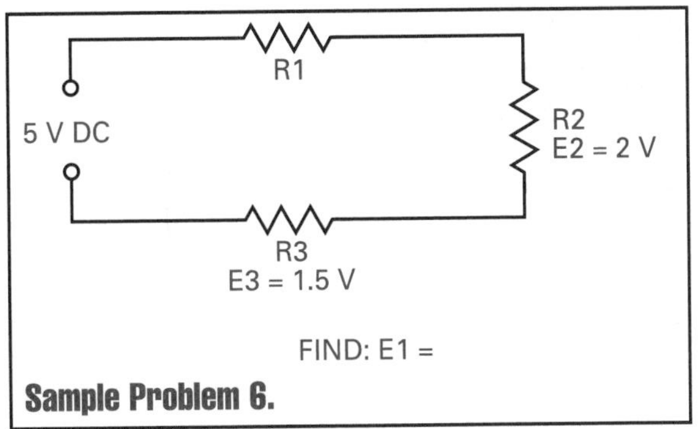

Sample Problem 6.

will change. However, the individual voltage drops will still add up to the source voltage.

Problem 6: Figure 7-5B is the same as Fig. 7-5A except that R has a resistance of 5 Ω instead of 1 Ω. Calculate the voltage drop for each resistor.

Solution:

1. Find the total resistance.

$R_t = R_1 + R_2 + R_3 = 5\ \Omega + 2\ \Omega + 3\ \Omega = 10\ \Omega$

2. Calculate the total current.

$$I_t = \frac{E_t}{R_t} = \frac{12}{10} = 1.2\text{ A}$$

3. Determine the voltage drop across each resistor.

Voltage drop across R_1: $E = IR = 1.2 \times 5 = 6.0$ V
Voltage drop across R_2: $E = IR = 1.2 \times 2 = 2.4$ V
Voltage drop across R_3: $E = IR = 1.2 \times 3 = 3.6$ V
Total voltage drops = 12 V

Notice that if any one of the resistor values changes in a circuit, the current and all the voltage drops change. But the sum of the individual voltage drops across the resistors, however, will always equal the applied voltage.

VOLTAGE DIVIDER

A series **voltage divider**, or network, is used to supply different values of voltage from one energy source. A simple voltage divider is shown in Fig. 7-6. In this circuit, a voltage of 12 V is applied to three resistors in series. The total resistance of this circuit limits the current to 1 A. The voltage drops are as follows:

Total current: $I = \frac{E}{R} = \frac{12}{2+4+6} = \frac{12}{12} = 1\text{ A}$

Voltage drop across CD: $E = I \times R = 1 \times 6 = 6$ V
Voltage drop across DE: $E = I \times R = 1 \times 4 = 4$ V
Voltage drop across EF: $E = I \times R = 1 \times 2 = 2$ V
Total voltage drop CF: $= 12$ V
Voltage drop across CE: $E = I \times R$
$= 1 \times (6 + 4) = 1 \times 10 = 10$ V

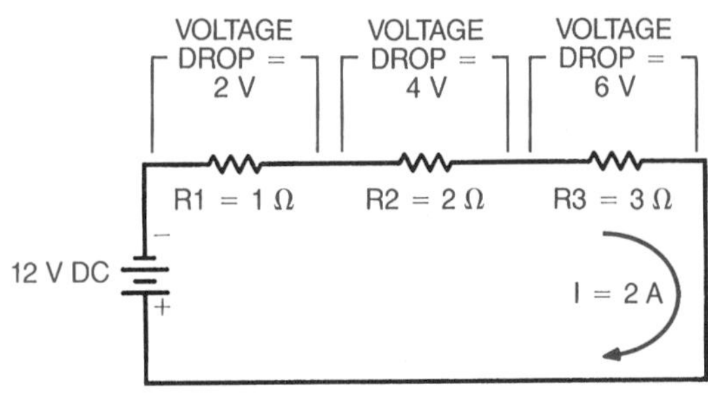

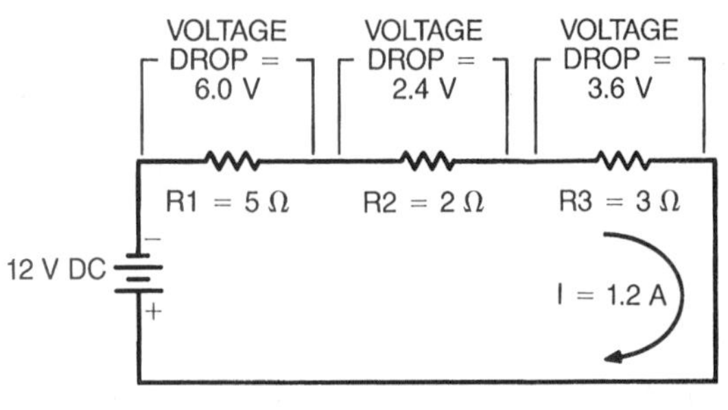

Fig. 7-5. Voltage drops: (A) diagram for Problem 5; (B) diagram for Problem 6.

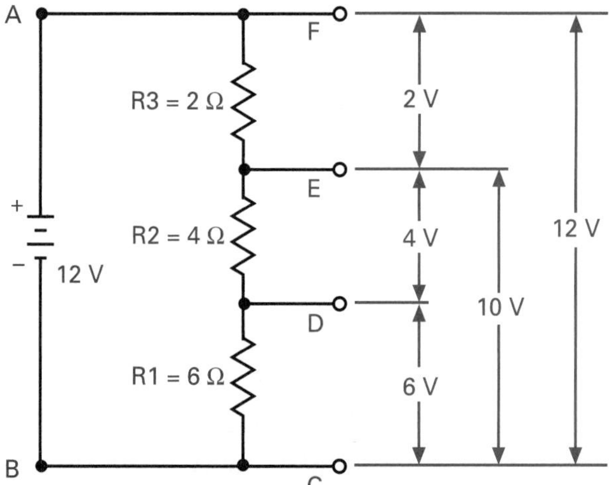

Fig. 7-6. A voltage-divider circuit.

Aiding Voltage Sources

Energy sources such as cells and batteries are often connected in series. These connections affect the total voltage and the total current. Cells connected in series produce a higher voltage than one cell does. The positive terminal of one cell is connected to the negative terminal of the next cell.

When cells are connected in this way, the total voltage is equal to the sum of the individual voltages (Fig. 7-7) that are called **aiding voltage sources**. Although voltage is increased, the current capacity of the cells in series is equal only to the cell with the lowest capacity. Batteries are cells, usually connected in series, so that individual cells aid each other.

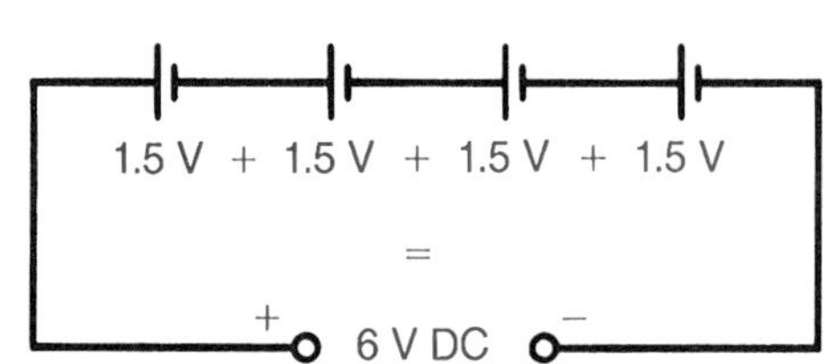

Fig. 7-7. Cells connected in series (aiding voltages).

Design a voltage divider that uses one of its outputs to supply a second series circuit using 1.5 A of current.

OPPOSING VOLTAGE SOURCES

In some circuits, the energy sources are connected so that the polarity of the connected terminals oppose each other (Fig. 7-8). When the voltages oppose each other, the resulting voltage to the load becomes the difference between the two source voltages that are called **opposing voltage sources**. In Fig. 7-8, the 70-V battery has the higher energy level but is reduced by the opposing 30-V battery.

Problem: What is the total voltage applied to the load in Fig. 7-8?

Solution:

$$E_t = E_2 - E_1 = 70\text{ V} - 30\text{ V} = 40\text{ Vdc}$$

The polarity is negative (-) at point A and positive (+) at point B because the 70-V battery overcomes the 30-V battery.

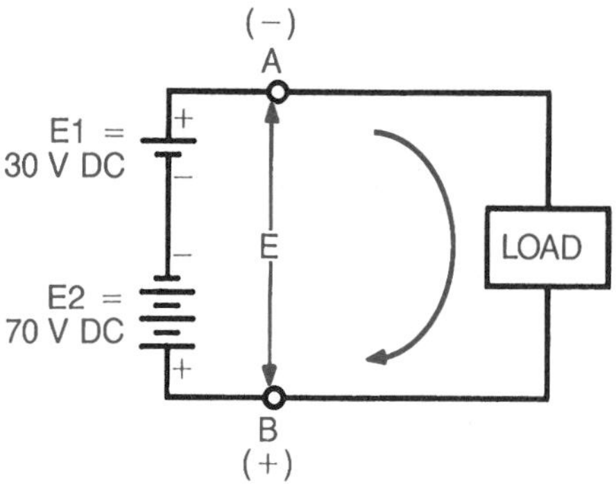

Fig. 7-8. Batteries connected in series (opposing voltages).

Chapter *Review*

Summary

- A series circuit is an electrical circuit that has only one path through which electrons can flow.
- Series circuits have energy sources, control devices, conductors, and loads.
- Resistance in a series circuit determines, along with the voltage, the current flow in a circuit. Ohm's law is the deciding formula. Therefore, the higher the resistance in a series circuit, the lower the current flow. The higher the voltage is, the higher the current will be.
- This current is steady throughout the series circuit and causes voltage to be dropped across each resistance. The sum of the voltage drops across all the resistors in a circuit is equal to the voltage available from the source. This is called Kirchhoff's voltage law.

Review Main Ideas

Review this chapter's main ideas by writing, on a separate sheet of paper, the word or words that most correctly complete the following statements:

1. A complete series circuit has four basic parts: (1) the energy source, (2) the controlling switch, (3) the conductor, and (4) the ______.
2. An open circuit, or open loop, means that the circuit is ______.
3. A terminal that connects two or more elements of a circuit is called a ______.
4. The total resistance of a series circuit is equal to the ______ of all the resistors in the circuit.
5. There is the same amount of ______ in all parts of a series circuit at any time.
6. In a series circuit, the ______ voltage is equal to the ______ of the voltage drops across the loads in the circuit.
7. A series network of resistors used to provide different voltages from a single source of energy is called a ______.
8. When cells are connected in series with the voltages aiding, the total voltage of the combination is equal to the ______ of the individual voltages.
9. When cells are connected in series with the voltages opposing, the resulting voltage to the load is equal to the ______ between the two source voltages.

Chapter *Review*

Apply Your Knowledge

1. Draw a series circuit with a 24-Vdc source and four resistors whose sum is 27 kΩ. What is the current in this circuit? How many different circuits can you design that have a total of 27 kΩ, but have no two resistance values the same?
2. Using one of the circuits you designed above, determine the voltage drops across the resistors.
3. Calculate the total source voltage of a circuit that has four 1.5-V batteries connected as aiding voltage sources.
4. Draw a voltage divider that has a 24-Vdc source and two equal outputs.
5. Explain what you think will happen when, instead of connecting source batteries as aiding (positive to negative), they are connected as opposing sources (negative to negative).

Make Connections

1. **Communication Skills.** Present an oral report to the class on Kirchhoff's voltage law. Explain the significance of the law to the analysis of series circuits.
2. **Mathematics.** Design a voltage divider network for equal voltages without load. Add a load across the lowest voltage. Then redesign the divider to bring the voltages back into equality.
3. **Science.** Write a report on the chemical action of a battery. Explain why the polarities must be connected negative to positive when connecting several batteries in series.
4. **Technology Skills.** Construct a model of a flashlight using 1.5 v cells arranged to satisfy the voltage requirements of a standard 4.5 Vdc bulb.
5. **Social Studies.** Prepare an oral or written report on Kirchhoff and what led him to discover the concepts behind his voltage law.

Answers to Sample Problems

1. 880 Ω
2. 13,000 Ω
3. 3 MΩ
4. 0.00114 A
5. E_1 = 6 Vdc; E_2 = 4 Vdc
6. E_1 = 1.5 Vdc

Chapter 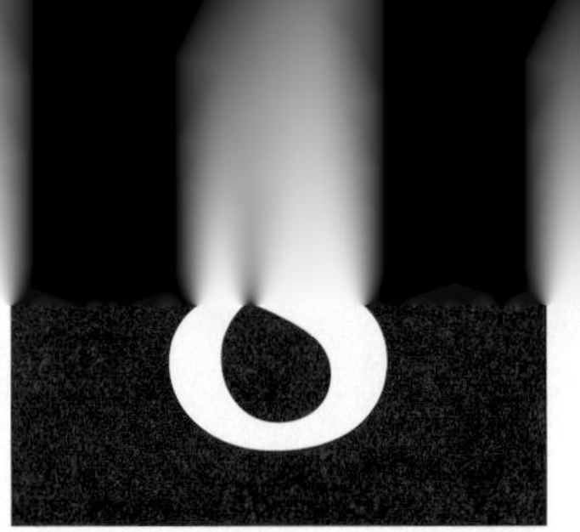

Parallel Circuits

Terms to Study

booster battery

branch circuit

equivalent resistance

Kirchhoff's current law

multiple connection

overload

parallel circuit

shunt connection

total current

OBJECTIVES

After completing this chapter, you will be able to:

- Draw and label a parallel circuit and indicate the polarity at each resistor.
- Define the term *overloaded circuit*.
- Use the three parallel-resistance formulas.
- Calculate the equivalent resistance of a parallel circuit.
- Calculate the total resistance, the total current, and the branch current in a parallel circuit.
- Connect batteries in parallel to increase current capacity.

In the previous chapter you learned that series circuits obey certain fundamental laws. Parallel circuits also obey certain laws. In this chapter you will learn how Ohm's law and Kirchhoff's law are used to determine how a parallel circuit functions.

PARALLEL CIRCUITS

This chapter explains the relationship between voltage, current, and resistance in parallel circuits and presents basic parallel-resistance formulas. The action of current as it enters and leaves junction points is explained. Circuit problems are solved in a step-by-step manner, allowing you to closely follow the mathematical process. You will also use the concept of "equivalent resistance" to solve parallel circuits and you will learn how to connect batteries in parallel to increase the current capacity of an energy source.

Sketch a parallel circuit of several loads in your home. See if you can name the loads.

In a **parallel circuit**, the loads are connected "across the line," meaning that they are connected between the two conductors that lead to the energy source (Fig. 8-1). The loads and their connecting wires often are called the **branch circuits**. Parallel connections are also called **multiple connections** and **shunt connections**.

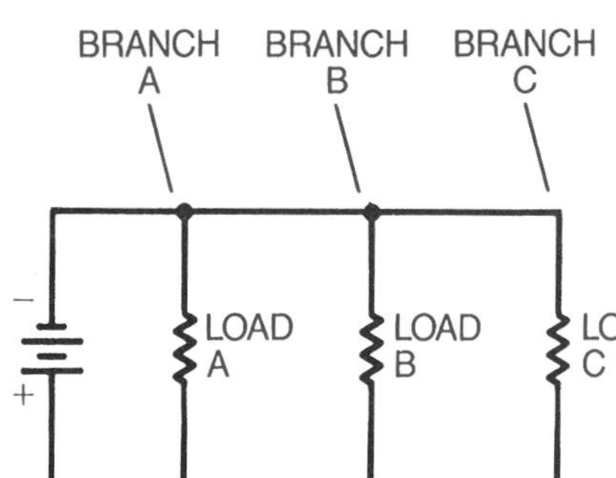

Fig. 8-1. A parallel circuit with three branches.

In a parallel circuit, the loads operate independently. Therefore, if one of the branches is disconnected, turned off, or burned open (as in an electric light bulb), the remaining branches can continue to operate.

Examples of Parallel Circuits

Sketches of various parallel circuits are shown in Fig. 8-2. Although the drawings look very different, all the loads are connected "across the line." This means that the energy source provides the same voltage to the loads. In Figs. 8-2F and 8-2G, the energy source is not shown. The terminals, however, provide the connecting point for the energy source. Notice that each load (resistance) can function independently since the voltage source is common to all loads. Figure 8-2A shows many more details about the circuit, including the direction of current flow in the branches.

Voltage and Current

Note that all the branches of a parallel circuit have the same voltage as the source (Fig. 8-3). The total current, however, is distributed among the branches. The amount of current in any branch can be found by using the Ohm's law formula $I = E/R$. The lower the resistance in any branch, the greater the current. Thus, while all the loads in a parallel circuit have the same voltage, they may or may not have the same amount of current. In a house wiring system, for example, lights and small appliances that operate at the same voltage require different amounts of current. Therefore, they are connected in parallel (Fig. 8-4).

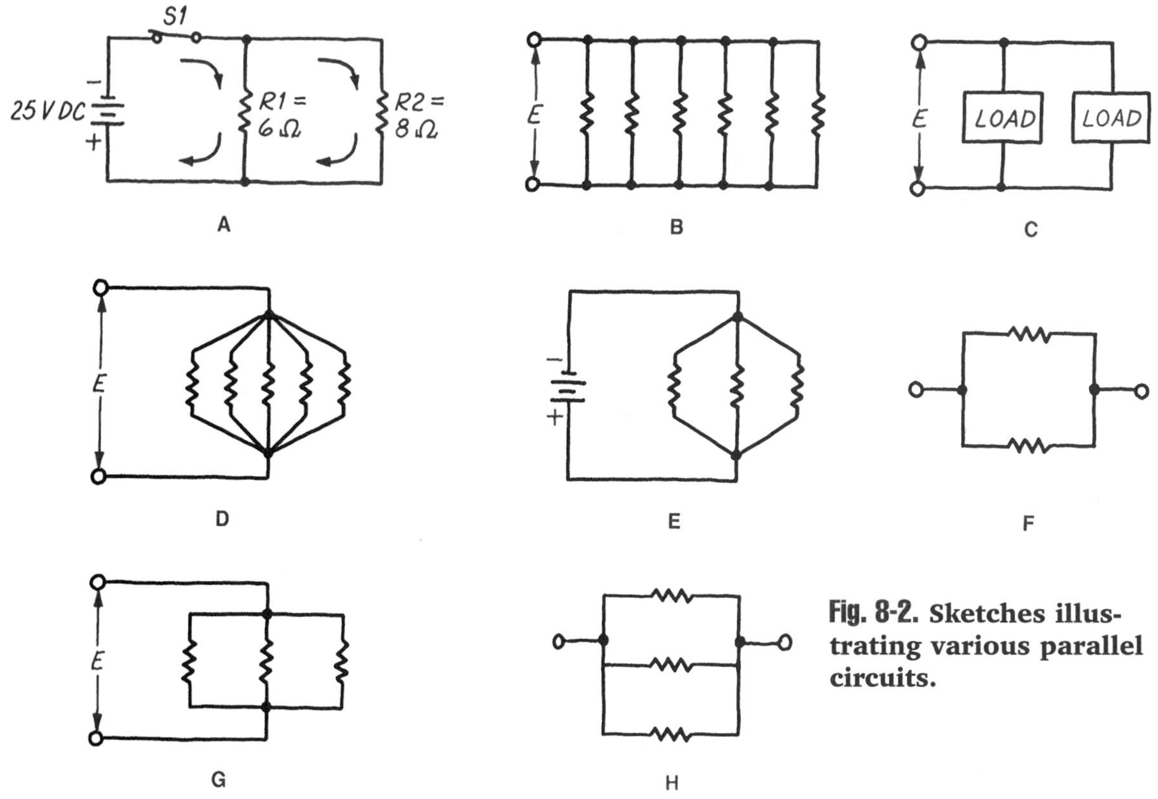

Fig. 8-2. Sketches illustrating various parallel circuits.

Fig. 8-3. The voltage is the same across all the branches in a parallel circuit.

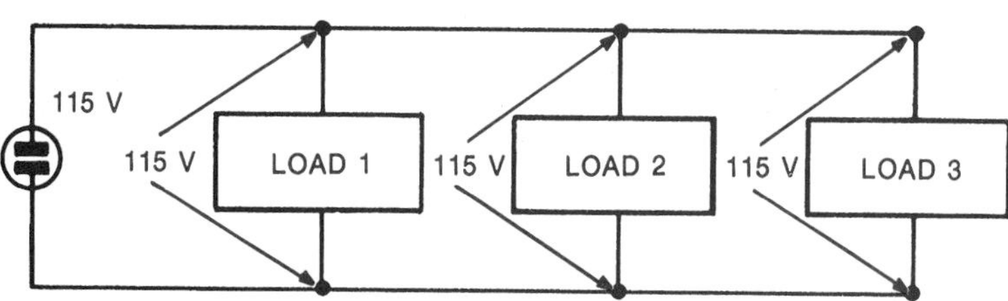

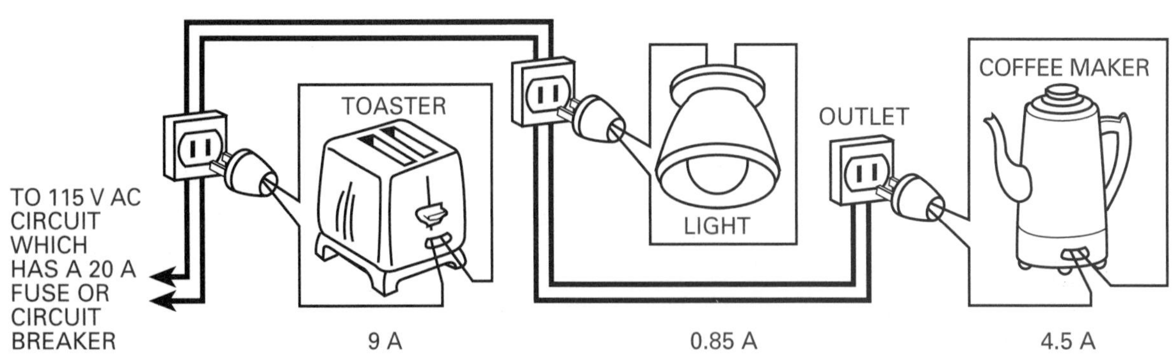

Fig. 8-4. Parallel circuits are used for the operation of household appliances, lights, and other electrical devices.

Fuses and Circuit Breakers

The **total current** delivered to a parallel circuit is the sum of the branch circuit currents. This is why a fuse in a household circuit burns open, or blows, when too many of the lights or appliances on a circuit are turned on. An **overload** occurs when the current increases until the total current is greater than the ampere rating of the fuse or circuit breaker. It is at this point that the fuse blows or the circuit breaker trips.

Residential Examples

Suppose, for example, that the circuit in Fig. 8-4 has a 20-A fuse. What will happen if another toaster is connected to the circuit? The 20-A fuse will blow. The fuse will open because the circuit is overloaded. Remember, *the total current delivered to a parallel circuit is equal to the sum of the branch circuits*. In Fig. 8-4, the current in each branch is:

Toaster	=	9.00 A
Light	=	0.85 A
Coffee maker	=	4.50 A
Total	=	14.35 A

The total current in Fig. 8-4 is 14.35 A. This current is well within the range of the 20-A fuse. If another toaster is connected into the circuit, there will be four branches. The current in each branch is:

Toaster 1	=	9.00 A
Toaster 2	=	9.00 A
Light	=	0.85 A
Coffee Maker	=	4.50 A
Total	=	23.35 A

With the second toaster connected in the circuit, the total current is 23.35 A. This circuit is now in an overload condition. The circuit is fused for 20 A and the load requires 23.35 A. The fuse, therefore, will burn out.

Ohm's Law in Parallel Circuits

Whenever the voltage is the same across each branch of a circuit, Ohm's law can be used to determine the current in that branch of the circuit (Fig. 8-5). For example, the current in branch 'A' is:

$$I = \frac{E}{R} = \frac{6}{2} = 3 \text{ A}$$

For branch B, the current is:

$$I = \frac{E}{R} = \frac{6}{3} = 2 \text{ A}$$

total current is:

$$\begin{aligned} I_t &= I_{\text{branch A}} + I_{\text{branch B}} \\ &= 3 + 2 \\ &= 5 \text{ A} \end{aligned}$$

In parallel circuits, therefore, the current in that part of the circuit is equal to its voltage across that part of the circuit divided by the resistance of that part of the circuit.

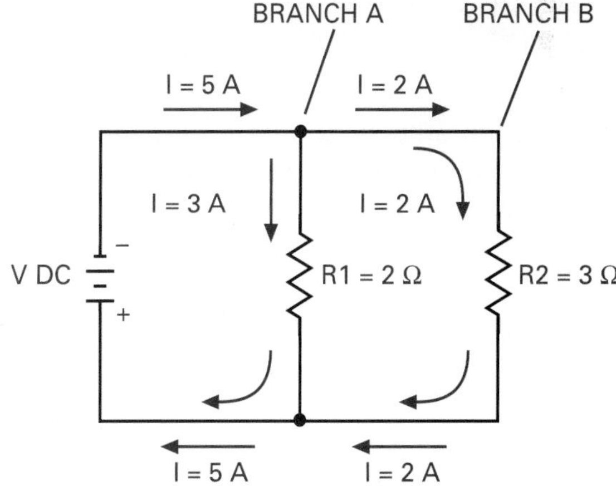

Fig. 8-5. The total current in a parallel circuit is equal to the sum of the currents in each branch.

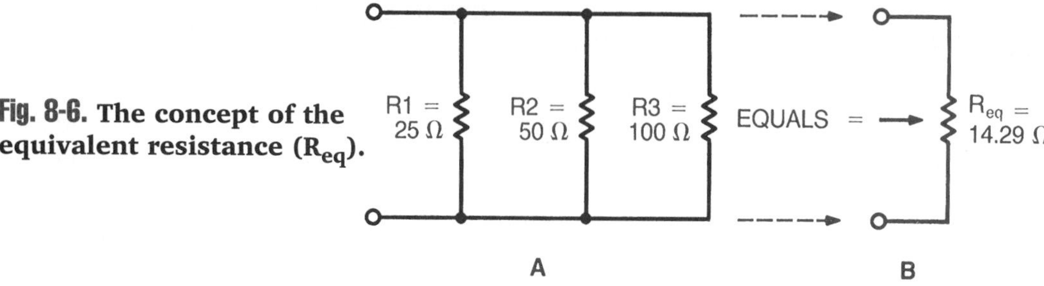

Fig. 8-6. The concept of the equivalent resistance (R_{eq}).

Resistance in Parallel Circuits

As branches are added to a parallel circuit, there are more paths through which current can flow. Therefore, total resistance to current decreases. The total resistance of a parallel circuit is always less than that of the branch with the lowest resistance (Fig. 8-6). The three resistances (25 Ω, 50 Ω, and 100 Ω) have an *equivalent resistance* of 14.29 Ω (Fig. 8-6B) when they are connected in parallel.

Equivalent resistance (R_{eq} is the electrical equivalent of the circuit components it represents. It is convenient to use in analyzing parallel and more complex circuits. If three resistances in Fig. 8-6 were connected in a series circuit, the total resistance 175 Ω (25 Ω + 50 Ω + 100 Ω = 175 Ω). The total, or equivalent, resistance is quite different connected in a series circuit than in a parallel circuit. Some parallel circuit problems that illustrate this difference follow.

1. Refer to the sketch you drew in the last Brain Booster. Determine which resistance formula would be the best one for you to use when determining the total resistance of your home circuits.

2. Assign resistance values to your circuit and solve for total resistance.

PARALLEL RESISTANCE FORMULAS

All parallel resistance combinations can be calculated to find an equivalent total resistance value by using the following *general formula.*

1. If the resistances are not all of equal value,

$$R_t = \frac{1}{\frac{1}{R_1} + \frac{1}{R_2} + \frac{1}{R_3} + \ldots + \frac{1}{R(n)}}$$

Under some conditions, special case formulas can be used to calculate equivalent total resistance. These can only be used when the circuit exactly matches the case, as in problems #2 and #3 below.

2. If there are two unequal resistor values,

$$R_t = \frac{R_1 \times R_2}{R_1 + R_2}$$

Where R_t = total resistance

R_1 and R_2 = parallel resistances

3. If all the resistances are equal in value,

$$R_t = \frac{\text{value of one resistance}}{\text{number or resistances}}$$

SOLVING THE PARALLEL RESISTANCE FORMULAS

To solve for total resistance in parallel circuits, you must first choose a formula. Remember that the general formula can be used for all parallel resistive circuits, but the special case formulas are used only when the circuit matches the formula. The following five examples illustrate the use of these formulas.

NOTE: In this chapter, there are sample problems, drawn freehand, which relate to the information in the text. These problems are for you to practice what you have learned. Use a separate sheet of paper to work out the problems. The answers to the problems are at the end of this chapter.

Problem 1: Three resistors with values of 2 Ω, 2.5 Ω, and 10 Ω are connected in parallel (Fig. 8-7). Find the total resistance.

Solution: Since there are more than two values of unequal resistance, use formula 1, general purpose formula.

NOTE: This solution supports the law that the total resistance of a parallel circuit is always lower than the branch with the least resistance. (See also Fig. 8-6.)

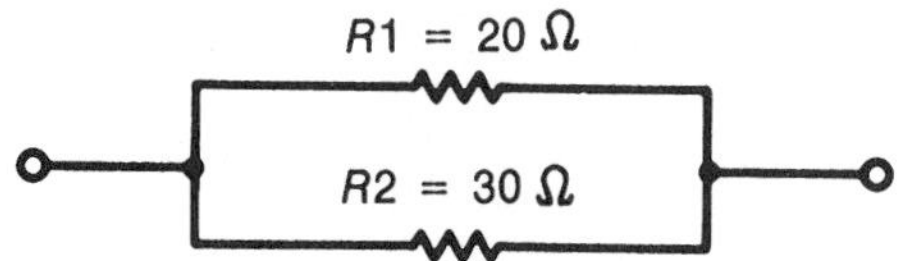

Fig. 8-7. Diagram for Problem 1.

FASCINATING FACTS

Adding more branches in parallel wiring lowers the total circuit resistance, but increases the total current. This may seem to be a contradiction. However, this phenomenon is similar to what happens when people leave a crowded theater that has several exits. Although there is resistance to movement at each exit, a larger number of exits allows a larger overall rate of movement.

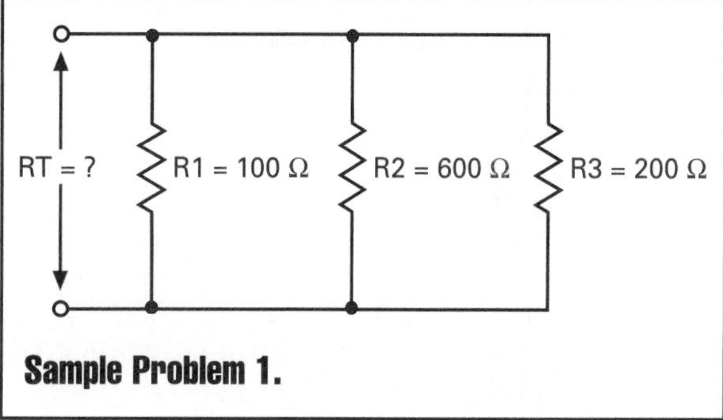

Sample Problem 1.

Problem 2: Two resistors with values of 20 Ω and 30 Ω are connected in parallel (Fig. 8-8). Calculate the total resistance.

Solution: Formula 2 is the easiest one to use in this case.

$$R_t = \frac{R_1 \times R_2}{R_1 + R_2} = \frac{20 \times 30}{20 + 30} = \frac{600}{50} = 12\ \Omega$$

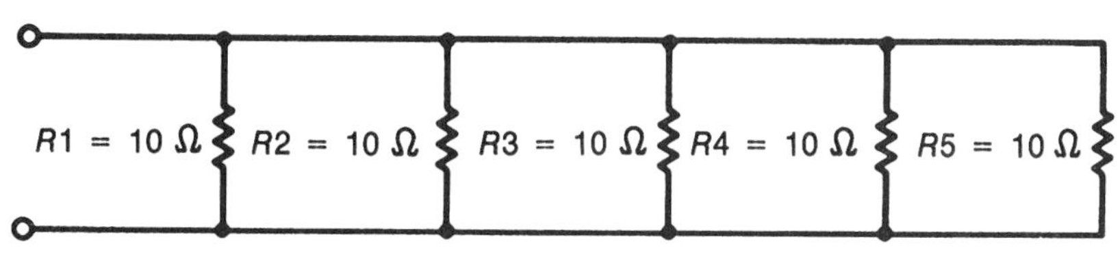

Fig. 8-8. Diagram for Problem 2.

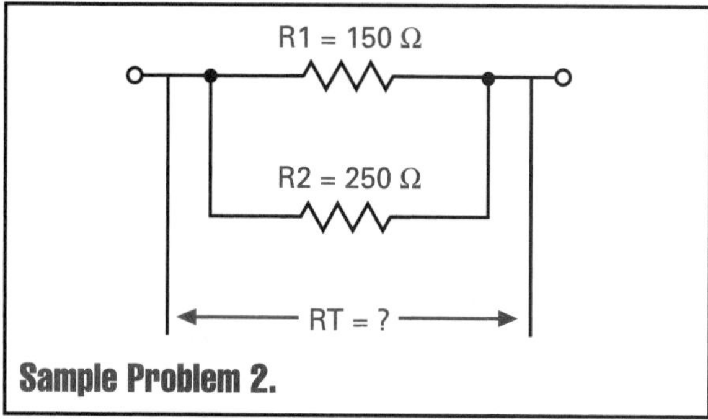

Sample Problem 2.

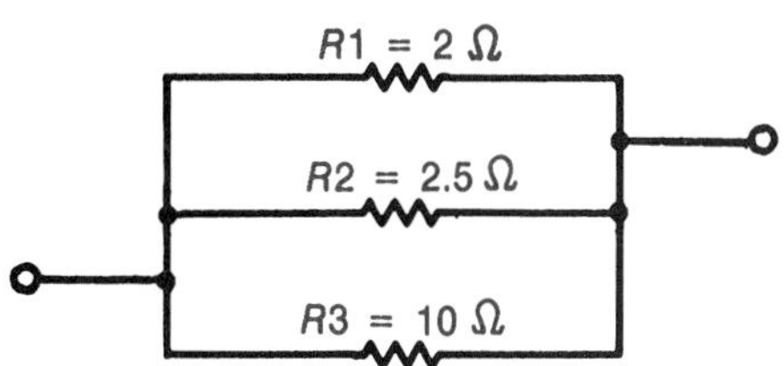

Fig. 8-9. Diagram for Problem 3.

Problem 3: Five resistors with a resistance of 10 Ω are connected in parallel (Fig. 8-9). Calculate the total resistance.

Solution: Use formula 2 because the resistors all have the same value.

$$R_t = \frac{\text{value of one resistance}}{\text{number or resistances}} = \frac{10}{5} = 2\ \Omega$$

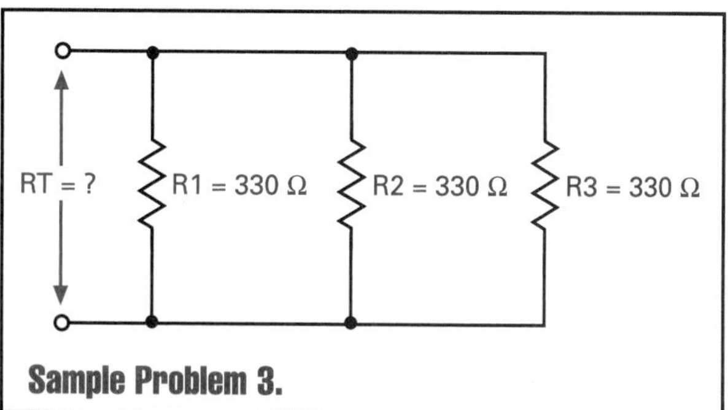

Sample Problem 3.

Problem 4: An engineer receives a circuit board with the circuit shown in Fig. 8-10. What resistance must be connected in parallel with that circuit board to obtain a total resistance of 1,482 Ω?

Solution: Use formula 1.

1. Use the *general purpose formula* to solve for R_3.

$$R_t = \frac{1}{\frac{1}{R_1} + \frac{1}{R_2} + \frac{1}{R_3}}$$

$$\frac{1}{R_t} = \frac{1}{R_1} + \frac{1}{R_2} + \frac{1}{R_3}$$

$$\frac{1}{R_3} = \frac{1}{R_t} - \left(\frac{1}{R_1} + \frac{1}{R_2}\right)$$

$$\frac{1}{R_3} = \frac{1}{1{,}482} - \left(\frac{1}{4{,}700} + \frac{1}{22{,}000}\right)$$

$$\frac{1}{R_3} = 0.0006747 - (0.0002127 + 0.0000454)$$

$$\frac{1}{R_3} = 0.0004166$$

$$R_3 = \frac{1}{0.0004166}$$

$$R_3 = 2400.38 = 2.4\ \text{k}\Omega$$

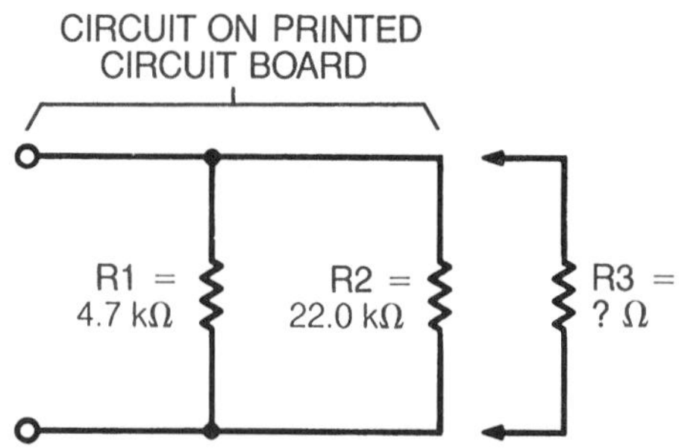

Fig. 8-10. Diagram for Problem 4.

2. Verify the results by substituting the calculated resistance (2.4 kΩ) for R_3.

$$R_t = \cfrac{1}{\cfrac{1}{R_1} + \cfrac{1}{R_2} + \cfrac{1}{R_3}}$$

$$= \cfrac{1}{\cfrac{1}{4{,}700} + \cfrac{1}{22{,}000} + \cfrac{1}{2{,}400}}$$

$$= \frac{1}{0.0002127 + 0.0000454 + 0.0004166}$$

$$= \frac{1}{0.0006747} = 1482.4 = 1{,}482\ \Omega$$

NOTE: This verifies the results. The calculated resistance of 2.4 kΩ connected in parallel with 4.7 kΩ and 22 kΩ equals 1,482 Ω.

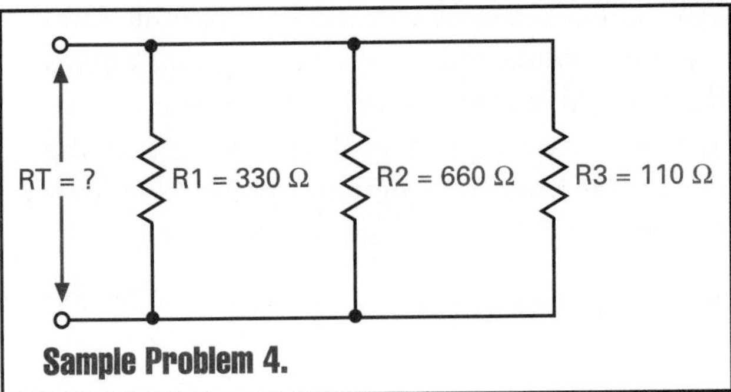

Sample Problem 4.

Problem 5: Three resistors valued at 50 Ω, 100 Ω, and 1,000 Ω are connected in parallel (Fig. 8-11). Calculate the total resistance, the total current, and the branch currents.

Solution:

1. The solution to this problem involves several steps. First, determine the total resistance in the circuit. Then, find the total and individual branch currents. Use formula 3.

$$R_t = \cfrac{1}{\cfrac{1}{R_1} + \cfrac{1}{R_2} + \cfrac{1}{R_3}}$$

$$= \cfrac{1}{\cfrac{1}{50} + \cfrac{1}{100} + \cfrac{1}{1{,}000}}$$

$$= \frac{1}{0.02 + 0.01 + 0.001}$$

$$= 32.3\ \Omega$$

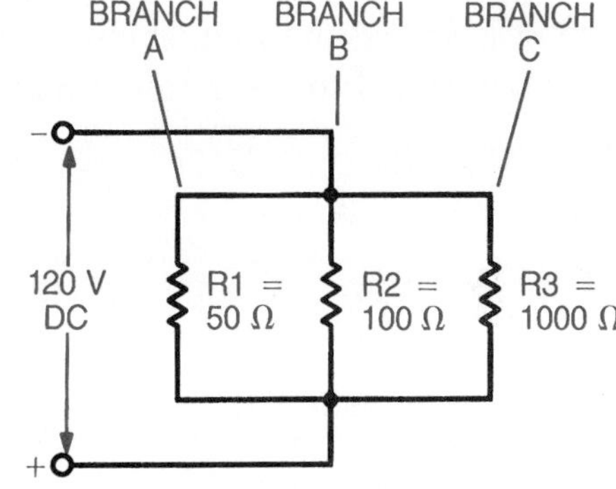

Fig. 8-11. Diagram for Problem 5.

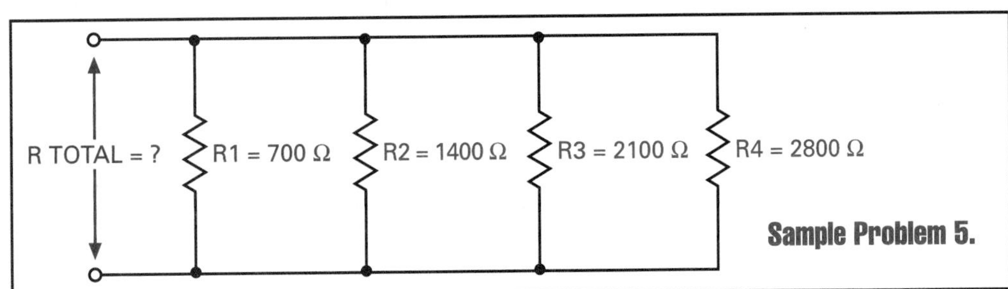

Sample Problem 5.

2. Calculate the total current in the circuit. Since the voltage and total resistance are known, use Ohm's law.

$$I_t = \frac{E_t}{R_t} = \frac{120}{32.3} = 3.72 \text{ A}$$

3. Calculate the individual currents in each branch circuit. Use Ohm's law for each part or branch in the circuit.

$$I_{\text{branch A}} = \frac{E}{R_1} = \frac{120}{50} = 2.4 \text{ A}$$

$$I_{\text{branch B}} = \frac{E}{R_2} = \frac{120}{100} = 1.2 \text{ A}$$

$$I_{\text{branch C}} = \frac{E}{R_3} = \frac{120}{1{,}000} = 0.12 \text{ A}$$

$$\text{Total} = 3.72 \text{ A}$$

4. Compare the total current of the individual branch circuits with step 2. Notice that the total current is the same. This supports another law: the sum of individual branch currents is equal to the total current in a parallel circuit (Fig. 8-12).

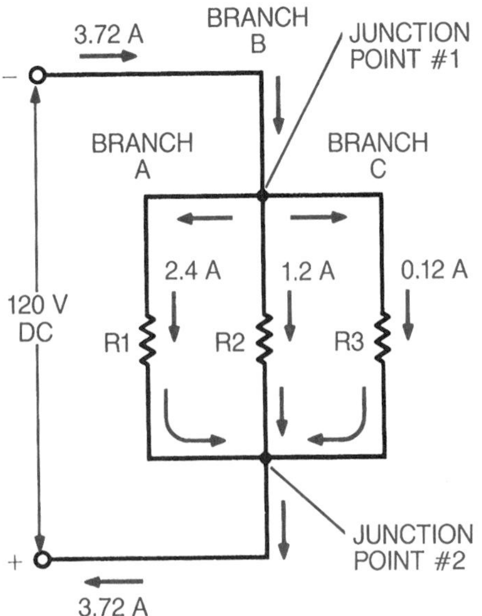

Fig. 8-12. The direction of current in the branch circuits from Problem 5. Notice that the current flowing from a junction point equals the current flowing to that junction point.

1. **Determine the branch currents of your circuit and test them using Kirchhoff's current law.**
2. **Determine the amount of current available to a half-charged car battery when it is connected in parallel to a booster battery with 100 A of current capacity.**

CURRENT AT JUNCTION POINTS

In the study of series circuits, you learned that the sum of the voltage drops is equal to the applied voltage. This is known as Kirchhoff's voltage law. **Kirchhoff's current law**, equally important, states that the total current flowing out of a junction point equals the current flowing into that junction point. This is illustrated in Fig. 8-12. Notice that the current flowing from junction point #1 equals the current flowing to junction point #2.

CELLS IN PARALLEL

Cells or batteries (as well as other types of voltage sources) can be connected in parallel to get more current.

LINKS

Read ahead to Chapter 16 to learn more about other voltage sources.

Rubidium

Gustav Kirchhoff, with Robert Bunsen, discovered the element rubidium in 1860. They named the element after the red lines in its spectrum, which were visible with a spectroscope. Today, rubidium is used in making certain catalysts and photoelectric cells.

Metallic rubidium is silvery white and very soft. One of the most active of the alkali metals, it instantly tarnishes when exposed to air and spontaneously ignites to form rubidium oxide. It also reacts violently with water. Rubidium melts at about 102°F and boils at about 1267°F. In its general chemical behavior, rubidium's performance resembles the alkali metals sodium and potassium.

Of the elements found in Earth's crust, rubidium ranks twenty-third in abundance. However, it is found only in small amounts in certain mineral waters and in many minerals usually associated with other alkali metals. It is also found in small quantities in tea, coffee, tobacco, and other plants. Trace quantities of the element may be necessary for living organisms.

With a parallel connection, the total voltage of the combination is equal to the sum of the voltages of all the branches (Fig. 8-13). The current-delivering capacity of the combination, however, is equal to the sum of the current capacities of all the cells or batteries.

Using a booster battery is an interesting and useful example of connecting batteries in parallel. The **booster battery** is an additional battery temporarily connected in parallel to an automobile battery that won't deliver enough current to start a car. The voltage applied to the automobile electrical system is still 12 V, but the booster battery will supply enough extra current to turn the cranking motor and start the engine. The plus terminal of the booster battery is connected to the plus terminal of the automobile's battery, and the two negative terminals are also connected. Using booster batteries is called "jump starting."

SAFETY TIP

Eye protection should be used when connecting a booster battery because battery acid can be splashed or an explosion can occur if the cables are improperly connected.

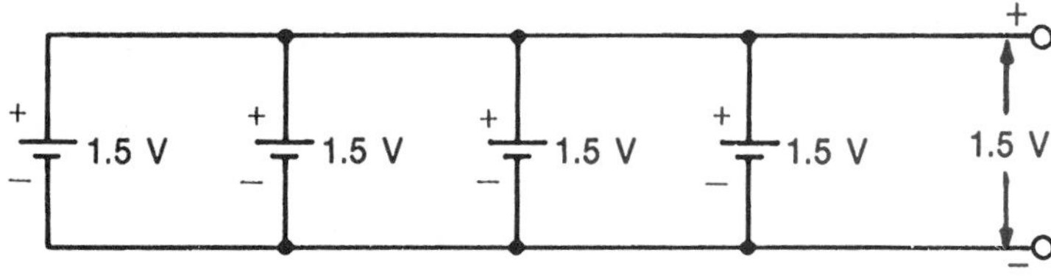

Fig. 8-13. The total voltage obtained when cells are connected in parallel is equal to the voltage of the individual cells.

Chapter *Review*

Summary

- Parallel circuits have branch circuits in which loads are connected "across the line." These loads operate independently. Therefore, each load will continue to operate if another load fails.
- Although the branch loads share a voltage source, they have different current demands. Voltage sources such as batteries can be wired in parallel to boost the current available to a circuit in order to meet those demands. If the current demand of a load becomes too great, however, it will blow a fuse or trip a breaker.
- The amount of current in a circuit depends on the total resistance of its loads.

Review Main Ideas

Review this chapter's main ideas by writing, on a separate sheet of paper, the word or words that most correctly complete the following statements:

1. In a parallel circuit, the loads are connected between the two ______ that lead to the source of energy.
2. All branches of a parallel circuit have the same ______ as the source.
3. The total current delivered to a parallel circuit is equal to the ______ of the ______ currents.
4. The total resistance of a parallel circuit is always ______ than that of the branch with the lowest resistance.
5. The formula for finding the total resistance of two parallel resistances having unequal values is ______.
6. The formula for finding the total resistance of two or more resistances that are equal in value and connected in parallel is ______.
7. The general formula for finding the total resistance of two or more resistances not all of equal value and connected in parallel is ______.
8. Cells and batteries are connected in parallel to get more ______.
9. When cells of equal voltage are connected in parallel, their combined voltage is equal to the sum of the voltages of ______.

Chapter Review

Apply Your Knowledge

1. Draw a parallel circuit with a battery source of 100 V and three loads: 20 Ω, 100 Ω, and 1 kΩ.
2. Using the general formula, determine the total resistance of this circuit.
3. Determine the branch and total currents of the circuit.
4. Draw a circuit with three 25-A sources in parallel. What is the total current available to the circuit?
5. Explain how to connect a booster battery to your car.

Make Connections

1. **Communication Skills.** Write a report comparing and contrasting series and parallel circuits.
2. **Mathematics.** Select several household appliances or power tools in your home. Determine which can be safely operated simultaneously on a 15 A 120 Vac circuit.
3. **Science.** Demonstrate to your class how adding resistors to a parallel circuit affects the required current from the power supply.
4. **Technology Skills.** Design a residential electrical system that does not use parallel connections for room receptacles. Analyze the result of your design. Evaluate whether it is an improvement over the parallel system used today.
5. **Social Studies.** Explain to the class why fuses were necessary when the parallel circuit was established as a standard for electrical appliances.

Answers to Sample Problems

1. 60 Ω
2. 93.75 Ω
3. 110 Ω
4. 73.33 Ω
5. 336 Ω

Chapter 9

Series-Parallel Circuits

Terms to Study

combination circuit

electrical system

network

series-parallel circuit

OBJECTIVES

After completing this chapter, you will be able to:

- Define the term *network*.
- Calculate the total resistance of two loads in series with a parallel circuit of two unequal loads.
- Determine the total resistance of a network.
- Calculate the total current of a series-parallel circuit, the voltage drop across each load, and the current flow in each.
- Perform the procedure that determines the resistance, voltage, and current in series-parallel circuits.

The previous two chapters discussed circuit loads that were connected in either series or parallel. This chapter discusses ways in which series and parallel circuits are connected together.

Do you know that the major circuit assemblies we use every day can be reduced to individual, simpler circuits? Do you know that a complicated circuit can be simplified as several circuits that are easier to understand? In this chapter, you will learn how to simplify complex resistance circuits using what you have learned from the previous chapters and in class.

SERIES-PARALLEL CIRCUITS

When connected together, they form circuits called **series-parallel circuits**, or **combination circuits**. Several interconnected series-parallel circuits are often called a **network**, or **electrical system**. Most electrical circuits are combination circuits. Automobiles, televisions, radios, and computers have many combination circuits that make up their electrical systems.

Figure 9-1 is an example of a series-parallel circuit. Load 1 is in series with a parallel circuit of loads 2 and 3. Notice that the current must first pass through load 1. Then the current separates into two paths at junction point A where it enters the parallel circuit. The two paths join again at junction point B where the current leaves the parallel circuit to return to the energy source. This illustration represents only one way in which series-parallel circuits are connected.

LINKS

If you need to review information on parallel circuits, refer to Chapter 8.

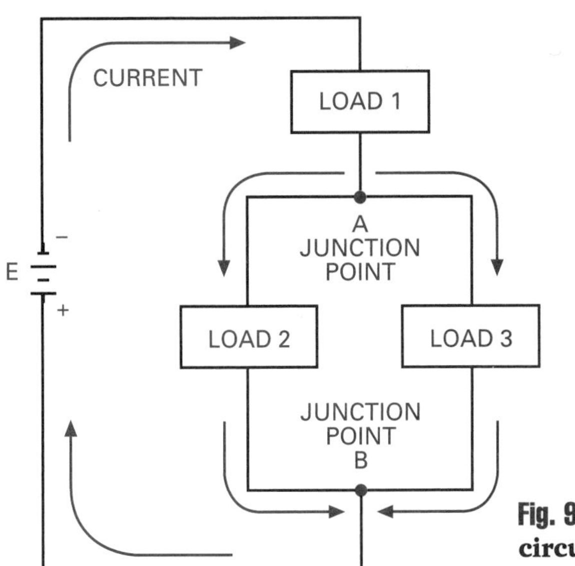

Fig. 9-1. **A series-parallel, or combination, circuit.**

Examples of Series-Parallel Circuits

Figure 9-2 shows several sketches of series-parallel circuits. Some are simple, others complex. These sketches show how the loads, represented by the symbol for resistance, are connected. In every sketch, one or more loads are in series with parallel loads. You may wish to compare the similarity and differences of these sketches with those shown in the two previous chapters. When you want to convey more complete information about a series-parallel circuit, draw your sketch as shown in Fig. 9-2A. This sketch contains basic information about the circuit. All the sketches, however, are meaningful and are often used by technicians and engineers to quickly convey information about series-parallel circuits.

Resistance

Let us take a close look at the series-parallel circuit shown in Fig. 9-3. In this circuit, loads 1 and 4 are in series with the combination of loads 2 and 3. Loads 2 and 3 are a parallel circuit.

FASCINATING FACTS

Often, circuits are added to the other devices in electrical equipment as safety mechanisms. In an elevator, for example, electric circuits are completed by contact points in the shaft doors on various floors and in the car gates. Completing the circuits in this manner prevents car operation when the gates and doors are open.

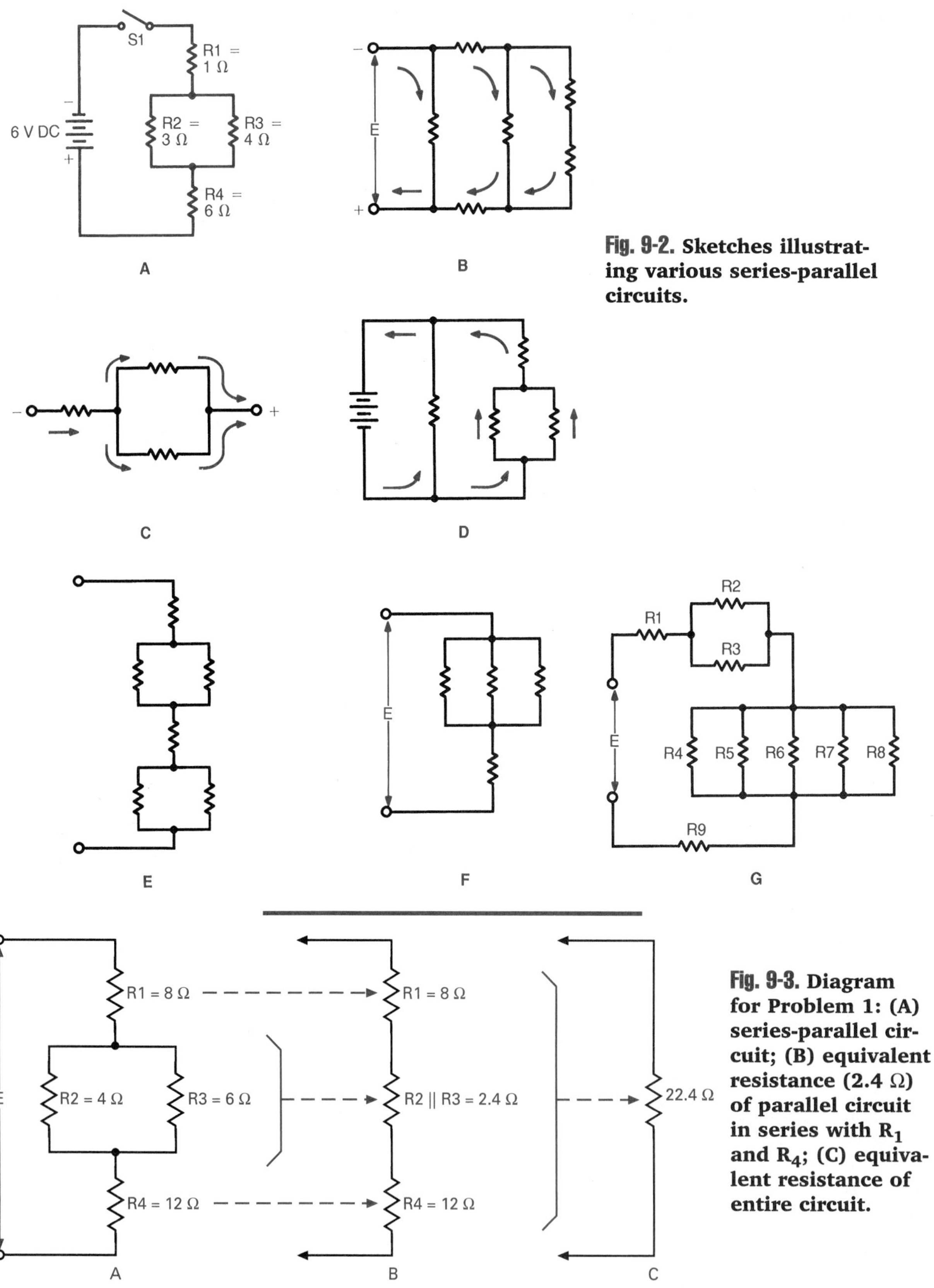

Fig. 9-2. Sketches illustrating various series-parallel circuits.

Fig. 9-3. Diagram for Problem 1: (A) series-parallel circuit; (B) equivalent resistance (2.4 Ω) of parallel circuit in series with R_1 and R_4; (C) equivalent resistance of entire circuit.

Wheatstone Bridge

An accurate measurement of resistance can be made with a galvanometer in a circuit called a Wheatstone bridge. This was named after the British physicist Sir Charles Wheatstone (1802-1875). Although invented by Samuel Christie (1784-1865), the circuit bears Wheatstone's name because he was the first to use it for measuring resistance in electric circuits.

This circuit consists of three known resistances and an unknown resistance connected in a diamond pattern. A dc voltage connects two opposite points of the diamond, and a galvanometer bridges the other two points. When all four resistances bear a fixed relationship to each other, the currents flowing through the two arms of the circuit will be equal, and no current will flow through the galvanometer. By varying the value of one of the known resistances, the bridge can be made to balance for any value of unknown resistance. The unknown resistance can then be calculated from the values of the other resistors.

Problem 1: What is the total resistance of the circuit shown in Fig. 9-3A?

Solution: To find the total resistance of a series-parallel combination, it is helpful to consider the individual parts that make up the total circuit. In this series-parallel circuit, for example, the parallel circuit is considered first.

1. Find the total resistance of the parallel part of the circuit made up of R2 and R3 (Fig. 9-3A).

$$R_t = \frac{R_2 \times R_3}{R_2 + R_3} = \frac{4 \times 6}{4 + 6} = \frac{24}{10} = 2.4\ \Omega$$

Since the parallel part of the circuit has an equivalent resistance of 2.4 Ω, you can now redraw the circuit showing the 2.4 Ω in series with R_1 and R_4 (Fig. 9-3B). This completes the first step in solving the problem.

2. Find the total resistance of the circuit. Use the redrawn circuit shown in Fig. 9-3B.

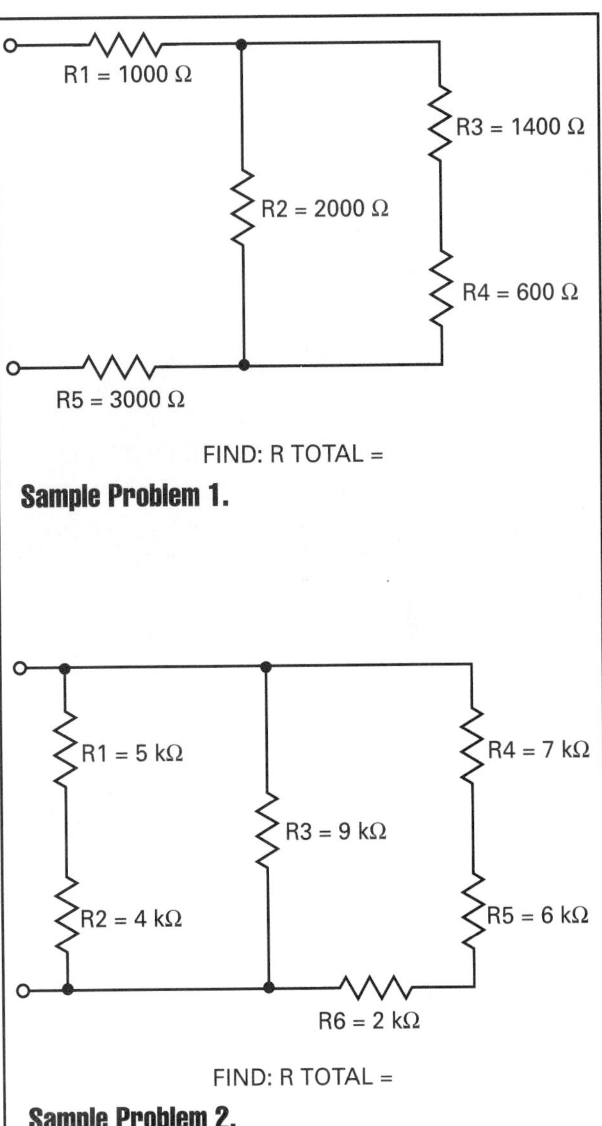

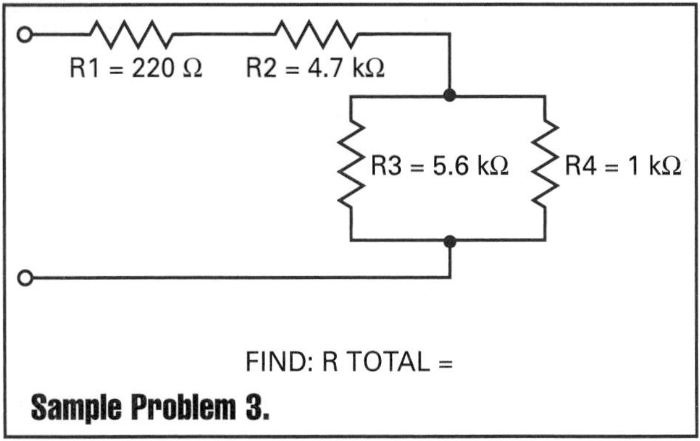

Sample Problem 3.

$$R_t = R_1 - R_2 \parallel R_3 + R_4$$
$$= 8 + 2.4 + 12$$
$$= 22.4\ \Omega$$

NOTE: In the following formula the mathematical symbol || means "parallel to." In step 1, the equivalent resistance of R_2 || R_3 is 2.4 Ω.

Figure 9-3 shows the steps followed in calculating the total resistance of this series-parallel circuit. In summary, first, analyze carefully each series-parallel circuit to determine its individual parts. Second, consider each individual part as a unit. In this circuit, the parallel circuit was considered as a separate unit. Third, combine the individual parts to form a simple series equivalent circuit.

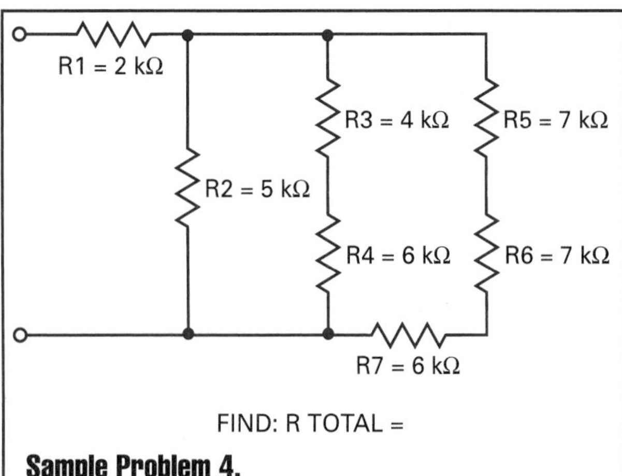

Sample Problem 4.

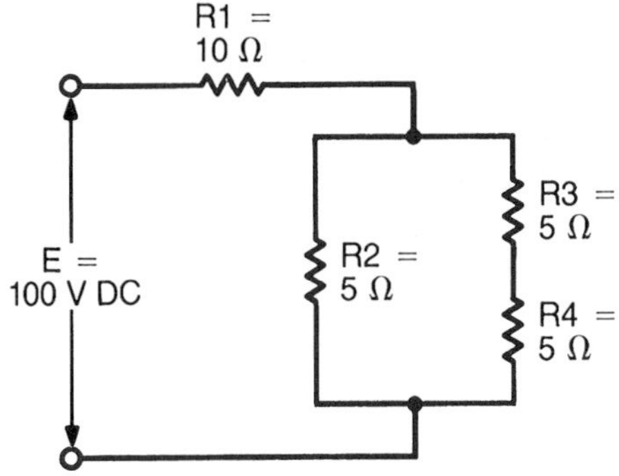

Fig. 9-4. Diagram for Problems 2 and 3.

Problem 2: Four resistors are connected as a series-parallel circuit (Fig. 9-4). Resistance R_1 is in series with a parallel circuit. The parallel circuit consists of two branches. One branch of the parallel circuit is made up of resistance R_2, and the other branch is made up of resistances R_3 and R_4. Find the total resistance.

Solution:

1. Since R_3 and R_4 are in series with each other in one branch of the parallel circuit, determine the total resistance R_{t1} of this series "string."

   $$R_{t1} = R_3 + R_4 = 5\ \Omega + 5\ \Omega = 10\ \Omega$$

2. Determine the total resistance R_t of the parallel part of the circuit.

   $$R_{t2} = \frac{R_2 \times R_{t1}}{R_2 + R_{t1}} = \frac{5 \times 10}{5 + 10} = \frac{50}{15} = 3.3\ \Omega$$

3. Since the parallel part of the circuit has an equivalent resistance of 3.3 Ω, calculate the total resistance R_{t3} of the series-parallel circuit.

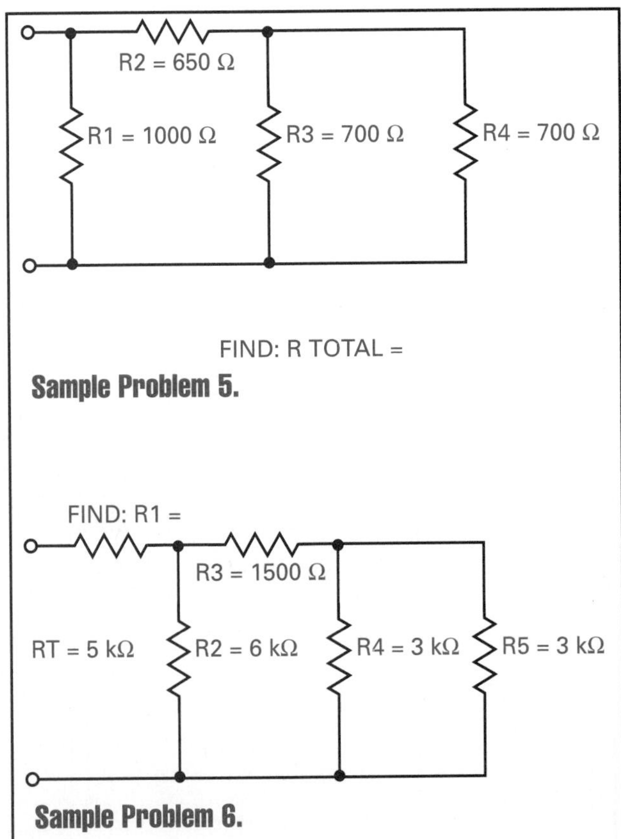

Sample Problem 5.

Sample Problem 6.

$$R_{t3} = R_1 + R_2 \parallel R_{t1} = 10 + 3.3 = 13.3\ \Omega$$

Voltage and Current

The previous problems illustrate the process of determining the resistance of a series-parallel circuit. The next problems explain how Kirchhoff's voltage and current laws are applied in series-parallel circuits. You already know that Kirchhoff's voltage law in series circuits states that the sum of the voltage drops in a circuit must equal the source voltage. This is also true with series-parallel circuits. However, the voltage across any parallel branch must be considered as one of the voltage drops in the total circuit. The current in these combination circuits also follows Ohm's law if the proper voltage and resistance are considered for that part of the circuit. Let us now examine the behavior of current and voltage in series-parallel or combination circuits.

Problem 3: Refer to the series-parallel circuit in Fig. 9-4. Find the total current in the circuit, the current in each resistance, and the voltage drops across each resistance.

Solution:

1. Since the total voltage is given and the total equivalent resistance (as calculated in Problem 2) is known, use Ohm's law formula to determine the total current.

$$I_t = \frac{E_t}{R_t} = \frac{100}{13.3} = 7.5\text{ A}$$

2. Calculate the voltage drop across R_1. Since the total current is known and R_1 is in series with the parallel circuit, you can use the Ohm's law formula for *this part of the circuit.*

$$E = IR = 7.5 \times 10 = 75\text{ V}$$

3. Find the voltage drop across the parallel circuit. We know from Kirchhoff's voltage law that the sum of the voltage drops around the circuit is equal to the applied voltage. Since the voltage drop across R_1 is 75 V, the voltage drop across the parallel circuit is 100 V minus 75 V or 25 V. The following calculation verifies that the sum of the voltage drops equals the applied voltage.

$$\begin{aligned} E_t &= IR_1 + I[R_2 \parallel (R_3 + R_4)] \\ &= 7.5 \times 10 + 7.5 \times 3.3 \\ &= 75 + 24.75 \\ &= 99.75\text{ V} \end{aligned}$$

The voltage drop calculated across the series and parallel circuit is 99.75 V. This is approximately equal to the applied voltage. The slight difference is because numbers were rounded during the calculations.

4. Since the voltage (25 V) across the parallel circuit is known, find the current in resistances R_2, R_3, and R_4. Use Ohm's law for calculating the current for this part of the circuit.

$$I = \frac{E}{R_2} = \frac{25}{5} = 5.0 \text{ A}$$

$$I = \frac{E}{R_3 + R_4} = \frac{25}{10} = 2.5 \text{ A}$$

$$\text{Total} = 7.5 \text{ A}$$

The current in the parallel branch with resistance R_2 is 5 A. The current in the parallel branch with resistances R_2 and R_3 is 2.5 A. Since R_2 and R_3 form a series "string," the current is the same in each resistance. Notice that the total current that passes through the parallel part is 7.5, the calculated total current for the entire circuit. The current in the various parts of this circuit is shown in Fig. 9-5. Notice that the current separates at junction point A and joins again at point B. The same amount of current flows out from point B as flows into point B. This verifies Kirchhoff's current law: the sum of the currents entering a point equals the sum of the currents leaving that point.

5. Calculate the voltage drops across the series string R_3 and R_4 within the parallel circuit. We know that the total voltage across the parallel circuit is 25 V and the current in the resistance string is 2.5 A. Use Ohm's law to find the voltage drops across this part of the circuit.

$$E = IR_1 = 2.5 \text{ x } 5 = 12.5 \text{ V}$$

$$E = IR_2 = 2.5 \text{ x } 5 = 12.5 \text{ V}$$

$$\text{Total} = 25.5 \text{ V}$$

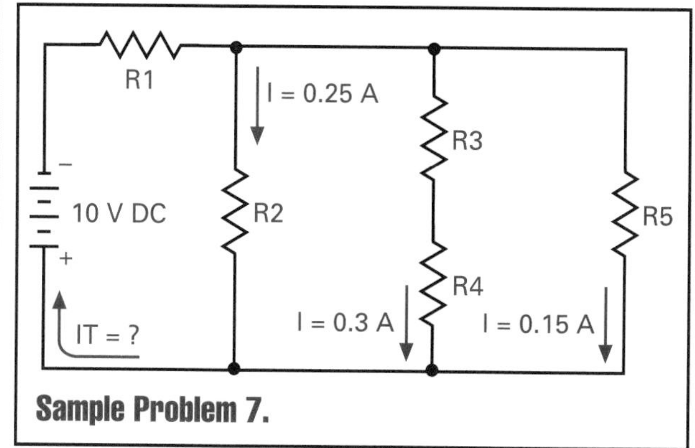

Sample Problem 7.

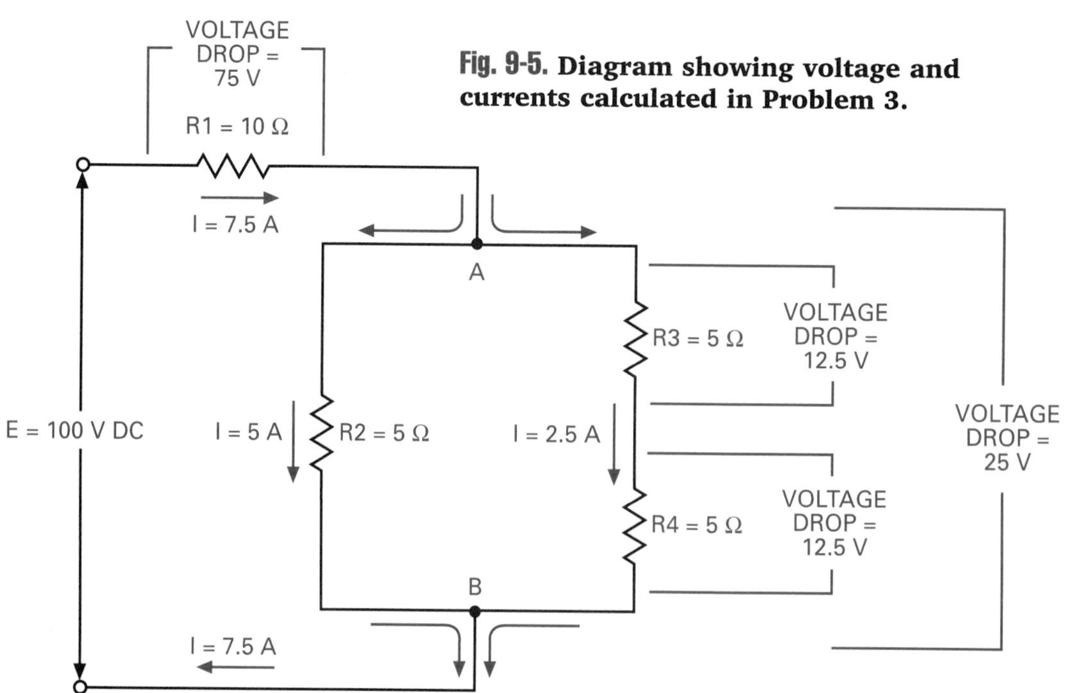

Fig. 9-5. Diagram showing voltage and currents calculated in Problem 3.

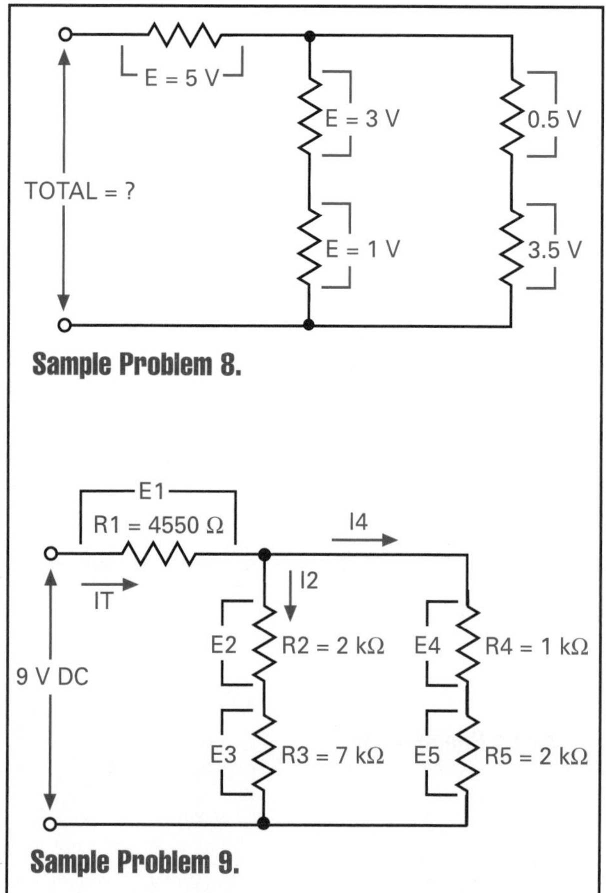

Sample Problem 8.

Sample Problem 9.

Think about the appliances you have used today. Which of them would have series-parallel circuits?

ANALYZING COMPLEX CIRCUITS

In analyzing complex series-parallel circuits, it is often convenient to rearrange the network into an equivalent circuit so that the series-parallel sections are easily recognized. This is illustrated in the next problem.

Problem 4: Find the total resistance and total current of the circuitry in Fig. 9-6A.

Solution:

1. Analyze the circuit shown in Fig. 9-6A and redraw it to more clearly show the series and parallel parts of the total circuit (Fig. 9-6B).

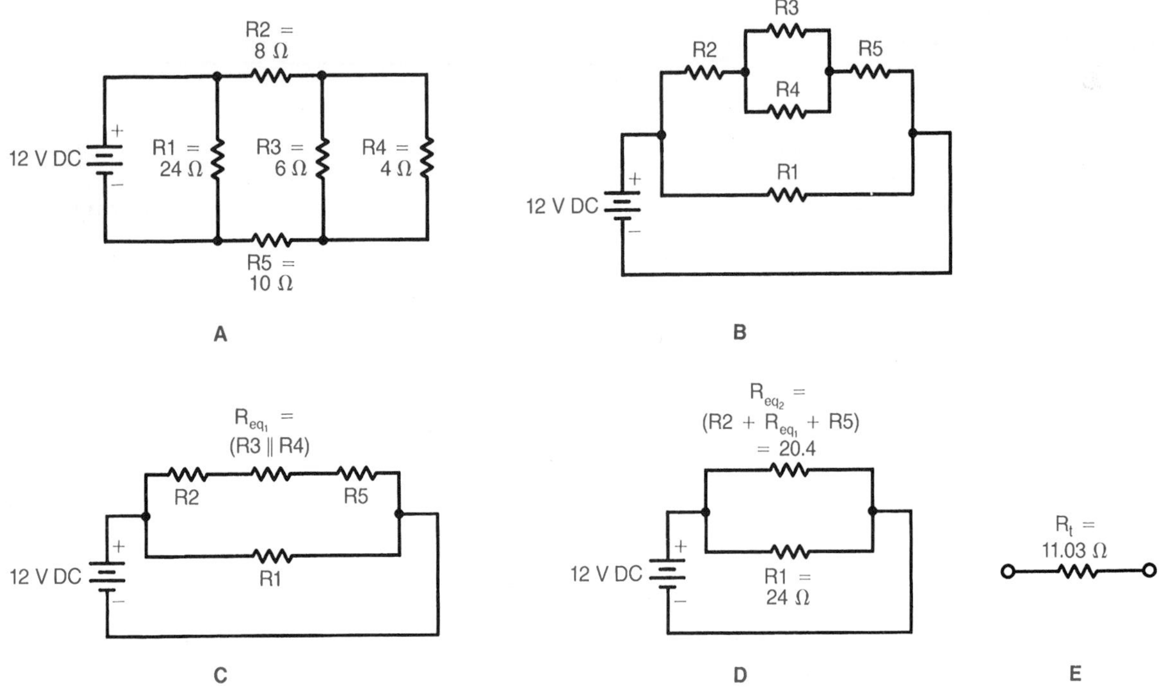

Fig. 9-6. Diagrams for Problem 4.

2. Determine the equivalent resistance (R_{eq1}) of the parallel circuit consisting of R_3 and R_4.

$$R_{eq1} = \frac{R_3 \times R_4}{R_3 + R_4} = \frac{6 \times 4}{6 + 4} = \frac{24}{10} = 2.4\Omega$$

3. The circuit now can be illustrated as shown in Fig. 9-6C. Calculate the total equivalent distance (R_{eq2}) of the upper branch.

$$R_{eq2} = R_2 + R_{eq1} + R_5 = 8 + 2.4 + 10 = 20.4\ \Omega$$

4. The circuit can now be redrawn as a simple parallel circuit (Fig. 9-6D). Find the resistance of this parallel circuit.

$$R_t = \frac{R_{eq2} \times R_1}{R_{eq2} + R_1} = \frac{20.4 \times 24}{20.4 + 24} = \frac{89.6}{44.4} = 11.03\ \Omega$$

The total resistance of the circuit can be represented by a single resistance equal to 11.03 Ω (Fig. 9-6E).

5. Find the total current in the circuit. Use Ohm's law.

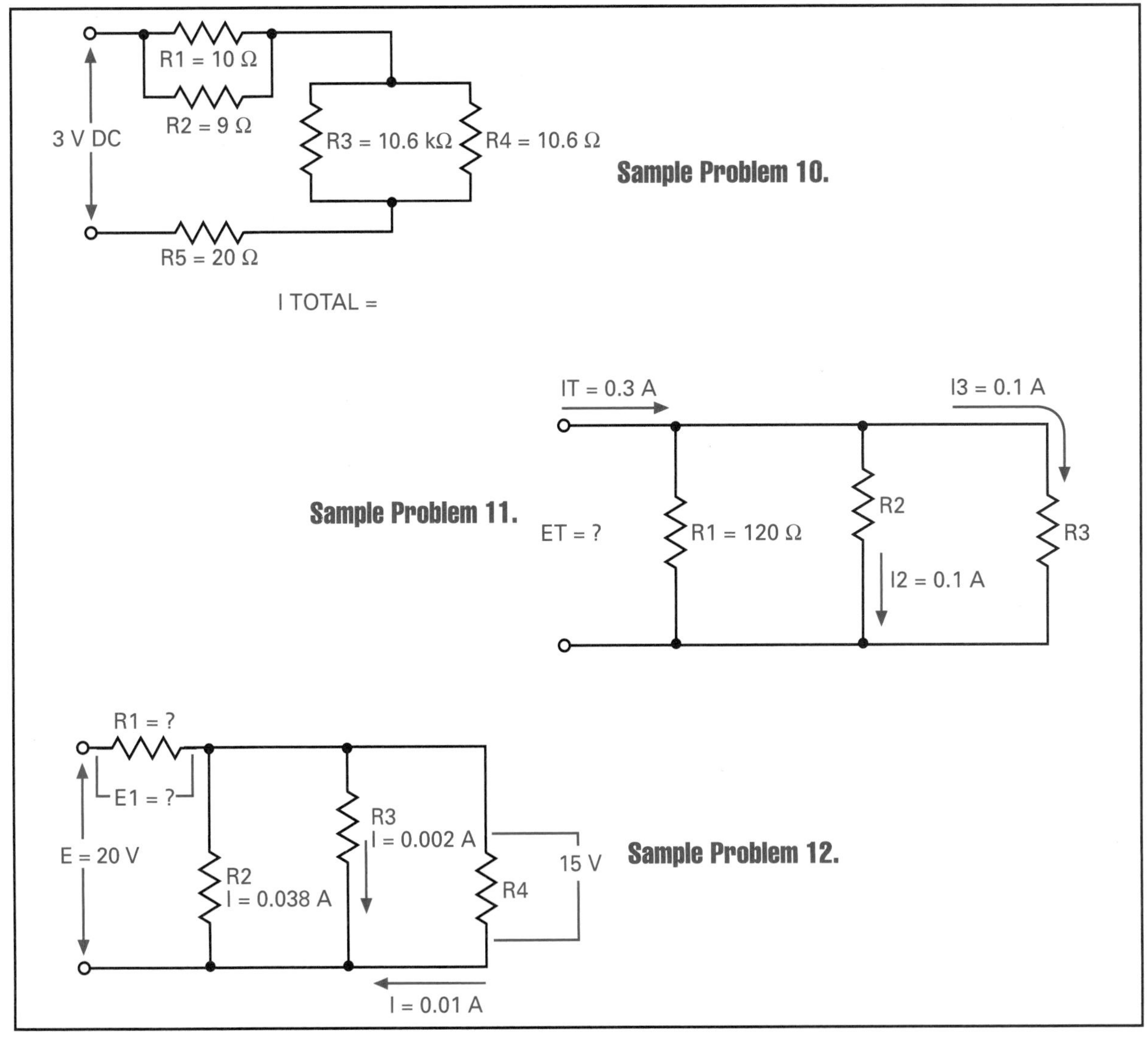

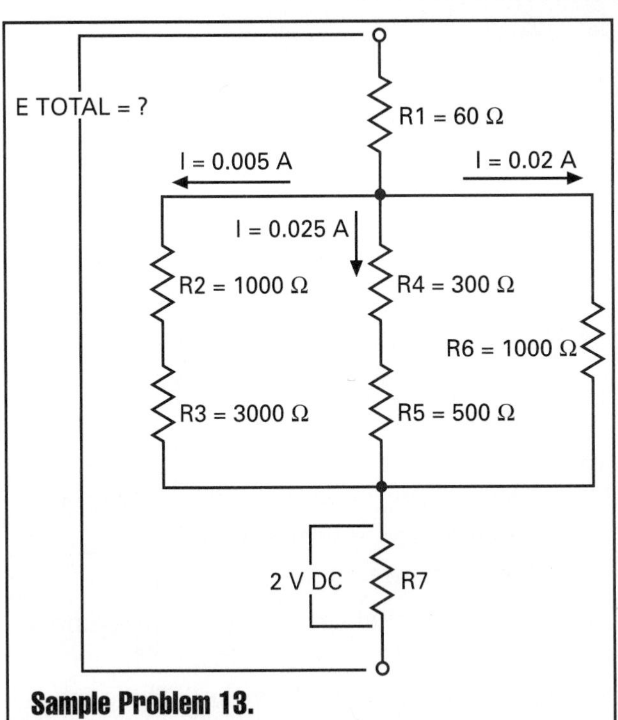

Sample Problem 13.

$$I_t = \frac{E_t}{R_t} = \frac{12}{11.03} = 1.09 \text{ A}$$

To determine the total resistance of a complex network:

1. Identify the loads of the network.
2. Combine parallel loads as equivalent resistances to create equivalent series circuits.
3. Determine the total resistance of the series circuit.

Chapter Review

Summary

- Series and parallel circuits can be combined to form series-parallel, or combination, circuits. These combination circuits can then be combined to form networks, or electrical systems.
- To analyze complex series-parallel circuits it is often easier to simplify the network by reducing it to equivalent circuits. Once simplified, the circuit currents and voltages can be found by using Kirchhoff's voltage and current laws.
- To analyze a network properly, however, it is necessary to know its total resistance, or impedance.

Review Main Ideas

Review this chapter's main ideas by writing, on a separate sheet of paper, the word or words that most correctly complete the following statements:

1. Several interconnected series-parallel circuits are called a ______, or electrical system.
2. A series-parallel circuit is often referred to as a ______ circuit.
3. Combination circuits have some loads connected in series and some in ______.
4. In solving series-parallel circuits, the resistances in the parallel circuits are first reduced to their ______.
5. In a series-parallel circuit, if two of the resistors form a series "string" with the energy source, the current is the ______ in each resistance.
6. A series-parallel circuit has two resistances in series with a parallel circuit. The parallel circuit has three branches. There are a total of ______ voltage drops in the circuit.
7. A series-parallel circuit is shown in Fig. 9-7. In solving the problem, the equivalent resistance of parallel circuit ______ should be determined first.

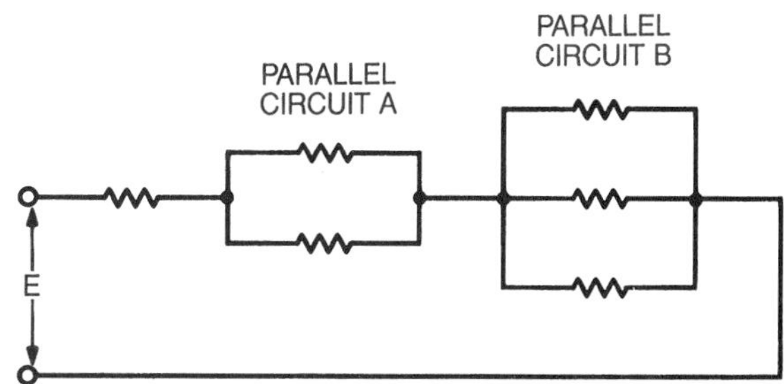

Fig. 9-7. Diagram for Review Main Ideas question 7.

8. The sum of the currents leaving a junction point ______ the sum of the current entering that junction point.

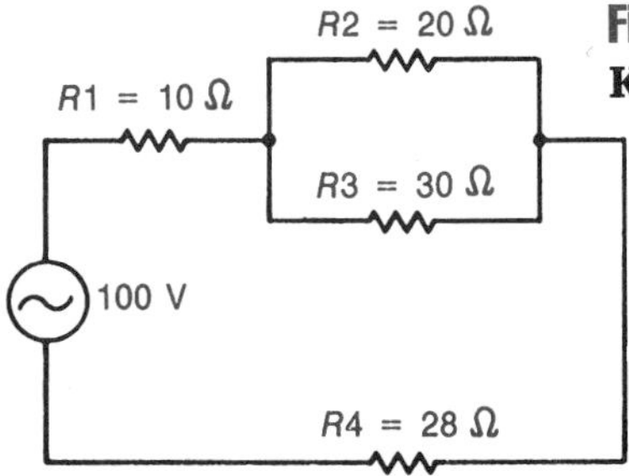

Fig. 9-8. Diagram for Apply Your Knowledge questions 1, 2, and 3.

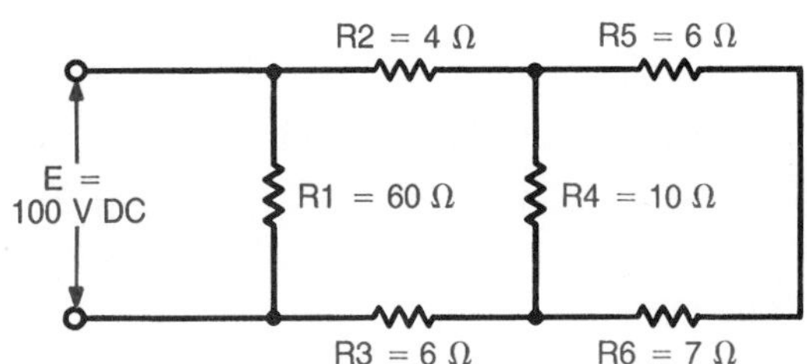

Fig. 9-9. Diagram for Apply Your Knowledge question 4.

Apply Your Knowledge

1. Find the equivalent resistance of the circuit in Fig. 9-8.
2. Find the total current of the circuit in Fig. 9-8.
3. Determine the voltage drop across each load in Fig. 9-8.
4. A series-parallel circuit is shown in Fig. 9-9. Find the voltage across R_6.
5. List the three steps in determining the total resistance of a complex network.

Make Connections

1. **Communication Skills.** Using a sample problem from this chapter, describe the process to follow in solving complex series-parallel circuits for total resistance, current, and voltage.
2. **Mathematics.** Select a complex circuit from the sample problems in this chapter. Develop a single equation that will solve for total resistance.
3. **Science.** Using 1/4 watt resistors, demonstrate how series-parallel combinations can be used to increase the power rating of resistive circuit elements.
4. **Technology Skills.** Obtain a schematic diagram of an electrical device such as a television set, a toaster, or an electric clothes dryer. Identify the parts in the diagram that are considered loads. Indicate the combinations of loads that make up series, parallel, and series-parallel circuits.
5. **Social Studies.** Explain to the class why it would be better to use parallel lighting circuits in the home than a combination of series-parallel circuits.

Answers to Sample Problems

1. 5,000 Ω
2. 3,000 Ω
3. 5,768.5 Ω
4. 4857 Ω
5. 500 Ω
6. 3,000 Ω
7. 0.7 A
8. 9 V
9. R_t = 6,800 Ω; I_t = 0.0013 A; E_1 = 6 V; I_2 = 0.33 mA; E_2 = 0.67 V; E_3 = 2.31 V; I_4 = 0.001 A; E_4 = 1 V; I_5 = 0.001 A; E_5 = 2 V
10. 0.1 A
11. 12 V
12. E = 5 V, R = 100 Ω
13. 25 V

Chapter 10

Capacitors

OBJECTIVES

After completing this chapter, you will be able to:

- Differentiate between the terms capacitance and capacitor.
- Specify the capacity of fixed and variable capacitors.
- Figure the time constant in a circuit that has resistance and capacitive reactance.
- Explain how to connect capacitors in series and in parallel.
- Show how to charge and discharge capacitors in a circuit with resistance.

Terms to Study

capacitance

capacitive reactance

capacitor

counter electromotive force

dielectric

electrostatic field

farad

impedance

plates

working voltage

Can we really harness electric energy and use it later? This can be done. In fact the storage of electrical energy is a key feature of many circuits and devices. This chapter explains how capacitors make it possible to accumulate a source of electrons to use when needed.

CAPACITOR ACTION

Capacitance is the ability of a circuit or a device to store electric energy. Although, almost all alternating-current circuits have capacitance, it may or may not be desirable. In circuits where a specific value of capacitance is needed, a device known as a *capacitor* may be used. The basic form of a **capacitor** is two conductors called **plates** separated by insulating material called the **dielectric.**

Capacitors perform several different circuit functions. Since they do not provide a continuous conducting path for electrons, they are used to block a direct current. An alternating current, however, can still flow through a capacitor. Capacitors are used with resistors and coils to filter, or smooth out, a varying direct current. Capacitors, along with resistors, form timing circuits that control other circuits. In radio and television receivers, capacitors are used with coils to form tuning circuits. These circuits let us choose the stations to which we want to listen. Capacitors are also very important components of oscillator circuits. These produce high-frequency alternating voltages used in timing circuits for most electronic equipment.

The action of a simple capacitor is shown in Fig. 10-1. When the plates are connected to a battery, electrons from the plate connected to the positive terminal of the battery move to the battery. This causes the plate to become positively charged. At the same time, an equal number of electrons are repelled to the other plate by the negative terminal of the battery. That plate becomes negatively charged. This produces a voltage across the plates (Fig. 10-1A). Because of the dielectric separating the plates, electrons cannot move directly from one plate to the other through the circuit that connects them. This flow of electrons in the wires connecting the capacitor to the battery continues until the voltage across the plates is equal to the voltage of the battery. When this happens, the capacitor is fully charged. An **electrostatic field**, which stores energy, exists between the plates of a charged capacitor.

FASCINATING FACTS

A relatively rare metal, tantalum, is used in capacitor production. At ordinary temperatures, this metal offers excellent resistance to nearly all acids. Tantalum capacitors store the largest amount of energy per unit volume. They are used extensively in miniaturized electrical circuitry.

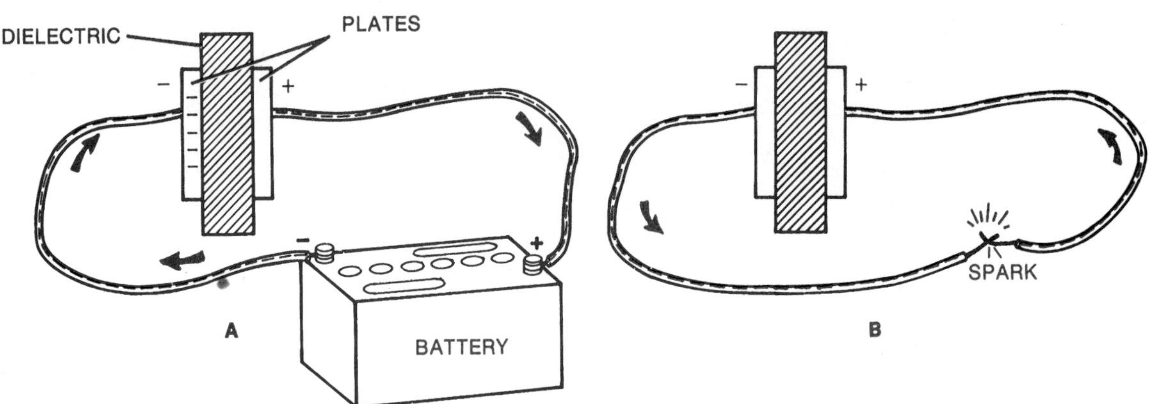

Fig. 10-1. Charging and discharging actions of a capacitor: (A) charging; (B) discharging.

When the battery is taken out of the circuit, the plates hold the charge. If the plates are now connected, electrons will flow through the circuit in the opposite direction. They will flow from the negatively charged plate to the positively charged plate (Fig. 10-1B). When the voltage across the plate decreases to zero, the flow of electrons stops. Now the capacitor is fully discharged.

Action in a Direct Current Circuit

The voltage across the plates of a fully charged capacitor is equal to the voltage of the battery to which it is connected. The capacitor voltage is also of the opposite polarity of the battery voltage. Because of this, a fully charged capacitor blocks current in a circuit (Fig. 10-2). The blocking action of a capacitor is used in many electronic circuits.

Action in an Alternating Current Circuit

When a capacitor is connected to a source of alternating voltage, the polarity of the applied voltage changes each half cycle. As a result, the capacitor is alternately charged, discharged, and recharged. It has the opposite polarity during each cycle (Fig. 10-3). However, there is no current flow through the dielectric that separates the plates.

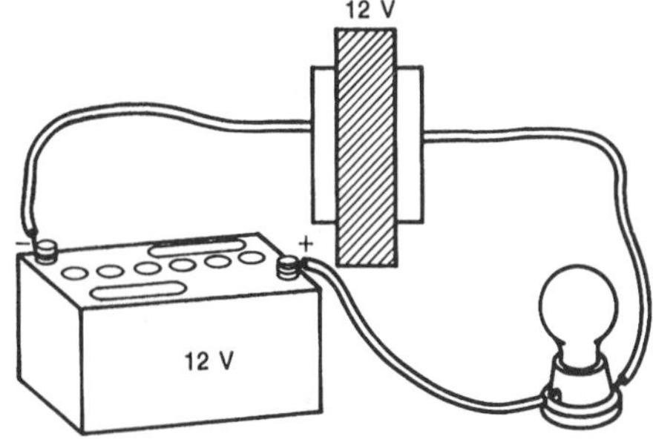

Fig. 10-2. The voltage across a fully charged capacitor is equal in value and polarity to the applied voltage. Thus, no current will flow in the circuit.

Capacitive Reactance

A capacitor allows the alternating current in a circuit to flow, but it does oppose the current. The opposition of a capacitor to the flow of alternating current is called **capacitive reactance.** The symbol for capacitive reactance is X_C. Its unit of measure is the ohm. The formula for figuring capacitive reactance is:

$$X_C = \frac{1}{2\pi f C}$$

where X_C = capacitive reactance in ohms
f = frequency in hertz
C = capacitance in farads
π = 3.14 (pi), a constant

Capacitive reactance is inversely proportional to both frequency and capacitance. This means that when the capacitance of a circuit is increased, the capacitive reactance decreases. The capacitive reactance also decreases when the frequency of the current in a circuit increases. In tuning circuits, capacitors determine the frequency at which the circuits operate.

Impedance

The total opposition to current in a circuit containing a combination of resistance and capacitive reactance is called **impedance**. It is also measured in ohms. The letter symbol for impedance is Z.

CAPACITANCE

The capacitance of a capacitor—the ability to store electric energy—depends on three things (1) the area of its plates—the greater the area, the greater the capacitance; (2) the spacing, or distance, between plates—the closer the spacing, the greater the capacitance; and (3) the dielectric material. The ability of a dielectric material to oppose electrostatic lines of force is its *dielectric constant*. Dielectric materials are air, Mylar, mica, and certain ceramic materials (Fig. 10-4). Other dielectric materials are paper, glass, and polystyrene.

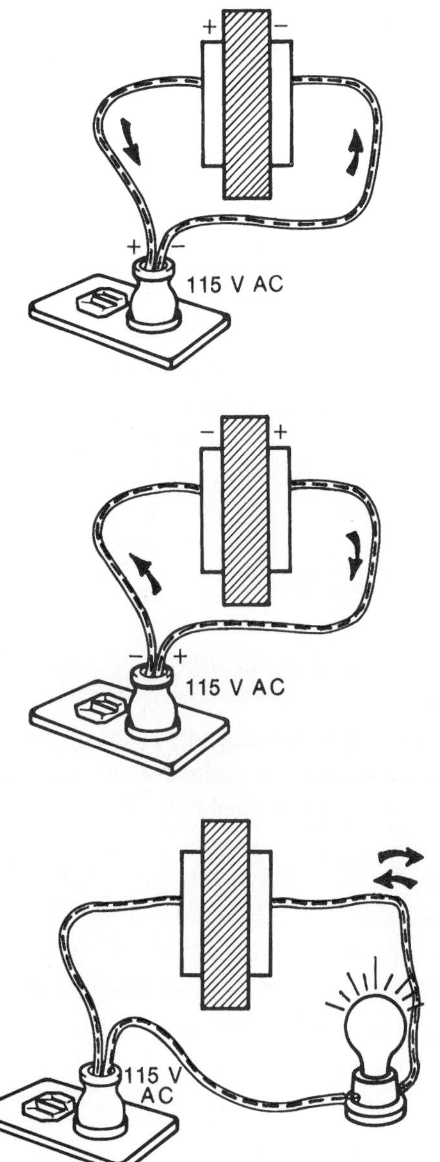

Fig. 10-3. Action of a capacitor in an ac circuit.

LINKS

Another component of impedance, called *inductive reactance*, is discussed in Chapter 12.

Unit of Capacitance

The basic unit of capacitance is the **farad**, abbreviated F. It is named in honor of Michael Faraday, an English scientist who studied electricity and lived from 1791 to 1867. A capacitor has a capacitance of one farad when an applied voltage that changes at the rate of one volt per second produces a current of one ampere in the capacitor circuit.

The capacitance of capacitors is most often given in microfarads (μF) or in picofarads (pF). One microfarad is equal to one-millionth of a farad. One picofarad is equal to one-millionth of a microfarad, or one trillionth of a farad.

Working Voltage

In addition to capacitance, fixed capacitors are also rated in terms of their working voltage

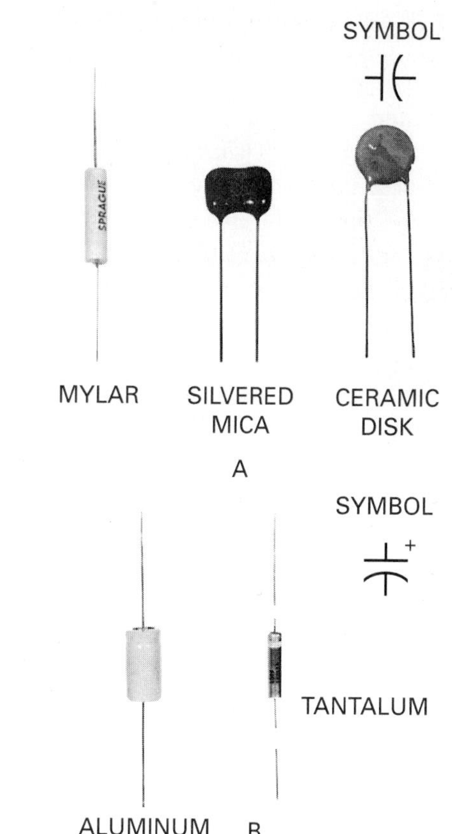

Fig. 10-4. Fixed capacitors: (A) nonelectrolytic; (B) electrolytic.

(WVdc). This is given in volts. **Working voltage** is the largest value of direct voltage that can safely be applied to a capacitor. If a higher voltage is applied, the dielectric breaks down and begins to conduct current. When this happens in a nonelectrolytic capacitor, the capacitor is said to be *leaking* and it must be replaced.

Draw two capacitors. Make one with larger plates and one with more dielectric. Specify which of the two has more capacitance. Explain why.

KINDS OF CAPACITORS

Capacitors are made in different ways. They are divided into two general classes: *fixed capacitors* and *variable capacitors*. Fixed capacitors have a specific single value of capacitance. Fixed capacitors are either *nonelectrolytic* or *electrolytic* which refers to the structure of the dielectric.

Michael Faraday

When asked what was his greatest discovery, the famous scientist Sir Humphry Davy answered, "Michael Faraday."

At age fourteen, Faraday was apprenticed to a bookbinder and read all the scientific books in the shop. He took careful notes when attending Davy's lectures, and sent them to Davy, asking for a job. Impressed by the boy's enthusiasm, Davy hired Faraday as a laboratory assistant.

Michael Faraday made several notable contributions to chemistry and electricity. His discoveries often led to other great developments. When he discovered hydrocarbon benzene in 1825, he became the father of an entire branch of organic chemistry.

In 1831 he found that moving a magnet through a coil of wire produced a current. In this way, he discovered electromagnetic induction. From this breakthrough, the electric generator was developed. Faraday's work in electromagnetism led James Maxwell to a theory that linked electricity, magnetism, and light.

In 1833, Faraday formulated the laws of electrolysis. Late in his career he discovered the rotation of the plane of polarization of light in a strong magnetic field. The invention of the radio was an indirect result of Faraday's work.

Electrolytic Capacitors

SAFETY TIP

Electrolytic capacitors use only polarized direct current and can, and often do, explode when an alternating or opposite voltage is applied to them.

The dielectric of electrolytic capacitors is a thin film of oxide. It is formed by electrochemical action directly on a plate of metal foil. The other plate is a paste electrolyte, either borax or a carbon salt. The very thin dielectric provides a large capacitance within a much smaller volume than would be possible with other materials.

Many electrolytic capacitors have more than one section. Some have two or more individual capacitor units within a single case. Multi-section capacitors have a positive lead for each of the capacitor units. They have a single negative lead common to all the capacitor units (Fig. 10-5).

The typical electrolytic capacitor used in electronic circuits is polarized. It must be connected into a circuit according to the plus and minus markings on the case (Fig. 10-6). Otherwise, the capacitor will overheat. This happens because of excessive leakage of current through the dielectric. Enough gas may be

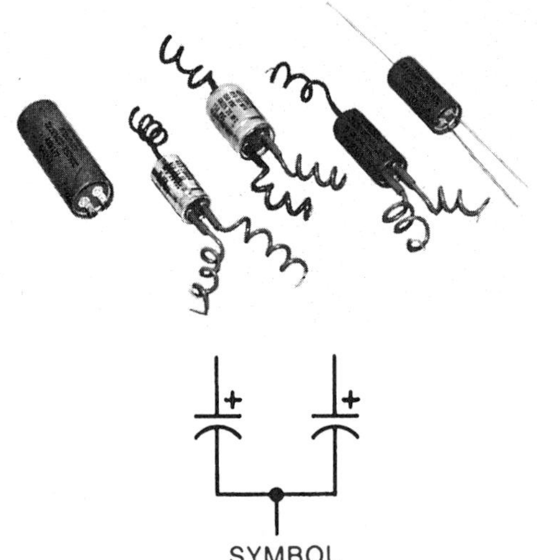

Fig. 10-5. Examples of dual, or two-section, electrolytic capacitors.

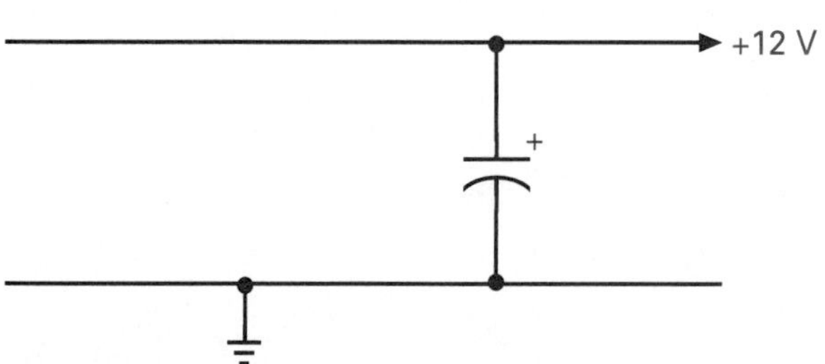

Fig. 10-6. The polarized electrolytic capacitor must be connected into a circuit according to its polarity markings.

produced to cause the capacitor to explode. Nonelectrolytic capacitors are not polarized. They can be connected into a circuit without regard to polarity.

Electrolytic capacitors are used in rectifier circuits to smooth out varying direct currents. They are also often used as passing capacitors in band-pass and band-stop filters. These block some ac currents and pass others.

Although electrolytic capacitors made from an aluminum foil are very popular, the tantalum electrolytic has very low leakage. It can be made much smaller than the aluminum electrolytic capacitor for the same capacity and working voltage. However, for extremely large values of capacitance for filtering power supplies, the aluminum electrolytic is generally preferred.

Capacitors with a capacitance of not more than one microfarad are generally nonelectrolytic capacitors. The capacitances of electrolytic capacitors usually range from one microfarad to hundreds of thousands of microfarads. Capacitance values and working voltages are usually printed on the body of the capacitor or are shown by means of a color code.

Ceramic Capacitors

Ceramic capacitors have disk-shaped or tubular dielectrics. These are made of ceramic materials such as titanium dioxide and barium titanate. The plates are thin coatings of a silver compound deposited on each side of the dielectric. Capacitances ranges from 5 pF to 0.05 μF. Working-voltage ratings exceed 10,000 V.

Silvered-Mica Capacitors

A silvered-mica capacitor is made of a thin sheet of mica (the dielectric) with coatings of a silver compound on each side (the plates). Silvered-mica capacitors have capacitance values that range from 1 to 10,000 pF. The working voltage is about 500 V.

Variable Capacitors

Variable capacitors are used primarily to tune circuits. This involves giving the circuits the least opposition to currents with a certain frequency (Fig. 10-7). One kind of variable capacitor, called a trimmer, is made of two metal plates separated by a sheet of mica dielectric. The space between the plates can be adjusted with a screw. As the distance between the plates is increased, the value of capacitance decreases.

Another kind of variable capacitor is made of two sets of metal plates separated either by air or by sheets of mica insulation. One set of

Fig. 10-7. **A trimmer capacitor.**

plates, the *stator assembly*, does not move. It is insulated from the frame of the capacitor, on which it is mounted. The other set of plates, the *rotor assembly*, is connected to the shaft. It can be turned. The rotor plates can move freely in or out between the stator plates. Thus, the capacitance of the capacitor can be easily adjusted from the lowest (plates apart) to the highest (plates together) value.

Mylar Capacitors

A Mylar capacitor is made of thin metal-foil plates and a Mylar dielectric. These are cut into long, narrow strips and rolled together into a compact unit. Capacitance values range from 0.001 to 1 microfarad. Working-voltage ratings are as high as 1,600 V (Fig. 10-8).

CAPACITORS IN CIRCUITS

Capacitors can be connected in either series or in parallel. The relationship between capacitors in a circuit is similar to the relationship between resistors.

CAPACITORS IN PARALLEL

Capacitors are often connected in parallel. This is done to get the capacitance value that cannot easily be obtained from one capacitor (Fig. 10-9A). The working voltage of the combination is equal to the lowest individual working-voltage rating. The total capacitance is equal to the sum of the individual capacitance values. Example: The formula for calculating two capacitors having the capacitance of 0.03 μF, and 0.02 μF is:

$$C\ (\text{total}) = C_1 + C_2 + \ldots C_n$$
$$= 0.03 + 0.02$$
$$= 0.05\ \mu F$$

Capacitors in Series

Capacitors can also be connected in series (Fig. 10-9B). When capacitors are connected in series, the applied voltage is divided among them. This is similar to resistors in series. There is a voltage drop across each capacitor. The total capacitance of capacitors connected in series is less than that of the smallest capacitor.

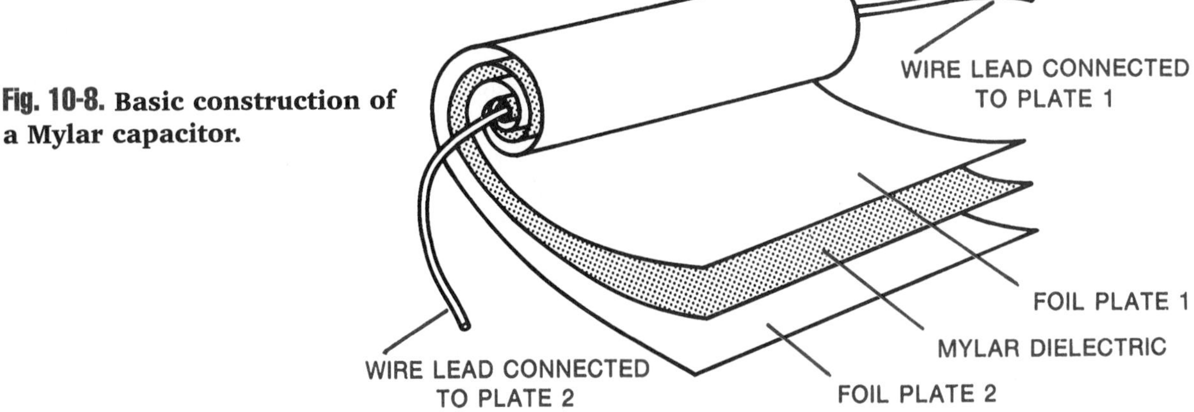

Fig. 10-8. **Basic construction of a Mylar capacitor.**

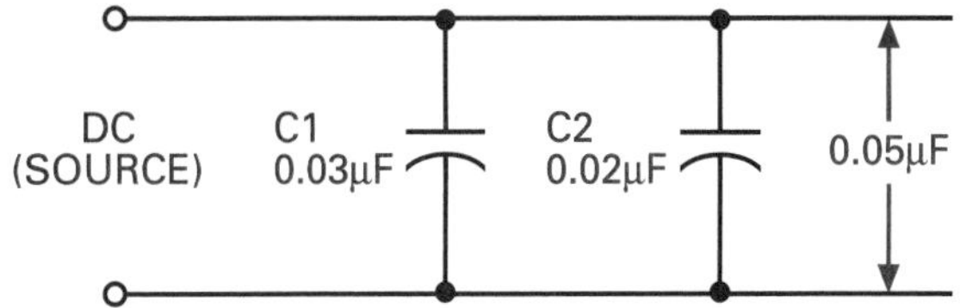

Fig. 10-9A. When capacitors are connected in parallel, the total capacitance is equal to the sum of the individual capacitance.

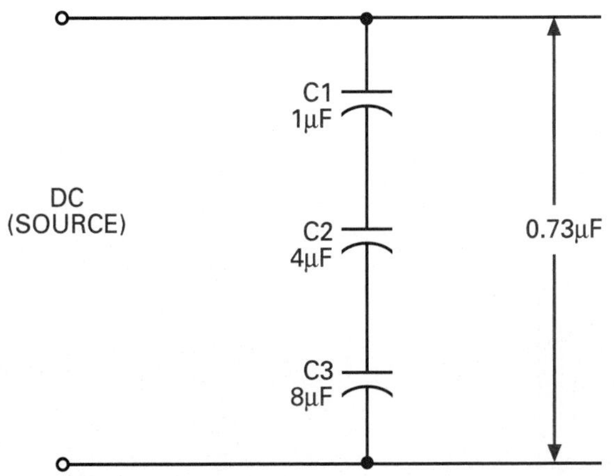

Fig. 10-9B. When capacitors are connected in series, the total capacitance is less than that of the smallest capacitor.

Example: The formula for calculating the capacitance of these capacitors connected in series having values of 1 μF, 4 μF, and 8 μF is:

$$\text{C Total} = \frac{1}{\frac{1}{C_1} + \frac{1}{C_2} + \frac{1}{C_2} + \cdots C_n}$$

$$= \frac{1}{\frac{1}{1} + \frac{1}{4} + \frac{1}{8}}$$

$$= \frac{1}{\frac{8 + 2 + 1}{8}}$$

$$= \frac{1}{\frac{11}{8}}$$

$$= \frac{8}{11} = 0.73\mu F$$

CHARGING CAPACITORS IN A CIRCUIT

Unlike a dry cell, a capacitor does not generate its own voltage. However, the voltage developed across the plates of a charged capacitor can be used as a voltage source. One important characteristic of a capacitor is that it can be charged. That charge can be stored for a time and then discharged through a load when needed.

Charging and Discharging Capacitors

The process of charging and discharging a capacitor can be illustrated in Fig. 10-10. At the instant the switch is closed between positions 1 and 2, electrons are transferred from one plate of the capacitor to the other (Fig. 10-10A). This causes a current (*I*) to flow in the circuit. At this same instant, the voltage (*E*) across the capacitor (*C*) is at a minimum since the voltage drop (*IR*) across *R* is at a maximum (Fig. 10-10B). As the current flow continues, the capacitor gradually becomes charged and a voltage begins to develop across the capacitor. This voltage opposes the battery voltage that is applied to the circuit. The opposing capacitor voltage is called a **counter electromotive force** (cemf). As a result, the current in the capacitor also decreases. Figure 10-10B illustrates the rise in voltage (*E*) and the corresponding decrease in current (*I*) during the process of charging a capacitor.

If the switch is closed from between positions 1 and 3, the capacitor immediately begins to discharge through the resistor (Fig. 10-10C). Since the voltage (*E*) across the capacitor is at maximum, the initial discharge current (*I*) is

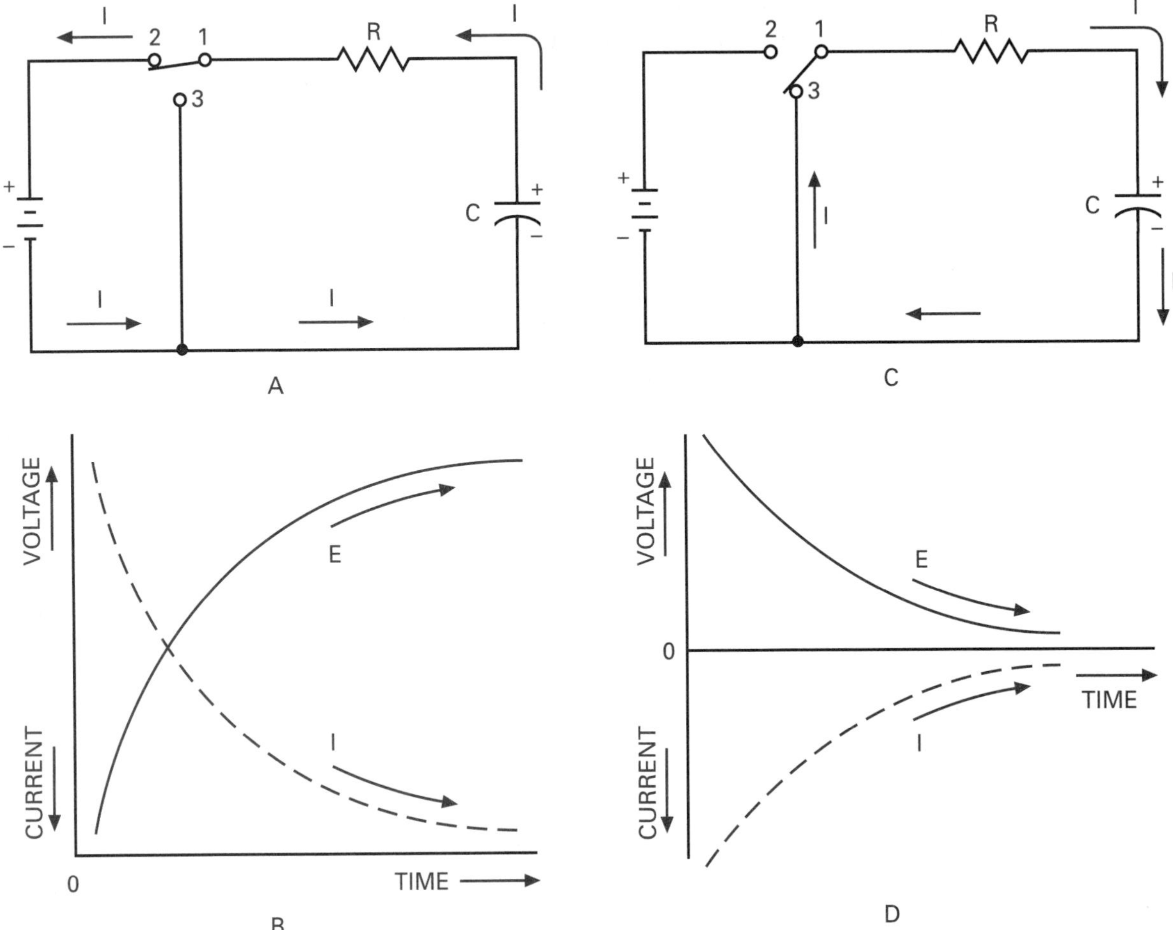

Fig. 10-10. The effect of capacitance on voltage and current in a dc circuit.

also at maximum (Fig. 10-10D). As the capacitor continues to discharge, the current gradually decreases until both the current in the circuit and the voltage reach zero.

The rate at which a capacitor charges and discharges can be controlled by varying the resistance in the capacitor circuit. The time needed for the voltage across a charging capacitor to reach 63.2 percent of the applied voltage is known as the time constant of the circuit. After five time-constant periods, the voltage across the capacitor will be very near 100 percent of the applied voltage. The time constant formula is:

$T = R \times C$

where T = the time constant in seconds
R = the resistance in ohms
C = the capacitance in farads

Charge and Discharge Rates

Capacitor charge and discharge rates can be used to control many circuits, including electronic timers and flashers. The circuit in Fig. 10-11 is a good example. In this circuit, the variable resistor $R1$ is used to change the time constant. During the first time constant, the capacitor will charge to 63.2 percent of 100 V, or 63.2 V. This may be enough voltage across the neon lamp to ionize the neon gas, causing it to conduct current. At low voltages, the neon

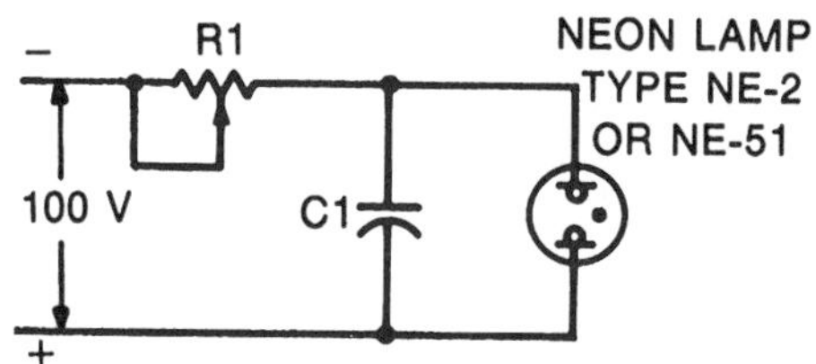

Fig. 10-11. Lamp-flasher circuit.

acts as an insulator, so no current passes through it.

If 63.2 V ionizes the lamp, the capacitor will quickly discharge through the lamp. There will be a flash of light. Now the capacitor will again begin to charge. The cycle is repeated. The time between flashes depends on the time constant of the circuit. A smaller value of resistance will let the capacitor charge more quickly. Flashes will occur more often. A larger resistance or capacitance will lengthen the time constant. Flashes will occur less often. This kind of circuit is often called a *relaxation oscillator*.

Capacitor Safety

Many capacitors, such as the electrolytic capacitors used in high-voltage rectifier circuits of television receivers, keep their charges for some time after the charging voltages have been removed. There could be dangerous voltages across these capacitors for many hours after their circuits have been turned off. These could give a person a bad electric shock or badly damage test instruments.

To prevent this, all capacitors in high voltage circuits should be discharged before the circuits are handled in any way. This can be done safely by briefly shorting the terminals of the capacitors with an insulated wire. A 10,000K Ω resistor can be connected in series with the wire to reduce sparking at the terminals.

SAFETY TIP

Do not attempt to perform this procedure without classroom instruction.

CAPACITOR DEFECTS

Fixed capacitors often become shorted because of a direct contact between their plates. This happens when the dielectric becomes worn or burns through because of age or too high of a working voltage. A shorted capacitor must be replaced. The capacitor not only may work improperly, but it also may let too much current pass through other components in its circuit. A capacitor can, but seldom does, become open. This happens when one of its leads becomes disconnected from its plate. Electrolytic capacitors can become defective if the chemical materials inside dry out. Such a capacitor has much less capacitance and should be replaced.

A variable capacitor becomes defective because of shorted plates. This often happens because the plates are bent while the capacitor is being installed or while its circuit is being repaired. A shorted variable capacitor can often be repaired by simply bending the touching plates back into place.

When replacing capacitors, it is best to replace them with a capacitor having the same capacitance or tolerance and type of dielectric and temperature characteristics. You may find the replacement capacitor to be physically smaller than the original capacitor. This is because the trend in recent years is to manufacture capacitors so that although the capacitance is increased, the package is smaller.

Write a short essay that explains the different ways in which a capacitor can fail.

Capacitor Color Codes

Capacitor values are often printed directly on the body of capacitors. The values may also be shown by a color code. The numbers represented by colors in this code are the same as those used in the resistor color code. Unless otherwise stated, capacitance values given by the color code are in picofarad units.

How the color code is used with three kinds of nonelectrolytic capacitors is shown in Fig. 10-12. As with the resistor color code, the multiplier color gives the number of zeros that are to be added to the first two numbers of the capacitance value. The temperature coefficient color in each of the codes shows the degree to which the capacitance will change with an increase of temperature.

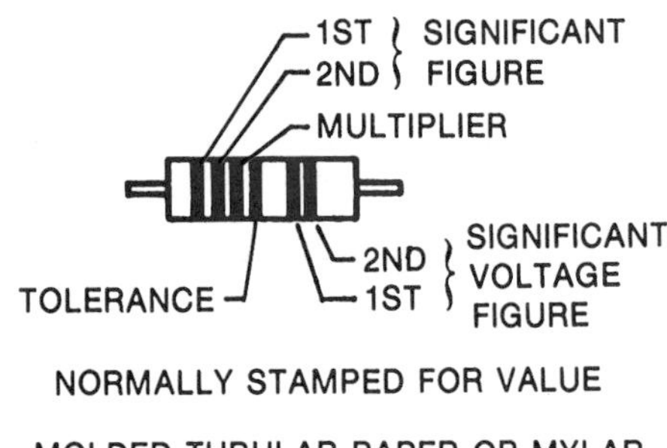

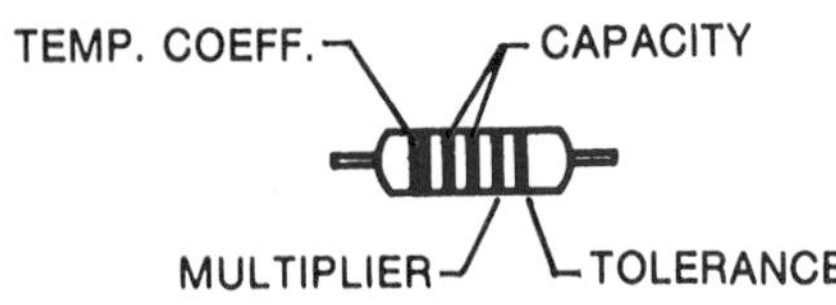

Fig. 10-12. How the capacitor color code is used.

Testing Capacitors

Capacitors can also be tested for short circuits with an ohmmeter. The ohmmeter is connected across the terminals of the capacitor being tested. If the capacitor is shorted, the ohmmeter will show continuity or a very low resistance. Leaky or shorted capacitors should be replaced. When testing a capacitor that is wired into a circuit, you should always take it out of, or isolate it from, the circuit (Fig. 10-13). This will keep the ohmmeter from also measuring the resistance of the circuit in parallel with the capacitor.

Nondefective nonelectrolytic capacitors have a very high resistance or no continuity between their terminals. Electrolytic capacitors have a lower resistance between their terminals. This is true even if they are in good condition. The amount of resistance will depend on the capacitance and working-voltage rating. With practice, you can soon learn how to tell if a capacitor is defective by testing it with an ohmmeter. If such a test is indefinite, the capacitor should be tested with a capacitor tester.

An ohmmeter applies a voltage to the component or part of a circuit to which it is connected. As the capacitor charges from the voltmeter battery, the resistance shown on the

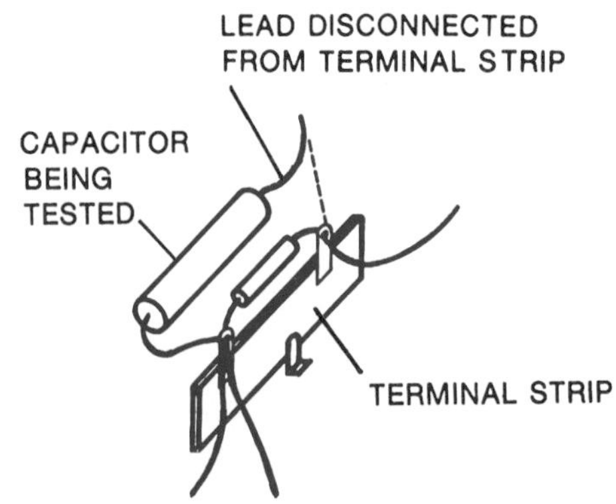

Fig. 10-13. Isolating a capacitor for test purposes.

meter will gradually increase until it shows as an open circuit. Under most circumstance, this indicates a working capacitor. To keep from damaging an electrolytic capacitor while testing it with an ohmmeter, make sure that the ohmmeter voltage is not greater than the working-voltage rating of the capacitor. It is also important to connect the ohmmeter to the capacitor in the correct polarity: positive terminal (+) to positive lead (red) and negative terminal (-) to negative lead (black).

To test a variable capacitor for short circuits, connect an ohmmeter to its terminals. Then slowly turn the shaft as far as it will go in both directions. If the capacitor is shorted, the ohmmeter will show direct continuity at that point.

Accurate measurements of capacitors can be made with a capacitor analyzer (Fig. 10-14). This type of meter performs two functions. It measures both capacitance and inductance. It is called a *Z meter* because it measures the total impedance (*Z*) of a circuit or component being tested. This chapter discusses how the meter measures capacitance. You will learn in Chapter 12 how the meter is used to measure inductance.

Capacitor Analyzers

The capacitor analyzer, such as the type shown in Fig. 10-14, can provide measurements (within 1 percent) of the rated capacitance of the capacitor. The following steps represent the general procedure to determine the capacitance of a capacitor.

1. Remove the power from the circuit if the capacitor to be checked has one end removed from the circuit but the other end still connected to the circuit. Whenever possible, completely remove the capacitor from the circuit.
2. Connect the cord of the meter to a properly grounded 115-Vac outlet.
3. Turn on the meter power switch.
4. Connect the test leads to the capacitor to be tested. Polarity of the test leads is important only if you are checking a polarized capacitor, such as an electrolytic capacitor. The red lead should be connected to the positive terminal of a polarized capacitor.
5. Depress the VALUE button under the capacitor section of the push button switch.
6. Read the value of the capacitor on the front panel digital readout. The value of the capacity will be in microfarads (μF) if the LED in front of the μF indicator is lighted. The capacity is in picofarads (pF) if the LED in front of the pF indicator is lighted. (The indicators are in the upper right-hand portion of the meter panel.)

Fig. 10-14. Capacitor analyzer: "Z meter" capacitor-inductor analyzer.

The foregoing procedures illustrate the ease with which this meter can be used to measure the capacitance of a capacitor. Many other tests and measurements of capacitors can be performed by this meter. It can, for example, detect leakage, dieletric absorption (the inability of a capacitor to discharge to zero), and intermittent capacitors.

LINKS

Refer to Chapter 12 for more information on inductance.

Chapter Review

Summary

- Capacitance is the ability of a circuit or a device such as a capacitor to store electric energy.
- When a battery is connected to a capacitor, the capacitor charges by developing an electrostatic charge between its plates that is equal to the battery voltage. This capacitor voltage will produce a current that flows (discharges) when the battery is disconnected. Because capacitors hold a charge, it is important to understand how to safely discharge them before handling a circuit.
- Capacitors can develop defects, and they can either open or short. A capacitor can be tested with either an ohmmeter or a capacitor analyzer for proper operation. It is important to replace a defective capacitor with one of equal working voltage and capacitance.

Review Main Ideas

Review this chapter's main ideas by writing, on a separate sheet of paper, the word or words that most correctly complete the following statements:

1. Capacitance is the ability of a circuit or a device to store ______ energy.
2. The basic form of a capacitor is two ______ separated by the ______.
3. A capacitor is charged by removing ______ from one of its plates and adding ______ to the other plate.
4. The voltage across the plates of a fully ______ capacitor is equal to the voltage of the battery to which it is connected.
5. The energy stored in a charged capacitor is in the form of an ______ field between its plates.
6. A fully charged capacitor ______ direct current from flowing in a circuit.
7. The opposition of a capacitor to the flow of alternating current in a circuit is called ______. This opposition is measured in ______. Its letter symbol is ______.
8. The total opposition to current in a circuit containing a combination of resistance and capacitive reactance is called ______. This total opposition is measured in ______. Its letter symbol is ______.
9. The ability of a dielectric material to oppose electrostatic lines of force is its ______.
10. A capacitor has a capacitance of one farad when an applied voltage that changes at the rate of one volt per ______ produces a current of one ______ in the capacitor circuit.
11. The most common units of capacitance are the _____ and the ______.
12. One ______ is equal to one-millionth of a microfarad.
13. The ______ voltage of a capacitor is the highest dc ______ that can be safely applied to a capacitor without it breaking down and conducting current.

14. The two general classes of capacitors are ______ capacitors and ______ capacitors.
15. An ______ capacitor is used to smooth out the varying dc output of a rectifier circuit.
16. The typical ______ capacitor is polarized. It must be connected into a circuit according to the ______ and ______ markings on the case.
17. Three common kinds of fixed nonelectrolytic capacitors are ______ capacitors, ______ capacitors, and ______ capacitors.
18. The primary use of variable capacitors is to ______ circuits.
19. When capacitors are connected in parallel, the total capacitance of the combination is equal to the ______ of the individual capacitance values.
20. The time needed for the voltage across a charging capacitor to reach 63.2 percent of the applied voltage is known as the ______ of the circuit.
21. The time constant formula is T = R x C, where T = the time constant in seconds, R = ______, and C = ______.
22. Nondefective nonelectrolytic capacitors have a ______ resistance or no continuity between their terminals. Electrolytic capacitors have a ______ resistance between their terminals.
23. For safety, charged high-voltage electrolytic capacitors should be ______ before the circuits are handled in any way.

Apply Your Knowledge

1. Draw a Mylar capacitor.
2. Define impedance and discuss how it is determined in a circuit.
3. Draw, label, and determine the impedance of a circuit of four series capacitors with values of 20 μF, 30 μF, 40 μF, and 80 μF.
4. Discuss the differences between capacitors connected in parallel and those connected in series.
5. Briefly explain the process of charging and discharging a capacitor.

Make Connections

1. **Communication Skills.** Write a succinct statement that defines the term *capacitance*.
2. **Mathematics.** Explain to your instructor how you would verify that one time constant is the time it takes to reach a charge level equal to 63.2% of the applied voltage in a circuit.
3. **Science.** Research the chemical action that takes place in an electrolytic capacitor. Summarize your research in a short essay.
4. **Technology Skills.** Break apart a damaged polarized capacitor and a damaged nonpolarized capacitor of the same size. Examine them carefully. Report your findings to the class.
5. **Social Studies.** Using your school or local library, research the effect of the radio on family life in the United States in the 1930s.

Chapter 

Magnetism

Terms to Study

degaussing

domains

electromagnetism

ferrite

flux

magnetic field

magnetism

magnetizer

permeability

relay

retentivity

saturation

solenoid

OBJECTIVES

After completing this chapter, you will be able to:

- Explain the domain theory of magnetism.
- Draw a bar magnet with its associated magnetic field and label the poles and the direction of the lines of force within the magnetic field.
- Define *degaussing*.
- Define *electromagnetism*.
- Describe the difference between a magnetic field produced by direct current and one produced by alternating current.
- Explain the operation of a solenoid.

Have you ever been sailing? Ships are very dependable, but fog makes it difficult to navigate them. Fortunately, most water craft have magnetic compasses on board. These compasses provide sailors reference points used to steer. This chapter discusses this form of energy called magnetism. You will learn how people use magnetism in many useful ways.

ATTRACTION AND REPULSION

Certain metals attract other metals (Fig. 11-1) through an attraction called **magnetism**. Materials that have magnetism are called magnets. Some magnets are found naturally in metallic ores. Others are manufactured. Magnets that keep their magnetism for a long time are called *permanent magnets*.

Magnetism is produced as a result of electrons spinning on their own axes while rotating about the nuclei of atoms (Fig. 11-2). In magnetic materials, there are certain areas, called **domains**, where atoms align. When this happens, most of their electrons spin in the same direction (Fig. 11-3A). Magnetization usually results in two *magnetic poles* being formed at the ends of the magnet. These are called the *north* and the *south poles*. In the absence of magnetization, the domains within a material are not aligned (Fig. 11-3B). The electrons then spin in all directions.

Magnetism is also produced as free electrons move through a conductor as current. This important relationship between electricity and

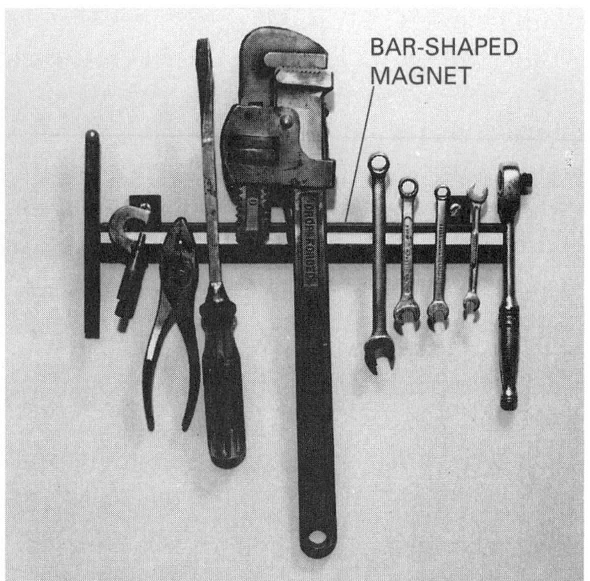

Fig. 11-1. Metal tools held in place by a magnetic tool holder.

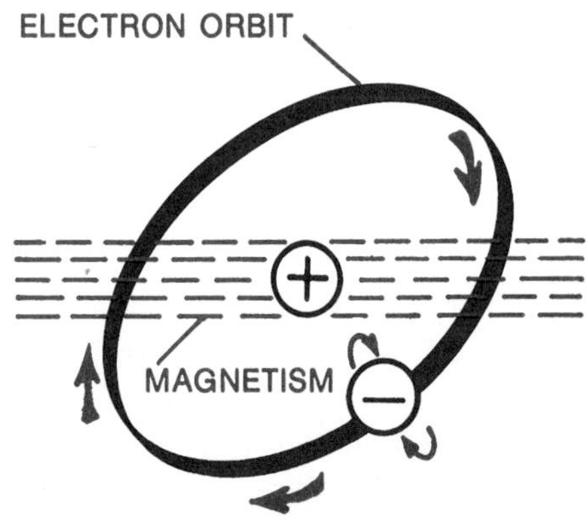

Fig. 11-2. Electrons spinning on their own axes while revolving about their nucleus produce magnetic fields.

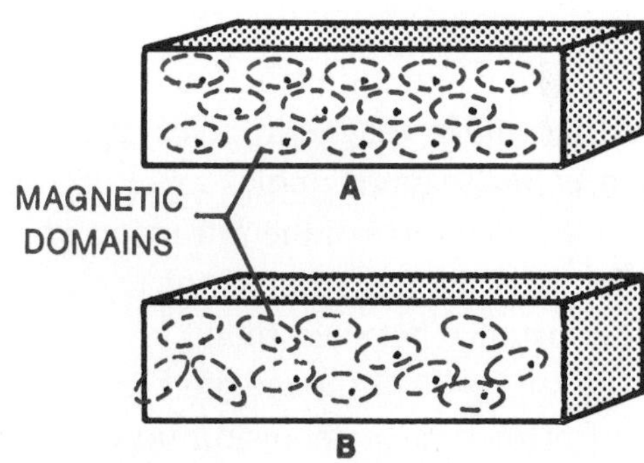

Fig. 11-3. Magnetic domains: (A) aligned; (B) not aligned.

FASCINATING FACTS

One advanced application of magnetism is the magnetic levitation, or maglev, train. Rather than using wheels or steel rails, the maglev system uses coils in the surface of the track to create a magnetic field that lifts the vehicles and propels them forward. In the late 1980s only short test systems had been built in Germany and Japan. Successful experimental runs were first made in the early 1990s using trains powered by environmentally friendly natural gas.

Because magnetism reduces friction, these trains may reach speeds of more than 200 miles per hour.

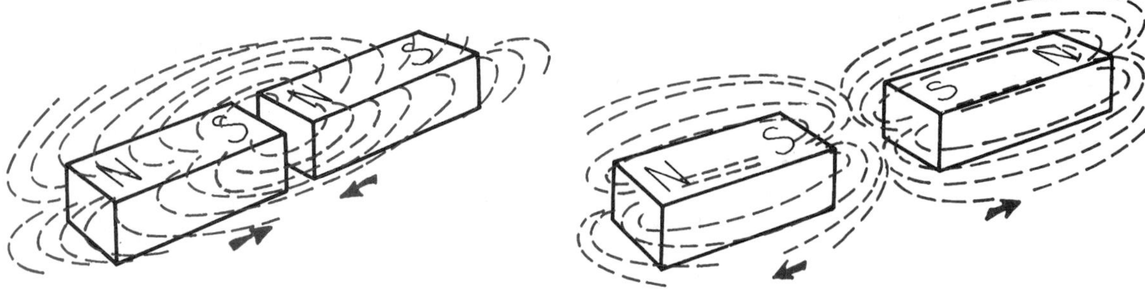

Fig. 11-4. The law of magnetic attraction and repulsion.

magnetism is known as electromagnetism, or the magnetic effect of current. It is used in the operation of many different kinds of circuits, products, and devices.

The law of magnetic attraction and repulsion states that *unlike* magnetic poles attract each other and *like* magnetic poles repel each other (Fig. 11-4). The force of the attraction or repulsion depends on the strength of the magnets and the distance between them.

As the strengths of the magnets increase, the force of attraction or repulsion between them also increases. The force of attraction or repulsion also increases as the distance between the poles of the two magnets decreases. If the distance between the poles of two magnets is reduced by one-half, the force of attraction between the magnets is increased four times. If the distance between the unlike poles doubled, the force of attraction between the magnets decreases to one-fourth the original value. Another way to state this relationship is to say that the force of attraction or repulsion between the poles of two magnets is inversely proportional to the square of the distance between them.

MAGNETIC FIELDS

The energy of a magnet is in the form of a **magnetic field**. This field surrounds the magnet. The magnetic field is made of invisible *lines of force*, or **flux**. Outside the magnet, the lines of force run from the north pole to the south pole (Fig. 11-5). Magnetic lines of force form an unbroken loop that do not cross each other. The strength of a magnetic field is directly related to the number of domains aligned during magnetization. Magnet **saturation** occurs when all the domains are aligned and the magnetic field becomes as strong as possible.

Gather magnetic and nonmagnetic materials from your classroom, workshop, or home. Touch each of them with a common magnet. Make lists of those that are magnetic and those that are not magnetic.

Magnetic and Nonmagnetic Materials

A *magnetic material* is one that is attracted to a magnet and can be made into a magnet. Among these materials are steel and the metallic elements iron, nickel, and cobalt. These are commonly called *ferromagnetic elements*. *Ferro* refers to iron and other materials and alloys that have certain properties similar to those of iron.

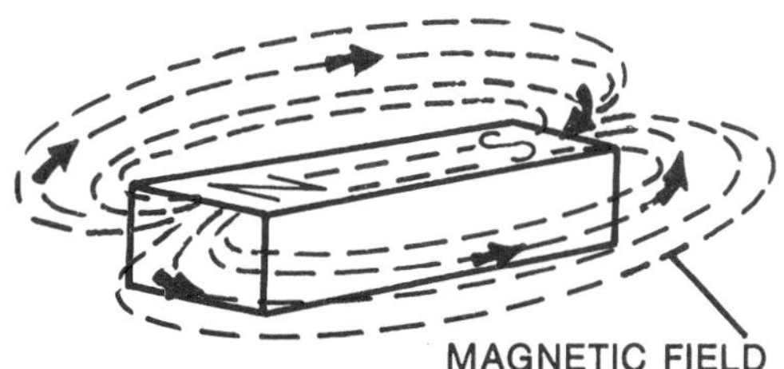

Fig. 11-5. The magnetic field surrounding a bar-shaped magnet.

Permanent magnets were originally natural magnets made of an iron ore called magnetite, or lodestone (Fig. 11-6). During the twelfth century, it was learned that a lodestone needle, when suspended so that it was free to turn, would point to the Earth's magnetic poles. Thus, it could serve as a magnetic compass. Lodestone is no longer used commercially as a magnetic material. Now, magnetic materials most often used for making modern permanent magnets are metallic alloys and compounds. These are usually combinations of iron oxide and certain other elements.

Metallic alloys and compounds have high **retentivity**, or are able to hold a magnetic-domain alignment for a long time. They also have high **permeability**, or are able to conduct magnetic lines of force easily. The magnetic permeability of a material is measured by comparing its ability to conduct magnetic lines of force with the ability of air to do so. Air, a vacuum, or other nonmagnetic substances are assumed to have a permeability of one.

Fig. 11-6. Magnetite, commonly known as lodestone, is a natural magnetic material.

Metals that are not attracted by a magnet and that cannot be made into a magnet are *nonmagnetic materials*. Copper, aluminum, gold, silver, and lead are nonmagnetic. Most nonmetallic materials such as cloth, paper, porcelain, plastic, and rubber are also nonmagnetic.

Alloys

One of the most common magnetic alloys is alnico. This alloy is made of aluminum, nickel, cobalt, and iron. Permanent alnico magnets come in different shapes and grades. The are used in such products as motors, generators, loudspeakers, microphones, and meters (Fig. 11-7). Other magnetic alloys are Permalloy (either nickel and iron; or cobalt, nickel, and iron), Supermalloy (nickel, iron, molybdenum, and manganese), and platinum-cobalt.

Ceramic Materials

Some hard ceramic magnetic materials are called **ferrites**. They are made by first grinding a combination of iron oxide and an element such as barium into a fine powder. The powder

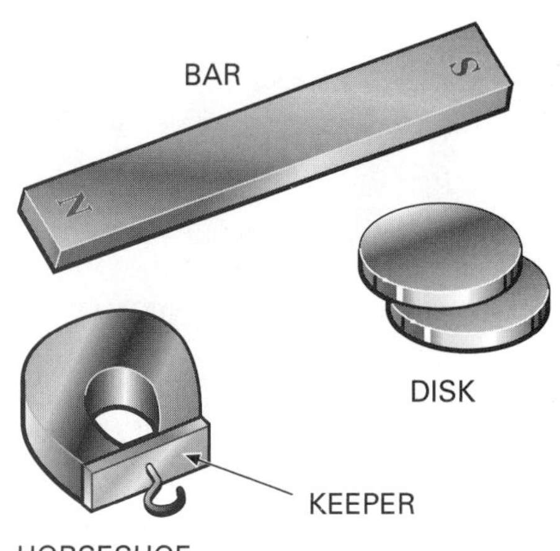

Fig. 11-7. Alnico permanent magnets. The soft-iron keeper placed across the poles of the horseshoe magnet helps the magnet keep its full strength for a longer period of time.

mixture is then pressed into the desired shape and baked at a high temperature. This produces a magnetic material that is very efficient and, unlike alloy materials, has a high electrical resistance.

Ceramic magnets are formed into many shapes (Fig. 11-8). Permanent magnets of this kind are used as latches on refrigerator doors. Such a magnet is in the form of a plastic strip filled with barium ferrite and bent to fit the shape of the door.

Magnets can be made by stroking a magnetic material in one direction only. Using a household magnet and a carpenter's nail, stroke the nail with the magnet until the nail attracts a paper clip or other small piece magnetic material.

Fig. 11-8. Ceramic permanent magnets can be made in many shapes and sizes.

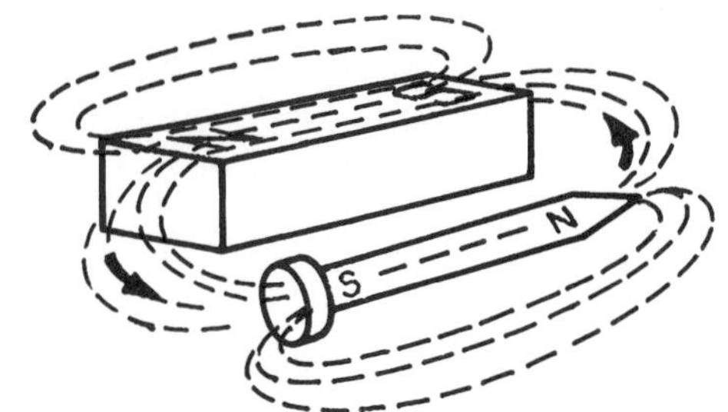

Fig. 11-9. Magnetic induction.

MAGNETS

In order to make a magnet, energy in the form of a magnetic field must be applied to a magnetic material. This process is called *magnetic induction*.

An example of magnetic induction is shown in Fig. 11-9. A nail is placed parallel to and near a permanent magnet. Since steel is more permeable than air, the lines of force from the permanent magnet pass easily through the nail. This causes some of the domains in the steel to align. Thus, the magnetic induction process produces magnetic poles at the ends of the nail. Note that the magnetic polarity of the nail is opposite of the permanent magnet. The nail thus becomes a magnet and is able to attract objects made of magnetic materials.

Commercially, permanent magnets are made with a **magnetizer**. Electromagnetic magnetizers use electric coils as the energy source (Fig. 11-10). The object to be magnetized is placed over the metal cores around which the coils are wound. When the switch is turned on, a direct current passes through the coils. This current produces a strong magnetic field that magnetizes the object quickly. The type of magnetism produced by the current is called **electromagnetism**.

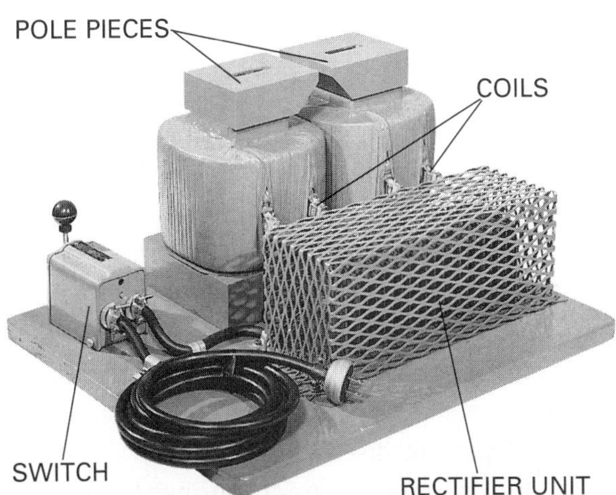

Fig. 11-10. An electromagnetic magnetizer.

Electromagnetic Fields

An *electromagnetic field* is the magnetic field produced by current that forms circles around the conductor. The larger the amount of current, the stronger the field. The direction of the magnetic field depends on the direction of the current (Fig. 11-11). If the current is a steady direct current, the magnetic field is constant in polarity and in strength. An alternating current produces a magnetic field that reverses in polarity and changes in strength. This is called a *moving magnetic field.*

Maxwell Field Equations

James Clerk Maxwell (1831-1879), a Scottish physicist and mathematician, believed electricity could cause waves in the air that would penetrate all substances. In 1864, Maxwell published a mathematical theory describing how the disturbance caused by moving electric charges or moving magnetic lines of force would travel. Maxwell said the disturbance would be treated as a wave motion with a speed of 3 x 10^8 meters per second—the experimentally determined speed of light.

By calculating the velocity of electromagnetic waves, Maxwell found that their speed was the same as that of light waves. He also concluded that light waves are electromagnetic in nature. Maxwell expressed all the fundamental laws of light, electricity, and magnetism in a few mathematical equations, commonly called the Maxwell field equations.

These equations were long considered a fundamental law of the universe. However, they have since been found not applicable to the phenomena governed by quantum theory, wave mechanics, and relativity.

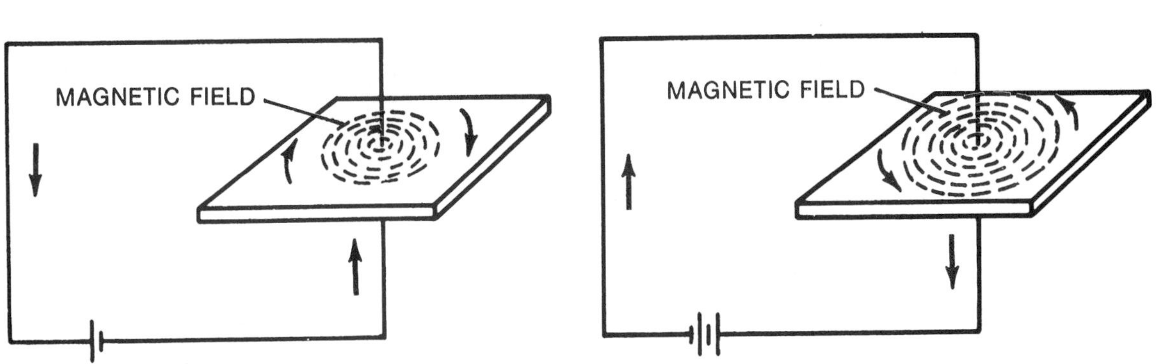

Fig. 11-11. The magnetic field produced by current in a conductor.

Electromagnets

To concentrate the magnetic field produced by current flowing in a wire, the wire is wound into a coil. When this is done, the magnetic fields around the turns of the coil are added together. This increases the magnetic strength of the coil. A coil wound in this manner is called a *solenoid* (Fig. 11-12). A **solenoid** has magnetic poles and a magnetic field with the same properties as those of a permanent magnet. If a solenoid is energized with alternating current, its magnetic polarity reverses with each reversal of the direction of the current.

Iron-core Electromagnets

An electromagnet can be made if magnetic wire is wound around an iron core. Because iron is more permeable than air, more lines of force can pass between the poles of the electromagnet to produce a stronger magnetic field (Fig. 11-13). *Residual magnetism* is the magnetism that remains in aligned domains after the power is removed from an electromagnet. Soft iron has little residual magnetism because soft iron has a low *retentivity*, or ability to stay magnetized.

The strength of an electromagnet is given by a unit called *ampere-turns*. This is the product of the coil current times the number of turns of wire with which the coil is wound. The ampere-turns unit is useful for comparing the strengths of electromagnets that operate at the same voltage.

Electromagnets of different sizes and strengths are used in a large number of devices. These include electric motor brakes, buzzers, horns, relays, circuit breakers, and electric clutches. As you will learn in later chapters of this book, electromagnetism or a combination of electromagnets and permanent magnets is used in many electronic products. Electromagnets are, for example, used in transformers, generators, loudspeakers, microphones, solenoid valves, and audio equipment for recording and playback.

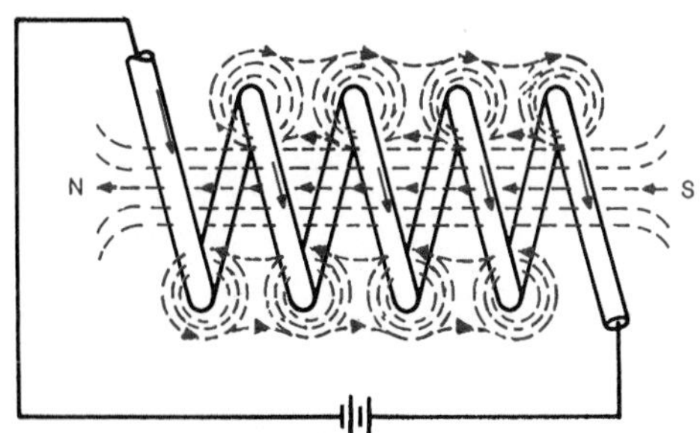

Fig. 11-12. By winding wire in the form of a coil, or solenoid, the magnetic field can be concentrated in a small area.

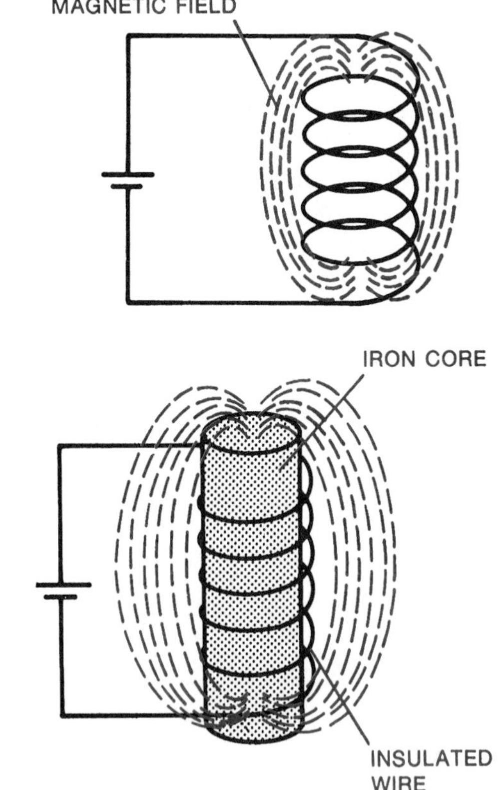

Fig. 11-13. The use of a soft-iron core increases the strength of an electromagnet.

Build a solenoid simulator with nonelectric parts such as a tube or other coil, an armature, and contacts.

SOLENOIDS

An example of a solenoid is shown in Fig. 11-14A. In this assembly, a plunger, or *armature* (movable piece), made of iron or steel is placed partly inside the coil. When the coil is energized, the armature is drawn further into it. With the right coupling, the movement of the armature can be used to control things such as valves, switches, relay contacts, and clutch mechanisms. A typical example of this kind of control is found in *relays*, which are discussed later in this chapter. (Fig. 11-14B).

Door-Chimes

The door chime is another common use of the solenoid. In this device, an iron or steel armature with a rubber or plastic tip is put inside the coil (Fig. 11-15A). When the coil is energized, the armature is attracted upward into it. This causes the tip to strike the chime bar and produce a sound (Fig. 11-15B).

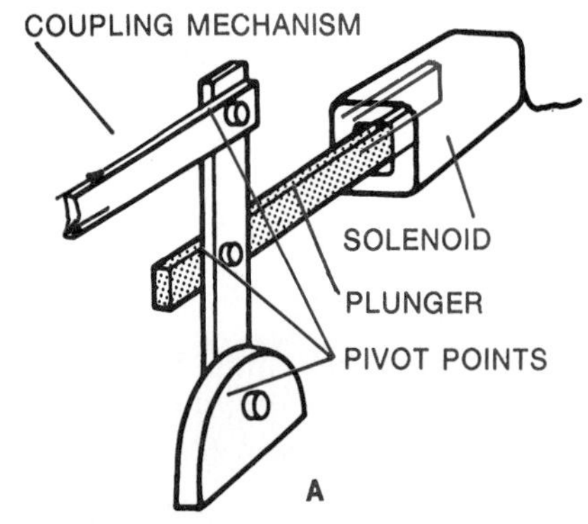

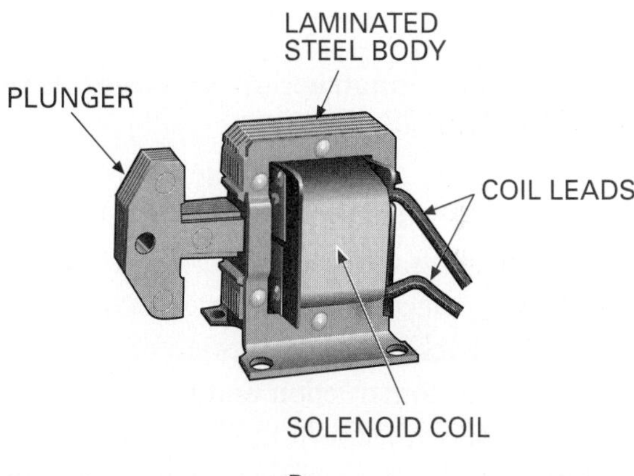

Fig. 11-14. Solenoid-control mechanism: (A) basic construction and operation; (B) typical solenoid assembly.

Fig. 11-15. Operation of a door chime.

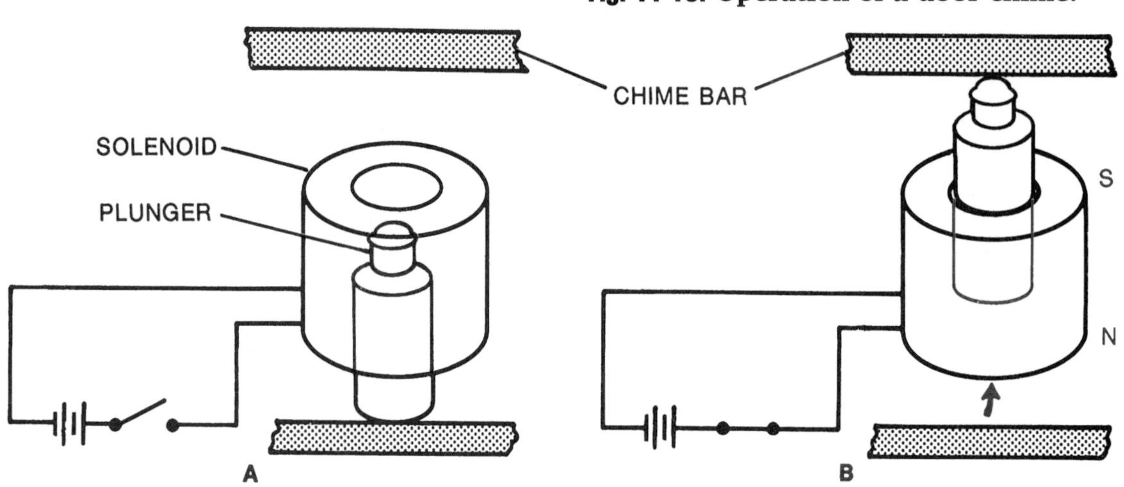

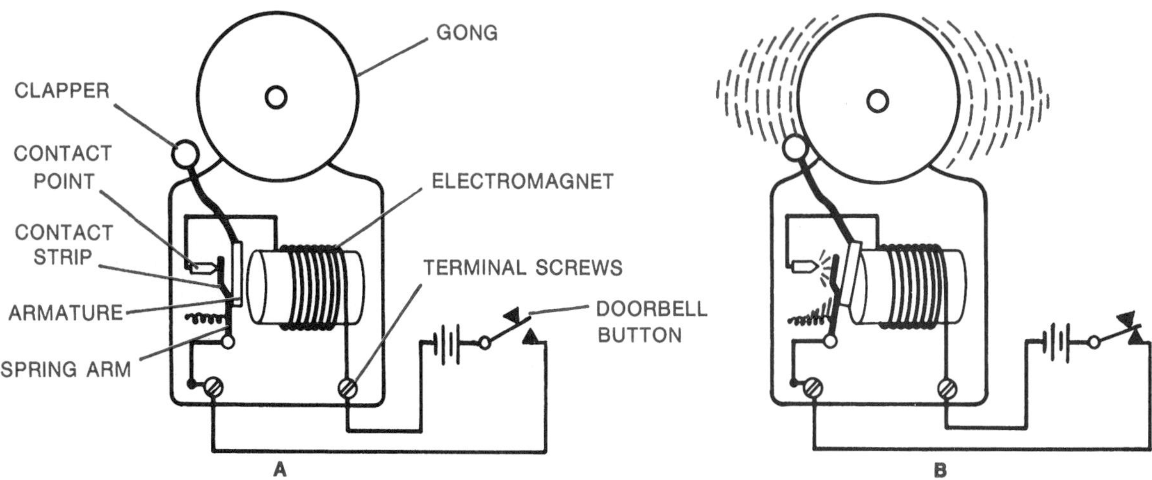

Fig. 11-16. Basic construction and operation of a doorbell.

Electric doorbells, and some kinds of horns, are an interesting group of *vibrating devices*. In these devices, electromagnetism is used to produce a rapid back-and-forth motion of an armature. The basic construction and operation of a typical doorbell are shown in Fig. 11-16.

Before the doorbell button is pressed, the contact strip is held against the contact point by the spring tension (Fig. 11-16A). When the button is pressed, the circuit is completed and current flows through the coil. The electromagnet then attracts the armature to the position shown in Fig. 11-16B. This causes the armature to move away from the contact point. The circuit is now turned off, and the electromagnet is deenergized. The spring is then able to pull the armature back to the contact point. This action is repeated very rapidly as long as the button is pressed. The *clapper* attached to the armature strikes the gong each time the armature is pulled toward the electromagnet. This produces a continuous ringing sound. In a buzzer, the sounds are produced as the armature strikes against one end of the electromagnet core.

Relays

A **relay** is a magnetically operated switch that opens or closes (makes or breaks) one or more of the contacts between its terminals (Fig. 11-17). As with mechanical switches, the action of relays is described by the number of lines (poles) that are controlled and the number of contacts (throws) each pole can make. The relay in Fig. 11-17 controls one line (single-pole) and can touch either of two contacts (double-throw).

The basic operation of a single-pole single-throw relay is shown in Fig. 11-18A. When the switch in the relay circuit is closed, the electromagnet is energized. It thus attracts the armature to the fixed contact point. There is now continuity between terminals 1 and 2, and the

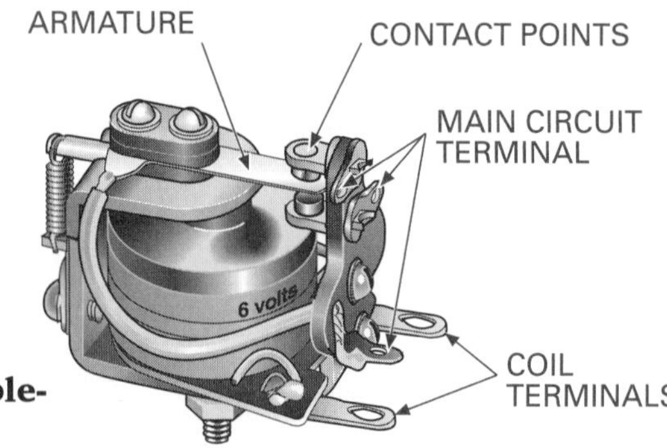

Fig. 11-17. Single-pole—double-throw relay.

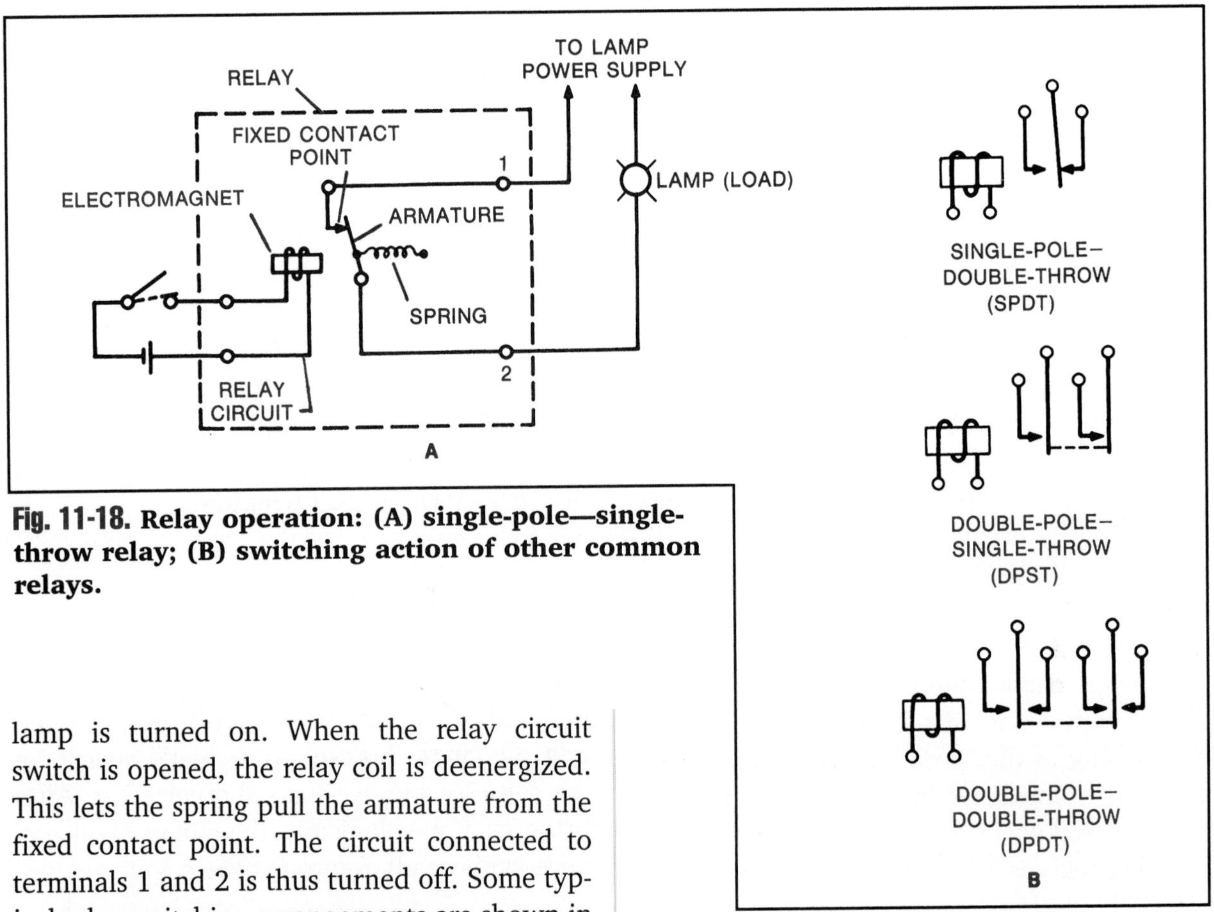

Fig. 11-18. Relay operation: (A) single-pole—single-throw relay; (B) switching action of other common relays.

lamp is turned on. When the relay circuit switch is opened, the relay coil is deenergized. This lets the spring pull the armature from the fixed contact point. The circuit connected to terminals 1 and 2 is thus turned off. Some typical relay switching arrangements are shown in Fig. 11-18B.

A relay can control a large load current at a high voltage with a small relay-energizing current at a low voltage. An example of this is shown by the circuit of Fig. 11-19. Here the relay is controlled by a 12 Vdc battery. The current in the control circuit, which consists of the battery and the relay coil is much less than the motor circuit. However, since the motor circuit that is controlled by the relay has a different source of energy, the current in that circuit can be and is quite large.

Ratings

General-purpose power relays are rated by: (1) the operating-voltage and frequency of the relay coil, (2) the current rating of the coil, and (3) the current and voltage ratings of its contacts. These indicate the largest safe load current the relay can control.

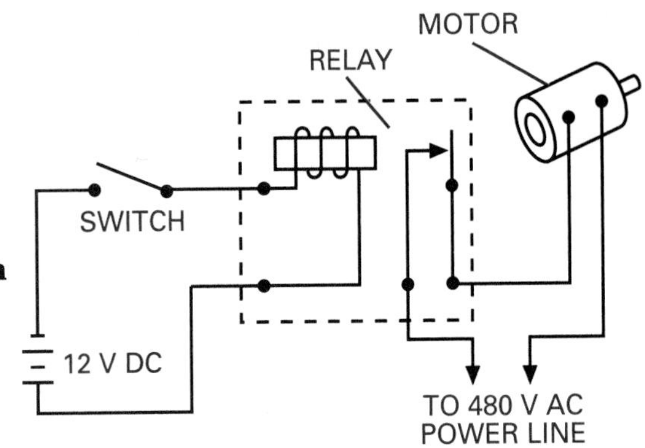

Fig. 11-19. Low voltage is used to control a motor circuit.

Contacts

The contacts of a relay are often described as being normally open (N.O.) or normally closed (N.C.). Normally open contacts are those that are separated when the relay is not energized. Normally closed contacts are those that are closed, or in contact, when the relay is not energized.

Reed Relays

The switching assembly of a typical magnetic-reed relay, or reed switch, is made of ferromagnetic reeds enclosed in a sealed glass tube (Fig. 11-20). In a complete relay assembly, the tube is placed in a coil. When the coil is energized, the reeds come together as a result of magnetic attraction. This kind of relay is very sensitive, which means that it can be operated with a very small amount of current.

In another kind of reed relay, the opening and closing of the reeds are controlled with a permanent magnet. The magnet is brought near the reed enclosure. When used in this way, the reed relay is sometimes called a magnetic *proximity switch*.

Magnetic Circuit Breakers

The magnetic *circuit breaker* is a device that protects a circuit from too much current. In one kind of circuit breaker, the coil of an electromagnet and two contact points are connected in series with one wire of a circuit (Fig. 11-21).

When the current is more than the ampere rating, or the trip size, of the circuit breaker, the electromagnet becomes strong enough to attract the armature. This moves and latches the contact point in the open position, and the circuit is broken. A brightly colored indicator pops out to show that the circuit breaker has been tripped by an overload. This makes it easy to locate the overloaded circuit. Before resetting the circuit breaker, you should search for the cause of the overload and correct it. After being reset mechanically, the circuit breaker is once again ready to protect the circuit.

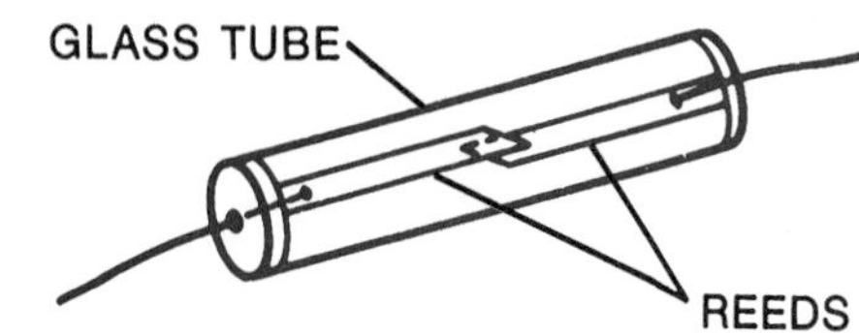

Fig. 11-20. Magnetic-reed relay.

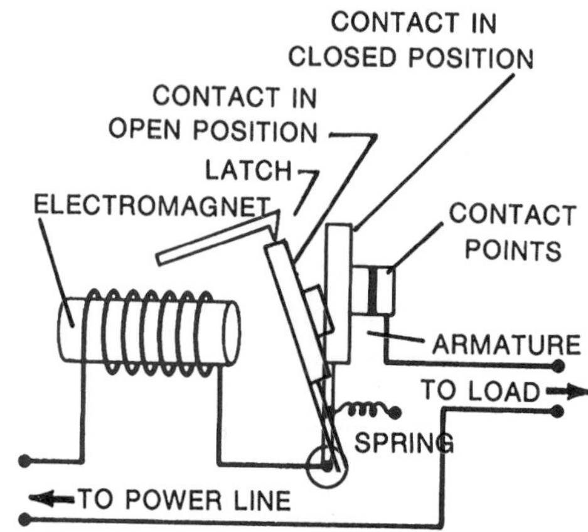

Fig. 11-21. Basic construction and operation of a magnetic circuit breaker.

MAGNET SHIELDING

Permeability is used widely to protect, or shield, certain components and instruments from the ill-effects of magnetic fields. In many instrument circuits, the effects of stray magnetic fields can cause significant errors in measurement.

A basic form of magnetic shield is shown in Fig. 11-22. With this type of shielding, a cover made of a highly permeable ferromagnetic material is placed over a circuit component. As a result, any external magnetic field that may be near the component is bypassed (Fig. 11-23).

ELIMINATING MAGNETISM

Removing magnetism from an object is called demagnetizing, or **degaussing**. For example, wristwatches made of magnetic materials will not keep the correct time after becoming magnetized. If metal-cutting tools such as drills and reamers become magnetized,

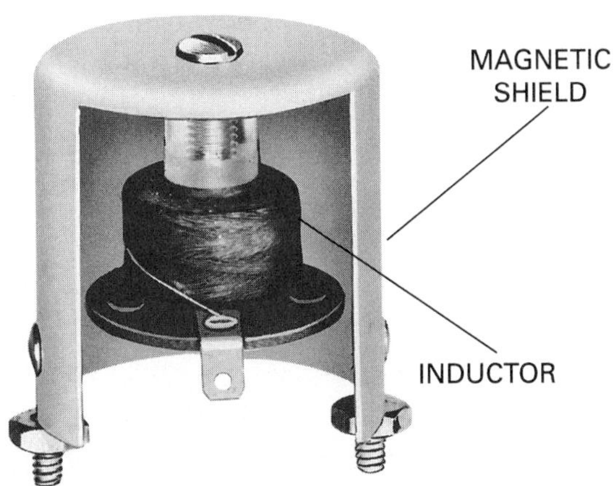

Fig. 11-22. cutaway view of a magnetic shield used to protect an inductor from the influence of external magnetic fields.

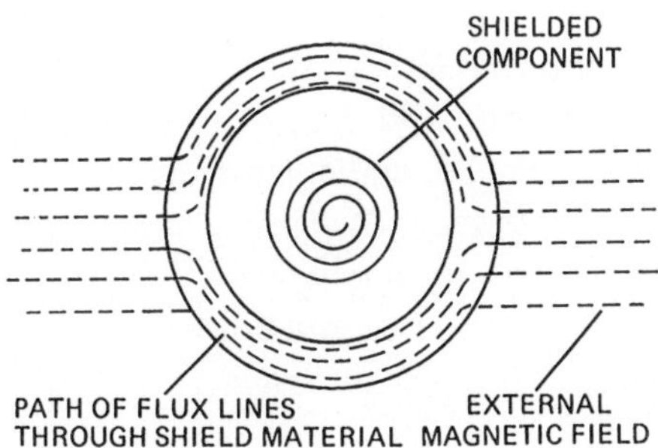

Fig. 11-23. Principle of magnetic shielding.

1. Using two magnets and magnetic and non-magnetic material, find the shielding materials that prevent the magnets from attracting or repelling each other.
2. Using the nail you magnetized earlier, hit it with a hammer until its magnetism is gone. After hitting the nail, can you detect any residual magnetism?
3. List at least five uses of magnets other than those explained here. Can you think of any new ways to use magnetism?

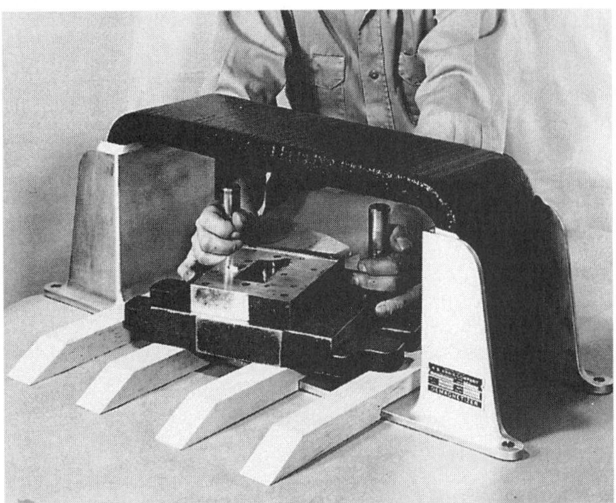

Fig. 11-24. Using a demagnetizer to remove the magnetism from a metal tool-holding device.

they will attract metal chips and filings. This causes the tools to become dull quickly.

In its simplest form, a demagnetizer is a coil of insulated wire through which alternating current passes. The object to be demagnetized is put within the coil. Then the object is slowly removed from the coil. The effect of the changing magnetic field disturbs the alignment of the domains and thus eliminates the magnetism from the object (Fig. 11-24).

A permanent magnet can also be demagnetized or greatly weakened by heating. It can also be demagnetized by striking it sharply with a metal object such as a hammer. In both cases, the domain alignment within the magnet is severely disturbed.

SAFETY TIP

Do not allow an A.C. coil to remain on after the object has been removed from its hollow core. A "coreless" coil will quickly burn open.

MAGNETS IN USE

Because Earth is a magnet and contains magnetic rock, magnets are one of the oldest applied technologies on earth. They have been used to steer ships, move objects, and to run equipment. Earth's magnetic poles are near the geological North and South Poles. There is evidence in the ocean floor that the polarity of these poles, has in ages past, reversed.

Compasses

A *magnetic compass* is used to find Earth's geological north. A small permanent magnet, the needle, is mounted on a pivot point so it is free to turn. The poles of the magnet are attracted by the magnetic poles of the Earth. Because of this, the needle of the compass points in a general north-to-south direction (Fig. 11-25).

The true magnetic poles of the Earth are actually about 1,500 miles (2414 km) from the geographic poles. Therefore, a compass needle

Magnetic Resonance Imaging

Magnetic resonance imaging (MRI), also called nuclear magnetic resonance (NMR), is a technique that involves subjecting atomic nuclei in specific organs of the human body to very strong stationary magnetic fields and then observing how they absorb very high frequency radio waves. This procedure is possible because specific nuclei in the water and fats of the body have characteristic magnetic behavior. The MRI has distinct advantages in medicine. It is relatively hazard-free, noninvasive, and is capable of sensitively differentiating between normal and diseased or damaged tissues.

By the late 1980s, MRI was considered superior to most other techniques in providing images of major organs such as the brain, liver, kidneys, spleen, pancreas, and breast. MRI images can reveal tumors, blood-starved tissues, and other diseased or damaged conditions of the body. Though the technique presents no health hazards, it cannot be used on patients with cardiac pacemakers or certain other metal-containing implants.

Fig. 11-25. The needle of a magnetic compass is attracted by the magnetic poles of the Earth. The north pole of a magnet is so named because it points to the Earth's north magnetic pole.

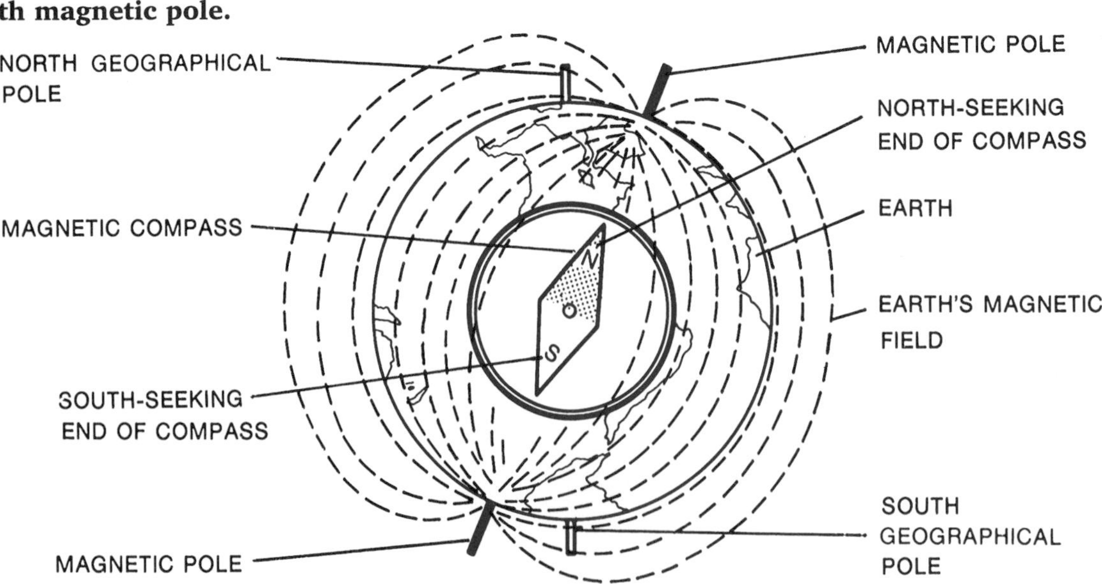

FASCINATING FACTS

Physicists working with magnets made of superconducting materials in the mid-1990s were able to produce a magnetic field 78,000 times stronger than the Earth's magnetic field.

does not point to the geographic north and south. The difference between a compass reading and the true direction is known as the *angle of declination*. This angle must, of course, be considered in finding truly geographic directions with a magnetic compass.

Defect Locators

Magnetic-particle inspection is a process that uses magnetism to find defects on the surfaces of metal parts. In this process, the object is magnetized and then coated with an oil or other liquid that contains fine particles of iron. The edges of defects such as cracks and scratches become magnetic poles to which the particles are attracted. This causes the particles to form a pattern that can easily be seen (Fig. 11-26). Tests of this kind are called *nondestructive testing*.

Future Uses of Magnetism

Scientists and engineers are constantly exploring new ways to use magnetism. The possibility of using the magnetic energy high above the surface of the Earth is now being studied. Many scientists believe that magnetism may someday be used to counteract the force of gravity. These ideas and others about uses of magnetism in the field of electronics are creating a new scientific frontier in the field of electronics.

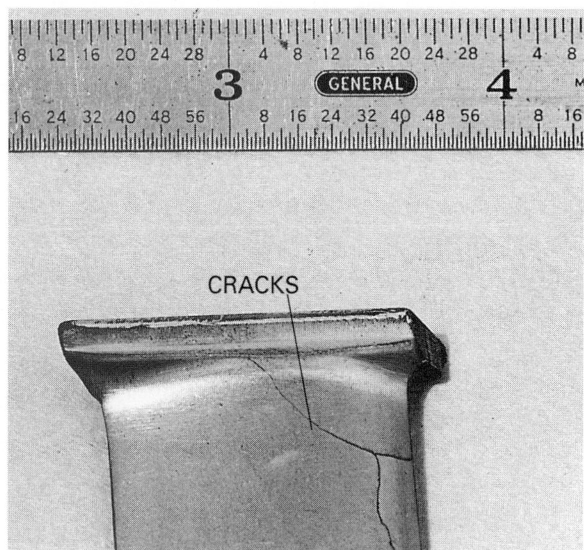

Fig. 11-26. **Cracks in the turbine blade of a jet aircraft engine discovered by magnetic particle inspection.**

Chapter *Review*

Summary

- Certain metals attract other metals. This attraction is called magnetism and is found only in magnetic materials such as those that contain iron or iron alloys.
- Magnetism is produced as a result of electrons spinning on their own axes while rotating about the nuclei of atoms.
- Magnetism is also produced as free electrons move through a conductor as current. This important relationship between electricity and magnetism is known as electromagnetism, or the magnetic effect of current.
- The law of magnetic attraction and repulsion states that unlike magnetic poles attract each other and like poles repel. The force of this attraction or repulsion is directly proportional to the distance between the magnets.
- Magnets form a magnetic field that is made of invisible lines of force called flux. These lines of force travel from north to south outside of the magnet and from south to north inside it.
- Magnetism is sometimes undesirable. If it is, it can be either shielded or eliminated.

Review Main Ideas

Review this chapter's main ideas by writing, on a separate sheet of paper, the word or words that most correctly complete the following statements:

1. Magnetism is produced as electrons ______on their own axes while rotating about the nuclei of atoms.

2. In a permanent magnet, atoms in certain areas called ______ are aligned.

3. Unlike magnetic poles ______ each other, while like magnetic poles ______ each other.

4. The force of attraction or repulsion between the poles of two magnets is inversely proportional to the ______ of the distance between them.

5. A magnetic field is made of invisible lines of ______, which are from the ______ pole to the ______pole of a magnet.

6. A magnet is said to be ______ when all its domains have been aligned.

7. Three metallic elements that are highly magnetic are ______, ______, and ______.

8. The first permanent magnets were natural magnets made of an iron ore called ______, or ______.

9. A material has a high ______ when it is able to keep its magnetic-domain alignment for a long time.

10. The ______ of a material is measured by comparing its ability to conduct magnetic lines of force with air's ability to do so.

11. Permanent alnico magnets are used in products such as ______, ______, ______, and ______.

12. Some ceramic magnetic materials are called ______.

13. Applying energy in the form of a magnetic field to a magnetic material is called ______.
14. The electromagnetic field is produced by current that forms ______ around the conductor.
15. A direct current produces a magnetic field that is ______ in polarity and strength.
16. An alternating current produces a magnetic field that reverses in ______ and ______.
17. If a solenoid is energized with alternating current, its magnetic polarity reverses with each reversal of the ______ of the current.
18. The magnetism remaining in the aligned domains of an electromagnet after the power has been removed is called ______.
19. Electromagnets are used in devices such as ______, ______, ______, ______, and ______.
20. The general-purpose power relay is rated in terms of ______, ______, and ______.
21. Normally open relay contacts are those that are separated when the relay is not ______.
22. External magnetic fields can be bypassed using a magnetic ______ made of high-permeability ferromagnetic material.
23. Removing magnetism from an object is sometimes called ______.
24. A nondestructive test called magnetic-particle inspection can be used to find ______ in the surfaces of metals.

Apply Your Knowledge

1. Explain why adding turns of wire to an electromagnetic coil causes the strength of the magnet to increase.
2. Draw a solenoid. Include the coil, armature, and energy source.
3. List three ways to demagnetize a magnet.
4. Explain how a relay works.
5. List six uses for electric magnets.

Make Connections

1. **Communication Skills.** Investigate and write a short report on the process of degaussing a large ocean-going vessel.
2. **Mathematics.** Write the formula that mathematically describes that the attraction and repulsion between two magnets is inversely proportional to the square of the distance between the magnets. Explain the formula to the class.
3. **Science.** Design a demonstration for the class to show the magnetic effect caused by current from a battery on a coil of wire that has an air core.
4. **Technology Skills**. Bring a magnetic stud finder to class. Demonstrate its use.
5. **Social Studies.** Write a report specifying the reasons you believe it was important to locate the geographic North Pole instead of the South Pole.

Chapter 12

Electromagnetic Induction and Inductance

Terms to Study

eddy current

henry

inductance

inductors

Lenz's law

short

transformer

OBJECTIVES

After completing this chapter, you will be able to:

- State Lenz's law of induced voltage and currents.
- Calculate the current delivered to a load from a transformer.
- Define inductance.
- Calculate inductive reactance.
- Describe the magnetic field produced by alternating current.
- Define and calculate an *RL* time constant.

The term *induction* means to cause something to happen. You will learn in this chapter how electromagnetic induction and inductance cause many things to happen electrically. You will learn how motion is needed to produce voltage by electromagnetic induction. You will also learn about moving magnetic fields and the operation of transformers.

LENZ'S LAW

Motion is needed to produce a voltage by electromagnetic induction. The magnetic field must move past the wire, or the wire must move past the magnetic field. When the magnet is moved near a wire, its magnetic field cuts across the wire and produces a voltage across the ends of the wire. If the ends of the wire are connected to a circuit, the voltage causes current to flow through the circuit (Fig. 12-1A). A voltage that is produced in this way is called an induced voltage. A current produced by an induced voltage is called an induced current. This important relationship between electricity and magnetism is called electromagnetic induction. When the magnet is moved in the opposite direction, the polarity of the induced voltage changes and the direction of the current in the circuit also changes (Fig. 12-1B).

A Russian physicist, Heinrich F. E. Lenz (1804 - 1865), discovered that when a current is induced by moving a conductor across lines of magnetic force, the induced current in the conductor creates a magnetic field that opposes the lines of magnetic force that produced it. He also found that the induced current and its own–magnetic field change direction depending upon which way the conductor is moved in the magnetic field. These important discoveries are known as **Lenz's law**.

Figure 12-2 illustrates these relationships in a single conductor. In Fig. 12-2A, the conductor is shown moving downward through the lines of magnetic force of a permanent magnet. The arrows on the conductor indicate the direction of the induced current. The arrows around the conductor indicate the direction of the magnetic field caused by the induced currents. Notice that the conductor's magnetic field opposes the permanent magnet's magnetic field (Fig. 12-2B). Work, therefore, must be done to overcome the magnetic effects of the induced currents. If the conductor is moved upward through the magnetic field, the directions of the induced current and its magnetic field reverse. Again, work must be done to overcome the magnetic effects of the induced currents.

One of the most important uses of electromagnetic induction is in *alternators* and other kinds of *generators*. These machines supply the electric energy that is distributed to our homes by electric power companies. Electromagnetic induction is also used to operate many other kinds of machines and devices. These include motors, transformers, microphones, heat treating processes, and tapeplayer playback heads.

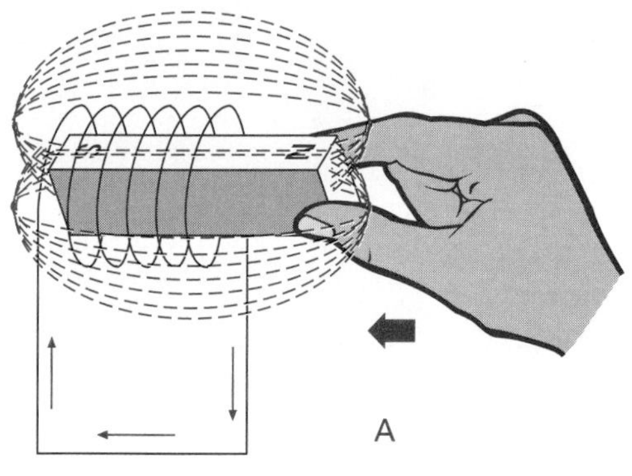

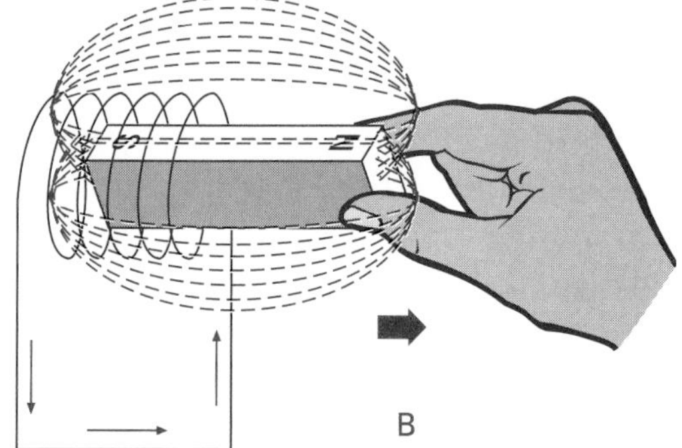

Fig. 12-1. Electromagnetic induction.

LINKS

You will learn in Chapter 15 how mechanical energy is used to do work needed to generate electricity.

Fig. 12-2. The effects of Lenz's law: (A) direction of the induced current and its magnetic field; (B) enlarged view of the opposing magnetic fields of the conductor and the permanent magnet.

Think of different ways a magnet can be moved near a wire. For instance: back and forth, by hand, or mechanically.

MOVING MAGNETIC FIELD

As you have learned, electromagnetic induction happens only when there is relative motion between a magnetic field and a conductor. In a generator, this motion is produced by turning its *rotor* assembly or its *field-coil* assembly with an engine of some kind. In a device such as a transformer that does not have any moving parts, a moving magnetic field is produced as the result of changes in the direction and value of the current.

Alternating Current

The magnetic field produced by an alternating current is shown in Fig. 12-3A. As the current increases, the magnetic field moves away from the conductor. As the current decreases, the magnetic field moves toward the conductor. A moving magnetic field such as this can produce electromagnetic induction.

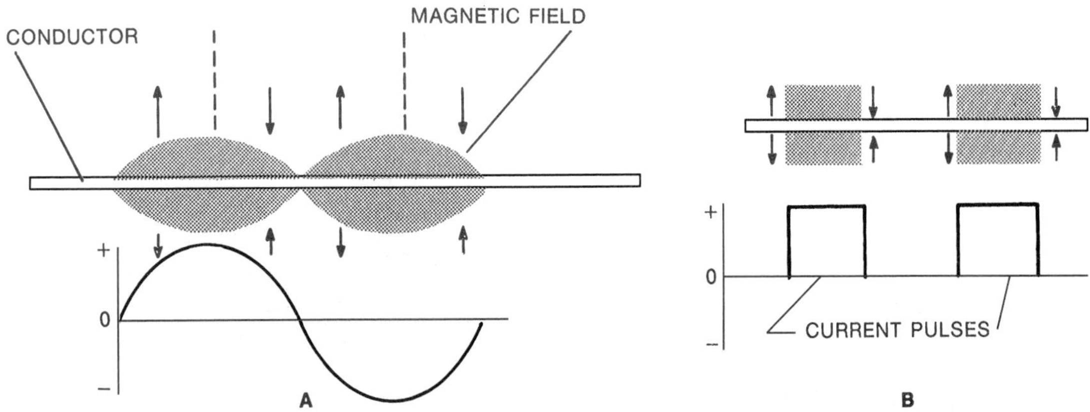

Fig. 12-3. Moving magnetic fields: (A) field produced by an alternating current; (B) field produced by a pulsating direct current.

Direct Current

A steady direct current does not produce a moving magnetic field. For this reason, such a current cannot produce electromagnetic induction. When a steady direct current must be used with a device such as a transformer, the current must be broken up into pulses. A pulsating direct current produces a changing magnetic field, which can cause electromagnetic induction (Fig. 12-3B).

Magnetic Air Gap

The magnetic field of a permanent magnet that does not move can be made to vary in strength. This is done by changing the permeability of its *air gap*, or the space between its poles. The air gap is changed by moving a soft-iron armature within it.

As the armature moves farther into the air gap, the permeability of the gap increases. This allows a stronger magnetic field to exist between the poles of the magnet (Fig. 12-4A). As the armature moves away from the air gap, the permeability of the gap decreases. When this occurs, the strength of the magnetic field also decreases (Fig. 12-4B). This method of changing the strength of a magnetic field is used in a magnetic phonograph cartridge.

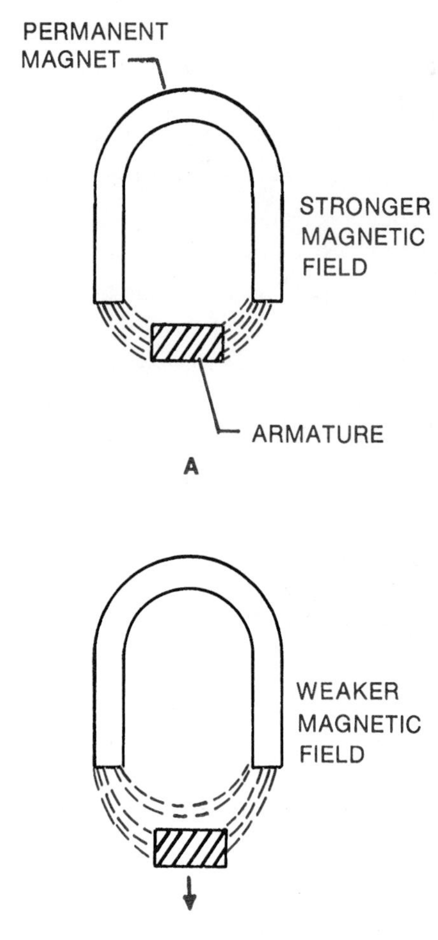

Fig. 12-4. Moving magnetic field produced by changing the position of a soft-iron armature within a magnetic air gap.

Consult your local power company. Do they use generators or alternators to produce electricity?

TRANSFORMERS

The word *transform* means "to change." A **transformer** is used to change the value of voltage or current in an electrical system (Fig. 12-5). If it reduces a voltage, it is called a *step-down* transformer. If it increases a voltage, it is called a *step-up* transformer. Some transformers do not change the value of the voltage. These are called *isolation* transformers. Isolation transformers are used when electrical equipment is electrically separate from the power line.

Transformers are used to change the value of voltage in many kinds of circuits. For example, they are used to change the power line voltage of 120 Vac to a lower voltage used in rectifier circuits. These circuits are used to change alternating household current into direct current for electronic products. The ignition coil in an automobile ignition system is a step-up transformer that supplies a high voltage to the spark plugs. Step-down transformers are used in systems that distribute electric energy from power plants to buildings. They are the most common type used.

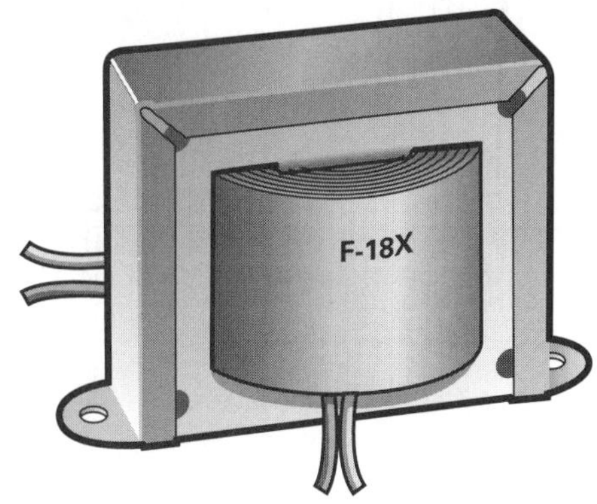

Fig. 12-5. A small audio transformer.

The operation of a simple transformer is shown in Fig. 12-6. Here the one winding, or coil, of the transformer is connected to the source of energy. This is called the *primary winding*. The other is electrically insulated from the primary winding and is connected to the load. This is called the *secondary winding*. Notice that there is *no direct electrical connection* between the primary and secondary windings.

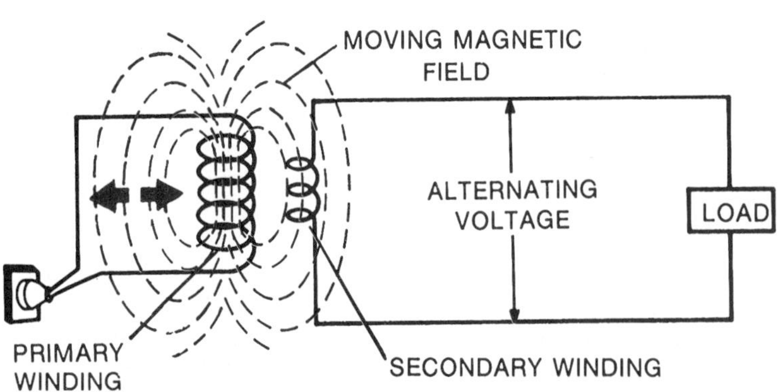

Fig. 12-6. Operation of a simple transformer.

The alternating current in the primary winding produces a moving magnetic field. This induces a voltage in the secondary winding. As a result, energy is transferred from the primary winding to the secondary winding. This is an example of what is known as *mutual induction*.

Polarity of Induced Voltage

The polarity of the voltage produced by electromagnetic induction depends on the direction in which magnetic lines of force cut across a conductor. Since a transformer operates with alternating– or pulsating direct currents, the magnetic field around the primary winding expands and collapses as it cuts across the secondary winding—first in one direction, then in the other. Therefore, the voltage induced across the secondary winding of any transformer is an alternating voltage.

Transformer Cores

The transformer in Fig. 12-6 is an *air-core transformer*. The coils of transformers operated with either alternating or pulsating direct currents are wound around *iron cores*. Despite their names, these cores are usually made of thin sheets of steel to which a small amount of silicon is added. These sheets are called laminations. Two kinds of cores and windings commonly used with iron-core transformers are shown in Fig. 12-7. Because the laminations are stacked together to form the core of these transformers, they are called *stacked-core transformers*.

This iron core allows a stronger magnetic field to exist in a transformer. This, in turn, permits a greater amount of energy to be transferred from the primary to the secondary winding.

Transformers and Toxic Waste

In 1977 the Environmental Protection Agency banned environmental discharge of PCBs (polychlorinated biphenyls), and in 1979 completely banned the manufacture of PCBs. They are toxic and tend to accumulate in animal tissues. PCB buildup in human tissues has been linked to cancer, birth defects, and other disorders. However, a significant quantity of PCBs remained in use, largely as heat dissipators in capacitors and transformers. In addition, the environment was still contaminated by PCBs discharged before the ban or dumped illegally or accidentally spilled afterward.

In the 1980s efforts to detoxify or destroy the PCBs remaining in the environment included attempts to remove the chlorine and the use of certain microorganisms which could degrade simple PCBs.

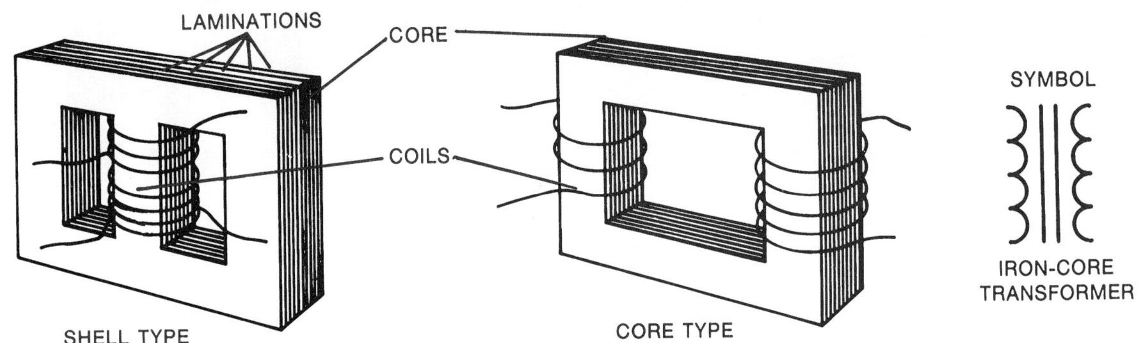

Fig. 12-7. Types of cores used with iron-core transformers.

A laminated, as opposed to solid, core is used to lessen the amount of energy wasted in the form of heat. Some of this heat is produced by **eddy currents** which are currents induced in the core of a transformer. Energy lost in the transformer core is known as *hysteresis loss*. This is the heat produced within the core as its magnetic domains try to align themselves with the changing magnetic field produced by alternating current. Another energy loss is from *copper losses*. This is the heat energy lost when current flows in the primary and secondary windings of a transformer.

Value of Induced Voltage

The value of the voltage induced across secondary winding depends on the number of wire turns it has compared with the number of primary-winding turns. This is called the *turns ratio* of the transformer. If the secondary winding has half as many turns as the primary winding, the voltage will be stepped down to one-half the primary-winding voltage (Fig. 12-8A). If the secondary winding contains twice as many turns as the primary winding, the voltage will be stepped up to twice the primary-winding voltage (Fig. 12-8B). This relationship can be expressed by the formula:

$$\frac{\text{secondary voltage}}{\text{primary voltage}} = \frac{\text{secondary turns}}{\text{primary turns}}$$

Problem: An experimenter has a 115-Vac transformer that has 725 turns of wire on the primary coil. Neglecting transformer losses, how many turns of wire are needed on the secondary winding to produce 10 Vac?

Solution: Use the preceding formula, and let x = secondary turns.

$$\frac{\text{secondary voltage}}{\text{primary voltage}} = \frac{\text{secondary turns}}{\text{primary turns}}$$

$$\frac{10}{115} = \frac{X}{725} = 115_x = 10 \times 725 = 115_x = 7250 = {}_x = 63 \text{ turns}$$

Do you use transformers at home? List some uses.

POWER TRANSFORMERS

A step-down transformer that supplies the voltage needed in electronic equipment is often called a *power transformer* (Fig. 12-9). A power transformer used in a rectifier circuit is often called a *rectifier transformer*.

Some power transformers have more than one secondary winding. Each of these windings is electrically insulated from the other. This

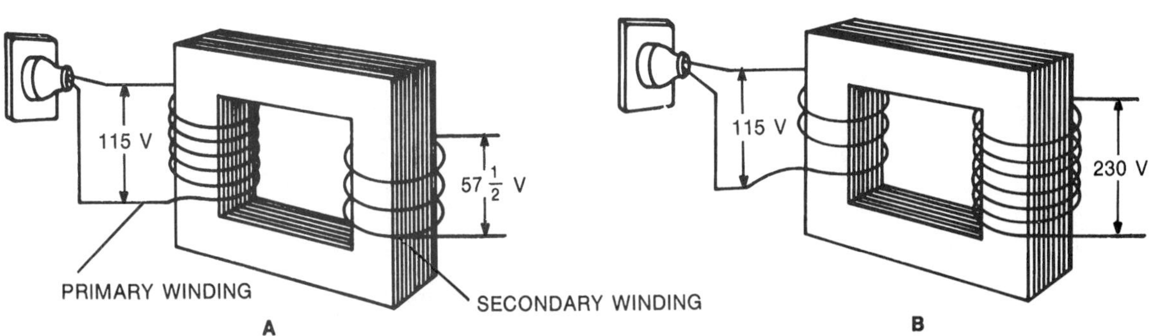

Fig. 12-8. The voltage induced across the secondary winding of a transformer depends on the winding turns ratio of the transformer. (A) Step-down transformer; (B) step-up transformer.

Fig. 12-9. **Power transformer.**

makes it possible to obtain different values of output voltages (Fig. 12-10) from connections called *taps*. Transformers are commonly found in manufacturing applications.

Voltage and Current

Although a power transformer can change voltage, it cannot deliver a greater amount of energy to a load from its secondary winding than is supplied to the primary winding. In fact, the energy delivered by the secondary winding is always less than that supplied to the primary winding. This is because of eddy currents and copper losses within the core and windings.

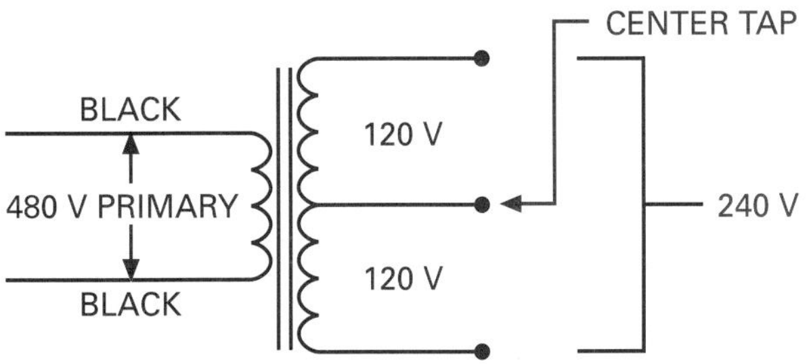

Fig. 12-10. **Schematic of a power transformer with a center tap.**

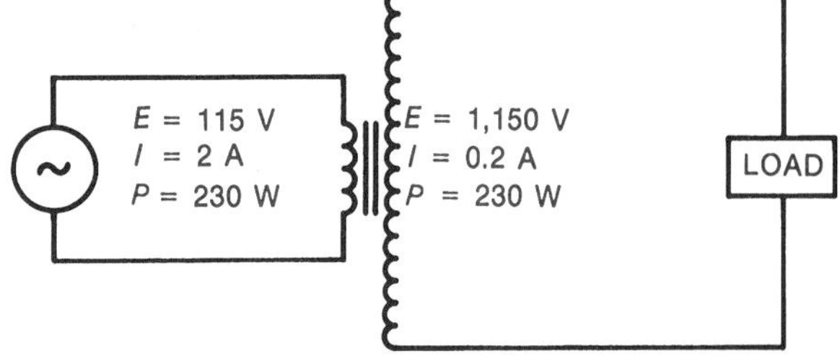

Fig. 12-11. **Relationship between voltage and current in a transformer.**

As an example, suppose that the primary winding of a transformer operating at 115 V has a current of 2 A. If the secondary winding has 10 times as many turns of wire as the primary, the voltage across the secondary winding will be stepped up to 115 x 10, or 1,150 V (Fig. 12-11).

Since $P = EI$, the wattage of the primary winding is equal to 115 x 2, or 230 W. Disregarding power losses within the transformer, the wattage of the secondary winding will also be 230 W. The largest current that the secondary winding can deliver to a load is then equal to

$$I = \frac{P}{E}$$

$$= \frac{230}{1150} = 0.2\text{A}$$

Therefore, as the voltage is increased 10 times, the current in the secondary winding is decreased 10 times, or from 2 A to 0.2 A.

Likewise, if a transformer steps down voltage, the largest current that can be delivered to a load by the secondary winding will be greater than the current in the primary winding. This condition must exist if the wattage ($P = EI$) of the secondary winding is to equal the wattage of the primary winding.

Power Transmission Lines

The relationships between input and output voltages and currents in transformers are

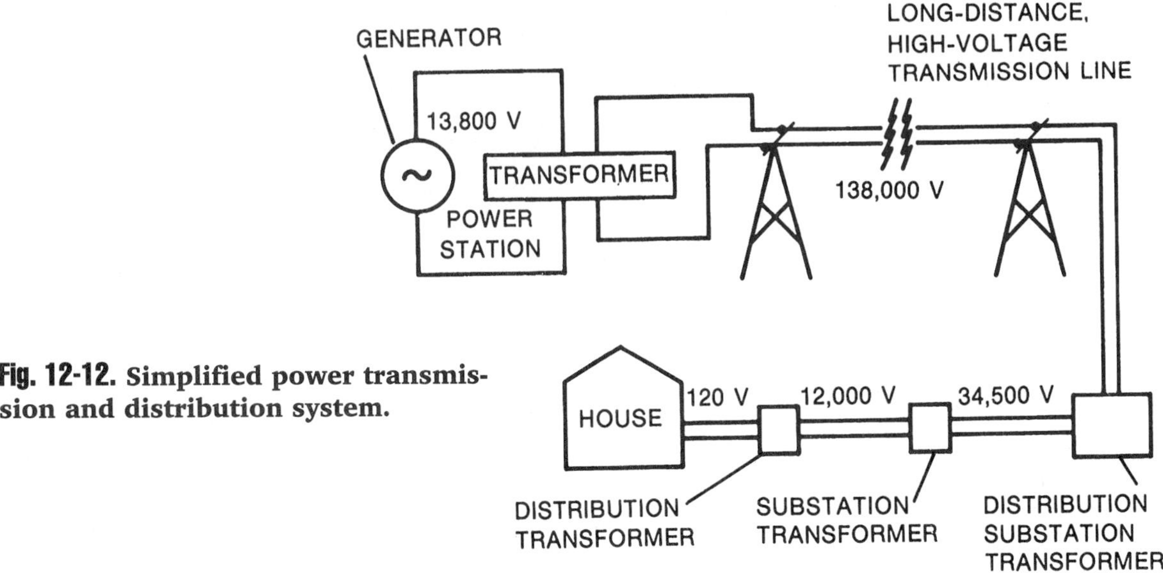

Fig. 12-12. Simplified power transmission and distribution system.

shown by *transmission lines*. Such lines deliver electric energy from power generating plants to homes and businesses (Fig. 12-12). Here, a transformer at the power station steps up the alternating voltage produced by the generator. Because of this very high voltage, it is possible to deliver the required amount of energy from the plant to the users along the long-distance transmission line using a high voltage and low current. This low current, in turn, lessens the power losses within the transmission line. It also permits the use of smaller conductors.

The transformers used with this power transmission system step down the high voltage of the transmission line to much lower voltages. Distribution transformers then further step down these voltages. Because the voltages are stepped down, the total current available to the users is many times greater than the current present in the transmission line. However, the total power is never greater than that originally produced by the generator. Actually, because of line losses, the available power is much less than that generated.

When an alternating current passes through a single coil, a moving magnetic field is produced around it. This action induces a voltage across the coil itself by means of electromagnetic induction. This is called self-inductance, or simply *inductance*. The voltage generated by self-inductance is always of a polarity that opposes any change in the value of the current through the coil.

Consider the circuits shown in Fig. 12-13. When the alternating voltage applied to these circuits increases in value during the first quarter of its cycle, the current through the coil also increases. This produces an expanding magnetic field around the coil (Fig. 12-13A). The voltage now induced across the coil is of a polarity that tends to oppose the change (increase) of the current through the coil.

When the applied voltage begins to decrease during the second quarter of its cycle, the magnetic field around the coil starts to collapse (Fig. 12-13B). As a result, the polarity of the voltage induced across the coil reverses. This voltage is now of a polarity that once again tends to oppose the change (decrease) of the current through the coil.

As the current moves through the second half of its cycle, the polarities of the applied

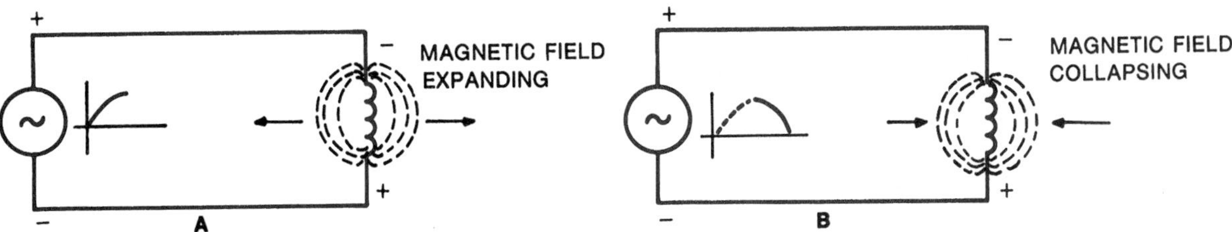

Fig. 12-13. **Self-inductance.**

voltage and the induced voltages reverse. However, the induced voltage continues to be of a polarity that tends to oppose any change in the value of the current. Because of this, the induced voltage also opposes the applied voltage that causes the current changes. Therefore, the voltage induced by self-inductance is called a *counter voltage* or a *counter electromotive force* (cemf), or **inductance**.

The unit of inductance is the **henry**. It is named after Joseph Henry, a nineteenth-century American teacher and physicist. A coil has an inductance of one henry when a current that changes at the rate of one ampere per second causes a voltage of one volt to be induced across it. The letter symbol for inductance is *L*.

FASCINATING FACTS

Although Michael Faraday is credited with the discovery of electromagnetic induction because of his earlier published writings, Joseph Henry independently found the principle of induced currents. Henry was officially recognized by the 1893 International Congress as the discoverer of self-induction. Therefore, the unit of measure for inductance is named the henry in his honor.

LINKS

You may wish to review Chapter 4. It discusses frequency, cycle, and hertz.

Inductive Reactance

Inductive reactance is the opposition to a changing current in a coil that is caused by the inductance of the coil. The unit of inductive reactance is the ohm. Its letter symbol is X_L The value of inductive reactance depends on the inductance of the coil and the frequency of the current that passes through it. The formula for inductive reactance is:

$$X_L \text{ (in ohms)} = 2\,P = fL$$

where f = the frequency of the current in hertz
L = the inductance of a coil in henrys
P = 3.14 (pi), a constant

This formula shows that inductive reactance is directly proportional to both frequency and inductance. As the frequency increases, the inductive reactance will increase. The inductive reactance will also increase if the inductance of a circuit is increased.

When inductive reactance and resistance are both present in a circuit, their total opposition to current is called impedance. The unit of impedance is the ohm. Its letter symbol is *Z*.

What is a substation? Consult your power company to find out where the one is for your house. See if you can follow the transmission lines to your neighborhood. Do this visually from the ground. Do not go near the wires.

EFFECT OF INDUCTANCE IN A DC CIRCUIT

Since the counter electromotive force (cemf) induced across an inductive component always opposes the applied voltage, the current changes in an inductive circuit lags behind the applied voltage whenever there is a change in the value of the current.

The current responses in a typical series *RL* circuit are illustrated in Fig. 12-14. Responses that occur when voltage is applied or removed are not instantaneous. The straight portion of the curve between the current-increasing and current-decreasing responses indicates the steady state of the circuit. During this time, no cemf is induced across the coil. Therefore, the current and the voltage are together because the current is limited only by the resistance of the circuit, which does not change with current direction.

Time Constant

The time required for the current in an inductive circuit to rise to its maximum magnitude (I_{max}) after the operating voltage has been applied depends upon the circuit's inductance and resistance. With a fixed amount of resistance in the circuit, the time period will

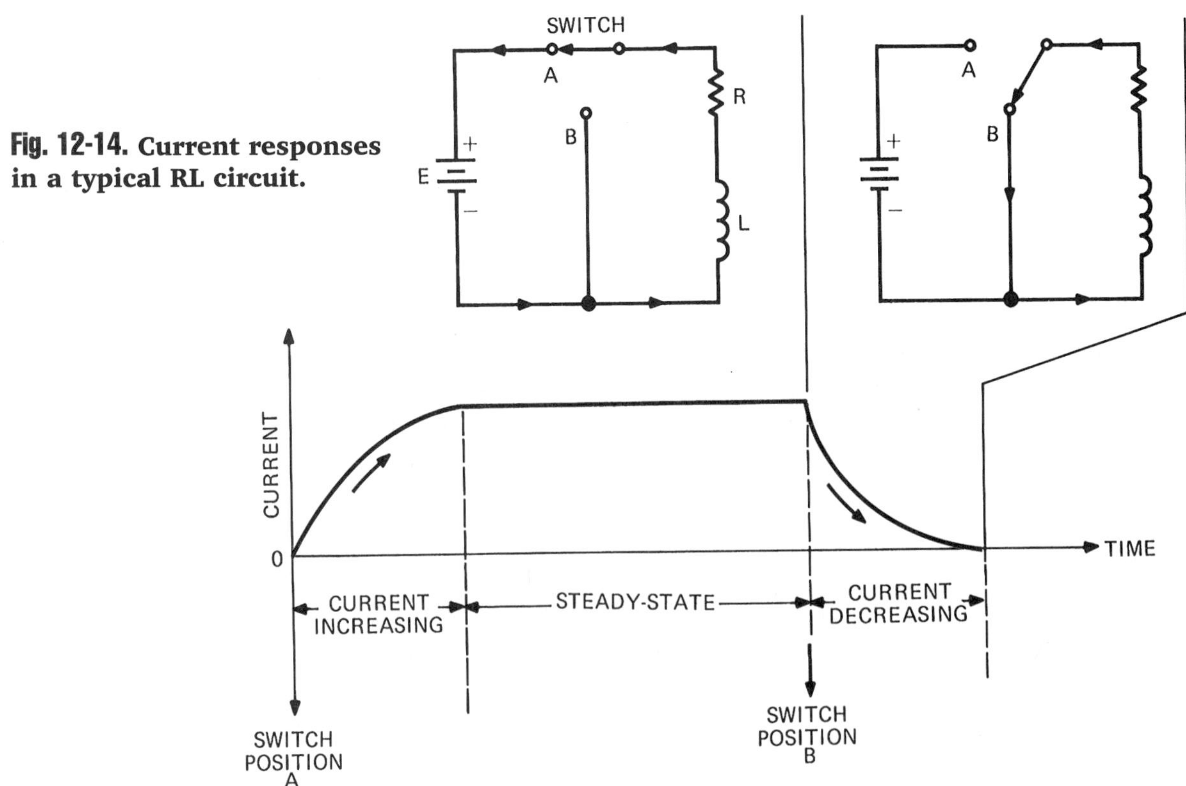

Fig. 12-14. Current responses in a typical RL circuit.

increase if the inductance increases. This occurs because the greater the inductance, the greater the magnitude of the cemf that is present to oppose the change in current. If the inductance of the circuit remains constant and resistance increases, the period of time required for attaining maximum current decreases. This is because the increase in resistance reduces the current flow in a circuit. Therefore, less cemf is generated to oppose the change in current.

The time required for the current in a series *RL* circuit to increase to 63.2 percent of its maximum (steady state) value is known as the *RL* time constant of the circuit. It is expressed as the formula,

$$t = \frac{L}{R}$$

where t = time constant, in seconds
L = inductance, in henrys
R = resistance, in ohms

The time constant also governs the rate at which a current decreases as the applied voltage is removed from a circuit (if a "discharge" path is provided by the remaining components of the circuit). Thus, if the current in a given circuit increases to 63.2 percent of its maximum in 0.25 second, it will also decrease by 63.2 percent of its maximum in 0.25 second.

Voltage Characteristics

During the time that the current in a series *RL* circuit is increasing, there are three different voltages present in the circuit. These voltages are: the applied voltage, the voltage drop across the resistance, and the cemf generated across the inductive component.

High Induced Voltages

When the current in an inductive circuit is increasing, the magnitude of the cemf induced across any coil in the circuit cannot become greater than the applied voltage. However, when the switch in such a circuit is opened, the resistance of the circuit suddenly spikes toward infinity. As a result, the *RL* time constant becomes extremely small, and the current decreases toward zero rapidly. This in turn causes a voltage many times greater than the applied voltage to be induced across the coil. Therefore, a spark can often be observed between a switch and other contacts when an inductive circuit is opened suddenly.

INDUCTORS

Inductors are coils that are wound to have a certain amount of inductance. Inductance depends on four things: (1) the number of turns of wire with which the coil is wound, (2) the cross-sectional area of the coil, (3) the permeability of its core materials, and (4) the length of the coil.

Iron-core inductors are sometimes called *choke coils* or *reactors*. They often are used to filter, or smooth, the output current from a rectifier circuit (Fig. 12-15). Inductors are also used with capacitors in radio and television tuning and in oscillator circuits.

1. **How many seconds will it take for a 10-L inductor and a 5-W resistor in a dc circuit to reach 63.2% of its steady state?**
2. **Does an inductor with one hundred turns have more, or less, inductance than one with fifty turns?**

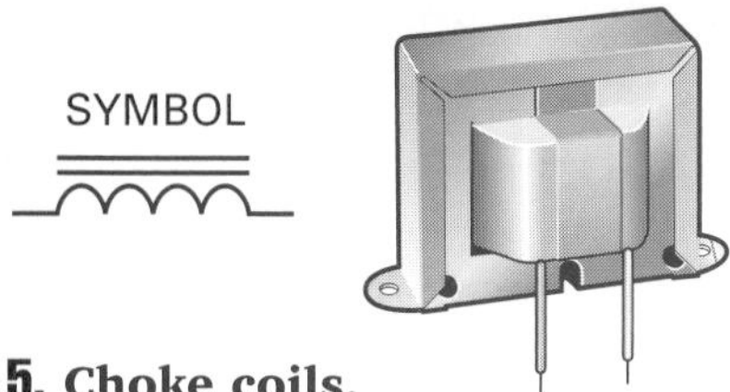

Fig. 12-15. **Choke coils.**

TOROIDAL TRANSFORMERS

Toroidal transformers are now used more frequently in electric circuits than ever before (Fig. 12-16A). The major advantages of these transformers over stacked-core types are: (1) less hum, (2) stray magnetic fields, (3) lower weight, and (4) smaller size.

The term *toroid* is used to define the geometry of the doughnut-shaped magnetic core. The core, made of powdered iron or silicon steel, has no air gap. Therefore, stray magnetic fields are minimized. They are also easy to mount on a circuit board (Fig. 12-16B). The wires of the primary winding are insulated from the core and wound around it. The primary winding is wound first. The secondary winding, which is insulated from the primary, is wound and distributed evenly on top of the primary winding. The toroidal transformer, like the stacked-core type, can have several secondary windings.

Toroidal transformers and inductors are used in industrial control equipment, computers, audio amplifiers, and power supplies.

Why do toroidal transformers hum less?

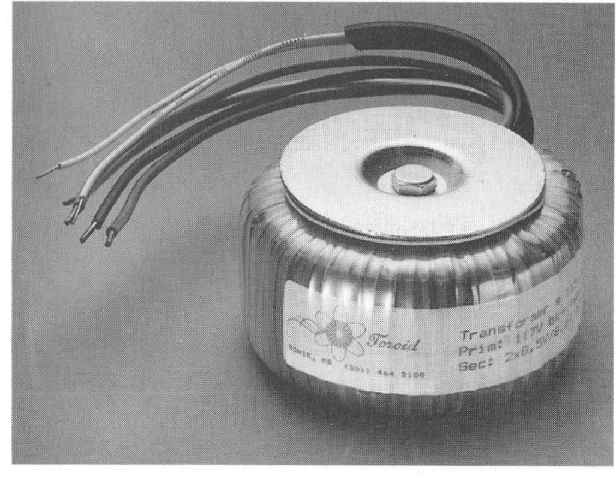

A

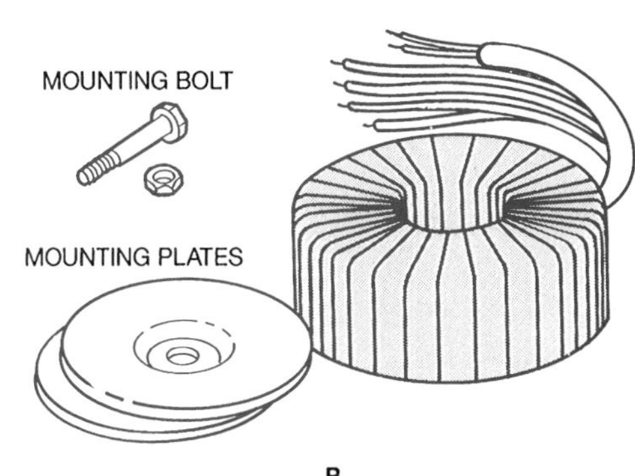

B

Fig. 12-16. **Toroidal transformer: (A) toroidal power transformer; (B) single bolt and mounting plates removed from transformer.**

TESTING TRANSFORMERS AND INDUCTORS

The most common defects of transformers are: (1) burned open windings caused by too much current, (2) **shorts** (wires touching that should be insulated from each other) between adjacent turns of wire, (3) windings grounded or shorted to the core, and (4) shorts between the primary and secondary windings. The first three of these defects also occur in inductors.

Continuity Tests

The different windings of a transformer can be identified easily by making *continuity tests*. A continuity test also can be made to see if there are any open windings or to see if there are any shorts between windings or between a winding and the core.

Shorts between the turns of a transformer or an inductor winding are usually found with an ohmmeter. Such shorts reduce the impedance of a transformer and cause windings to overheat even when they are operating at normal current.

Continuity tests of a transformer or an inductor should be made only after the device has been disconnected from its circuit. If this is not done, other components in the circuit may interfere with the test results.

Voltage Tests

After testing for continuity, the secondary winding should be tested for voltage. To do this, it is necessary to apply voltage to the primary winding. This should be done only after checking to see that bare ends of secondary-winding leads are not in contact with each other, ground, or other component. Where very high secondary voltages are expected, a reduced primary voltage may be used and the voltage ratio determined. For example, a transformer with a 120-V primary and a 600-V secondary winding has a voltage ratio of 1 to 5.

After you have found the voltage of each secondary winding, label the leads for future reference.

SAFETY TIP

Do not conduct any testing without having been instructed by a licensed technician or qualified instructor. Power transformers often have one very high voltage winding called a "wild leg." Therefore, extreme care must be taken when applying the rated primary voltage. Never touch any lead of any electrical component while making checks. Always treat every electrical component as if it is "live."

Measuring Inductance

The digital meter discussed previously in Chapter 10 (see Fig. 10-15) performs two functions. It measures both capacitance and inductance. It is called a *Z meter* because it measures the total impedance of a circuit's components (capacitors, resistors, and inductors) being tested. This chapter discusses how the meter measures inductance.

The following steps represent the general procedures to determine the inductance of an inductor (coil).

1. Turn the power off by disconnecting the ac line cord to the equipment under test.
2. Wait 5 minutes for the circuit capacitors to discharge.
3. Disconnect the inductor leads from the circuit.
4. Connect the cord of the meter to a properly grounded 117-V ac outlet.
5. Turn the meter power switch on.
6. Connect the test leads to the coil or transformer to be tested.
7. Depress the VALUE button under the inductor section of the push button switch.
8. Read the value of inductance of the coil or transformer on the digital display. The LED will light in front of the *m*H if the value is in micro-henrys or in front of the mH if the value is in milli-henrys.

How do you think a resistor in series with an inductor will affect its *Z* reading?

Chapter *Review*

Summary

- Lenz's law describes the relationship between magnetism and current.
- Motion, magnetism, and a conductor must be present for induction to take place.
- Induction through transformers is used to change the value of voltage in electrical systems. They can either step up or step down voltage. Transformers cannot change power, or energy, values.
- Inductive reactance is the opposition to a changing current in a coil. This reactance is caused by the coil's inductance, or its ability to self-induce voltages in its own coil.
- The most-used formulas to design circuits with resistance and inductance are the formulas for inductive reactance and the *RL* time constant.

Review Main Ideas

Review this chapter's main ideas by writing, on a separate sheet of paper, the word or words that most correctly complete the following statements:

1. If a magnet is moved near a wire, its ______ cuts across the wire and produces a voltage across the ends of the wire.
2. Lenz's law states that when a ______ is induced by moving a conductor across lines of magnetic force, the induced current creates a magnetic field that ______ the lines of magnetic force that produced it.
3. One of the most important uses of electromagnetic induction is in alternators and other kinds of ______.
4. The magnetic field produced by an alternating current can be thought of as a ______ magnetic field.
5. If a transformer is operated with direct current, the current must be broken up into ______.
6. The transformer is used to ______ the value of a voltage or current in an electrical system.
7. If a transformer reduces voltage, it is called a ______ transformer. If it increases voltage, it is called a ______.
8. The winding of a transformer that is connected to a source of energy is called the ______ winding.

9. The voltage induced across the secondary winding of any transformer is an ______ voltage.
10. The thin sheets of steel from which transformer cores are made are called ______.
11. A laminated transformer core is used to lessen the amount of energy that is wasted in the form of ______.
12. A step-down transformer that supplies the voltage needed in electrical equipment is called a ______transformer. If this type of transformer is used in a rectifier circuit, it is called a ______ transformer.
13. Although a power transformer can change voltage, it cannot deliver a ______ amount of energy to a load from its secondary winding than is ______ to the primary winding.
14. The voltage induced across the coil itself by means of electromagnetic induction is an example of ______-inductance.
15. The voltage induced by inductive reactance is called ______.
16. Coils that are wound so as to have a certain amount of inductance are called ______.
17. Iron-core inductors used to filter the output current from rectifier circuits are called ______ coils or ______.
18. A ______ transformer is designed in such a way that it has no air gaps.

Apply Your Knowledge

1. If a transformer has a primary voltage of 240 V and turns of 3:1, what is its secondary voltage if it is a step-up transformer? What is its secondary voltage if it is a step-down transformer?
2. What is the inductive reactance of a coil with 5 henrys of inductance in a 60-hertz (frequency) circuit?
3. Calculate the *RL* time constant of a circuit with 1 KΩ of resistance.
4. List the four most common defects of transformers and inductors.
5. Explain why continuity testing is performed on transformers.

Make Connections

1. **Communication Skills.** Write a short synopsis of the life of the Russian physicist Heinrich F. E. Lenz. Identify where he lived and the institution where he worked.
2. **Mathematics.** An experimenter has a 115-Vac transformer that has 725 turns of wire on the primary coil. Neglecting transformer losses, calculate how many turns of wire are needed on the secondary coil to produce 25 Vac.
3. **Science.** In a written report, discuss the causes of heat loss in transformers.
4. **Workplace Skills.** Obtain a transformer that has several secondary taps. Show the class how each secondary winding is identified.
5. **Social Studies.** On a clear day, the electric power goes off in your home. You learn that the electric power is also off in several other homes on your block. Identify the steps to follow to correct the situation. Identify the possible cause of the problem.

Chapter 

Circuit Diagrams and Symbols

OBJECTIVES

After completing this chapter, you will be able to:

- Define the following circuit diagrams: pictorial, schematic, block, and architectural.
- Draw a schematic wiring diagram from a pictorial diagram.
- Draw from memory seven graphic symbols used in electrical and electronic diagrams and label each symbol with the proper class designation letter.

Terms to Study

ANSI

block diagram

chassis ground

class designation letters

computer-generated diagrams

diagram

earth ground

graphic symbols

pictorial diagram

scale

schematic diagram

Suppose you want to tell a friend how to get to a soccer field several blocks away. What's the best and most convenient way to do this? Should you just tell him or her where the field is located? Or, would it be clearer and easier for the friend to look at a sketch with lines drawn on paper that represent city blocks? A rectangular shape could represent the soccer field. The sketch would be a plan to follow to find the soccer field. An electric diagram, like the sketch, is a plan to follow when building a circuit.

PICTORIAL DIAGRAMS

Once you have developed a circuit, you should plan how to build it. In electrical and electronics work, the basic part of such a plan is usually a diagram. A **diagram** shows how different parts are connected to form a circuit. Developing the diagram is an important step in designing circuits. A diagram guides those who maintain and repair the circuit, and serves as a record for those who need to copy, study, or change the circuit.

This unit discusses several common kinds of diagrams. It also offers much useful information about reading and drawing diagrams.

If you build an electrical product from a do-it-yourself kit, one of the items included in the construction manual will be a pictorial diagram. A **pictorial diagram** is a sketch of circuit parts showing how the parts are connected and where they are to be located within the assembly (Fig. 13-1). The pictorial diagram also shows the wiring devices, such as terminal ends and sockets that are used. The pictorial diagram is usually drawn to **scale**, meaning that the *relationships* of the sizes and locations of the parts are accurate whether they are shown full size, smaller, or larger.

Pictorial diagrams are easy to follow. For this reason, they are used with do-it-yourself kits and in manufacturing to show what the finished products look like.

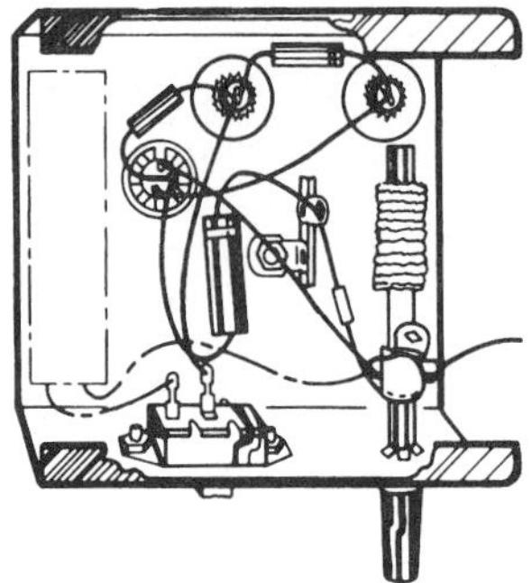

Fig. 13-1. Pictorial diagram of a one-transistor radio.

FASCINATING FACTS

During World War II, defense plants employed millions of unskilled workers. To help them in their work, new techniques were developed to show mechanical diagrams more simply than with conventional methods. Rather then being drawn to scale, drawings were made in true or exploded scale, showing all components in proper size and proper relative positions. Stacks of transparent drawings representing a complex device's complete assembly were bound as books. These simplified drawings are now widely used as supplements to conventional mechanical drawings.

One disadvantage of a pictorial diagram is that it does not give clear electrical information about the circuit. It simply shows what the circuit will look like after it is put together. Such diagrams do not show the electron path or the way the parts relate to one another electrically. Pictorial diagrams usually take too much time to prepare and use too much space.

Draw a floor plan diagram of your room in your house. Include the doorways, furniture, windows, electrical outlets, and lighting fixtures. Draw it to scale.

SCHEMATIC DIAGRAMS

Schematic diagrams are the standard way to communicate information about electric and electronic circuits. In these diagrams, **graphic symbols** represent components. These letters, drawings, or figures stand for something else. In electricity and electronics, a graphic symbol shows the connecting points of a part in a circuit. Because the symbols are small, diagrams are more compact than pictorial diagrams. The symbols and related lines show how the parts of a circuit are connected and how they relate to each other.

Using Graphic Symbols

Examine the schematic diagram of the two-transistor radio circuit shown in Fig. 13-2. If you read this diagram from left to right, you will see the components (or parts) in the order in which they convert radio waves into sound energy. The antenna (at the left) collects the radio waves (energy) which the circuit converts to electronic signals. The headset (at the right) converts the electronic signals into sound. By using the diagram, you can trace the operation of the circuit from beginning to end. Because of this, engineers and technicians who design, construct, troubleshoot, and maintain electrical circuits use them often.

Identifying Components

As shown in Fig. 13-2, the components on a schematic diagram are also identified by **class designation letters**. These letters include *R* for resistors, *C* for capacitors, and *Q* for transistors. Table 13-A shows the class designation letters for selected items. The letters are further combined with numbers, such as R_1 R_2, and R_3, to distinguish different components of the same kind. The numerical values of components are often shown directly on the schematic diagram

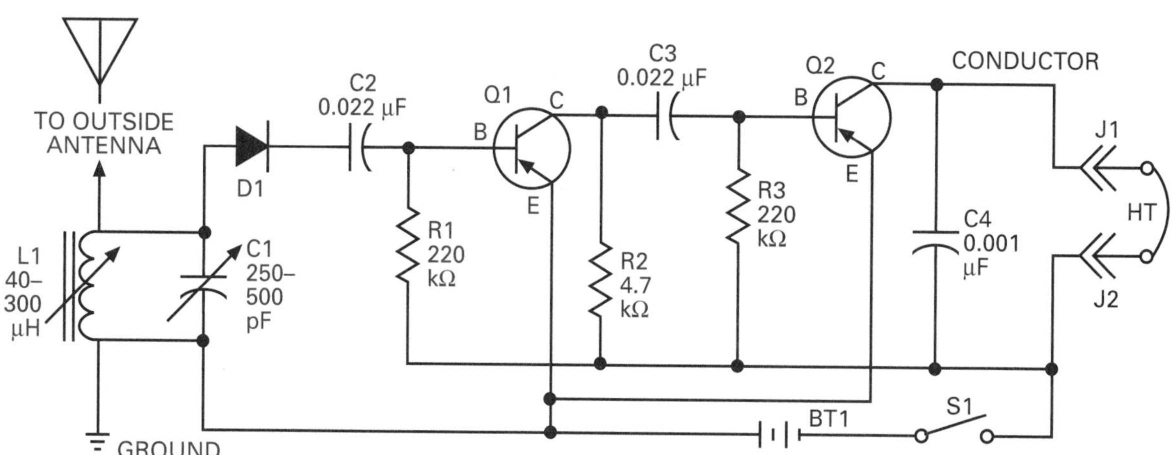

Identification of components

BT1	Battery
C1	Variable capacitor (the arrow crossing the symbol indicates the value of the component can be varied.)
C2, C3, and C4	Fixed capacitors
CR1	Crystal diode
HT	Double headset
J1 and J2	Insulated tip jacks-connectors
L1	Antenna coil (variable ferrite-core inductor)
Q1 and Q2	p-n-p transistors
R1, R2, and R3	Carbon-composition resistors
S1	Single-pole—single-throw switch

Conductor symbols

Conductors crossing but not connected.

Conductors electrically connected (the dot is often omitted when the connection is obvious).

NOTE: The letters B, C, and E near the transistor symbols indicate the Base, Collector, and Emitter leads of the transistors.

Fig. 13-2. Schematic diagram of a two-transistor radio.

Table 13-A Selected Class Designation Letters

Designation Letter(s)	Class of Items
AR	Amplifier (other than rotating) Repeater
BT	Battery Solar cell
C	Capacitor
CB	Circuit breaker
D or CR	Crystal diode Metallic rectifier Varactor
DS	Alphanumeric display device Light-emitting solid state device Signal light
F	Fuse
HT	Earphones Electrical headset Telephone receiver
L	Inductor
M	Motor
Q	Semiconductor controlled switch Phototransistor (three terminals) Transistors
R	Potentiometer Resistor Rheostat
S	Contactor Disconnecting device (switch) Switch Telegraph key Thermostat
T	Transformer

with their component identifications. If they are not given in this way, they will be stated in the parts list in the notes that come with the diagram. Very often, symbol letters are shown with slanted letters, called *italics*, on diagrams.

Other Graphic Symbols

A number of other common symbols used with schematic diagrams are shown in Fig. 13-3. One of the organizations responsible for standardizing these and other symbols is the American National Standards Institute (**ANSI**).

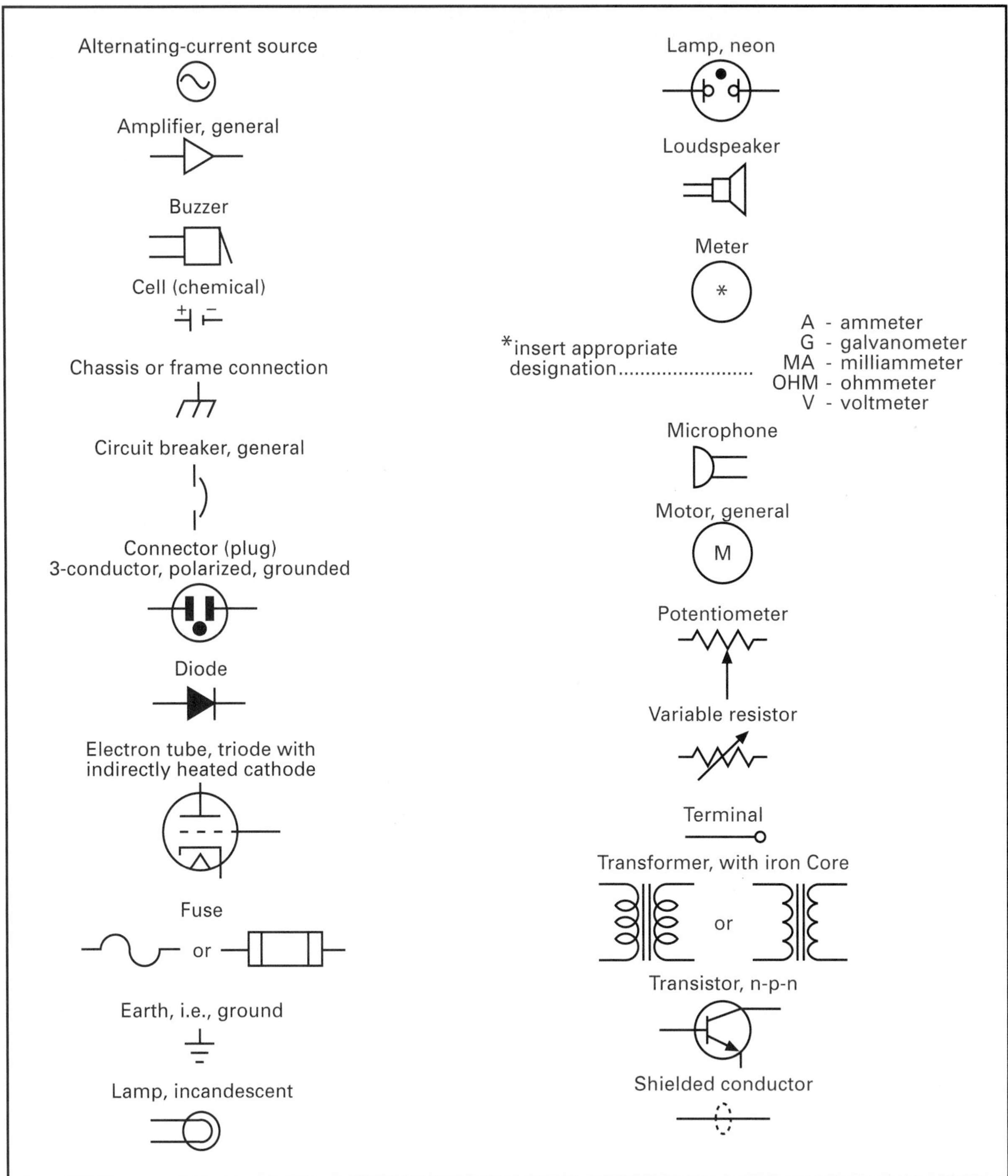

Fig. 13-3. Common symbols used on schematic diagrams. Other symbols are shown throughout the book.

Complete listings of the symbols used in electricity and electronics are available from this organization. Standards used in other technical fields are also available from the ANSI.

Symbols for *earth ground* and *chassis ground* connections are commonly found on schematic diagrams. These symbols serve as common reference points in a circuit. **Chassis ground** stands for a common circuit return to a structure, such as the frame of an air, space, or land vehicle that is not electrically connected to earth. If the circuit, however, has a direct connection to the earth or a body of water, additional information should be placed adjacent to the symbol. The word *earth* adjacent to a ground symbol, for example, indicates an earth ground. An **earth ground** connection is made by fastening a wire to a metal rod that is driven into the earth. The symbol for a chassis, or frame, connection also stands for a common circuit return. This symbol represents an electrical connection to a chassis, a common wire, or a common area on a printed circuit board. This connection, however, may be at a higher voltage level with respect to the earth or structure in which this chassis frame (or printed circuit board) is mounted. In complex circuits, such as those for television and radio, many connections have a common return. The use of these symbols makes it possible to show all common connecting points without cluttering the diagram with extra lines. Water pipe grounds are no longer admissible by all codes.

SAFETY TIP

Chassis grounds are connected to earth ground in appliances and motor-driven equipment. This is done as a safety measure in case a motor lead shorts to the chassis. If a motor lead shorts to the chassis and the chassis is not "grounded," personnel who are not grounded might touch the frame and suffer a severe electrical shock. However, if a motor lead shorts to a grounded frame, a circuit breaker will trip or a fuse will blow. This will stop the motor and protect both people and equipment from unwanted current.

Reading the Schematic Diagram

Being able to read schematic diagrams is important in understanding electrical circuitry. Most circuits you may want to study, build, or repair are probably illustrated by such a diagram. Reading a diagram is not as hard as it may first appear. With some practice, you will find that referring to a schematic is convenient and time saving.

To learn how to read schematic diagrams, begin by looking at the pictorial and the schematic diagrams of the same circuit shown in Fig. 13-4. Read from left to right. Find and identify the same components in both. Trace the wire connections. Some connections in the pictorial diagram may be covered by other components. Try to find these wires in the schematic diagram.

Unlike a pictorial diagram, a schematic diagram does not show where the parts or connecting wires are located on the chassis. Also, note that the schematic diagram does not show any wiring devices such as terminal ends or sockets. How to use any other wiring devices depends on the sizes and shapes of the components, available space, and on the electrical characteristics of the circuit.

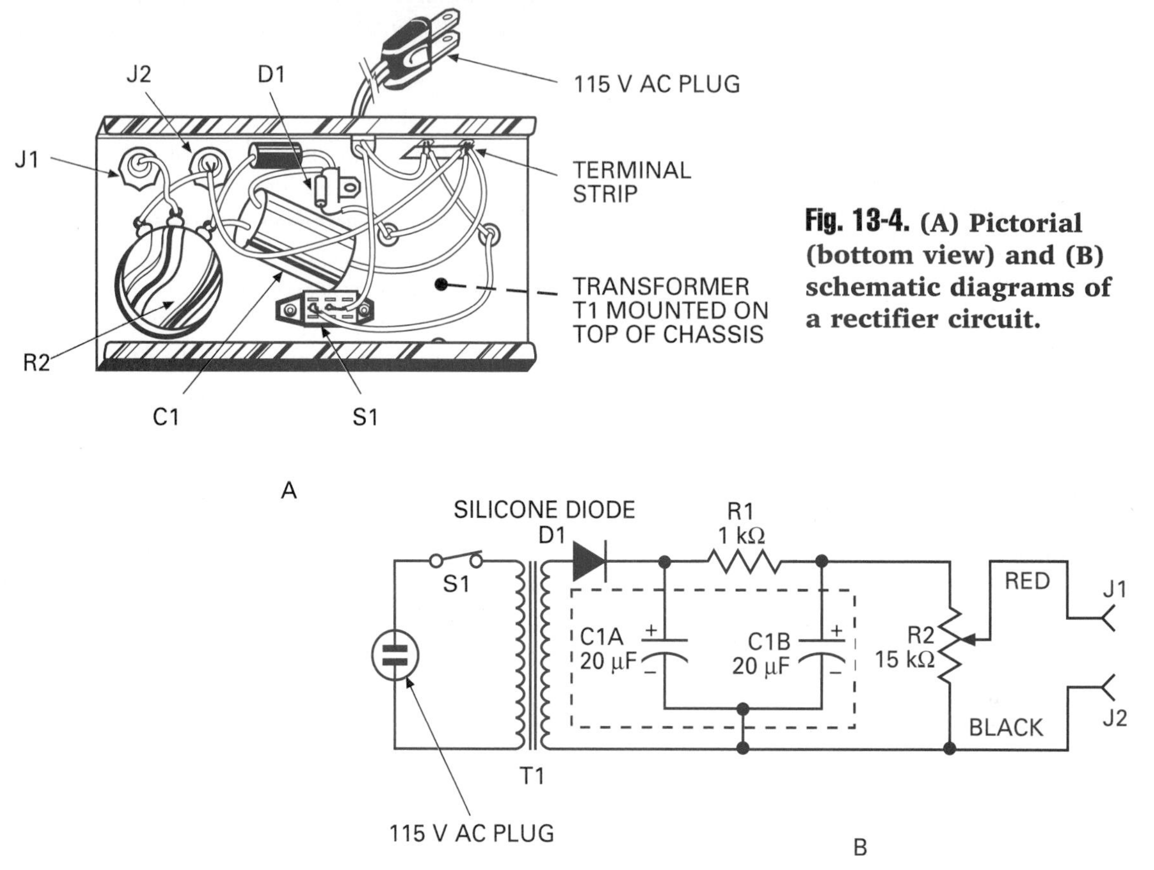

Fig. 13-4. (A) Pictorial (bottom view) and (B) schematic diagrams of a rectifier circuit.

Wiring a Schematic Diagram

To wire a circuit from a schematic diagram, you must know how to read the diagram. You also need to know about wire, about making connections, and about wiring tools and devices.

LINKS

Chapter 20 provides details on wiring materials, tools, and processes. For now, you should concentrate on understanding and drawing the various types of circuit diagrams discussed in this chapter.

Drawing a Schematic Diagram

Sometimes you will wire a circuit without using a schematic diagram. At other times, you may have to work on an existing circuit for which there is no diagram. In such cases, you need to draw a schematic diagram by looking at the wired circuit. The following suggestions will help you make a neat, readable diagram:

1. Use standard symbols for all components. If there is no standard, make a symbol for your use, but note its meaning on the diagram and in a separate symbol list.
2. Locate the symbols on the page so the lines showing the connecting conductors will not be too close together. Use as few line crossings and bends as possible (Fig. 13-5A).

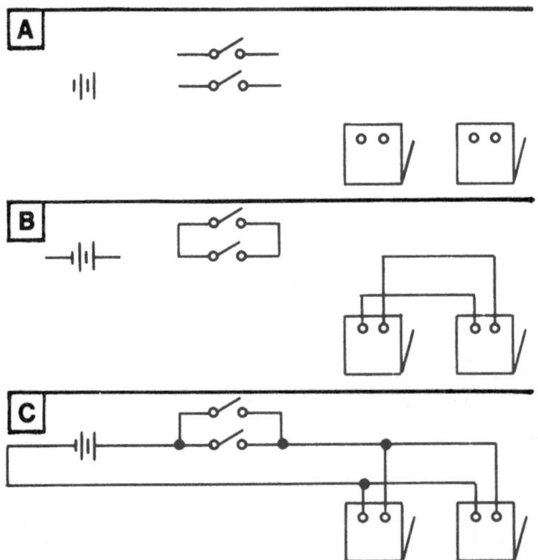

Fig. 13-5. Drawing the schematic diagram.

3. Do not draw the symbols so near each other that they crowd the diagram and make it hard to read.
4. Use unruled paper for schematic diagrams to keep from confusing drawn lines with the lines printed on the paper.
5. After drawing the symbols, connect them with straight, vertical, and horizontal lines (Fig. 13-5B and 13-5C).
6. Use the dot symbol when it is necessary to show where conductors are electrically connected.

OTHER ELECTRICAL DIAGRAMS

While the pictorial diagrams and schematic wiring diagrams are very common, there are other types of drawings to show electrical wiring. These are wiring (connection diagrams), block diagrams, architectural floor plan diagrams, and computer circuit diagrams.

Block Diagrams

A **block diagram** is a convenient way to quickly give information about complex electrical systems (Fig. 13-6). Each block represents electrical circuits that perform specific system functions. The functions the circuits perform are written in each block. Where appropriate, block diagrams are often combined with graphic symbols. Block diagrams usually read from left to right with arrows indicating current direction. Wires and individual components are not shown on block diagrams. The *power supply* block has arrows pointing toward the other blocks. This means that its function is to supply electrical power to all the circuits. The power supply block is sometimes omitted in block diagrams because it is assumed that all electrical circuits need power to operate.

Architectural Floor Plan Diagrams

Architects, electrical designers, and contractors use floor plan diagrams to show where the parts of a building's electrical system are to be

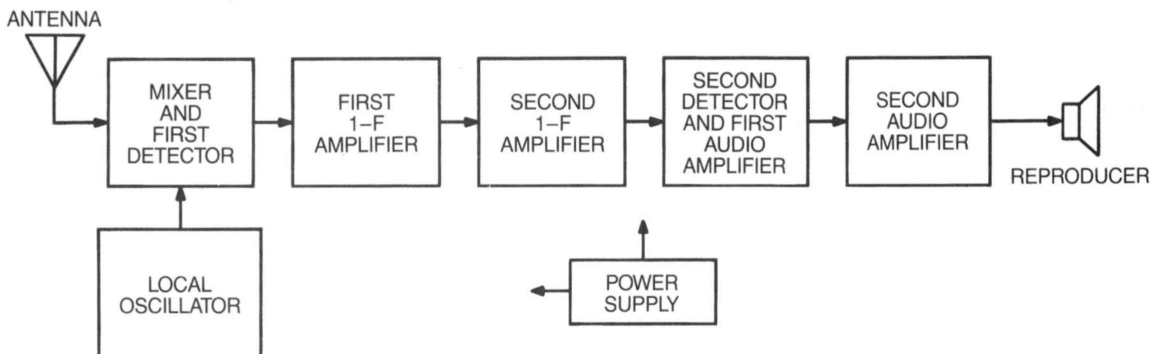

Fig. 13-6. Block diagram of a superheterodyne radio receiver.

located. These include receptacle outlets, switches, lighting fixtures, conduit, and other wiring devices. (These are shown with symbols in Fig. 13-7). A basic floor plan diagram sometimes omits the conduit system and associated wiring. These are shown, however, on the contractor's working drawings. It is from these drawings that the electrician does the wiring job.

Computer-generated Diagrams

Using a computer software program to design electrical circuits is a common practice in industry. Computer programs can easily plot horizontal and vertical lines to represent wires connected to graphic symbols in **computer-generated diagrams**. A library of many types of symbols representing transistors, flip-flops, gates, timers, LED displays, resistors, capacitors, transformers, integrated circuits, etc., are available to the designer. With a click of the mouse, the designer can drag and drop a graphic symbol to its proper location on the diagram. If a symbol is not available in the library, the designer can add a standard symbol to the library. The designer, however, can also develop a symbol to meet his or her special need. Any non-standard symbol, however, must be noted in the drawing legend.

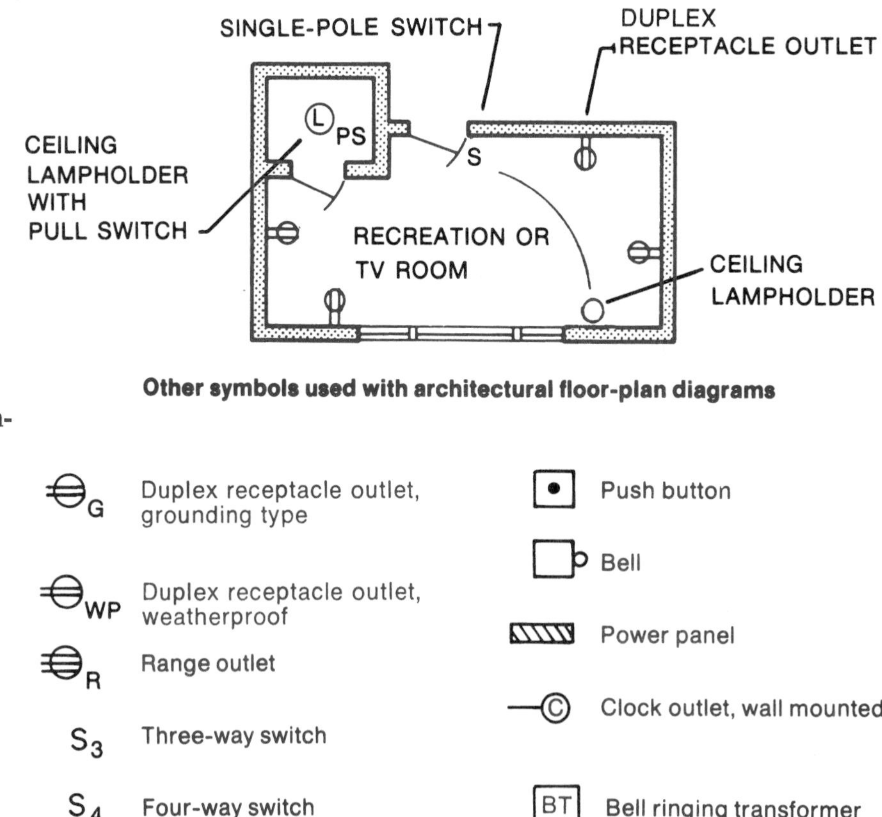

Fig. 13-7. **A simple architectural floor-plan diagram and some standard symbols used on these diagrams.**

Some sophisticated computer software programs can automatically layout printed circuit boards based on a circuit design. Other software programs include, for example, circuit simulation, waveform analysis, and circuit measurement.

Compare the time and quality of a finished schematic diagram drawn by hand with a schematic printed using a computer schematic program.

CAD and CAM

Many drafters use computer-aided design (CAD) to prepare diagrams because it is quick and versatile. CAD programs allow a drafter to manipulate and test models on computer screens until they obtain an appropriate balance of features, including ease of production and reasonable cost.

Computer diagrams can be produced in two or three dimensions. CAD can produce extremely accurate drawings and quickly make small or large changes. However, computers with more powerful microchips and a great deal of memory are necessary to handle the strong programs and large amount of data.

Once a diagram is complete, it is stored. If necessary, a reproduction is made on drafting film inserted into a plotting or printing device connected directly to the computer. When CAD diagrams are produced, they can be transmitted to machinery that make the individual parts for assembly. This system of transmittal and manufacture is called computer-aided manufacturing (CAM). CAM engineers use computer modeling to determine the best overall manufacturing procedures. The CAD/CAM system is particularly helpful in designs that repeatedly use the same features.

Chapter Review

Summary

- A pictorial diagram is a sketch of circuit parts that shows the wiring devices and is drawn to scale.
- Schematic diagrams are the standard way to communicate electrical information. A schematic diagram uses graphic symbols that stand for electrical components or connections. Standard component diagrams are determined by ANSI, the American National Standards Institute.
- Other types of electrical diagrams are wiring, block, computer generated, and architectural floor plan diagrams. These are used by engineers and technicians to design, troubleshoot, and maintain electrical circuitry and equipment.

Review Main Ideas

Review this chapter's main ideas by writing, on a separate sheet of paper, the word or words that most correctly complete the following statements:

1. A ______ is a sketch of circuit parts showing how the different parts are connected and where they are located.
2. The pictorial diagram is usually drawn to ______.
3. On a schematic diagram, the components are represented by ______.
4. You read most schematic diagrams from ______ to ______.
5. Using a schematic diagram makes it possible to trace the ______ of a circuit from beginning to end.
6. Examples of class designation letters used to identify components on a schematic diagram are ______ for resistors, ______ for capacitors, and ______ for transistors.
7. ANSI stands for the ______.
8. In drawing a schematic diagram when there is no standard symbol for a ______, make your own symbol and note its meaning on a ______.
9. The dot symbol is used to show where conductors are electrically ______.

10. Architectural floor-plan diagrams show where the parts of a building's ______ system are to be located.
11. A ______ diagram is a convenient way to quickly give information about a complex electrical system.
12. Computer software programs can easily plot lines to represent ______.

Apply Your Knowledge

1. Describe a class designation and list those for the following:
   capacitor
   amplifier
   battery
   transistor
   motor
   fuse
   transformer
2. Look at the pictorial diagram in Fig. 13-4. Cover the schematic diagram with a piece of paper. Draw your own schematic wiring diagram from the pictorial diagram.

Make Connections

1. **Communication Skills.** Write a short paragraph about an activity that you enjoy, such as fishing, boating, playing basketball, or roller blading. Then, describe the activity through the use of drawings.
2. **Mathematics.** Show the class how the pictorial diagram in Fig. 13-1 can be increased in size according to a specific scale or proportion.
3. **Science.** Describe symbolically the element copper.
4. **Workplace Skills.** Draw a circuit that can be used to individually control a front door bell and a back door buzzer in a residential home. The circuit will be powered by a 12 Vdc battery. Label all components with the proper class letter.
5. **Social Studies.** Identify six different fields of knowledge. Indicate at least three different symbols used in each field.

Section 3

Producing Electricity

Chapter 

Chemical Cells and Batteries

Terms to Study

anode

battery

capacity

cathode

cell

dry cell

electrode

electrolyte

hydrometer

primary cell

secondary cell

shelf life

voltage rating

voltaic chemical cell

wet cell

OBJECTIVES

After completing this chapter, you will be able to:

- Define the terms *cell* and *battery*.
- Describe the basic chemical action of a cell.
- State the correct way to connect a battery charger to a battery.
- Define the term *shelf life* as it applies to dry cells.
- Decide when a dry cell should be discarded and replaced with a new one.
- Use a hydrometer and other battery testers to determine the charge of a lead-acid automobile battery.
- Describe the chemical action during discharge and charge of a lead-acid automobile battery.
- Explain what is meant by the term *low-maintenance battery*.

Have you ever been a spelunker? This term refers to people who explore caves or caverns for recreation or enjoyment or to gain scientific knowledge. If you have ever been deep into a cave, one thing you remember is the complete darkness. One piece of equipment a cave explorer carries is a portable light powered, most likely, by a battery.

BASIC ACTION OF A CELL

The chapter you are about to study discusses chemical cells and batteries. You will learn some tips on how to test a cell or battery to decide whether it is charged fully or should be discarded. You will learn also how cells are connected to increase their capacity to power a device such as a light bulb for long periods of time.

A **voltaic chemical cell** is a combination of materials used to change chemical energy into electric energy in the form of voltage. The words *cell* and *battery* are often used to mean the same thing. However, this is not technically correct. A **cell** is a single unit. A **battery** is formed when two or more cells are connected in series or in parallel (Fig. 14-1).

A chemical cell is made up of two electrodes in contact with a substance in which there are many ions. An **electrode** is a terminal that conducts electric current. As you learned in Chapter 3, an ion is a charged atom. One substance that has many ions is an **electrolyte**, which is a solution made with acids, bases, or salts. Salt water, for example, is an electrolyte. Electrolytes are also good conductors of electricity.

The chemical actions that produce voltage are complicated. Studying how a very simple cell works will help you understand the basics of how chemical cells work.

In the cell, the electrolyte breaks down, or *ionizes*, to form positive and negative ions (Fig. 14-2A). At the same time, chemical action causes the atoms in one of the electrodes to ionize. This causes electrons to be deposited on the electrode. Positive ions move from the electrode into the electrolyte solution. This creates a negative charge on the electrode and leaves the area near it positively charged (Fig. 14-2B). An **anode**, however, is a negatively charged electrode.

SYMBOL
POSITIVE TERMINAL
WEATHERPROOF LANTERN BATTERY

Fig. 14-1. A battery consists of two or more cells connected together.

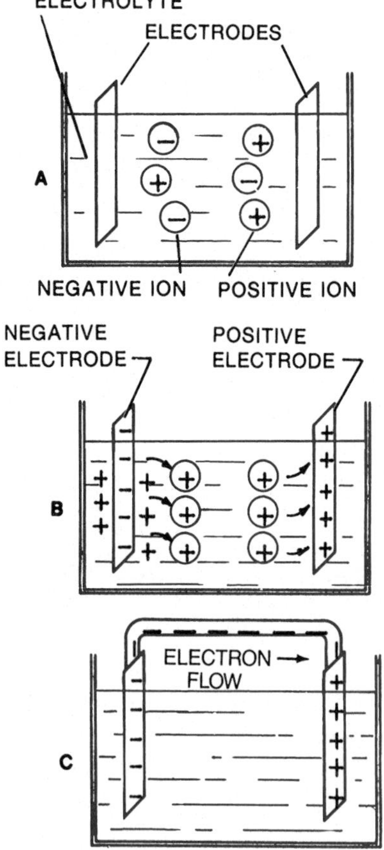

Fig. 14-2. Basic action of a chemical voltaic cell.

Some of the positive ions produced because of the electrolyte's ionization move to the other electrode. At this electrode, these ions combine with electrons. Since this process removes electrons from the electrode, this electrode becomes positively charged. A positively charged electrode is sometimes called a **cathode**.

The chemical action described above causes the electrodes to have opposite charges. One electrode is positive and the other is negative. These opposite charges cause a voltage between the electrodes. If a wire is connected between the electrodes of the cell, excess electrons from the negative electrode pass through the wire and into the positive electrode (Fig. 14-2C). This current continues until the materials in the cell become chemically inactive.

The electrolyte of a cell may be liquid or a paste. A **wet cell** contains a liquid electrolyte. A **dry cell** contains a paste electrolyte.

Chemical batteries are not the only type of batteries available. Think about other types of batteries, such as solar-powered batteries. Under what conditions are chemical cells and batteries superior to these other types? What factors might determine which type of battery should be used?

CHARACTERISTICS OF DRY CELLS

The characteristics of cells differ depending on the job they are intended to do. However, almost all chemical cells work in similar ways.

Primary and Secondary Cells

Primary cells are those that cannot be *recharged*. They cannot be returned to good condition after their output voltage drops too low. **Secondary cells** can be recharged. During recharging, the chemicals that provide electric energy are restored to their original condition. This is done by passing direct current through a cell in a direction opposite to the direction of the current that the cell delivers to a circuit (Fig. 14-3).

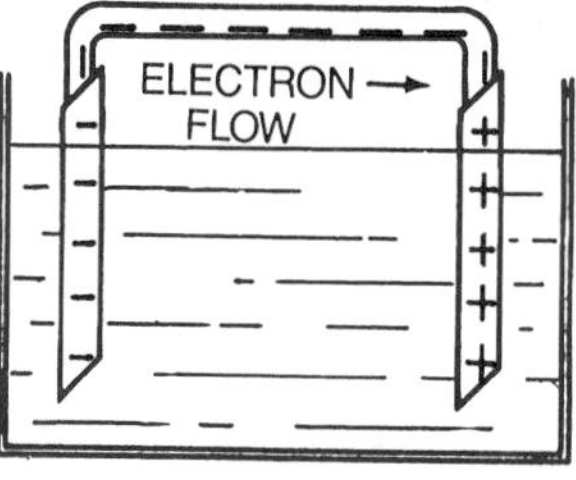

Fig. 14-3. Direction of current flow when a cell is being recharged.

A cell (or battery) is recharged by connecting it to a battery charger in "like-to-like" polarity as shown in Fig. 14-4. The charger is a rectifier circuit that can produce a variable output voltage. Many battery chargers have a voltmeter and an ammeter that show the charging voltage and current.

Voltage and Current of a Cell

The **voltage rating** of a cell is the voltage that the cell produces when it is not connected to a circuit. This is called the *open-circuit voltage*. The open-circuit voltage depends on the materials from which the cell is made.

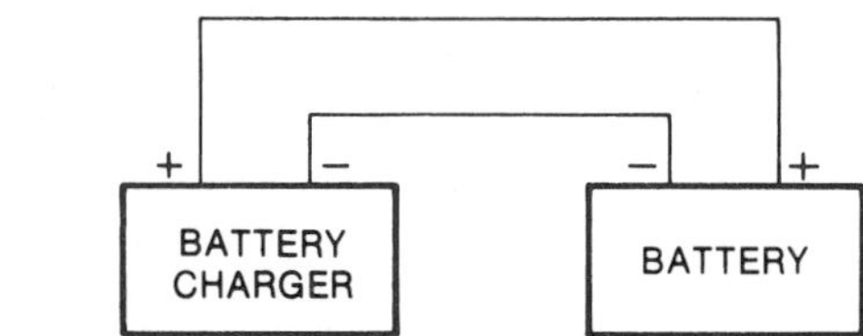

Fig. 14-4. Battery charging circuit.

LINKS

For more information about rectifier circuits, see Chapters 17 and 18. For information about voltmeters and ammeters, see Chapter 22.

LINKS

You may wish to review chapters 7 through 9. These chapters give details on connecting cells in series and in parallel.

The **capacity** of a cell is its ability to deliver a given amount of current to a circuit. A cell's capacity depends on two factors. The first is the amount and condition of the electrolyte in the cell. The second is the size of the electrodes. When two cells have the same kinds of electrolyte and electrodes, the larger cell can usually deliver more current for a longer period of time than a smaller cell.

As mentioned previously, cells can be connected either in series or in parallel. When you connect cells in series, a higher voltage results. When you connect cells in parallel, a greater current results.

Shelf Life

All dry cells, even when not in use, lose energy to some extent. The **shelf life** of a cell is that period of time during which the cell can be stored without losing more than about 10% of its original capacity. A cell loses capacity for two reasons. First, its electrolyte dries out over time, reducing the available ions. Also, chemical reactions occur that change the materials within the cell. Since heat speeds both of these processes, you can lengthen shelf life by keeping a cell in a cool place.

Consider what you know about how voltage is created in chemical cells. When does a cell become chemically inactive? What is the difference, at this point, between the two electrodes?

TYPES OF DRY CELLS AND BATTERIES

Several different types of cells and batteries have been designed over the years.

Carbon-Zinc Cells and Batteries

The carbon-zinc cell is one of the oldest and most widely used primary dry cells. Its electrolyte is a paste of ammonium chloride and zinc chloride dissolved in water. These and other chemicals are contained in an area of the cell known as the *bobbin* (Fig. 14-5).

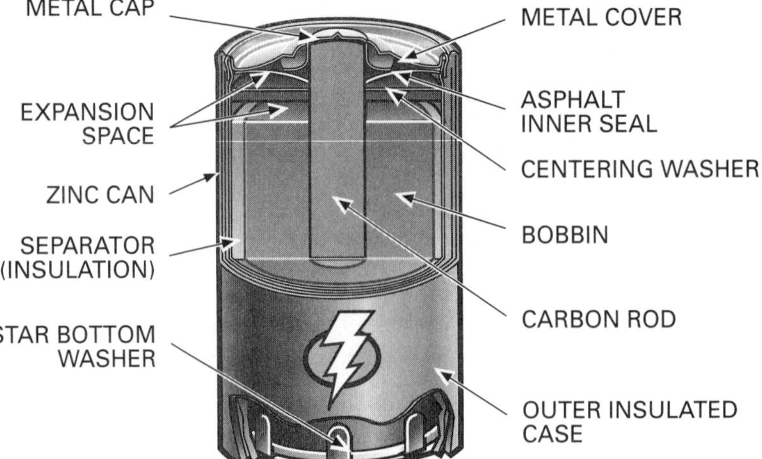

Fig. 14-5. Principal parts of a typical carbon-zinc dry cell.

The negative electrode in a carbon-zinc cell is a zinc can. The positive electrode is a mixture of powdered carbon and a black mineral called *manganese dioxide.* The carbon decreases the resistance of the electrode. A solid carbon rod passes through the center of the mixture. The rod provides a good electrical contact between the positive electrode and the positive terminal of the cell.

Polarization

During the operation of a carbon-zinc cell, the carbon rod becomes coated with hydrogen gas due to a process known as *polarization.* The manganese dioxide removes the hydrogen gas from the cell. In this case, the manganese dioxide serves as a *depolarizing agent.* As a result, the cell gives much better service.

Sizes and Voltage of Carbon-Zinc Cells

Carbon-zinc cells are available in several sizes (Fig. 14-6). They can have an open-circuit voltage of 1.5 to 1.6 V.

There are several kinds of carbon-zinc batteries. The most common have voltages of 3, 4.5, 6, 9, 13.5, 22.5, and 45 V (Fig. 14-7). In some batteries, the cells are cylindrical. In others, they are flat.

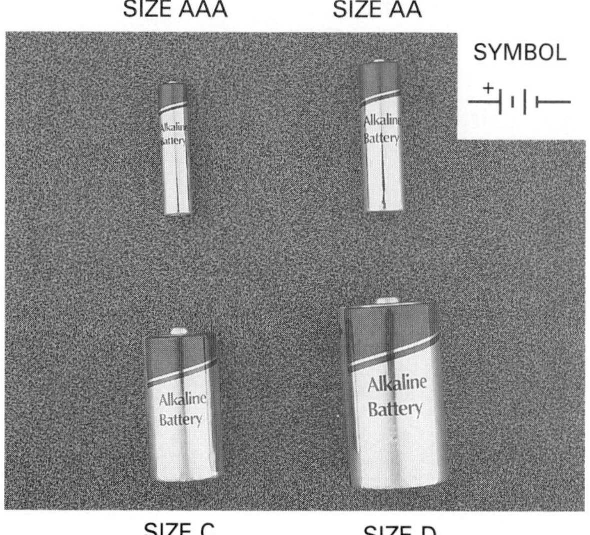

Fig. 14-6. Common sizes of carbon-zinc cells.

Fig. 14-7. Heavy-duty lantern battery. The heavy-duty lantern battery shown here has a strong metal case with four large 1½-V cylindrical batteries connected in series to provide 6 V.

Operating Efficiency

Ordinary carbon-zinc cells and batteries provide the most efficient service when they are used for short periods of time at relatively low currents. This method allows the cells and batteries to stay polarized.

Alkaline Cells

Alkaline cells can be either primary or secondary (rechargeable). Primary alkaline cells are similar to the secondary cells and have the same open-circuit voltage.

The secondary alkaline cell was a major advance in portable energy sources. In this dry cell, the electrolyte is potassium hydroxide. The anode is zinc. The cathode is manganese dioxide (Fig. 14-8).

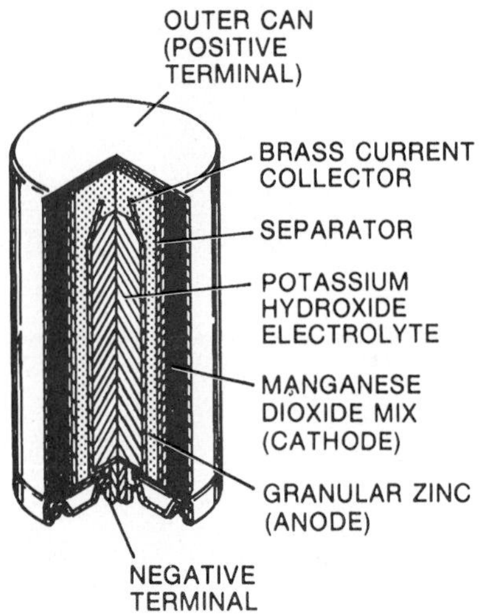

Fig. 14-8. Principal parts of a rechargeable alkaline cell.

Secondary alkaline cells have an open-circuit voltage of 1.5 V. They commonly come in AA, C, and D sizes. They can be used as direct replacements for carbon-zinc cells of the same sizes. The most common secondary alkaline batteries have voltages of 4.5, 7.5, 13.5, and 15 V.

Both the cells and the batteries come in a fully charged condition. Before recharging alkaline cells or batteries, it is a good idea to allow them to discharge fully. For best efficiency, recharge them according to the manufacturer's guidelines.

Alkaline cells last longer than carbon-zinc cells of the same size when used in the same way. Both primary and secondary alkaline cells (and batteries) give good service when used with high current loads. For this reason, they are widely used as sources of energy in radios, television sets, tape recorders, cameras, and cordless appliances.

Silver-Oxide Cells

Silver-oxide cells have now replaced mercury cells (14-9). The use of mercury cells is being phased out.

Fig. 14-9. Silver-oxide cells.

Nickel-Cadmium Cells

The nickel-cadmium cell was originally developed in Europe as a secondary wet cell for use in automobiles. In the secondary nickel-cadmium dry cell, the electrolyte is potassium hydroxide. The anode is nickel hydroxide, and the cathode is cadmium oxide.

The open-circuit voltage of nickel-cadmium dry cells is 1.25 V. These cells come in several sizes, including the common AA, C, and D cells. Others have the flat button shapes needed for use in such items as watches and hearing aids (Fig. 14-10). The most common nickel-cadmium batteries have voltages of 6, 9.6, or 12 V.

Nickel-cadmium cells are rugged. Because they were originally intended as automobile cells, they work well under extreme conditions of shock, vibration, and temperature.

Lithium Batteries

Lithium batteries have excellent mechanical and electrical properties. They have high-energy density and a long shelf life. They are also

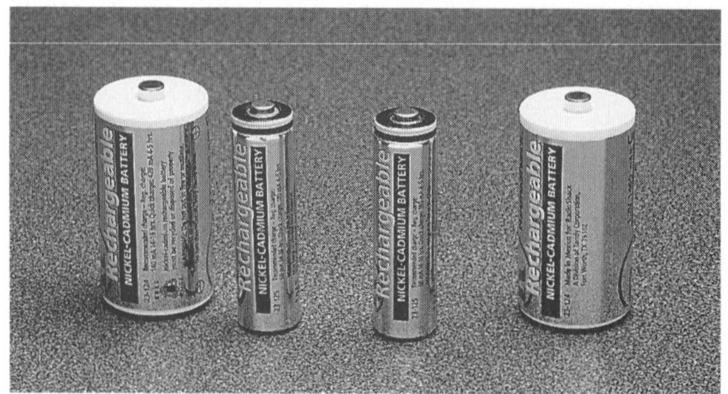

Fig. 14-10. Nickel-cadmium cells.

Fig. 14-11. **Lithium batteries are manufactured in a variety of shapes to meet industrial needs.**

extremely reliable. Because of these advantages, they are ideally suited as power sources for the long-term supply of microelectronic circuitry such as watches, calculators, and photo devices.

Lithium batteries are manufactured in button or cylindrical shapes (Fig. 14-11). Lithium (chemical symbol Li) is a very light metal. It has a silvery luster but quickly tarnishes when exposed to moist air. It is an active metal and gives up its electrons easily to form lithium ions (Li^+). One type of lithium battery contains lithium and manganese dioxide (Li/MnO_2). The voltage of this type of lithium battery is 3.0 V. Button cells have an available capacity range of 30 to 200 mAh. The capacity of cylindrical cells ranges from 160 to 1400 mAh. The Li/MnO_2 cell uses lithium for the anode. The active cathode material is a specially prepared, heat-treated form of MnO_2. The electrolyte contains lithium perchlorate in an organic solvent. Figure 14-12A shows the construction of a button cell; Fig. 14-12B shows a cylindrical cell.

Contact a local Environmental Protection Agency. Find out the proper disposal method for each of the five types of dry cell batteries discussed in the text. Be sure to dispose of batteries properly.

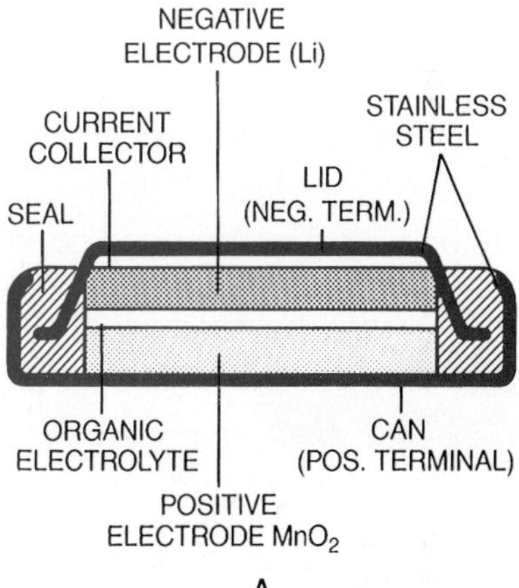

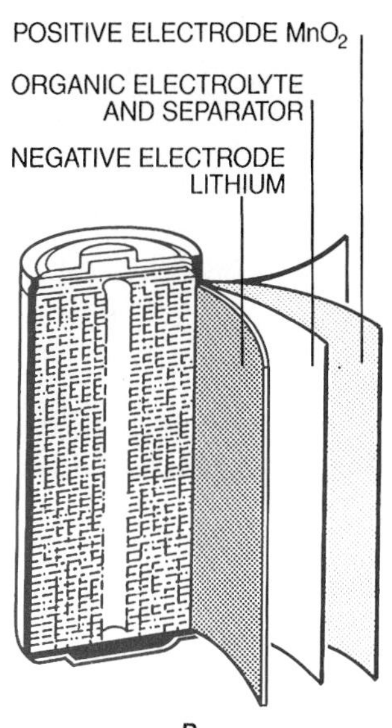

Fig. 14-12. **Lithium batteries: (A) construction of an Li/MnO2 button cell; (B) construction of an Li/MnO2 cylindrical cell.**

TESTING DRY CELLS AND BATTERIES

To check the condition of a dry cell or battery, measure its voltage when connected to a load. If the value of the voltage is less than 80% of the open-circuit voltage, the cell or battery should be replaced.

The voltage test must be made with the load connected and the current flowing. The load current produces a voltage drop across the internal resistance of the cell or battery. If a dry cell or battery is not in good condition, its internal resistance is high. This is generally because the electrolyte has dried out. A relatively large internal voltage drop occurs, so the voltage at the terminals is greatly reduced.

If you tested the cell or battery under an open-circuit condition, you would see only a very small internal voltage drop. As a result, the voltage across the terminals of the cell or the battery could be close to the rated voltage, even if the battery is not in good condition.

Describe how to test a dry cell battery for voltage.

WET CELLS AND BATTERIES

The most common wet cells used today are those used in lead-acid secondary storage batteries in automobiles, farm and garden tractors, motorcycles, and marine engines. This section discusses lead-acid technology. In a fully charged lead-acid cell, the electrolyte is a solution of water and sulfuric acid. About 27% of the total volume is acid. The active material on the brown positive plates (electrodes) is lead peroxide. The active material on the gray negative plates is pure lead in a spongelike form.

The lead-acid cell has an open-circuit voltage of a little more than 2 V. In the typical automobile battery, six cells are connected in series to produce a total voltage of 12 V (Fig. 14-13).

Chemical Action

As a lead-acid cell discharges, some of the acid leaves the electrolyte. The acid combines with the active material on both plates (Fig. 14-14). This chemical reaction changes the material on both plates to lead sulfate.

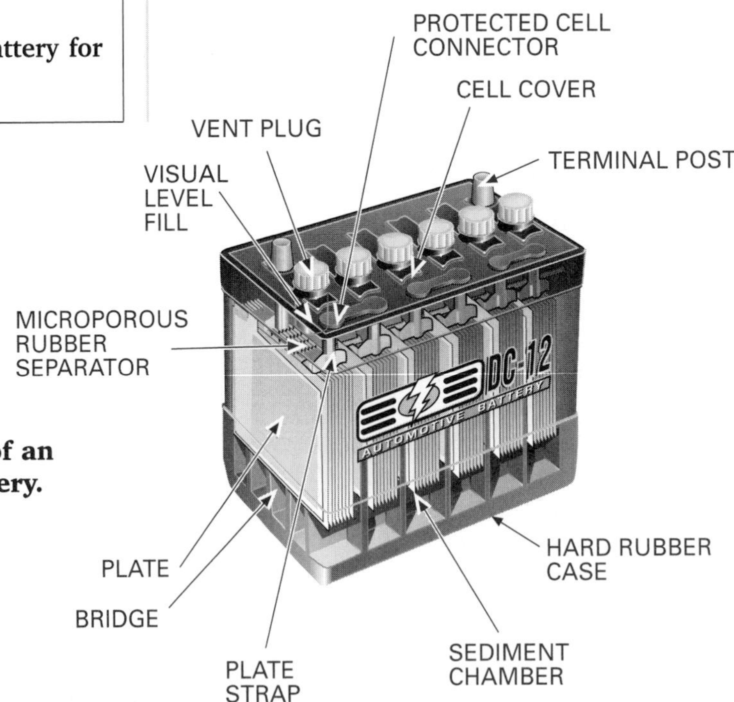

Fig. 14-13. **Principal parts of an automobile lead-acid battery.**

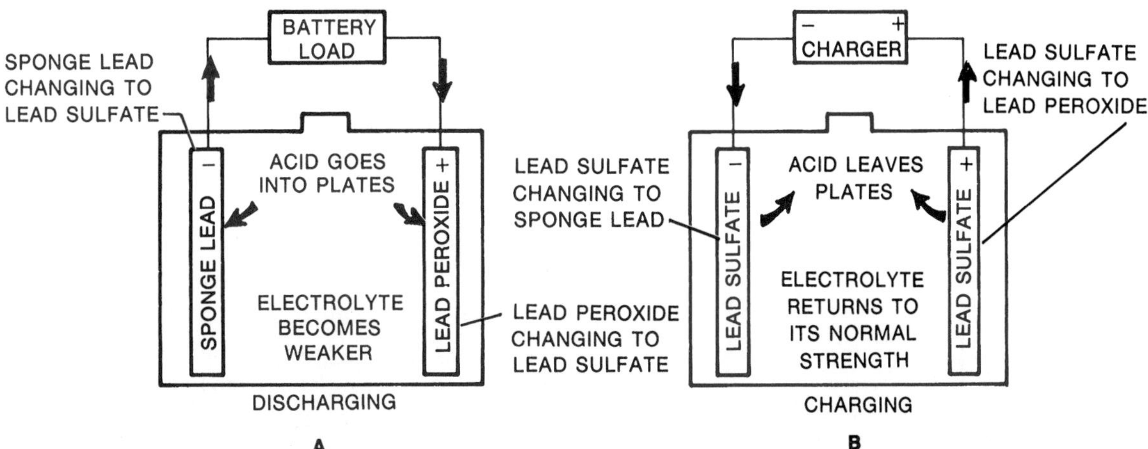

Fig. 14-14. **Chemical actions of a lead-acid cell: (A) discharging; (B) charging.**

When the cell is being charged, the reverse action takes place. Now the acid that was absorbed by the plates returns to the electrolyte. The active materials on the plates change back to their original combination of lead peroxide and lead.

Testing Lead-Acid Cells and Batteries

Although most automobile batteries are now considered "maintenance free," farm and garden tractor, motorcycle, and marine batteries still need to be maintained. Wet cells such as lead-acid cells are usually tested using one of several types of floating devices. These devices measure the specific gravity of the electrolyte in a lead-acid cell. The *specific gravity* of a liquid is its weight as compared to that of an equal volume of pure water. Since sulfuric acid is heavier than water, the specific gravity of the electrolyte in a lead-acid cell decreases as the cell discharges. Therefore, by measuring the specific gravity of the electrolyte, you can determine the charge condition of lead-acid cells.

Hydrometers

One of the most common lead-acid battery testing devices is the **hydrometer**, (Fig. 14-15) a device used to measure the specific gravity of the electrolyte in a lead-acid cell.

SAFETY TIP

The sulfuric acid used in a battery is a highly corrosive chemical and can cause severe burns. When using a hydrometer, do not allow any of the electrolyte to touch your skin or clothing. In case of contact, immediately wash the affected area with large quantities of soap and water.

To use a hydrometer, insert the end of the hydrometer into the electrolyte. Then squeeze the rubber bulb and release it slowly. The resulting suction causes the electrolyte to be drawn into the glass tube. The float in the hydrometer rises within the electrolyte to a level that depends on the specific gravity of the electrolyte. After the float has settled to its level, read the specific gravity from the markings on the float (Fig. 14-16).

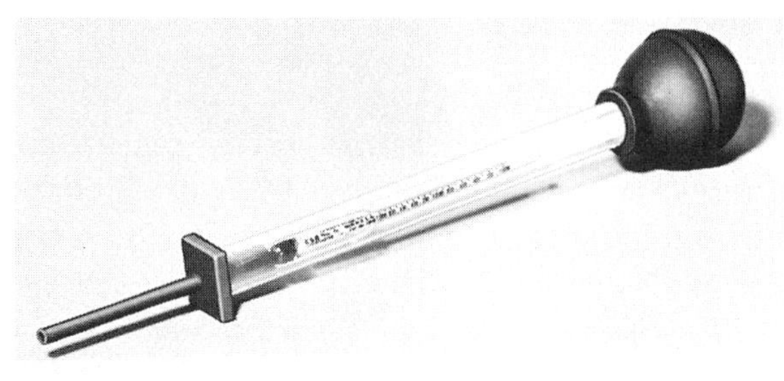

Fig. 14-15. **Hydrometer.**

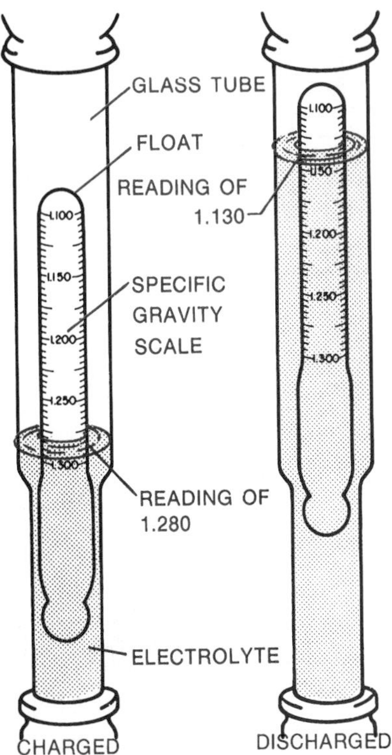

Fig. 14-16. Reading a hydrometer scale when a battery is fully charged and when it is discharged.

Since the cells are connected in series, a single defective cell causes the entire battery to be defective. Therefore, you must test each cell of the battery.

The specific gravity of the electrolyte in a fully charged automobile battery is about 1.280. As the cell discharges, the specific gravity gradually decreases to about 1.130. At that point, the cell is fully discharged.

FASCINATING FACTS

The hydrometer can determine a liquid's specific gravity, density, or weight per unit volume. Beyond its application in testing batteries and radiators, a hydrometer may also be used to determine the richness of milk, the percentage of sugar in sugar solutions, and the strength of saline solutions.

Other Battery Testers

Another type of battery tester uses floating balls to help individuals decide the status of the charge in a lead-acid battery. The floating balls work on the same scientific principle as the float in a hydrometer. For example, one type of tester has five balls to show the level of charge in the battery. To use the tester, draw the electrolyte up in the glass tube. If no ball floats, the battery is discharged. If one ball floats, the level of charge is 25%. If two balls float, the level of charge is 50%. If three balls float, the level of charge is 75%. If four balls float, the level of charge is 100%. If five balls float, the battery is overcharged.

Rating

The capacity rating of a lead-acid battery is usually given in ampere-hours (Ah) for a specific discharge period. The most commonly used rating system is based on the Society of Automotive Engineers (SAE) 20-hour rate. This means that a new battery rated at 100 Ah should deliver 5 A of current continuously for 20 hours and maintain at least 1.75 V per cell.

Dry Charge

Some lead-acid batteries are shipped from the manufacturer as dry-charge batteries. The plates of a dry-charge battery are in the condition of a fully charged battery. This means that one plate is lead and the other is lead peroxide. To prepare the battery for use, sulfuric acid of the proper specific gravity is added to each cell. The wet-cell battery is then ready to be used.

Battery Maintenance

Although modern automobile batteries are maintenance-free, other batteries, such as those for tractors, must be maintained. The following tips will help you keep a lead-acid automobile battery in good condition.

- Keep the battery cells properly filled at all times.

The First Battery

Allesandra Volta (1745-1827) and Luigi Galvani (1737-1798) conducted the first important experiments in electrical currents. In 1775, Volta devised the electrophorus, an instrument that produced charges of static electricity. Static electricity was now available for scientific study. The electrical unit known as the volt was also named in his honor.

- Make sure the terminals are clean and free of corrosion.
- Make sure the battery-cable clamps are connected solidly.
- Check the specific gravity of the electrolyte often.

Reasons for Battery Discharge

In general, lead-acid batteries show a low charge for one of three reasons. First, faulty wiring in the electrical system may be causing the battery discharge. Second, a defect in the generator or regulator may be allowing the battery to discharge. Finally, a loose or greasy alternator (fan) belt can cause the alternator to run more slowly, allowing the battery to discharge. Be sure to fix these problems.

Charging and Overcharging

Do not overcharge a lead-acid battery. This will weaken the electrolyte and may cause serious damage to the cell plates.

SAFETY TIP

Hydrogen, an explosive gas, is released from lead-acid batteries while they are being charged. For this reason, batteries should be charged in a well-ventilated area, away from open flames or sparks.

Storing a Lead-Acid Battery

If you must store a lead-acid battery, first make sure that it is fully charged. The plates of a stored battery that is partly or fully discharged soon become coated with a sulfate compound. This compound hardens and causes the plate material to become chemically inactive. A partly or fully discharged battery will also freeze at a temperature of about 20° F (6.7° C). Never store a battery on concrete.

Low-Maintenance Batteries

A low-maintenance battery usually has a low rate of water loss. This reduces routine maintenance such as adding water.

Most types of low-maintenance batteries produce very little gas at normal charging voltages. This reduces the rate of water loss. Some of these batteries have venting systems, but water cannot be added.

New developments in the design and manufacture of the battery plate grids that hold the active materials have helped make low-maintenance batteries possible. In older automobile batteries, antimony is used as an alloy in the plate grids to improve their strength. The presence of antimony, however, increases the rate of water loss and self-discharge. In low-maintenance batteries, metals such as cadmium and strontium are substituted for the antimony. These metals provide the necessary strength, but reduce gassing and self-discharge.

Although the newer batteries require little or no maintenance, you should always carefully follow the manufacturer's instructions in using and recharging these batteries.

Explain the chemical composition of a lead-acid battery.

Chapter *Review*

Summary

- A voltaic chemical cell changes chemical energy into voltage. Two or more cells connected in a series or in parallel form a battery.
- A chemical cell consists of two electrodes in contact with either a dry or wet electrolyte. Opposite charges between the cell's anode and cathode cause voltage to occur.
- The various dry cells and batteries differ in the voltage they produce, capacity, shelf life, and whether they can be recharged.
- To test a dry cell or battery's condition, determine whether the value of its voltage when connected to a load measures less than 80% of the open-circuit voltage. If so, replace the battery.
- The most common wet cell is the lead-acid type used in automobiles. Each cell is tested by measuring the specific gravity of the electrolyte with a hydrometer.

Review Main Ideas

Review this chapter's main ideas by writing, on a separate sheet of paper, the word or words that most correctly complete the following statements:

1. A cell that changes chemical energy into electric energy in the form of voltage is called a ______ cell.
2. A ______ is formed when two or more cells are connected.
3. A chemical cell is made up of two ______ in contact with a substance in which there are many ions.
4. A cell in which the electrolyte is a liquid is often called a ______ cell. A cell in which the electrolyte is a paste is called a ______ cell.
5. Cells that cannot be recharged are called ______ cells. Cells that can be recharged are called ______ cells.
6. A cell or a battery is recharged by passing current through it in a direction ______ to the direction of the current that the cell delivers to a circuit.
7. A battery is recharged by connecting it to a battery charger in _______ polarity.
8. The open-circuit voltage of a cell depends on the ______ from which it is made.
9. The cell's capacity depends on the amount and condition of its ______ and the size of its ______.
10. The ______ of a cell is the period of time during which it can be stored without losing more than 10% of its original capacity.
11. The open-circuit voltage of carbon-zinc cells ranges from ______ to ______ V.
12. The open-circuit voltage of alkaline cells is ______ V.
13. Both primary and secondary alkaline cells give good service when used with ______ current loads.
14. Nickel-cadium dry cells have an open-circuit voltage of ______ V.

15. Batteries containing lithium and manganese dioxide have a nominal voltage of ______ Vdc.
16. In checking a dry-cell battery, if the value of the voltage is less than ______, the battery should be replaced.
17. The float in a hydrometer rises within the electrolyte to a level that depends on the ______ of the electrolyte.
18. Since the cells in a lead-acid battery are connected in series, a ______ defective cell causes the ______ to be defective.

Apply Your Knowledge

1. Describe the construction of a simple chemical cell and explain how it produces voltage.
2. What advantages do alkaline cells provide compared to carbon-zinc cells?
3. Do products such as calculators and cellular phones use wet-cell or dry-cell batteries? Explain why the other type is not appropriate.
4. What is the advantage of shipping lead-acid batteries from the manufacturer as dry-charge batteries?
5. Think about the properties of a low-maintenance battery. Why is there no way to add water to a low-maintenance battery?

Make Connections

1. **Communication Skills.** Visit a store that specializes in providing equipment for outdoor activities. Discuss with the manager the types of electrical flashlights and lanterns that are available for individuals who hike in the wilderness or in caves. Report orally your findings to the class.
2. **Mathematics.** Obtain a rechargeable nickel-cadmium battery for 9-V applications. Fully recharge the battery according to the specifications on the battery. Connect the battery to an appropriate load. On a chart, periodically record the current and voltage while the battery is connected to the load until the battery's voltage is 10% of the open-circuit voltage. Report your findings to the class.
3. **Science.** Review information related to the robotic spacecraft, called Cassini, that is expected to study the planet Saturn in 2004. Write a report on the device that is designed to produce electrical power for the spacecraft.
4. **Technology Skills.** Research the process of recycling an automobile battery. Explain to the class how you would establish and conduct a battery recycling program in your school.
5. **Social Studies.** Investigate the development of battery-powered vehicles in the United States. In a written report, cite two advantages and two disadvantages of these types of vehicles. Determine what impact the vehicles would have on the electric utility industry if produced in large numbers.

Chapter 15

Generators

OBJECTIVES

After completing this chapter, you will be able to:

- Explain the three factors that produce a voltage through generator action.
- Describe the differences between ac and dc generators.
- Compare the similarities and differences of slip rings and commutators.
- Contrast rotating-armature and rotating-field generators.
- Draw a graph depicting how a three-phase voltage is produced in an ac generator.

Terms to Study

alternator

armature

commutator

electric generator

excitation current

phase

sine wave

slip ring

Have you ever started something? In order to start something, you have to take action. This chapter discusses how to generate a voltage. As you might imagine, something has to do the work. You will learn what that something is in this chapter. This chapter discusses generators.More than 95% of the world's electric energy is supplied by generators with electromagnetic induction.

BASIC GENERATORS

The phrase *to generate* means "to produce." An **electric generator** is a machine that produces a voltage by means of electromagnetic induction. This is done by rotating coils of wire through a magnetic field or by rotating a magnetic field past coils of wire. The modern generator is a result of Michael Faraday and Joseph Henry's work in the early 1800s.

A continuous alternating voltage can be produced by rotating a coil of wire between the poles of a permanent magnet (Fig. 15-1). This is a simple generator. The coil is called the **armature.** The ends of the armature coil are connected to **slip rings** (Fig. 15-1) that are insulated from each other and from the armature shaft on which they are mounted. The brushes press against the slip rings. They make it possible to connect the rotating armature to an external circuit. The armature must be driven by mechanical force.

Generator Action

The value of the voltage induced by generator action at any instant depends on three things: (1) *the flux density of the magnetic field* (the greater the flux density, the greater the induced voltage); (2) *the velocity of the conductor motion* (induced voltage increases as the velocity of the conductor increases); and (3) *the angle at which a conductor cuts across flux lines* (the greatest voltage is induced when the conductor cuts across flux lines at a 90-degree angle).

The action of a single coil in producing a complete cycle of alternating voltage is shown in Fig. 15-2. If a load is connected across the terminals, an alternating current will flow through the circuit. With the coil in position 1 (Fig. 15-2), it is parallel to the flux lines. Therefore, no voltage is induced because the armature is not cutting across any flux lines. When the coil is in this position, no current flows through the load circuit.

As the armature moves from position 1 to position 2, it cuts across more and more flux lines. Thus, the voltage increases from zero to its highest value in one direction. This increase in voltage results in a similar increase in current. This is shown by the first quarter of the sine wave. At position 2, the coil is cutting flux lines at a 90-degree angle. This produces the highest voltage.

As it moves from position 2 to position 3 in the same direction, the armature cuts across fewer flux lines at angles less than 90°, causing the voltage to decrease. Consequently, the voltage decreases from its highest value to zero when the coil is parallel to the flux lines and not cutting across them. During this time, the

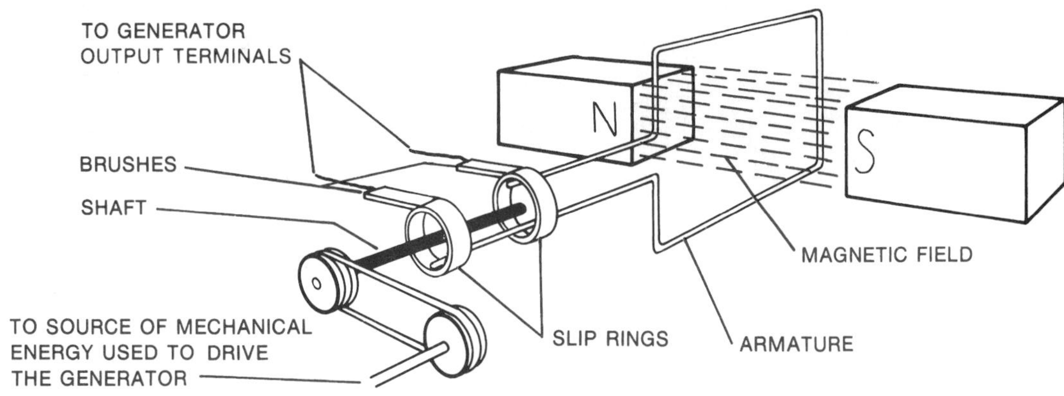

Fig. 15-1. **A simple one-coil ac generator.**

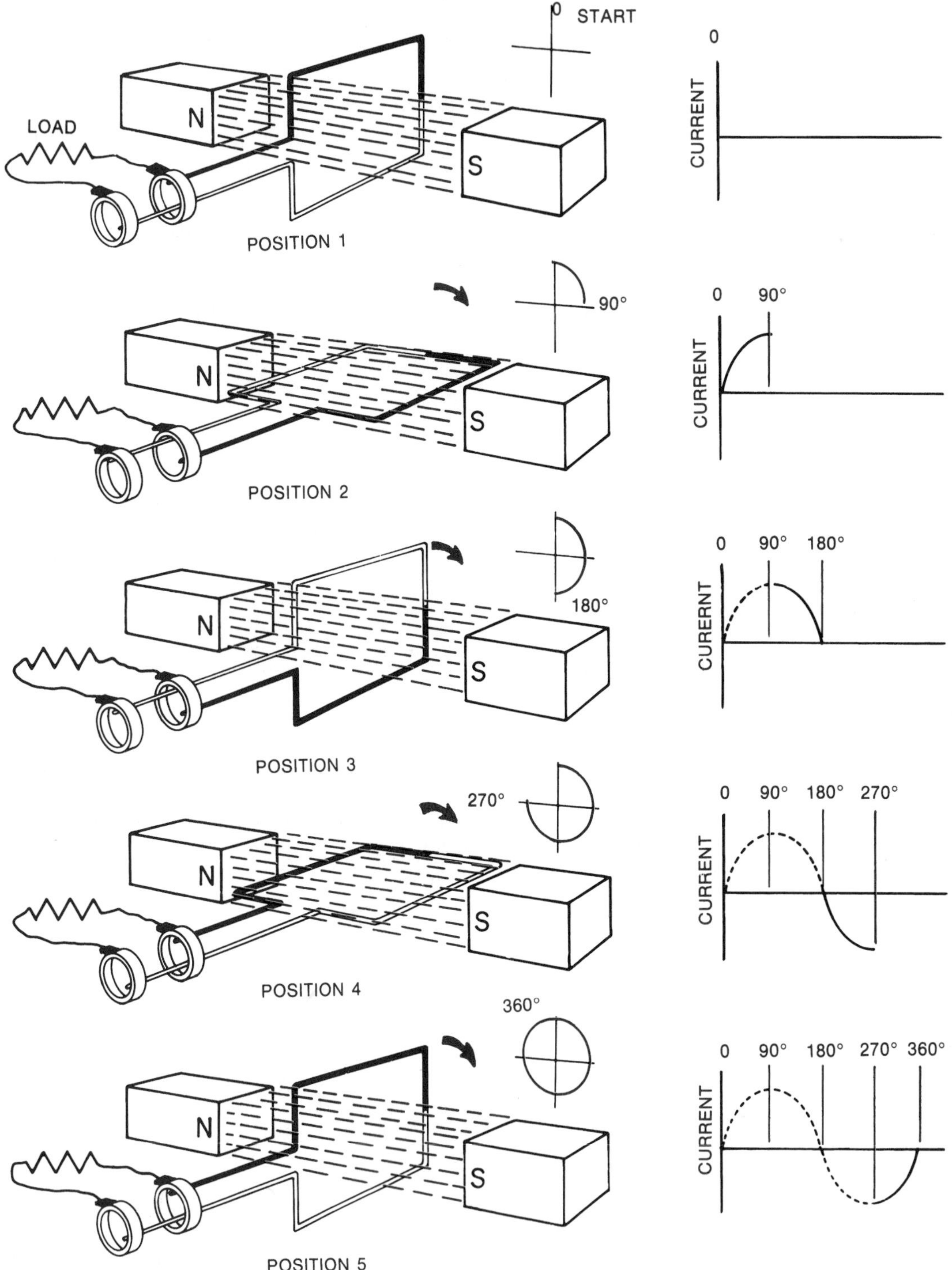

Fig. 15-2. **Generating one cycle of voltage with a single-coil ac generator.**

current also decreases to zero. This is shown by the second quarter of the sine wave.

As the armature rotates to position 4, each of its sides cuts across the magnetic field in the opposite direction. This changes the polarity of the voltage and the direction of the current. Once again, the voltage and current increase from zero to their highest values during the third quarter of the sine wave.

From position 4 to position 5, the armature rotates back to its starting point. During this time, the voltage and current decrease from their highest values to zero. This completes the cycle.

Voltage Waveform

Figure 15-2 shows the generation of a *single-cycle voltage waveform*. Notice that the coil completes one 360-degree revolution in the ac generator. This waveform can be expressed as a mathematical sine function within a specific time period. As the coil rotates, it is represented by the sine of the angle every one-quarter revolution, for example: 90 degrees, 180 degrees, 270 degrees, and 360 degrees. The shape of this waveform is often called a **sine wave**. The angles are referred to as *electrical degrees*.

Phase

If the load in an external ac circuit is purely resistive, the waveform produced by the voltage and current will rise and fall at the same instant. When this occurs, the voltage and current are said to be in **phase** with each other (Fig. 15-3A).

As you learned in previous chapters, circuits that contain capacitive or inductive reactance cause the current to be out-of-phase from the voltage. Capacitive reactance opposes a change in voltage. Thus, in an ac circuit that has only capacitive reactance, the current leads the voltage by 90 degrees. In this case, the current waveform is ahead in time compared to the voltage waveform (Fig. 15-3B). The opposite

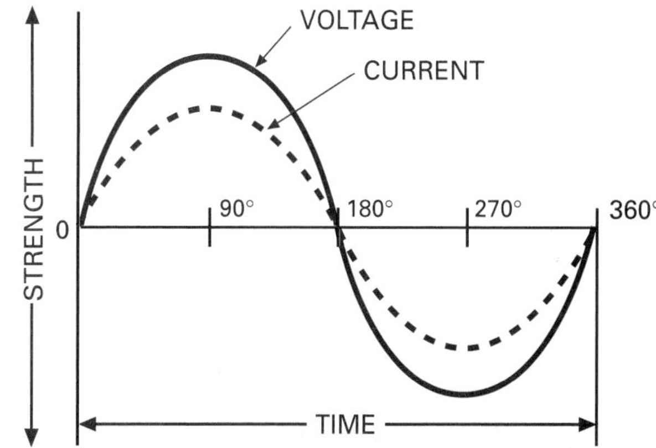

Fig. 15-3A. Voltage and current in phase with each other.

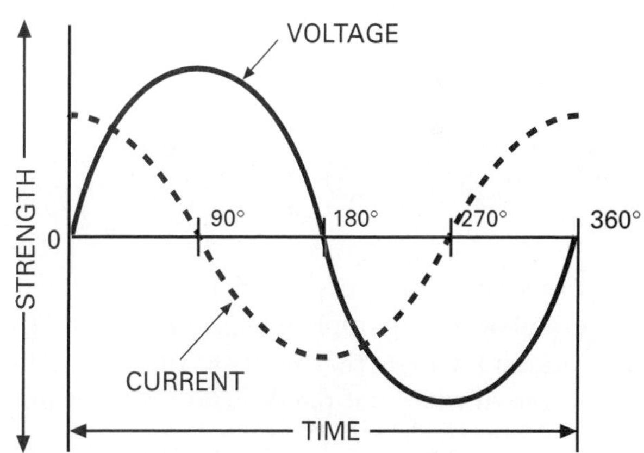

Fig. 15-3B. Current leads voltage by 90 degrees in a circuit with capacitive reactance.

occurs when the ac circuit has only inductive reactance. In this case, the inductive reactance opposes a change in current. Therefore, the current waveform lags the voltage waveform by 90 degrees (Fig. 15-3C). Thereby, the current and voltage are said to be out-of-phase from each other in an ac circuit because they occur at different points in time. In practical ac circuits, there is usually a combination of resistance, capacitive, and inductive reactance. This causes the current waveform to be out-of-phase from the voltage waveform by a value between zero and 90 electrical degrees.

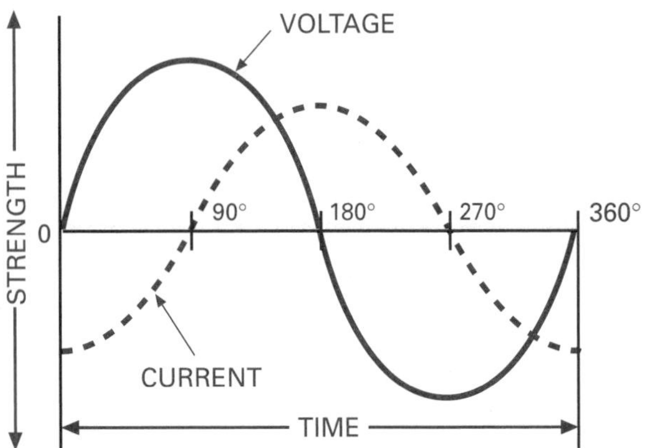

Fig. 15-3C. Current lags voltage by 90 degrees in a circuit with inductive reactance.

Draw your own version of a pictorial diagram of a generator. Include the lines of flux and arrows that indicate the rotation of the magnet and the direction of current flow.

DIRECT CURRENT GENERATORS

In a direct-current generator, the ends of the armature coil or coils are connected to a **commutator**. This device is needed to produce a direct current. In its basic form, a commutator is a ring-like device made up of metal pieces called *segments*. The segments are insulated from each other and from the shaft on which they are mounted.

The operation of a simple dc generator is shown in Fig. 15-4. In Fig. 15-4A, the armature coil cuts across the magnetic field. This motion produces a voltage that causes current to move through the load circuit in the direction shown by the arrows. In this position of the coil, commutator segment 1 is in contact with brush 1. Commutator segment 2 is in contact with brush 2.

Commutator Action and Direct Current

As the armature rotates a half turn in a clockwise direction, the contacts between the commutator segments and the brushes are reversed (Fig. 15-4B). Now segment 1 is in contact with brush 2. Segment 2 is in contact with brush 1. Because of this *commutating action*, the side of the armature coil in contact with either of the

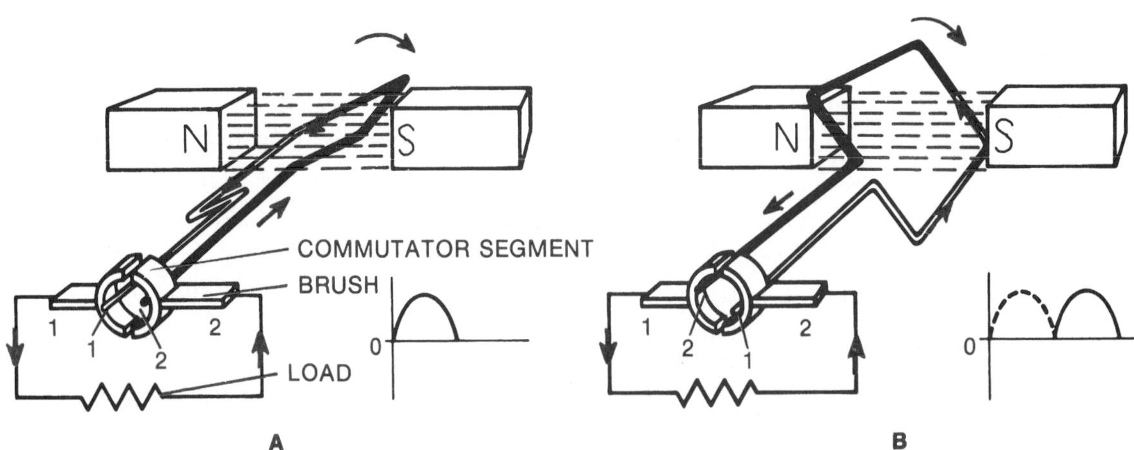

Fig. 15-4. Basic operation of a dc generator.

The Battle of the Currents

Thomas Edison (1847-1931) opened the Pearl Street electrical station in New York City on Sept. 4, 1882. This station contained six direct current (dc) generators of about 100 kilowatts, each of which could serve an area of about 1 square mile.

However, George Westinghouse (1846-1914) recognized an alternating current (ac) system's ability to distribute electricity over longer distances than possible with the direct-current system. Importing a polyphase ac system from Europe, Westinghouse purchased the patents of Nikola Tesla's (1856-1943) ac motor and hired him to improve and modify it for use. The ac system used alternating-current generators and step-up and step-down transformers to distribute electricity.

Once the new system was ready, advocates of dc power set out to discredit ac power. Public acceptance of ac power came soon after Westinghouse dramatically proved its advantages at the World's Columbian Exposition in Chicago (1893). Using incandescent and arc lighting, the fairground was set aglow with light. This usage marked the start of large-scale outdoor lighting and illuminated advertising signs. In 1895, using an ac system, the Westinghouse Electric Co. transmitted electricity 22 miles, from Niagara Falls to Buffalo, New York.

brushes is always cutting across the magnetic field in the same direction. Therefore, brushes 1 and 2 have a constant polarity. A direct voltage is applied to the external load circuit.

Practical Generators

The main parts of a commercial dc generator are shown in Fig. 15-5. In this generator, the armature is made up of coils of magnet wire, the conductors of which are placed in the slots

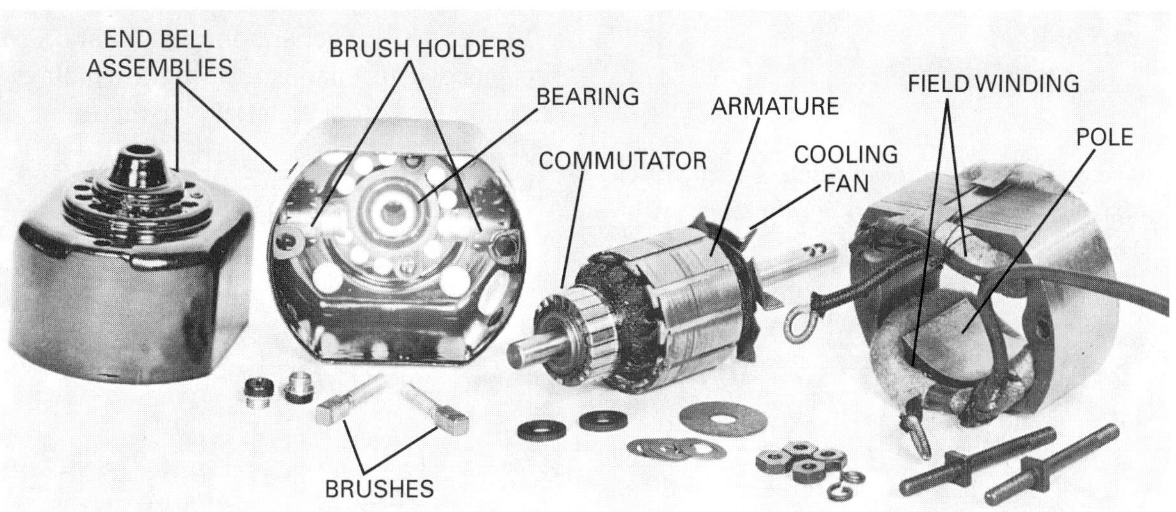

Fig. 15-5. Main parts of a commercial dc generator.

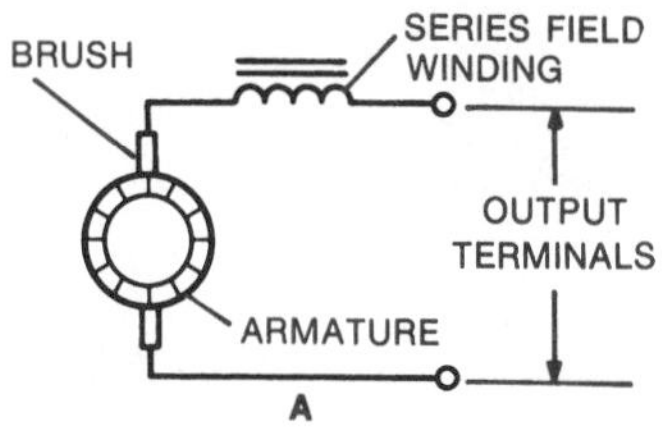

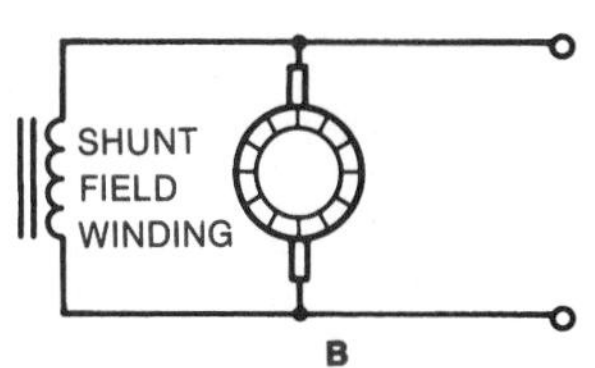

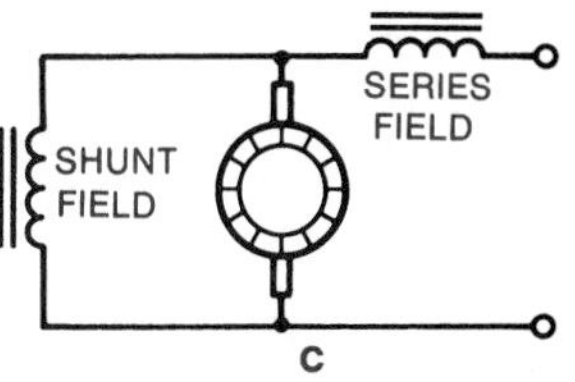

Fig. 15-6. Schematic diagrams of common dc generators: (A) series generator with field windings connected in series with the armature; (B) shunt generator with field windings connected in parallel with the armature; (C) compound generator containing both series and shunt field windings.

of the laminated armature core. The ends of the coils are connected to the commutator segments. The field windings are electromagnets that produce the magnetic field needed to operate the generator.

The direct current used to energize the field windings and start a generator is the **excitation current**. Once the generator develops an output, current is obtained from this output (Fig. 15-6).

Most small dc generators are driven by alternating-current motors or by gasoline engines. They are used in trains, in welders, for charging batteries, and for the operation of field telephone equipment.

List all the ways you can think of to provide excitation current to a generator.

ALTERNATING CURRENT GENERATORS

The rotating parts of large ac generators are called rotors. They are turned by steam turbines, hydro- (water-driven) turbines, or diesel engines. These generators produce the electric energy used in our homes and in industry.

FASCINATING FACTS

The tremendous volume of water that pours over Niagara Falls is used to generate power in six hydroelectric plants. The turbine generators of these plants develop a maximum of 5¾ million horsepower: 55% on the American side and about 45% on the Canadian side.

Small ac generators are usually driven by gasoline engines (Fig. 15-7). Such generators are commonly used to provide emergency power. Alternating current generators are also called **alternators**.

Excitation

In the small ac generator, the excitation current needed to energize the field windings is first obtained from a battery or the output of

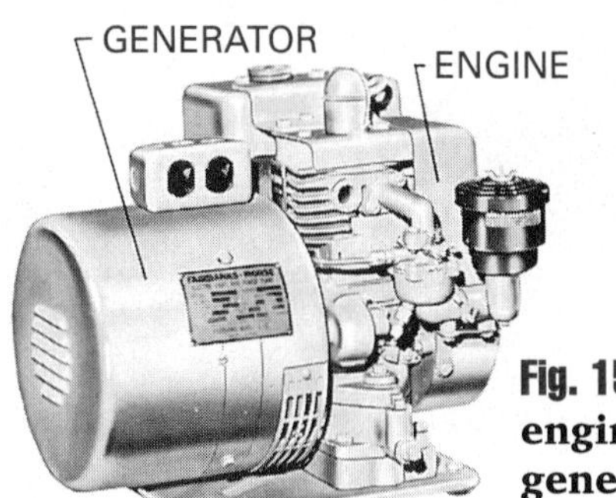

Fig. 15-7. A gasoline-engine-driven ac generator.

the generator itself. The excitation current must be direct, but the output current is alternating. Thus, the output current must first pass through a rectifier to be changed into direct current and used as excitation current. In some alternators, the rectifier circuit is located within the generator housing. In large ac generators, dc *exciter generators* produce the excitation current. This is mounted either on the shaft of the main generator or located nearby.

Rotating-Armature Generator

In small ac generators, the armatures are usually the *rotors*, or rotating parts. The rotors turn within magnetic fields produced by stationary field windings called *stators*. Rotors have collectors, or slip rings, in contact with carbon brushes (Fig. 15-8).

Fig. 15-8. Slip rings mounted on the armature assembly of a rotating armature ac generator.

Rotating-Field Generator

In rotating-field generators, armatures are stationary. They are made of winding conductors placed in the slots of the frame assembly (Fig. 15-9A). The field windings are wound around pole pieces on the rotor assembly and they are connected to slip rings (Fig. 15-9B). The excitation current passes to the field windings through the carbon brushes that are in contact with the slip rings. Generators of this kind are used in most large power-generating plants.

Frequency

The frequency of the current produced by an alternator depends on the speed of the rotor (either the armature or the field windings) and on the number of magnetic poles formed by the field windings. Power company generators in most parts of the United States produce a frequency of 60 cycles per second, or hertz. Special-purpose generators can have higher or lower frequencies.

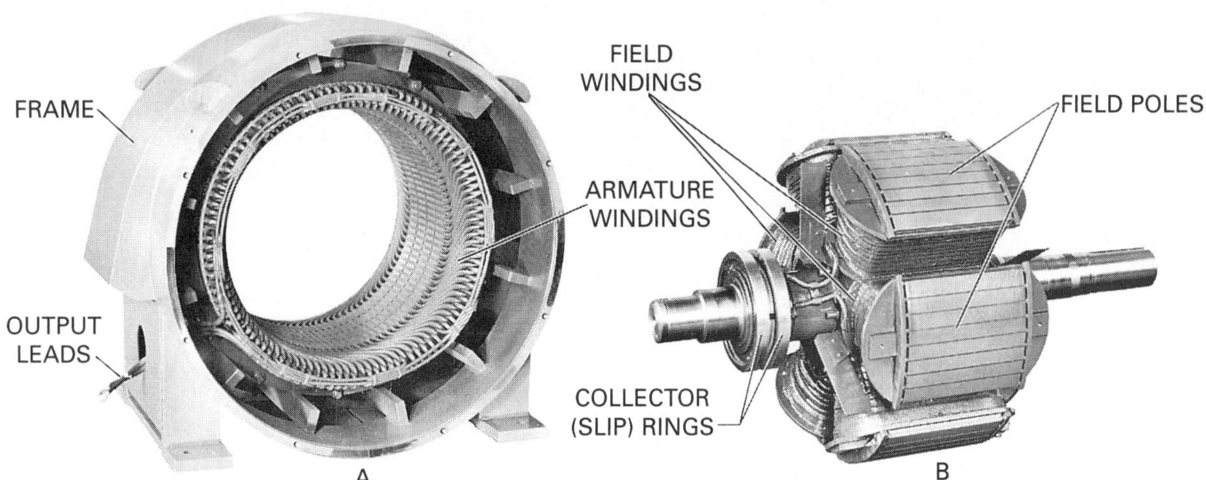

Fig 15-9. Principal parts of a rotating-field generator: (A) armature (stator); (B) field (rotor).

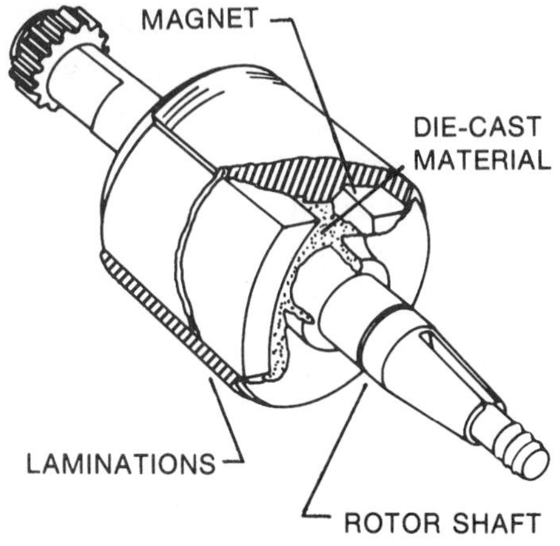

Fig. 15-10. **Rotor assembly of a typical magneto showing the bar-shaped permanent magnets embedded in the rotor.**

Voltage

The output voltage of a generator depends on the speed of the rotor, the number of armature coils, and the strength of the magnetic field produced by the field windings. Power-station generators usually have output voltages of 10,000 V or more.

MAGNETO

Magnetos are ac generators in which the magnetic fields are produced by one or more permanent magnets instead of electromagnets. In some magnetos, the permanent magnets are in the rotor assembly (Fig. 15-10).

A flywheel magneto is commonly used with small gasoline engines. One or more of the permanent magnets of this generator are mounted on the flywheel assembly. As the flywheel turns, a magnetic field cuts across a stationary coil. This is an ignition coil, and it is a *step-up transformer* (Fig. 15-11). The voltage induced across the secondary winding of the ignition coil is applied to the spark plugs.

THREE-PHASE GENERATORS

A generator with a single set of windings and one pair of slip rings produces a single wave of voltage. This is known as a *single-phase system*.

A *three-phase generator* has three separate sets of windings. Spaced 120 degrees from each other around the armature, one end of each winding is connected to a slip ring (Fig.

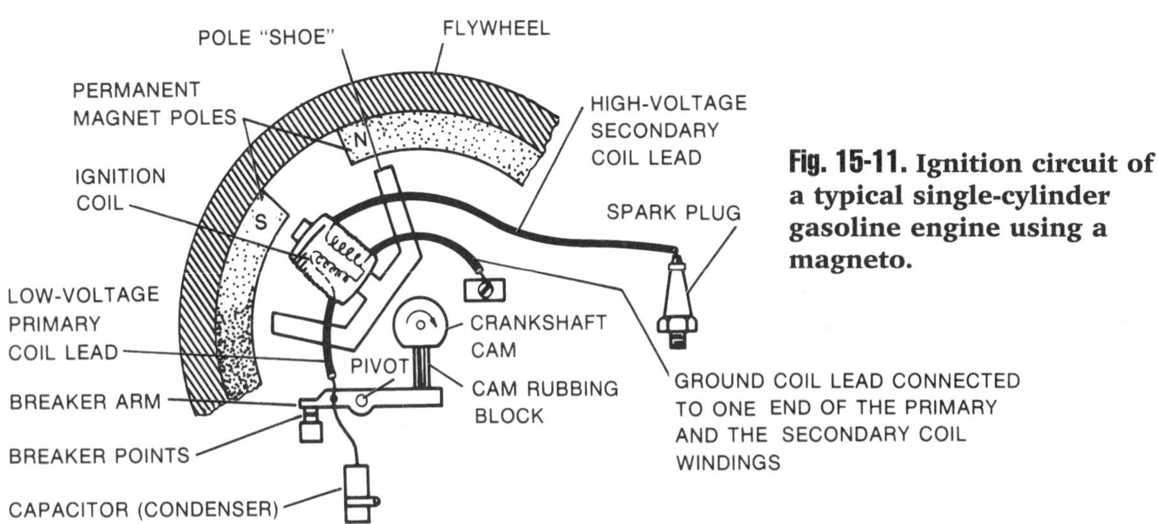

Fig. 15-11. **Ignition circuit of a typical single-cylinder gasoline engine using a magneto.**

THREE-PHASE GENERATOR

15-12A). In such a generator, each complete turn of the rotor produces three separate voltages that are 120 degrees out-of-phase from each other (Fig. 15-12B). These are applied to a load through a three-conductor power line. Three-phase power systems are commonly used in manufacturing applications to operate *induction motors*.

LINKS

For more information on motors, refer to Chapter 29.

1. **Draw a schematic diagram that includes:**
   - **The generator.**
   - **Transmission lines from the power plant.**
   - **The transformer at the sub-station.**
   - **The transformer on your block.**
   - **The wires into your house.**
2. **Obtain a drawing of an exploded view of a repair parts list of a gasoline-powered lawn mower or weed trimmer. Show the class how the flywheel assembly and ignition coil are arranged mechanically to generate a high voltage for the spark plug.**
3. **Contact your local school maintenance personnel and list the location and function performed by any three phase generators used in and around the school. Report your findings to the class.**

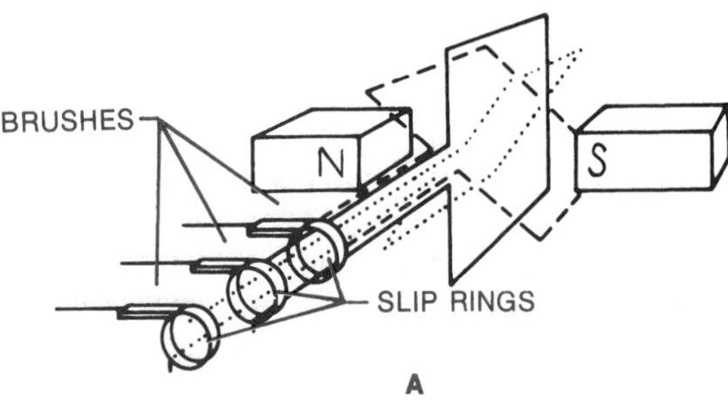

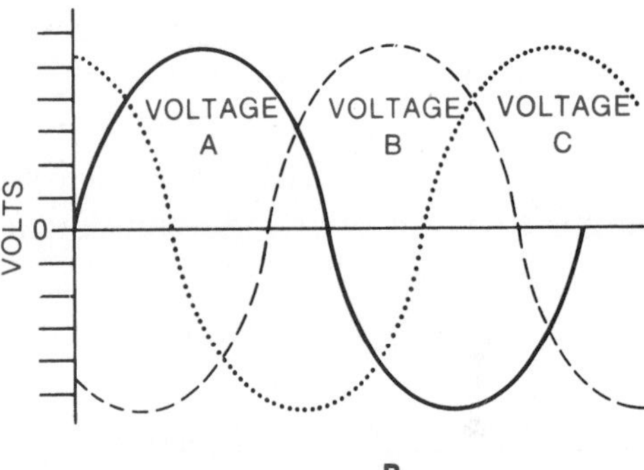

Fig. 15-12. Three-phase generator: (A) basic construction of a simple three-phase generator; (B) sine waveforms of a three-phase voltage.

Almost all power companies in this country use three-phase generators and three-phase power-distribution lines. Although industry uses all three phases, single-phase loads can be connected to one of the three conductors of the power line (if the voltages are correct) and to a fourth conductor, called a *neutral wire*.

LINKS

You may wish to read ahead in Chapter 16 (The Electric Power Industry). It provides more details on three-phase generators.

Chapter *Review*

Summary

- Basic generators produce voltage by rotating a coil of wire between the poles of a permanent magnet. The amount of voltage output depends on flux density, speed of rotation, and the angle at which the coil cuts across the lines of flux.
- The voltage output of a generator changes from zero to maximum twice during each rotation of the coil. The direction of current flow also reverses. The voltage pattern developed by each rotation of the coil is called a sine wave. Each sine wave is called a cycle. The number of cycles per second is called frequency and is measured in hertz. Each frequency output is a phase.
- Some generators produce three-phase voltage. Three-phase voltage is used in manufacturing to operate certain kinds of motors.

Review Main Ideas

Review this chapter's main ideas by writing, on a separate sheet of paper, the word or words that most correctly complete the following statements:

1. An electric generator produces a voltage by means of ______.
2. A generator can be operated by rotating coils of wire through a ______ or by rotating a ______ past coils of wire.
3. Voltage and current are said to be in ______ with each other when the waveform they produce rises and falls at the same instant.
4. A commutator is used in a ______ generator.
5. In its basic form, a commutator is a ringlike device made up of metal pieces called ______.
6. The direct current used to energize the field windings of a generator is called the ______ current.
7. Most small dc generators are driven by ______ or by ______.
8. Large ac generators are turned by sources of mechanical energy such as ______, ______, or ______.

9. Alternating-current generators are also called ______.
10. A rotating armature in an ac generator is equipped with rotors that have ______ or ______ rings that are in contact with ______.
11. The frequency of the current produced by the alternator depends on the speed of the ______ and on the number of magnetic ______ formed by the field windings.
12. ______ are ac generators in which the magnetic fields are produced by one or more permanent magnets.
14. A three-phase generator has three separate sets of ______.

Apply Your Knowledge

1. List the three factors that influence the value of generator output voltage.
2. Draw the following waveforms:
   a. Single phase
   b. Inductive—voltage versus current
   c. Capacitive—voltage versus current
3. Describe the difference between dc and ac generators.
4. Contrast rotating-armature and rotating-field generators.
5. Describe a three-phase system.

Make Connections

1. **Communication Skills.** Write a report describing the difference between generating electricity to produce a direct current and generating electricity to produce alternating current.
2. **Mathematics.** Explain what is meant by the sine function of angles between 0 and 90 degrees.
3. **Science.** Develop an experiment showing the effect that induced electromotive force has on the current in a circuit.
4. **Technology Skills.** Describe a practical procedure to reduce the effect of heat produced in an armature of a generator.
5. **Social Studies.** Using library facilities and/or the Internet, research the percentage of the world output of electricity produced by rotating generators and the percentage of electricity produced by other methods. Develop a chart showing the results of your investigation.

Chapter 10

The Electric Power Industry

OBJECTIVES

After completing this chapter, you will be able to:

- Identify four major problems created by a country using large amounts of energy.
- Describe four sources of energy, other than fossil fuels, that can produce electric energy.
- Explain the process of generating electricity in each of the three main types of electric power plants.
- Identify the main elements of a transmission system used to carry electric energy from power plants to places where it will be used.

Terms to Study

atomic fission

cogeneration

electrolyte

fossil fuel

geothermal energy

grid

photovoltaic cell

reactor

solar array

transmission lines

We all use electric power in many ways and have become accustomed to it being ready when we need it. Can you remember the last time you were without electricity? Our society depends on a constant supply of electrical energy. It is a system that supports society and can not be removed without causing major changes and inconveniences, as well as economic disaster.

GENERALLY USED ENERGY SOURCES

This chapter discusses how our electrical system is constructed and how energy in one form is changed to electrical energy so that it can be delivered through the system to you.

Modern power plants consist of three main kinds: (1) fossil fuel plants, (2) nuclear plants, and (3) hydroelectric plants. All power plants use large alternating-current generators, or alternators, to produce electric energy. This energy is delivered to users by transmission networks, or **grids**. The generators are usually driven by turbines, often called *prime movers*. These turbines are turned by the energy of steam pressure (in the case of fossil fuel and nuclear plants) or by moving water (in the case of hydroelectric plants). The basic principle of a simple turbine is shown in Fig. 16-1. Smaller generators are usually driven by gasoline or diesel fuel engines.

Fossil Fuel Power Plants

Fossil fuel plants were the earliest type of power plant and are still most common. In these plants, water is heated in a boiler to create steam (Fig. 16-2). The fuel used to heat the water can be either coal, oil, or natural gas (called **fossil fuels**). In some electric plants, the steam is heated to 1000° F (538° C) and is under a pressure of 3600 pounds per square inch (24,800 kilopascals (kPa). This steam is directed at the turbine rotor (Fig. 16-3). A series of blades or buckets on the turbine rotor converts the steam jet force into rotary motion. After the energy of the steam is used in the turbine, the steam is condensed into water and continually recycled to the boiler. The generator turns at a constant speed even though the load requirements on the generator change. As more electric energy is needed, more fuel is burned. More steam is thus made available to be converted into mechanical energy and then into electric energy. This process is an *energy-conversion system*. Some high-efficiency steam-electric plants use as little as three-quarters of a pound (0.34 kg) of coal to generate 1 kWh of electric energy. The efficiency of steam plants has been increased significantly by a system called **cogeneration**. In such a system, the unused steam and hot water discharged by the turbine are used for air conditioning, heating, cooking, and other commercial and industrial applications before being recycled to the boiler.

Fig. 16-2. A large steam-electric plant.

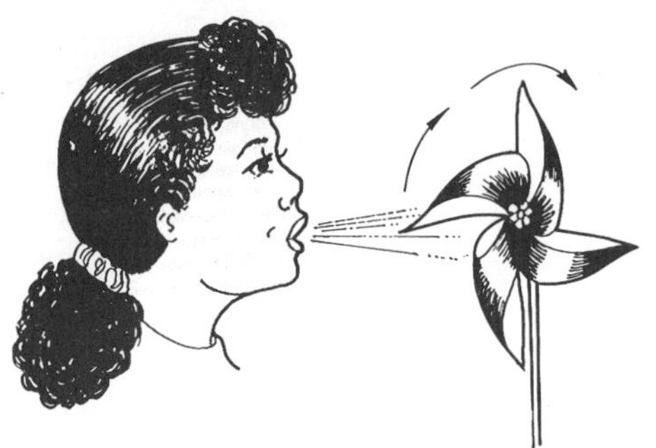

Fig. 16-1. Principle of a turbine. Air pressure forces the vanes (blades) of the pinwheel (rotor) to rotate.

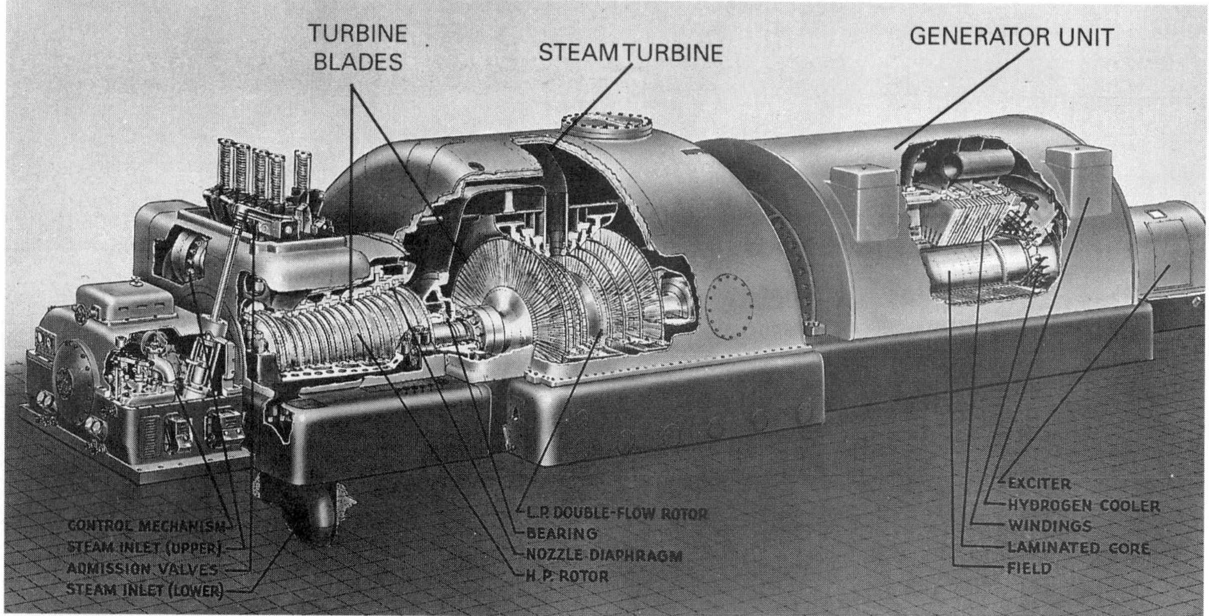

Fig. 16-3. A large steam turbine-generator unit. The turbine blades are mounted on the high-pressure (HP) rotor and the low-pressure (LP) rotor.

Hydroelectric Power Plants

In a hydroelectric power plant, water from a river, lake, or reservoir is directed against the blades of a turbine by gravity. The water pressure causes the shaft of the turbine to rotate. Then this shaft turns the rotor of a generator (Fig. 16-4).

A number of hydroelectric power plants in the United States are operated by the federal government. The dams built to control water flow in these plants are also used to control floods. (Fig. 16-5).

Nuclear Plants

The fuel used in a nuclear power plant is uranium. An atomic particle, the neutron, strikes the nucleus of a uranium atom and splits in a process known as **atomic fission**. Fission causes large amounts of energy and neutrons to be released. These neutrons split more uranium, and more energy and neutrons are released, and so on. The result is a *nuclear chain reaction*. In atomic bombs, this chain

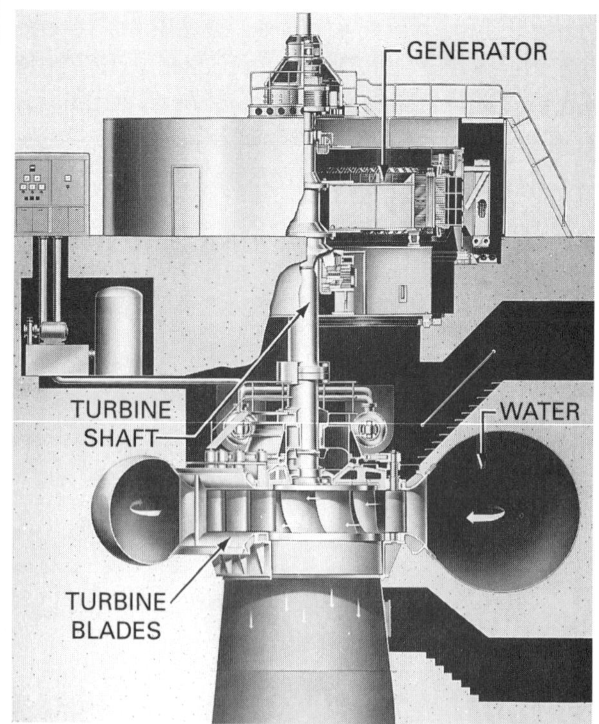

Fig. 16-4. A vertical hydroelectric generator.

Fig. 16-5. Aerial view of Shasta Dam and Shasta Lake—a large hydroelectric power plant near Redding, California.

reaction releases tremendous amounts of heat energy almost instantly. However, it can be controlled to release energy slowly.

In a nuclear power plant, a controlled amount of heat is produced by nuclear chain reactions in a **reactor**. This heat boils water, and the steam produced is used to operate turbines. These turbines are similar to those in a steam-electric power plant.

The main elements of a modern nuclear power plant are shown in Fig. 16-6. Note that in this kind of atomic electric plant, the water is cycled directly from the main circulating pump to the reactor vessel. The water surrounding the reactor is heated to about 600° F (316° C) and is changed into steam at a pressure of 1000 pounds per square inch (6900 kP). The water used is demineralized and thus

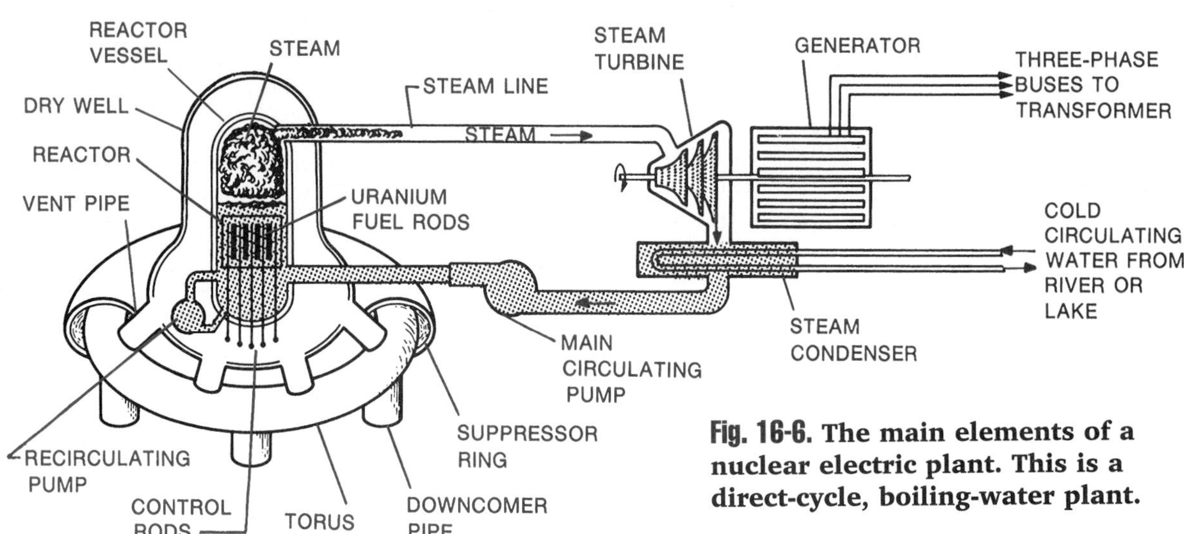

Fig. 16-6. The main elements of a nuclear electric plant. This is a direct-cycle, boiling-water plant.

helps to slow the neutrons, causing them to hit a larger number of atoms. Recirculating pumps also help to regulate fission. The pump reduces the amount of bubbling that occurs in the water when steam is produced. The fewer the bubbles, the greater the rate of the atomic reaction.

Reactor and Reactor Vessel

The reactor vessel in some plants can be 65 ft (20 m) long and weigh nearly 700 tons (635 metric tons). The vessel is supported in an upright position, partly filled with water, and contains the uranium fuel rods. As the fuel rods emit neutrons, some of the atoms split and release heat energy.

Other rods of neutron-absorbing metal are used to control the rate of the chain reaction in the reactor. The control rods are placed between the fuel rods. Moving the control rods in and out of the reactor regulates the amount of heat. By controlling the heat from 115 tons (104 metric tons) of uranium in the core of the reactor, a plant can produce electric power equal to that obtained from 6 million tons (5.4 million metric tons) of coal.

Steam Generator

The steam pressure generated at the top of the reactor vessel is directed through pipes to the turbine blades, making the blades spin. The turbine shaft turns the generator. In some modern plants, the generators convert over 800,000 hp (597,000 kW) into electricity. When the electricity is produced, it is transformed and distributed over the *grid*, a network of transmission lines connected to other power plants and consumers.

Steam Condenser

After the steam is used in the turbine, it is condensed to water in a *steam condenser*. There, large amounts of cold water are used to speed the condensation of the steam to water. The water is then recycled through the system by the main circulating pumps. Since this plant converts the water around the reactor into steam, it is called a *direct-cycle, boiling-water plant*. Because large amounts of cold water are needed by the condenser, plants that use steam turbines are usually found near large rivers, lakes, or reservoirs. Some condensers use 250,000 gallons of water per minute (15.8 cubic meters per second) to condense the steam from the turbines.

Safety

Many safeguards are used in nuclear power plants because they harbor dangerous levels of radiation. For example, in Fig. 16-6, a dry well surrounds the entire reactor vessel. Should a break occur in the reactor vessel, the steam and water would rush out into the dry well. The steam and water would then go through the vent pipes into the *torus*, the doughnut-shaped base of the reactor vessel, and the downcomer pipes. These pipes are placed around the torus with their openings under water. The water helps condense the steam, thus reducing the pressure. The large suppresser ring that surrounds the dry well has a chamber that acts as a cushion for any quick changes in pressure. This protects the entire system from damage. If water is removed from the reactor vessel, the reactor will automatically shut down.

One of the major advantages of nuclear power plants is that they are operated with fuel that is available in practically unlimited quantities. Wisely using this energy, with coal, oil, and natural gas, will help to meet the energy needs of the future. It is very important, however, to safely generate nuclear energy.

Nuclear power plants are massive and complex structures. They are designed to withstand earthquakes and tornadoes. They have many special safety devices and systems that help protect the plant, the plant personnel, and the public in the event of a serious accident.

However, in spite of these precautions, an accident did occur on March 28, 1979, at the Three Mile Island nuclear power plant. This plant is located near Middletown, Pennsylvania, on the Susquehanna River. The accident raised questions in the minds of many people about the use of nuclear energy to generate electricity.

On April 26, 1986, a more serious explosion of a reactor occurred at the Chernobyl nuclear power plant near Kiev, in the Ukraine. Experiments conducted with the control rods and the cooling water turned off resulted in a build up of pressure that was greater than the containment chamber was designed to withstand. This may have caused the explosion. Unfavorable winds carried the radioactive particles across much of the northern hemisphere. Because of the potential public health threat, nuclear power plants in the United States are operated under the general direction of the *Nuclear Regulatory Commission* (NRC).

TRANSMISSION SYSTEM

To carry electric energy from power plants to the places where it will be used takes large transmission and distribution systems (Fig. 16-7). The main parts of these systems are discussed on the next page.

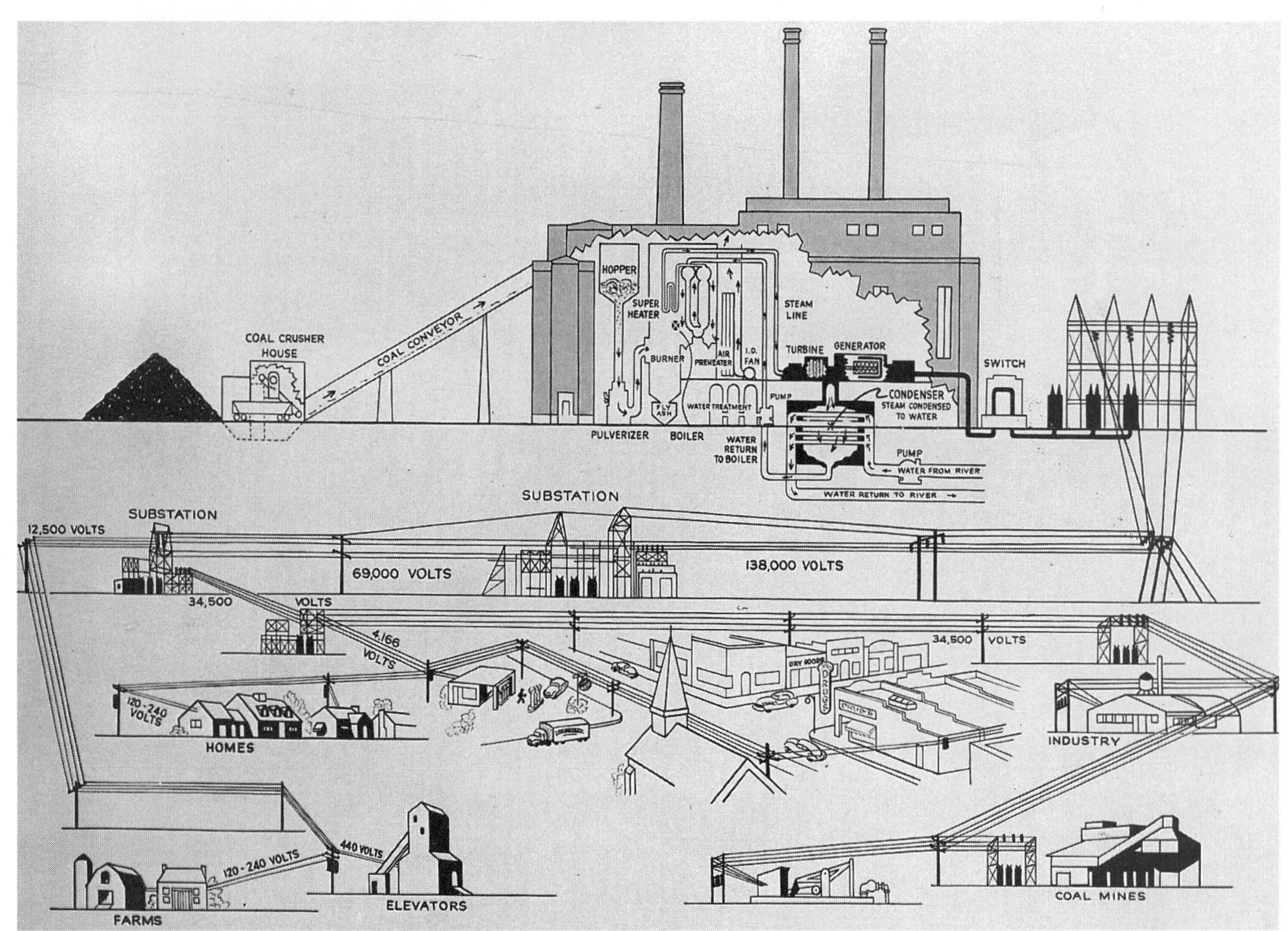

Fig. 16-7. An overview of power distribution.

Contact local OSHA authorities and discover how safety regulations are enforced in a nuclear power plant.

Transmission Lines

Transmission lines, overhead wires supported by high towers, transmit electric energy from power plants. The center strands of transmission lines are made of steel to give them strength. The outer strands are made of aluminum because of its lightness and ability to carry current. The wires are insulated from the towers by porcelain insulators to prevent the loss of electric energy. Figure 16-8 shows line workers stringing wires for a high-voltage (250,000 V) transmission line. In some cases, especially in cities, wires run underground in ducts.

Extensive laboratory and usability testing have proven the underground transmission cable, shown in Fig. 16-9, to be successful. The

Fig. 16-8. High-voltage transmission tower: (A) high-voltage insulators; (B) conductors.

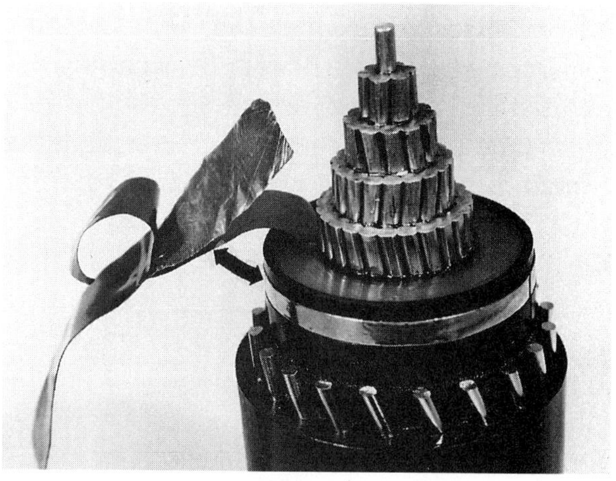

A

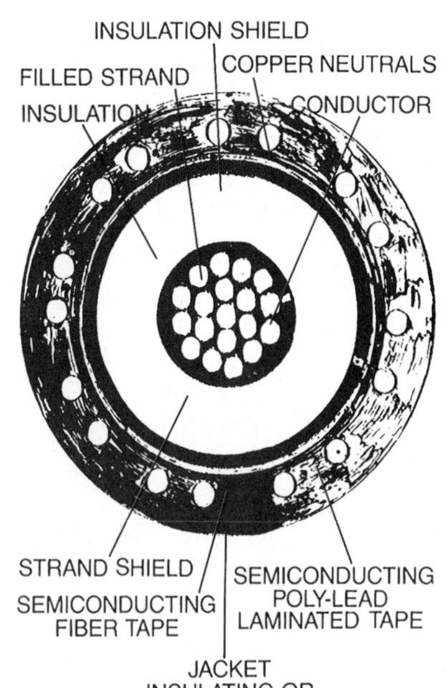

B

Fig. 16-9. Underground electric transmission cable: (A) cutaway view; (B) cross-sectional view of the core.

double-tipped arrow in Fig. 16-9A points to a lead-foil, plastic-laminate, insulating layer of the underground 35-kV cable against an actual cutaway. This type of design prevents water from leaking into the cable. This leakage problem is known as "watertreeing," and can eventually ruin a power cable.

Transformers

In an alternating-current system, energy loss across transmission lines is reduced because transformers make it possible to raise the ac voltage to very high values. Some transmission systems use 750,000 V. These higher voltages allow the same level of electric power made available at a lower current. This results in less energy loss, smaller transmission cables, and higher efficiency. Transformers make it possible to transmit electricity economically over distances of 200 to 300 miles (322 to 483 km). In addition to stepping up, or raising, the voltage for long-distance transmission, transformers step down, or lower, the voltage to the requirements of the load.

Figure 16-10 shows two transformers connected in parallel fed by isolated *buses*, or bars of copper, directly from a three-phase alternator. Each of the three phases of the alternator is isolated from the other. Each bus is separated by insulators in the center of a pipe-like protector. The high-voltage three-phase leads (Fig. 16-10) go directly to a substation.

Substation and Distribution Networks

Figure 16-11 shows a *substation*, or distribution center. Substations step down the voltage for industrial and home uses. They distribute electric energy conveniently to various loads and make it easy to isolate system trouble.

Fig. 16-10. Two transformers connected in parallel.

Fig. 16-11. Substation.

Fig. 16-12. Close-up view of substation switchgear.

Circuit Breakers

A *circuit breaker* is an automatic electric switch. It opens by itself whenever an overload causes excessive current. This protects the transmission lines and other parts of the circuit from high currents. Because of the high currents and voltages involved (as high as 34.5-kV and 1,200 A), the circuit breakers are placed in oil or another arc-suppressing medium such as a vacuum.

Special devices are used to minimize the arc formed when the breakers open under load. Smaller circuit breakers operate in air. They are called *air circuit breakers*. Figure 16-12 shows two circuit breakers installed in a substation. Above the circuit breakers is a disconnect switch. Maintenance personnel can open this switch from the ground and thereby completely disconnect the load from the source. This allows work on the system to be done safely.

SAFETY TIP

It is important to know that older circuit breakers and transformers contain oil that is harmful to both the environment and humans. See a local Environmental Protection Agency office for special instructions on handling polychlorinated biphenyl (PCB).

Contact your local EPA and learn how to safely handle polychlorinated biphenyl.

ELECTRIC POWER COMPANIES

Electric power companies have many departments and working groups. They provide efficient service, maintain facilities, and plan the improvements and expansions needed to satisfy the demand for energy. One major factor influencing these services is the weather.

Power Outages

Hurricanes, tornadoes, downbursts, hail, and heavy, wet snow cause power outages. To respond to these weather conditions, some electric power companies use early warning computer systems. These systems monitor the damage and provide information for supervisors to direct repair crews. One electric power company, for example, had to repair a large portion of their distribution system when two fierce storms, only two days apart, hit their area (Fig. 16-13).

Priorities

Usually the power company can tell when an outage occurs because customers call. They will report, for example, that a tree is down and they have no lights. The power company can also tell because the power demand to a disconnected service area drops dramatically. This important information is fed into the computer system and analyzed. Repairs are made depending on the need. Hospitals and fire departments receive highest priority since human life can be endangered when power is interrupted.

Improved Customer Service

After repairs are completed, evaluation helps improve the company's response to future power outages. This also provides information on how to improve customer service. Based on the studies, depleted electrical supplies and equipment are reordered to meet future storm damages. Broken wires and cables, transformers, and other damaged equipment are examined for their salvage value. Materials, such as copper and aluminum wire, are picked up by a commercial scrap dealer for recycling.

Make a list of all the functions of a power company. Be certain to consider services such as billing and environmental awareness advertising.

Fig. 16-13. **Overhead Line Crews restoring electric power due to storm damage.**

OTHER ENERGY SOURCES

Energy sources are an important part of technology. Technology involves using scientific ideas to make machines that can do work. Countries such as the United States can maintain social and technical standards only by using large amounts of energy. This use of energy, however, creates several problems. These include: (1) depleting (using up) fuel supplies, (2) polluting our environment, (3) depending on external sources for fuel supplies, and (4) spending large amounts of money on imported fuels.

Electric energy is one of the cleanest forms of energy. It is easy to transport. It can be produced in large quantities at a reasonable cost. But producing electricity is damaging to our environment and ways to produce electricity in an environmentally "friendly" manner are being sought.

Today the production of most electric energy begins with heat energy. As mentioned earlier in this chapter, fossil fuels, such as coal, oil, and natural gas are burned to produce heat. The water is then heated and changed to steam. The steam is then used to power turbines that drive large generators.

However, the world's supplies of coal, oil, natural gas, and fissionable materials are limited. Thus, other sources of energy must be used. Some of these are sunlight, wind, ocean heat, wave motion, and *organic* (animal or plant) materials. These sources are particularly useful for producing electricity.

Photovoltaic

The use of sunlight as a source of energy to generate electric power has many advantages. Sunlight is *inexhaustible* (cannot be used up) and is available everywhere. Using sunlight does little damage to the environment and produces few wastes. Solar, or **photovoltaic cells**, generate electricity directly from sunlight and use no fuel. Changing sunlight into electricity may help solve the long-term energy problems of the United States.

An Opportunity in Crisis

By the late 1970s, the United States was consuming more than one third of the world's supply of natural gas and petroleum to meet three-fourths of its energy needs. About one half of the petroleum used was imported. In addition, oil-exporting countries increased oil prices by more than 1,500% between 1970 and 1980. An energy crisis had begun.

It was apparent that natural gas and petroleum resources might soon be depleted, and energy conservation became an important government policy. In 1977 President Jimmy Carter established the Department of Energy and proposed a comprehensive energy program. Passed by Congress in 1978, the program included measures to discourage energy consumption, especially of natural gas and petroleum. It provided incentives for alternative energy sources. An Energy Security Act, to develop synthetic fuels, was passed in 1980. The energy crisis of the 1970s had shifted the country's focus from fossil fuels to developing and using alternative sources of power.

Solar Cells

These cells are usually made of a layer of *silicon*, a common nonmetallic element, that coats a metal base known as the substrate. The silicon is treated chemically to make *p-type* and *n-type* forms. The area where these two forms touch each other is called a *p-n junction*.

LINKS

You will find a more in-depth discussion about p-n junctions in Chapter 17.

FASCINATING FACTS

The energy efficiency of most photovoltaic cells is about 7 to 11%. Only that percentage of radiant energy is converted to electric energy. Above the atmosphere, the solar radiation intensity is only about 125 watts per square foot and even less at the EarthÕs surface. Therefore, very large solar panels are necessary for large load requirements. Consequently, photovoltaic cells have thus far been used primarily for very low-power applications such as calculators, watches, and cameras.

A cross section of a silicon solar cell is shown in Fig. 16-14. The top layer of p-type silicon is very thin. Light energy passes through it and reaches the p-n junction. The light energy gives energy to the electrons in the p-type layer. This causes the electrons to move across the p-n junction and into the n-type layer. The n-type layer then becomes negatively charged. A voltage is produced across the layers on each side of the p-n junction.

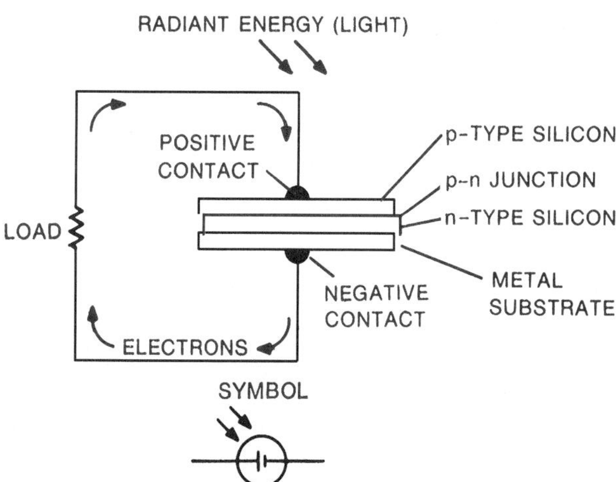

Fig. 16-14. Cross-sectional view of a silicon solar cell.

The amount of voltage produced by a single solar cell is quite small. For this reason, many solar cells are often connected to form a *solar panel*. One very important use of solar panels is to provide energy for charging batteries and operating circuits in space vehicles.

Figure 16-15 shows a **solar array**, or several solar panels connected together, at Lake Hoare

Fig. 16-15. Experimental water purifier powered by solar energy.

in the Dry Valley region of Antarctica. This array, built by NASA (National Aeronautics and Space Administration) researchers delivers 1.5 kilowatts of power that provides for the living needs, as well as power for the laboratory at that site. Other uses of smaller solar arrays include powering crossing signals on the Alaska railroad, generating power in locations where power is not available and supporting emergency telephones installed on the nation's highways.

Wind

Energy from the wind has been used in the United States for many years. The windmill, for example, is a machine that converts wind energy into useful work. In the rural United States, windmills were often used to pump water. Only recently have windmills been seriously considered as another means of generating electricity.

Figure 16-16 illustrates several wind-driven generators located on farmland near the Altamont Pass in California. Surveys indicate that the wind energy in this area averages 700 to 1,200 W per square meter for the eight-month windy season. A wind farm uses many wind generators in areas where there are strong and persistent winds. The wind farm industry is new but, is growing in importance because wind energy is abundant and environmentally sound. Single wind-driven generators, however, may be installed for residential or business use. Looking toward the future, one state has a goal to produce ten percent of its electricity from wind energy by the year 2000. If this capacity was currently available, it would result in a direct savings of 40 million barrels of oil a year.

More details about wind-driven generators are shown in Fig. 16-17. This illustration shows workers installing a three-blade airplane-shaped propeller on one type of wind-driven generator. The blades of this wind-driven generator are 16.6 ft (5.1 m) in length. The blades, generator, and associated equipment are mounted on a tower up to 60 ft (18.3 m) high. It takes a "cut-on" wind speed of ten miles per hour (16.1 km/h) to begin generating electricity. At a wind speed of 60 miles per hour (96.6 km/h), the wind generator will turn off itself. This wind speed is known as the "cut-off" wind speed. These wind-driven generators are designed, however, to withstand a wind speed of 125 miles per hour (201.2 km/h). They are rated at 95 kW at a wind speed of 40 miles per hour (64.4 km/h). The output of these wind generators is fed into the local utility system.

Fig. 16-16. In the future, wind farms may be a common sight over land areas where there is a strong and persistent wind.

Fig. 16-17. Workers installing airplane propeller-like blades on a wind-driven generator.

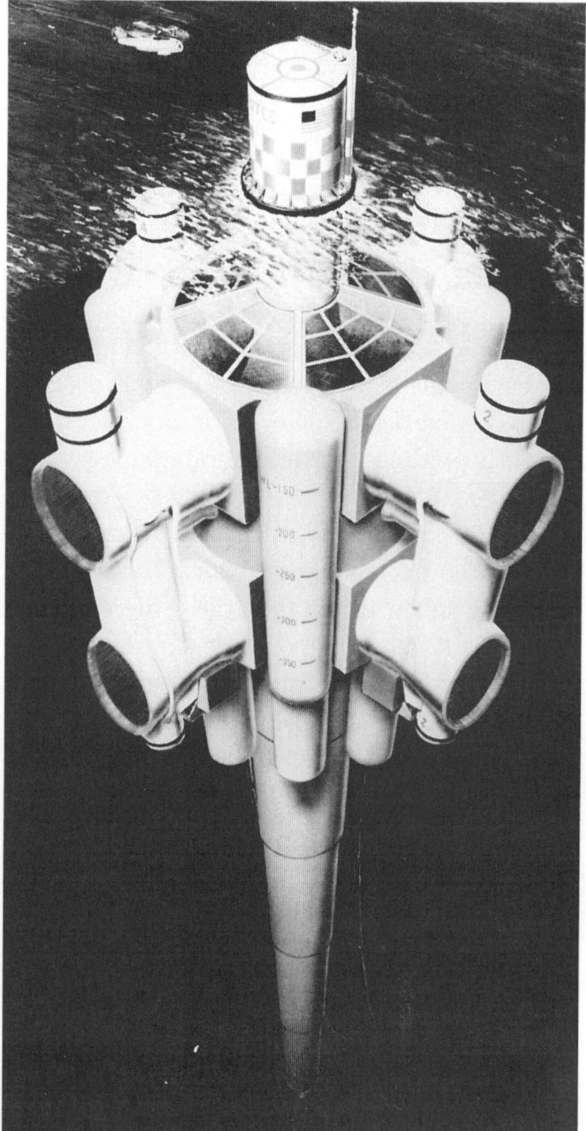

Fig. 16-18. Artist's concept of a floating structure for an ocean wave motion energy conversion system.

Ocean Heat

Another possible way to generate electricity is to use *thermal* (heat) energy from the ocean. This involves using ocean water to collect and store energy from the sun. A system for doing this might take the form of a large floating structure anchored to the ocean floor (Fig. 16-18). This system uses the temperature difference between the heated upper layers of water and the colder layers below. The warmer water can change a liquid such as ammonia into a gas. That gas pressure can then be used to drive a turbine. An electric generator can be connected to the turbine and the gas. After passing through the turbine, it can be changed back into a liquid by the colder water.

However, an ocean thermal energy conversion system would operate at low temperatures and pressures. Very large amounts of ocean water would be needed and power would need to be transmitted to land by underwater cables where they would connect with regular power lines. Large numbers of such systems would need to be built, and would probably be located close to major population areas along coastlines.

Geothermal Energy

In areas of the world where geologic activities, such as magma chambers contained within volcanic features are close to the surface of the earth, the heat produced can be used to generate electricity. Some geothermal fields are currently producing several hundred megawatts of power. Typically, in this process, water is passed through pipes that have been placed close to the hot geologic activity. This water is turned to steam which then turns the turbines. This provides power to operate generators.

Emerging countries with less developed electrical systems and available geologic formations are making use of this resource to meet the electric needs of their countries. It is estimated that one megawatt of power will service the needs of 1000 households and that **geothermal energy**, or energy supplied from the Earth's heat, can replace 150,000 barrels of oil annually. The savings experienced from using alternative energy sources can make money available for developing other areas of the local economy and for furthering environmental concerns.

Wave Motion

Another source of energy being considered is the motion of ocean waves. One way of doing this uses an anchored *buoy* (float). A buoy for using wave motion to produce electricity has two parts. The lower section is the part that floats. The upper part is shaped like a ball with its bottom open. Waves cause water to move up and down in this upper part. This forces air in and out through a tube containing a rotor with windmill-like blades. The air moves the blades of the rotor, which are designed to always turn in the same direction. Connected to the shaft of the rotor is an electric generator (Fig. 16-18). Such a system could be an inexpensive way of generating electricity for countries with ocean shoreline.

Biomass

Organic material that can be changed into usable fuel is another possible source of energy. Such materials include agricultural wastes from crops, animals, and sewage. These can be changed into methane gas, alcohol, and oil. Recent developments in this technology have made it possible to produce enough energy to supply all the electricity demands of some rural states. Some farmers are now growing energy crops, such as corn for ethanol, to use as fuel. Also included are certain species of trees and grasses, naturally rich in the fuels with the high BTU levels needed to produce electricity.

Design an independent power system for your home. Use at least three alternate power sources and do not include a power company.

High-Altitude Transmission of Power

Figure 16-19 shows an artist's idea of one kind of satellite power station, a Space-Based Power Conversion System. This system is being studied by NASA. Such a system can be placed in a fixed *orbit* (path) around the Earth. There, it can collect pollution-free energy from the sun. This energy can be transmitted to the Earth in the form of *microwaves*. These are ultra high-frequency radio waves. The receiving station on Earth can then change the microwaves into electricity. Located high above the Earth, this system would not be affected by clouds or night. Thus, it could operate continuously.

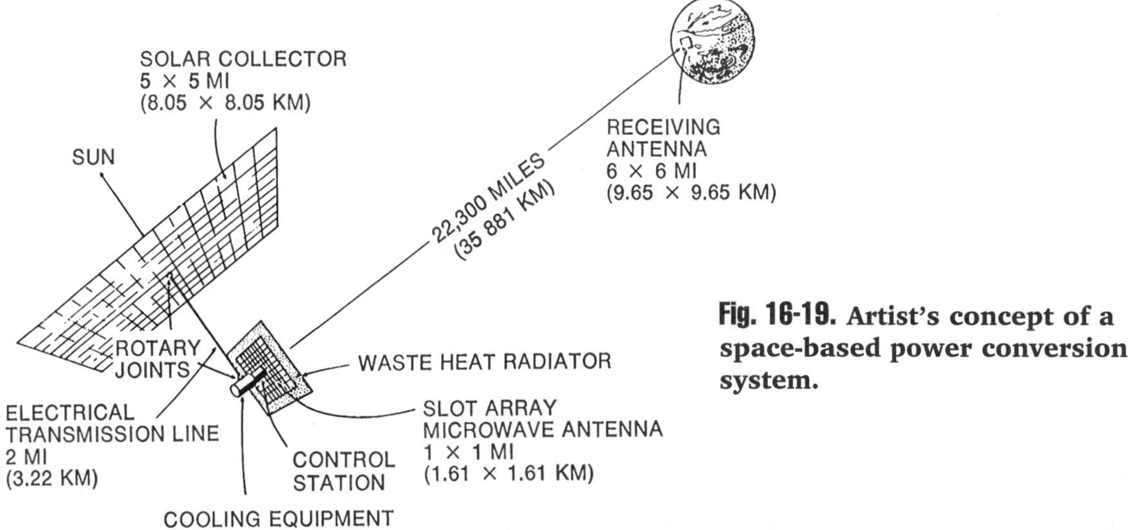

Fig. 16-19. **Artist's concept of a space-based power conversion system.**

Fuel Cells

In a *fuel cell*, reactions between oxygen and a fuel cause chemical energy to be changed directly into electric energy. The fuels most commonly used in such cells are hydrogen and methane.

The basic construction of a hydrogen-oxygen fuel cell is shown in Fig. 16-20. In this cell, hydrogen in either liquid or gas form is supplied to the cathode, or negative electrode. The hydrogen then spreads through the **electrolyte** which is an ionized liquid capable of conducting current. This causes electrons to be released from the electrolyte. These electrons are deposited on the cathode, causing it to become negatively charged. The electrons then move through the load to the anode, or positive electrode.

A very common use of fuel cells is in space vehicles. Other land-based uses are as temporary portable electric sources. By changing the combination of electrolyte and reactive chemicals, different types of fuel cells can be produced. The efficiency ratings of these vary, but a fuel cell can be expected to convert 45% of the inputted energy to electricity and also provide 35% of the converted energy to usable heat for on-site purposes. With only a 20% energy loss, the fuel cell is becoming very competitive for local electric production.

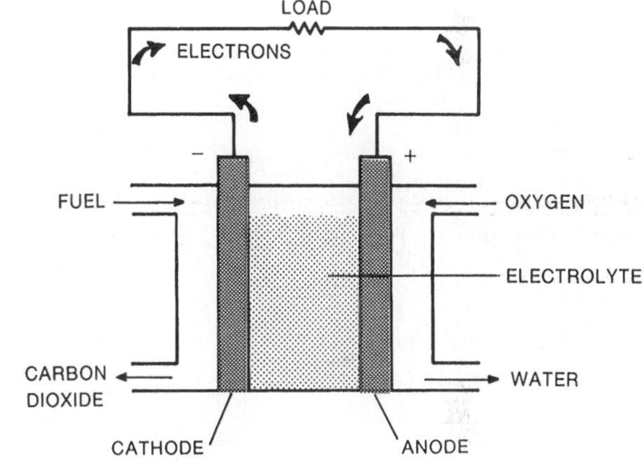

Fig. 16-20. **Basic operation of a fuel cell.**

Magnetohydrodynamic (MHD) Conversion

Another device that could generate a large amount of electricity is the *magnetohydrodynamic* (MHD) *converter*. MHD converters are based on two principles. The first is that gases can be ionized by high temperatures, thus made into good conductors of electricity. The second is the principle of electromagnetic induction.

Figure 16-21 shows the basic elements of the MHD converter. Gas is heated to a temperature of about 5000°F (2760°C). This ionizes the gas, which is then forced through a strong magnetic field. The electrical force that results makes contact with electrodes at the top and bottom of the gas stream. Current flows in the external circuit. The gas moves to a condenser unit and then back to the heat source to complete the cycle. The ionized gas thus takes the place of metal conductors.

MHD converters have few mechanical moving parts. New materials will allow them to withstand the tremendous heat and pressures needed. Someday it may be possible to design and build MHD converters with outputs of more than 1 million watts. These will be smaller, lighter, more efficient, and more reliable than equivalent turbine-generator combinations. Figure 16-22 is a flow diagram that shows the operation of an MHD converter.

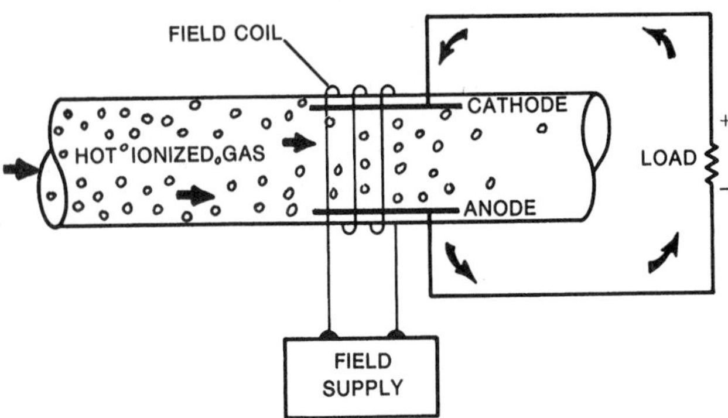

Fig. 16-21. Basic elements of a magnetohydrodynamic (MHD) converter.

Thermionic Energy Conversion

Another system for changing heat directly into electricity is the *thermionic energy converter*. It consists of two electrodes in a vacuum. The *emitter* electrode is heated to produce free electrons. The *collector* electrode, at a much lower temperature, receives the electrons released by the emitter.

Fig. 16-22. Flow diagram of a magnetohydrodynamic (MHD) electric power converter.

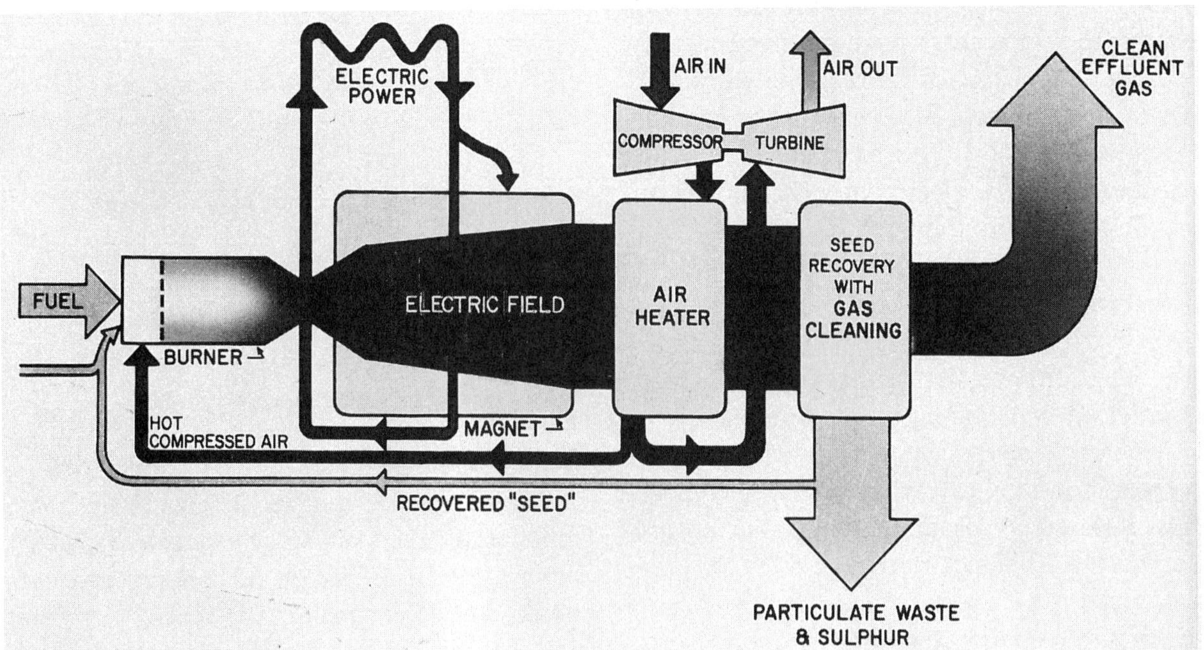

The *coaxial thermionic energy converter* shown in Fig. 16-23 has a heat source in the center of the converter. This will heat the emitter, which will then give off electrons. The electrons then move to the collector and on to the external load.

An alternating current can be produced directly in a thermionic converter. This is done by applying a small *modulating* (varying) signal to a *grid electrode* located between the emitter and the collector. Large thermionic converters could someday replace ordinary turbine-driven generators.

LINKS

Look in Chapters 27 and 32 to find further discussions about modulating signals and the effect of grid electrodes.

Fig. 16-23. Coaxial thermionic energy converter.

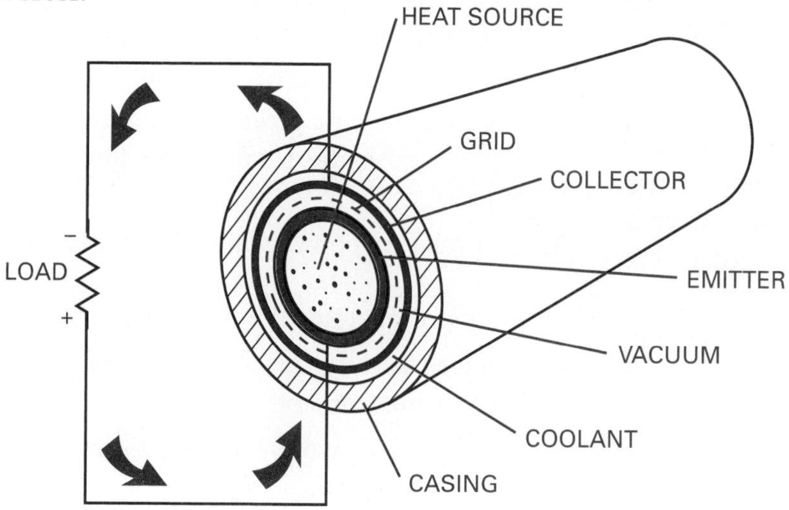

Chapter *Review*

Summary

- There are three main types of power plants in the United States: fossil fuel, nuclear, and hydroelectric.
- Once the electricity is produced, it is distributed to a grid, or network of transmission lines shared by consumers and other power companies.
- Large countries with enormous power demands face four major problems: (1) depleted fuel supplies, (2) pollution of the environment, (3) dependence on external sources for fuel, and (4) high costs for imported fuels.
- There are a number of alternate energy sources for producing electricity. Examples are solar arrays, wind-powered generators, ocean heat, geothermal energy, ocean waves, biomass, high altitude power transmission, fuel cells, magnetohydrodynamic converters, and thermionic converters.

Review Main Ideas

Review this chapter's main ideas by writing, on a separate sheet of paper, the word or words that most correctly complete the following statements:

1. The three main kinds of electric power plants are ______, ______, and ______.
2. In a hydroelectric power plant, gravity and ______ cause the shaft of the turbine to rotate.
3. In a nuclear power plant, chain reactions in a ______ heat water, and the ______ produced operates the turbines.
4. Voltages as high as ______ V have been used in transmission systems.
5. ______ step-down the voltage for industrial and home use.
6. A ______ switch in a substation makes it possible for workers to completely disconnect the load from the source.
7. ______ is an inexhaustible source of energy.
8. Devices that change sunlight directly into electricity are called ______.
9. A ______ has "cut-on" and "cut-off" wind speed requirements.

Chapter *Review*

10. Thermal energy uses ______ to generate electricity.
11. Energy supplied from the Earth's heat is called ______ energy.
12. A potential source of energy which could be inexpensive for countries with ocean shoreline is ______.
13. A satellite power station placed in a fixed orbit around Earth can transmit energy to Earth in the form of ______.
14. A device that converts chemical fuel energy directly into electric energy is the ______.
15. A device that changes heat energy into electric energy by using a hot ionized gas or electromagnetic induction is the ______ converter.
16. A system for changing heat directly into electricity is called a ______.

Apply Your Knowledge

1. Explain how a steam turbine operates.
2. Draw a flow diagram of how water is processed in a nuclear power plant.
3. Describe a transmission system. Include the circuit breakers, transformers, disconnect switches, and loads.
4. List how four alternate energy sources discussed in this chapter prevent pollution or preserve resources.
5. Describe four problems caused by a nation using large amounts of energy.

Make Connections

1. **Communication Skills.** Prepare a report for your class on the job opportunities in the electric utility industry in your area. Include information such as job classifications, employment locations, average wages, and education requirements.
2. **Mathematics.** Have each of your classmates find out how many kilowatthours their household used for a selected month. Average the usage for your class. Find out how many households there are in your community. Calculate the residential power requirements for that month for your community.
3. **Science.** Using photovoltaic cells, construct an array that will produce sufficient electric energy to charge a 1.5 Vdc rechargeable battery. Conduct experiments to determine that charge rate under several lighting conditions. Report your findings to the class.
4. **Technology Skills.** Using a permanent magnet motor, construct a generator. Determine how much input energy is needed to light a 4-watt 6-Volt bulb to a normal level.
5. **Social Studies.** Prepare a written report on how electricity is generated for your use. Discuss what is being done to protect the environment from the by-products of the electricity generation process.

Section 4

Semiconductors and Integrated Circuits

Chapter 

Semiconductors and Diodes

OBJECTIVES

After completing this chapter, you will be able to:

- Draw and label a schematic diagram that uses a single diode to illustrate half-wave rectification of an ac current.
- Draw and label a schematic diagram using four diodes to illustrate full-wave rectification of an ac current.
- Test a diode for performance.
- Explain hole movement and electron flow.
- Describe a heat sink.

Terms to Study

amplifier

base

bridge

collector

covalent bond

diode

doping

emitter

hole

transistor

Alternating current (ac) is delivered by the electric companies to homes, schools, factories, and stores. However, electronic equipment operates on direct current (dc). How is the energy changed to the appropriate current? The component that almost single-handedly changes the current from ac to dc is the diode. In this chapter, you will learn how the diode works and how it changes ac to dc. The diode circuit is the most important electronic component. It is the basis for all other semiconductors such as *transistors* and *silicon controlled rectifiers*.

SEMICONDUCTOR STRUCTURE

A *semiconductor material* is one that, in its pure state and at normal room temperature, is neither a good conductor nor a bad insulator. However, pure semiconductor materials are seldom used in electronics. Instead, **doping**, a process that adds small amounts of other substances such as arsenic, creates free electrons in the semiconductor through *covalent bonding*. Thereby, the semiconductor resistance is made much lower than that of the pure semiconductor, or *intrinsic*, materials.

The most widely used semiconductor materials are germanium and silicon. They are used in solid-state diodes and transistors. A **diode** is a semiconductor that, unlike the electron tube, is made as a solid unit. It does not have a glass envelope or a filament (Fig. 17-1).

The atoms of germanium and silicon have four valence electrons in the outermost shell (Fig. 17-2). In these materials, the atoms are arranged in what is called a *crystal-lattice structure*. In a **covalent bond**, an atom shares each of its valence electrons with those of nearby atoms. This forms what are called *electron-pairs* between the atoms (Fig. 17-3).

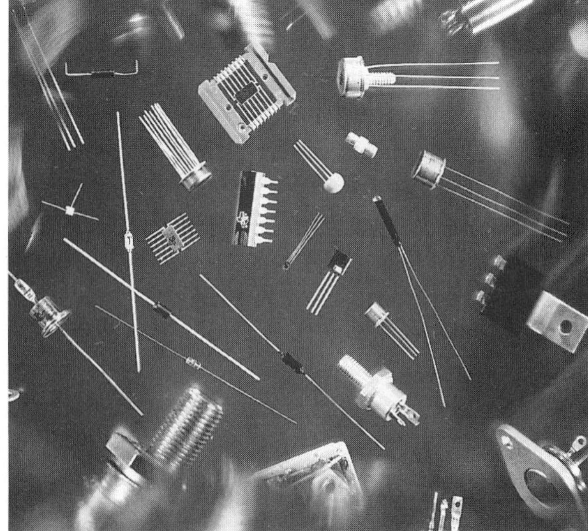

Fig. 17-1. Semiconductor solid-state devices.

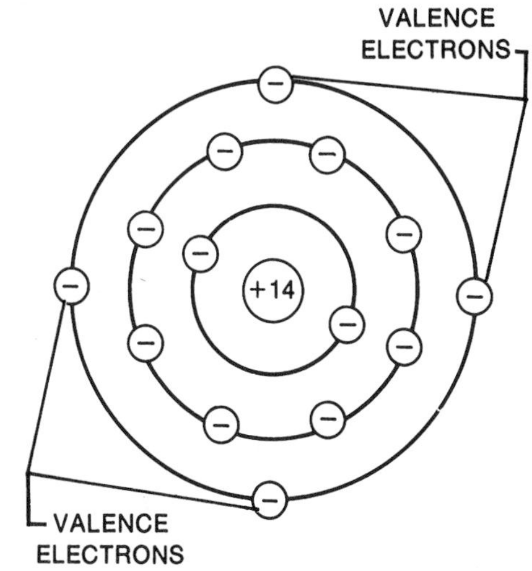

Fig. 17-2. Representation of a silicon atom.

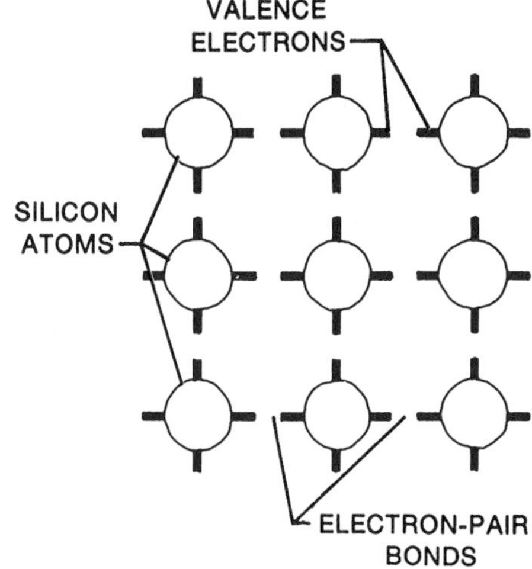

Fig. 17-3. Electron-pair, or covalent, bonds in the crystal-lattice structure of pure silicon.

Semiconductor Materials: N-Type

One impurity with which a pure semiconductor material, such as silicon, is often doped is arsenic. Arsenic has five valence electrons, four of which form electron-pair bonds with the valence electrons of each silicon atom. One valence electron of each arsenic atom is then left unattached and it becomes a free electron

Silicon

Silicon *is the second most abundant element, making up about 28% of the Earth's crust. It occurs only in combined forms such as silicon dioxide (in quartz and sand) and silicate rocks and minerals.*

Commercially, pure silicon is prepared by removing the oxide through a reaction with a carbon-based substance such as coke in electric furnaces. Some silicon is obtained through a reaction with aluminum. Pure silicon is a hard, dark-gray solid with a metallic luster. Its crystalline structure is the same as carbon's diamond form. Therefore, they share many chemical and physical properties.

Only when doped does pure silicon yield electrons for a current. Containing all the components of an electronic circuit, tiny chips of silicon can be connected to form complex, yet compact circuits. Since silicon plays such a significant role in microprocessors, the region in California famous for computer technology is called ***Silicon Valley****.*

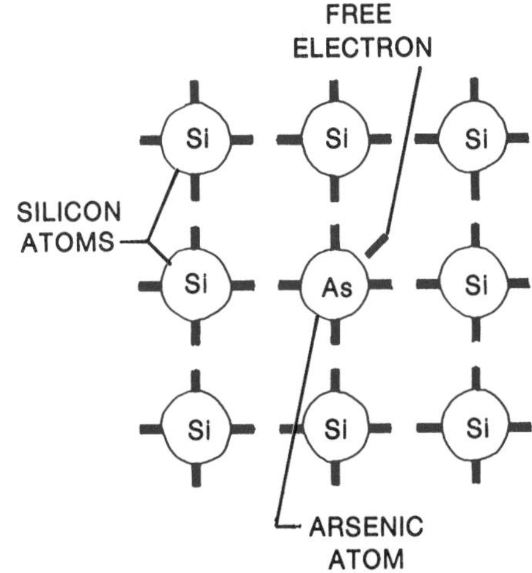

Fig. 17-4. Free electron resulting from the addition of arsenic to silicon.

(Fig. 17-4). Since the silicon then contains free electrons, or *negatively* charged particles, it is called an *n-type semiconductor material*. Because the arsenic donated a free electron to the silicon, it is called a *donor impurity*.

Semiconductor Materials: P-Type

Pure silicon can also be doped with an impurity such as indium, which has only three valence electrons. In this case, the valence electrons of each indium atom form electron-pair bonds with the valence electrons of three silicon atoms. This leaves one valence electron of a nearby silicon atom without an electron-pair bond. Consequently, there is an electron lacking in the crystal-lattice structure of the silicon. A **hole** is this void in the crystal-lattice structure caused by doping with indium or another acceptor impurity (Fig. 17-5A).

If voltage is applied to the silicon, an electron from an electron-pair bond can gain enough energy to break the bond and move into the hole. This creates a new hole where the electron had been. This hole is then ready to accept another electron that has broken its electron-pair bond. As this continues, the hole is said to *move* through the silicon, as shown in Fig. 17-5B.

Since a hole represents a missing electron, it is considered a *positively* charged current carrier. Therefore, semiconductor materials containing many holes are called *p-type materials*. Indium is then called an acceptor impurity because each atom of indium can accept an electron from an electron-pair bond.

Draw a crystal-lattice structure of silicon doped with arsenic. Illustrate clearly the valence and free electrons.

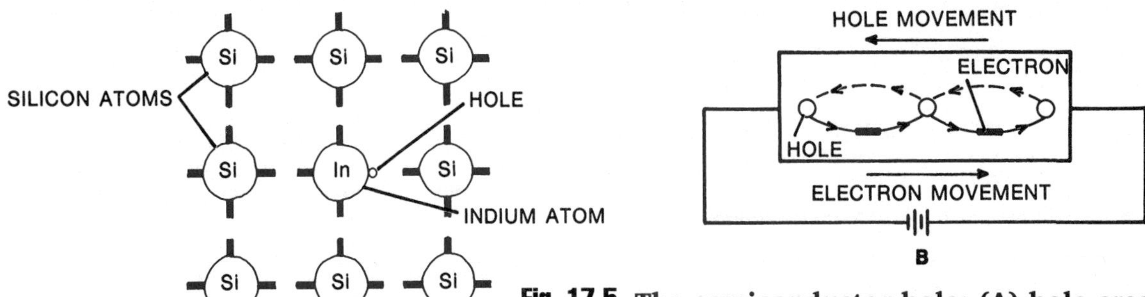

Fig. 17-5. The semiconductor hole: (A) hole created when indium is added to silicon; (B) movement, or drift, of holes and electrons through a p-type material.

P-N JUNCTION DIODES

The basic solid-state diode is a combination of p- and n-type semiconductor materials. Through a chemical process, these are made as a single unit known as a *p-n junction* (Fig. 17-6).

Depletion Regions

When a p-n junction is formed, free electrons within the n section spread across the junction and fill the holes near the junction in the p section. At the same time, holes from the p section cross the junction and combine with free electrons in the n section near that junction (Fig. 17-6A). As a result, the atoms near the junction of each section ionize. This action reduces the number of n-section electrons and p-section holes. This area where the process of ionization occurs, is called a *depletion region*.

Potential Barriers

Because there are electron and hole combinations within the depletion region, atoms in the p section gain electrons and become negative

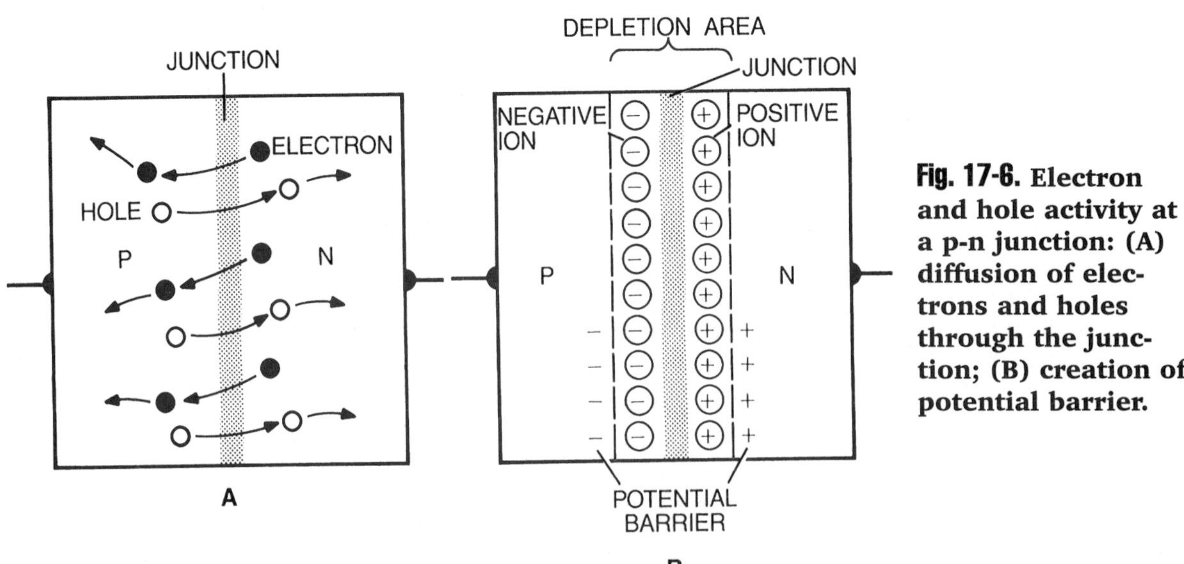

Fig. 17-6. Electron and hole activity at a p-n junction: (A) diffusion of electrons and holes through the junction; (B) creation of potential barrier.

ions, while atoms in the n section lose electrons and become positive ions. This condition creates a small voltage across the junction that is referred to as a *potential barrier* (Fig. 17-6B). Because of the polarity of the potential barrier, there is no further significant movement of the electrons from the n section into the p section or of holes from the p section into the n section.

Write a paragraph that describes how a potential barrier is created.

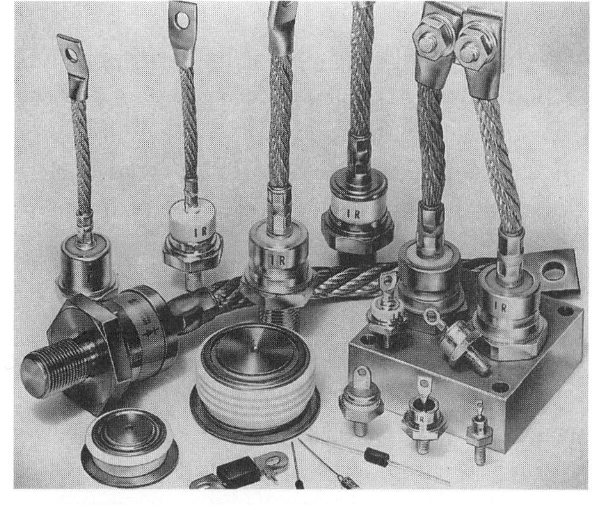

A

HOW DIODES WORK

The common solid-state diode is made of silicon. It works by acting as a gate. A *gate* lets current pass through it in one direction but not in the other. The polarity and value of the voltage applied to a diode determines if the diode will conduct current. The two polarities of voltage applied to diodes are known as *forward* and *reverse biases.*

Diodes are enclosed in protective cases of various shapes with lugs or leads for making the needed circuit connections (Fig. 17-7A). The two electrodes of the diode are the *anode* and the *cathode*. The anode is the p-type material in a diode. The cathode is the n-type material in a diode. The anode and cathode terminals of a diode are identified in several different ways, as shown in Fig. 17-7B.

The graphic symbol for a diode is shown in Fig. 17-10, and the cathode (-) and anode (+) are marked. The class designation for a diode is CR, as illustrated in Fig. 17-12. Each diode in a circuit is given a unique number following the CR label. This gives the reader a way of identifying all individual diodes in a circuit.

Most diodes are identified by a number-letter code, such as 1N, which is followed typically by 2, 3, or 4 numbers or number-letter combinations. This is the number that specifies the device as a diode. Some manufacturers use their own numbering system, and these must then be cross-referenced to find the specifications.

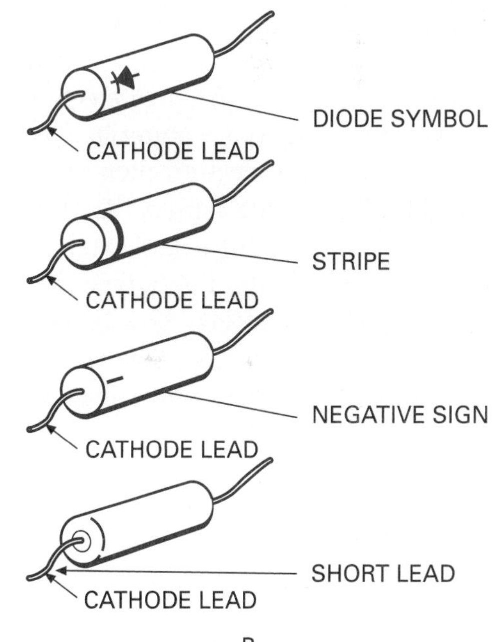

B

Fig. 17-7. Silicon rectifier diodes: (A) typical diodes; (B) methods used to identify the cathode lead or terminal of small-size diodes.

Forward Biasing

A diode is forward-biased when the positive terminal of a voltage source, such as a battery, is connected to its anode and the negative terminal is connected to its cathode (Fig. 17-8A). Under this condition, the positive terminal of the battery repels holes within the p-type material toward the p-n junction. At the same time, free electrons within the n-type material are repelled toward the junction by the negative terminal of the battery. When holes and electrons reach the junction, some of them break through it (Fig. 17-8B). Then holes combine with electrons within the n-type material. Electrons combine with holes within the p-type material.

Each time that a hole combines with an electron or an electron combines with a hole near the junction, an electron from an electron-pair bond within the p-type material (anode) breaks its bond. This electron then enters the positive terminal of the battery. At the same time, an electron from the negative terminal of the battery moves into the n-type material (cathode) of the diode (Fig. 17-8C). This produces a negative-to-positive flow of electrons in the external circuit to which the diode is connected.

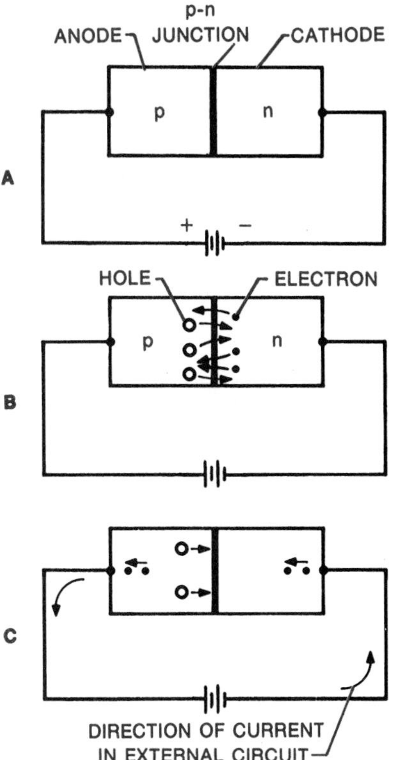

Fig. 17-8. Diode circuit with forward bias.

Reverse Biasing

A diode is reverse biased when its anode is connected to the negative terminal of the battery and its cathode is connected to the positive terminal of the battery (Fig. 17-9A). In this condition, holes within the p-type material are attracted toward the negative terminal of the battery and away from the p-n junction. At the same time, free electrons within the n-type material are attracted toward the positive ter-

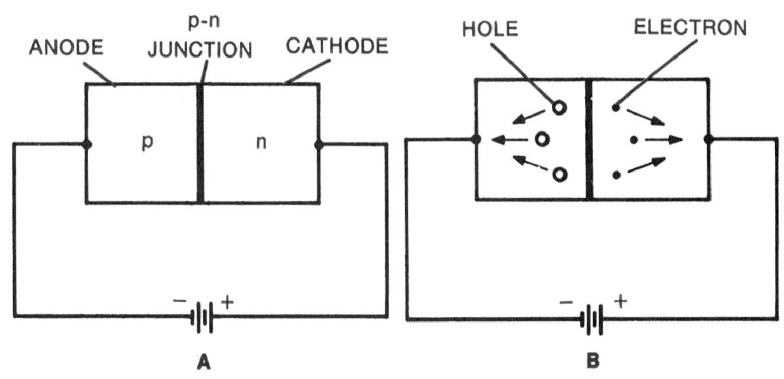

Fig. 17-9. Diode circuit with reverse bias.

minal of the battery and away from the junction (Fig. 17-9B). This action causes the diode to present a high resistance to current. Under this condition, no current will flow in the external circuit.

Draw and correctly label a diode. Include the p- and n-materials, holes, and free electrons.

RECTIFYING

Most power companies produce alternating current. One of the main reasons for this is that alternating current can be sent over long distances very efficiently. Many processes and devices, however, can be operated only with direct current. For example, transistors and certain control devices need direct current. If circuits containing these devices are to be operated from ordinary outlets, they must use rectifier circuits. *Rectifier circuits* change alternating current to direct current. Diodes are used as rectifiers in battery chargers and in other power-supplies that provide direct current. They also are used in rectifier circuits in radios, television receivers, computers, and other electronic and electrical products.

Half-Wave Rectifier Circuits

When a diode is connected to a source of alternating voltage, it is alternately forward biased and then reverse biased during each cycle. Therefore, when a single diode is used in a rectifier circuit, current passes through the circuit load during only one-half of each cycle of the input voltage (Fig. 17-10). For this reason, the circuit is called a *half-wave rectifier*.

Outputs

The outputs of half-wave rectifier circuits are pulsating direct currents. Such currents can be used in some circuits. However, they produce a loud humming in radios, television sets, and amplifiers. To eliminate or reduce this humming, the pulsating direct current is *filtered*, or smoothed.

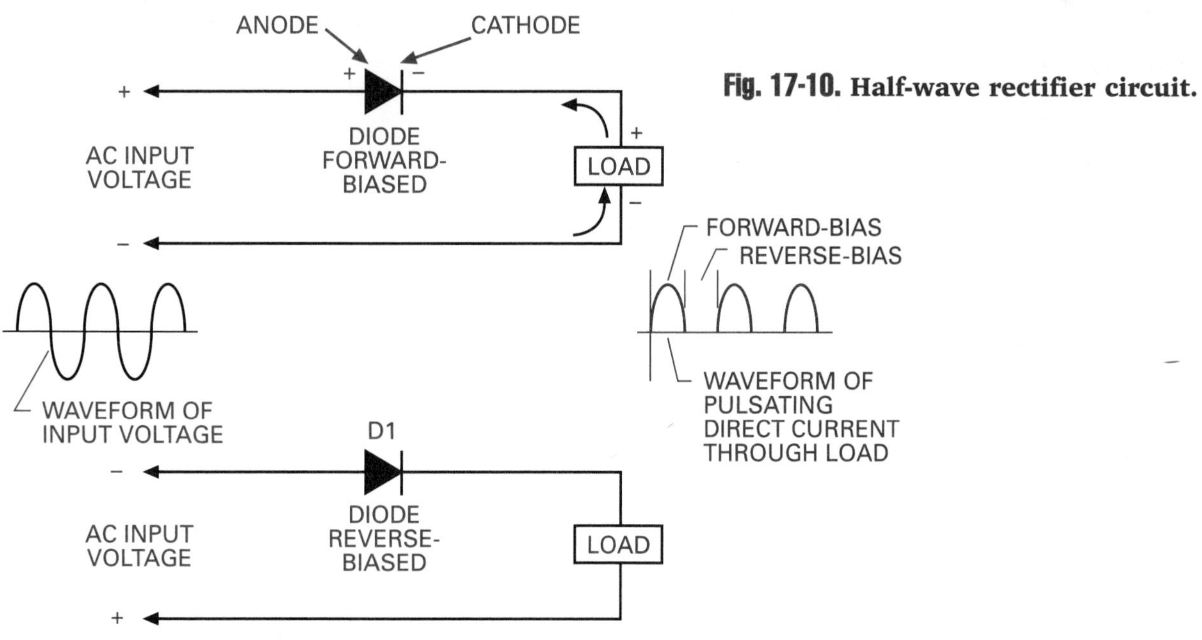

Fig. 17-10. **Half-wave rectifier circuit.**

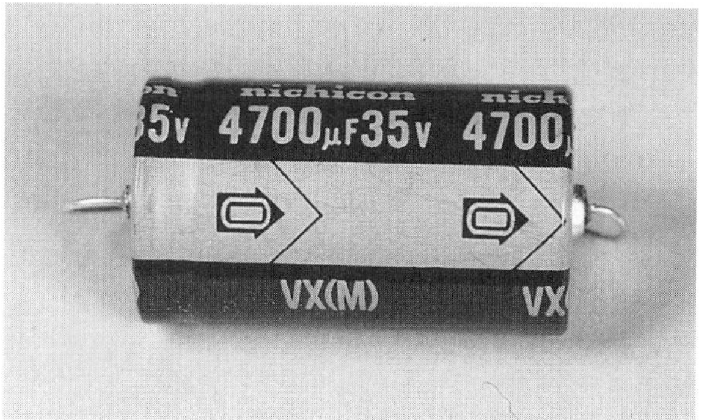

Fig. 17-11. **Electrolytic capacitor.**

Filtering

The basic filtering component of rectifier circuits is an electrolytic capacitor (Fig. 17-11). The action of such capacitors, in half-wave circuits, is shown in Fig. 17-12.

During the time that the diode is forward biased, voltage builds up across the output circuit. The capacitor becomes charged. As the voltage decreases, the capacitor begins to discharge through the resistor R . When the diode is reverse biased, it ceases to conduct. Meanwhile the capacitor keeps discharging through R . When the diode is forward biased again, the capacitor, which was not fully discharged, is brought back to full charge. It then begins the discharge cycle again. Thus, the output voltage and current are smoothed.

Full-Wave Rectifier Circuits

A full-wave-rectifier circuit is one that rectifies the entire cycle of an applied voltage. A basic full-wave-rectifier uses two diodes. The action of these diodes during each half cycle of the applied voltage is shown in Fig. 17-13. The diodes may be individual units, or they may both be in a single package (Fig. 17-14).

Transformers used in these circuits must have center-tapped secondary windings. The dc output voltage of the rectifier circuit depends mainly on the voltage across the secondary winding of the transformer, although other circuit components such as output resistors and filter capacitors will affect the output as well.

Full-wave rectifiers have steadier output voltages than half-wave rectifiers do. This is because full-wave rectifiers produce a pulse of voltage in the output circuit during each half cycle of the applied voltage. After filtering, the load can be quite smooth.

Bridge Rectifiers

A **bridge** is a full-wave rectifier that uses four diodes in specific configuration. The diode action during each half cycle of the applied

Fig. 17-12. **Filtering action of an electrolytic capacitor in a half-wave-rectifier circuit.**

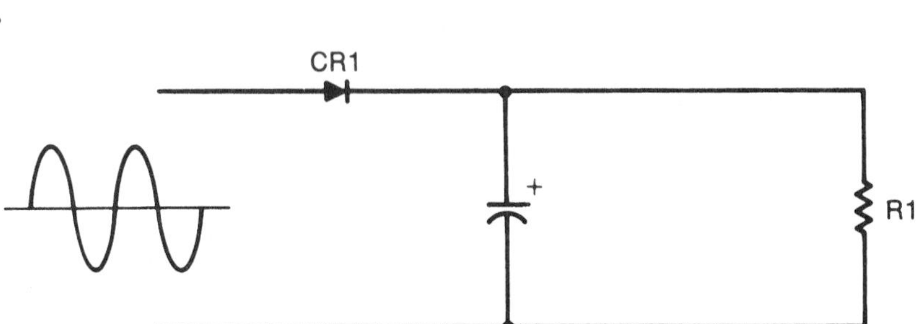

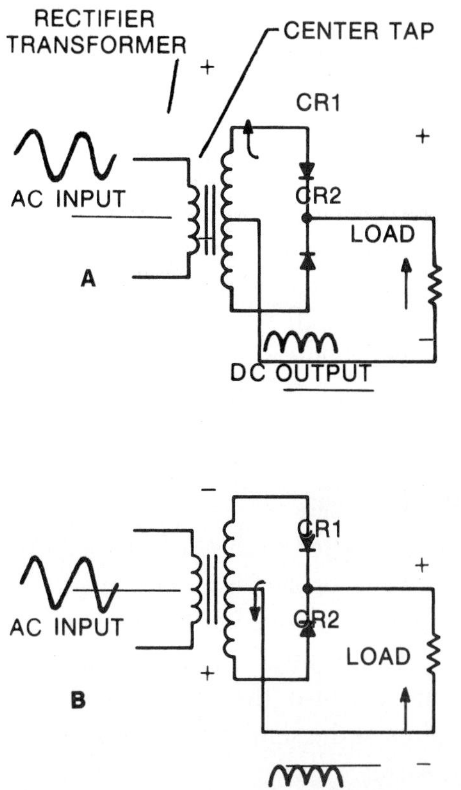

Fig. 17-13. **Basic two-diode, full-wave-rectifier circuit: (A) diode CR1 conducting, diode CR2 not conducting; (B) diode CR2 conducting, diode CR1 not conducting.**

alternating input voltage is shown in Fig. 17-15. Note that the rectifier transformer used in a bridge circuit does not have a center-tapped secondary winding. The diodes may be individual units or packaged into a single unit (Fig.

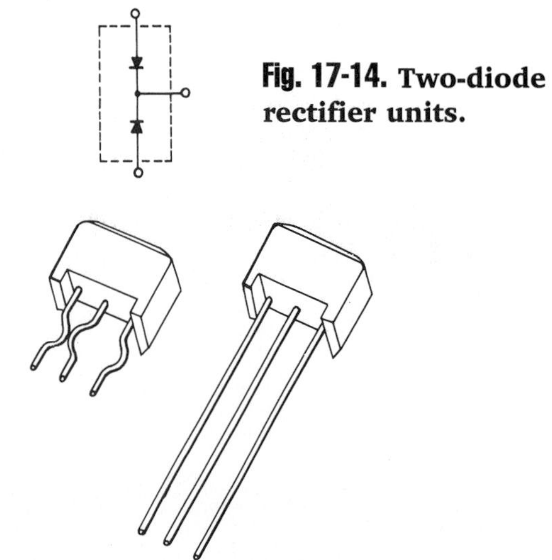

Fig. 17-14. **Two-diode rectifier units.**

17-16). A filter is also needed to remove the pulses from this rectified dc.

Special Diodes

In addition to rectifying ac signals, there are many different applications for diodes such as regulating dc outputs and controlling ac. Diodes used for these applications are zener diodes, TRIACs and DIACs.

Zener diodes are designed to regulate dc voltages by operating in reverse-bias mode. They conduct when the reverse-bias *breakdown voltage* is reached. This is the reverse voltage that the diode can block. Once it conducts, the output voltage regulates at that level and compensates for load current and input voltage level changes.

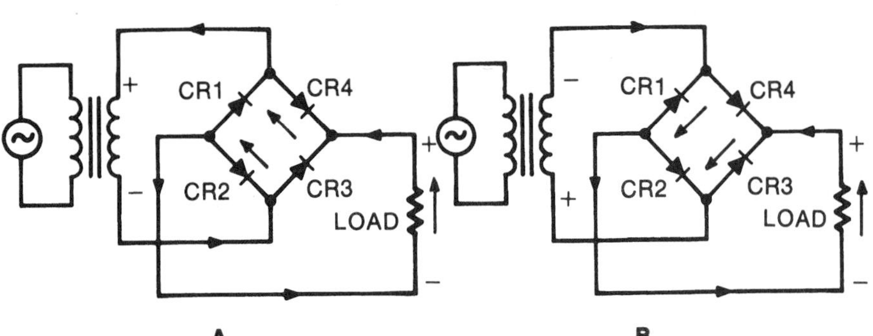

Fig. 17-15. **Bridge rectifier circuit: (A) diodes CR2 and CR4 conducting, diodes CR1 and CR3 not conducting; (B) diodes CR1 and CR3 conducting, diodes CR2 and CR4 not conducting.**

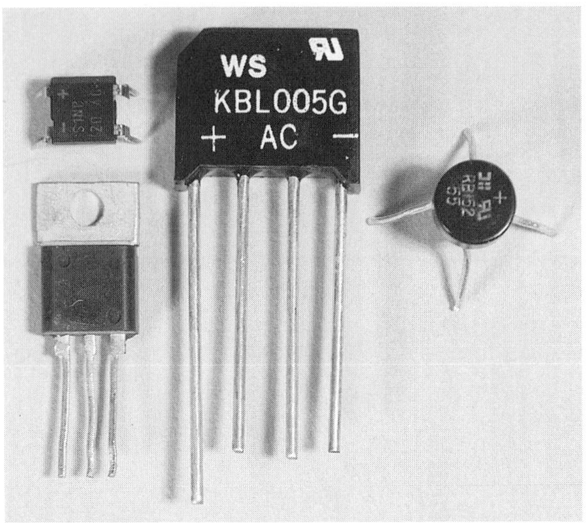

Fig. 17-16. Four-diode bridge rectifier units.

TRIACs and DIACs are ac devices used to control current flowing in both directions. TRIACs can be turned off and on with a *gate*; and DIACs are *triggered,* or caused to conduct, by reverse bias voltage.

LINKS

Regulators are covered more fully in Chapter 18.

DIODE RATINGS

Rectifier diodes are rated by the maximum current that they can conduct safely and the maximum forward and reverse voltages that can be applied to them. The maximum reverse voltage is called the peak inverse voltage (PIV), or sometimes the peak reverse voltage (PRV). If more than the maximum reverse-bias voltage is applied, the diode can be badly damaged. The ratings for diodes are found in various specification manuals published by semiconductor manufacturers.

DIODE TESTING

When a diode becomes defective, it is almost always open or shorted. Each of these conditions generally occurs because overheating damages the semiconductor material. This happens because the atomic and molecular structures in the semiconductor become in serious disarray.

Ohmmeter testing of a diode is based on forward and reverse bias. When the negative terminal of the ohmmeter power supply is connected to the diode's cathode, the diode is forward biased. Its resistance will be relatively low (Fig. 17-17A). If the positive terminal of the ohmmeter is connected to the diode's cathode, the diode is reverse-biased. Its resistance will be relatively high (Fig. 17-17B).

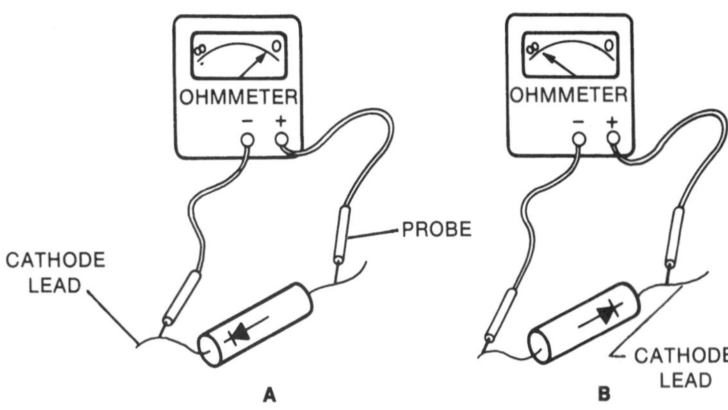

Fig. 17-17. Testing a diode with an ohmmeter.

1. **Draw and label the following three waveforms: a) half-wave rectified, b) dual-diode, full-wave rectified, and c) four-diode, full-wave rectified.**
2. **Determine the PIV and PRV ratings of five diodes. Consult a library for specification manuals.**
3. **Explain what causes a shorted or open diode.**

To test a diode with an ohmmeter, connect the meter to the diode. Then note the resistance of the diode. Next, reverse the connections of the ohmmeter to the diode. If the resistance of the diode is much greater in one direction than in the other, you can assume that the diode is in good condition. If the test shows direct continuity or the same amount of resistance in both directions, the diode is defective and must be replaced. Some meters have a diode testing setting that includes an audio indication of continuity.

TRANSISTORS

Transistors have changed how electronic products look and work. Portable disc players or televisions would not be possible without transistors. **Transistors** are semiconductor devices used to control current and amplify input voltages or currents (Fig. 17-18). In a transistor radio, for instance, transistors are used to amplify the very small signal existing across the antenna coil. Thus, a much larger signal, strong enough to operate the radio's loudspeaker, is created.

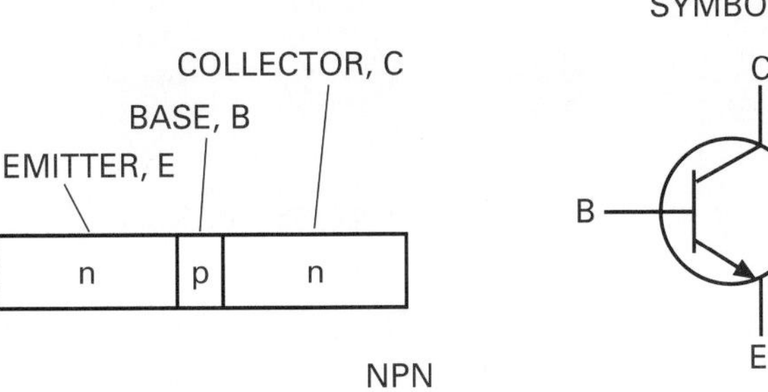

A

Fig. 17-18. **Transistor construction. (A) NPN transistor; (B) PNP transistor; (C) typical construction of germanium alloy PNP power transistor.**

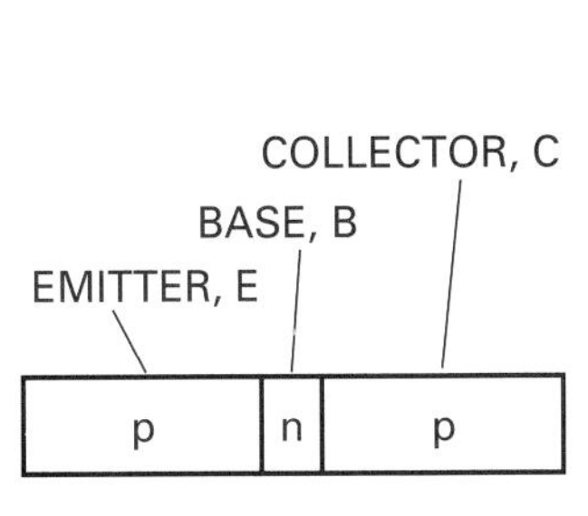

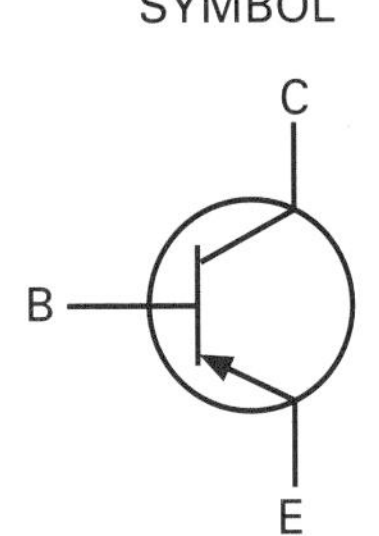

B

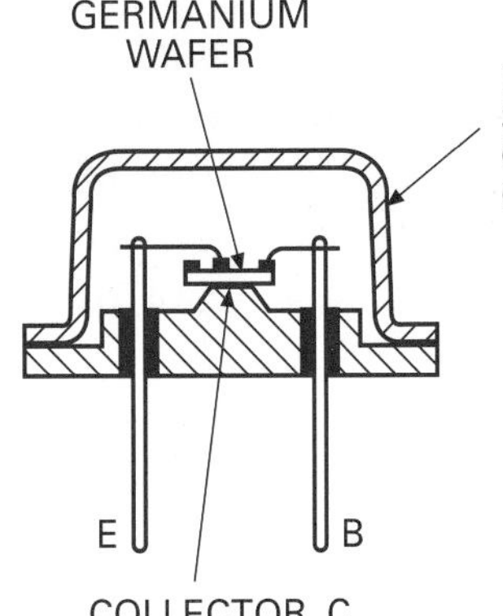

C

FASCINATING FACTS

Three Bell Telephone Laboratories researchers, John Bardeen, William Shockley, and Walter Brattain invented the transistor in 1947. A fellow employee, John Pierce, originally named this accomplishment the transresistor. However, he settled on transistor because it was easier to say. Bardeen, Shockley, and Brattain shared the Nobel prize in 1956 for their research on semiconductors and the discovery of the transistor effect.

Kinds of Transistors

The two most common kinds of transistors are *NPN* and *PNP transistors*. They are often called *bipolar* transistors because their operation depends on the movement of holes and electrons. These transistors are made by combining n- and p-type materials. The materials are arranged as two diodes connected in a "back-to-back" fashion. This arrangement forms three regions called the emitter, the base, and the collector. The **emitter** is the common conductive region to both the base and the collector. The **base** (to emitter) is the input junction and the **collector** (to emitter) is the output junction. These regions are identified by the symbols E, B, and C in Fig. 17-18. Transistor regions are joined to wire leads or pins, which connect the transistor to the circuit (Fig. 17-18C).

Transistors enclosed in metal cases often have a fourth lead known as the shield lead. This lead is internally attached to the case and connected to a common point in a circuit. The metal case shields the transistor from nearby electrostatic and magnetic fields.

Symbol Interpretation

There is a convenient way to remember whether the symbol for a junction transistor represents the NPN or the PNP type. Note in which direction the emitter arrow is pointing. If the arrow points away from the base, it can be thought of as "Not Pointing iN." The symbol thus represents an NPN transistor. If the arrow points toward the base, it can be thought of as "Point iN Please." This symbol thus represents a PNP transistor (see Fig. 17-18B).

Identification

Most transistors are identified by a number-letter code, such as 2N, followed by a series of numbers as, for example, 2N104, 2N337, and 2N2556. Other transistors are identified by a series of numbers or by a combination of numbers and a letter, such as 40050, 40404, and 4D20.

Transistor Manuals

Transistor identification codes do not indicate whether the device is the NPN or the PNP

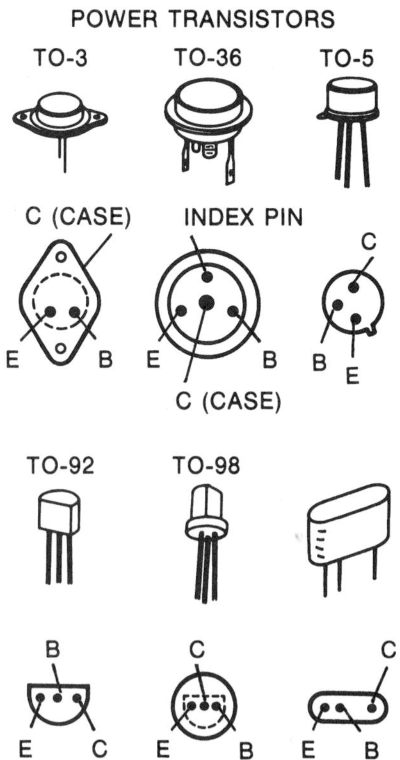

Fig. 17-19. Common NPN and PNP transistor cases and a baseview identification of leads.

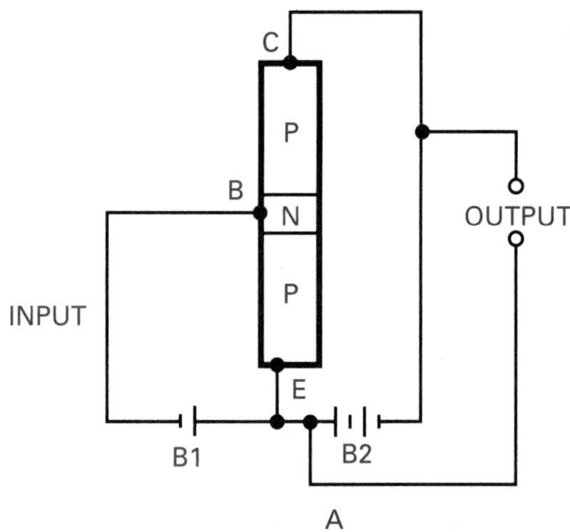

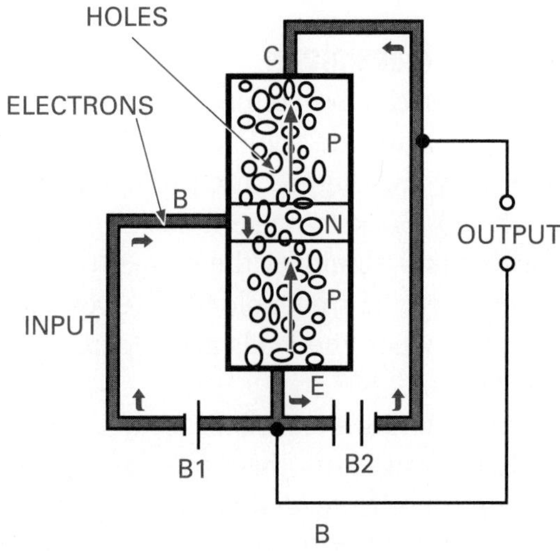

Fig. 17-22. A common-emitter PNP transistor amplifier circuit; (A) circuit diagram; (B) electron flow.

Draw PNP and NPN transistors. Label all their parts and illustrate the biasing direction at the base-emitter junctions.

AMPLIFIER CIRCUITS

A simple one-transistor amplifier circuit is shown in Fig. 17-23. In this circuit, the bias voltage applied to the PNP transistor is supplied by battery B1. Resistor R_1 is a current limiter that controls the base to emitter current. As in any transistor amplifier circuit, the base-collector p-n junction is reverse-biased.

Input Signals

The input signal from an audio-frequency source, such as a microphone, is the input. It is coupled to the base of the transistor through capacitor C_1. As the input-signal voltage varies in value and reverses in polarity, it aids and opposes the forward bias applied to the base. This causes the current in the input circuit to vary directly with the audio signal. The output current varies in the same way.

As the current in the output circuit changes in value, the voltage drop across output resistor R_2 changes in value. The battery, transistor, and R_2 form a series circuit. Therefore, as the output current increases, the voltage drop across resistor R_1 increases. When the current

Fig. 17-23. Schematic diagram of a single transistor, common-emitter amplifier circuit. The transistors are identified by the letter Q.

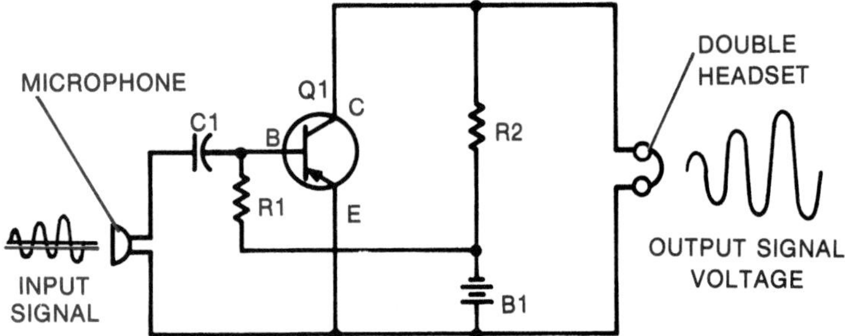

in the output circuit decreases, the voltage drop across R_1 decreases. The output voltage increases. The changes in the output-signal voltage are greater than those to the input-signal voltage. Thus, the input signal has been amplified. The changing output voltage is then applied to the headset. There, it is changed into sound waves.

Operating Points

The input signal voltage in Fig. 17-23 is ac, and, therefore, it reverses the bias of the base-emitter junction during each negative half of the sine wave. Since the emitter-base junction is like a diode, it will only conduct in one direction. The input is rectified by this junction and half of the signal is lost. To eliminate this problem, a steady current is passed through the emitter-base circuit. The level of this current is determined by the value of R_1 and the voltage source at B1 (Fig. 17-23). The value of R_1 determines the base current level so the negative half cycle of the incoming signal is less than the operating point. Then no rectification occurs and the entire input signal voltage is amplified, as shown by the output signal voltage (Fig. 17-23). Operating points are also called the quiescent or Q points.

WORKING WITH TRANSISTORS

Transistors can be damaged. Too much heat can permanently damage the crystal-lattice structure of the material. Great effort has been made by engineers to cool electronic devices during their operation. Service people also need to be cautious when soldering transistor circuits.

Heat Sinks

Transistors that conduct large amounts of current are often mounted on *heat sinks* to keep them from overheating (Fig. 17-24). Not only does heat damage semiconductors, but semiconductor resistance changes in direct proportion to its temperature. Fluctuations in transistor temperatures can also lower the reliability of an electronic device if temperature is not carefully controlled. A heat sink absorbs heat from a transistor and dissipates it, or throws it off, more quickly than the transistor itself can. This lets the transistor work at higher currents and lower temperatures.

Connecting Transistors

Transistors can be connected in two ways. Either their leads are soldered to circuit terminals, or they are plugged into transistor sockets (Fig. 17-25). Sockets make it easier to replace transistors and eliminate the danger of a transistor overheating from soldering. Overheating can happen when the leads are soldered to the circuit.

When soldering a transistor lead, it is best to use a soldering iron that does not produce more heat than is needed to do the job. A soldering iron rated at 30 to 50 W is usually hot enough when using a lead-alloy solder. A heat sink should always be attached to a transistor lead that is being soldered. The jaws of long-nose pliers or other tool, such as a hemostat, can be used as a heat sink (Fig. 17-26). A tran-

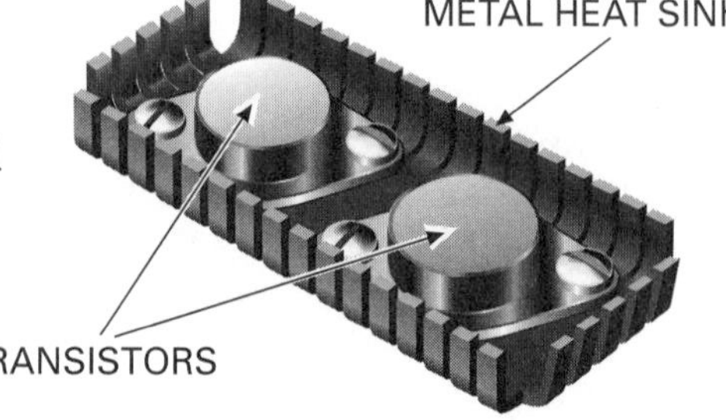

Fig. 17-24. Power transistors mounted on a heat sink.

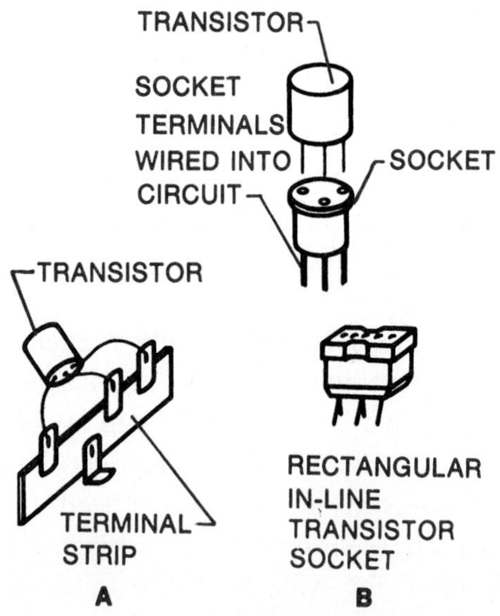

Fig. 17-25. Methods of connecting transistors: (A) leads soldered to circuit; (B) transistor plugged into socket.

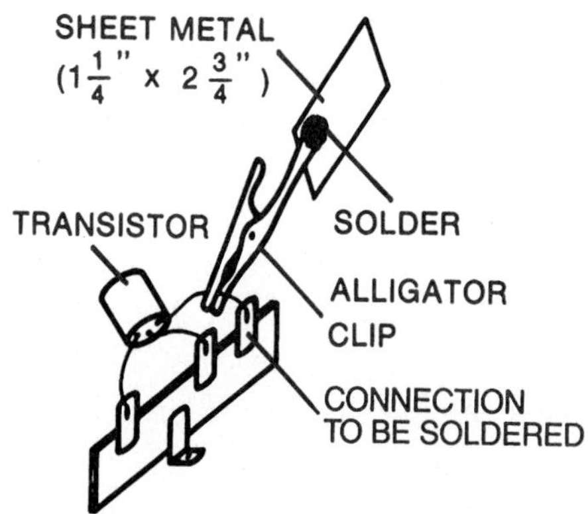

Fig. 17-26. Using a simple heat sink made of a small piece of sheet metal and an alligator clip.

sistor should always be removed when a socket terminal is being soldered or unsoldered.

Bias Voltages

Never put a transistor into a circuit until you are sure the right value of bias voltage will be applied to its terminals. As the base-emitter bias voltage in either NPN or PNP transistors moves toward the collector voltage supply, the transistor is turned on. As the base voltage moves away from the collector voltage, the transistor is turned off. A base-emitter voltage, for example, is established usually midway between the cut-off and maximum region of operation. For amplifier operation, the base-emitter dc bias is approximately 0.3 V for germanium transistors and 0.7 V for silicon transistors. Any excessive voltage applied to the p-n junction of a transistor will cause the transistor to conduct more current than it can safely handle. A transistor should never be taken out of or put into a live circuit because damaging surges of current can then pass through the transistor.

1. **Why does the base-emitter junction conduct in only one direction?**
2. **Choose one of the three transistors chosen in the previous Brain Booster. Determine which reverse polarity voltage will destroy it. Use the manufacturer's specifications.**

Correct Polarities

Transistors can be damaged if high enough reverse bias voltages are applied to them. The danger of such damage is much less with common-emitter circuits. Nevertheless, it is always a good idea to check carefully to see that any transistor is properly connected before applying voltage to the circuit. This is particularly true with circuits that are being used for the first time.

TESTING TRANSISTORS

Transistors are often defective because they have been overheated. Overheating is caused by too much current flowing through the base-to-emitter diode section, the base-to-collector diode section, both these sections, or from emitter to collector. As with diodes, overheating seriously disrupts the crystal-lattice structure. This may result in either an open or a shorted transistor.

The junctions of transistors can be tested with an ohmmeter. If the transistor being tested is wired into a circuit, it should be isolated from the circuit by disconnecting its base lead. With all the source voltages disconnected, the ohmmeter, adjusted to a low or medium range, is connected between the base and the emitter. The resistance reading is noted. The ohmmeter leads to the base and to the emitter are then reversed. If the resistance between the base and the emitter is significantly higher in one direction than in the other, the base-emitter diode section of the transistor can be assumed to be good (Fig. 17-27). To test the base-collector diode section, the same procedure is followed with the ohmmeter connected to the base and to the collector. As a final test, the ohmmeter is connected between the emitter and the collector with the source removed and one leg disconnected.

SAFETY TIP

Always disconnect all voltage sources when making ohmmeter tests.

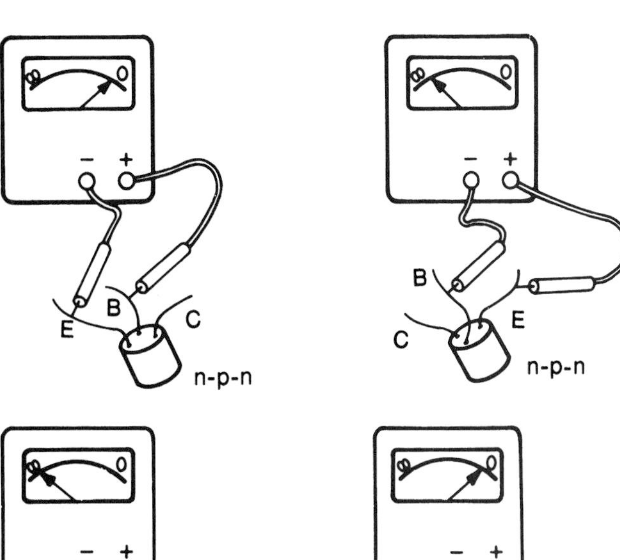

Fig. 17-27. Testing NPN and PNP transistors with an ohmmeter.

If any of these tests show direct continuity, the transistor probably has a short circuit. If any of the tests show infinite resistance, the transistor is open. If there is some doubt about the accuracy of the ohmmeter test, recheck the transistor against a new, identical transistor.

SAFETY TIP

Some ohmmeters on the R x 1 range, can supply enough voltage to damage some transistors. Also, some ohmmeters supply enough voltage on the highest range to break down transistor junctions.

Some meters have a transistor testing setting. At this setting, the meter reads the actual voltage across a junction while the meter is on that scale setting. The ohmmeter battery is used as the source so the circuit source must still be disconnected. See any meter manual that has this option for further instruction. Transistor testers are also available from most electronic testing equipment suppliers.

Explain why a transistor must have all three junctions tested during a performance analysis.

Chapter *Review*

Summary

- Pure semiconductor material is neither a good nor a poor conductor. Resistance of the semiconductor can be altered by adding small amounts of other materials, a process called doping. The doping creates surplus electrons or holes. These electrons and holes move through the semiconductor material, causing it to conduct electricity.
- Germanium and silicon are the most frequently used semiconductor materials. The most commonly used doping materials are arsenic and indium. Arsenic creates an n-type material that has excess free electrons, and indium creates a p-type material that has excess holes.
- The basis for all semiconductor devices is the combination of n-type and p-type materials. Excess holes and electrons flow across the junctions of these combinations at different rates, depending upon the bias voltage applied to them. The amount of bias voltage applied determines the amount of current that flows through the device.
- The most common and basic semiconductor device is the diode. Diodes are a p-n junction through which current will flow only in one direction.
- Diode configurations are used to build many types of electronic devices, such as zeners, TRIACs, DIACs, and transistors.

Review Main Ideas

Review this chapter's main ideas by writing, on a separate sheet of paper, the word or words that most correctly complete the following statements:

1. ______ is a process that adds small amounts of other substances to pure semiconductor materials.
2. The most widely used semiconductor materials are ______ and ______.
3. Atoms of germanium and silicon semiconductors have ______ valence electrons in the outermost shell.
4. In a ______, an atom shares each of its valence electrons with nearby atoms.
5. An n-type semiconductor contains many ______ electrons.
6. The void in a crystal-lattice structure caused by doping is called a ______.
7. A hole is considered a ______ charged current carrier.
8. A ______ semiconductor material contains many holes.
9. The two electrodes of a typical semiconductor diode are called the ______ and the ______.
10. A diode is ______-biased when the positive terminal of a voltage source is connected to its anode and the negative terminal is connected to its cathode.

11. The maximum reverse-bias voltage that can be applied to a rectifier diode is its ______ rating.
12. N-p-n and p-n-p transistors are often called ______ transistors because their operation depends on the movement of holes and electrons.
13. The three regions of an n-p-n or p-n-p transistor are the ______, the ______, and the ______.
14. The arrow on the emitter of a p-n-p transistor symbol points ______ the base.
15. A transistor amplifier circuit in which the emitter is common to both the input and output circuits is called a ______ circuit.

Apply Your Knowledge

1. Write a brief essay describing how doping affects silicon and germanium.
2. Describe the operation of a common solid-state diode.
3. Explain the importance of heat sinks.
4. Draw a bridge rectifier.
5. Discuss how a transistor amplifies an electronic signal.

Make Connections

1. **Communication Skills.** Write a report comparing and contrasting the use of 1, 2, and 4 diodes to change ac to dc.
2. **Mathematics.** Obtain a set of specifications for a rectifier diode and for standard heat sinks. Using the data from these manuals, calculate and graph the current limit increase for the diode with three sizes of heat sinks added to the diode.
3. **Science.** Research the chemistry used to change pure silicon to semiconductor-grade N and P material. Report your findings to the class.
4. **Workplace Skills.** Design and construct a circuit to test selected diodes for function and quality at designed characteristics.
5. **Social Studies.** Research the effect the semiconductor has had on society through changes in communication. Report your findings to your class.

Chapter 18

Integrated Circuits and Special Devices

Terms to Study

gate

integrated circuit

modulated signal

operational amplifier

optoelectronics

Peltier Effect

photodiode

pyrometer

silicon-controlled rectifier

thermistor

thermocouple

thyristor

OBJECTIVES

After completing this chapter, you will be able to:

- Explain how silicon-controlled rectifiers operate.
- Describe how monolithic integrated circuits are built.
- Compare the characteristics of TTL-type integrated circuits to MOS-type integrated circuit.
- List five applications of optoelectronics.

You have probably used electronic equipment that makes entertainment, education, or employment easier. Many of the parts discussed in this chapter are found in common electronic equipment. Many of them are used everyday. This chapter gives insight into how electronic parts work.

DETECTOR DIODES

In addition to semiconductor silicon diodes and junction transistors, there are other solid-state rectifying, current-controlling, light-detecting, and amplifying devices. These include special forms of rectifiers, regulators, field-effect transistors, and photocells. Another important solid-state product is the integrated circuit, or IC. An integrated circuit contains a large number of components reduced to tiny chips, or wafers, of certain semiconductor materials.

Detector diodes are semiconductor devices used in radio and television circuits to produce a rectifying action known as *audio detection*, or *demodulation* (Fig. 18-1). This separates audio signals (voice or music) from high-frequency carrier signals. Carrier signals are used in radio, television, and other communication systems. **Modulated signals** vary according to voices or music. Audio signals must be modulated so that a carrier signal can transport its information. These signals are mixed with or placed on the carrier signal at a broadcasting station.

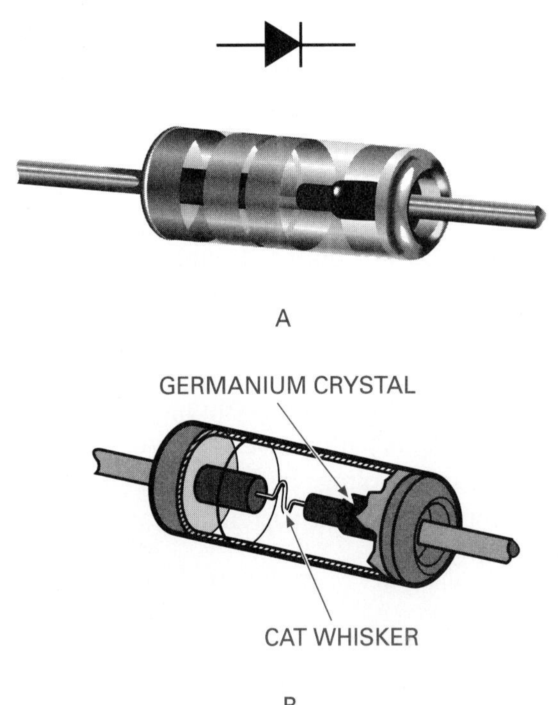

Fig. 18-1. Germanium "crystal" detector diode: (A) typical diode; (B) interior construction.

A basic amplitude-modulated (AM) detector circuit is shown in Fig. 18-2. In this circuit, the modulated carrier signal is applied to detector diode DI. This diode acts as a rectifier. As a result, a pulsating direct current passes through resistor R_1. The value of this current varies according to the variations of the modulated carrier signal. This causes a voltage that represents the audio-modulating signal to be present across the resistor and the output of the detector circuit. The capacitor, typically 100 pF, provides a low impedance to the higher carrier frequency and a high impedance to the lower frequency of the audio signal; therefore, the high frequency signal is separated from the audio signal by the capacitor.

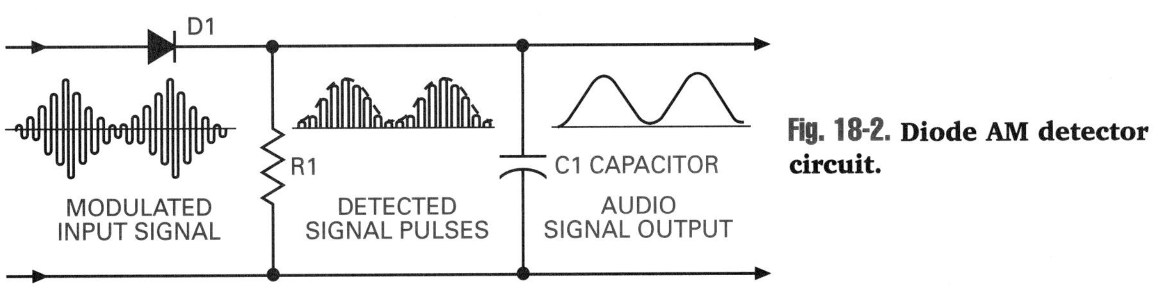

Fig. 18-2. Diode AM detector circuit.

REGULATORS

Line voltage to power supplies often fluctuates and loads do not operate at steady current demands. Therefore, power supply output voltages must be controlled, or regulated. This can be achieved with zener diodes (Fig. 18-3A).

Zener diodes control the output voltage of a power supply by operating in the reverse-biased mode. This occurs when the power supply's output voltage crosses over the zener's reverse-bias breakdown voltage. Because a zener's resistance decreases as current increases, the output voltage across the zener remains constant as the load current or input voltage fluctuates.

1. Define, in detail, a carrier frequency.

2. Draw and label the symbol for a zener diode. Indicate the direction of current flow.

3. Explain how an SCR is triggered.

THYRISTORS

The silicon-controlled rectifier (SCR) is one of a group of devices commonly called thyristors. **Thyristors** are used in switching or current-control circuits in which a trigger voltage is applied to their control electrode. Other thyristors are called TRIACs (*triode ac semiconductor*) and DIACs (*diode ac semiconductors*).

Silicon-Controlled Rectifiers

Silicon-controlled rectifiers (SCR) are four-layer semiconductor devices equipped with three external connections. These are the *anode, cathode*, and a control electrode called the gate (Fig. 18-3B). Like silicon diodes, silicon-controlled rectifiers conduct current in only one direction, from cathode to anode. However, applying a trigger voltage to the gate controls the exact time conduction begins.

An example of a silicon-controlled rectifier circuit is shown in Fig. 18-4. In this circuit, the amount of control voltage applied to the gate can be varied by adjusting variable resistor R . As the control voltage is varied, the silicon-controlled rectifier can be made to conduct early or late during the positive alternation. Early conduction produces the largest load current. Late conduction gives less load current. Because of this current-control feature, silicon controlled rectifiers are used in lamp dimming circuits and in speed controllers for motors.

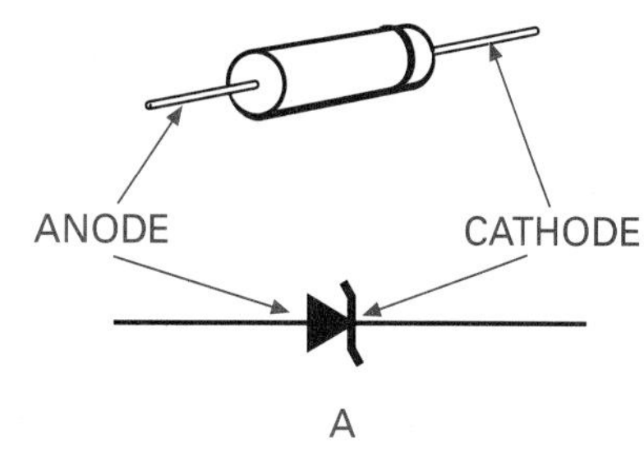

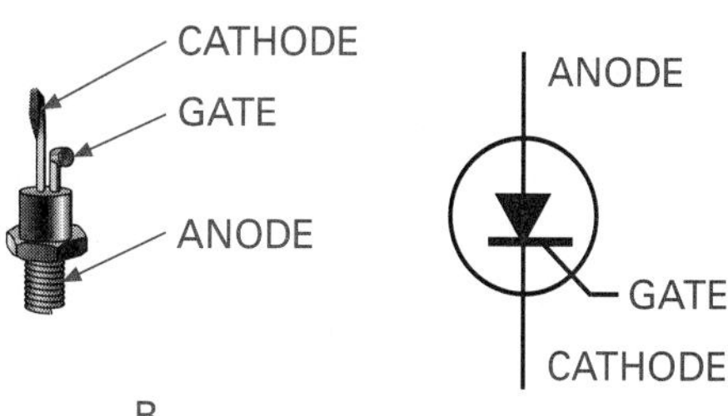

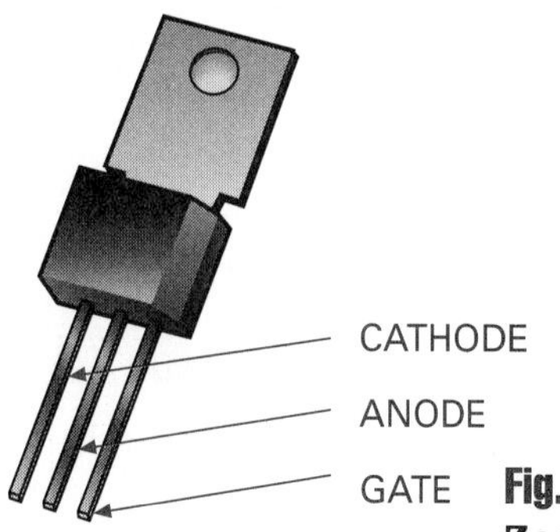

Fig. 18-3. Silicon controlled rectifiers: (A) Zener diode and symbol; (B) stud-mounted and symbol; and (C) plastic-case.

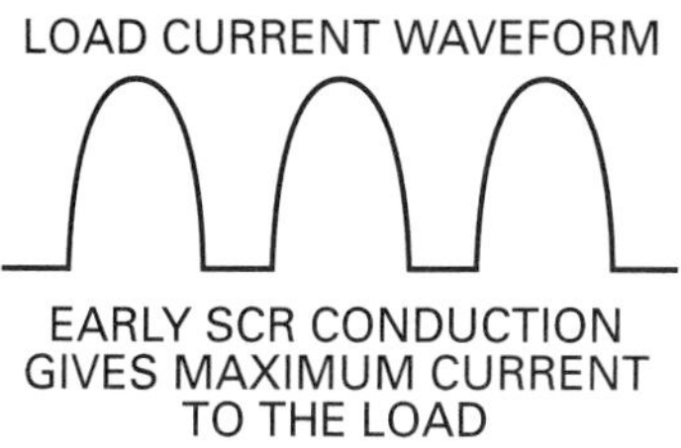

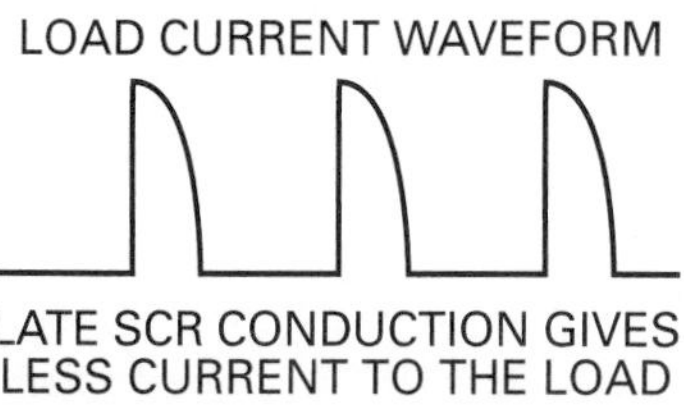

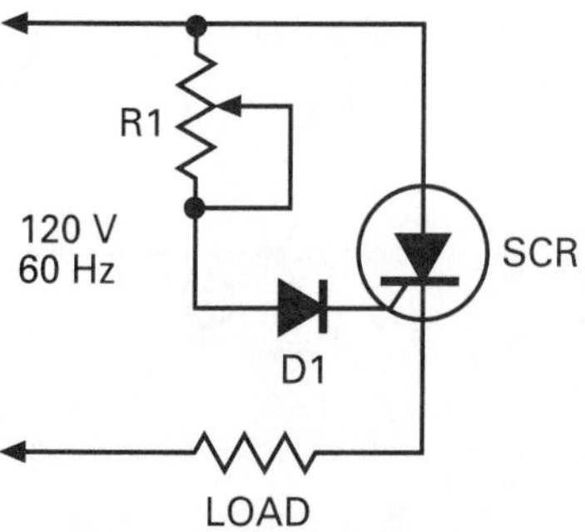

Fig. 18-4. **Simple silicon controlled rectifier current-control circuit.**

DIACs and TRIACs

TRIACs conduct ac current in both directions. They are triggered in the same way as an SCR, but can be used to switch an ac current. Because TRIACs are less sensitive to triggering current in one biasing direction, DIACs, which are doped equally in both directions, are used to gate them. When used in conjunction with a variable resistor and a capacitor, a DIAC/TRIAC combination can be used to control motor speed, heater temperature, or lighting (Fig. 18-5).

FIELD-EFFECT TRANSISTORS

In one common kind of *field-effect transistor* (FET), the control electrode is the gate. The **gate** is a region of p-type semiconductor material that forms a peninsula within a block of n-type semiconductor material. Two other electrodes, the *source* and the *drain*, are of n-type material (Fig. 18-6A). The symbol is shown in Fig. 18-6B.

When a direct voltage that makes the gate negative with respect to the source is applied to the transistor, an electrostatic field is created

Fig. 18-5. **Common DIAC/TRIAC circuit.**

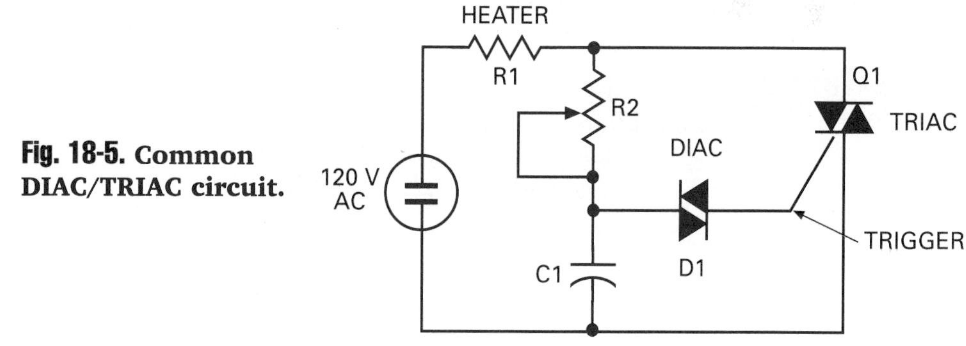

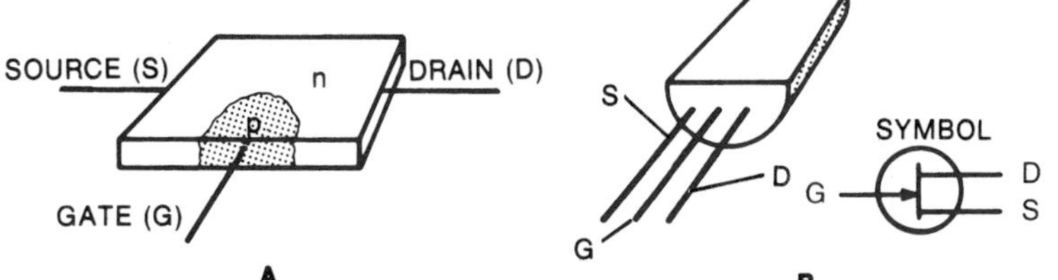

Fig. 18-6. **An n-channel field-effect transistor: (A) basic construction; (B) transistor in TO-92 case.**

about the gate. This field acts on free electrons within the n-type material and limits current flow through the transistor from source to drain. As the voltage between the gate and the source decreases, thereby making the gate less negative with respect to the source, the transistor conducts more current from source to drain.

An amplifier circuit with one field-effect transistor is shown in Fig. 18-7. In this circuit, the input signal causes variations of the voltage applied between the gate and the source. As a result, greater variations of voltage appear across the output circuit. Therefore, the transistor acts as a voltage amplifier. Field-effect transistors are used in several kinds of circuits, including those found in audio amplifiers, electronic timers, and test instruments. Other types of field-effect transistors are junction (JFET) and metal oxide semiconductors (MOSFET).

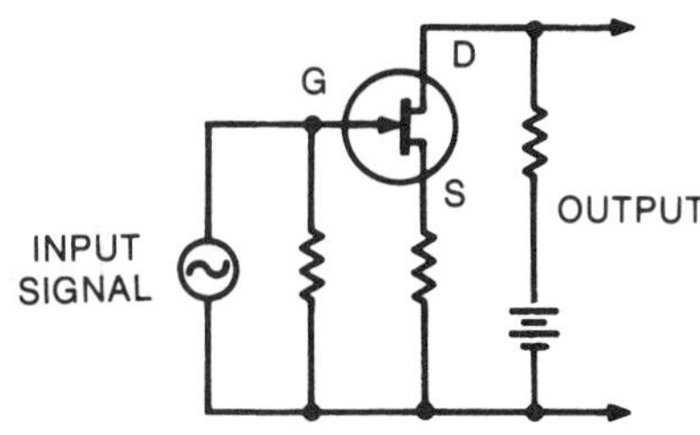

Fig. 18-7. Field-effect-transistor amplifier circuit.

Name the three electrodes of a field-effect transistor.

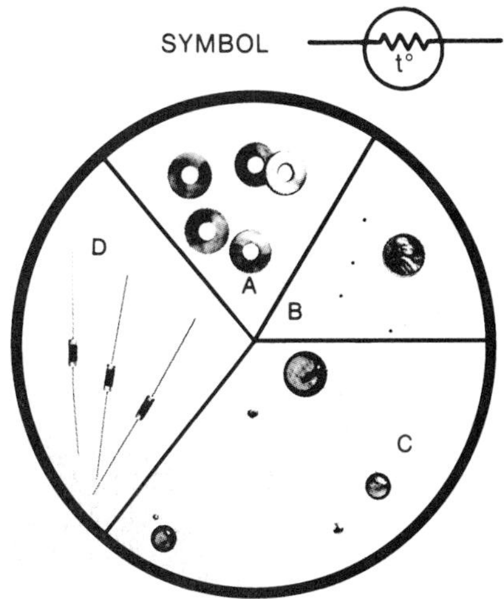

Fig. 18-8. Thermistors: (A) washer; (B) bead; (C) disk; (D) rod.

TEMPERATURE DEVICES

There are many different kinds of electronic temperature controllers available. Although there are also many differences between them, they all require some sort of temperature sensor. Three types of sensors are *thermistors*, *thermocouples*, and *resistance temperature devices* (RTDs).

Thermistors

Thermistors, or thermal (heat) resistors, are devices designed so the resistance decreases when temperature increases. They are made of compounds called *oxides*. Oxides are combinations of oxygen and metals, such as manganese, nickel, and cobalt. Thermistors come in various shapes, some of which are shown in Fig. 18-8.

Because thermistor resistances vary with temperature, they operate as heat-operated resistors. Thus, thermistors can be used as *heat sensors*. These are devices that convert temperature changes into corresponding changes in the value of current in a circuit.

An example of such a circuit used for temperature measurement is shown in Fig. 18-9. The thermistor is connected in series with an ordinary dry cell and an ammeter. As the temperature around the thermistor changes, the value of the current also changes. The meter scale can be calibrated, or marked off, in

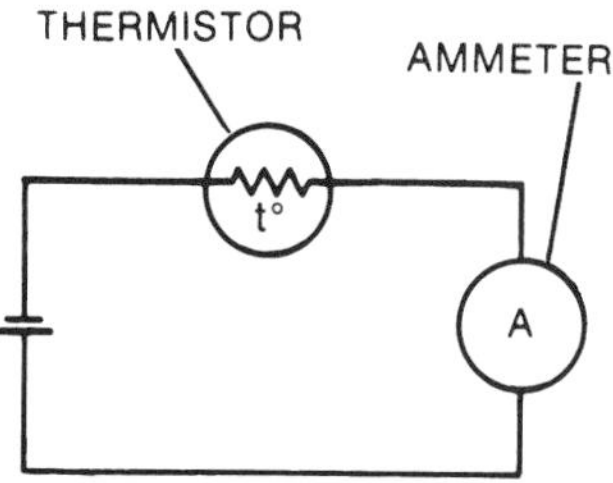

Fig. 18-9. A simple circuit using a thermistor as a heat-sensing or temperature measurement device.

degrees so that a direct temperature reading can be made.

Thermocouples

Thermocouples are solid-state devices used to change heat energy into voltage. They consist of two different metals joined at a junction (Fig. 18-10A). When the junction is heated, the electrons in one of the metals gain enough energy to become free. These electrons then move across the junction and into the other metal. These changes produce a voltage across the terminals of the thermocouple. Several combinations of metals are used to make thermocouples. These metals include iron and constantan, copper and constantan, and antimony and bismuth. A typical thermocouple is shown in Fig. 18-10B.

Pyrometers are meters designed to measure thermocouple voltages. When a thermocouple is in a furnace, and the temperature of the furnace increases, voltage produced by the thermocouple also increases. As a result, more current passes through the meter. The meter then indicates the increase of current as a higher temperature. Temperatures ranging from 2,700° to over 10,800 F (1500° to 6000 C) can be measured very accurately with pyrometers.

Resistance Temperature Devices

Some materials, such as platinum and silver, increase in resistance as their temperature increases. The changes in resistance are directly proportional to temperature changes and are used to measure temperature by a controller.

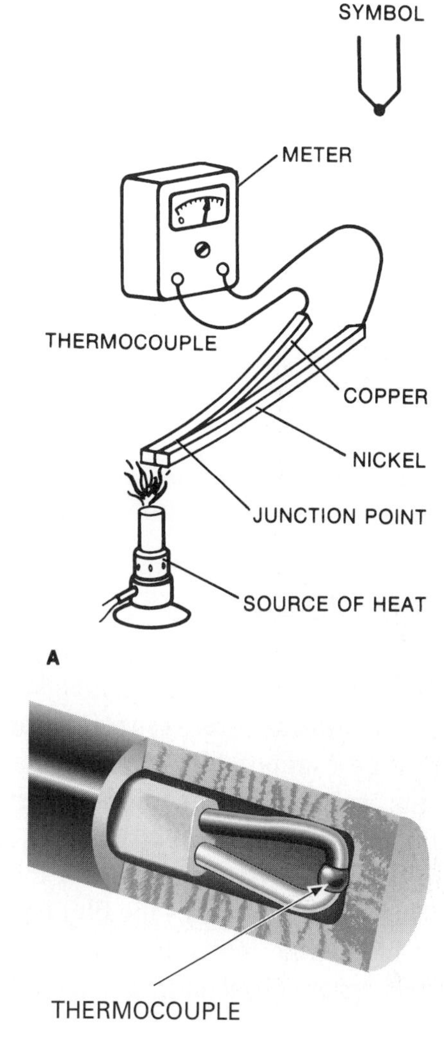

Fig. 18-10. The thermocouple: (A) basic construction and operation; (B) typical thermocouple.

Peltier Effect

If two semiconductor materials are connected in a loop to form two junctions and a voltage is applied, a change in temperature occurs at the junctions. One junction will increase in temperature and one will decrease. The **Peltier Effect** states that the polarity of the current flowing through the junctions will determine which junction will cool and which will heat. Named for Jean Charles Peltier, a French physicist, this principle is used in such applications as cooling individual integrated circuits in confined spaces and delicate electronic equipment.

What does a pyrometer measure?

INTEGRATED CIRCUITS

A new way to build circuits, **integrated circuits** (ICs) contain components such as diodes, transistors, and resistors that are formed into a common block of base material called the substrate. This process forms a *monolithic*, or single, block of material. They can also be formed on the surface of the substrate. This process produces the film-type IC.

In the monolithic integrated circuit, diodes and transistors are formed into the substrate by a diffusion process. This process causes impurities to spread into given areas of the substrate. This creates p-type and n-type regions within the substrate. These regions then make up the sections of the diodes and transistors (Fig. 18-11).

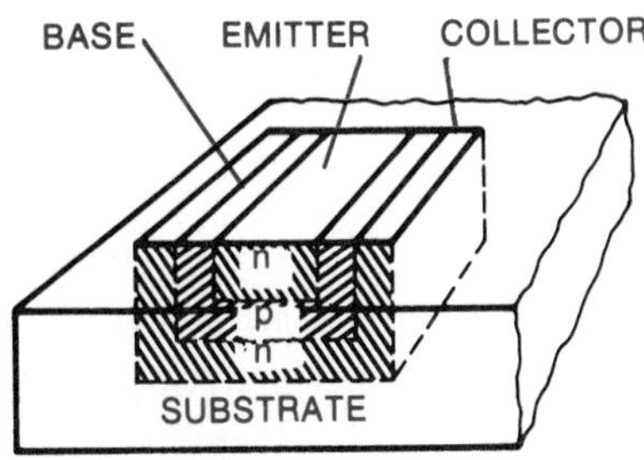

Fig. 18-11. Basic fabrication of a monolithic NPN transistor.

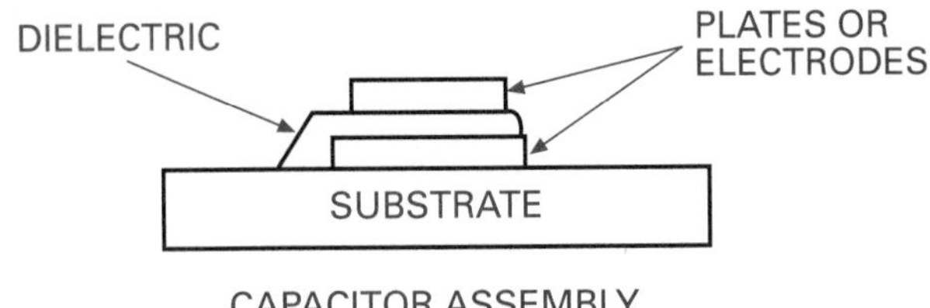

Fig. 18-12. Thick-film integrated circuits.

In the thick- or thin-film integrated circuit, components such as resistors and capacitors are formed by placing or depositing certain materials on the substrate (Fig. 18-12). Hybrid integrated circuits are made of components of both monolithic and the film types.

An integrated-circuit chip is shown in Fig. 18-13. Circuit assemblies of this size or smaller may have dozens of components. For this reason, integrated circuits have the important advantages of being very small and light. This is often referred to as *microminiaturization*.

In its complete form, an integrated-circuit assembly is packaged in some kind of enclosure (Fig. 18-14). Each of these packages may contain the components of a specific circuit such as an amplifier, or a combination of circuits.

FASCINATING FACTS

Jack Kilby (born 1923), an American engineer working for Texas Instruments, Inc., designed the first integrated circuit in 1958. Kilby selected silicon for this circuit because once doped, it creates pathways that electricity can flow through. The development of the integrated circuit greatly increased computing power. It was the first step in continuing miniaturization of all electronic devices.

Fig. 18-13. Integrated circuit.

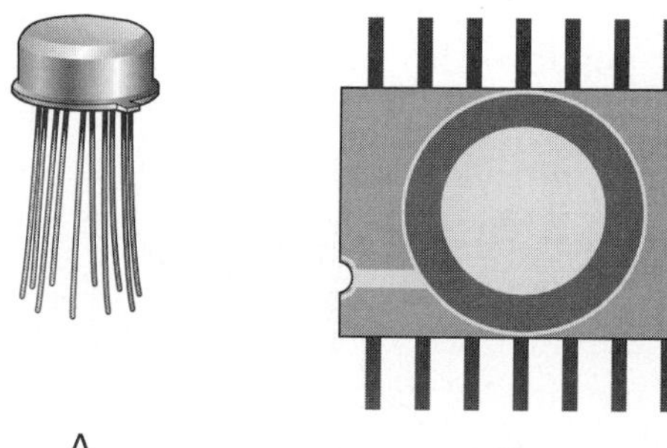

Fig. 18-14. Integrated-circuit packages: (A) TO-5 case; (B) dual inline case; (C) flat-pack.

Types of Integrated Circuits

There are several types, or families, of integrated circuits available today. Some of these types were developed to meet the need for high-speed switching circuits, called *logic gates*, in computers.

Families of integrated circuits include *resistor-transistor logic* (RTL), *diode-transistor logic* (DTL), and *transistor-transistor logic* (TTL). Of these three types, the TTL has been the most useful because of its versatility and use in logic gates. The TTL-type IC is known as the "74 series." Examples of the 74 series integrated circuit are the 7400N and the 74LSO4N. These ICs are used in computers. The functions performed and the descriptions of these and other integrated circuits are available in various manufacturers' specification sheets and catalogs. Because of the method of manufacturing the TTL-type ICs, the supply voltage is critical to its operation. The TTL supply voltage should be maintained between 4.5 and 5.5 dc.

LINKS

Logic gates and digital circuits are discussed in Chapter 24, "Digital Electronic Circuits."

MOS/CMOS

Two other common types, or families, of integrated circuits are the *metallic-oxide semiconductor* (MOS) and the *complementary-metal-oxide semiconductor* (CMOS). These integrated circuits are *unipolar* (in contrast to the *bipolar*) TTL types. Bipolar TTL transistors have two charge carriers (holes and electrons). MOS transistors are manufactured either as p-types with positively charged carriers (holes) or as n-types with negatively charged carriers (electrons). Since only one type of charge carrier is used, these are called *unipolar*.

Figure 18-15 illustrates the graphic symbols and connections of one MOS transistor. A very thin piece of n-type silicon has two p-type regions diffused into forming two p-type junctions in the n-type substrate. In the MOS transistor, one of the p-type regions is called the source; the other is called the *drain*. The source and drain are manufactured with a very small gap between them. Over the gap, an insulating layer of silicon oxide is formed. The gate is the aluminum layer formed on top of the insulating layer. Notice in Fig. 18-15 that there are no direct connections between the gate and the junctions formed by the drain and source. The conductivity between the source and drain of the MOS transistor is controlled by the voltage on the gate.

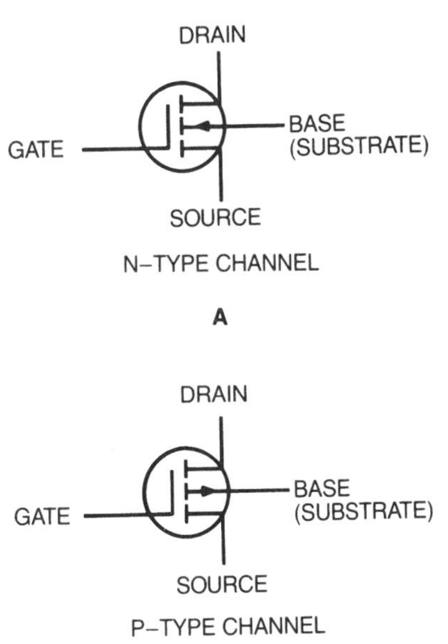

Fig. 18-15. Metallic-oxide semiconductors: (A) n-type channel and graphic symbol; (B) p-type channel with symbol.

When both the p-type channel and n-type channel MOS transistors are manufactured on the same integrated circuit chip, the CMOS family emerges. Some of the advantages of the MOS/CMOS integrated circuit include lower power consumption, smaller size as compared to bipolar transistors, and greater capacity to operate at voltages from 3 to 15 Vdc.

Although the MOS/CMOS integrated circuits have several advantages, they must be handled carefully to prevent damage from electrostatic charges to the internal electrodes. Conducting plastic foam is used to provide an external conducting path for the pins until the IC is soldered in the circuit. Once the IC is soldered, any damaging electrostatic charges are discharged through the resistances in the circuit.

One of the main advantages of MOS and CMOS integrated circuits is that large-scale integration (LSI) can be achieved. This has made possible the manufacture of complex electronic circuits within a single IC package, such as a *microprocessor* (Fig. 18-16). A microprocessor is an LSI circuit that can perform many functions. Microprocessors are used on some automobiles to monitor selected functions: the time of day, current rate of gas consumption, inside and outside temperature, and battery condition. The microprocessor is also used in automobiles to control the air-fuel mixture and engine timing.

LINKS

The microprocessor is discussed more fully in Chapter 25, "Fundamentals of Computers."

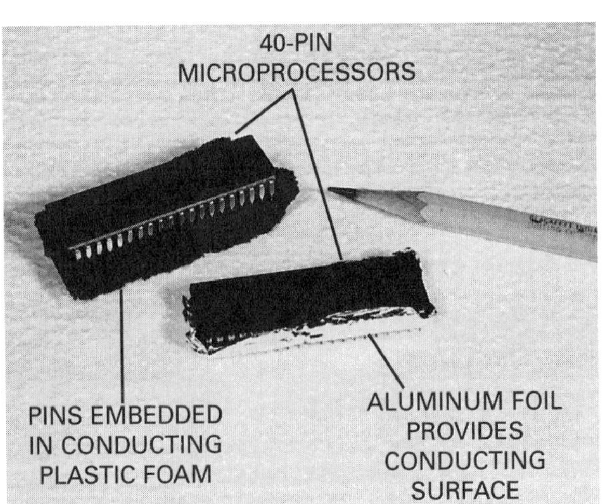

Fig. 18-16. Microprocessors. (The pencil provides comparison for size of integrated-circuit packages).

There are other types of integrated circuit families, and in the future, newer types will no doubt be developed. A more in-depth discussion, however, of integrated circuits is beyond the scope of this text.

Describe the construction of a MOS.

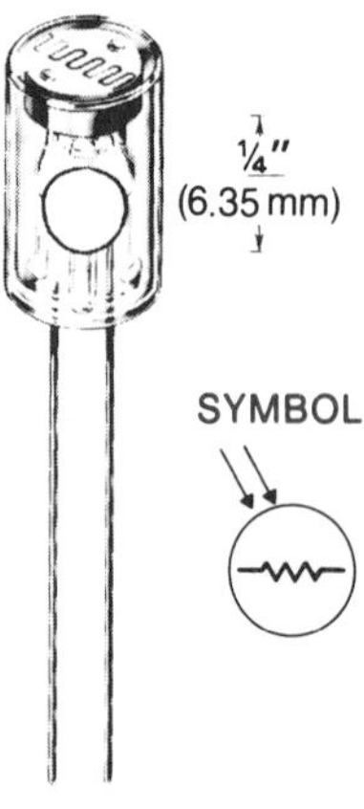

Fig. 18-17. Cadmium sulfide photoconductive cell.

OPTOELECTRONICS

Optoelectronics is a field of study which combines optics (properties of light) and electricity. An *optoelectronic device* is a device that has both optic and electronic parts.

The field of optoelectronics is increasing in importance. Optoelectronic components, for example, are being used in the photographic industry, in infrared remote-control systems, in displays on instrument panels, and in signal transmission. Optoelectronic devices are also used in systems that control production, regulate equipment, and position machine tools for processing materials. The following sections discuss various optoelectronic components.

Photoconductive Cell

A typical cadmium sulfide photoconductive *cell* is shown in Fig. 18-17. Cadmium sulfide is a semiconductor compound consisting of cadmium and sulfur.

As light strikes the active material of the cell, some of the valence electrons in atoms of the material gain enough energy to escape from their parent atoms. They become free electrons. As the intensity of the light increases, more and more free electrons are made available. The resistance between the cell terminals decreases.

How a photoconductive cell is used as a light-sensing device is shown in Fig. 18-18. As light strikes the photocell, its resistance decreases. This allows more current to flow in the circuit. The ammeter reading will always be directly related to the intensity of the light.

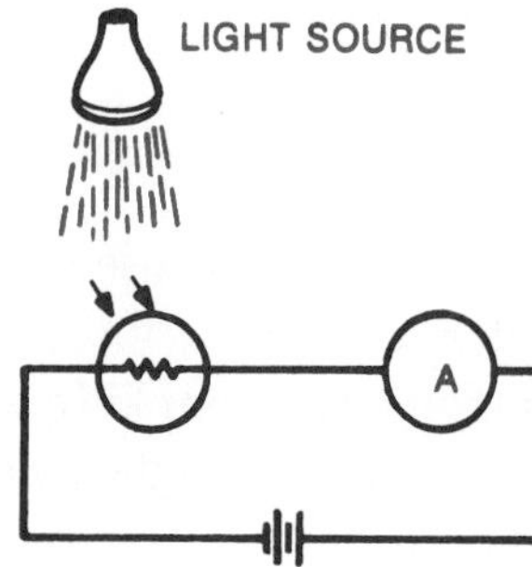

Fig. 18-18. This circuit is an example of how a photoconductive cell can be used as a light-sensing device.

Photoconductive cells are also used in systems that include light sources and relays. In such systems, when the path between the light source and the photocell is interrupted, the current in the relay circuit decreases. The relay, which acts as an on/off switch, is used to control counters, alarm systems, inspection or supervision equipment, and other devices. Photocells are frequently used in lighting circuits to automatically turn on lights when it gets dark.

Light-Emitting Diodes

Light-emitting diodes (LEDs) are commonly found on many electronic devices such as indi-

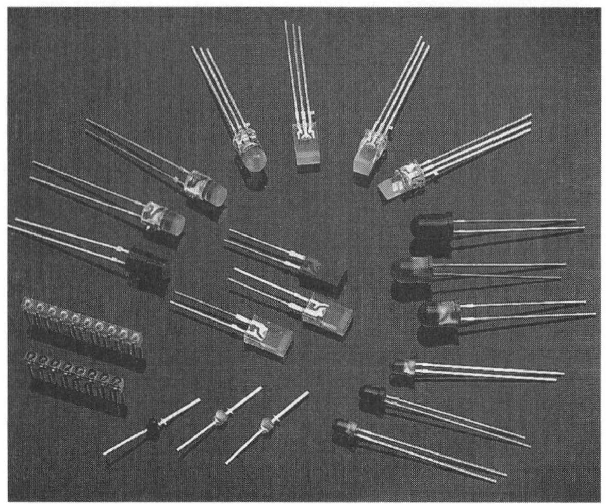

Fig. 18-19. **LED lamps.**

Fig. 18-20. **Infrared emitters.**

cating lamps (Fig. 18-19). They are used, for example, to indicate if a switch is closed or open, or if a voltage is high or low. An LED is a p-n junction that emits or gives out light when the junction is forward biased. The color of the light depends upon the materials used. Some LEDs emit a light that is red, green, yellow, or blue. Others emit a light that is invisible to the eye, such as an *infrared light*. LEDs that can emit invisible light are called *infrared emitters* (Fig. 18-20). Since the light is invisible, these emitters are used in combination with photodetectors in security systems, presence detectors in manufacturing, and remote control devices.

In an LED, the material used that emits a red light is gallium arsenide phosphide (GaAsP). For a green light, gallium phosphide (GaP) is used. The LED is encapsulated, or embedded, where the light is emitted in a plastic package that is shaped into the form of a lens. Single lamps, or arrays, may be used individually or stacked together to form lines of any length. Arrays are used often on meters to indicate which meter scale is being used. Some LEDs have a colorless lens with two-color operation, for example, red and green. These LEDs have two diodes and three leads, one of which is a common cathode.

Typical LEDs use very little electric power. Available ratings for one red miniature LED are power dissipation (85 mW); forward voltage (2.4 Vdc); forward current (35 mA); viewing angle (100); and a dominant red wavelength of 645 nanometers (nm). This wavelength is within visible range. The wavelength normally visible to the human eye is approximately 400 to 700 nm. Specification sheets for every type of LED should be examined carefully so that maximum ratings are not exceeded.

Photodiodes

Photodiodes are semiconductor devices that detect the presence of light (Fig. 18-21). The p-n junction is designed so that light can easily fall upon the light-sensitive, or radiant, area. When light strikes a reversed-biased p-n junction, charge carrier pairs (holes and electrons) are generated. The resulting current is called a *photocurrent* and is proportional to the intensity of the light striking the radiant sensitive surface. A silicon photodiode can detect light when it is reversed-biased. It can also operate as a photovoltaic cell. As a photovoltaic cell, it has a high open-circuit voltage, and, therefore, is used often in camera exposure meters.

Fig. 18-21. Photodiodes.

The junction is often encapsulated in a plastic package with a round, flat glass lens exposing the radiant sensitive area. The types shown in Fig. 18-21 are easily mounted on printed circuit *boards*, and they are compatible with devices using integrated circuits. Photovoltaic cells can be designed to receive visible as well as invisible light. Photodiodes are used in light-controlled regulating equipment, such as twilight switches, fire detectors, perforated tapes, and in signal transmission.

As an example, some of the characteristics of one type of silicon photodiode are maximum power dissipation (250 mW); reverse voltage (7 Vdc); photosensitivity to wavelength of 550 nm (within the visible range); and an open-circuit voltage of 390 mV.

Phototransistors

Phototransistors are similar in operation to photodiodes except they have built-in amplifiers. They are over 100 times more photosensitive than comparable photodiodes. Phototransistors not only respond well to the light of incandescent lamps but are also effective when combined with an infrared-emitting diode. Figure 18-22 illustrates this arrangement. Through carefully controlled manufacturing techniques, the emitter's light is focused in a narrow beam, as in a flashlight. This beam can be pointed at the radiant sensitive area within a phototransistor. The resulting photocurrent is a function of the base diode current and the current amplification factor of the transistor. The infrared emitter and phototransistor combination is used as a beam interrupter in security and counting systems.

Optical Couplers

The integrated circuit devices shown in Figure 18-23 are called *optical couplers*, or *isolators*. An optical coupler has two components in a single IC package. One component is a light source such as a light-emitting diode; the other is a photodetector. Together, they form an optically coupled pair separated by a transparent insulator. The insulator provides very high isolation

Lasers

The photodiode and laser work together to play back information contained on a CD, CD-ROM, or videodisc. As the disc spins, a very narrow laser beam focuses on the series of depressions, or pits, and flat areas, called lands, arranged on the disc in a spiral pattern spaced about one micron apart. (A micron equals one millionth of a meter, or three millions of a foot.) Each pit represents a digital pulse from the original signal. When the laser hits a pit, the light scatters and is not reflected to the photodiode detector. When the laser strikes a land, the light reflects to the photodiode through a semitransparent mirror and lens.

The result is a series of light flashes corresponding to the pulses of the original digitized signal. The photodiode converts these laser light pulses into digitized "on" and "off" electrical signals. It then sends them to other circuits that recreate the original sound, text, or images. These are then sent to a CD player, computer, or television.

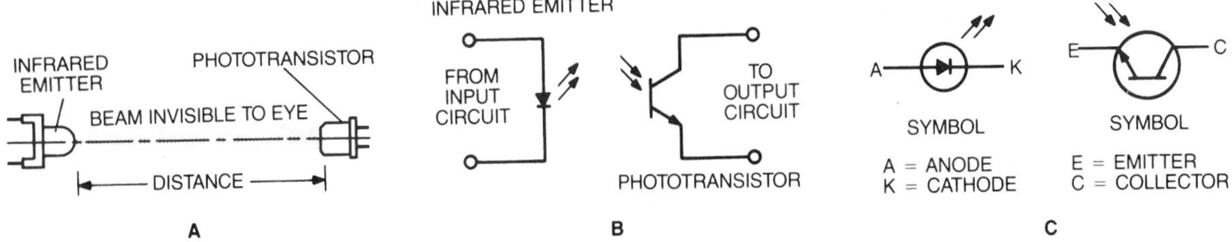

Fig. 18-22. Using an infrared emitter and phototransistor: (A) a narrow infrared beam focused on a phototransistor; (B) circuit diagram; (C) graphic symbols.

Fig. 18-23. Optocouplers.

breakdown voltage, for example, 2,800 to 6,000 V. Because of this, they are especially suited for interfacing, or connecting, circuits that need to be isolated from each other. Connecting a circuit that operates at a high voltage to a circuit that operates at a low voltage is a common way to use these devices. The optical coupler replaces relays and transformers.

The material for the light source (transmitter) is often gallium aresenide (GaAs). The detector (receiver) is generally a phototransistor, but other semiconductors can be used. Figure 18-24 illustrates the application symbol and circuit diagram of one type of optical coupler.

Package and Pin Configurations

Figure 18-25 illustrates the specifications of a dual-channel phototransistor optoisolator. It has an optically coupled pair using gallium arsenide infrared LEDs and silicon NPN phototransistors. The drawings illustrate the standard method of providing detailed information on integrated dual-in-line packages (DIPs). These particular devices provide 100% minimum current transfer ratio, are compatible with the 7400 series IC, have an isolation breakdown voltage of 7,500 V, and have a cou-

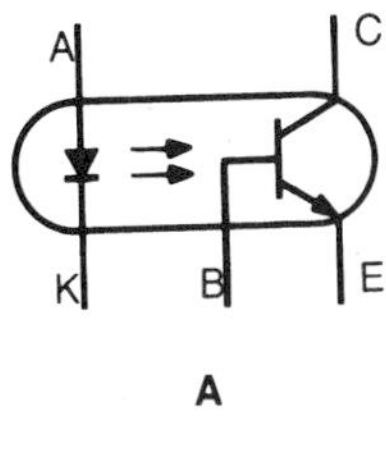

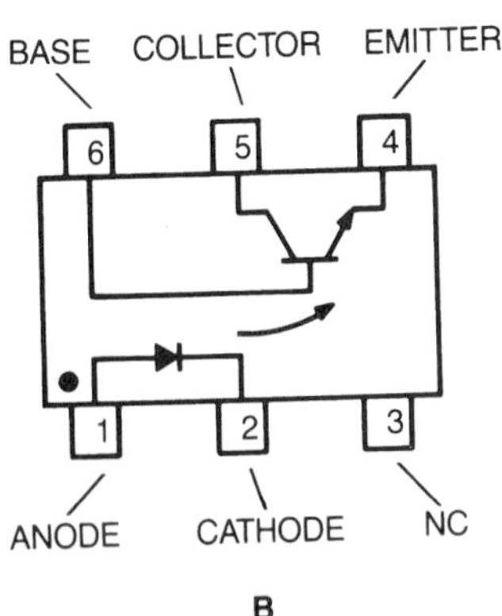

Fig. 18-24. Optocoupler or photon-coupled isolator: (A) symbol; (B) circuit diagram (top view) pin configuration. The dot identifies where pin 1 is located.

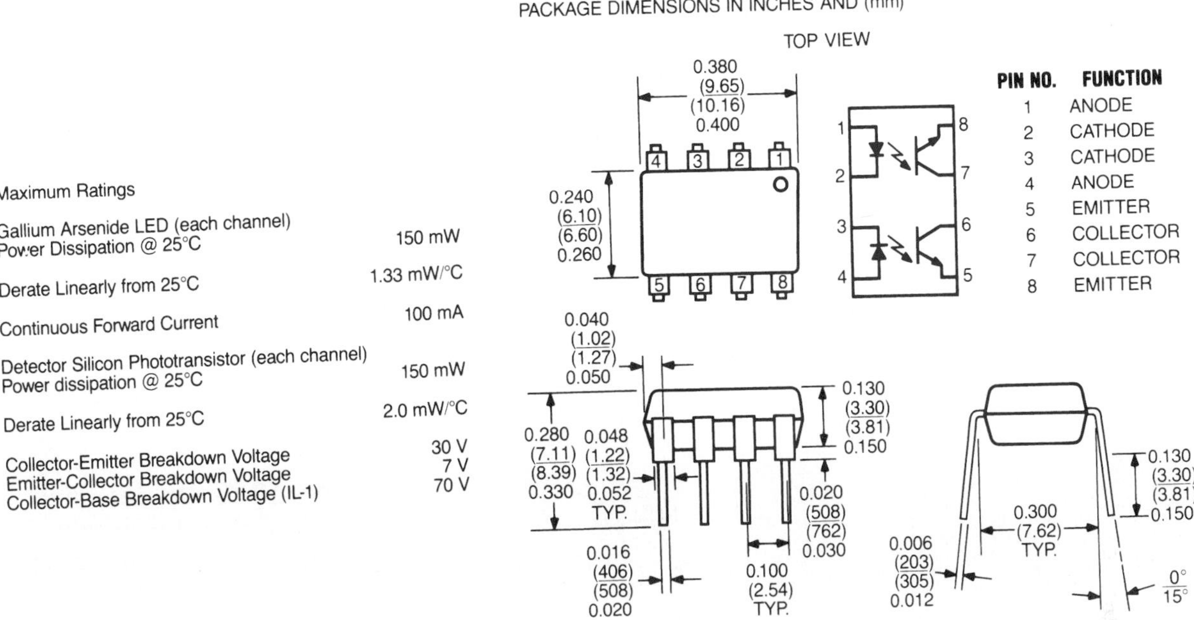

Fig. 18-25. Specifications for the ILD-2 dual-channel phototransistor optoisolator.

pling capacitance of .5 pF. They can be operated in a single channel or dual-circuit arrangement without *cross-talk*, or interference from another channel. The *dot*, or small circle, on the top view identifies pin 1 for both the 8-pin dual and the 16-pin quad package. Other methods, such as a notch, are used to identify the location of pin 1 on a DIP.

Describe how a photoconductive cell operates in a circuit.

OPERATIONAL AMPLIFIERS

Operational amplifiers (op amp) are versatile integrated-circuit devices that can perform mathematical operations, such as addition, subtraction, and integration. Operational amplifiers also function as comparators, oscillators, and linear amplifiers. Op amps are high-gain amplifiers. Open-loop gains of 100,000 to 250,000 are not uncommon. They also have very high input impedance, for example, 100 billion ohms is possible. Op amps are designed for external circuit elements.

Figure 18-26 illustrates an internal view of an op amp. The standard TO-99 package is shown with its dimensions and pin identification (Fig. 18-26B). The pin configuration and functional diagram are shown in Fig. 18-26C. Notice that the amplifier input pins have inverting (pin 2) and noninverting (pin 3) inputs. The output of the single op amp is connected to pin 6. The power supply is connected to pins 7 and 4. Most op amp applications

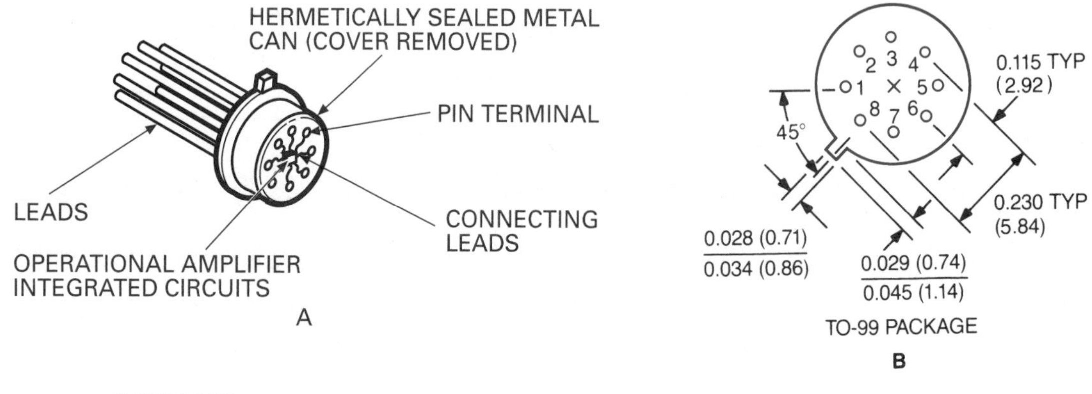

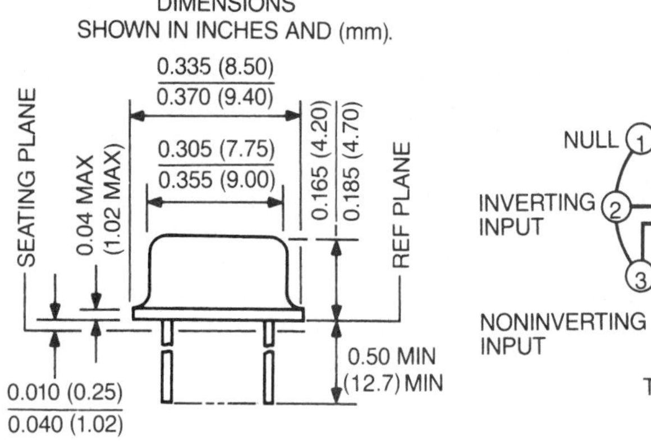

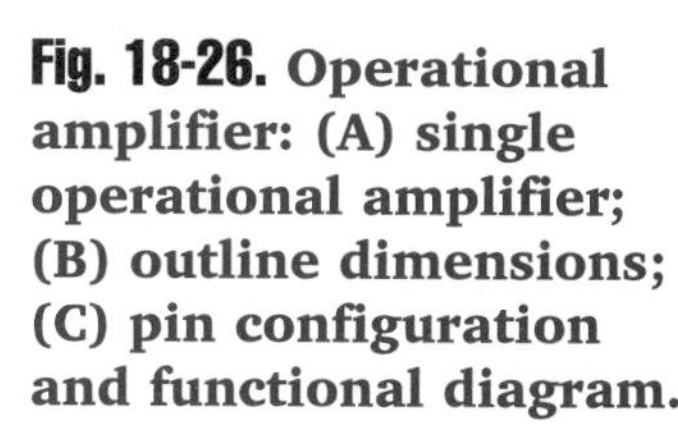

Fig. 18-26. Operational amplifier: (A) single operational amplifier; (B) outline dimensions; (C) pin configuration and functional diagram.

require a dual power supply, ±12 V, for example. The total supply current for an op amp is low, around 2.5 mA. The high open-loop gain ensures high linearity.

Inverting Input

The inverting input is identified with a negative (-) polarity sign. The noninverting input is identified with a positive (+) polarity sign. These polarities should not be confused with the power-supply voltages. If the input signal is connected to the inverting input and the noninverting input is connected to the ground, there is a phase reversal at the output. With this type of circuit arrangement, the operational amplifier functions as an inverting amplifier (Fig. 18-27). Resistor R_3 provides some feedback voltage from the output to the inverting input. This is known as *negative feedback voltage* because the op amp inverts the phase of the input signal. Although negative feedback reduces the gain somewhat, it improves the stability of the op amp's operating characteristics.

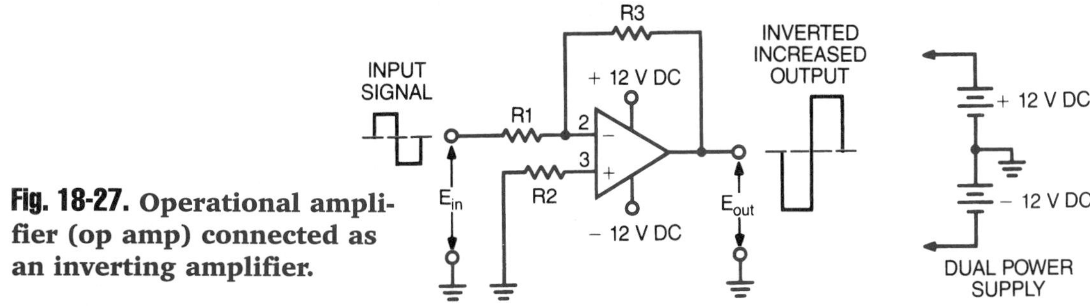

Fig. 18-27. Operational amplifier (op amp) connected as an inverting amplifier.

Noninverting Input

If the input signal is connected to the noninverting input, the output signal changes in phase (in step) with the input signal. When the op amp is connected in this circuit arrangement, the output voltage follows the input voltage.

State the advantage of negative feedback to an op amp.

Chapter *Review*

Summary

- In addition to silicon diodes and junction transistors, there are many other solid-state devices that rectify or control current, control temperature, detect light, or amplify.
- Detector diodes separate modulated audio signals from carrier signals.
- Regulators use special diodes, such as zener diodes, to control the voltage output of a power supply.
- TRIACs, DIACs, and SCRs are in a family of electronic devices called thyrisitors. TRIACs and DIACs are used together to form a gate circuit with which to control ac current. SCRs are also gated to control the amount of current flow in a circuit, usually to a motor.
- Field-effect transistors (FETs) are also turned off and on by a gate current. The source and drain connections serve as conducting electrodes.
- Among the devices used to measure or control temperature are thermistors, thermocouples, and resistance temperature devices.
- Another important solid-state product is the integrated circuit, or IC. An integrated circuit contains a large number of tiny components encased in a chip. Of the three types of ICs, TTL (transistor-transistor logic) is most often used because of its versatility and use in logic gates.
- Other types of semi-conductors are optoelectronics devices and operational amplifiers. Optoelectronics devices can be light-driven or light-emitting. Some examples are photoconductive cells, light-emitting diodes, photodiodes, phototransistors, and optical couplers. Operational amplifiers are commonly used in circuits that increase a voltage or current, such as volume and speed control circuits.

Review Main Ideas

Review this chapter's main ideas by writing, on a separate sheet of paper, the word or words that most correctly complete the following statements:

1. Separating an audio signal from a high-frequency carrier signal is called ______.
2. A detector diode acts as a ______.
3. The electrodes of a silicon controlled rectifier are the ______, the ______, and the ______.
4. Silicon controlled rectifiers are part of a group of devices called ______.
5. The electrodes of a field-effect transistor are the ______, the ______, and the ______.
6. In a field-effect-transistor amplifier circuit, the input signal causes variations of the voltage applied between the ______ and the ______.
7. A thermistor can be used in a circuit to operate as a ______ controlled resistor.

8. A thermocouple is a solid-state device used to change ______ energy into ______.
9. In a monolithic integrated circuit, transistors and diodes are formed into a substrate by a ______ process.
10. Several types, or ______, of integrated circuits are available today.
11. The TTL-type IC supply voltage should be maintained between ______ and ______ dc.
12. A ______ is an LSI circuit that can perform many functions.
13. A field of study that combines optics and electricity is called ______.
14. Photoconductive cells, light-emitting diodes, ______, phototransistors, optical couplers, and ______ belong to a group called optoelectronic devices.
15. A photoconductive cell is used as a ______device.
16. A ______ is a p-n junction that emits or gives out light when the junction is forward biased.
17. The wavelength normally visible to the human eye is approximately ______ to 700 nm.
18. ______ are semiconductor devices that can detect the presence of light.
19. A phototransistor is over ______ times more photo sensitive than a comparable photodiode.

Apply Your Knowledge

1. Draw a schematic diagram of a basic silicon controlled rectifier circuit.
2. Describe the construction of a thermocouple.
3. Describe the differences between a TTL-type and MOS-type integrated circuits.
4. Explain the difference between a light emitting diode and a photodiode.
5. List the advantages of optical couplers.
6. Identify the main characteristics of an operational amplifier.

Make Connections

1. **Communication Skills.** Select one of the devices discussed in this chapter. Report to the class on its use in consumer electronic devices.
2. **Mathematics.** Refer to Fig. 18-2. Assume a standard AM intermediate frequency (455 kHz) and a 400 Hz audio modulating signal. Calculate the capacitive reactance of the capacitor (100 pF) to the two signals. Explain why the two signals split using the results of your calculations as proof.
3. **Science.** Research the fundamental principle that operates in the Peltier Effect. Report your findings to the class.
4. **Technology.** Prepare a report for your class comparing and contrasting the use of visible and IR devices in industrial and commercial devices. Determine why each type of device is appropriate for a specific use.
5. **Social Studies.** Prepare a report for your class on the development of the integrated circuit and how it has affected the consumer electronics industry.

Section 5

Wiring and Soldering

Chapter 19

Wires and Cables

OBJECTIVES

After completing this chapter, you will be able to:

- Calculate the circular mil area of a round wire.
- Identify by name four common types of wire used for electrical and electronic wiring jobs.
- List three common types of communication cables.
- Define the term interface.
- Identify five items that should be included to label and document an assembled communications cable.

Terms to Study

ampacity

audio cable

cable

circular mil

coaxial cable

cord

grounding wire

heater cord

lamp cord

service cord

This chapter discusses the wires and cables that are used as conductors in electrical and electronic circuits. These wires and cables also carry electrical power and information to you at home, school, or work. The next time you are out walking, look up at the wires and cables on the utility poles. Did you know that these cables are for electrical power, telephone, cable television, and other services?

WIRE SIZE

Wires used in electrical and electronics work are usually round, and made of soft, annealed copper. They may be solid or stranded, and bare or insulated (Fig. 19-1A). Stranded wire is made of many smaller wires laid together. It is more flexible than solid wire and, therefore, less likely to break if frequently bent or twisted. Several insulated wires contained within a single covering form a cable (Fig. 19-1B). Some kinds of flexible cables, especially those used with appliances and lamps, are called cords.

An American wire gage (AWG) number tells the size of bare, round wire. This is based on its diameter. The larger the AWG number, the smaller the wire (Fig. 19-2). The American wire gage is also known as the Brown and Sharpe (B&S) wire gage.

Wire Gage

The AWG number and wire diameter can be found with a wire gage (Fig. 19-3). To use this gage, put the uninsulated or stripped wire to be measured into the slot that gives the snuggest fit without binding (Fig. 19-4). Then read the AWG number of the wire at the bottom of the slot on one side of the gage. The diameter of the wire is given on the other side.

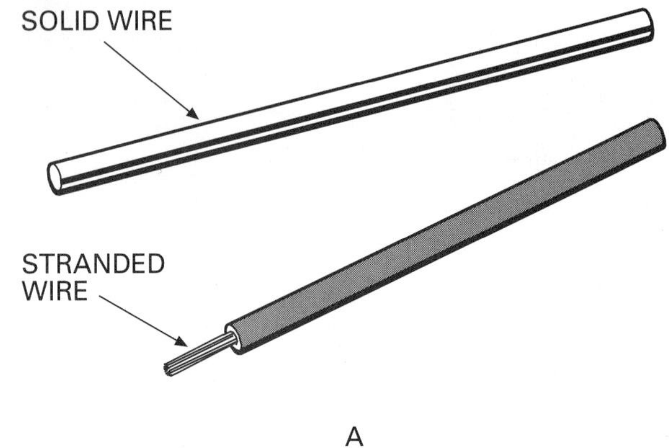

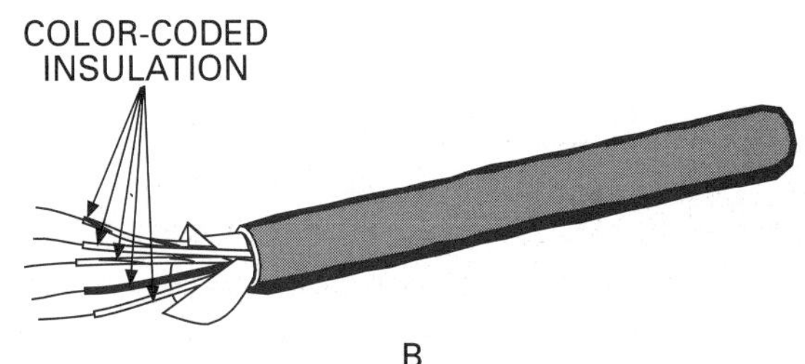

Fig. 19-1. Wires and cable.

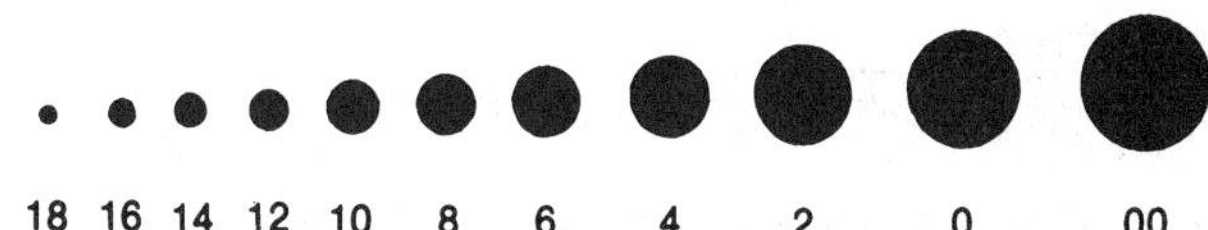

Fig. 19-2. Actual American wire gage sizes.

FASCINATING FACTS

Any material that can be drawn out or hammered thin can be made into wire. Most wires are made of steel, copper, aluminum, or their alloys. Modern techniques have extended the number of materials useful for wire. For example, gold can be drawn into very fine wires for the cross hairs of optical instruments and other uses. One grain of gold (0.002 ounce, or 0.065 gram) can be drawn into a wire 0.00002 inch (0.00005 centimeter) in diameter and more than one mile (1.6 kilometers) long.

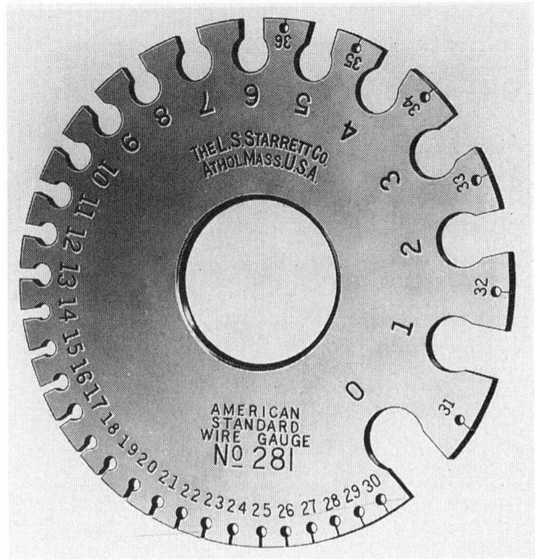

Fig. 19-3. American standard wire gage.

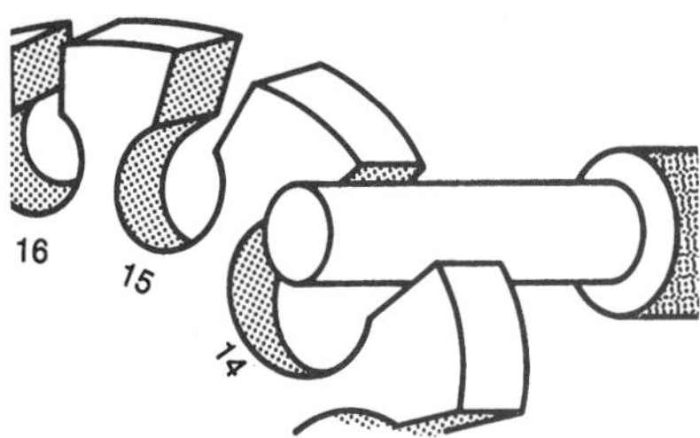

Fig. 19-4. **Using the wire gage.**

Circular Mil Area

The cross sectional area of round wires is often stated in terms of circular mils. One *mil* is one-thousandth (0.001) of an inch (0.0254 mm). One **circular mil** is equal to the area of a circle with a diameter of 0.001 in. (0.0254 mm). To find the circular mil area of a wire, square its diameter, given in mils. Thus, a number 20 AWG wire (closest standard size—0.80 mm), which has a diameter of 0.03196 in (0.81 mm), has a circular mil area of 31.96^2, or about 1,022 circular mils.

SAFE CURRENT-CARRYING CAPACITY

The safe current-carrying capacity of a wire is that amount of current in amperes that the wire will conduct without overheating. The smaller the gauge of a wire, the higher its resistance will be. This means the more current that flows, the greater the power use that will be in a smaller wire. The greater the power usage, the higher the temperature. Therefore, the larger the cross-sectional area of a wire, the greater its current-carrying capacity will be. The safe current-carrying capacity of a wire also depends on its insulating material. The safe current-carrying capacity of a wire is sometimes called the **ampacity** of the wire.

Wire Table

Information about some of the physical and electrical characteristics of wires can be found in a wire table (Table 19-A). Other wire tables give the safe current-carrying capacities of wires insulated with different materials.

Although the largest wire size in Table 19-A is shown as 0000 (also written as 4/0 and pronounced "four-ought"), larger sized wires are also commonly used. Wires larger than 4/0 are generally used in transmission and distribution systems and in building electrical installations. These larger wires are measured in thousands of circular mils, such as 250 and 500 MCM, rather than in the AWG number systems.

COMMON WIRES

Many different kinds of wires are used for electrical and electronic wiring jobs. Each of these is best suited for a particular purpose because of its size, insulation, or composition. Some wires are *tinned*, or covered with a thin coat of solder. This prevents corrosion and makes it easier for one wire to be soldered to another.

Hookup Wire

Hookup wire is used mostly for wiring the components in a circuit. It is either solid or stranded copper wire with a plastic or Teflon insulation. The most common sizes range from number 18 to number 24 AWG (1.00 to 0.50 mm). Some large wiring jobs are hooked up or connected with a wire harness. The wire in the harness is pre-shaped, tied, and wrapped to meet the needs of various applications (Fig. 19-5).

BRAIN BOOSTERS

1. **How many circular mils is a wire with a diameter of 55.00 mm?**
2. **Using Table 19-A, determine the resistance for 4/0 wire.**

Table 19-A. Wire Table for Standard Annealed Copper

AWG (B&S) Gage	Standard Metric Size (mm)	Diameter in mils	Cross-sectional Area		Ohms per 1,000 ft at 20°C (68°F)	Lbs per 1,000 ft	Ft per lb
			circular mils	square inches			
0000	11.8	460.0	211,600	0.1662	0.04901	640.5	1.561
000	10.0	409.6	167,800	0.1318	0.06180	507.9	1.968
00	9.0	364.8	133,100	0.1045	0.07793	402.8	2.482
0	8.0	324.9	105,500	0.08289	0.09827	319.5	3.130
1	7.1	289.3	83,690	0.06573	0.1239	253.3	3.947
2	6.3	257.6	66,370	0.05213	0.1563	200.9	4.977
3	5.6	229.4	52,640	0.04134	0.1970	159.3	6.276
4	5.0	204.3	41,740	0.03278	0.2485	126.4	7.914
5	4.5	181.9	33,100	0.02600	0.3133	100.2	9.980
6	4.0	162.0	26,250	0.02062	0.3951	79.46	12.58
7	3.55	144.3	20,820	0.01635	0.4982	63.02	15.87
8	3.15	128.5	16,510	0.01297	0.6282	49.98	20.01
9	2.80	114.4	13,090	0.01028	0.7921	39.63	25.23
10	2.50	101.9	10,380	0.008155	0.9989	31.43	31.82
11	2.24	90.74	8,234	0.006467	1.260	24.92	40.12
12	2.00	80.81	6,530	0.005129	1.588	19.77	50.59
13	1.80	71.96	5,178	0.004067	2.003	15.68	63.80
14	1.60	64.08	4,107	0.003225	2.525	12.43	80.44
15	1.40	57.07	3,257	0.002558	3.184	9.858	101.4
16	1.25	50.82	2,583	0.002028	4.016	7.818	127.9
17	1.12	45.26	2,048	0.001609	5.064	6.200	161.3
18	1.00	40.30	1,624	0.001276	6.385	4.917	203.4
19	0.90	35.89	1,288	0.001012	8.051	3.899	256.5
20	0.80	31.96	1,022	0.0008023	10.15	3.092	323.4
21	0.71	28.46	810.1	0.0006363	12.80	2.452	407.8
22	0.63	25.35	642.4	0.0005046	16.14	1.945	514.2
23	0.56	22.57	509.5	0.0004002	20.36	1.542	648.4
24	0.50	20.10	404.0	0.0003173	25.67	1.223	817.7
25	0.45	17.90	320.4	0.0002517	32.37	0.9699	1,031.0
26	0.40	15.94	254.1	0.0001996	40.81	0.7692	1,300
27	0.355	14.20	201.5	0.0001583	51.47	0.6100	1,639
28	0.315	12.64	159.8	0.0001255	64.90	0.4837	2,067
29	0.280	11.26	126.7	0.00009953	81.83	0.3836	2,607

Table 19-A. Wire Table for Standard Annealed Copper (Cont'd.)

AWG (B&S) Gage	Standard Metric Size (mm)	Diameter in mils	Cross-sectional Area		Ohms per 1,000 ft at 20°C (68°F)	Lbs per 1,000 ft	Ft per lb
			circular mils	square inches			
30	0.250	10.03	100.5	0.00007894	103.2	0.3042	3,287
31	0.224	8.928	79.70	0.00006260	130.1	0.2413	4,145
32	0.200	7.950	63.21	0.00004964	164.1	0.1913	5,227
33	0.180	7.080	50.13	0.00003937	206.9	0.1517	6,591
34	0.160	6.305	39.75	0.00003122	260.9	0.1203	8,310
35	0.140	5.615	31.52	0.00002476	329.0	0.09542	10,480
36	0.125	5.000	25.00	0.00001964	414.8	0.07568	13,210
37	0.112	4.453	19.83	0.00001557	523.1	0.06001	16,660
38	0.100	3.965	15.72	0.00001235	659.6	0.04759	21,010
39	0.090	3.531	12.47	0.000009793	831.8	0.03774	26,500
40	0.080	3.145	9.888	0.000007766	1049.0	0.02993	33,410

Fig. 19-5. Making a wire harness.

Magnet Wire

Magnet wire is solid wire. It is usually round and is insulated with a thin, tough coating of a varnish-like, plastic compound. This kind of wire is often referred to as *plain enamel (PE) magnet wire*. The most common sizes range from number 14 to number 40 AWG (1.60 to 0.08 mm). It is called magnet wire because it is used for winding the coils in electromagnets, solenoids, transformers, motors, and generators (Fig. 19-6).

Bus Bar Wire

Bus-bar wire is solid, tinned, and uninsulated copper wire. Usual sizes range from number 12 to number 30 AWG (2.00 to 0.25 mm). It is used as a common terminal for several wires (Fig. 19-7).

Fig. 19-6. Magnet wire used in a fan motor.

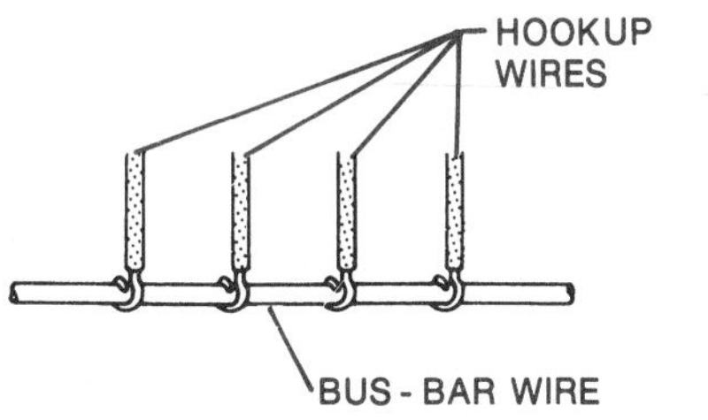

Fig. 19-7. Bus-bar wire used as a common connecting point in a circuit.

Test-Lead Wire

Test-lead wire is a very flexible, stranded wire. It is insulated with rubber or with a combination of rubber and other materials. The most common sizes are number 18 and number 20 AWG (1.00 and 0.80 mm). It is used for making the test leads of instruments such as ohmmeters that endure heavy use.

Antenna Wire

Radio antenna wire is stranded wire that may be either bare or insulated. Each strand is usually made of Copperweld, which is wire with a copper-coated steel core. This gives the wire high tensile strength, keeping it from sagging when a length of it is suspended between supports.

LINKS

For the use of wire in residential wiring, refer to Chapter 28.

Television Lead-In Wire

The most common transmission line used to connect a television to an antenna or to cable service is 75-Ω coaxial cable. **Coaxial cable** has a center conductor of Copperweld wire covered by a polyethylene insulation of a very specific diameter. A braid of copper wire is wrapped around the center insulation and then plastic insulation covers the entire assembly. The braided wire serves as a shield for the high frequency signals commonly carried by this type of cable (See Fig. 19-10C). Coaxial cable provides the correct impedance match between the line and the signal input to the set. The impedance of the wire is established by the gage of the conductors, the insulation used, and the spacing between the conductors.

Another type of antenna lead-in wire is called *twin-lead*, or two-conductor antenna wire. The wires are most often made of Copperweld insulated with polyethylene plastic. They are usually covered by an outer jacket, also made of polyethylene (Fig. 19-8A). The outer jacket allows precise spacing between the wires. This spacing makes the impedance of this wire 300-Ω. Twin-lead wire is also used to connect a frequency modulation (FM) radio receiver with its antenna, and also makes a very effective FM antenna.

Some television receivers have two input connectors. These connectors allow both the 300-Ω and 75-Ω transmission lines to be used when needed. It is important to have good impedance matching between the receiver and the transmission line so that the maximum

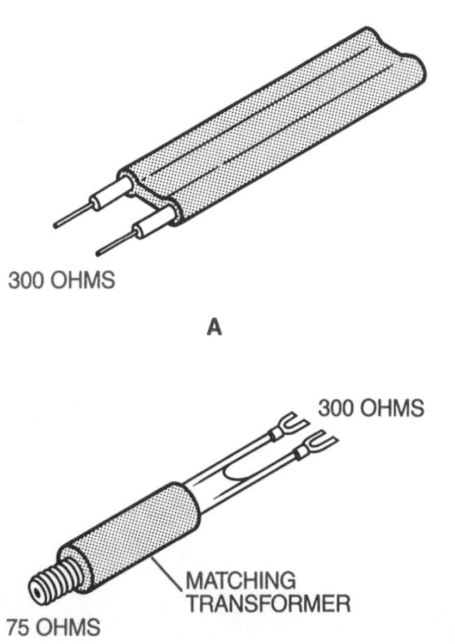

Fig. 19-8. (A) Television twin-lead lead-in wire; (B) indoor-outdoor matching transformer, reversible.

amount of signal energy received from the antenna can be transferred. It is often necessary to connect a 75-Ω transmission line to a 300-Ω input on a television receiver or vice versa, but a *matching transformer* can match the different impedances. A matching transformer has two connections: a cable connector for the 75-Ω coaxial cable and two metal screws for connecting the 300-Ω twin-lead line (Fig. 19-8B).

Grounding Wire

Grounding wire is usually a wire that connects metal cabinets and certain parts of circuits, such as those in a house wiring system, to ground rods. It will be either solid or stranded, and either green or bare.

Romex

Romex is used to wire most homes. It has two copper conductors that are insulated and one, for ground, that is bare. These wires are molded into a flat plastic ribbon that is easy to install in household wiring devices. There are special Romex stripping and connecting tools.

List the wires you need to use if you are wiring a desk lamp, a household range, and a television antenna.

ELECTRICAL CORDS

Cords are flexible and are used to connect electrical and electronic equipment and appliances to outlets. Thus, they are able to withstand the bending and twisting that occur as portable products are used.

Lamp Cord

Lamp cord is made with two parallel, stranded wires. Each wire is separately covered with rubber or plastic, and the two are joined into what looks like a single unit. (Fig. 19-9A). The wires in a lamp cord are usually number 16 or number 18 AWG (1.25 or 1.00 mm). Lamp cord is used with products such as portable lamp fixtures, light-duty appliances, radios, and television sets.

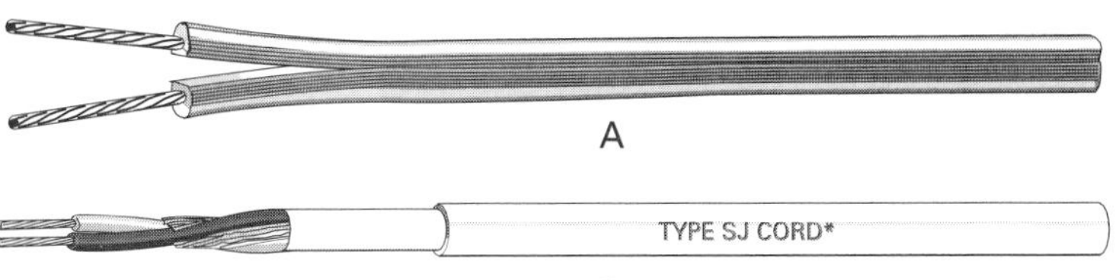

Fig. 19-9. Electrical cords: (A) rubber-insulated lamp cord; (B) service cord.

Service Cords

Service cords have two or three stranded, bare-copper wires. These wires are each insulated and put together in a round outer jacket made of rubber or plastic (Fig. 19-9B). The wires usually range in size from number 10 to number 18 AWG (2.50 to 1.00 mm). The most common service cords are listed by the Underwriters' Laboratories as types S, SV, and SJ, which represent the jacket ratings for voltages and environments. Service cords are also rated by conductor size and number of conductors. For instance, 14/3 cord has three number 14 conductors. These cords are most often used with portable power tools and medium- to heavy-duty appliances. Service cords often have a grounding conductor that has green insulation.

SAFETY TIP

Care must always be taken when working with and disposing of older high-temperature insulation. It is often asbestos and is hazardous to both the environment and to humans.

Heater Cord

Heater cords are insulated with materials that can withstand high temperatures. Such cords are used with heat-producing appliances and with soldering irons and guns. Heater-cord wires are stranded and are usually number 16 or number 18 AWG (1.25 or 1.00 mm).

One common kind of heater cord looks much like ordinary rubber-insulated lamp cord. However, its insulation is made of a thermosetting plastic compound that can withstand high temperatures.

Go to your local library and find, in the National Electrical Code, the ampacity differences between wires and cords.

COMMUNICATION CABLES

A **cable** is a bundle of insulated wires through which electric current can be sent. Cables are most often used for connecting components or products that are some distance

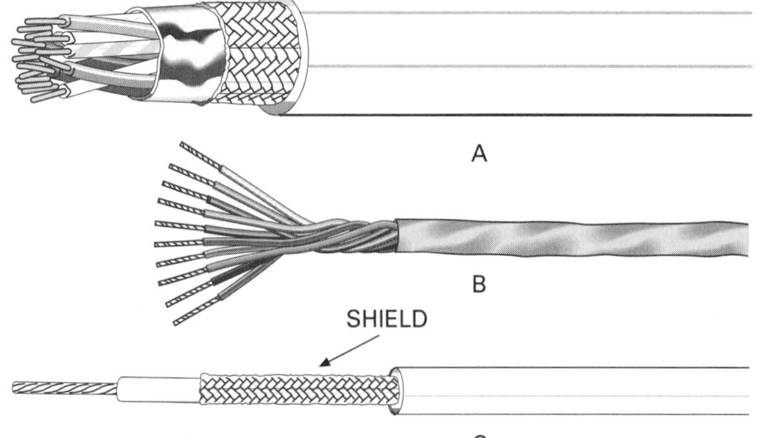

Fig. 19-10. **Communication cables: (A) shielded cable; (B) audio cable; (C) coaxial shielded cable.**

apart. Cables may be unshielded or shielded. A *shielded cable* is made up of one or more insulated wires. These are often contained within a sheath of copper wires woven together to form a braid (Fig. 19-10A). This braid, or shield, protects the inner wire or wires from undesirable external electric or magnetic fields. It is placed directly beneath the outer covering jacket of a shielded cable. It may or may not serve as one conductor of the cable.

Audio Cable

Audio cable is cable that is commonly used in telephone circuits, in intercommunications circuits, and in circuits connecting the loudspeakers of public-address systems to amplifiers. This cable is made up of two or more solid or stranded wires. These are color-coded, insulated, and placed in an outer jacket (Fig. 19-10B). Wire sizes range from number 20 to number 24 AWG (0.80 to 0.50 mm).

Coaxial Cable

Coaxial cable is cable in which the conductors in the cable have a common center, or axis. Most coaxial cable is made of one stranded or solid copper wire insulated with polyethylene and placed in the center of a copper-braid shield. The outer jacket of the cable is vinyl-plastic insulation (Fig. 19-10C). The braid is one conductor of the cable. Coaxial cable is used, for example, to connect a microphone to an audio amplifier.

PLUG IN TO Science

Underwater Cables

Ocean cables must be insulated against ocean water's high conductivity. They must also be armored against damage by objects such as anchors. Other assaults come from tides and icebergs. Even the tiny sea animal, the teredo (or the ship worm), can bore through cable insulation. Underwater cables must also be strong enough to withstand the pull on them as they are laid.

In spite of the challenges underwater cables face, their useful life can range from 30 to 40 years.

Computer Cables and Connectors

The point of connection between a device and the attached cables is called an *interface*. The use of proper cables and connectors is essential to interface a computer with its peripherals. These external devices are connected by cables to make up the total computer system (Fig. 19-11).

Because computers generate high frequencies, they must be manufactured to reduce the effects of electromagnetic interference (EMI) and radio frequency interference (RFI). The rules of the Federal Communications Commission set limits on how much *"noise"* or interference a computer system can legally emit. Since the rules cover the entire system,

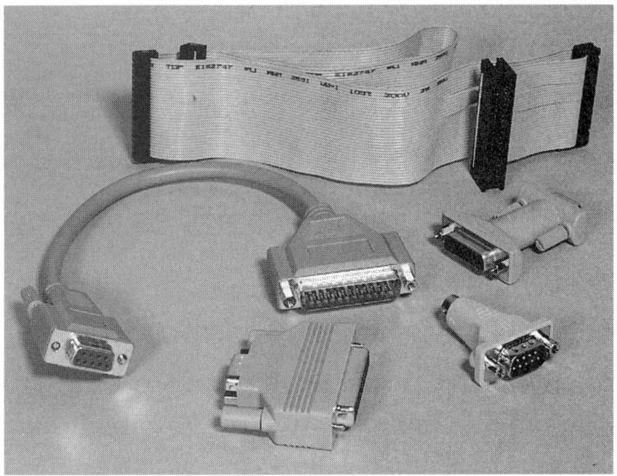

Fig. 19-11. Common connectors for computers and peripherals.

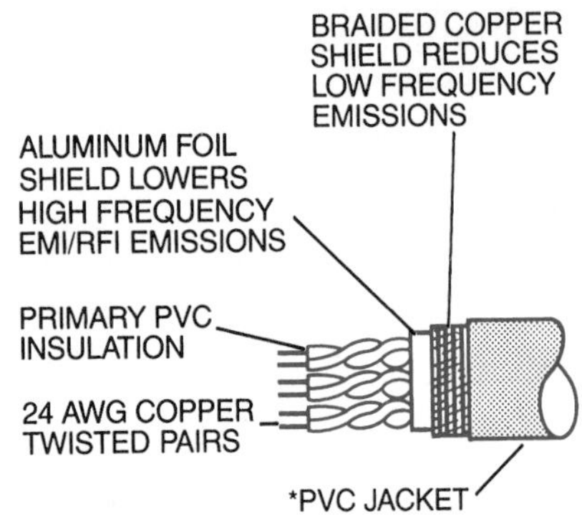

Fig. 19-12. Metallic shields reduce electromagnetic and radio interference.

metallic shields and metallic hoods are used on the cables and connectors that interface the peripherals with the computer.

Figure 19-12 illustrates the details of one type of a six-wire, twisted-pair cable. This type of cable minimizes interference (EMI/RFI) because of its double layer of metallic shielding. Some types of shielded, twisted-pair cables also use special insulation on the wires and the cable jacket. This makes them suitable for applications that extend beyond the standard 50-ft limit between work stations and peripherals. Before building a cable to install a computer, terminal, or peripheral equipment at distances greater than 50 ft, make sure it offers both EMI/RFI shielding and low capacitance.

CABLE FABRICATION TECHNIQUES

One of the most common types of cable connections used with computers and associated peripherals is the 25-pin D connector. In assembling such a cable, you should strive to obtain good electrical contact between the pins and the conductors and to minimize interference (EMI/RFI). Two general methods are used to connect a cable to a D-type connector. One method uses *pin insertion*; the other, soldering.

The pin insertion method allows you to service or change the pin locations in the connector. Special pin insertion and extraction tools are required, however. Male and female pins are used. The first task is to determine the number of conductors needed. Although the connector (solder or pin insertion type) has a fixed number of pins, not all pins are needed for a specific installation. To reduce cost, purchase a cable with the number of conductors required.

Figure 19-13 shows how individual male pins are connected to the conductors of a cable. Wire strippers are used to cut back the outer sheath 1 to 1.5 in. (25.4 to 38.1 mm). Each conductor in the cable is stripped to one-eighth to three-sixteenth in. (3.2 to 4.8 mm) from its end. The stripped conductors will have either a male or female pin or socket crimped to it. Each pin has two crimp locations, called *flags*. The first crimp makes the electrical connection only with the conductor. The second flag crimps the insulation for strain relief, or to put pressure on the insulation instead of on the pin. Special crimping tools are used to crimp the flags so that the connections are secure. Once the necessary crimps are made, a pin insertion tool is used to insert the pins in the correct shell holes. Male pins should be inserted only into male shells and female pins into female shells (Fig. 19-14).

The last step in fabricating a cable is to attach the hood. The hood has a plastic or

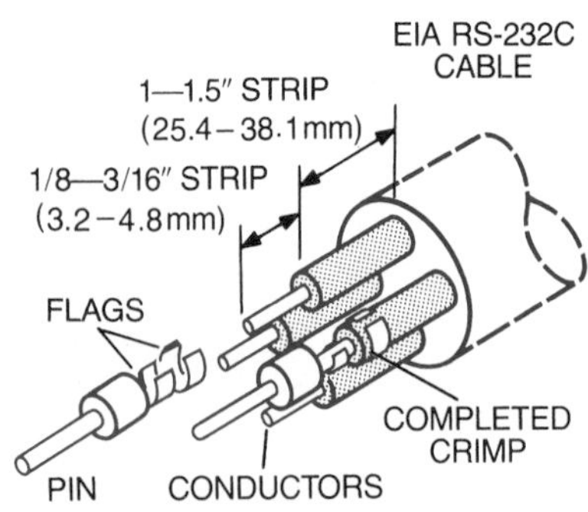

Fig. 19-13. Crimping a male pin to a conductor in a cable.

Fig. 19-14. **Assembling a D connector.**

metal cover with two or three cover screws (with or without nuts). Most hoods have a strain relief screw. The completed cable shell assembly should be positioned between the two hood pieces and the cover screws inserted and partially screwed into place. Before the cover screws are tightened down, the device attachment washers and screws and the relief screw should be attached. After this is done, the cover screws can be tightened.

The soldering method uses a connector that has the pins embedded firmly within the shell (body) of the connector. The opposite end of each pin has a cup-shaped slot. Wire cutters or strippers are used to cut back the outer sheath of the cable 1 to 1.5 in. (25.4 to 38.1 mm). Strip away about three-sixteenth in. (4.8 mm) of the insulation from the ends of each conductor. Solder the ends of each conductor into the cup-shaped slot of the proper pin. Care must be used not to short-circuit the connections by using too much solder. The hood is then assembled as mentioned previously.

One of the most overlooked steps in cable assembly is the proper labeling and documenting of the assembled cable. The cable should have a stick-on label that indicates where the pin documentation is located so that anyone can identify the electrical signals associated with each pin. An illustration of a typical master drawing is shown in Fig. 19-15. A male connector is indicated by the arrows pointing out; a female, by the arrows pointing in. The drawing should also identify the cable length, wire size, number of conductors, crossovers, type of connector (male, female, or both), jumpers, and any special information.

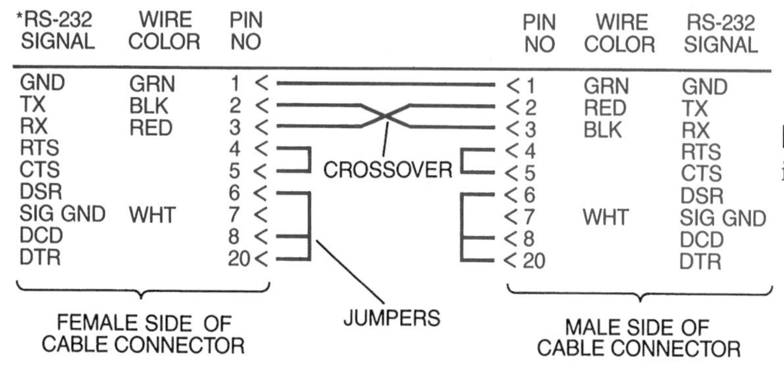

Fig. 19-15. **Example of documenting an assembled cable.**

Chapter Review

Summary

- Wires, cords, and cables are sized by an American wire gage number—the bigger the number, the smaller the wire.
- The size of a conductor determines its ampacity, or current-carrying capacity. The smaller the wire (or the higher the wire gage number) the less ampacity a conductor has.
- There are many different types of cables, cords, and wires. Some of the most common are magnetic, bus bar, test lead, antenna, television lead-in, and grounding wires. Common cords are lamp, service, and heater cords. Communication cables often used are audio, coaxial, and computer cables.
- Grounding wires are available in cables and cords and are either bare or have green insulation. The shield on cables is often connected to ground and is either copper or aluminum, often braided, and bare.
- There are special techniques and tools associated with cable fabrication. They are designed to insert, extract, and crimp pins and sockets, strip insulation, and solder. It is important to remember to label the connector according to its documentation and use after it is assembled.

Review Main Ideas

Review this chapter's main ideas by writing, on a separate sheet of paper, the word or words that most correctly complete the following statements:

1. Stranded wire is more ______ than solid wire.
2. The abbreviation for the ______ is AWG.
3. The larger the AWG number, the ______ the wire.
4. The AWG number and a wire's diameter can be found with a ______.
5. The circular-mil area of a wire is found by ______ the diameter of the wire expressed in ______.
6. The safe current-carrying capacity of a wire is that amount of ______ in amperes the wire will conduct without ______.
7. Wires are often tinned to prevent ______.
8. Copperweld wire consists of a ______ core that is coated with ______.
9. A type of television antenna lead-in wire is known as a ______ or two-conductor wire.
10. ______ connect metal cabinets and certain parts of circuits to ground rods.
11. Some kinds of flexible cables used to connect equipment and appliances to outlets are called ______.
12. Heater cords are often insulated with a material that can withstand ______.

13. A combination of insulated wires through which electric current can be sent is called a ______.
14. Some cables have shields to protect the internal wire from undesirable external ______ or ______ fields.
15. In a coaxial cable, all the conductors have a common ______.
16. Cables and connectors are essential to ______ a computer with its peripherals.
17. Metallic shielding and hoods are used on cables and connectors to reduce the effects of ______ on a computer.
18. Two common methods of connecting the wires of a cable to a connector are the ______ method and ______.

Apply Your Knowledge

1. Explain how to find the circular mils of a wire.
2. List four types of wire used for electronic and electrical wiring jobs.
3. List five items included in labeling and documenting communications cable.
4. Explain the necessity of computer interfaces.

Make Connections

1. **Communication Skills.** Find a place in your community where cables are supported by poles. From the ground, observe and identify the types of cables on the pole. Report your findings to the class.
2. **Mathematics.** Use a standard wire gauge to measure the diameter of five wires of consecutive gauge sizes. Using an equal length of wire from each wire sample, calculate the volume of wire in each sample. Compare the amount of copper in each sample. Relate this to its current carrying capability.
3. **Science.** Using the periodic table, compare the characteristics of ten conductive elements. Report to the class on how they compare with respect to valence and conductivity.
4. **Workplace Skills.** Develop and construct a device that will enable a worker to identify individual wires, end to end, in a multiconductor cable without removing the outer casing.
5. **Social Studies.** Research the installation of the first trans-Atlantic cable. Report to the class on the importance of this event.

Chapter 20

Wiring Tools and Devices

OBJECTIVES

After completing this chapter, you will be able to:

- Identify three common types of pliers by their proper names.
- Describe three types of wire strippers.
- Identify and explain the use of banana plugs and jacks, tie plugs and jacks, and open-circuit and closed-circuit phone jacks.
- Identify and explain the purpose of at least six coaxial cable connectors and adapters.
- Specify five different types of insulating materials and explain the use of each.
- List five methods of marking wires.
- Explain the processes of wire wrapping and unwrapping.
- List the advantages of wire wrapping over soldering connections.

Terms to Study

alligator clips

breadboard

discrete circuits

Fahnestock clips

grommets

ground wire

hand wiring

lug

polarized plug

terminal strips

Underwriters' knot

wire nuts

wire strippers

wire wrapping

Tools have been an important part of the development of civilization. In fact, the use of tools is one factor anthropologists use to classify cultures. Think what life would be like if we did not have tools. Could you have the comforts and conveniences that you enjoy each day?

WIRING TOOLS

New tools are being invented every day. Some tools are made by workers to do a very specific job or meet a need. Others have a more general purpose and are made in large quantities for sale to the general public. Have you ever needed to improvise a tool to do a very specific job? If so, think about the process you went through to make that tool and why you made it the way you did. This chapter discusses the various tools needed to work efficiently with wiring.

Hand wiring involves using wires to connect the terminals of sockets and other devices. Integrated circuits and printed circuits have greatly lessened the need for wiring by hand. However, hand wiring is still used in manufacturing and in repairing and servicing many products. To do hand wiring correctly, you must know about wires. You must also be familiar with the tools and the wiring devices used with circuits. Such knowledge will help you do new wiring as well as repair jobs in a neat, efficient way.

To do hand wiring correctly, you must use the proper tools. The tools used most commonly in wiring include several types of pliers and wire strippers.

Pliers

Pliers are essential wiring tools. You have probably seen or used some kind of pliers in the past. Because they are used for many different purposes, there are many different kinds of pliers. The three kinds discussed below are generally used when working with wires.

Diagonal-Cutting Pliers

Diagonal-cutting pliers are for cutting wires only (Fig. 20-1A). They should never be used as a gripping tool or to notch sheet metal. This can damage the cutting jaws. The jaws can also be damaged if they are used to cut hardened iron or steel wires. As with other kinds of pliers, diagonal-cutting pliers are sized according to their length. The common sizes are 4, 5, and 6 in. (102, 127, and 152 mm).

Longnose Pliers

Longnose pliers are used to grip and bend small wires. They sometimes have wire-cutting jaws (Fig. 20-1B). The jaws of these pliers are relatively weak. They may be bent out of shape if too much pressure is used to grip objects. Common sizes are 5 and 6 in. (127 and 152 mm).

Because of the shape of the *nose,* or gripping jaw, longnose pliers are very useful for bending wires into a loop. This is often done to fasten the end of a wire to a screw terminal.

Some pliers have gripping jaws that are even longer and thinner than those on standard longnose pliers. These pliers are called *needlenose pliers*. These pliers are often used to work with wires in tight places, especially when the wire is very thin or must be handled delicately.

All-Purpose Pliers

All-purpose pliers are very handy to have around. Because of the shape of the jaws, they can be used as a nut wrench, stripper, cutter, or crimper (Fig. 20-1C). Common sizes are 5 and 6 in. (127 and 152 mm).

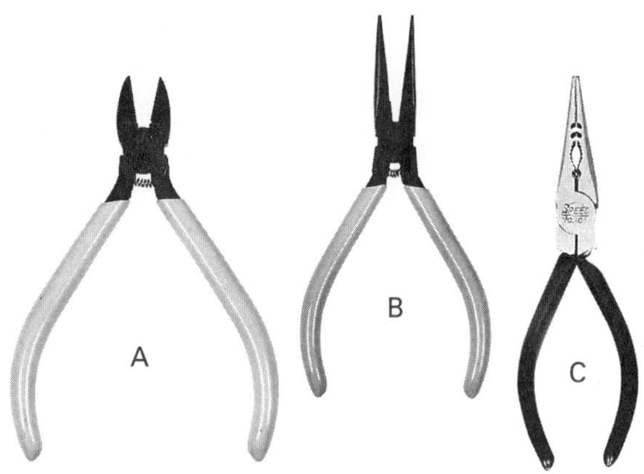

Fig. 20-1. Pliers: (A) diagonal-cutting, spring-return pliers with insulated grips; (B) longnose, spring-return pliers with insulated grips; (C) all-purpose pliers.

Wire Strippers

Wire strippers are tools used to remove insulation from wires. The insulation must be removed before wires can be joined or connected. The following types of wire strippers are used in hand wiring.

Basic Wire Strippers

A simple wire stripper is shown in Fig. 20-2A. To use this tool, press the blades together until they cut through the insulation of the wire. Then pull the tool away from the wire to remove the insulation. When using this kind of wire stripper, be careful not to press the blades together more than is needed to cut through the insulation. Otherwise, the blades may cut into the wire and weaken it.

Another common kind of wire stripper is shown in Fig. 20-2B. This tool has stripping notches that help prevent cutting into the wires accidentally. Each notch is designed to strip a specific wire size. To keep from damaging the wires, be sure to use the right size stripping notch.

Automatic Wire Stripper

To use an automatic wire stripper, put the wire into the right size notch on the cutting jaws. Then squeeze the handles. This causes the jaws to separate. The cutting jaws cut through the insulation. At the same time, the clamping jaws pull the insulation from the wire (Fig. 20-2C). Automatic wire strippers are very useful for removing the tough plastic of fabric insulation from wires. Since they do not require any pulling, they are also a good tool to use with wires that have one end connected to a terminal.

Other Stripping Methods

Thermal wire strippers use heat to remove insulation. These wire strippers have adjustable, electrically heated "jaws" that melt through the insulation. The insulation of some wires, including some magnet wires, can be removed using the heat of a soldering iron without further stripping or cleaning of any kind.

Some wires are insulated with enamel or varnish. These wires are often stripped chemically with a paint solvent.

Using Pliers as Wire Strippers

With practice, you can learn to use the wire-cutting jaws of pliers as a wire stripper, particularly on wires with a thin plastic insulation. To do this, first put a wire between the jaws. Then apply enough pressure to the handles to allow the jaws to cut through the insulation. Then pull the pliers away from the wire to remove the insulation (Fig. 20-3A). Be sure to use just the right amount of pressure so that the wire itself will not be nicked or cut.

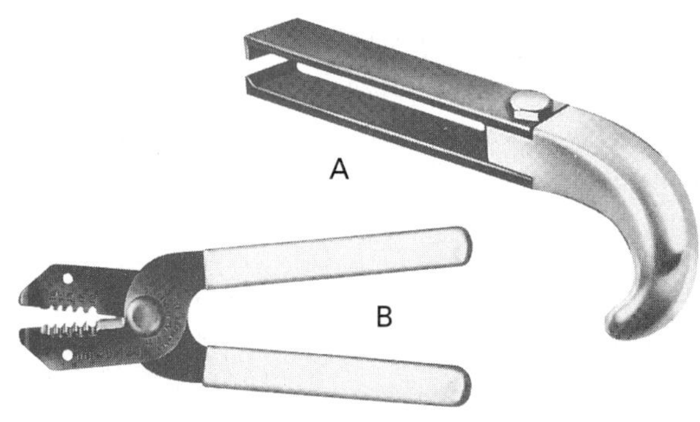

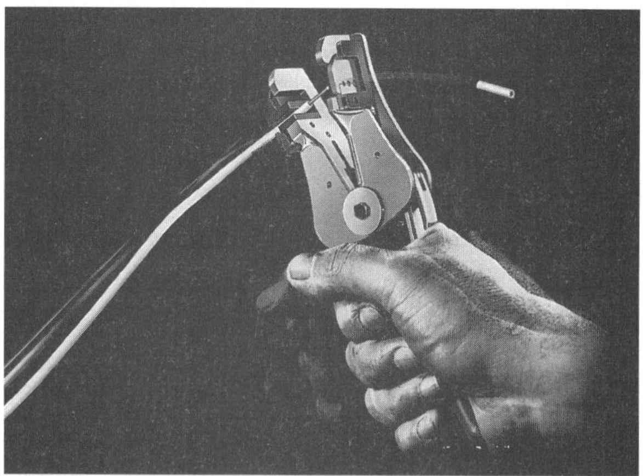

Fig. 20-2. Wire strippers: (A) basic wire stripper; (B) notch-type wire stripper; (C) using an automatic wire stripper.

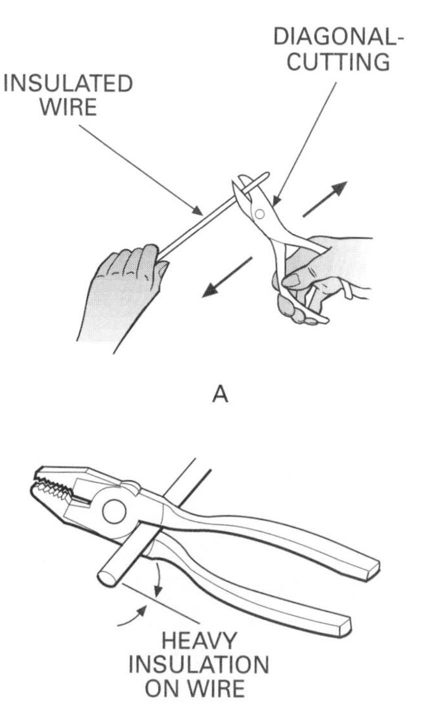

Fig. 20-3. Using pliers as wire strippers: (A) using diagonal-cutting pliers; (B) crushing the insulation on a heavily insulated wire before stripping it.

To use pliers to strip a wire covered with a heavy plastic or fabric insulation, you must often crush the insulation first. With most pliers, you can do this by inserting the wire between the handles just behind the pivot point and by applying enough pressure (Fig. 20-3B). You can then either pick the insulation off the wire, or remove it using the cutting jaws.

Have you ever tried to connect headphones to a radio, tape player, or CD player—only to find that the plug doesn't fit the jack? Think of ways to overcome this problem. How would you do it?

FASCINATING FACTS

The Matsushita Electrical Industrial Company was founded around 1918 solely to make electric plugs. The company later expanded into other electrical products, including lights, radios, motors, and batteries. After World War II it made major appliances. In the 1970s, the company marketed the VHS system for videocassette recorders. In the 1980s, it further diversified into automation and semiconductors. Today, selling under brand names such as Panasonic and Quasar, this company is the largest maker of electronics products in the world.

PLUGS, JACKS, AND CONNECTORS

Electrical plugs provide a convenient way to connect devices to a circuit. To connect devices using electrical plugs, you insert the plug into a jack or a receptacle wired to the circuit.

Attachment Plugs

Attachment plugs are used with lamp and service cords to connect electric appliances and devices to electrical outlets. On many products, molded attachment plugs are used (Fig. 20-4).

Plugs with Screw Terminals

For some heavy-duty applications, attachment plugs are fastened to a cord by pressure contacts or screw terminals. Attachment plugs with screw terminals are wired as shown in Fig. 20-5.

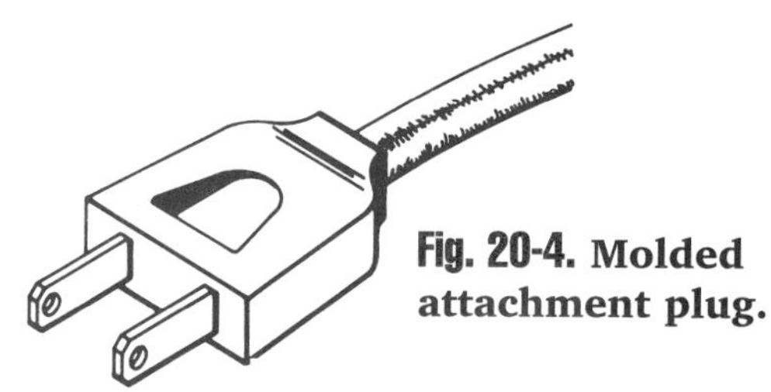

Fig. 20-4. Molded attachment plug.

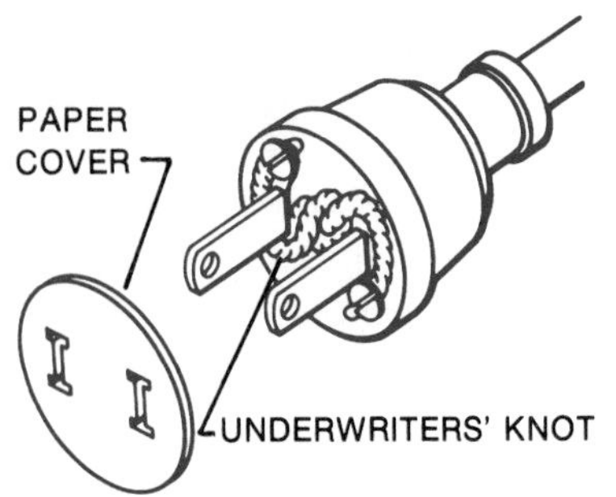

Fig. 20-5. **Attachment plug with screw terminals. The Underwriter's knot is covered with a fiber cover after the wires have been attached.**

To wire attachment plugs with screw terminals, first remove about ½ in. (12.7 mm) of the insulation from each wire of the cord. Then twist the bare end of each stranded wire tightly so that the strands form a compact bundle. Apply solder to the tips to keep them from separating. Otherwise, a loose strand might cause a short circuit. Wrap each wire around a terminal screw in a clockwise direction. This helps keep the wire under the screwhead while you tighten the screw.

When space permits and when an attachment plug does not have a cord clamp, tie the two wires in an **Underwriters' knot** as shown in Fig. 20-6. This will keep the wires from pulling loose from the screw terminals if the cord is pulled.

Polarized Plugs

On some attachment plugs, one prong is wider than the other (Fig. 20-7). This is referred to as a **polarized plug** because the difference in prong size ensures that the connection will be made in only one way. Receptacles are wired so that the neutral connection to the electrical system is made to the wider prong and the line connection is made to the smaller prong.

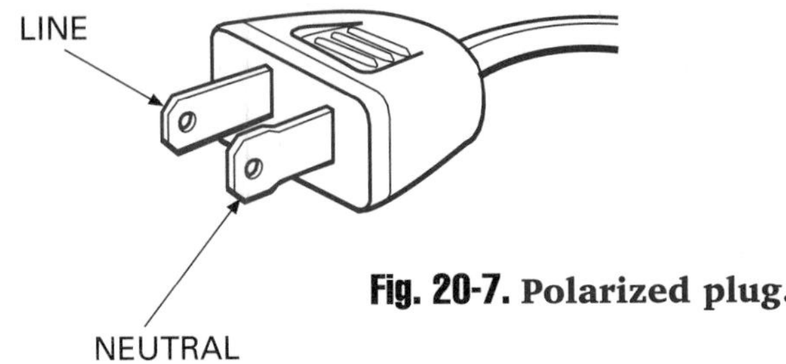

Fig. 20-7. **Polarized plug.**

The difference in prong size is a safety feature. The neutral connection of the plug cable or cord is often made to the metal frame (chassis) of an appliance. If an operator accidentally reversed a non-polarized plug when inserting it, the chassis would have 120 volts ac to ground. If the operator touched the metal chassis and was grounded, a shock situation could occur. The difference in prong size eliminates the possibility of connecting an appliance in an unsafe way.

Grounded Plugs

Modern attachment plugs have three prongs. The third prong is connected to the green wire, called the **ground wire.** The ground wire provides for safe grounding (Fig. 20-8). The third prong is offset from the others and is often "U"

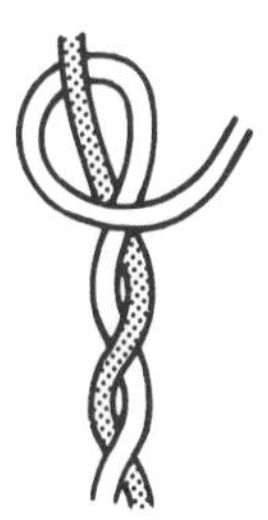

Fig. 20-6. **Tying an underwriter's knot.**

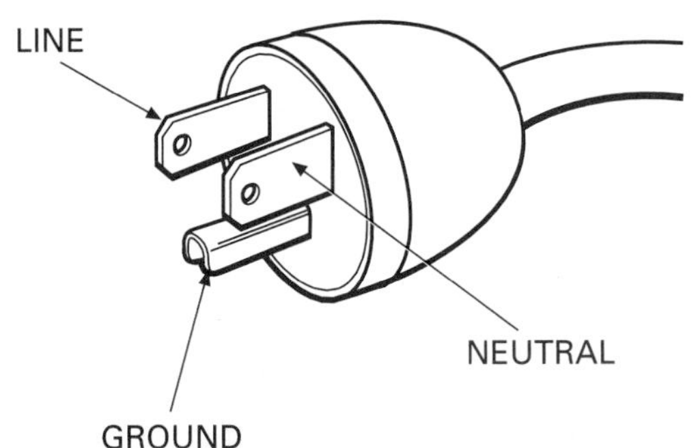

Fig. 20-8. **Grounded plug.**

LINKS

For more detailed information about wiring devices and equipment, read ahead to Chapter 28.

shaped or round. The other end of the ground wire is connected to a frame or metal chassis. If an electrical failure occurs in an appliance, tool, or other device while it is operating, the current is conducted safely to ground.

Other Plugs, Jacks, and Connectors

A variety of other types of plugs, jacks, and connectors are used for specific wiring purposes. Basic descriptions of some of the major types are discussed below.

Banana Plugs and Jacks

Banana plugs and jacks are commonly used for connecting test leads to instruments and for other connections (Fig. 20-9). They are especially useful for making quick or temporary connections. Wire leads are held in the plugs by set screws, pressure, or soldering.

Tip Plugs and Jacks

Tip plugs and jacks are used mostly with test instruments (Fig. 20-10). Because they are small, these devices are often used in spaces where other kinds of connecting devices do not fit.

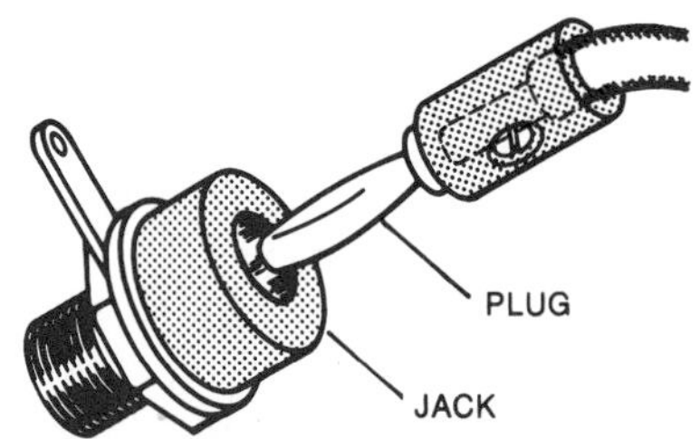

Fig. 20-9. **Banana plug and jack.**

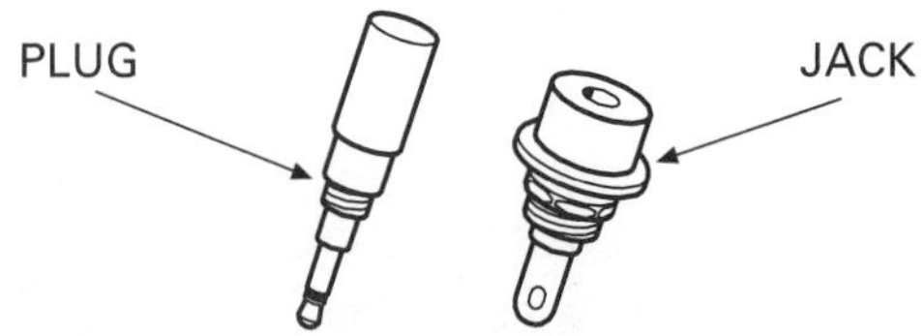

Fig. 20-10. **Tip plug and jack.**

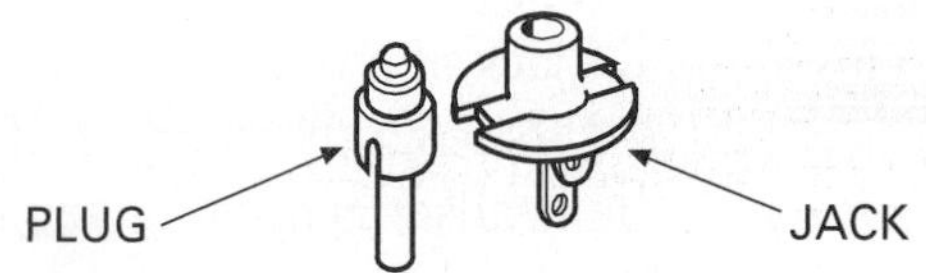

Fig. 20-11. **Phono plug and jack.**

Phono Plugs and Jacks

Phono plugs and jacks got their name from their original purpose. They were used to connect phonograph cartridge output to the input of an amplifier (Fig. 20-11). Now they are often used to make other audio-line connections, such as those needed by today's home audio systems. These devices are usually connected to shielded cables by soldering.

Phone Plugs and Jacks

Standard phone plugs and jacks are used for connecting headsets to radio receivers and audio amplifiers (Fig. 20-12). The standard plug prong and the jack socket of these devices are ¼ in. (6.35 mm) in diameter.

Miniature phone plugs and jacks are used for the same purposes. They are also similar in

Fig. 20-12. Phone plug and jack.

construction. However, the prong and socket are only ⅛ in. (3.18 mm) in diameter. These plugs are used in small audio devices such as headphones for portable CD players.

Phone jacks may be either open-circuit or closed-circuit (Fig. 20-13A). The closed-circuit jack allows a circuit to be completed through the jack contacts when a plug is not inserted into it. If, for example, a headset is plugged into a closed-circuit jack of a radio receiver, the circuit is completed through the headset. If, on the other hand, the headset is not plugged into the jack, the radio circuit is completed to the loudspeaker (Fig. 20-13B).

Connectors and Adapters

Various types of connectors are used at the ends of cables to connect the cables to equipment. Since different types of equipment require different connectors, adapters are often needed to form a proper connection. This section describes several types of connectors and adapters.

Microphone Connectors

Microphone connectors are used to connect the coaxial cables of microphone cords to amplifiers (Fig. 20-14). They are also commonly used for connecting leads to various kinds of test instruments.

Coaxial Cable Connectors and Adapters

Figure 20-15 illustrates a variety of coaxial cables, connectors, and adapters available for use today. These devices make it possible to connect with the cables needed for various

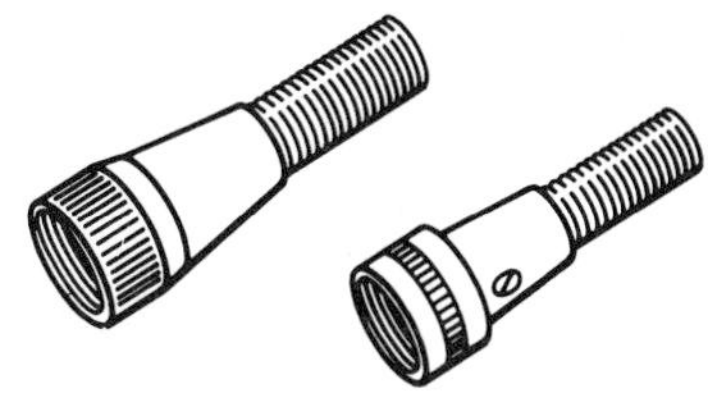

Fig. 20-14. Microphone connectors.

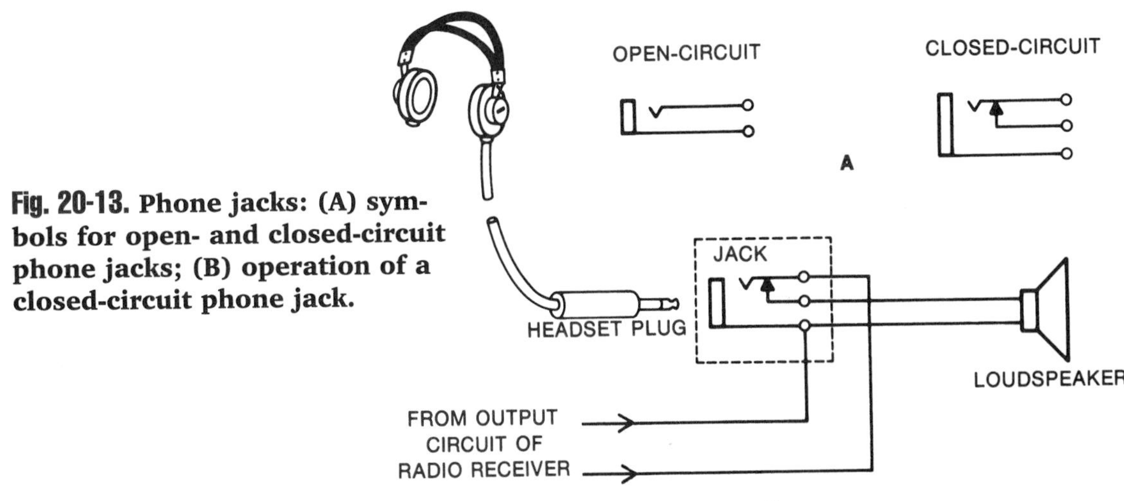

Fig. 20-13. Phone jacks: (A) symbols for open- and closed-circuit phone jacks; (B) operation of a closed-circuit phone jack.

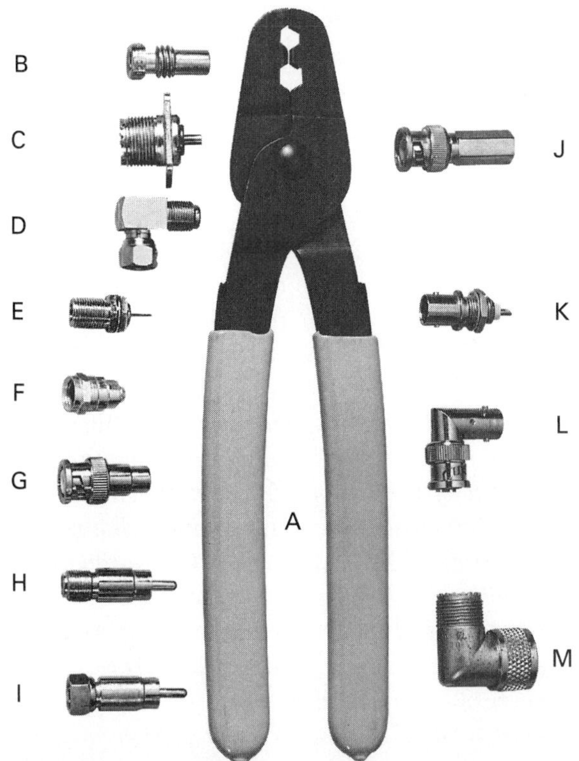

Fig. 20-15. **(A) Heavy-duty hex crimping tool; (B) through (M) selected coaxial cable connectors and adapters.**

videocassette recorders, television receivers, and FM equipment. For example, a radio frequency (rf) plug and a chassis mount connector are used with coaxial cables for citizens band (CB) communications. High-grade coaxial cables are used extensively for amateur radio, CB, TV, and FM equipment.

Coaxial cables are available to meet various electrical specifications. They are made from low-loss polyethylene dielectric with a tough outer jacket. All conductors are made of copper. It is essential that the type of connector and cable match properly.

The items illustrated in Fig. 20-15A through M are:

A. *Heavy-duty hex crimp tool*. This tool is used to crimp connectors to a coaxial cable. No soldering is necessary.

B. *Cable reducer*. Reducing adapters make it possible to use cable of a different diameter with a larger, standard-size plug.

C. *Chassis mount connector*. This type of connector is commonly found on amateur and citizens band radio equipment.

D. *Right-angle adapter*. This adapter is used on coaxial cables for television sets and FM radios. The adapter is especially useful when space is limited at the rear of a chassis.

E. *Antenna chassis connector*. This type of connector is often used on television sets and FM radios to connect antennas to the equipment.

F. *Coaxial connector*. This connector is crimped onto a coaxial cable with a hex crimping tool. It is used extensively to connect coaxial cables to television sets. It is relatively easy to install and does not need soldering.

G. *RCA phono jack to male BNC connector*. This type of connector makes it possible to join two different types of connections. BNC connectors are very secure because of the coupling mechanism. The coupling mechanism has spiral ramps that engage in projections to lock the mating halves together. A phono plug can be pressed firmly into the jack at the other end to form a connection.

H. *Female F connector to RCA phono plug*. The two ends of this type of coaxial connector are different. One end has threads to connect with a male F connector secured on a coaxial cable. The other end is pressed firmly into a mating socket.

I. *Male F connector to RCA phono plug*. This type of connector is used to connect a coaxial cable to a phono plug.

J. *Male BNC solderless connector*. This type of connector has a locking connector on one end and a solderless connector on the other end.

K. *Female BNC connector*. This type of connector is used on a chassis to provide a connector for a male BNC plug. It provides a very secure connection that can withstand considerable vibration.

L. *Right-angle adapter, BNC male to BNC female*. This type of adapter provides a convenient method to make a secure connection where space is limited between cables terminated with BNC connectors.

M. *Right-angle adapter*. This adapter is especially useful to connect an antenna transmission line to a transmitter where space is limited.

Wire Nuts

Wire nuts are devices that make solderless connections in house wiring systems, motors, and various appliances (Fig. 20-16A). Wire nuts come in different sizes. The size needed depends on the number and sizes of wire to be connected. Figure 20-16B shows two wires being inserted into a wire nut.

Coaxial Cable

Coaxial cable's ability to carry many conversations on a single line made dramatic advancements in the telephone industry possible. However, the use of coaxial cable for long-distance telephone connections has steadily dropped with increasing technological advances. In the United States in 1950 about 25% of the 112 million long-distance circuit miles were carried by coaxial cable. (Circuit miles are the number of calls that can be carried simultaneously multiplied by the length of connections. For example, a network with 10,000 miles of connections with each line capable of carrying 10,000 calls would have a capacity of 100 million circuit miles.)

By the mid-1980s, the total long-distance circuit miles increased to nearly 600 million. However, the number of coaxial cables used for this purpose dropped to 4%. Microwave relay stations assumed almost 84%. At this time, satellites were just beginning to make significant contributions to domestic long-distance telecommunications in the United States. Although satellites already carried a large portion of international telephone and data communications, the remainder was carried mainly by undersea coaxial cables.

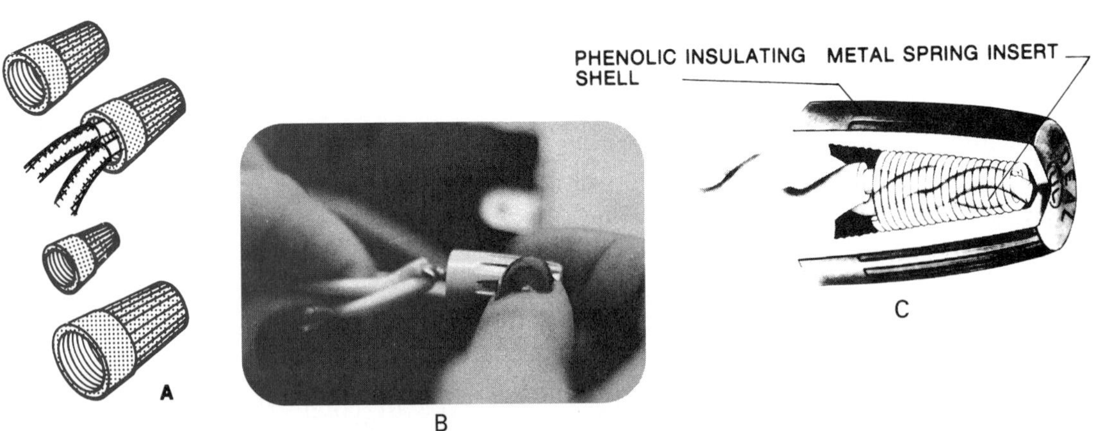

Fig. 20-16. **(A) Wire nuts come in various sizes. (B) Strip the ends of the wires and insert them into the wire nut as shown. (C) When you twist the wire nut, a metal spring inside the wire nut holds the connection firmly in place.**

To use a wire nut, strip about ⅜ in. (10 mm) of insulation from the ends of the wires to be connected. Put the wires into the wire nut. Then firmly screw the nut over the wires (Fig. 20-16C).

Clips

Various kinds of clips are used for making temporary wire connections to circuits. **Alligator clips** have spring-loaded jaws with teeth to hold a wire securely (Fig. 20-17A). They allow a firm wire connection to be made to a terminal point. Because they can be installed and removed quickly, they are used with instrument test leads and other temporary wiring systems.

Fahnestock clips have a spring lever. Fahnestock clips are often used for making temporary connections to experimental models.

To use a Fahnestock clip, push the lever down and insert the wire the clip hasp (Fig. 20-17B). For a good connection, you should strip and thoroughly clean the end of the wire before you insert it.

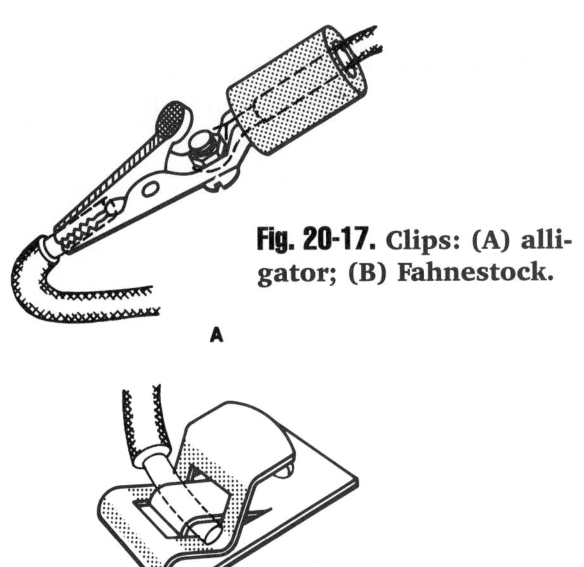

Fig. 20-17. Clips: (A) alligator; (B) Fahnestock.

Terminal Strips and Lugs

Terminal strips consist of solder lugs and a mounting lug or lugs, usually fastened to a thin phenolic strip (Fig. 20-18A). A **lug** is a metal connector that provides a convenient terminal for connecting wires and components. The connections can be soldered (solder lugs) or screwed to another surface (mounting lugs). See Fig. 20-18B. Whenever possible, soldered connections should be made on terminal strips.

Terminal lugs, or wire lugs, are used to connect wires to screw terminals. These lugs come in solder and solderless versions (Fig. 20-19A and B). Solderless lugs are fastened to the ends of wires with a crimping, or terminal, tool, as shown in Fig. 20-19C.

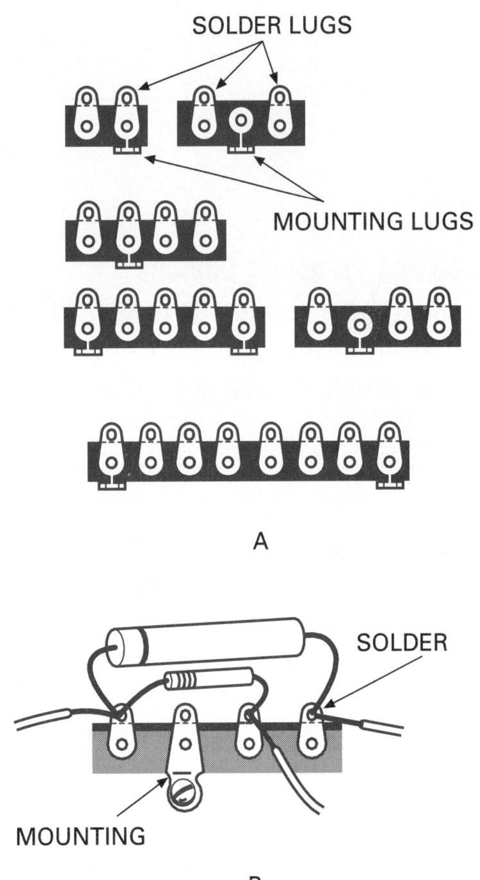

Fig. 20-18. (A) Terminal strips; (B) using a terminal strip to make soldered connections.

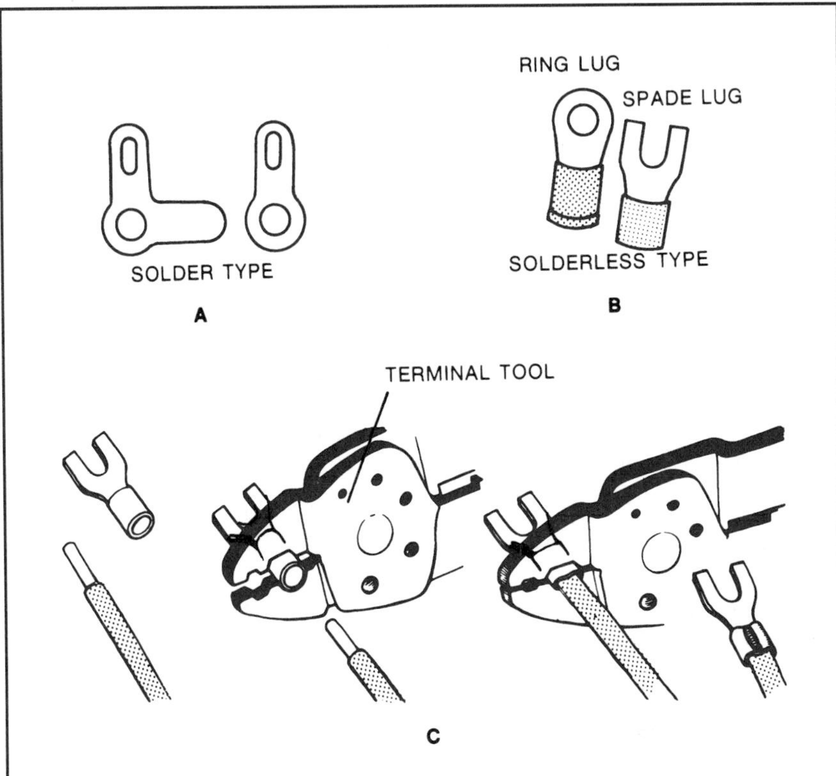

Fig. 20-19. **Terminal or wire lugs: (A) solder lugs; (B) solderless lugs; (C) fastening a solderless lug to the end of a wire using a crimping tool.**

Clamping and Strapping Devices

Clamping and strapping devices hold wires and cables in place. They make the job look neater and lessen the physical strain on individual wires.

- *Cable clamps* are made of metal or plastic (Fig. 20-20A). They are used to secure wires and cables to the surfaces of metal, wood, or plastic.
- *Insulated staples* are a form of wire or cable clamp with an insulating saddle. They are used to secure wires and cables to wood surfaces (Fig. 20-20B).
- *Plastic tie straps* are used to secure a bundle of wires and cables (Fig. 20-20C). Wires and cables are also commonly tied or laced together with heavy lacing string, as shown in Fig. 20-20D.

List five types of connectors. Describe in detail one of them and explain its purpose(s).

BREADBOARDS

In the early days of radio, a flat wooden board, termed a **breadboard** because of its appearance, was used as a base for mounting components. Fahnestock clips, held to the board with wood screws, provided a convenient way to make temporary connections. This made it easy to experiment with different circuits. Modern breadboards have many forms.

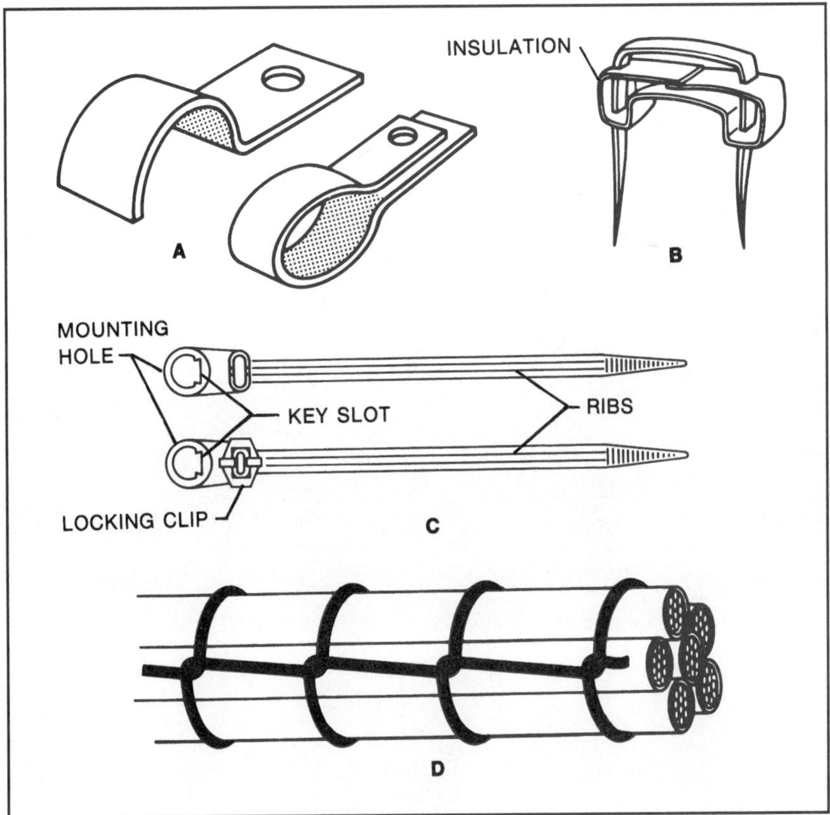

Fig. 20-20. **Clamping and strapping devices: (A) cable clamps; (B) insulated staple; (C) plastic tie straps; (D) laced cables.**

Solderless Connections

The breadboard shown in Fig. 20-21 contains flat spring clips that hold wires tightly to make quick, solderless connections. Short lengths of ordinary solid hookup wires are used to make the connections. The wire size can vary between no. 22 and no. 33 AWG.

In most breadboards today, five holes are placed in a row. This row is called a *tie point*. Any lead pushed into a hole is connected automatically with all the other holes in the row. Therefore, five connections can be made in any row of five holes.

To use this type of breadboard, strip about ⅜ in. (10 mm) of insulation from the ends of the wire. Put the bare wire lead into a hole in the board. Connect the other end to an appropriate tie point to make the electrical connection.

Note that in the illustration both discrete and integrated circuits are connected to the solderless breadboard. **Discrete circuits** are made up of separate components connected by leads. *Integrated circuits* contain many interconnected components in a single block of material.

LINKS

Review Chapter 18 for more detailed information about integrated circuits.

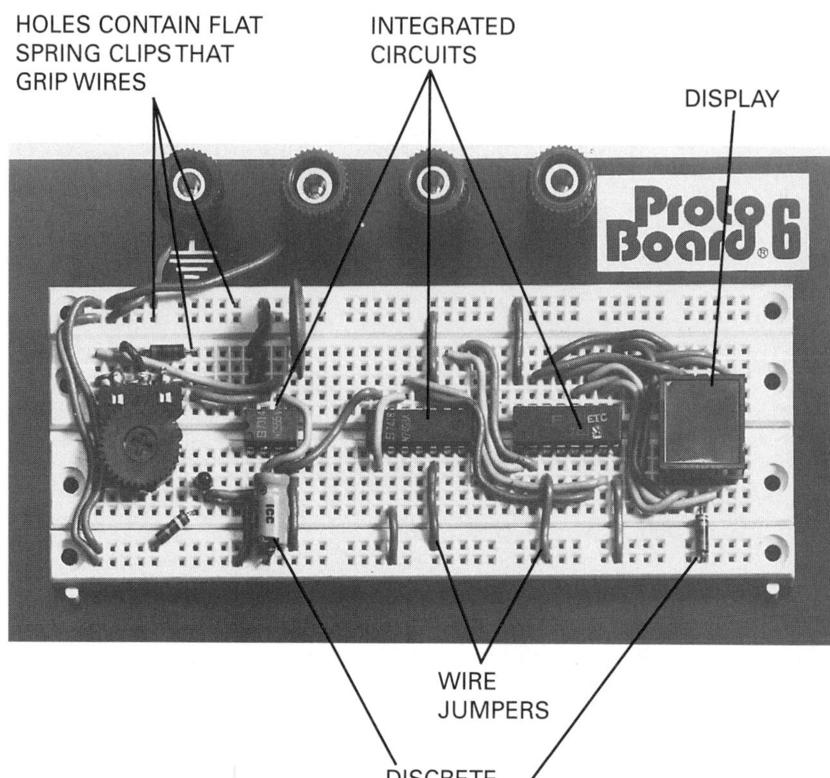

Fig. 20-21. This solderless breadboard has been used to create a decade counter circuit with display.

Soldered Connections

Figure 20-22 shows a prepunched circuit board. This board is made of a phenolic material and comes in several sizes. Prepunched circuit boards make a convenient base for mounting and wiring circuit components. These circuit boards are often used as breadboards for experimental work. Sometimes a special pin is pushed into a hole, as shown in Fig. 20-23. These pins form convenient contact points for soldering the leads of components.

Manufacturers' surplus printed circuit boards make another very useful and inexpensive type of breadboard. You can find them in many radio and electronic supply stores. Many of these boards do not have components soldered in place. If they do, you can desolder the components and remove them. Remove the components carefully so they can be used again.

Although the circuit on the board is not the one you want for your experimental circuit, you can modify it. Figure 20-24 shows a breadboard made from a surplus printed circuit board. This breadboard circuit is an experimen-

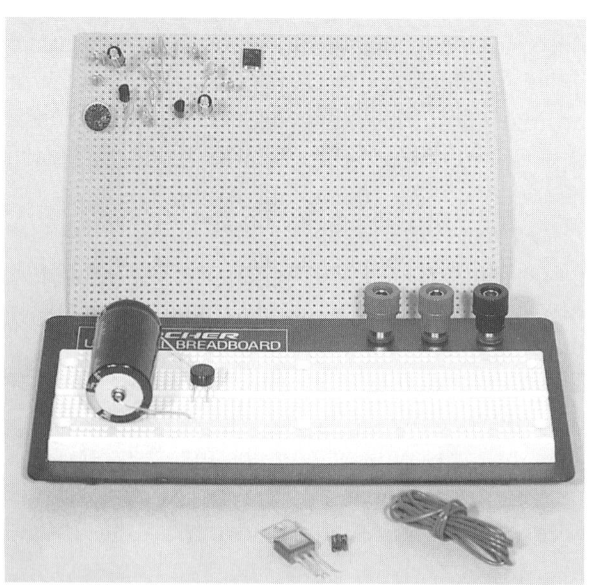

Fig. 20-22. Components mounted on a prepunched circuit board.

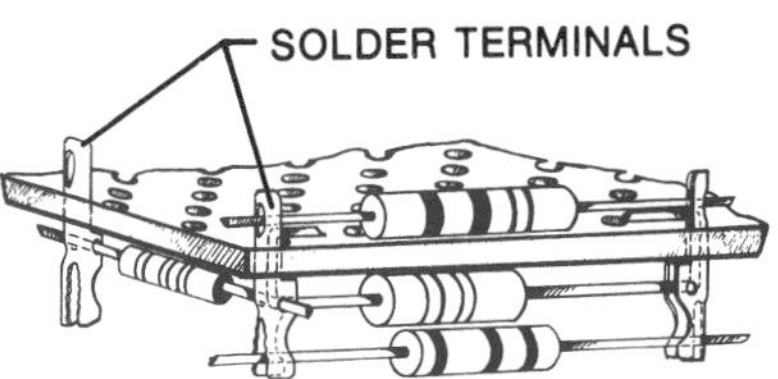

Fig. 20-23. **Using push-in terminals that need to be soldered.**

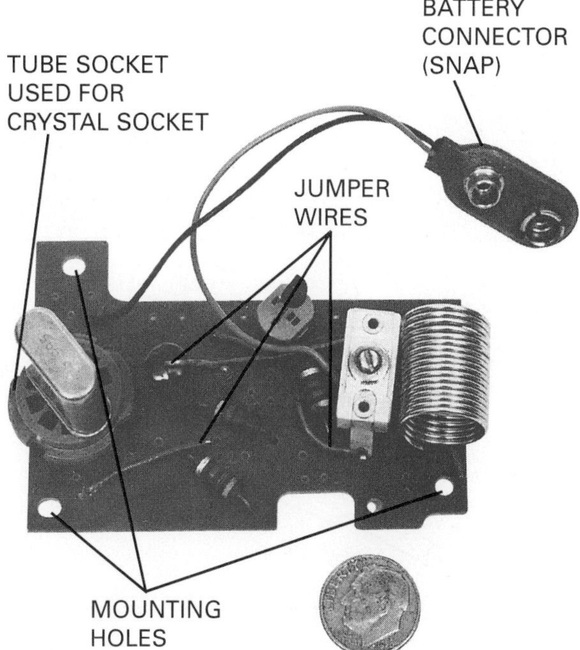

Fig. 20-24. **Breadboard circuit of an rf oscillator mounted on a surplus printed circuit board.**

tal rf oscillator. The components are soldered to any convenient printed circuit. Even the predrilled mounting holes can be used to mount the circuit. Note that a tube socket is used instead of a regular crystal socket. Jumper wires are used to modify the previously printed circuit to meet the needs of the new circuit. Using surplus printed circuit boards and old components saves time and money. Also, it avoids any need to etch the circuits.

Advantages of Breadboard Circuits

Breadboards allow technicians and electronic engineers to design, build, and test circuits quickly. Breadboard circuits can be used to make industrial *prototypes*, or first models, of circuits to see if they will work as planned. After being checked out completely, a prototype is rearranged as needed before the finished product is made.

Breadboards can help you build circuits more quickly, more easily, and less expensively. They also can help you test and improve your circuits.

As you know, using power cords with frayed or missing insulation is dangerous. Without buying a new cord, how could you repair the cord so that it can be used safely once again?

INSULATING TAPES, DEVICES, AND MATERIALS

Insulation is used to prevent undesirable electrical contacts. *Electrical tape* is made of plastic, rubber, or cloth, usually 3/4 in. (19 mm) wide (Fig. 20-25). Plastic tape is thinner than rubber or cloth tape. It is also more adhesive and has a higher dielectric strength. Electrical tape made of black, coated cloth is generally called *friction tape*.

Tubing

Plastic insulating tubing, or *spaghetti*, provides a very convenient way to insulate wires (Fig. 20-26). The size of insulating tubing is usually specified by its inside diameter and the largest wire size that can fit into it.

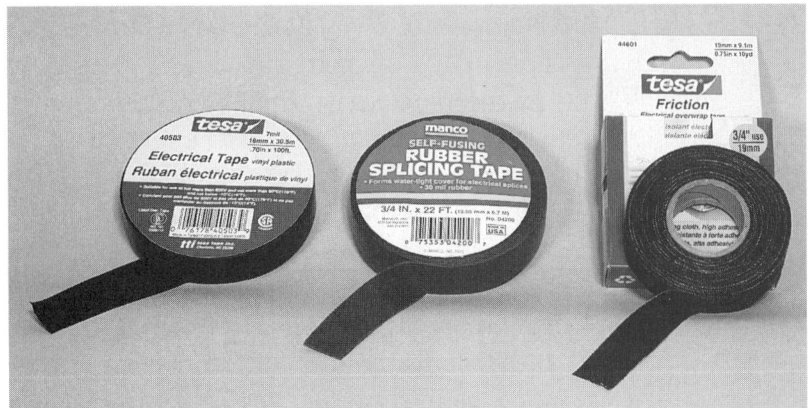

Fig. 20-25. Various types of electrical tapes.

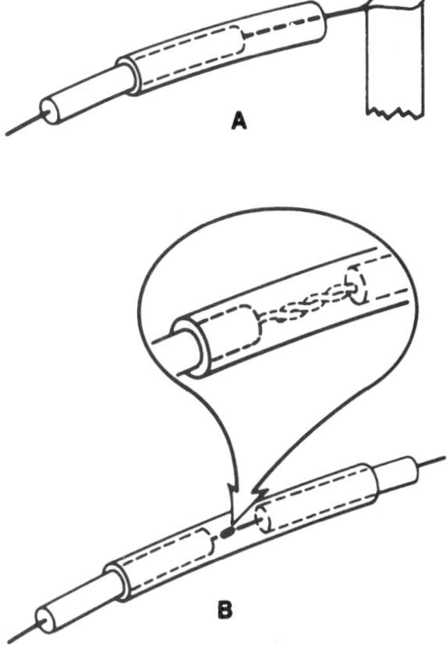

Fig. 20-26. (A) Using insulating tubing to cover exposed wiring at a terminal point; (B) using the tubing to cover a splice.

Heat-shrinkable tubing is also used to insulate wire connections. This tubing fits around a connection. When you apply heat, it shrinks to half its original diameter.

One way to apply the necessary heat is to hold a 40-W soldering iron about ½ in. (12.7 mm) away from the tubing. You can also use a hair blow-dryer to shrink the tubing. Heat-shrinkable tubing is available in sizes 1/16 to ½ in. diameter (1.6 to 12.7 mm).

Grommets

Grommets made of rubber or plastic are used to protect wires and cords that must pass through holes in metal (Fig. 20-27). The size of the grommet is measured by the diameter of the mounting hole into which it will fit snugly.

Fiber Washers

Fiber washers insulate devices such as jacks and mounting screws from metal surfaces (Fig. 20-28A). Fiber shoulder washers insulate the device from the inside surface of the mounting hole. Flat fiber washers are often used with shoulder washers to completely insulate the jack or screw. Figure 20-28B shows shoulder and flat fiber washers being used to insulate a banana plug from a metal chassis. The size of a shoulder washer is usually given by the inside and the outside diameters of its shoulder.

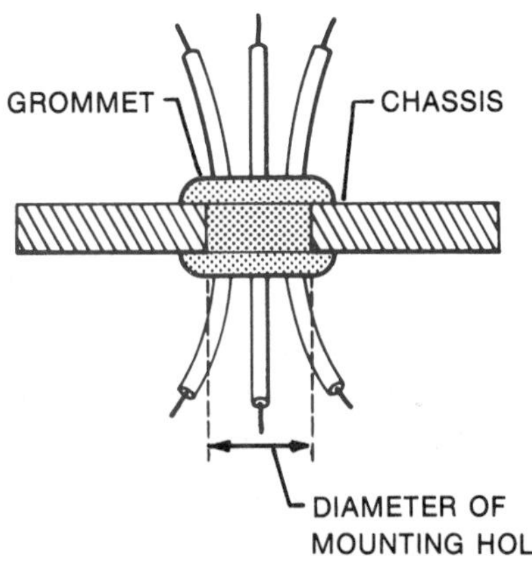

Fig. 20-27. A grommet being used to protect wires passing through a hole in a chassis.

Fig. 20-28. **(A) Types of fiber washers; (B) use of shoulder and flat washers to insulate a banana jack.**

Insulating Papers, Cloths, and Varnishes

Rag and kraft papers of various thicknesses and shapes are often used as insulation in machines such as motors and generators (Fig. 20-29). Insulating papers and insulating cloths made of grass fiber or varnished cotton, silk, or linen are also used to insulate the windings of transformer coils from the metal cores around which they are wound. Varnished linen is known as *varnished cambric.*

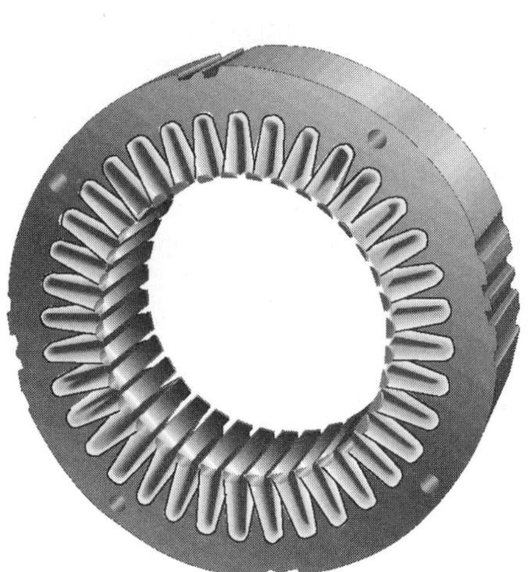

Fig. 20-29. **Insulating paper used to line the winding slots in the laminated stator of an electric motor.**

Insulating varnishes are also used to strengthen the insulation of magnet wires and to repair defective insulation on these and other wires. These varnishes are applied by dipping, spraying, or brushing. Some insulating varnishes can be air-dried. Others must be baked at a temperature of 220° to 270°F (104° to 132°C) for several hours.

Some equipment and other devices include many different wires. For someone who doesn't understand how they are identified, it may take a while to figure out the purpose of each wire. Under what conditions might it be important to be able to identify the purpose and proper connection of wires?

MARKING WIRES

Many wires, individual or in cables, are color-coded. For further identification, *wire markers* can be used to identify wires, cables, and terminals in wiring systems (Fig. 20-30A). They make it possible to find and trace wires in a circuit quickly. Marked wires and terminals are also very useful in testing certain circuits and in replacing the wires in them. Other ways of marking wires are shown in Fig. 20-30B.

WIRE WRAPPING AND UNWRAPPING

Wire wrapping makes an electrical connection by tightly coiling a wire around a metal terminal. This process is used because soldering is slow and has other disadvantages. For example, soldering uses heat. This can be dangerous both to a worker and to the components being soldered. Also, a soldered connection takes up a lot of space. Modern electronic circuits are very compact, with little space available. In addition, unsoldering a connection can be difficult. Wire wrapping and unwrapping meet the needs of many electronic circuits.

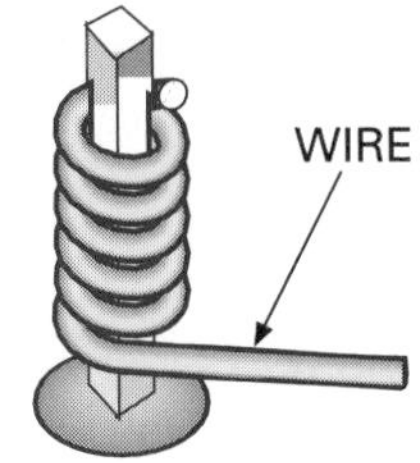

Fig. 20-31. Terminal post for wire wrapping.

Terminal Post

To make a wire-wrapping connection, you will need a terminal post that has at least two sharp edges. Most terminal posts have four sharp edges (Fig. 20-31). Terminal posts used in wire wrapping are generally square, with each side 0.025 x 0.025 in. (0.64 x 0.64 mm). Insulated solid copper wire from no. 26 to no. 32 AWG (0.40 to 0.20 mm) is commonly used for wire wrapping.

Tools for Wrapping and Unwrapping Wire

Figure 20-32A shows a package containing a wire-wrapping tool, an unwrapping tool, and four spools of wire. Fig. 20-32B illustrates four

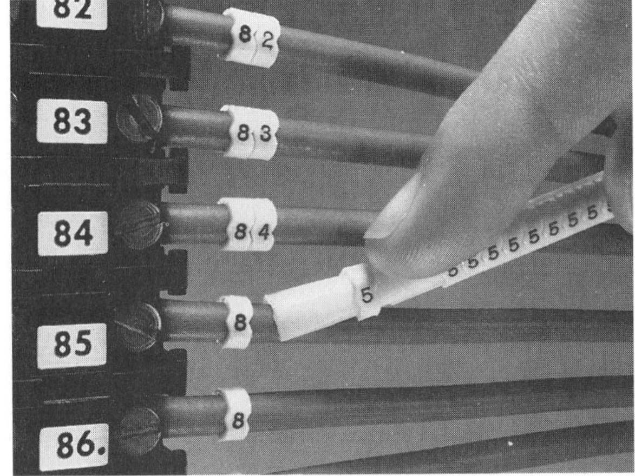

A

B

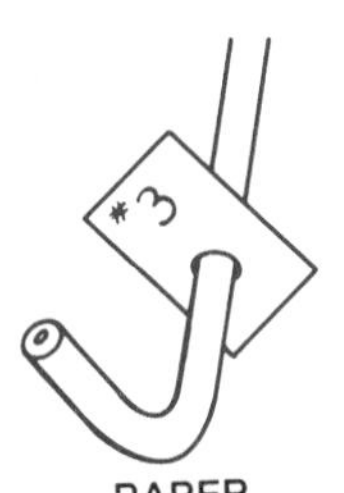

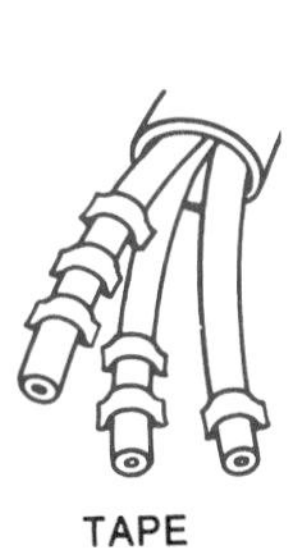

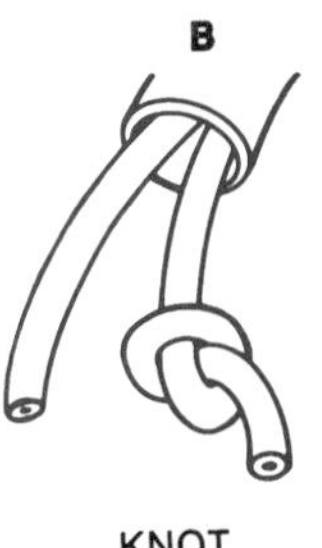

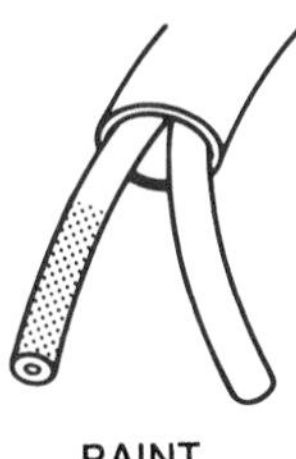

Fig. 20-30. Marking wire: (A) applying a clip-sleeve wire marker; (B) other practical methods of marking wires.

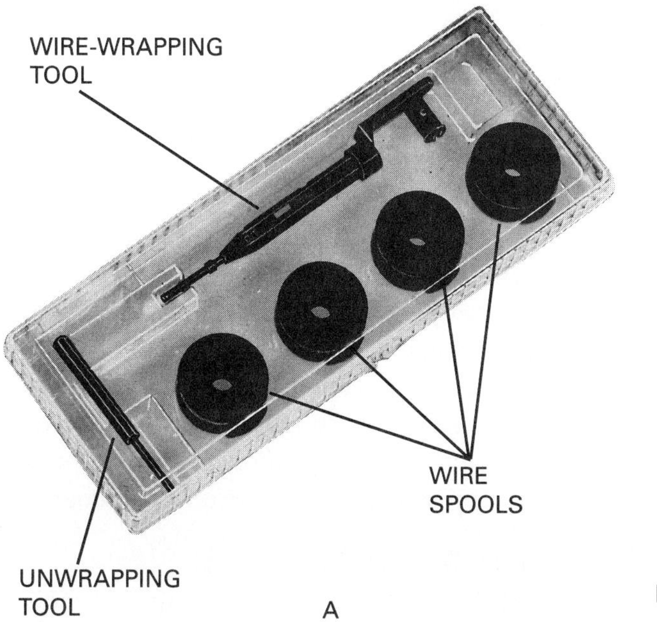

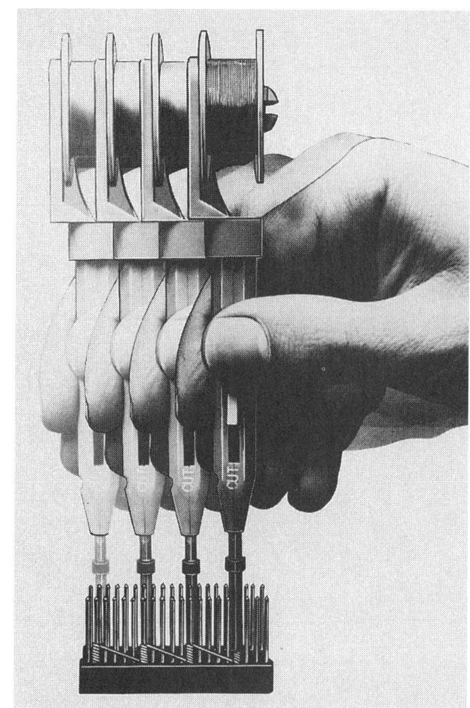

Fig. 20-32. (A) Wire wrapping and unwrapping tools; (B) using the wire wrapping tool.

wire-wrapped connections made to four terminal posts. Each wire wrap is made separately. Then the tool is moved to the next terminal post without cutting or stripping the wire.

The wire-wrapping tool shown in Fig. 20-32 uses a solid copper wire with a special insulation that can be pierced by the sharp corners of terminal posts. Therefore, bare copper wire from no. 26 to no. 32 AWG (0.40 to 0.20 mm) is commonly used for wire wrapping.

Using a Wire-Wrapping Tool

When using a hand wire-wrapping tool, be sure the tool is designed to wrap around the terminal posts being used. To use the wire-wrapping tool, move the slide to the "wrap" position. Then feed the wire through the center hole in the body of the tool, through the eye, and through the wrapping bit (Fig. 20-33). Allow about ¼ in. (6 mm) of wire to stick out of the wrapping bit.

Put the tool over the terminal post. Rotate the tool clockwise about 8 to 10 turns. Keep a firm, but gentle, pressure on the tool while turning it. While you turn the tool, the wire is pulled down from the spool and wrapped tightly around the terminal. To make a point-to-point connection, lift the tool off the first terminal and put it over the next terminal. Repeat the clockwise turning to make another connection. To cut the wire, move the slide to the "cut" position. Keep turning the tool in a clockwise direction. This cuts the wire close to the terminal.

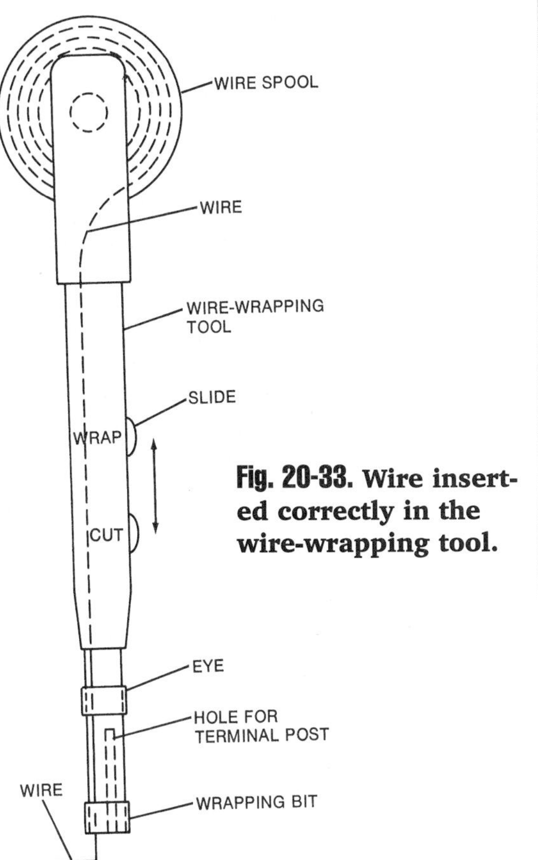

Fig. 20-33. Wire inserted correctly in the wire-wrapping tool.

Unwrapping Wire

To unwrap a connection, put the unwrapping tool over the wire-wrapped terminal post so that it engages the post. Turn the unwrapping tool in a counterclockwise direction. This frees the wire from the terminal post. Thus, the connection is easily removed. Because wire is nicked by wrapping, you should throw it away after unwrapping it. Use only new wire for wrapping.

Tools for wrapping and unwrapping wire make it easy to change connections in a circuit. Therefore, these methods are often used to develop prototype circuits on breadboards.

Other Wire-Wrapping Tools

A variety of other wire-wrapping tools and unwrapping tools are available. Figure 20-34 shows a fast-action tool.

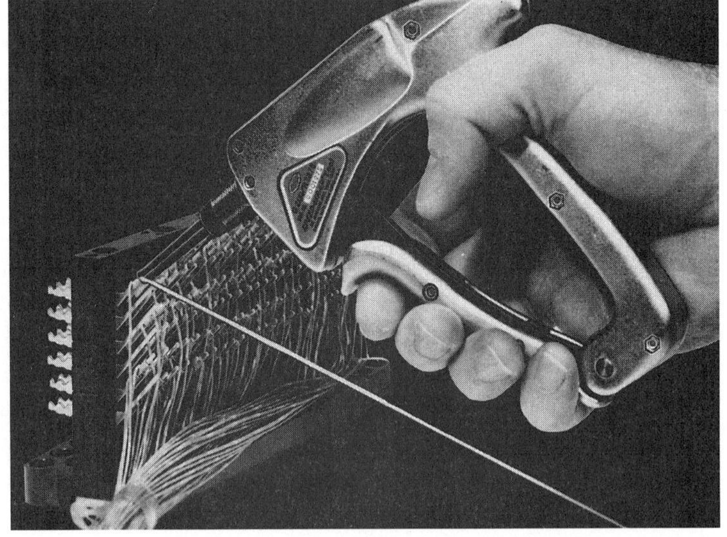

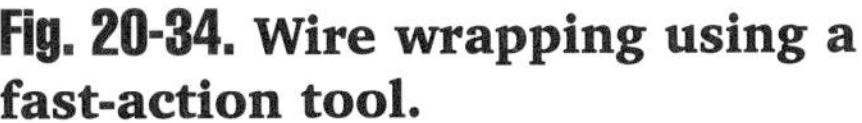

Fig. 20-34. Wire wrapping using a fast-action tool.

With some wire-wrapping tools, you must strip the ends of the wire before wrapping. Usually, about 1 in. (25 mm) of insulation is stripped away. Figure 20-35 shows a wire-wrapping tool that slits the insulation on the wire just before wrapping it. This saves time by removing the need to strip the wire before using it.

Advantages of Wire Wrapping

Wire wrapping has several advantages over other forms of connections. The following list describes a few of these advantages.

- A wire-wrapped connection can be removed as easily as it is made.
- The tools are simple to use and require little training.
- Wire wrapping can be done in tight spaces.
- Wire-wrapped connections are dependable.
- Wire-wrapped connections are as durable as most connections.
- Wire wrapping is economical.

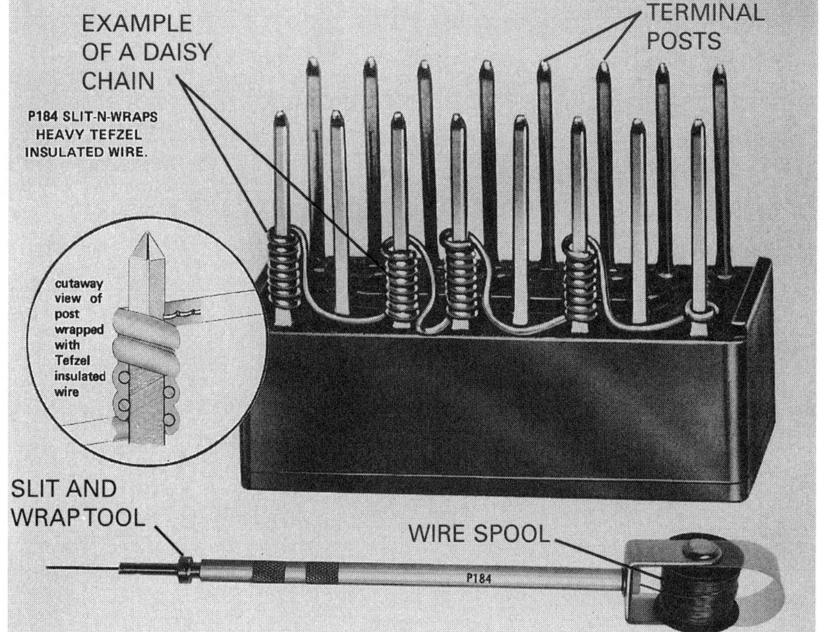

Fig. 20-35. Wire wrapping with a tool that slits the insulation before wrapping.

Chapter *Review*

Summary

- To work efficiently with electrical wiring, you must have the proper tools and use the right techniques.
- Electricians and electronics technicians work with cords and cables that have many different kinds of connectors. These may include plugs and jacks, adapters, wire nuts, clips, or other specialized connectors.
- Most new circuits are the result of many experiments. Breadboards provide the means to experiment with a new circuit until it works properly.
- Other considerations in working with wiring include insulation and the identification of wires or sets of wires. Various techniques and products are available to help you insulate and identify wiring.

Review Main Ideas

Review this chapter's main ideas by writing, on a separate sheet of paper, the word or words that most correctly complete the following statements:

1. Diagonal-cutting pliers should be used only for ______ wires.
2. The size of pliers is given by their ______.
3. An ______ plier can be used as a nut wrench, stripper, cutter, or crimper.
4. The ______ is a tool used to remove insulation from wires.
5. Attachment plugs are fastened to a cord by ______ or ______.
6. The ______ knot is used to keep wires from being pulled loose from the screw terminals if the cord is pulled.
7. The ______ phone jack allows a circuit to be completed through the jack contacts when a plug is not inserted into it.
8. A ______ connects a coaxial cable to television sets.
9. A ______ nut is a device used for making solderless connections in house wiring systems, motors, and various appliances.
10. ______ are fastened to the ends of wires with a crimping, or terminal, tool.
11. Plastic tie strips are used to secure a ______ of wires and cables.
12. ______ circuits are made up of separate components connected by leads.

13. A ______ is used by technicians and electronic engineers to design, build, and test circuits quickly.
14. Electrical tape made of black, coated cloth is called ______.
15. Plastic insulating tubing, or ______, provides a convenient way to insulate wires.
16. Wires and cords that pass through holes in metal are protected by a ______.
17. Insulating varnishes are ______ or are dried by being ______.
18. ______ makes an electrical connection by tightly coiling a wire around a metal terminal.
19. In making an electrical connection, ______ is slow, uses heat, takes up space, and can be difficult.

Apply Your Knowledge

1. Why is an Underwriters' knot used in some attachment plugs?
2. Explain the difference between an open-circuit and a closed-circuit phone jack. Describe an appropriate application for each.
3. Explain why the handles of pliers and wire strippers used in electrical wiring are insulated.
4. In this chapter, you read that alligator clips are often used for instrument test leads and temporary connections because they are easy to install and remove. Describe another application for alligator clips. Be specific.
5. This chapter instructs you to use terminal strips whenever possible in creating permanent circuits. Yet breadboards allow many more connections and are faster and easier to use. Why should you use terminal strips instead of breadboards for permanent electrical circuitry?

Make Connections

1. **Communication Skills.** Write a report on the development of wiring tools.
2. **Mathematics.** Select a pair of long-nose pliers and a pair of short-nose pliers. Calculate and graph the mechanical advantage for each at several equal increments from the tip to the pivot.
3. **Science.** Select a tool. Report on how it applies the principle of mechanical advantage.
4. **Technology.** Select a standard hand tool used in wiring circuits. Redesign it to improve its ease of use.
5. **Social Studies.** Report to the class on the role that tools play in the development and growth of a societal structure.

Chapter 21

Soldering

Terms to Study

cold solder joint

eutectic solder

flux

heat sink

soft soldering

solder

soldering

soldering gun

soldering iron

tinning

OBJECTIVES

After completing this chapter, you will be able to:

- Define *soldering*.
- Specify the proportion of alloys used to produce a low-melting-point solder that melts between 360° and 370°F (182° and 188°C).
- Select the proper size soldering iron for soldering power cords and for soldering electrical components to a printed circuit board.
- Describe the construction features of a switchable dual-heat soldering iron with replaceable tips.
- Explain why acid flux is not used when soldering copper wires.
- Perform the following: tin a wire, solder two wires together, solder wires to lug terminal strips, and solder a microphone cable to a phono plug.

Soldering is a technique that requires practice to do well. Have you ever tried to make a soldered electrical connection? After you read this chapter, you should obtain the needed soldering tools and materials, as well as a pair of safety glasses, and practice soldering on spare and junk parts. When you have gained some experience with soldering, you will be ready to make solder connections on projects.

SOLDER AND SOLDER FLUX

Soldering is the process of joining metals using another metal that has a low melting point. The metal used in this process is called solder. In electricity and electronics, the metals being joined are usually copper wires, lugs, terminal points, and so on. Electrical soldering is called **soft soldering**. Its main purpose is to make a good electrical contact between the soldered surfaces.

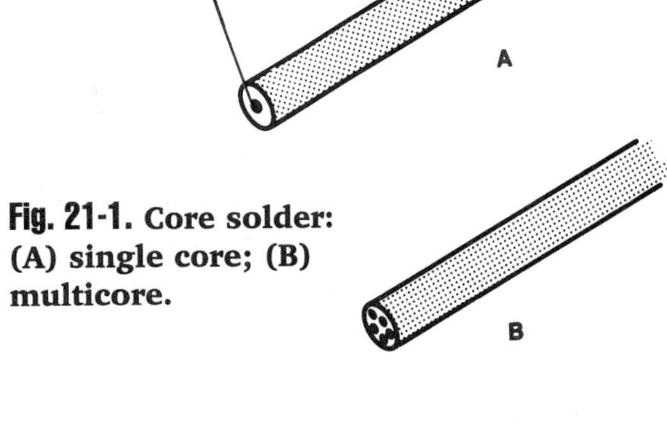

Fig. 21-1. Core solder: (A) single core; (B) multicore.

PLUG IN TO *Industry*

Brazing

A process called brazing *is also widely used to join two metal surfaces. The distinction between brazing and soldering is based on the metallurgical character of the joint and the temperatures necessary to form the joint. The braze, sometimes called hard solder, melts at temperatures above 800°F (430°C), and often contains a large percentage of copper.*

Brazing is accomplished by various methods. In mass production, for example, furnaces uniformly heat the braze on assembly pieces in a protective atmosphere or vacuum. In salt-bath brazing, the parts of an assembly are immersed in a bath of molten salt that melts the braze, protects the parts, and acts as a flux. In dip brazing, the parts are dipped in a bath of molten braze. For single, nonrepetitive operations, a joint is heated with infrared heating, or with a gas, oxyacetylene, or oxyhydrogen torch applied to the joint.

Since brazing depends on capillary action to distribute the filler between the pieces, the distance between parts must be very small, from 0.001 to 0.005 inch (0.025 to 0.13 millimeter). As with soldering, surfaces must be precisely machined and carefully cleaned. Brazing also requires a flux, frequently borax. While the process can join almost all metals, it is most commonly used with alloys of copper, silver, aluminum, nickel, and zinc.

The solder used for electrical and electronic soldering is usually in the form of a wire. This wire-shaped solder has one or more cores of rosin **flux** (Fig. 21-1). When the solder melts, the flux flows over the surfaces to be soldered. It acts as a cleaner to remove any oxides coating the surfaces. These oxides are produced when oxygen in the air combines chemically with the metal. They must be removed, because solder will not stick to surfaces covered with oxides.

An acid flux should never be used when soldering copper wires. Acid causes copper to corrode. This produces a weak solder joint and a high resistance between the connected wires.

Composition and Melting Point of Solder

Soft solder is an alloy of tin and lead. It is usually 40% tin and 60% lead; 50% tin and 50% lead; or 60% tin and 40% lead. Of these three, the 60%-tin/40%-lead solder has the lowest melting point. It ranges from about 360° to 370°F (182° to 188°C). A solder with a low melting point makes it possible to solder with less heat. Using lower heat lessens the danger of damaging components or insulation. This is very important in connecting semiconductor components.

Eutectic solder is 63% tin and 37% lead. It has the lowest melting point of any tin-lead combination: 360°F (182°C).

Size

Solders are specified by diameter as well as by composition. Solder is commonly available in diameters from 0.010 in. to 0.125 in. (0.25 mm to 3.18 mm) in as many as nine increments. Choose a diameter to match the amount of solder needed for the connection. Smaller connections require less solder.

Think of the items you have used today. Which of them might have been manufactured using the soldering process?

SOLDERING IRONS

A **soldering iron** is a tool used to melt solder. In an electric soldering iron, current passes through a heating element to produce heat. The heating element is usually a coil wound with high-resistance wire (Fig. 21-2).

The amount of heat the soldering iron produces is proportional to the wattage rating of the heating element. For electrical and electronic work, soldering iron ratings range from 25 to 100 W. Heavy-duty soldering irons, used for soldering very large wires or for soldering wires to large objects, may have a wattage rating of 250 W or more. A general rule is to choose a soldering iron that does not produce more heat than is needed for the job.

Soldering Iron Holders

A heated soldering iron can cause a fire if not handled properly. Thus, you should be very careful when using any soldering iron. Always keep a heated soldering iron in a holder or support while not using it (Fig. 21-3). Carelessly placed irons may char or burn work surfaces. Holders enclosed with a perforated metal guard give the best protection against burning flesh or clothing.

Tips

Soldering iron tips come in several different shapes (Fig. 21-4). They are usually made of copper. However, some are nickel-plated or iron-clad to reduce corrosion.

Most soldering irons have replaceable tips. The tips either screw into the barrel socket or are held in place by a set screw. When putting a tip into a soldering iron, make sure that it is fully inserted into the socket of the iron.

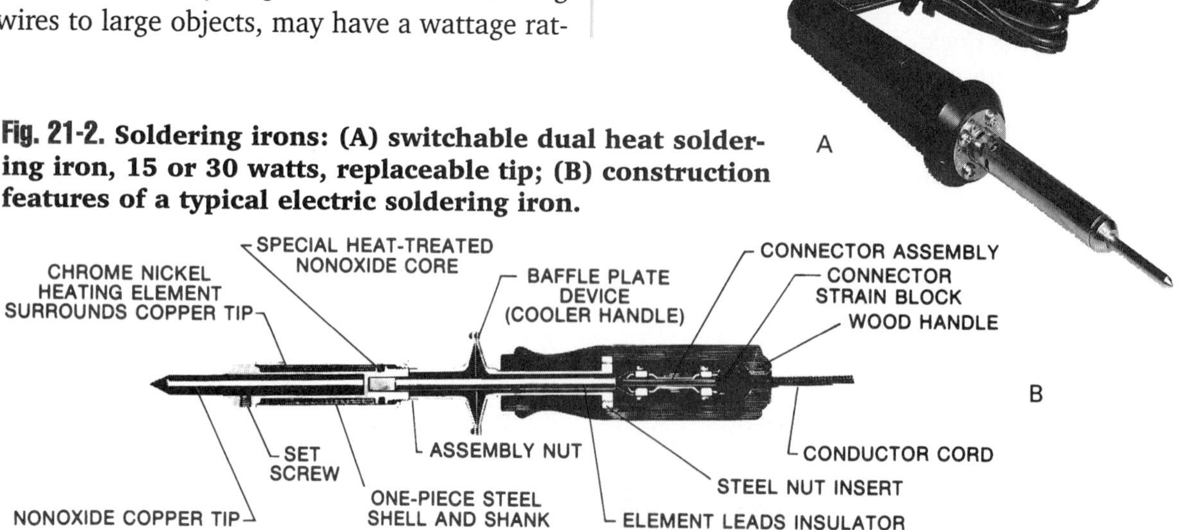

Fig. 21-2. Soldering irons: (A) switchable dual heat soldering iron, 15 or 30 watts, replaceable tip; (B) construction features of a typical electric soldering iron.

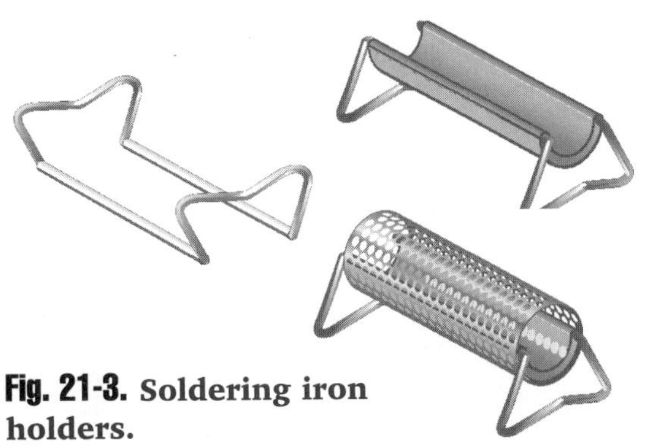

Fig. 21-3. **Soldering iron holders.**

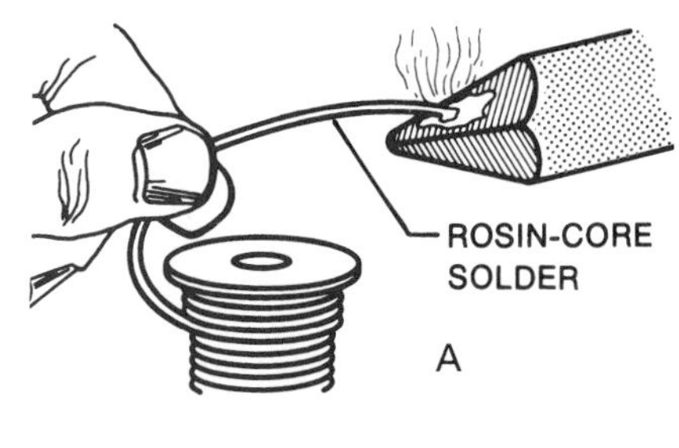

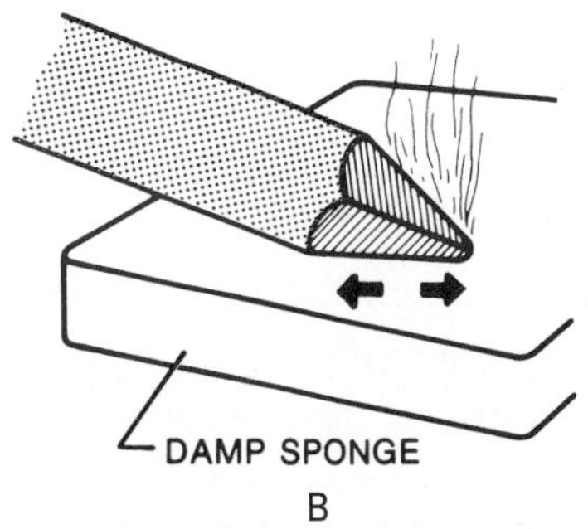

Fig. 21-5. **Preparing the soldering iron tip for use: (A) tinning; (B) removing excess solder.**

After a period of time, soldering iron tips become coated with a thick layer of oxide. The oxide reduces the amount of heat that is transferred from the iron to the tip. It also causes the tip to stick in the iron. To keep these things from happening, remove the tip from the soldering iron after it has been used for some time. Clean it with fine sandpaper or emery cloth before putting it back into the iron.

Tinning and Cleaning

Before being used for the first time, a soldering iron tip should be tinned. **Tinning** is the process of melting a thin coat of solder over the surfaces of the tip (Fig. 21-5A). This allows the most heat to be conducted from the tip to the surfaces being soldered. After tinning a tip, remove any extra solder from it by carefully rubbing it over a slightly damp sponge (Fig. 21-5B).

After being used for some time, the faces, or flat surfaces, of the tip of a soldering iron become coated with the burned residue of solder flux. This coating lowers the amount of heat delivered to the surfaces being soldered. When using a soldering iron, always keep the tip clean and shiny by rubbing it over a damp sponge. However, never clean a plated tip with a file or with a wire brush. This can damage its protective coating.

SOLDERING GUNS

A **soldering gun** is a solder-melting tool that has a step-down transformer. The transformer applies a low voltage across the tip. The cur-

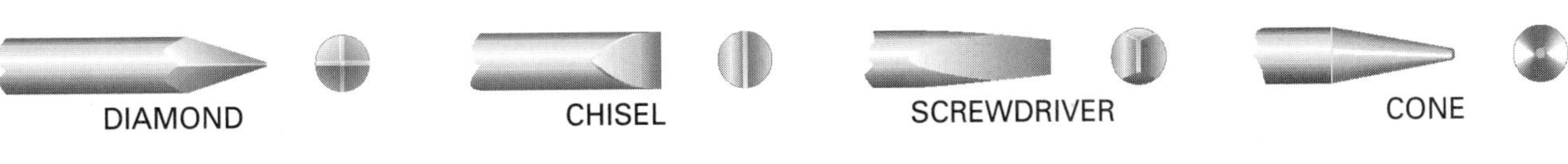

Fig. 21-4. **Soldering iron tips.**

FASCINATING FACTS

An electrolytic tinning method is commonly used to make the tinplate for a tin can. Despite its name, a tin can, if not made completely of aluminum, actually contains only 0.5 to 2% tin as the coating over a steel base. The tin coating on some cans may be as thin as 0.000015 of an inch (0.0004 of a millimeter).

rent passing through the tip produces enough heat to melt solder (Fig. 21-6). The gun is controlled by an on/off trigger switch. The switch often has two trigger positions, allowing the soldering gun to be operated at two heat levels. The wattage ratings for the two levels could be, for example, 100 and 140 W.

When you depress the switch to the first trigger position, the tip becomes heated to soldering temperature in a few seconds. To solder, touch the tip to the metal. Feed a little solder to the tip to release the flux. Then apply solder to the work until the solder flows freely. Withdraw the gun immediately and release the trigger.

The second trigger position produces more heat for heavier jobs. Most soldering guns have one or more lamps that light up when you press the trigger. The lamps illuminate the surfaces being soldered and show that the iron is on.

Soldering guns have replaceable tips that come in a variety of shapes. You should tin and clean them using the same methods you use with soldering iron tips.

A soldering gun with an open tip, such as the one shown in Fig. 21-6, should never be used for soldering semiconductor devices such as transistors and integrated circuits. This is because a strong magnetic field surrounds the tip. This field can seriously disturb the atomic arrangement within semiconductor devices, damaging or ruining them.

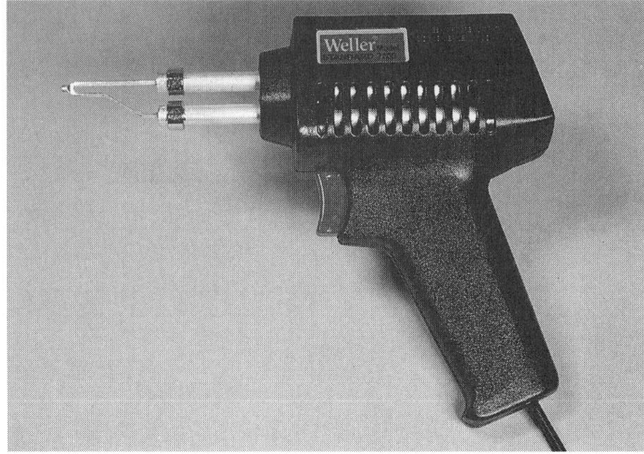

A

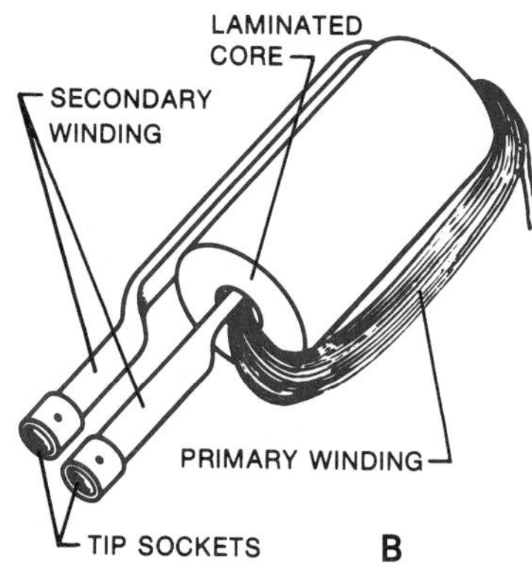

Fig. 21-6. **(A) A typical soldering gun; (B) construction details.**

Electric soldering guns allow more intricate soldering. Think about the difficulty of soldering a complex job with a soldering iron.

PREPARING WIRES FOR SOLDERING

For most soldering jobs, about ½ in. (13 mm) of insulation should be removed from the wire. Wires from which insulation has just been removed are usually clean enough for soldering. However, bare wires such as capacitor and

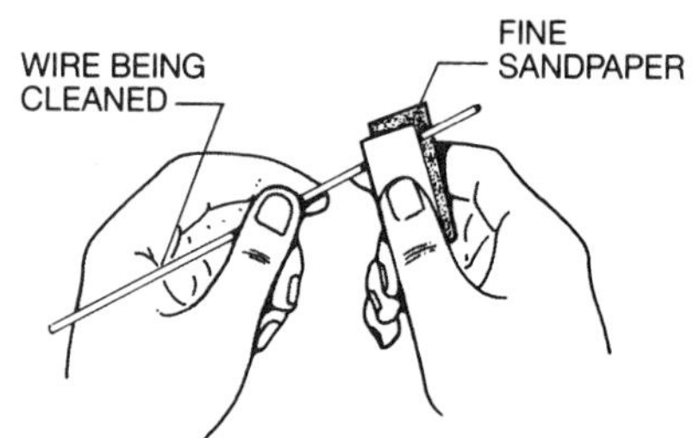

Fig. 21-7. **Cleaning a wire with sandpaper.**

resistor leads may be coated with oil, dirt, or an oxide coating. You should clean these wires with an ink eraser, fine sandpaper, or the blade of a knife (Fig. 21-7). When cleaning any wire, be careful not to nick the wire or to remove too much copper.

After you have removed the insulation from a stranded wire, twist the strands together to form a firm bundle. This keeps the loose strands from making contact with other wires and causing a short circuit.

Some magnet wires are coated with an insulation that burns off as the wires are soldered. To remove the insulation on all other magnet wires, use fine sandpaper, a knife blade, or a chemical solvent.

Tinning Wires

Bare copper wires should always be tinned before soldering. This allows quicker and more effective soldering with less heat. You can often solder pretinned wires more quickly if you retin them just before soldering.

You can tin a wire quickly by leaving the soldering iron in the holder, as shown in Fig. 21-8. To do this, first melt a drop of solder on the soldering iron tip. Put the wire in the solder. Melt a small amount of additional solder on the tip, turning the wire until all its surfaces are thinly coated with solder.

Most terminals and lugs to which wires are soldered are pretinned. They do not need more tinning. However, before soldering you should tin any bare surface to which a solder connection will be made.

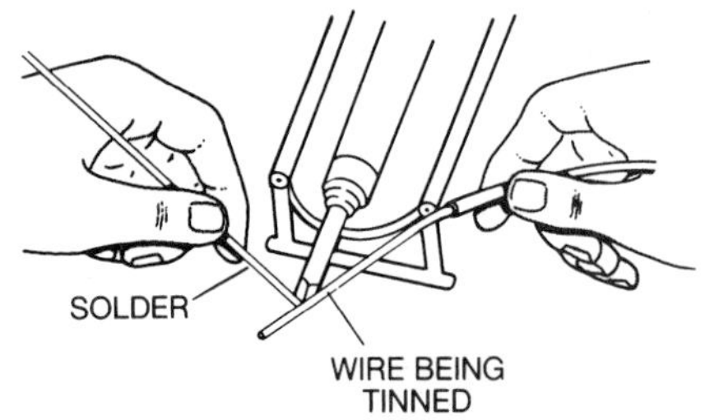

Fig. 21-8. **Tinning a wire using a soldering iron in a holder.**

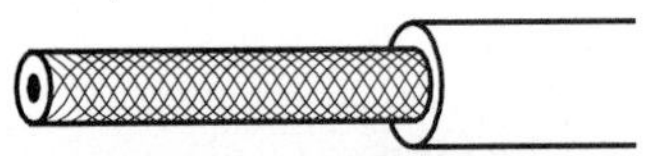

1. REMOVE APPROXIMATELY 1½ IN. (38.1 mm) OF OUTER JACKET.

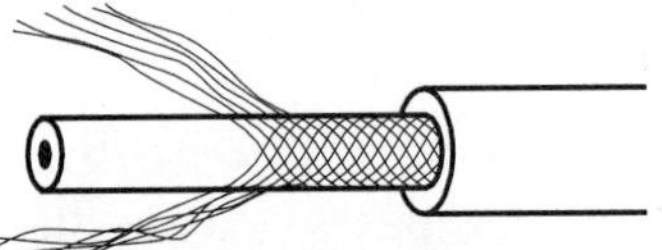

2. UNBRAID THE SHIELD WITH A SCRIBER OR OTHER SHARP-POINTED TOOL.

3. TWIST UNBRAIDED PORTION OF SHIELD TOGETHER AND REMOVE A SHORT LENGTH OF THE INSULATION THAT COVERS THE CONDUCTOR.

Fig. 21-9. **Preparing shielded coaxial cable for soldering.**

Preparing Coaxial Cable

Figure 21-9 shows how to prepare shielded coaxial cable for soldering when the braid shield must be connected into a circuit. When working with coaxial cable, be careful not to damage the insulation between the shield and the inner conductor. This is particularly important when you are soldering a wire lead to the shield. If you apply too much heat to such a joint, the insulation may melt enough to cause a short circuit between it and the inner conductor.

SOLDERING TECHNIQUES

Some soldering techniques apply to all soldering jobs. For example, you should always handle a soldering iron or gun with care to keep from burning the insulation or damaging the components. Burned or charred insulation does not look good. In addition, it may cause a short circuit or ground.

Other soldering techniques are unique to the type of soldering job you are performing. The following text explains techniques for three specific soldering jobs.

Soldering Wires to Lugs

Before soldering a wire to a lug, fasten it to the lug mechanically to hold it in place (Fig. 21-10A). If the wire is the lead of a semiconductor device such as a diode or transistor, attach a heat sink to it (Fig. 21-10B). A **heat sink** absorbs or dissipates excess heat that could otherwise damage the semiconductor device.

Next, melt a drop of solder on the tip of the soldering iron or gun. Press the tip to one side of the joint (Fig. 21-10C). The melted solder allows more heat to be conducted from the tip to the joint. Briefly heat the joint. Then press one end of a length of solder wire to the other side of the lug until a small amount of the solder has melted (Fig. 21-10D). Remove the solder wire and let the melted solder flow thoroughly into the joint. Hold the tip in place a while longer. Then remove it from the joint. After the joint has cooled, remove the heat sink.

Use only enough solder to cover all surfaces of the joint without hiding the shape of the joint (Fig. 21-11). This enables you to inspect the joint. It will also keep extra solder from causing a short circuit with nearby wires or terminals.

Do not move or otherwise handle a newly soldered wire until the solder has cooled. If you move a warm connection, the solder will suddenly become less shiny. A dull-colored solder joint indicates that not enough heat was used. Such a joint, commonly called a **cold solder joint**, does not provide a good electrical contact.

Soldering Wires to Hollow Terminals

The technique for soldering a wire to a hollow terminal, such as a phono plug, is shown in Fig. 21-12. To solder, first melt a small amount of solder on the tip of the soldering iron or gun.

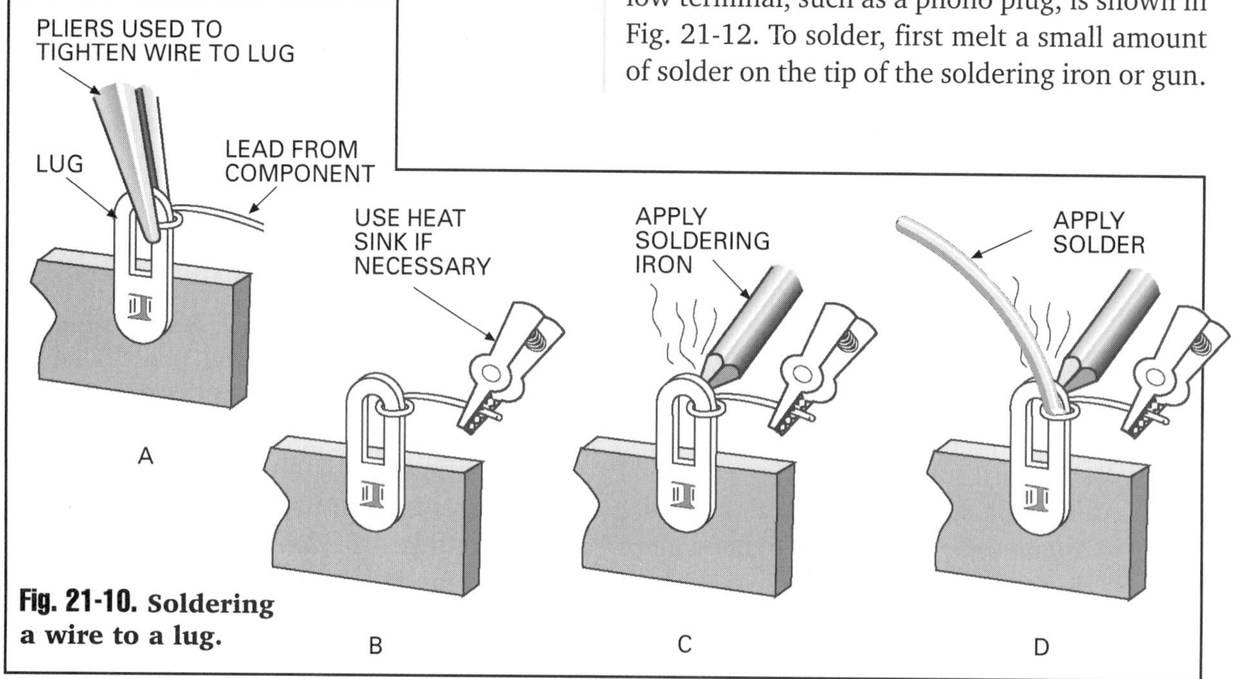

Fig. 21-10. Soldering a wire to a lug.

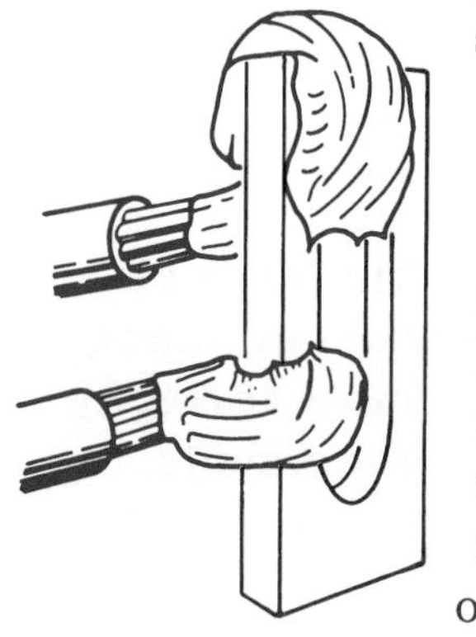

Fig. 21-11. Solder joints showing the correct amount of solder to be used.

Hold the tip of the plug in place until the solder has been drawn up into it. Then add some more solder if needed. Cut or file away any extra solder or wire that extends beyond the tip of the plug.

Soldering Wires Together

Wires are usually soldered at terminal points, such as lugs. However, it is sometimes necessary to solder two wires together without a lug. Before soldering, twist or hook the wires together tightly (Fig. 21-13A). Then solder them together, using only the amount of solder needed to make a good connection.

Make the solder joint as short as possible. Cover the joint with insulating tubing or other insulation (Fig. 21-13B). The tubing insulates the joint and makes the job look neater.

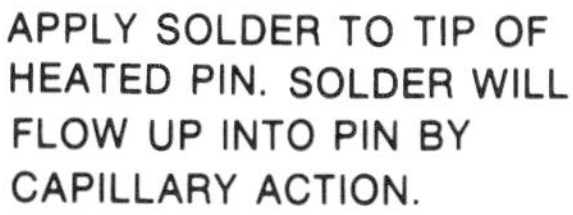

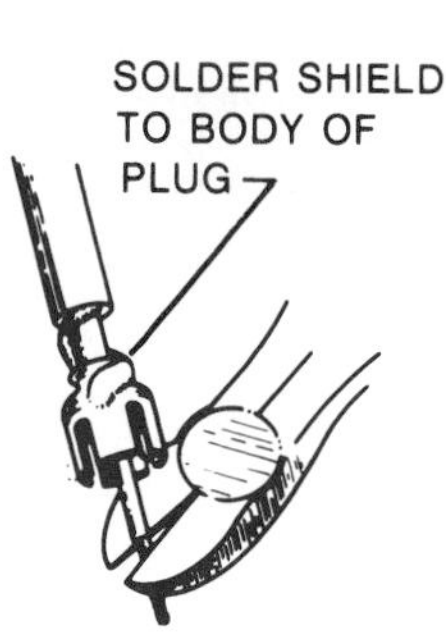

Fig. 21-12. Soldering a wire to a hollow terminal such as that of a phono plug.

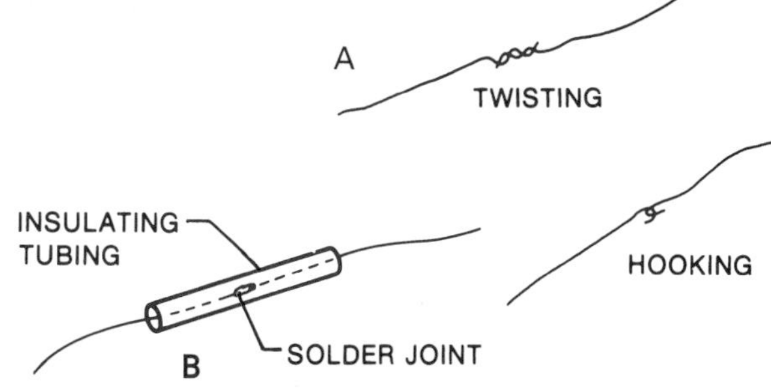

Fig. 21-13. Soldering two wires together: (A) twist or hook the wires together tightly to provide mechanical strength before soldering; (B) cover the solder joint with insulating tubing.

Soldering Integrated Circuits

Components of an integrated circuit are often small and fit closely together on a circuit board. When soldering these components, you must concentrate on each soldered joint to prevent short circuits (Fig. 21-14).

LINKS

You may wish to refer to Chapter 23, "Testing, Troubleshooting, and Repair," for information on testing and troubleshooting.

Fig. 21-14. Soldering components on a circuit board.

Chapter *Review*

Summary

- Soldering is the process of joining metals using another metal. The solder used for electrical and electronic soldering is usually a wire that has one or more cores of rosin flux.
- Soldering irons and guns are used to melt solder. The amount of heat produced by these tools is proportional to the wattage rating of their heating element.
- Tips for soldering irons and guns come in several shapes and are usually made of copper.
- Wires must be prepared for soldering. This may involve removing insulation, cleaning, twisting strands together, or tinning.
- Some soldering techniques apply to all soldering jobs. Others are unique to the type of soldering job you are performing.

Review Main Ideas

Review this chapter's main ideas by writing, on a separate sheet of paper, the word or words that most correctly complete the following statements:

1. Soldering is the process of joining metals using another metal that has a ______.
2. Solder flux is used as a cleaner to remove any ______ coating the surfaces to be soldered.
3. Never use ______ flux when soldering copper wires.
4. Soft solder is generally an alloy of ______ and ______.
5. The melting point of 60/40 solder ranges from about ______ to ______ F.
6. Solder is commonly available in diameters from ______ to ______ in as many as nine increments.
7. The amount of heat a soldering iron produces is proportional to the ______ rating of the heating element.
8. In doing a soldering job, it is always best to choose a soldering iron that does not produce more ______ than is needed for the job.
9. A soldering-iron or -gun tip is tinned by coating its surfaces with ______.
10. A soldering gun with an open tip should never be used for ______ devices such as transistors and integrated circuits.

11. Bare copper wires should always be ______ before soldering.
12. Before soldering a wire to a lug, fasten the lug ______ to hold it in place.
13. A ______ absorbs excess heat that could otherwise damage a semiconductor device.
14. A solder joint to which enough heat has not been applied is called a ______ solder joint.
15. Always handle a soldering iron or gun with care to keep from burning the ______ or ______ the components.

Apply Your Knowledge

1. How is the solder used for electrical connections different from other types of solders? Why does it need to be different?
2. Explain why soldering irons and guns are made in a variety of electrical sizes.
3. Suppose you are using a soldering iron and notice that the soldered joints are dull-looking, even though you applied heat for an appropriate amount of time. What might cause this? What could you do to prevent it?
4. Explain why it is often a good idea to retin pretinned wire.
5. Explain how to solder wire to lug terminal strips.

Make Connections

1. **Communication Skills.** Prepare a demonstration on proper soldering techniques. Discuss the materials needed to solder electronic components to printed circuit boards.
2. **Mathematics.** Devise a method to determine the length of wire solder in a one-pound spool of 60/40 resin core solder, diameter #22 AWG.
3. **Science.** Devise an experiment to compare the melting points of 60/40 resin core solder with eutectic solder 63/37.
4. **Workplace Skills.** Demonstrate the use of acid core or liquid flux and resin core solder to show why acid flux is not used on electronic parts.
5. **Social Studies.** Report to your class on the dangers of lead base solder when used in areas where human contact and ingestion are possible (utensils, drinking water pipes, and metal plates). Emphasize proper handling of solder.

Section 6

Testing and Troubleshooting

Chapter 22

Measuring Electrical Quantities

OBJECTIVES

After completing this chapter, you will be able to:

- Identify common meters.
- Explain how to operate a digital multimeter.
- Describe how a voltmeter is connected to an electric circuit.
- Illustrate how an ammeter is connected into an electric circuit.
- Explain the operation of a dual-trace oscilloscope.
- Explain how to use signal generators.

Terms to Study

ammeter

dual-trace oscilloscope

frequency counter

function generator

multimeter

ohmmeter

oscilloscope

signal generator

virtual instrumentation

voltmeter

In earlier chapters you learned about electrical quantities, such as voltage, current, and resistance. Your teacher may have used instruments called meters to measure the quantities as they were being explained. Technicians and engineers use these instruments to measure the various quantities to decide how a circuit operates. An understanding of how meters and instruments work will help you study electronic principles.

METERS

In order to "see" electricity, instruments called meters are used to measure electrical quantities such as voltage, current, resistance, and power. This chapter discusses three types of meters: digital, analog, and virtual. To indicate various electrical quantities, digital meters use numbers, or digits, on a display. Analog meters use moving pointers over scales, and virtual meters use computer software to produce meter images on a screen. These are used in designing, manufacturing, and servicing all kinds of electrical and electronic products.

Voltmeters

A **voltmeter** is a testing device used to measure voltage. The meter test leads or wire connections are attached to the tested circuit in parallel with the device that is expected to have a voltage drop. Various levels of voltage can be measured, from microvolts to megavolts, when the proper instruments are used. Measurement of very large levels of voltage require specially insulated equipment to ensure operator safety. Voltmeters are available as instruments for specific applications, or as one of the several selectable functions of a *multimeter*. **Multimeters** are meters that can measure current, voltage, and resistance. Panel voltmeters are single purpose instruments (Fig. 22-1). They can be mounted in an area for a technician to easily observe the reading and use the information to operate a system. *Digital multimeters* (DMM) that measure voltage are used when individual tests of circuits are needed to diagnose a problem. The meters shown in Fig. 22-2 are a handheld portable model (A) and a bench model (B).

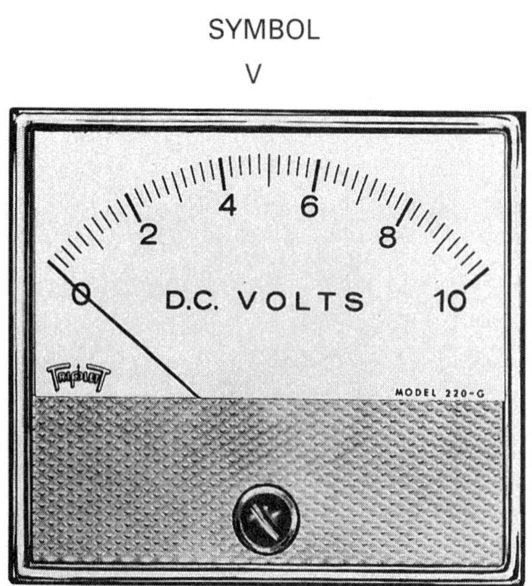

Fig. 22-1. **A dc voltmeter.**

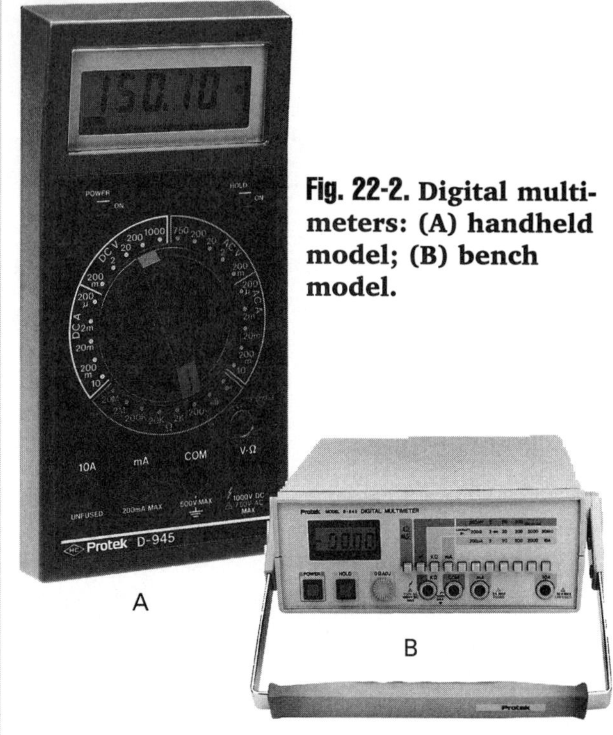

Fig. 22-2. **Digital multimeters: (A) handheld model; (B) bench model.**

Ammeters

An **ammeter** is a testing device used to measure current. Microammeters measure current in millionths of an ampere, and milliammeters measure in thousands of an ampere (Fig. 22-3). A special type of ammeter called a *galvanometer* detects and indicates the direction of extremely small amounts of current. As with voltmeters, ammeters can be built as individual instruments or panel meters. They can also be part of multimeters or digital displays. Multimeters use a single meter movement or

Fig. 22-3. Panel-mounted milliammeter.

digital display but different internal circuitry to measure voltage, current, and resistance. Panel ammeters measure either direct or alternating current. Many ammeters measure the voltage drop across a fixed resistor called a *shunt*. The voltage drop changes as current changes, therefore, the meter will reflect current measurements.

Ohmmeters

Electrical resistance is measured with an **ohmmeter**. The digital display or movement of ohmmeters is the same as that in voltmeters and ammeters. Ohmmeters can be either analog or digital and part of an individual unit or a multimeter (Fig. 22-4).

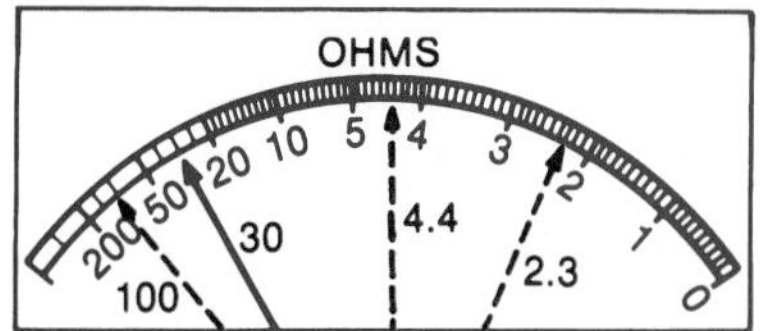

Fig. 22-4. Reading the ohms scale on an analog volt-ohm-milliammeter (VOM).

An ohmmeter circuit requires a source of energy such as a cell, a battery, or both. Contained in the case of the meter, the voltage of the cell provides a standard voltage which sources a current. By measuring this current and applying Ohm's Law, the resistance is calculated and displayed. The voltage on the cells or batteries of an ohmmeter should be measured periodically. This measurement should be at or slightly above the rated voltages on the cells or batteries. With this voltage, the ohmmeter will operate properly and resistance readings will be accurate.

A *megohmmeter* is used to test materials with high resistances. Equipment cases and wire insulation are examples. A high voltage output from the megohmmeter (meggar) is needed to measure these resistances. The source of energy in the megohmmeter is sometimes a hand-cranked generator that provides this high voltage.

Illustrate a shunt-type ammeter. Include in the illustration a circuit diagram and explain how the voltage reading reflects current.

DIGITAL METERS

Digital meters display a measured value on a numeric display with fixed increments. These meters sample a signal and convert it from an analog quantity into digital form. The device that does this is called an analog-to-digital converter or ADC.

LINKS

Analog-to-digital converters are discussed more fully in Chapters 26 and 27.

The input to a DMM has field-effect transistors as the first amplifier for the test signal. This gives the DMM a very high input impedance or resistance. To measure voltage in a circuit, the meter is connected in parallel to the device being measured. Remember that the total resistance of a parallel circuit is reduced when a circuit is added in parallel. The high input impedance of the DMM, therefore, has little effect on the test circuit because the current sample being taken from the test circuit is very small.

Digital meters have replaced many of the analog meters used to measure electrical quantities for three main reasons. First, the digital meter is easy to read. Second, the measurements are more accurate and more quickly accommodate today's electronic circuits. Third, the purchasing cost of a digital meter is much less than that of an analog meter.

Digital meters are classed by the number of digits on their displays. Generally the display will have 3 ½, 4 ½ or 4 ¾ digits plus a positive (+) and negative (-) sign. The left-most digit is the fractional digit and displays the character "1" or is off. The + or - sign shows the polarity of the value under test. Some meters have an analog bar graph which gives the user a second reference by which to observe and understand the measurement. The input is not a continuous sampling, but is done a fixed number of times per second. This is called the sampling rate. If a continuous sampling is used, the circuitry is fast enough that the display would be constantly changing and no reading would be possible.

Fig. 22-5. A digital multimeter.

Digital Multimeters

The *digital multimeter* shown in Fig. 22-5 is a sophisticated type of meter used by technicians and engineers. This meter is designed to measure volts, ohms, and milliamperes. Some of its features include a function selector, a power "on/off" switch, 3 and ¾ digit display with a polarity sign, and floating decimal point. The display is a liquid crystal type, which produces black numbers on a gray background. It has a test function for diode junction, as well as ac and dc voltage, ac and dc current, ohms, and other specialized items. This meter also has an RS232 standard interface to send measurement data to a personal computer (PC) for storage and analysis. Many models of digital meters have an audible tone for continuity tests. This feature allows the user to make such tests without having to look away from the device being tested. The meter operates well on a 9-volt battery and has a low voltage level indicator on the display to tell the user when it is time to change the battery.

dB Meters

Some meters are also equipped with a decibel (dB) scale. This is used to measure the sound output level of audio amplifiers for public address systems. The decibel unit is a comparison of the power level of an output signal to a known reference power level, usually the input power level of the device being rated. The dB value is the mathematical expression of how many times larger (+dB number) or smaller (-dB number) the tested signal is than the reference.

Clamp-on-Meters

Most meters must be connected directly into a circuit to get a reading. *Clamp-on meters* are designed to avoid this and are generally used to measure current and voltage. (Voltage can be measured with a set of test leads and wires.) Current is measured by opening the jaws, passing a single conductor under test into the space and closing the jaws around the conductor (Fig. 22-6). When a current passes through it, the magnetic field that surrounds the wire will be directly proportional to the quantity of current. The circuitry in the meter uses the magnetic field, determines the level of associated current, and displays a reading. Clamp-on meters make it easy to take current measurements in wires that cannot be easily disconnected from their terminals.

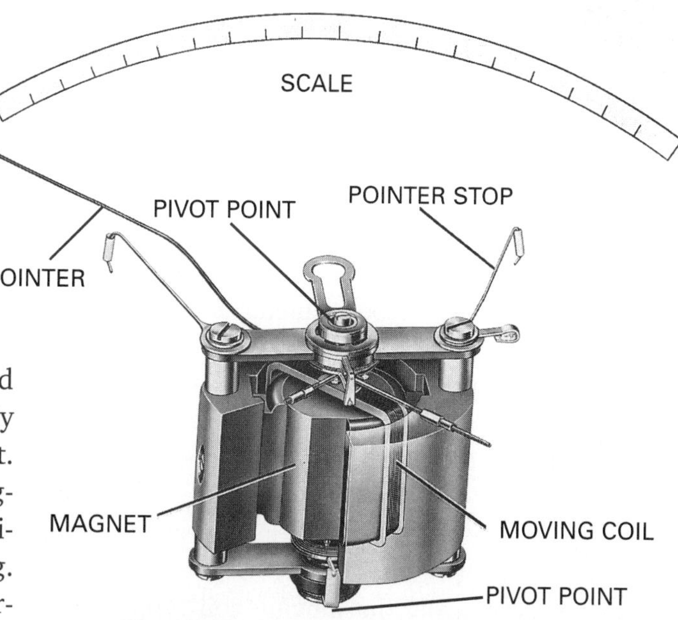

Fig. 22-7. **The permanent-magnet moving coil meter movement.**

BRAIN BOOSTER

Describe the uses of a multimeter. How do digital and analog multimeters differ?

Fig. 22-6. **A clamp-on meter.**

THE MOVING COIL METER

A typical *moving coil meter* works like a small electric motor and the magnetic action causes the meter pointer to move. The amount of this movement depends on the amount of current passing through the meter circuit. Although the pointer moves because of current, other electrical quantities, such as voltage and resistance, can be measured. The meter face can be marked to show the different electrical quantities. The arrangement of parts that cause the pointer to move is called the meter movement. Many meters use permanent-magnet moving coil movement (Fig. 22-7). The coil is wound around a frame mounted between pivot points so that the coil is free to turn. The pointer is fastened to the coil assembly. A small spring holds the coil so that the pointer is at zero on the meter scale. The ends of the coil are connected to the meter's stationary terminals.

When current passes through the coil, a small electromagnet is produced. The poles of the electromagnet are repelled by the poles of the permanent magnet. Since the movement of the coil and pointer depends upon current flowing, the pointer will move. From zero, it moves a distance determined by the value of the current. The amount of current needed to move the pointer to the highest reading on the meter scale is called the *full-scale deflection* current of the meter. This type of meter is called an analog meter. It is called an analog meter because it operates with a continuous variable quantity. In this meter, the continuous variable quantity is the current flowing through the coil of the meter.

USING METERS

All meters, especially the moving coil type, are delicate instruments. They must be used carefully to protect them mechanically and electrically. You must choose the correct range on a voltmeter or ammeter. This means that the instrument must be adjusted to safely handle the highest voltage or largest current being tested.

You should always learn how to operate any meter before trying to use it. One of the best ways to do this is to read carefully the meter's operation and application manual. This manual provides detailed instructions for using and reading the meter safely. Each of the instruments discussed in this chapter and those with which you will work are supplied with an instruction or user's manual from their manufacturer. Instruction manuals are very important and should be read before any instrument is used. The information contained in it includes the proper way to make connections, design specifications for measurements being made, and important safety instructions to follow. Your teacher probably has an instruction manual for each of the instruments you will be using in class. It is strongly suggested that this manual be read before using a new instrument, or as an aid when operational questions arise during use of any instrument.

A very important part of the instruction manual is the safety instruction section. Read the safety instructions carefully. Be sure you understand them before you operate any test instrument. Also listed in the manual is a specifications table that summarizes the characteristics of an instrument. Included in the listing is the power requirements, ranges of measurement, available output signals, input impedances for probes, battery type needed, and much more.

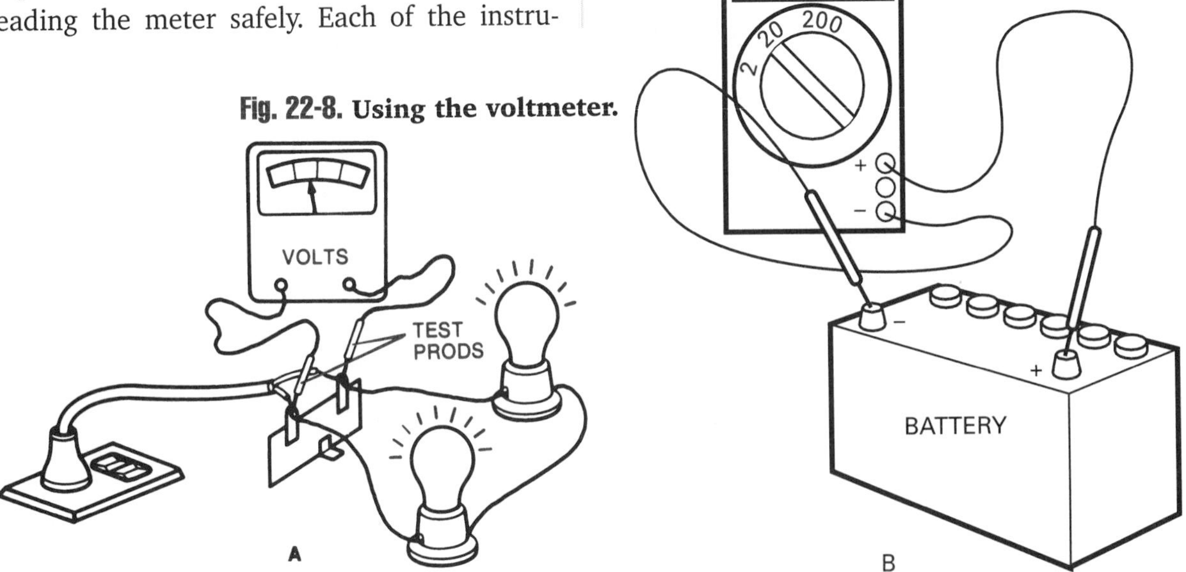

Fig. 22-8. Using the voltmeter.

Voltmeters and ammeters are connected to energized circuits or to circuits in which voltage will be applied. It is very important, therefore, to follow the safety rules while using them. Voltmeters and ammeters are often connected to a circuit with clips or screw terminals. It is always safer to turn off the power supply before making these connections. The circuit can then be turned on to make the measurement. Always check the test lead insulation for cracks or breaks that might connect you to the circuit you are measuring. Also, be careful not to touch bare circuit wires when using test leads to connect a meter to a circuit.

Using Voltmeters

Voltmeters are always connected in parallel to the device or circuit across which the voltage is to be measured. A dc voltmeter is a *polarized instrument*. This means that care must be taken to connect the test leads to the correct polarity, or + to + and - to - (Fig. 22-8). If this is not done, a digital meter will indicate a negative voltage, which might confuse the data. Also, the pointer of a moving coil meter will move in the wrong direction and damage the mechanism. As you have learned, alternating current reverses direction every half cycle and, therefore, ac does not have polarity. So, an ac voltmeter can be connected across two points under test regardless of polarity.

Fig. 22-9. **Using the ammeter.**

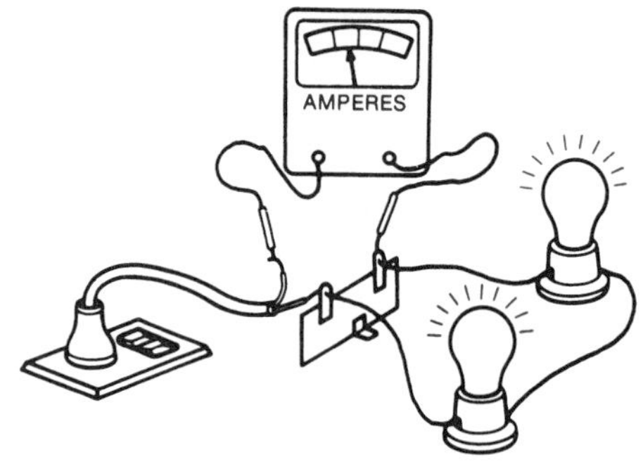

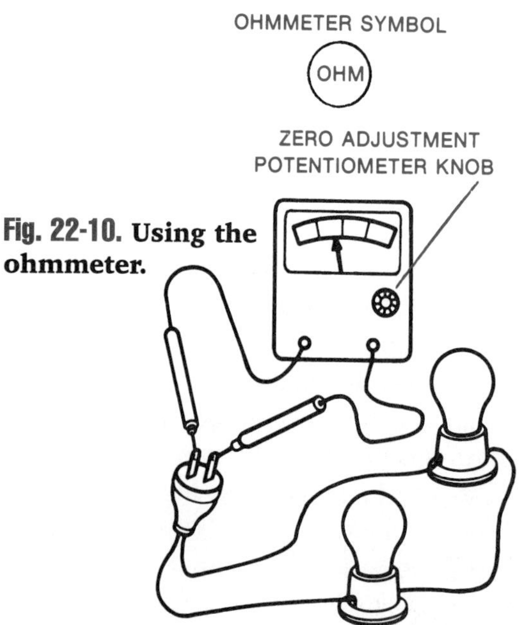

Fig. 22-10. **Using the ohmmeter.**

Using Ammeters

Ammeters are connected in series with the conductors and the load being tested (Fig. 22-9). They can be permanently damaged if connected into circuits with too much current. A dc ammeter should be connected into a circuit with the correct polarity. A digital ammeter which has been connected incorrectly will indicate a reverse polarity on the display. If a moving coil meter is connected in reverse, the pointer will move in the wrong direction and may be damaged. An ac ammeter can be connected into a circuit without regard for polarity.

Using Ohmmeters

Ohmmeters are connected in parallel with the terminals of the device or circuit to be tested (Fig. 22-10) The device or circuit must be disconnected from power and from any portion of the circuit that is not to be tested. False readings will result if other components are connected to the test circuit. Do not let the fingers of both hands touch the tips of the test leads while measuring resistance. If you do, the ohmmeter will measure the combined resistance of your body and the circuit being tested. Be very careful of this when measuring high values of resistance, as your body resistance is high and will affect the total resistance of another high resistance. A false reading can cause you to make a poor decision when troubleshooting a circuit.

Measuring Resistances

Typical ohmmeters have five or six ranges from which to take readings. If the test resistance is unknown, select a high range and take the measurement. If one or two digits appear on the display, the range is too high and should be adjusted at a lower range. If the range is too low, some indication will appear to tell you that the test value is above the selected range. This indication might be something like "OL" (over limit), which means the range must be changed to a higher one to get any reading. A display of 1.987 with the meter selector set to a 2K range indicates a resistance value of 1,987.

Continuity Tests

In addition to measuring resistance, an ohmmeter is used to make *continuity tests*. These determine if there is a continuous current path from one test point to another. Continuity tests are very useful for checking hidden conductors of coils and other devices in which the complete path cannot be seen. Continuity tests are commonly used for checking switches, transformers, and relays.

When making a continuity test, adjust the ohmmeter to its lowest range and connect it to the terminals of the device or to the point of the circuit being tested. If the display does not indicate a resistance reading, the device or the circuit being tested is open (Fig. 22-11A).

If, on the other hand, the ohmmeter indicates some resistance, there is continuity between the terminals (Fig. 22-11B). The amount of resistance will be determined by the characteristics of the device or the circuit being tested. When the resistance is very low, as shown when the meter display has a low number, such as .05 or .00, there is said to be direct continuity between the points under test.

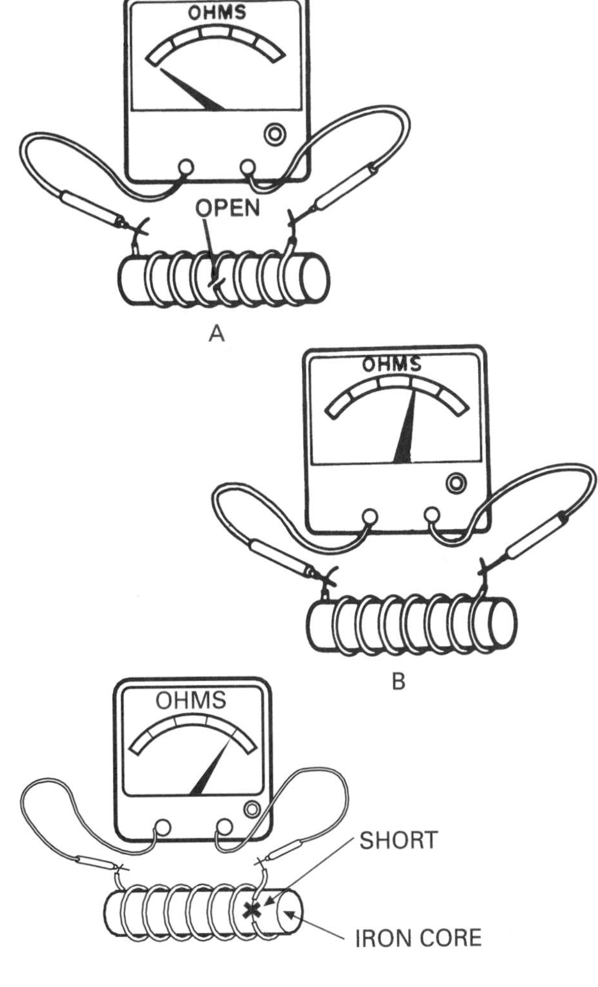

Fig. 22-11. Making a continuity test with an ohmmeter.

Some digital multimeters have an audio feature for continuity testing. With this feature, continuity tests can be made without looking at the display. This is useful for checking closely placed pins and sockets on cable ends. It is often necessary to test continuity from end to end on these cables to make sure that the proper pins are connected and not shorted. The tone allows the technician to test the circuit without looking away from the leads to the display and risk moving the test probe. Continuity tests are also very useful for checking shorts and grounds. In the test shown in Fig. 22-11C, direct continuity exists between one terminal of a coil wound with insulated wire and the core. In this case, the wire insulation has become defective at some point. There is a short circuit between the wire and the core. Since this condition might be dangerous or cause the circuit to work improperly, the coil should be replaced.

What is the single most important aspect to remember when using any kind of meter?

VIRTUAL INSTRUMENTATION

With the development of ever faster personal computers and ever-growing memory capacities, it is now possible to create software programs that provide technicians and engineers with a wide variety of computer based instrumentation. When computers perform the function of electronic instruments, it is called **virtual instrumentation**. Because of a computer's power, virtual instrumentation provides technicians and engineers with a masterful diagnostic tool. This allows users greater versatility when custom instrumentation is required. Engineers can design systems that will yield very specific data sets to meet very specific needs. The virtual instrument system can also be designed to mimic a standard instrumentation system composed of individual units. Data obtained by this system is in the computer ready for transmission via the Internet to interested researchers. This option is not available with standard bench instrumentation. Another advantage of the virtual system is that the user can update instead of purchasing new instruments. A disadvantage of this system is that it depends on a computer. If the computer isn't working, all virtual instrumentation capabilities are lost.

Some of the common instruments available in the virtual mode include digital multimeters, oscilloscopes, function generators, dynamic signal analyzers (DSAs), and data loggers. Each of these instruments is generated in computer software programs and appears on a screen. A keyboard and mouse, on screen buttons, knobs, and slider controls can be operated just like standard instrumentation. The power of a PC when added to a virtual instrumentation system increases the ability to process, display, and store data. The speed of PCs allows data acquisition in increments of microseconds, which can then be displayed in table form for analysis. Data and waveforms displayed on the screen can also be printed.

External test circuit connections are made to the virtual instrumentation software through a connection device and a *data acquisition (DAQ) interface* card (Fig. 22-12A) added to the host computer at one of the card connector slots. The external connector (Fig. 22-12B) is provid-

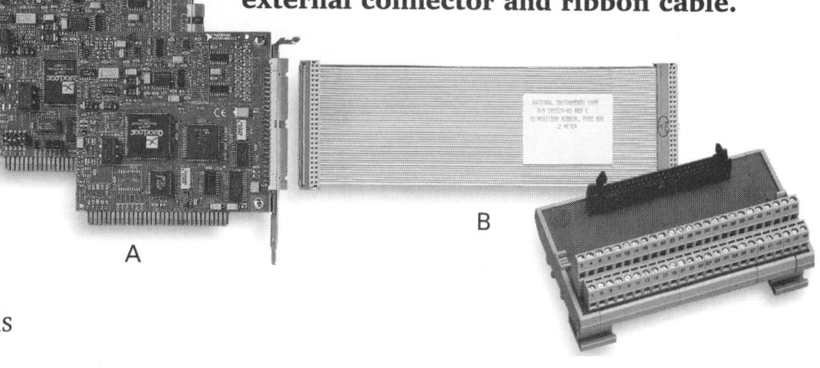

Fig. 22-12. Virtual instrumentation: (A) data acquisition (DAQ) interface card; (B) external connector and ribbon cable.

ed with a number of input and output terminals to allow the user to connect several devices to the DAQ card simultaneously. The software is then programmed to switch very rapidly between inputs and to assemble the data from each into a display.

Write a brief essay that describes the advantages of virtual instrumentation.

SELECTED VIRTUAL INSTRUMENTATION

The selected instrumentation that follows represents some of the possibilities using a computer's power. There are many more virtual instruments available to meet the advancing needs of the research and development industry. Beyond the virtual instrumentation already developed, software programs can meet specialized needs allowing users to create customized virtual instrumentation.

Virtual Multimeter

Digital multimeters are used to take voltage and current (ac and dc), resistance, and special measurements such as temperature and diode junction tests. The virtual DMM in Fig. 22-13 has input ranges which can be automatically or manually selected from 1 mV to 10 V.

Virtual Oscilloscopes

Figure 22-14 shows a virtual oscilloscope display. This instrument is capable of displaying wave patterns of circuitry with multiple channels and frequency bandwidths from dc to 1.6 megahertz. The voltage display has a range

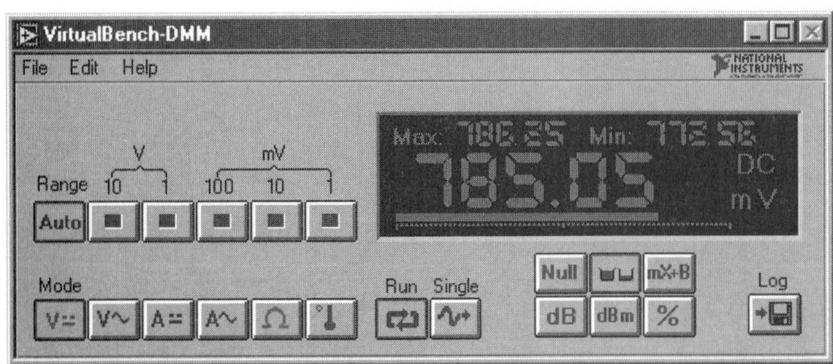

Fig. 22-13. **VirtualBench™ digital multimeter.**

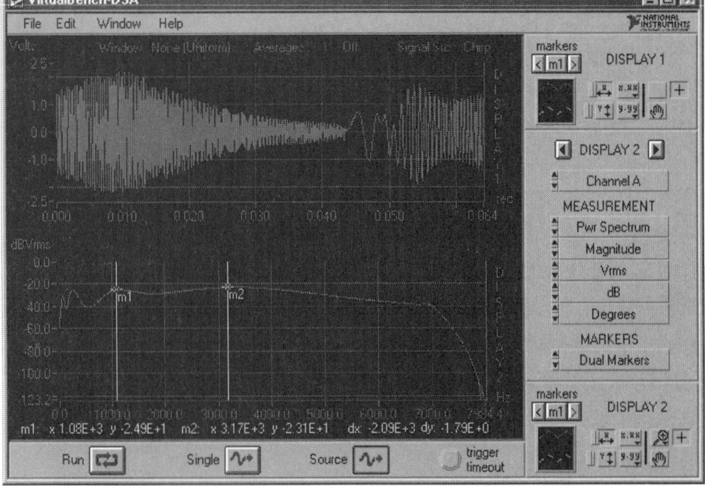

Fig. 22-14. **VirtualBench™ oscilloscope.**

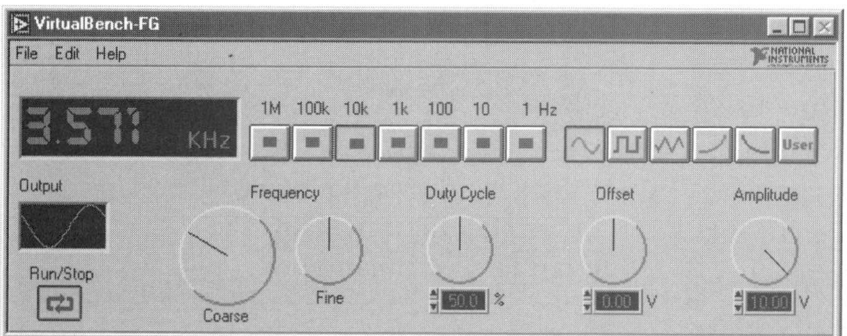

Fig. 22-15. **VirtualBench™ function generator.**

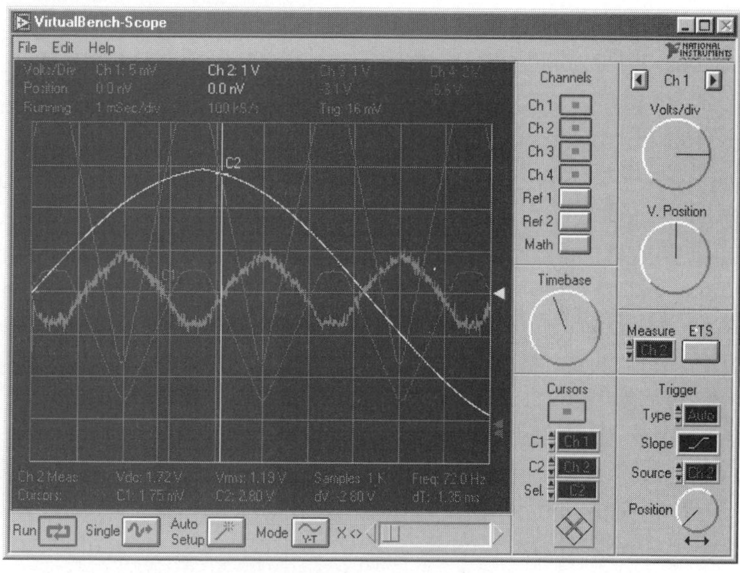

Fig. 22-16. **VirtualBench™ dynamic signal analyzer.**

from 1 mV/division to 10 V/division. The sweep rate of the display is variable from 10 nanoseconds per division to 100 milliseconds per division. By taking advantage of the mathematical powers of a host computer, this virtual oscilloscope can also do many calculations using data from the waveform. In this way, much more information is delivered on screen to the user. Accurate measurements of dc voltage, RMS voltages, maximum and minimum voltage, peak to peak, frequency can be made.

Virtual Function Generators

Function generators (Fig. 22-15) provide signal voltages in a variety of waveforms and shapes at very precise, selectable frequencies. In a virtual situation, a familiar image of a typical waveform generator is displayed on the screen. The controls are push buttons and knobs manipulated by the mouse. By placing the cursor over the desired control and adjusting it, waveforms and voltage levels are selected. The signals are sent to the external connector and then applied to the corresponding circuitry. Available waveforms include dc, sine, rectangular, triangle, sawtooth, noise, stairstep, and others. Because they are computer generated, custom waveforms can be created from equations by drawing with the mouse, or by combining different waveforms from different time periods.

Virtual Dynamic Signal Analyzers

Waveforms of extremely high frequencies or very short impulses are difficult to observe and analyze. However, the important information contained in all types of signals can be found with a *dynamic signal analyzer* (DSA) (Fig. 22-16) Some selected quantities that can be measured with a DSA are amplitude and power spectrums, frequency coherence, and impulse response. The user can also obtain data about harmonic distortion and frequency response. Pseudo random noise, chirp, and fixed sine waves can also be generated and injected into a test circuit, which can then be analyzed for responses. Data can be gathered continuously while the display updates specific values on the computer screen. Also, one set of data can be obtained when the instrument is triggered, for example, by a particular signal or voltage level. The triggered operation mode allows the capture and analysis of very short impulses and electrical events.

Virtual Data Loggers

It is often necessary to collect data over a period of time and then store it in a form to be used for later analysis. This is the function of a *data logger* (Fig. 22-17). This instrument is unusual to virtual instrumentation because it takes advantage of a computer's memory capacity. Data can be recorded over a period with the start-and-stop mode selected to operate on command or at set times. This data can then be displayed on the screen for visual analysis. It can also be presented in graphic form on the screen or printed on paper. This type of tool allows technicians to examine signal data that is too fast for human eyes to see. This instrument will also collect data that occurs over a long period where it would be impractical for a worker to manually collect the data.

Write a "letter" to your " boss" or " instructor" explaining why he or she should invest in virtual instrumentation.

BENCH INSTRUMENTS

Although virtual instrumentation is gaining popularity, most test benches have individual instruments. These instruments are used by engineers and technicians to analyze the internal operation of circuitry. The information obtained with these instruments, together with a thorough understanding of electronics, enables users to develop new circuitry, troubleshoot new circuits, and correct circuit problems. Common bench instruments are discussed in this chapter's conclusion.

Oscilloscopes

Oscilloscopes, or *scopes* as they are often called, are important test instruments. They are used in many kinds of design, service, and maintenance activities (Fig. 22-18). The main purpose of an **oscilloscope** is to show a waveform (Fig. 22-19), or picture, of a voltage signal. The amplitude of the wave represents various voltage values for the wave, and the horizontal axis of the screen is used to measure time between points on the wave. This makes it possible to test circuits by seeing if the waveforms of the circuit voltages are at the correct points and if key points are properly timed.

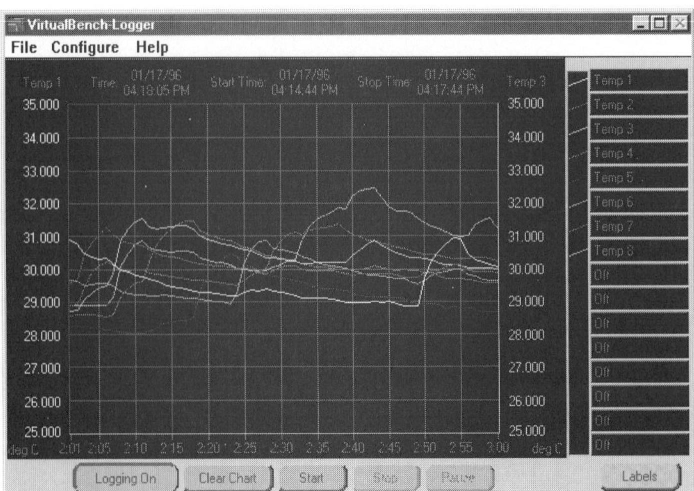

Fig. 22-17. VirtualBench™ data logger.

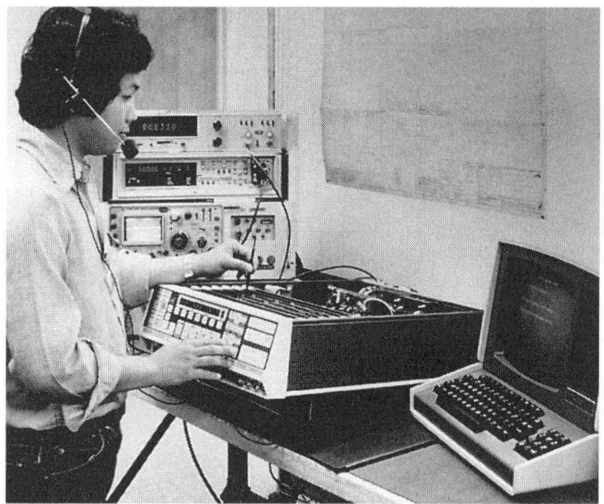

Fig. 22-18. **Student using an oscilloscope and other test equipment.**

Oscilloscopes

Oscilloscopes are used in medicine in patient monitoring systems. Biosensors attached to the body closely observe such activities as temperature, brainwaves, breathing rate, and heartbeat. This information is often displayed on oscilloscopes for doctors to analyze.

Electrical monitoring systems are most widely used for unstable newborn infants and seriously ill patients with heart and respiratory disease. These systems analyze the signal patterns taken from such machines as electrocardiograms. They relay the information to the hospital staff, usually as a visual display on an oscilloscope. Any variation in heart rate, blood pressure, or respiratory rate will set off an alarm at the nursing station.

The Cathode Ray Tube

The waveforms shown by an oscilloscope are displayed on the screen of a cathode ray tube (CRT). This is, in many ways, similar to the picture tube of a black-and-white television set. An electron-gun assembly in the neck, or narrow end of the tube produces a beam of electrons, sometimes called a *cathode ray* (Fig. 22-20A). When the electron beam strikes the phosphor coating on the screen of the tube, the phosphor *fluoresces*, or glows. If the beam does not move horizontally or vertically, a tiny spot of light will appear on the screen (Fig. 22-20B) Most oscilloscopes, called **dual-trace oscilloscopes**, can display two wave patterns at the same time. This system allows technicians to compare related patterns for amplitude and timing. Separate inputs and controls are provided for each of the beams and are referred to as channel 1 (Ch1) and channel 2 (Ch2), or A and B.

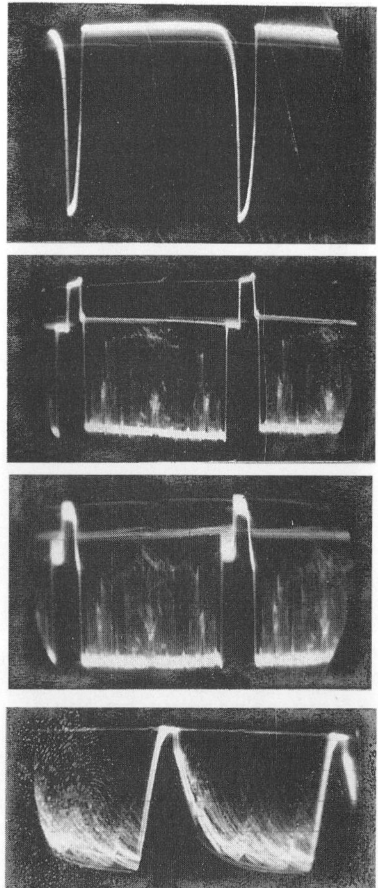

Fig. 22-19. **Oscilloscope displays showing voltage waveforms present at different points of a television circuit.**

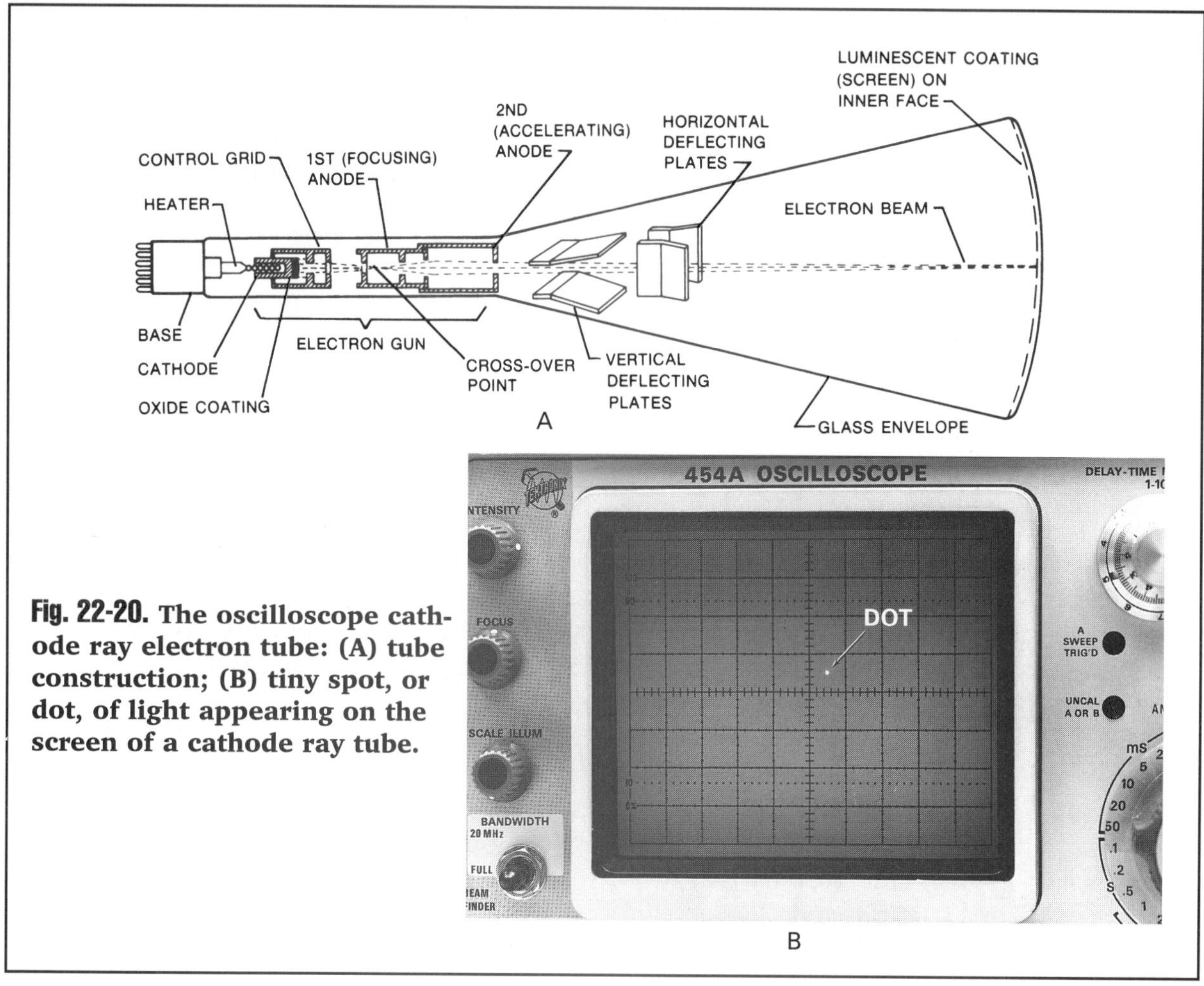

Fig. 22-20. The oscilloscope cathode ray electron tube: (A) tube construction; (B) tiny spot, or dot, of light appearing on the screen of a cathode ray tube.

Electrostatic Deflection

The electron beam of the cathode ray tube in an oscilloscope is moved by a process known as *electrostatic deflection.* Since the beam is made of electrons, or negative charges, it can be made to move by the electrostatic forces of attraction and repulsion. This is done by applying voltages to the horizontal and the vertical deflection plates in the tube (see Fig. 22-21). The amount of vertical deflection of the electron beam is proportional to the level of the tested voltage. The screen is divided into equal increments, usually centimeters (cm). The voltage value for each division is selected by a switch on the front panel. Examples of horizontal and vertical beam deflection are shown in Fig. 22-21A and Fig. 22-21B. If one horizontal and one vertical deflection plate are equally positive or negative at the same time, the beam will move diagonally across the screen (Fig. 22-21C).

Sweep Circuit

When the vertical gain control of an oscilloscope is turned all the way down and the horizontal gain control is turned all the way up, the electron beam *sweeps,* or moves, across the cathode ray tube in a horizontal direction. The beam spot moves so quickly that, because of a characteristic of human eyes known as *persistence of vision,* it appears as an unbroken line (Fig. 22-22).

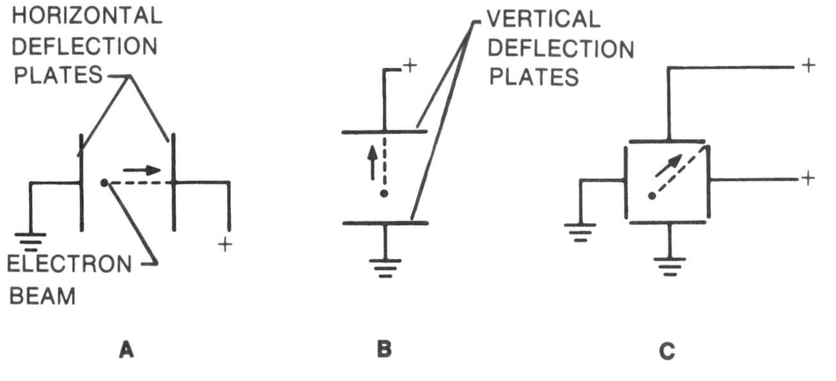

Fig. 22-21. Electrostatic deflection of the electron beam in a cathode ray tube: (A) horizontal; (B) vertical; (C) diagonal.

The horizontal sweep of the beam is produced by an oscilloscope circuit commonly called a *sweep generator*. This circuit generates a sawtooth-shaped output voltage when a timed trigger pulse is applied to the generator circuit. In this way, the sweep will begin exactly at the same point on each wave displayed. The pattern will appear to be constant and stationary, and therefore, it will be easy to observe. The main factor, however, is that the horizontal gradations on the screen will have a time period assigned by the sweep rate. This time period will range from several seconds to fractions of microseconds for each gradation. When the sweep voltage is applied to the horizontal deflection plates, the beam is moved across the screen from left to right. It is then very quickly brought back to the starting position. The speed at which this action is repeated is determined by the frequency of the sweep generator. The frequency of the generator is varied by switching the time base control. This is usually possible using a switch with indicators of the time period at each position, for example, *t/cm*. There is often a variable sweep control that adjusts the sweep rate. This control helps the operator synchronize a display. This control must be locked in the "off" position for the time base to be accurate.

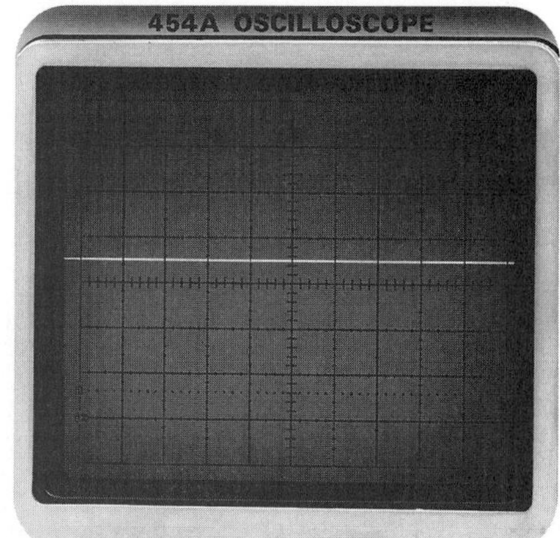

Fig. 22-22. Horizontal line, or trace, appearing on the screen of an oscilloscope cathode ray tube.

FASCINATING FACTS

The persistence of vision *phenomenon explains how the eyes receive the impression of motion pictures. When the eye's retina registers an image, the image persists from 1/50 to 1/25 of a second. A movie is, in fact, a rapid series of still pictures flashed on a screen, with about 1/60 of a second of complete darkness after each image. The eye's persistence of vision fills in the dark moment. It blends each picture perfectly with the one before it. Thus, with rapid motion of a series of images, our eye creates the same impression that true motion produces.*

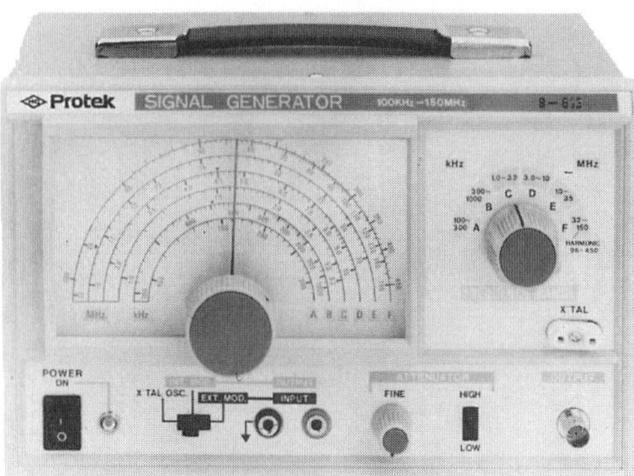

Fig. 22-23. **An RF signal generator.**

Signal Generators

Signal generators are test instruments that supply different voltage frequencies (Fig. 22-23). These voltages, or signals, can then be applied to certain circuits such as radio receivers. This is done to test and adjust the circuits for efficient operation.

Audio-Frequency and Radio-Frequency Signal Generator

An audio-frequency (AF) signal generator produces an output signal with a frequency range of about 10 Hz to 100 kHz. An audio-frequency generator produces sinusoidal, or *sine wave*, outputs. Many also produce square waves. Radio frequency (RF) signal generators have frequencies from about 100 kHz to 100 MHz.

Modulation

In typical radio frequency signal generators, signals from a 400-Hz oscillator circuit in the generator modulate the output signals. *Modulation* is to vary the amplitude or the frequency of a carrier wave. In this case, the carrier wave is the sine wave output voltage. When the amplitude is varied, it is called *amplitude modulation* (AM). When the frequency is varied, it is called *frequency modulation* (FM).

Because of its modulated outputs, the radio-frequency signal generator is very often used as a miniature radio transmitter for testing radio receiver circuits. In such testing, the signal generator is connected to certain parts of a radio circuit. If the circuit operates properly, the 400 Hz signal, or tone, with which the carrier signal is modulated is reproduced by the loudspeaker. Special kinds of radio frequency signal generators are also commonly used to test and adjust television receiver circuits.

Frequency Counter

A **frequency counter** is an instrument that can be used with oscilloscopes to study circuit frequencies. These instruments count the number of cycles present at the input leads for a short period of time and then display that count. This process is repeated periodically, and the display is updated.

Frequency counters, such as the one pictured in Fig. 22-24, have many uses. One application is to be sure that the transmitting equipment is operating properly and within Federal Communications Commission (FCC) regulations. Another use is to set local oscillators in radio receivers.

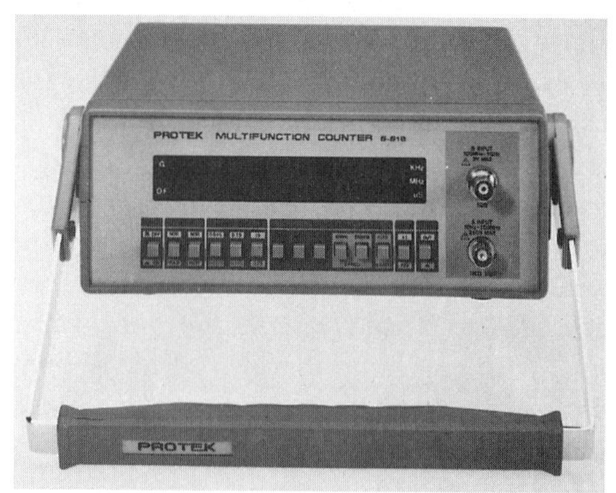

Fig. 22-24. **Wide-band frequency counter.**

Signal Tracers

Signal tracers are instruments that trace, or follow, a signal voltage through circuits such as those found in amplifiers and radio receivers. This is done by connecting different points of the circuit under test to the input of the signal tracer. If the circuit is operating properly, the signal tracer will indicate this by an audio output signal heard through a loudspeaker or headphone.

When a signal tracer is used with a modulated radio-frequency signal, it must have a detector circuit. This circuit separates the audio signals from the radio-frequency carrier. This separation is necessary since a signal-tracer loudspeaker will not respond to signals above the audio range of frequencies.

Describe the differences between a signal generator and a function generation.

USING ELECTRONIC TEST EQUIPMENT

The procedures given below are intended to provide basic information on operating an oscilloscope. The controls mentioned are those found on typical, general-purpose scopes. After learning to operate an oscilloscope in this way, you will be able to use it for various purposes. Many oscilloscopes have additional controls beyond those discussed here. Therefore, before operating an oscilloscope or other test equipment, you should carefully read its instruction manual. Basic information on the operation of a signal generator is also given in the following procedures. Notice that this is a drawing of a dual-trace oscilloscope. The following procedure applies to the operation of either channel.

Procedure

1. Turn on the oscilloscope. Adjust the following controls as stated:
2. *Vertical position*—adjust the control so that a horizontal trace, or line, appears through the center of the oscilloscope screen.
3. *Intensity control*—adjust so that the line can be clearly viewed without too much brightness. Then adjust the focus control so that the line is "clean," or well defined.
4. *Vertical input jacks*—connect an audio-frequency signal generator to these jacks on the oscilloscope. Set the output selector to square wave and adjust the frequency to 10 kHz. Set the output-level control to a midrange position.
5. *Sweep selector*—adjust to a range 10 microseconds per millimeter. Adjust the volt per millimeter control to a position that yields a wave that is about 3/4 of the screen's full height.
6. Slowly adjust the signal generator's output frequency and observe the oscilloscope's display. The display should be stable and expand or contract as the frequency is decreased and increased, respectively.
7. Adjust the signal generator to produce a sign-wave output. Observe these waveforms at different frequencies.
8. Connect a crystal microphone to the oscilloscope. Observe the voice waveforms generated by talking into the microphone.
9. Connect a 6-V battery in series with a door buzzer. Connect the vertical input leads of the oscilloscope to the buzzer terminals.
10. Adjust the necessary controls of the oscilloscope so that a clear pulsating-current waveform appears. This waveform is produced by the action of the buzzer contacts in making and breaking the circuit.

Chapter *Review*

Summary

- Instruments called meters are used to measure electrical quantities such as voltage, current, resistance, and power. In addition to meters, other instruments such as oscilloscopes, frequency generators, function generators, and signal generators are used to view and diagnose electrical circuits.
- Voltmeters measure voltage, ammeters measure current, ohmmeters measure resistance, and oscilloscopes view "pictures" of electrical signals. Voltmeters, ammeters, ohmmeters and most other instrumentation are all available in analog, digital, and virtual models.
- Every form of instrumentation requires both practice and instruction to use safely and correctly.

Review Main Ideas

Review this chapter's main ideas by writing, on a separate sheet of paper, the word or words that most correctly complete the following statements:

1. Voltage is measured with a ______.
2. Current is measured with an ______.
3. Meters that present the measured value in a numeric form are called ______ meters.
4. Direct-current voltmeters are ______ instruments whose test leads must be connected in the correct ______.
5. Ammeters are connected in ______ with one of the conductors and the load being tested.
6. If you touch the test leads while measuring, an ohmmeter will measure the combined ______ of your body and the circuit being tested.
7. A ______ test is used to determine if there is a continuous current path from one test point to another, especially where the complete path cannot be seen.
8. Computers can run software programs that have ______ instrumentation to make and record test measurements.
9. ______ provide signal voltages in a variety of waveforms and shapes at very precise, selectable frequencies.
10. Some quantities measurable with a dynamic signal analyzer are ______, power spectrums, ______, and ______.

11. The main purpose of an oscilloscope is to show a ______, or picture, of a voltage.
12. The horizontal axis on the oscilloscope represents ______ between points on the wave.
13. An oscilloscope that displays two wave patterns at a time is called a ______ oscilloscope.
14. A signal generator is a test instrument used to supply voltages of different ______.
15. Two general kinds of signal generators are ______ signal generators and ______ signal generators.
16. An instrument used to trace a signal voltage through circuits such as those in amplifiers and radio receivers is called a ______.

Apply Your Knowledge

1. Explain some of the special features available with a digital multimeter.
2. Draw a diagram showing how a voltmeter is connected to circuit points being tested.
3. Briefly describe how an ohmmeter is used for measuring the resistance of a device or of a circuit.
4. Explain the differences between an analog meter, a digital meter, and a virtual meter.
5. Explain the difference between amplitude modulation and frequency modulation.

Make Connections

1. **Communication Skills.** Research the method used by early investigators to measure the speed of light before the use of modern instrumentation. Report your findings to the class.
2. **Mathematics.** Compare the measurements from several different types of digital and analog meters (e.g., multimeters and panel meters) used to measure a known electrical quantity, such as voltage, current, or resistance. Determine the percentage of error for each instrument tested. Report your findings to the teacher.
3. **Science.** Investigate the basis for the determination of the measurement standard currently used for the volt. Report your findings to the class.
4. **Workplace Skills.** Research how instruments are used in the workplace by technicians and engineers to collect data when designing and testing new components.
5. **Social Studies.** Prepare a written report, two pages in length, that shows how digital instrumentation has affected the design and use of home appliances.

Chapter 23

Troubleshooting and Repair

OBJECTIVES

After completing this chapter, you will be able to:

- Identify the main differences between thru-hole and surface mount soldering of individual components.
- Identify how common sense can be used as a guide for troubleshooting.
- Explain why an oscilloscope is useful in troubleshooting.
- Identify and describe five types of electronic test equipment used in troubleshooting electronic circuits.

Terms to Study

desoldering

diagnostic software

logic levels

logic probe

oscilloscope

signal generator

solder wick

surface mount technology

thru-hole technology

troubleshooting

Even though electronic equipment is very reliable, there are times when it breaks down and needs to be repaired. If you have ever tried to make such a repair, you probably found that it takes a combination of knowledge and skills to do so. Professional electronic repair people have obtained the necessary knowledge and skill through school, training, and practice.

TROUBLESHOOTING PROCEDURES

Troubleshooting is the process of determining why a device or a circuit is not working. To do this efficiently, you must have a broad knowledge of how things work. Technicians who troubleshoot, repair, and service electronic equipment must have a variety of skills and knowledge. They must know what test instruments and procedures will help find a defect. They must also know what tools and materials are needed to repair it. Finally, they must have the ability to *diagnose*. This means that they must be able to recognize the symptoms of a defective device or circuit and then form a plan of action to repair it (Fig. 23-1). This chapter explains how to troubleshoot and repair electronic equipment.

The troubleshooting process for a circuit is often a systematic procedure. For example, the following are the steps for troubleshooting a radio circuit:

1. Visually check all components, wiring, and printed circuit boards for discoloration, burns, and breaks.
2. Check all batteries and replace any with low voltage.
3. Measure voltages at the leads of transistors for unusual voltage levels.
4. Disconnect all power sources.
5. Test the voice coil of the loudspeaker with an ohmmeter.
6. Check potentiometers with an ohmmeter and clean or replace as necessary.
7. Clean contacts of headphone jack.
8. Check detector diodes with an oscilloscope (reconnect power for scope testing) or ohmmeter; replace if necessary.
9. Test amplifier stages, especially the transistors and transformers.

In general troubleshooting, there are usually no definite rules to follow. In the typical case, technicians must apply common sense and their knowledge of theory, test instruments, tools, and materials to solve a problem.

BRAIN BOOSTER

Make a list of troubleshooting steps for a nonfunctioning series circuit that has one load, one switch, two resistors, and a single voltage source.

Fig. 23-1. A technician troubleshooting an electronic device.

USING COMMON SENSE

Often, successful troubleshooting is simply the result of using common sense. If an electronic or electrical circuit is dirty, smoking, hot, noisy, burnt, or obviously broken, then the problem is easily found. Four of the five human senses plus intuition are effective tools for troubleshooting and provide valuable clues to the source of the trouble.

Vision

Unless the probable source of difficulty is known, troubleshooting should begin with a thorough visual inspection. Loose connections, broken wires, poor solder joints, burned-out resistors, or even a blown fuse or tripped circuit breaker are found this way.

For example, if a floor lamp is not working, the bulb may be defective. If replacing the bulb does not correct the trouble, then the lamp switch, lamp socket, or plug might be defective. Perhaps if you had inspected the plug before replacing anything, you would have discovered that a wire was loose at a plug terminal screw. With this corrected, the lamp would have worked. Therefore, in this case, a thorough visual inspection would have saved time and money. In more complicated troubleshooting and repair jobs, even more time and money are wasted on unnecessary procedures than if a visual inspection would have been made first.

Vision also plays an important part in troubleshooting television circuits since the picture tube can be used as a test instrument. Many technicians are able to quickly localize defects in a television circuit by observing picture loss or distortions that appear in a picture or in a test pattern.

Smell, Touch, and Hearing

In addition to sight, the senses of smell, hearing, and touch are valuable troubleshooting aids. Overheated resistors, transformers, choke coils, motor windings, and insulation materials can very often be detected by their odor. Each condition produces a characteristic odor that experienced technicians can use to locate defective components. When overheating is noticed, the circuit should be turned off at once.

Overheated components can also be found by touch. This is especially true of transformers, choke coils, electrolytic capacitors, and transistors. If any of these components feels too warm, there may be a defect in the component or in the circuit wiring.

SAFETY TIP

Be certain to touch circuits only when the power has been disconnected and enough time has elapsed to discharge any charged capacitors. This is particularly important with television circuits that may also need to have capacitors discharged to ground.

Hearing is also used in troubleshooting loudspeaker circuits by using the loudspeaker as a test instrument. The lack of sound or the distortions of sound heard through it can indicate specific circuit defects. Experienced technicians can quickly locate defects using this procedure.

Electrical arcing sometimes produces a crackling noise that is easily heard. In high-voltage circuits such as those in television receivers, the arcing of current between certain wires and the chassis or between two wires produces a chirping noise. By tracing these noises, it is possible to quickly locate the source of the arcing.

Describe four ways in which the human senses aid troubleshooting.

ELECTRICAL TESTING

When a common sense inspection does not reveal a problem, it is time to make electrical tests. There are many different kinds of electrical equipment available for testing electronic and electrical circuits. This text, however, will discuss the most common diagnostic equipment: voltmeters, ohmmeters, oscilloscopes, logic probes, and software.

Voltmeters

After determining that all fuses, breakers, and switches are in good condition, the next step is to test the voltage at certain supply points. Begin with the incoming voltage and work down to the five volts for the processor, if any. This includes the power transformer primary and secondary, the power supplies, and other voltage sources. If the supply voltage is normal, it is good practice to then measure the voltages at the electrodes of the transistors. Unless these voltages are of the correct value, a circuit will fail to operate or will operate improperly.

The voltages that should be present between the electrodes of transistors or electron tubes and the common point of their circuit are usually shown on the schematic diagram of the circuit (Fig. 23-2). If you do not have a diagram, it is often helpful to look at the schematic diagram of another circuit in which identical (or similar) transistors are used, or consult the manufacturer's specification manual. Often, the correct voltages are printed on the board, transformer, or other device.

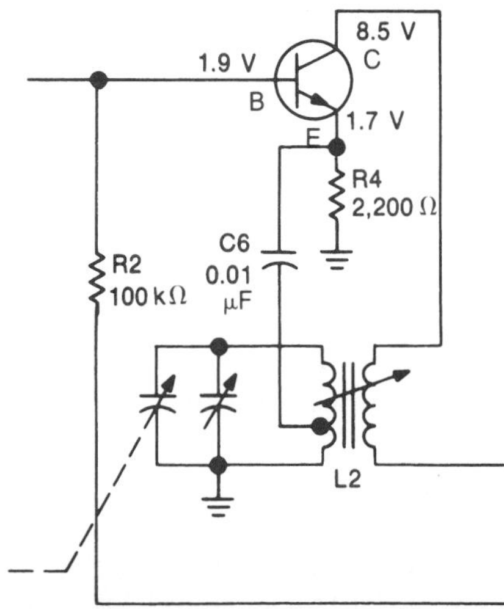

Fig. 23-2. Portion of a radio-receiver schematic diagram showing the dc voltages that should be present between the electrodes of a transistor and the common (ground) point of the circuit.

Ohmmeters

A lack of voltage or an incorrect voltage at a given point of a circuit is very often caused by a defective resistor, diode, or capacitor. To find this type of defect, it is helpful to test the resistance of that part of the circuit from which voltage is obtained for the electrode in question. In this way, it is possible to locate defective components that are preventing a circuit from operating. Be sure to disconnect all power sources before making any resistance tests.

Troubleshooting many devices and circuits involves continuity and resistance tests made with an ohmmeter. Continuity tests should be made wherever an open-circuit condition is suspected (Fig. 23-3). The test can also be used to check for shorts. It is important to remember that an ohmmeter should never be used on a

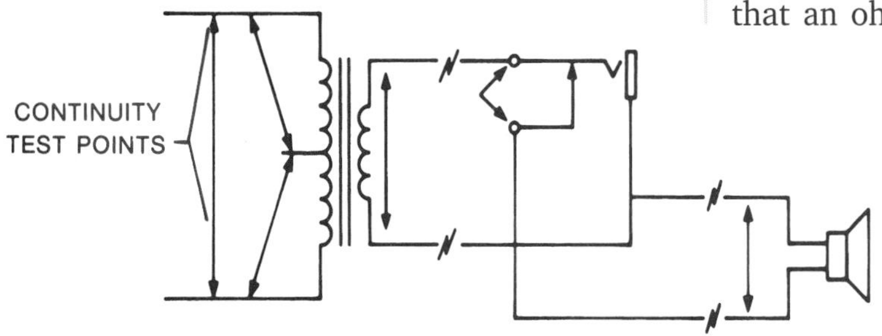

Fig. 23-3. Typical continuity test points used in checking the output transformer, headphone jack, and loudspeaker of a radio receiver.

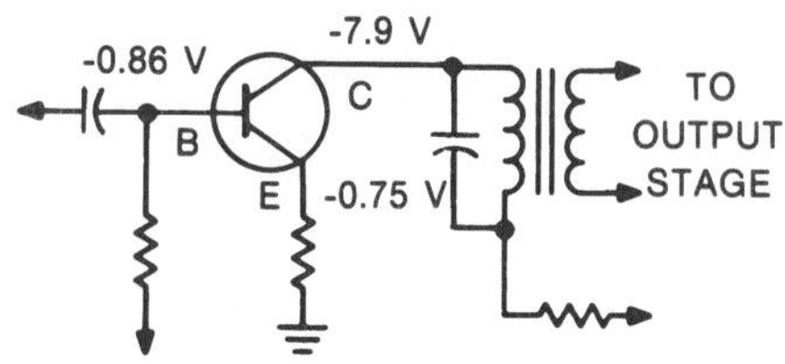

Fig. 23-4. Portion of a radio-receiver circuit showing the resistors that should be checked if correct voltages are not present at the electrodes of the transistor.

circuit to which a voltage is applied. Also, a component should be isolated by disconnecting one of its leads before continuity tests are done.

Oscilloscopes

Very often, components change characteristics enough to cause distortion in operation, but not enough to fail completely. An **oscilloscope** is a CRT device that will show deviations in wave patterns and help find problems. A **signal generator** is an electronic device that produces a variety of electronic control waveforms. By connecting one to the input of a circuit, such as in Fig. 23-4, technicians can use an oscilloscope to observe wave shapes as they pass through a circuit. If a signal is present at an input to a component and the correct output signal is not at the output, then the fault is usually isolated at that component. If a signal is seen on the oscilloscope at the base of the transistor and distorted at the collector, then the transistor is suspected.

At a local library, find out more about oscilloscopes. Write a brief essay describing the history of their development.

DIAGNOSING CIRCUIT TROUBLES

A circuit defect such as a broken connection, loose solder joint, or a worn-out component can usually be repaired or replaced simply. However, a repair involves further diagnosis of the reason or reasons for a defect. For example, a resistor, a diode, or a transistor can overheat and burn open because a second component in the circuit is defective and demands too much current. If the first component is replaced without further diagnosis, it is probable that the replacement component will also be damaged.

Suppose that the carbon-composition resistor in the rectifier circuit of an amplifier is burned out (Fig. 23-5). Further suppose that an inexperienced person immediately replaces the resistor only to find that the replacement resistor also quickly burns out. It is obviously necessary to diagnose the situation more carefully.

First, fundamental theory says that the resistor burned out because it was forced to carry excessive current. Because resistors fail open, it is reasonable to assume that excessive current is not caused by a defect in the resistor itself. The conclusion must be that the excessive current was caused by some other defect in the circuit.

In the circuit in Fig. 23-5, two electrolytic filter capacitors, C_1 and C_2, are connected across the rectifier output. We know that electrolytic capacitors can become shorted. If C1 is shorted, perhaps one or both of the diodes will be damaged. However, because of the circuit arrangement, this would not cause excessive current through the resistor. On the other

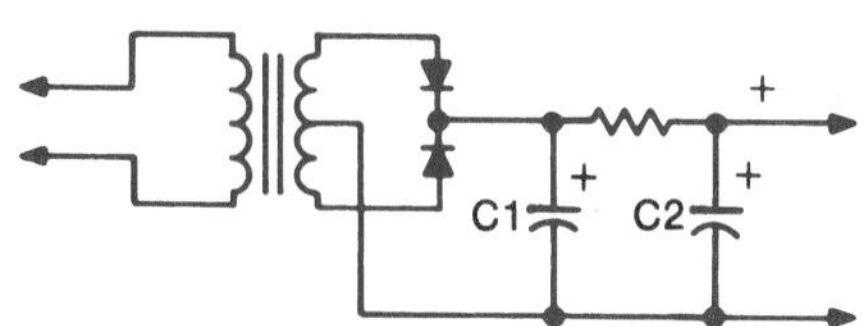

Fig. 23-5. Troubleshooting a rectifier filter circuit.

hand, if C2 is shorted, excessive current will definitely pass through the resistor. Therefore, the solution to the problem is to check and replace C2, if necessary, before replacing the resistor.

The situation just described is a very basic troubleshooting and repair job. It is also a good example of how knowledge of theory, practical experience, and a logical troubleshooting procedure can be used to solve an electrical circuit problem.

Logic Probes

Logic probes are test instruments used mostly to troubleshoot different families of *integrated circuits*, or ICs (Fig. 23-6). Logic probes detect pulses of voltage. In digital work, **logic levels** are these pulses of voltage, or *threshold levels*. Logic probes also store and display logic levels. These logic levels are displayed on *light-emitting diode* (LED) readouts.

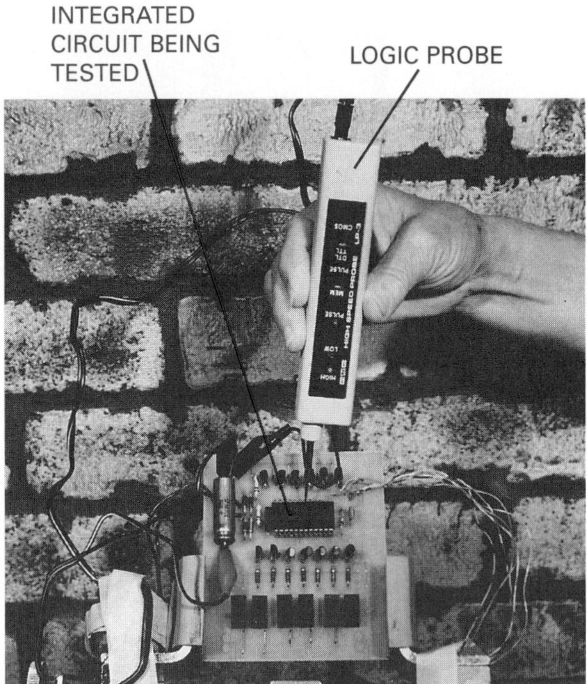

Fig. 23-6. Using a logic probe to check a digital-clock integrated circuit.

LINKS

You may wish to review various integrated circuits in Chapter 18, "Integrated Circuits and Special Devices."

By flipping a switch, logic probes can be set to detect threshold levels of 2.25 V (logic 1) and 0.8 V (logic 0) for a family of ICs known as DTL or TTL. These abbreviations refer to *diode-transistor logic* (DTL) or *transistor-transistor logic* (TTL or T^2L). Switching to another position sets the probe to detect threshold levels for a family of ICs known as CMOS/HTL. These abbreviations refer to *complementary metallic oxide semiconductors* (CMOS) or *high-threshold logic* (HTL). In CMOS/HTL ICs, threshold levels for digital operation are determined by the *applied voltage* (Vcc). When a threshold level is greater than 70 percent of Vcc, it is logic 1. When the threshold level is less than 30 percent of Vcc, it is logic 0. Some logic probes detect pulse widths as narrow as 10 nanoseconds at a frequency of 50 MHz. These features make the logic probe an important troubleshooting instrument. The power supply of a circuit under test also powers the logic probe. The following general procedures are suggested for using a logic probe:

1. Connect the clip leads to the power supply of the circuit.
2. Set the switch to the logic family to be tested (DTL/TTL or CMOS/HTL).
3. Set the MEMORY/PULSE switch to the PULSE position.
4. Touch the probe tip to a node, or terminal, on the IC to be analyzed. The three LEDs (high, low, and pulse) on the logic probe will give an instant reading of the signal activity at the node.

If the logic probe is to store a pulse, touch the probe tip to the node under test. Then move the MEM/PULSE switch to the MEM position. The next pulse will activate the PULSE LED.

Diagnostic Software Tools

The more complex a piece of equipment is, the more time-consuming it is to diagnose a problem. If a technician tests each circuit component in a computer until a fault is detected, much valuable time would be wasted. **Diagnostic software** is a set of computer programs that will troubleshoot computers. It has been developed for most types of computers and will electronically test the circuitry in question to determine the quality of operation. When a fault is found, the computer reports the results either on the computer screen or to a printer.

Data Communication Testing

When computing equipment sends data from one location to another, many controlling signals are sent along with the data. The presence and timing of these signals is very important. Figure 23-7 shows a test set that can be connected in series with a data cable. Once in place, a technician can use a logic probe or oscilloscope to test for the key factors in data transmission.

List each testing device listed above and match it to its purpose.

FASCINATING FACTS

Operators often describe a computer as having "a bug" when it is slower or less efficient than it should be. The term bug *was coined in the early 1940s, when programmers looking for the cause of a mysterious malfunction in one of the huge early computers discovered a moth in a vital electrical switch.*

REPAIRING PRINTED CIRCUIT BOARDS

Repair jobs commonly done on printed circuits are replacing defective components, resoldering, and repairing broken conductors. The most important rule for doing any printed circuit repair is not to use too much heat while soldering or desoldering, or removing solder from a joint. Too much heat will very often seriously damage conductors. Again, a light-duty soldering iron of 25 or 40 W is best.

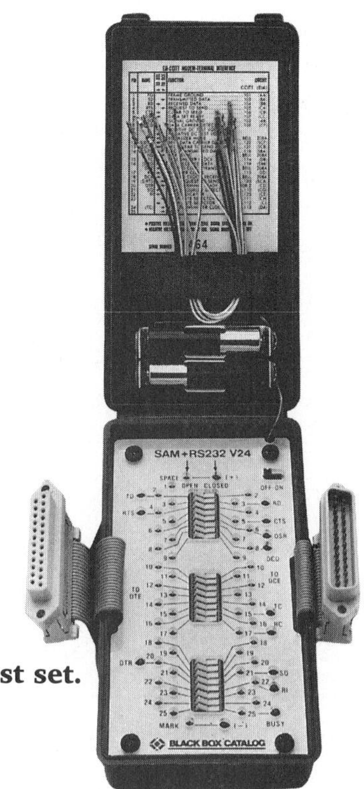

Fig. 23-7. **Data line test set.**

Conductive and Convective Heating

Hand held conductive heating devices generally fall in two categories: continuous heating devices and pulse heating devices. Continuous heating devices include soldering irons, thermal tweezers, and thermal pik devices. Convective heating methods are found in devices, such as semi-automatic bench top work stations, high-powered, hand-held, hot-air guns and jet hand pieces.

Most continuous heating devices have tips that can be tinned. These devices are used primarily on thru-hole component installation and removal. Pulse heated devices, such as resistance tweezers and other hand-held devices, produce heat directly in the tip or work with high current and low voltage. They are useful for installing and removing surface mount components.

Convective heating is used primarily with surface-mounted component installation and removal. Two purposes of convective heat are to prevent the rush of thermal shock to the board or component, and to complete the soldering task without direct mechanical contact. Also, it may be necessary to preheat PC circuit assemblies with great thermal mass to make them easier to solder.

Solder and Flux Fume Extraction

A major concern in industry is the exposure of workers to the fumes generated from solder and flux. There are basically two types of fume extractors available to separate fumes from workers. The first is commonly called "Tip Extraction" (Fig. 23-8). Fumes are removed from the tip of the soldering iron by mounting an extraction (vacuum) tube directly on the iron itself. These systems move small amounts of air so they do not affect the tip temperature.

Expressing Conduction Mathematically

Conduction is a method of heat transfer in opaque solids. If one end of a metal rod is heated, heat is conducted to the colder end. Exactly how heat is conducted in solids is not completely understood. It is believed that heat conduction is partially caused by the movement of free electrons in the solid, which transports energy when a temperature difference is applied. This theory helps explain why good electrical conductors also tend to be good heat conductors.

In 1882 French mathematician Jean Baptiste Joseph Fourier gave conduction a precise mathematical expression in what is known as Fourier's law of heat conduction.

The other option is an "Arm-Extraction" (Fig. 23-9). This technology uses extraction arms that are positioned for maximum extraction performance. Often these extraction arms are statically safe. These systems move large amounts of air through the system.

All fume extraction systems must use a High Efficiency Particulate Air (HEPA) filter to capture the tiny particles of solder and flux fumes. A gas filter, such as activated carbon, collects the gases present in solder and flux fumes. Most systems are easy to install and provide a safe and healthful workplace for technicians.

Fig. 23-8. Fume extraction right at the soldering iron tip.

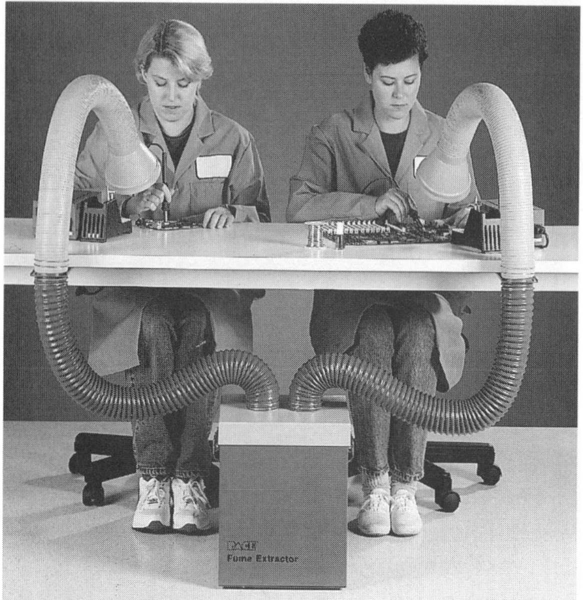

Fig. 23-9. **A high-performance fume extraction system for a workstation.**

LINKS

Review Chapter 21, "Soldering," for more specifics on correct soldering procedures.

What other health risks, besides fumes and particulate matter, can you think of that are associated with printed circuit repair?

TYPES OF PRINTED CIRCUIT BOARDS

Electronic components are often mounted on one side of a PC. In **thru-hole soldering**, the leads of some components are inserted through drilled holes and soldered to the printed circuit on the other side of the baseboard.

In **surface mount technology**, components can be mounted directly on the surface of the printed circuit boards without drilling holes. One advantage of this technology is that components can be mounted very close to each other. This makes it possible to have densely populated printed circuit boards. Because of this and other advantages, PC boards are being used in all kinds of electrical and electronic products. Various names are often given to describe electronic circuits connected to flat surfaces, such as video cards, adapter boards, accelerator boards, and mother boards.

Repairing Thru-hole Circuit Boards

Usually, the hardest part of replacing components is removing the defective components from the printed circuit board. This must be done very carefully to keep from damaging conductors.

One way to remove a component such as a fixed resistor or capacitor is to simply cut its leads as near to the body of the component as possible. The leads of the replacement component are then soldered to the wire stubs (Fig. 23-10). This method is particularly good when the soldered side of the board cannot be conveniently reached.

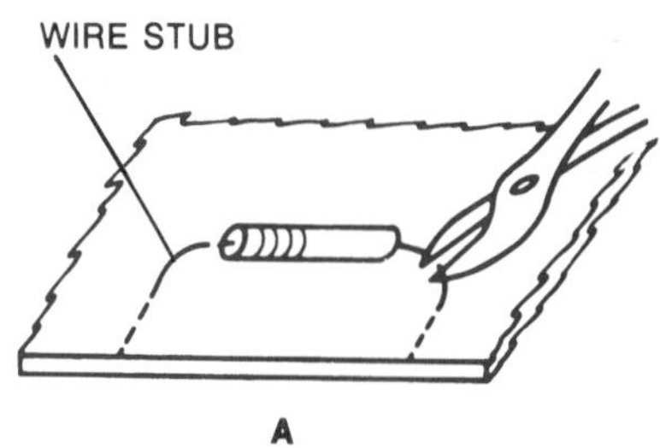

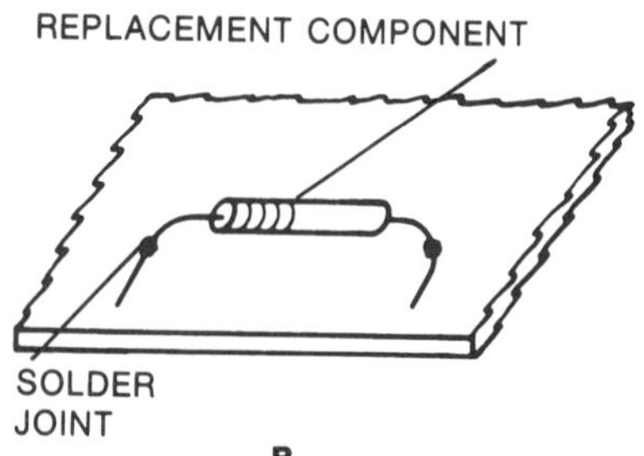

Fig. 23-10. **Replacing a defective printed-circuit component: (A) cut the leads of the defective component; (B) solder the leads of the replacement component to the wire stubs.**

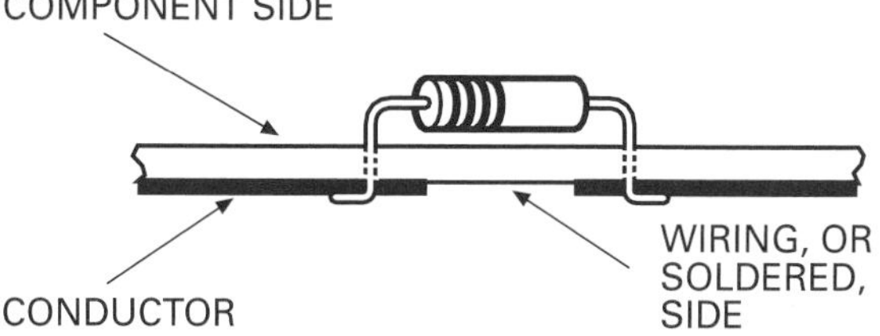

Fig. 23-11. Mounting components on a printed-circuit board.

When the soldered side of a printed circuit board can be reached, a component can be replaced by desoldering its leads and slowly pulling them from the mounting holes. The leads of the replacement component are then put into the mounting holes and soldered into place (Fig. 23-11). It may be necessary to melt and remove old solder around the mounting hole before putting in the leads. This method makes for a neater job.

Removing components such as potentiometers and multi-sectioned electrolytic capacitors that have fixed terminals is a more difficult job. **Desoldering** is the melting and removal of molten solder. In desoldering, the molten solder must be completely removed from all the terminal joints before the component itself can be taken out. Otherwise, a desoldered terminal joint will become resoldered while another joint is being heated. This problem can be solved by using a *desoldering tool* (Fig. 23-12). It consists of a soldering iron and a suction device. First, the solder on a joint is melted with the hollow-tip soldering iron. The molten solder is then removed by suction into the rubber syringe bulb. After all the terminals of a component have been desoldered in this way, the component can be easily taken out of the printed circuit board.

Solder wick, a solder-absorbing, braided copper ribbon, can also be used to desolder a joint. The ribbon is laid on the joint and heated. The solder is then pulled into the "wick" by the heat. The ribbon is removed from the joint before it cools.

Loose Solder Joints

Loose solder joints on a printed circuit board can often be repaired by reheating the joint without using more solder. If more solder must be put on a joint, this should be done carefully to keep any from flowing to a nearby terminal or conductor.

Broken Conductors

If the break in a conductor is a thin crack, it can usually be filled with solder (Fig. 23-13A). On some printed circuit boards, however, the conductors have protective coating that must be removed before the crack can be soldered. This can be done by lightly rubbing with fine sandpaper or by lightly scraping with a knife blade.

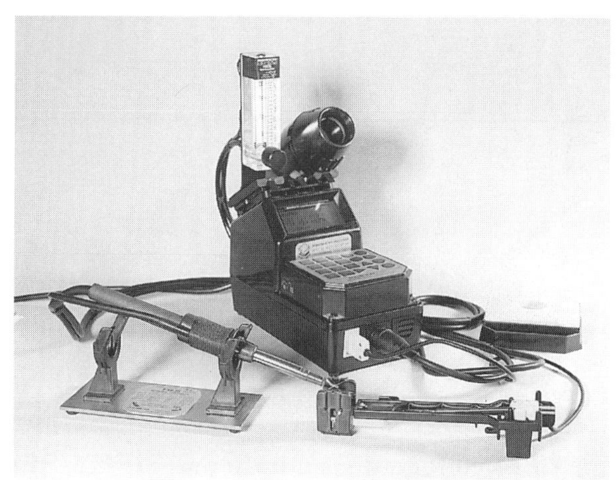

Fig. 23-12. A digital desoldering-resoldering tool. (*Edsyn, Inc.*)

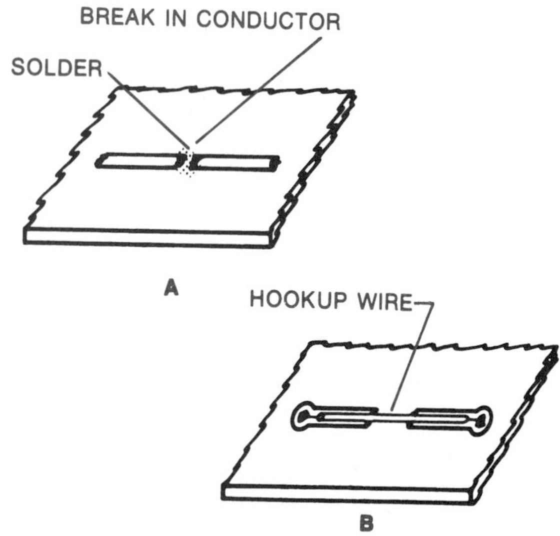

Fig. 23-13. Repairing broken conductors on a printed-circuit board.

A broken printed circuit conductor can also be repaired by "bridging" over the break with hookup wire connected to the conductor terminal points (Fig. 23-13B). This method is best when a piece of a conductor has been badly damaged.

Nondestructive Removal of Surface Mount Components

To remove surface mount components, it may be necessary to preheat the PC circuit board assembly. Heat must be applied in a rapid, controlled fashion to get complete, simultaneous reflow (melt) of all the soldered joints. Some components have as many as 100 pins or leads. The ThermoPik (Fig. 23-14A) can remove these Quad Flat Packs without damaging the board or surrounding components. Its special tips are designed to fit the shapes of the components. For instance, the Soldr-X-Tractor (Fig. 23-14B) has special tips which can remove Thin Quad Flat Packs (TQFPs) and Thin Small Outline Packages (TSOPs). These tips have thin walls to remove components from the most densely populated circuitry such as on pagers, cellular telephones, and computer adapter boards. Direct heat from the Soldr-X-Tractor tool insures rapid simultaneous reflow of solder from all the component leads at very low soldering temperatures. Both the Soldr-X-Tractor and ThermoPik use a vacuum to remove the component without damaging the leads. Selecting the proper removal tip will help to protect surrounding components. Other special soldering tools are the Solder Pen (Fig. 23-14C), ThermoTweeze (Fig. 23-14D), Dual ThermoPik (Fig. 23-14E), and the ThermoJet (Fig. 23-14F).

The bridge fill method of removing a Quad Flat Pack (QFP) is shown in Fig. 23-15A through Fig. 23-15E:

1. Remove any surface coating and clean the work of any contamination, oxides, or residues.
2. Install a vacuum cup on the vacuum pick-up tube of the hand piece.

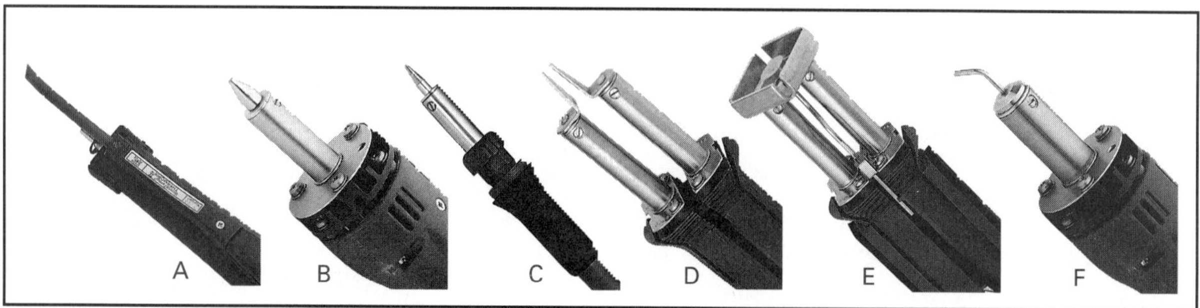

Fig. 23-14. Selected thru-hole and surface mount handpieces and tips.

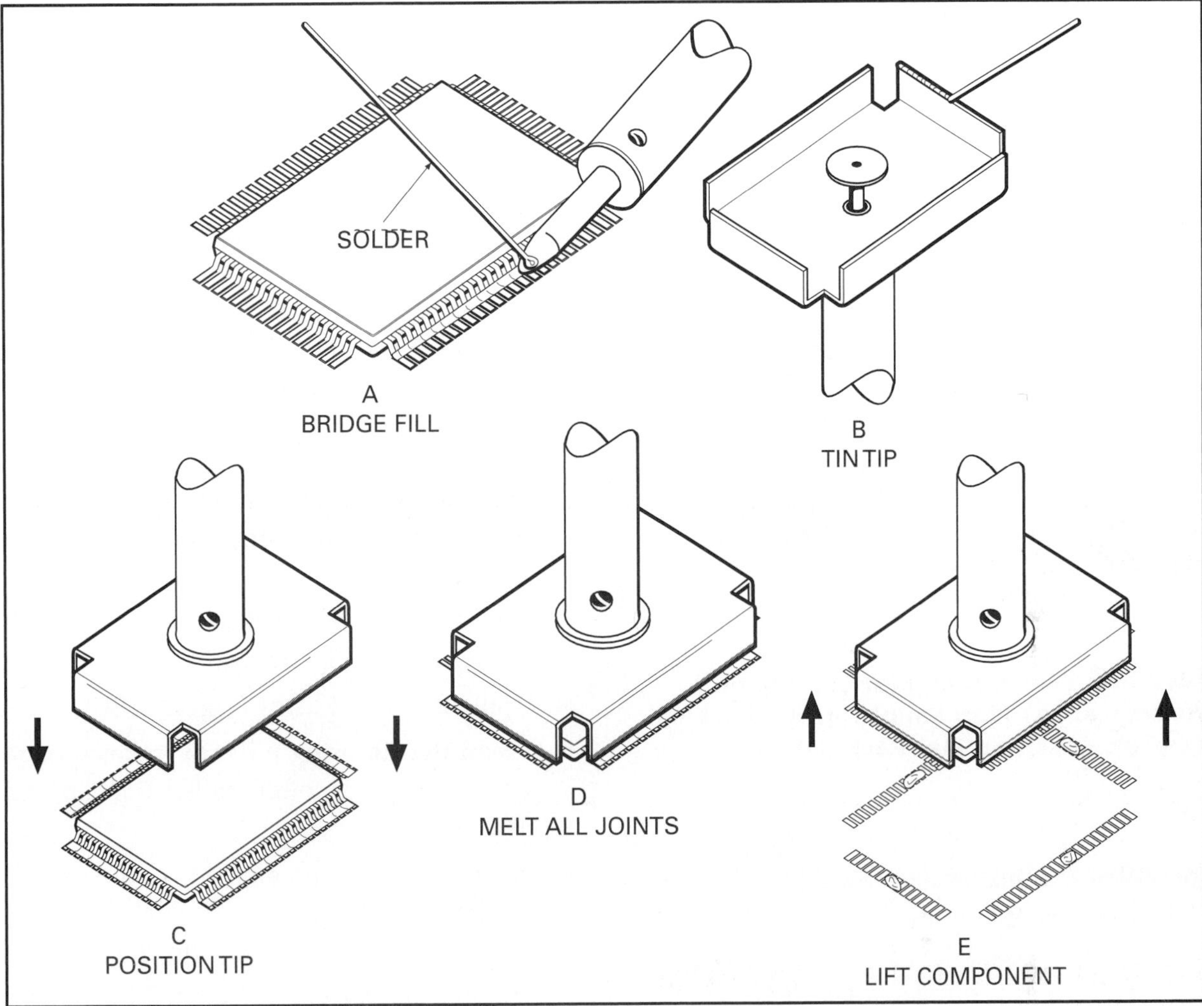

Fig. 23-15. Bridge fill method of removing a Quad Flat Pack component.

3. Set the tip temperature for 600° F (315° C).
4. Melt the solder to form a bridge fill using a Mini-Wave tip on the soldering hand piece.
5. Join all component leads with the solder bridge (Fig. 23-15A).
6. Select, install, and tin the bottom edges of the removal tip (Fig. 23-15B).
7. Gently lower the removal tip over the component to be removed, contacting all the leads (Fig. 23-15C).
8. Confirm that the solder has melted on all the joints (Fig. 23-15D).
9. Start the vacuum pump by pressing the switch on the hand piece and lift the component from the PC board (Fig. 23-15E).
10. Release the component on a heat resistant surface.
11. Prepare the lands for component replacement.

Rectangular component packages are shown in Fig. 23-16.

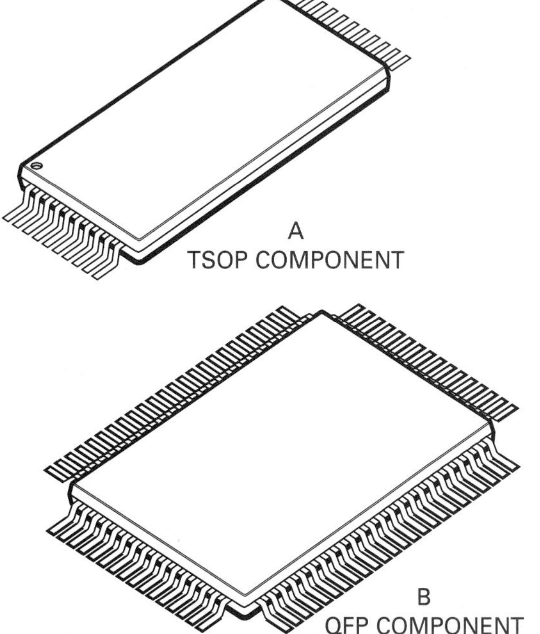

Fig. 23-16. Two rectangular component packages: (A) Thin Small Outline Package (TSOP); (B) Quad Flat Pack (QFP).

Installing Surface-Mount Components

It is important to remove any protective coating on the PC board. It is also necessary to thoroughly clean any surface areas where an electrical connection is to be made. This removes any oxide on the lands and other metal surfaces. Preheat the assembly if required.

Figure 23-17 shows a Mini-Wave solder tip. The tip has a concave elliptical cup, or well, that holds a miniature reservoir of solder. As the reservoir of solder moves over the component leads, it deposits the right amount of solder and heat at each lead and land. Safety is assured since low temperature is used and only the molten solder touches the leads and lands. This method is like an upside-down wave soldering process. A variety of tip sizes are available. Solder joint temperature should remain above the melting point of the solder alloy for the proper time to get a reliable connection. Avoid thermal and mechanical damage to component, board, adjacent components, and their joints.

Figures 23-18A through 23-18E show details on the Mini-Wave soldering method:

1. Set the soldering temperature to 600° F (315° C).
2. Position the component on the lands using your fingers, tweezers, or vacuum pick (Fig. 23-18A).
3. Apply the flux and tack solder on opposite corner leads. (Fig. 23-18B).
4. Apply flux to all leads (Fig. 23-18C).
5. Clean the tip using a damp, clean sponge on the Prep-Set, often called the Tip and Tool Stand.
6. Apply solder to the cupped portion of the tip (Fig. 23-18D).
7. Position Mini-Wave tip so the solder bead contacts the leads.
8. Slowly move the tip over one row of leads to form solder fills at each joint (Fig. 23-18E).
9. Refill the cupped portion of the Mini-Wave tip with solder as needed.

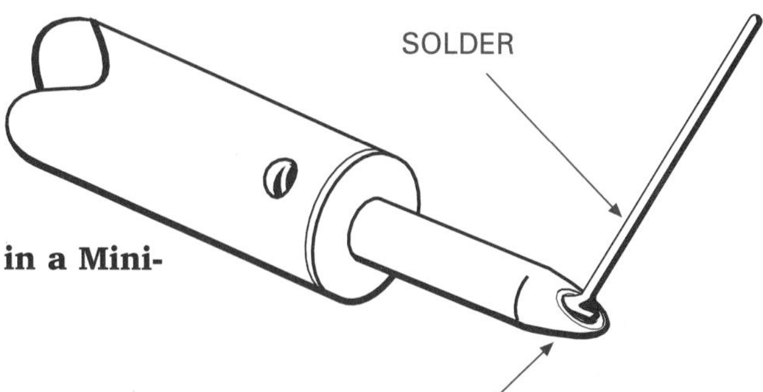

Fig. 23-17. Solder reservoir in a Mini-Wave tip.

10. Repeat the process on the remaining sides of the component.
11. Upon completion, tin the tip to extend the tip's life and return the hand piece to the Tip and Tool Stand.
12. Clean and inspect the soldered joints on the component.

BRAIN BOOSTER

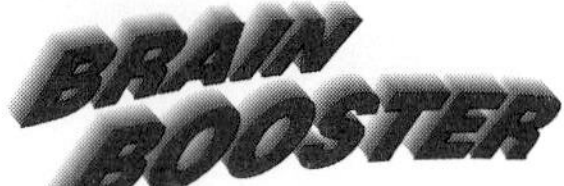

How is heat damaging to components? List three ways to avoid overheating a printed circuit board while making repairs to it.

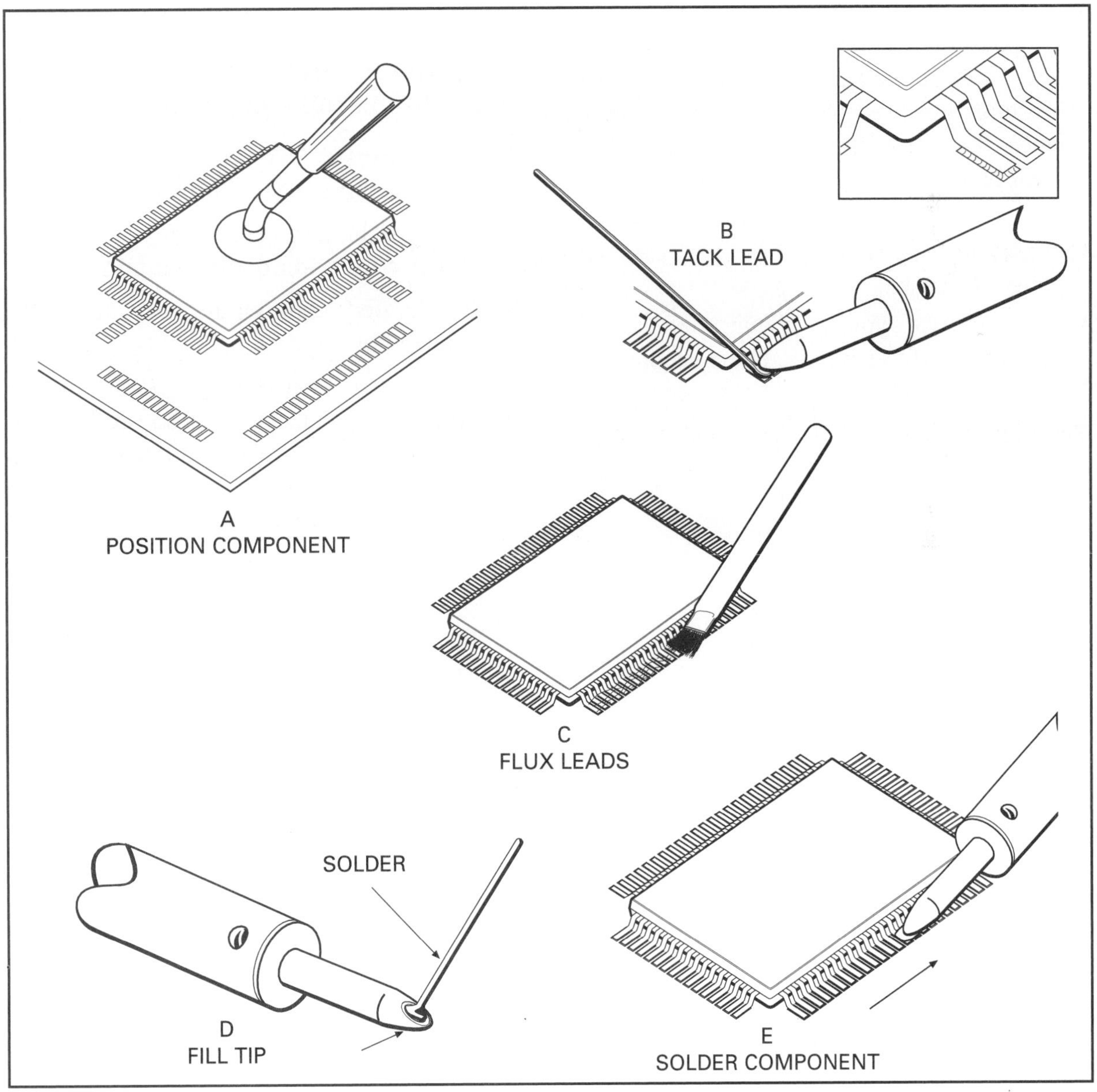

Fig. 23-18. **Installing a surface mount component using the Mini-Wave soldering method.**

Chapter *Review*

Summary

- Troubleshooting determines why a device or circuit is not working.
- Troubleshooting is usually a step-by-step procedure. Human senses and electrical testing tools are used to identify the problem. Once the problem has been identified, it must be repaired.
- During the repair of circuit boards, it is important to prevent overheating of the component or board.
- There are two types of circuit boards: thru-hole and surface mount. These require different soldering and desoldering procedures.

Review Main Ideas

Review this chapter's main ideas by writing, on a separate sheet of paper, the word or words that most correctly complete the following statements:

1. The process of determining why a device or a circuit is not working is called ______.
2. The five human senses plus ______ are effective tools in troubleshooting.
3. Overheated components can often be detected by ______.
4. In a voltmeter, the voltages that should be present between the electrodes of transistors or electron tubes and the common point of their circuit are often shown on ______ diagrams.
5. Repair jobs often involve further diagnosis of the ______ for defects.
6. The troubleshooting of many devices and circuits involves ______ and resistance tests made with an ohmmeter.
7. By connecting an oscilloscope to the input of a circuit, technicians can observe ______ as they pass through a circuit.
8. The oscilloscope is used to observe deviations in ______ patterns in circuits under test.
9. Integrated circuits can be analyzed with a test instrument called a ______.
10. CMOS refers to a family of integrated circuits called ______.

11. ______ is a set of computer programs that troubleshoot circuitry in computers.

12. ______ is used primarily with surface-mounted component installation and removal.

13. In removing components that have fixed terminals, molten solder must first be ______ from all terminal joints.

14. ______ can also be used to desolder a joint.

15. When part of a conductor has been badly damaged, it is best to ______ terminal points with hookup wire.

16. In installing surface-mount components, remove protecting coatings on the ______ and clean ______ where the connection will be made.

Apply Your Knowledge

1. Describe the differences between repairing surface mount and thru-hole circuit boards.
2. List the factors that should be considered before applying heat to a printed circuit board.
3. Describe fully the use of common sense in troubleshooting.
4. Explain resistance testing.

Make Connections

1. **Communication Skills.** Prepare a wall chart showing standard troubleshooting procedures.
2. **Mathematics.** Visit a local electronics repair shop. Find out what the itemized costs would be for typical repair of a television set. With this data, prepare a report for the class showing the itemized costs and total cost for the repairs.
3. **Science.** Research the chemical composition of solders and flux. Explain the scientific principle that causes molten solder to flow.
4. **Workplace Skills.** Observe the internal operation of a VHS VCR. Make a list of all functions and sequence of functions when the VCR is working normally. Compare this information with the observed operation of a malfunctioning VCR. Make a list of the differences you observe.
5. **Social Studies.** Obtain a list of job opportunities that require troubleshooting skills. Organize the list by the categories of industry (manufacturing, service, etc.) Do some jobs place greater emphasis on troubleshooting than others?

Section 7

Electronics in an Information Age

Chapter 24

Digital Electronic Circuits

Terms to Study

Boolean algebra

combinatorial circuit

comparator

digital electronics

flip-flop

full-adder

half-adder

inverter

logic gate

register

truth table

OBJECTIVES

After completing this chapter, you will be able to:

- Define *digital electronics*.
- Name three methods by which binary conditions or states can be achieved.
- Draw the symbol and truth tables for the following logic gates: AND, OR, NOT, and EXCLUSIVE-OR.
- Contrast the binary number system with the decimal system.
- Add binary numbers.
- Explain how bistable multivibrators (flip-flops) divide frequency.

A compact disc recording of your favorite group is performed with continuously varying sounds we recognize as music. However, the information needed to reproduce it in a CD player is recorded as digital data. The CD player can read this data and convert it to analog sound.

BINARY CONDITIONS

Digital electronics is a type of electronics that uses a technique that employs low voltage pulses to operate electronic equipment such as computers, stereos, and televisions. This technology has replaced many electronic functions previously performed by analog circuitry.

Images sent to earth from Mars by Sojourner in 1997 were sent as digital information. Why do you think it is so necessary to handle analog information, such as music and video images, as digital information? What other devices can you think of that handle information in digital forms? What do you think is in these devices that makes it possible to handle so much data in very short amounts of time?

Operating most digital circuits depends on a series of switching actions performed in a sequence to produce the desired result. In this chapter, you will learn about the basic switching actions that relate to logic circuits, adders, counters, and clocks. Other common applications of modern digital circuits include computers and other data processing systems, numerically controlled machine systems, and test instruments. The term *digital* is also generally used to describe those systems in which the output, or readout, is indicated by means of alphanumeric characters.

In many electronic circuit systems, signals with different voltage levels and frequencies produce the desired result, or output. *Signals* are voltages or currents. For example, a loudspeaker in an amplifier system produces a wide range of sounds when currents of many different levels and frequencies are applied to it. The phrase *digital electronics* describes those circuit systems which primarily operate with two different voltage levels or two other binary conditions. *Binary* means consisting of two parts.

The two different conditions, or states, by which digital circuits operate can be in several forms. They can, in their most simple form, consist of the opening and closing of a single-pole, single-throw switch (Fig. 24-1A). In this case, the closed-switch condition can be represented by logic 1 (high) and the open-switch condition by logic 0 (low). Likewise, the lighted lamp can be considered as being in the logic 1 condition when the switch is closed, while the unlighted lamp can be considered as being in the logic 0 condition when the switch is open.

A very common method of digital operation uses voltage pulses (Fig. 24-1B). Here, a positive pulse is represented by logic 1 and the absence of a pulse by logic 0. With a square-wave signal, the positive-going pulses can represent logic 1, and the negative pulses can represent logic 0 (Fig. 24-1C).

Another commonly used binary condition is obtained using two different voltage levels (Fig. 24-1D). In this case, a voltage that is positive with respect to a given point can be represented by logic 1 while a voltage that is less positive with respect to the point can be represented by logic 0.

LINKS

A more thorough discussion of the binary numbering system follows in Chapter 25, "Fundamentals of Computers."

Draw a ten-second waveform of load current that is switched ON for one second and OFF for one second. Assume a +5 Vdc power supply.

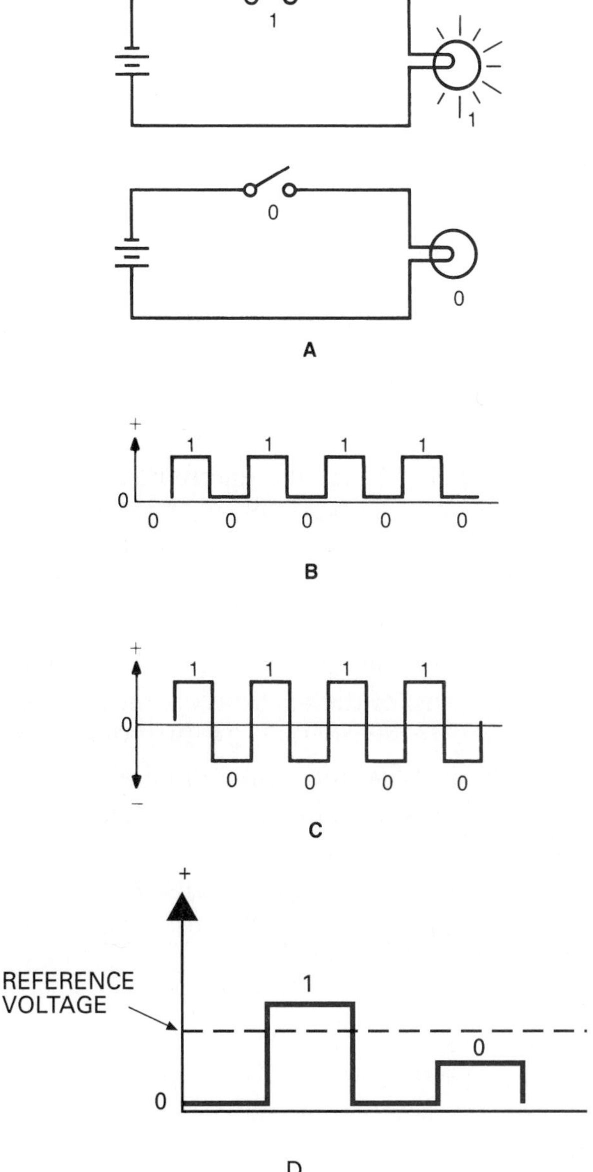

Fig. 24-1. **Common methods of producing binary conditions: (A) switch; (B) voltage pulses; (C) positive versus negative pulses; (D) reference voltage.**

LOGIC CIRCUITS

The term *logic*, when used with digital circuits describes circuits that provide definite, predetermined output conditions when one or more specific input conditions are satisfied. Circuits designed to operate in this manner are said to perform a *logic function*.

Input conditions of a logic circuit can be controlled with a combination of switches that are ideally suited for binary applications because they have ON (1) and OFF (0) actions.

In early logic gates, relays were very commonly used to perform switching functions. However, because of their slow operating speed and relatively large power requirements, relays have been replaced by switching devices such as transistors, diodes, silicon-controlled rectifiers, and other solid-state devices.

Integrated circuits containing many transistors are the most commonly used switching devices in digital logic gates. The reasons for this are that they are compact and capable of operating at extremely high speeds.

A basic PNP transistor circuit functions as an ON/OFF switch (Fig. 24-2). In this circuit, switch S1 controls the input signal (battery voltage).

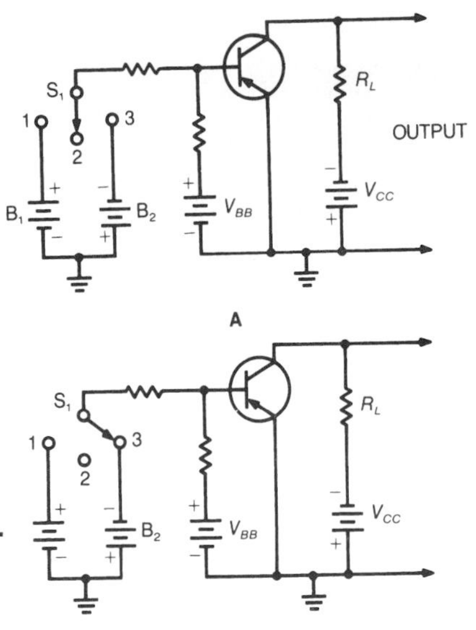

Fig. 24-2. **A transistor as a switching device: (A) transistor OFF; (B) transistor ON.**

With S1 open (position 2 closed), the emitter-base junction of the transistor is reverse-biased to the cutoff point (Fig. 24-2A). In this condition, the current in the collector, or output, portion of the circuit is at a minimum. The transistor "switch" can now be considered as being turned OFF, or in the low state.

When position #1 of S1 is closed, a positive voltage is applied to the base of the transistor. Since this voltage does not affect the reverse-bias condition of the base-emitter junction, the transistor output signal stays low.

The transistor is turned on by closing position #3 of S1 (Fig. 24-2B). In this condition, a negative voltage is applied to the base of the transistor. The negative voltage overcomes the reverse-bias applied to the base-emitter junction by V_{BB} because B2 is the higher voltage, and the transistor is "switched" to the conducting (ON or logic 1) state.

Write a brief essay describing the switching action of a transistor.

TYPES OF LOGIC GATES

Logic gates are electronic circuits that process data by comparing signal inputs logically, or in a predetermined manner. The three basic types of digital logic gates are AND gates, OR gates, and NOT gates. As with all logic gates, the output of each is binary, or in one of two states: ON (1) or OFF (0). Each of these output states is, in turn, produced by switching (ON or OFF) actions, or logic conditions, that are applied to the inputs of the logic gate. How each of the basic types of logic gates operates and an example of each are described in the following paragraphs.

AND Gates

Let us suppose that in a manufacturing plant, two workers are stationed at an assembly line. Also, suppose that in this system it is possible for the motor-driven, assembly-line belt to move only when both workers are ready and that it is desirable for each worker to be able to stop the belt.

AND gates can solve the problem in designing such controlled-output systems (Fig. 24-3A). Since switches *A* and *B* are in series, both of them must be on (1) in order to apply a voltage to the motor so it will run. This voltage, or output signal (*C*), can be represented by 1. If either switch *A* or switch *B* is OFF or in the 0 state, no voltage will be applied to the motor and it will stop. This output condition is then represented by 0.

Truth tables describe the operation of the circuit in terms of the relationship between its inputs and its output (Fig. 24-3B). The truth table indicates that the output of an AND gate is 1, if and only if, all inputs are 1. If any or all inputs are 0, the output is 0.

Comparators are circuits that use AND gates to compare quantities, conditions, or states to determine equality. For example, a single two-input AND gate can be designed so that its output is 1 only when two voltages reach a predetermined positive value. The output can then be used to activate an indicating device of some kind.

A basic AND gate using an NPN transistor is shown in Fig. 24-3C. Here switches *A* and *B* control the bias voltages that switch the transistor to provide either a high or low signal output across the emitter resistor. Notice that the truth table is also valid for this gate.

The operation of an AND gate is mathematically expressed by the equation $A \cdot B = C$. This can be read as "input *A* and input *B* equals output *C*."

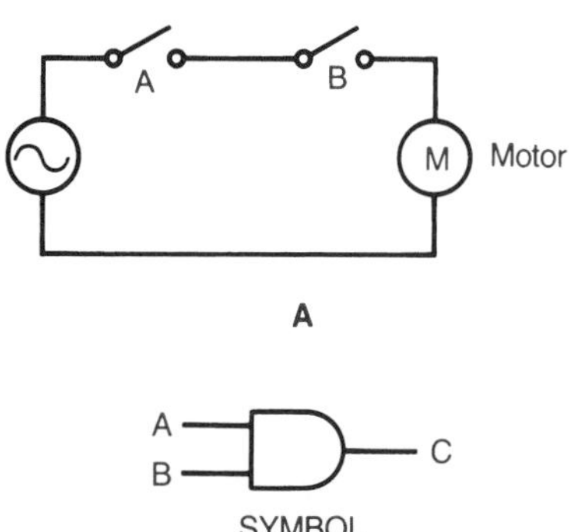

Truth Table

Inputs		Output
A	B	C
0	0	0
1	0	0
0	1	0
1	1	1

B

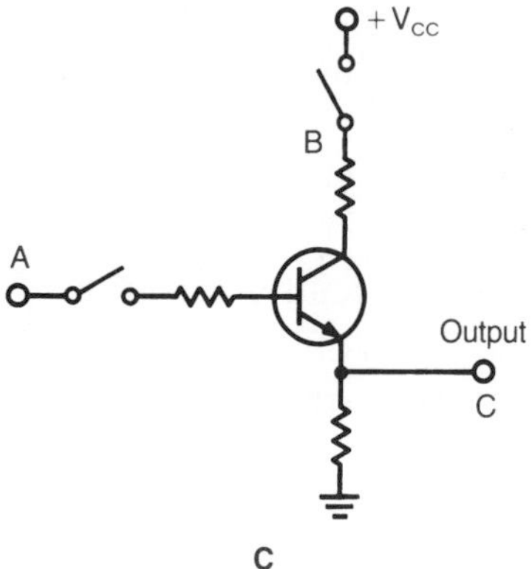

Fig. 24-3. AND logic gate: (A) motor control; (B) AND and its truth table; (C) transistor.

Logic Mathematics

Boolean algebra is the method of mathematics used to describe logic circuits. Although it looks like ordinary algebra, its symbols and operations are used differently.

OR Gates

A very practical example of a simple OR gate is the combination of push-button switches used in a residential, two-door signal system (Fig. 24-4A). In this gate, switches *A* and *B* are connected in parallel. As a result, the buzzer sounds (output *C* is 1) when either switch *A* or switch *B* is closed (1) or when both switches are closed. The buzzer will not sound (output is 0) when both switches are open, or in a 0 state. These relationships for the gate are shown by its truth table (Fig. 24-4B).

An example of a basic OR gate using an NPN transistor is shown in Fig. 24-4C. In this gate, switches *A* and *B* set the conditions by which the transistor switches to provide a 1 or a 0 output signal. The truth table also reflects the I/O (input to output) relationships of this gate.

The operation of an OR gate is often expressed by the equation $A + B = C$. This can be read as "input *A* or input *B* (or both) equals output *C*."

A New Approach to Reasoning

For centuries philosophers have studied logic, which is orderly and precise reasoning. However, in 1854 George Boole (1815-64), an English mathematician, argued that logic should be allied with mathematics rather than with philosophy. In An Investigation of the Laws of Thought, *he demonstrated logical principles with mathematical symbols rather than words. This approach bridged the gap between mathematics and logic. It introduced a new field of mathematical/philosophical study: symbolic logic.*

Symbolic, or modern, logic is different from classical logic in several ways. Symbolic logic uses symbols to represent the logical subject and predicate of a sentence and to signify the classes, members of classes, relationships of class membership, and class inclusion. It differs from classical logic in the assumptions made about the existence of things referred to in its universal statements. For example, the statement, "All A's are B's" is instead conveyed as, "If anything is an A, then it is a B." This approach does not assume that any A's exist.

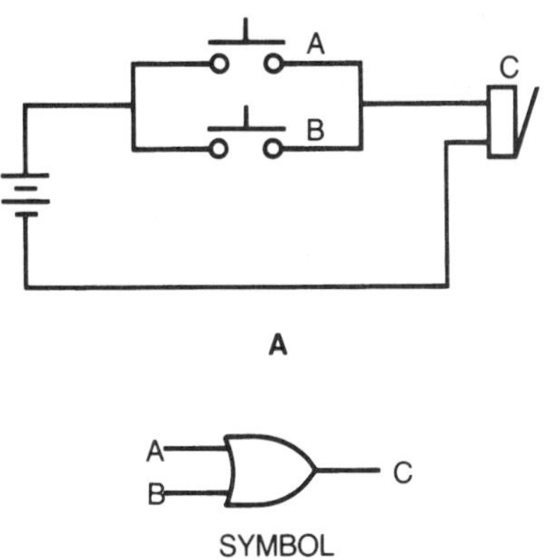

Truth Table

Inputs		Output
A	B	C
0	0	0
1	0	1
0	1	1
1	1	1

B

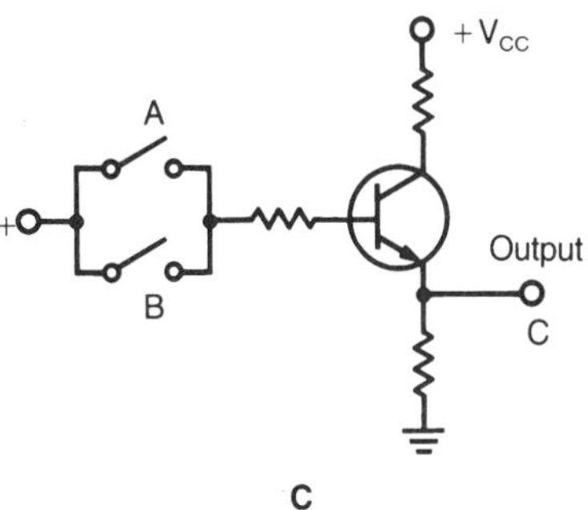

Fig. 24-4. **OR logic gate: (A) buzzer control; (B) truth table; (C) transistor.**

NOT Gates

The use of **inverters**, or *complementary gates*, is an important application of NOT gates that reverses signals. Such gates have only one input.

A very common example of a basic NOT gate is shown in the relationship between temperature and the operation of a gas furnace in a residential heating system. The control device in such a system is a thermostat that is affected by the binary conditions of a higher or a lower temperature. Let us assume that a higher temperature above the thermostat setting is a 1 and that a temperature below the thermostat setting is a 0.

If the temperature in a room rises above the setting of the thermostat, the thermostat opens (Fig. 24-5A). This turns off the output current to the relay that controls the gas valves, and the furnace goes low. If the temperature decreases below the setting of the thermostat (input 0), the thermostat closes (output 1), and the furnace begins to operate, or goes high.

A basic NOT circuit using an NPN transistor is shown in Fig. 24-5B. When the input voltage to the base of the transistor is high enough in the positive direction (1), the transistor conducts heavily. At this time, the output voltage, or the voltage across the transistor, is low (0).

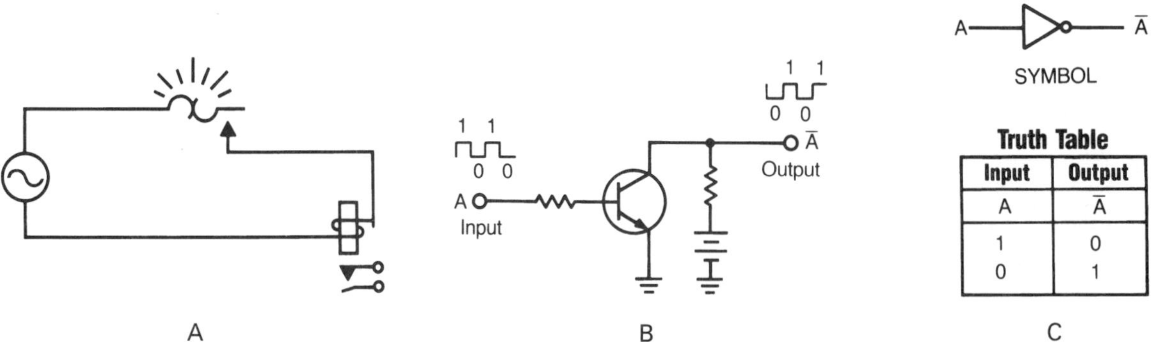

Truth Table

Input	Output
A	$\overline{A}$
1	0
0	1

Fig. 24-5. **NOT logic gate: (A) furnace control; (B) transistor; (C) truth table.**

When there is no input voltage or the input voltage is low (0), the transistor operates at cutoff, thus producing an output voltage which is high (or in the 1 state). A line over the letter designating one or more of the inputs or the output of a gate indicates an inverted, or complimentary state. This is referred to as a "not," and, therefore, $\bar{A}$ or $\bar{C}$ can be expressed as "not *A*" or "not *C*."

The operation of a NOT gate is mathematically expressed by the equation $A = \bar{A}$. This can be read as "input *A* equals output NOT *A*." The truth table for a NOT gate is shown in Fig. 24-5C.

NAND AND NOR GATES

Together, the three basic gates that have already been discussed can perform any logic function or operation. However, it is very often convenient to combine AND, OR, and NOT gates to simplify logic circuit structure. The two most commonly used combinations of these gates are NAND gates and NOR gates.

A NAND gate is a combination of an AND gate and a NOT gate (Fig. 24-6A). As indicated by its truth table, if the inputs to such a gate are in the same condition (both either 1 or 0), the output will be in the opposite condition. If the inputs are not in the same condition, the output will be 1.

A NOR gate is a combination of an OR gate and a NOT gate (Fig. 24-6B). The truth table for this gate shows that if both inputs are 1, the output will be 0 and if both inputs are 0, the output will be 1. When the inputs are not the same condition, the output is 0.

The operation of a NAND gate is mathematically expressed by the equation $\bar{A} \cdot \bar{B} = C$. The NOR gate is expressed by the equation $\bar{A} + \bar{B} = C$. The NAND equation can be read as "input not *A* and not *B* equals *C*," and the NOR equation can be read "not *A* or not *B* equals *C*."

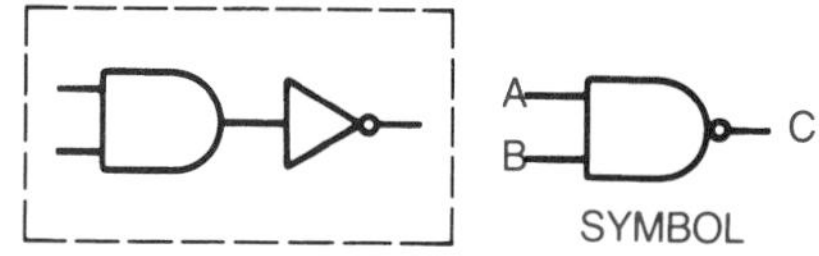

Truth Table

Inputs		Output
A	B	C
1	1	0
0	0	1
1	0	1
0	1	1

A

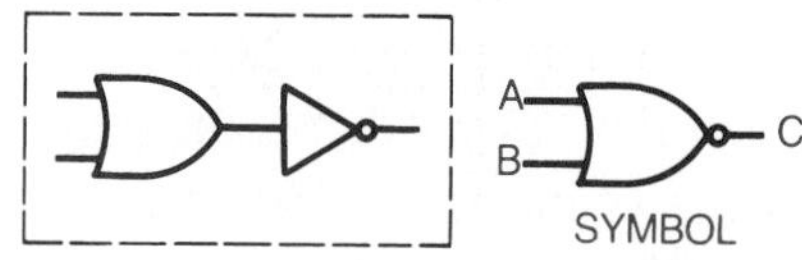

Truth Table

Inputs		Output
A	B	C
1	1	0
0	0	1
1	0	0
0	1	0

B

Fig. 24-6. NAND and NOR logic gates: (A) NAND gate; (B) NOR gate.

1. **Connect the outputs of two AND gates to the inputs of an OR gate. Label the inputs of the first AND gate *A* and *B*. Label the inputs of the second AND gate *C* and *D*. Determine the outputs of OR gate for the following conditions:**
   $A = 1, B = 0, C = 1, D = 1$
   $A = 0, B = 0, C = 0, D = 0$
   $A = 1, B = 1, C = 0, D = 0$
   $A = 1, B = 0, C = 1, D = 0$
2. **Compare and contrast NAND and NOR gates.**

EXCLUSIVE-OR (XOR) GATES

Exclusive-OR (XOR) gates are special gates. They are used when the logic desired is true for only one of two states at any given time. As shown in the truth table in Fig. 24-7, when any one input is logic 1, the output will also be logic 1. If the input logic has two logic 1s or two 0s, the output will be 0. An example of this is when you are playing a two-player video game. Only player 1 or player 2 can play the game at any given time. Both players cannot operate the game at the same time. Therefore, this is an exclusive OR situation. If there are more than two inputs to an XOR gate, an odd number of input 1s produces an output 1: an even number of input 1s produces a 0 output.

The operation of an XOR gate is mathematically expressed by the equation $A \oplus B = C$. The XOR equation is read as "input A exclusively or B exclusively equals C."

LOGIC CIRCUIT DESIGN

Suppose a circuit with the following characteristics is needed: output C is 1 if inputs A and B are both 1; C is 1 if inputs A and B are both 0; C is 0 for all other input combinations. Expressed logically, these statements are

(1) If $A = 1$ and $B = 1$, then $C = 1$

(2) If $A = 0$ and $B = 0$, then $C = 1$

(3) $C = 0$ if otherwise

To design a logic circuit with these characteristics, each requirement is first fulfilled individually. They are then combined to give the output. Requirement (1) can be fulfilled with a simple AND, $C = A \cdot B$, as shown in Fig. 24-8A. The second requirement needs a circuit that is 1 when both inputs are 0. This can be done with an AND circuit that has inverted inputs (Fig. 24-8B). If both inputs are 0, then the inputs to the AND circuit are 1, and hence the output is 1.

These two circuits can be fed into the OR circuit so that either circuit gives an output when the right inputs are present. A **combinatorial circuit** has this type of circuit arrangement (Fig. 24-8C).

A table of the first part of the problem can be created. First, all the possible combinations of inputs are listed. With two inputs, there are four combinations. For each set of inputs, the rules for the logic circuits determine the output C. With a specific set of inputs, intermediate logic values can be determined, and the output value is clear. This output value is listed in the truth table. A new set of inputs is then chosen and this is done for all combinations. The final truth table is shown in Fig. 24-8D. Note that the output C is 1 for the two input combinations 0,0 and 1,1 and 0 for any other. In addition, the third requirement in the problem is met automatically. This is often the case, but it should be checked.

The logic circuit design problem just described is a *coincidence circuit* and is a common circuit found in computers. Its output is 1 if its inputs are *coincident*, that is, either both 1 or both 0.

THE BINARY NUMBER SYSTEM

Representing any number electronically using digits of the decimal system (zero through nine) requires ten different circuit conditions or states—each representing a different digit. Although this can be done, it necessitates using complicated circuits consisting of tremendous numbers of components.

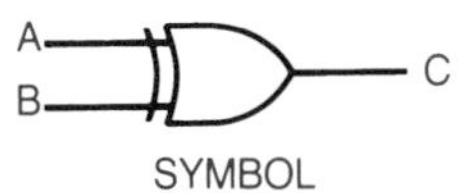

Truth Table

Input		Output
A	B	C
0	0	0
0	1	1
1	0	1
1	1	0

Fig. 24-7. Exclusive-OR (XOR) logic gate.

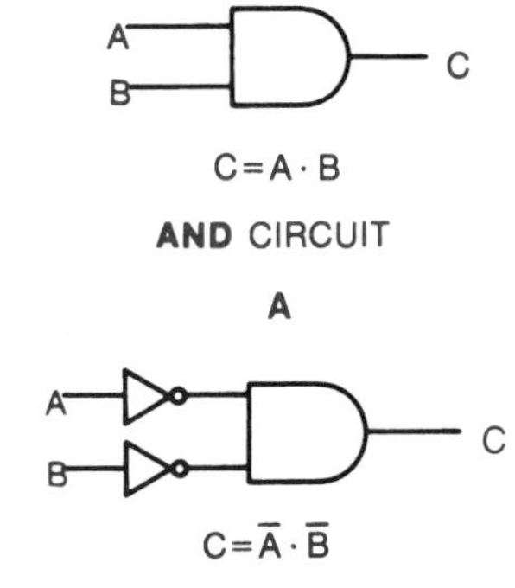

A
B
C

$C=\bar{A}\cdot\bar{B}$

AND CIRCUIT WITH INPUTS INVERTED USING **NOT** CIRCUITS

B

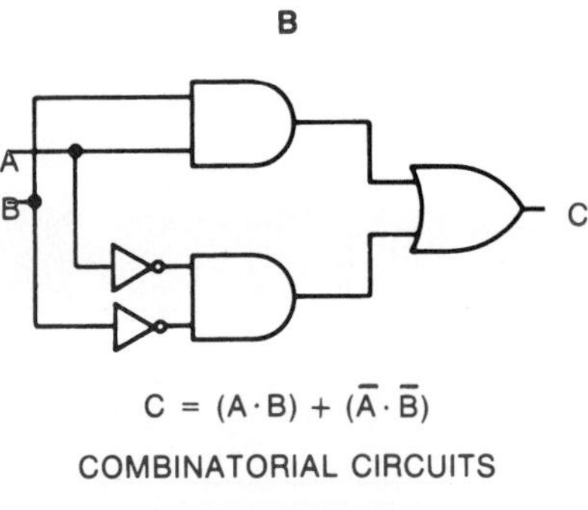

Truth Table

Inputs		Outputs
A	B	C
0	0	1
0	1	0
1	1	1
1	0	0

D

Fig. 24-8. Steps for designing a logic circuit: (A) the AND circuit; (B) the AND circuit, with inputs inverted using NOT circuits; (C) combinatorial circuit; (D) truth table for the conditions of the problem.

The logic circuits that have been discussed are binary in nature. This means that operating them depends on only two conditions, or states. These are very easily represented electronically by means of simple switching devices. In some digital circuits, however, there is a need to express decimal numbers or groups of numbers. Examples of these are circuits found in counters, calculators, and data processing systems such as computers.

The binary number system represents decimal numbers by combining binary digits (usually called bits), 1 and 0. Consequently, the mathematical operations of addition, subtraction, multiplication, and division can be performed with great speed by digital circuits.

The binary number system is based on the fact that any decimal number can be represented by a combination of numbers using the number base 2. For example,

$1 = 2^0$

$2 = 2^1$

$3 = 2^0 + 2^1$ (1+2)

$25 = 2^0 + 2^3 + 2^4$ (1 + 8 +16)

Several decimal numbers and their binary number equivalents are given in Table 24-A. This table shows that, reading from right to left, the decimal number 7 is equal to

One (2^0) + One (2^1) + One (2^2)

or 1 + 2 + 4

or, expressed as a binary number, 7 = 000111 = 111

(Leading zeros may be omitted.)

The binary equivalents of decimal numbers are used in calculators and computers because people prefer decimal numbers. Therefore, the two systems have to be translated, or converted back and forth, as people work with the digital equipment.

1. List three applications of an XOR gate.
2. Write an essay describing a comparator. List at least two specific applications.

As indicated by Table 24-A, each bit of a binary number represents some decimal number. This number equals 2 raised to some power. A binary number can be converted by adding its equivalent decimal numbers. As the table shows, each of the decimal numbers equals 2 raised to the power indicated by the column in which the corresponding bit appears. The important thing to remember when converting a binary number to a decimal number is that the powers of 2 begin with 2^0 in the right-hand column. The following example shows how to convert a binary, or base 2, number to a decimal, or base 10, number:

$$\begin{array}{ccccccccccl} 1 & & 0 & & 1 & & 1 & & 0 & & \text{binary} \\ \downarrow & & \downarrow & & \downarrow & & \downarrow & & \downarrow & & \\ 2^4 & & 2^3 & & 2^2 & & 2^1 & & 2^0 & & \\ \downarrow & & \downarrow & & \downarrow & & \downarrow & & \downarrow & & \\ 16 & + & 0 & + & 4 & + & 2 & + & 0 & = 22 & \text{decimal} \end{array}$$

Table 24-A shows that very large binary numbers can become complicated for humans, who are accustomed to the decimal number

Table 24-A Decimal Numbers and Their Equivalent Binary Numbers

Decimal Numbers	**2^5** (32)	**2^4** (16)	**2^3** (8)	**2^2** (4)	**2^1** (2)	**2^0** (1)
	Binary Numbers					
0	0	0	0	0	0	0
1	0	0	0	0	0	1
2	0	0	0	0	1	0
3	0	0	0	0	1	1
4	0	0	0	1	0	0
5	0	0	0	1	0	1
6	0	0	0	1	1	0
7	0	0	0	1	1	1
8	0	0	1	0	0	0
9	0	0	1	0	0	1
10	0	0	1	0	1	0
20	0	1	0	1	0	0
30	0	1	1	1	1	0
40	1	0	1	0	0	0
50	1	1	0	0	1	0

FASCINATING FACTS

To simplify computer programming, a low-level language called assembly language assigns a code to each machine-language instruction. For example, in binary code, the command to move the number 255 into a CPU register would be 0011111011111111. In assembly language, this might be written MVIA,OFFH. To the programmer this means Move immediately to register A the value FF (hexadecimal for 255). A program may include thousands of such codes, which are then translated into binary code.

system. For instance, to translate the number 1111111111111 from binary to decimal requires adding thirteen numbers. For this reason, *binary coded decimals*, or BCDs, are used to translate binary numbers from computerized equipment for people. A binary coded decimal uses binary numbers to represent each decimal digit. For example, the decimal number 325 is represented by the binary coded decimal 0011 0010 0101. The binary coded decimal equivalent of 1111111111111 is

1000 0001 1001 0001 =
8 1 9 1 =
8,191

Because it can be done at a glance, using binary coded decimals to translate large binary numbers is much simpler than adding: 4096 + 2048 + 1024 + 512 + 256 + 128 + 64 + 32 + 16 + 8 + 4 + 2 + 1 to equal 8191.

Explain the differences between the binary number system and the decimal number system.

DIGITAL ARITHMETIC

A vital part of operating many digital systems includes mathematical functions. As an example, the arithmetic unit of a computer system can add, subtract, multiply, divide, extract roots, and provide other means of processing information. In actuality, however, all mathematical operations are performed by adding circuits.

Addition of Binary Numbers

The rules for adding binary numbers are:

0 + 0 = 0
0 + 1 = 1
1 + 0 = 1
1 + 1 = 10

The 10 can be read as "one-zero," as "0 and carry 1," or as "sum 0 and carry 1." Examples of using these rules follow (decimal number equivalents are written in parentheses):

011 (3)	1101 (13)	110 (6)
010 (2)	1111 (15)	011 (3)
101 (5)	11100 (28)	101 (5)
	1110 (14)	

Half-Adders

Half-adders are digital circuits capable of adding two, one-bit binary numbers. The half-adder in Fig. 24-9 provides an interesting example of how to use ANDs, ORs, and NOT logic gates. The associated truth table gives the binary condition (0 or 1) at several points in the circuit when the various combinations of one-bit numbers are added.

Full-Adders

Full-adders are circuits that can add two bits and carry a bit from a prior adding circuit. A basic full-adder circuit is a combination of two half-adders plus another logic gate. By using many full-adders, it is possible to add large binary numbers.

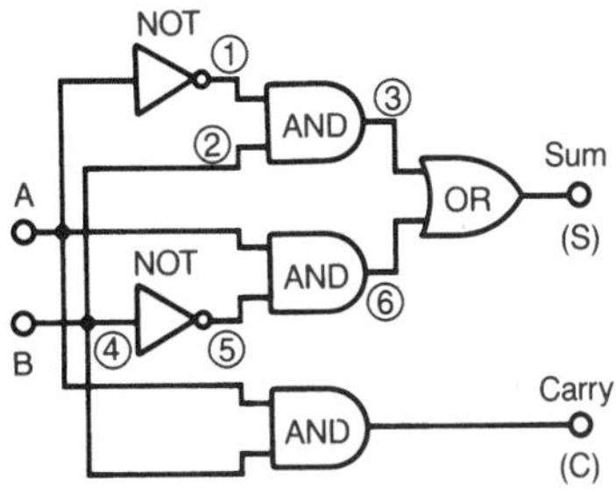

Truth Table

Inputs		Points in Circuit						Answer	
A	B	1	2	3	4	5	6	C	S
0	0	1	0	0	0	1	0	0	0
0	1	1	1	1	1	0	0	0	1
1	0	0	0	0	0	1	1	0	1
1	1	0	1	0	1	0	0	1	0

Fig. 24-9. A half-adder circuit and truth table.

Subtracting, Multiplying, and Dividing

Subtracting, multiplying, and dividing are all variations of adding. Subtracting is the inverse, or opposite, of adding; multiplying is repeated adding; and dividing is the inverse of multiplying. Because of these relationships, variations of adders also perform other mathematical operations.

REGISTERS AND COUNTERS

Registers are memory devices that store bits of data as words, or groups of bits. These devices can operate as simple storage vaults for data bits or can operate in such ways that change data by shifting it left or right. One very common use for a register is as a timer that works by counting predictable clock pulses from an oscillator. Another way registers are used is to count input pulses from a field device such as a limit switch.

Counters are one of the most common and useful digital circuit systems. In addition to counting, they can also divide frequencies. For example, a 100 Hz signal can be divided to 10 Hz or to 1 Hz by digital counting circuitry.

Bistable multivibrators, or **flip-flops** (FF), are the basic circuits used in most binary counters. These circuits have two stable output states and can be switched from one output state to the other. In T-type flip-flops, applying a trigger (T) voltage pulse to their inputs (Fig. 24-10) does this. Here 1 represents an output voltage pulse and 0 represents no output voltage. In the symbol for flip-flops, the not Q ($\bar{Q}$) at one of the outputs indicates that this output is always the inverse of the other (Q) output.

The *clock* is often the trigger voltage pulse applied to a flip-flop. When the pulse is a positive voltage, it is an *up-clock.* Conversely, a negative trigger pulse is a *down-clock.*

Each clock pulse causes a transition to occur in the output states of Q and $\bar{Q}$. When Q is high, $\bar{Q}$ is low. The outputs will remain at these states until another clock pulse causes a transition.

A *cascade* is a simplified counter circuit arrangement consisting of flip-flops connected together as the four T-types shown in Fig. 24-

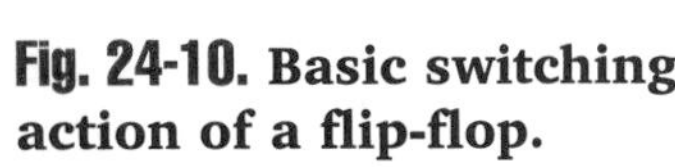
Fig. 24-10. Basic switching action of a flip-flop.

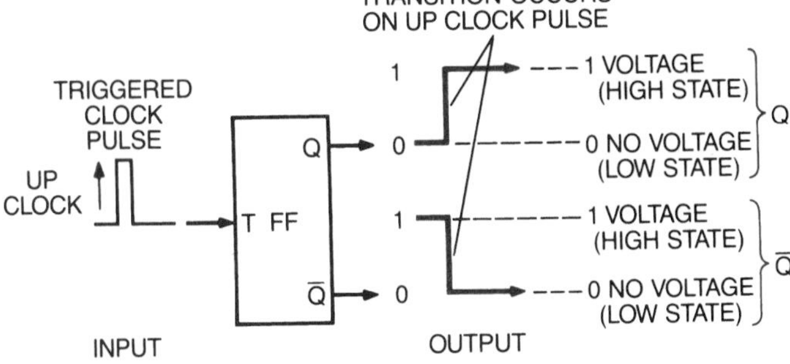

1. **Add the following binary numbers. Write the answers in decimal form.**
   1 + 1
   0 + 1
   1 + 1 + 1
   01 + 11 + 1011
2. **How many flip-flops will it take to count to 32?**

11. The $\bar{Q}$ outputs are connected to lamps that are the readout devices. The reset line allows a voltage pulse to be simultaneously applied to all the flip-flops. This pulse switches the flip-flops so that the $\bar{Q}$ outputs are at 0. When reset, the readout of the counter is 0000. In this circuit, assume that a positive trigger pulse (a 1 state) applied to any flip-flop will cause it to switch or change states.

Assuming the initial state of all the flip-flops is 1, when the first positive pulse is applied to the trigger T of FF#1, it switches this flip-flop, causing the $\bar{Q}$ output to go high, or become 1. Since, at this time the Q output of FF#1 is 0, the pulse does not affect FF#2. Lamp #1 now lights and the readout of the counter is 0001.

A second positive pulse applied to FF#1 causes this flip-flop to switch back to its original state. Now that the $\bar{Q}$ output of FF#1 is 0, its Q output is 1, and lamp #1 goes out. As the high output of FF#1 is applied to the trigger of FF#2, this flip-flop switches, causing its $\bar{Q}$ output to become 1 and its Q output to become 0. As a result, lamp 2 lights and the readout of the counter is 0010.

A third positive pulse applied to FF#1 causes it to switch, thus producing a $\bar{Q}$ of 1 and causing lamp #1 to light. The outputs of FF#2 remain unchanged since its input, the Q output of FF#1, is in the 0 state. Likewise, FF#3 is not affected because the Q output of FF#2 is also in the 0 state. In this condition of the circuit, lamps #1 and #2 are both lighted, and the readout of the counter is 0011.

By carefully "tracing" the operation of the counter beginning with the fourth pulse, the fourth through the fifteenth pulses applied to the trigger of FF#1 cause the binary readouts in Fig. 24-11. The maximum number of pulses that this counter can count is 15. The sixteenth input pulse automatically resets the counter to 0000 by setting all the flip-flop Q outputs to 1. By adding more flip-flops to the circuit, a counter can count additional pulses.

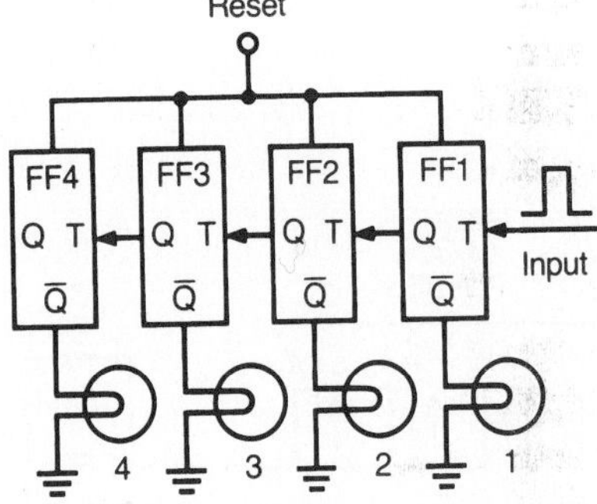

Binary Readout				Pulse Number
0	0	0	1	1
0	0	1	0	2
0	0	1	1	3
0	1	0	0	4
0	1	0	1	5
0	1	1	0	6
0	1	1	1	7
1	0	0	0	8
1	0	0	1	9
1	0	1	0	10
1	0	1	1	11
1	1	0	0	12
1	1	0	1	13
1	1	1	0	14
1	1	1	1	15
0	0	0	0	16

Fig. 24-11. A four flip-flop digital counter.

A counter such as the one described is called an *up counter* since it counts events in an upward sequence (1, 2, 3, 4, 5, etc.). Conversely, a counter that counts in a downward sequence is called a *down counter*.

FREQUENCY DIVIDERS

Figure 24-12A shows how a single flip-flop is used as a frequency divider. The first input pulse causes the Q output to go to a 1 state and the $\bar{Q}$ output to 0 state. The second input pulse causes the Q output to go to 0 state and the $\bar{Q}$ output to 1 state. The third input pulse causes the Q output to go to 1 state and the $\bar{Q}$ output to 0 state. The fourth pulse causes the Q output to go to 0 state and the $\bar{Q}$ output to go to 1 state. Notice that it takes four clock input pulses to cause the output Q to go to a 1 state twice. The Q output, therefore, effectively divides the input pulses by 2. When the Q output is connected to the input of another flip-flop as shown in Fig. 24-12B, the input pulses at each flip-flop are divided by two.

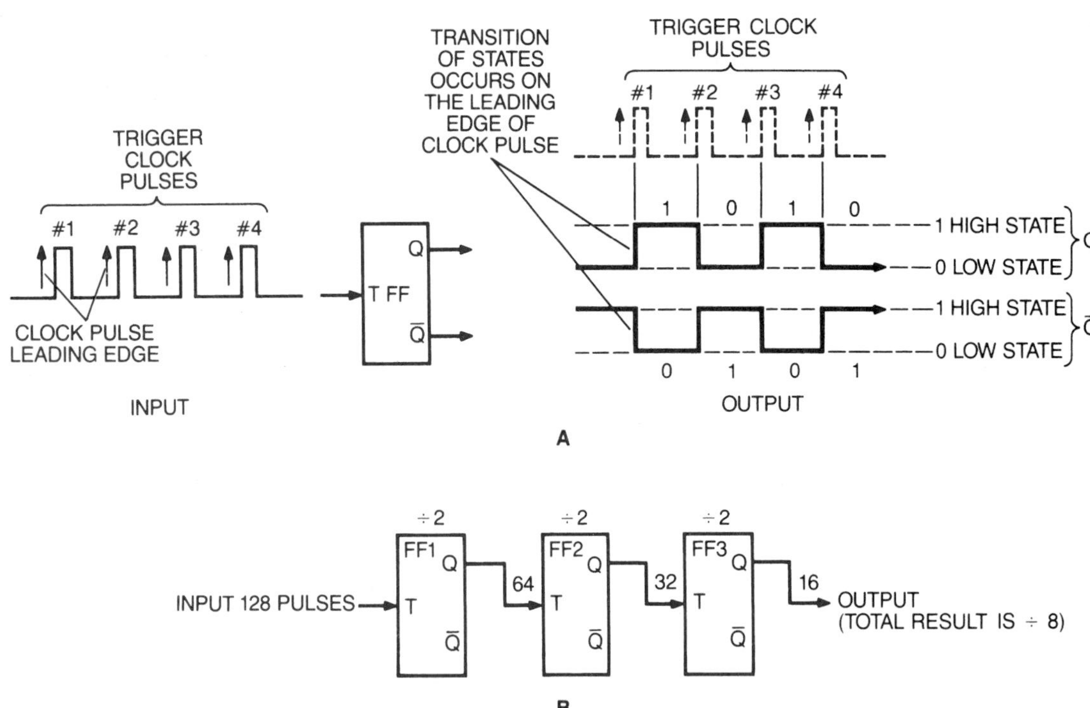

Fig. 24-12. Frequency divider: (A) basic operation; (B) flip-flops connected as a frequency divider.

Digital Clocks

Figure 24-13 shows a simplified block diagram of a digital clock that operates from 60 Hz power. This diagram does not, of course, show the several different kinds of circuits involved in the operation of the clock. It does provide an idea of how digital counters are used to reduce power line frequencies to accumulate seconds, minutes, and hours.

How many flip-flops are needed to divide a clock frequency of 1 MHz to operate a counter requiring 31.25 KHz?

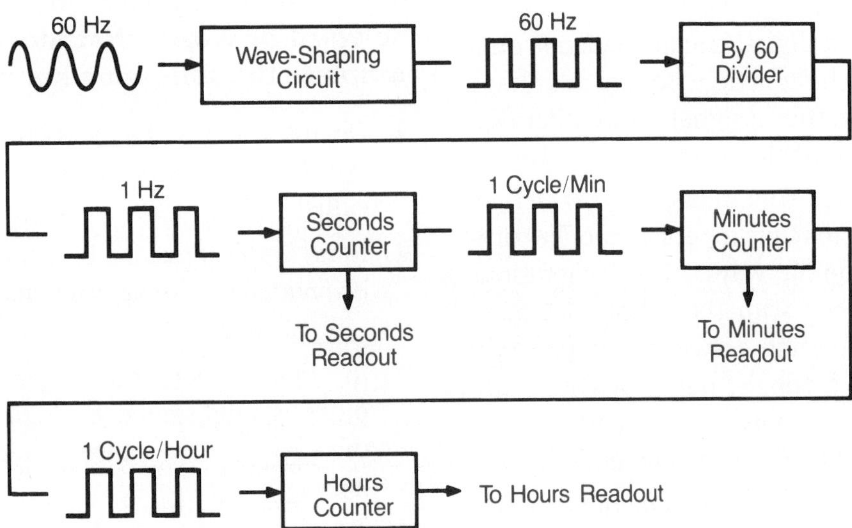

Fig. 24-13. Block diagram of a digital clock circuit system.

Chapter *Review*

Summary

- Digital electronics uses high and low voltage pulses to operate equipment such as computers, stereos, and televisions.
- Operating most digital circuits depends on a series of switching actions to create voltage pulses that produce a signal for another digital, or even an analog, circuit.
- The two different conditions, ON and OFF, by which digital circuits operate can be produced in different ways. These conditions can be produced with the simple action of opening and closing a switch, the presence or absence of a voltage pulse, or with a reference voltage against which a positive or negative voltage level is compared.
- Logic circuits perform logic functions to provide definite, predictable signal outputs according to the inputs. These circuits are constructed of common logic gates such as AND gates, OR gates, NOT gates, and others. The outcome of these gates is predicted using Boolean algebra.
- Because the binary system requires only two voltage states, it is the number system that a computer uses to perform mathematics. This is done using 1s and 0s, or high and low signals, that operate the transistors in logic gates such as full-adders and half-adders.

Review Main Ideas

Review this chapter's main ideas by writing, on a separate sheet of paper, the word or words that most correctly complete the following statements:

1. Circuit systems that work primarily with two fixed levels of voltage conditions are called _____.
2. In its simplest form, a binary condition can be created by the opening and closing of a ______ switch.
3. In digital operation, a positive voltage pulse is represented by logic ______ and the ______ of a pulse is represented by logic ______.
4. A binary condition can also be obtained by using two different ______.
5. Another name for logic circuitry is _____.
6. Four logic gates are the _____, _____, _____, and _____.
7. The input to a logic gate can have one of two states _____ or _____.
8. Truth tables describe the operation of the circuit in terms of the relationship between its ______ and its ______.
9. Comparators use ______ gates to compare quantities, conditions, or states to determine equality.
10. A residential, two-door signal system is an example of an ______ gate.

11. An ______ is an example of a NOT gate that reverses signals.
12. Representing numbers electronically in the ______ system requires complicated circuits consisting of large numbers of components.
13. The ______ number system, which represents decimal numbers by combining binary digits, performs ______ operations quickly.
14. The mathematics used to describe logic circuits is called _____.
15. The decimal number 9 can be expressed as _____ in the binary numbering system.
16. The addition of binary numbers (011 + 010) is equal to the binary number _____.
17. The basic circuit used in most binary counters is the bistable multivibrator called a ______.

Apply Your Knowledge

1. Name, describe, and write the symbols for three basic types of logic gates.
2. Draw a flip-flop circuit that counts to 100 in binary. Include lights that indicate the count.
3. Assuming a 60 Hz source, design a frequency divider that produces a pulse every second.
4. Illustrate the difference between a full-adder and a half-adder circuit.
5. Write the rules for adding binary numbers.

Make Connections

1. **Communication Skills.** Devise a system of communication between yourself and a friend using binary values to represent words. Demonstrate your system for the class.
2. **Mathematics.** Design a circuit consisting of logic gates that will handle the addition of 2 one-bit numbers. Build the circuit and present it to the class.
3. **Science.** Report on the reasons why there is so much emphasis on digital methods of transmission and reception of data, in all forms, as compared to analog methods.
4. **Technology Skills.** Research the type of transistors used for switches in digital logic design. Identify the parameters needed for a typical transistor used as a switch in an OR gate.
5. **Social Studies.** Write a report identifying the major issues between the cable companies and the satellite companies in providing TV services to consumers.

Chapter 25

Fundamentals of Computers

OBJECTIVES

After completing this chapter, you will be able to:

- Identify the three basic elements required in a computer system.
- Define terms used to describe computer memory.
- Convert binary data into hexadecimal and octal codes.
- Describe the function of the central processing unit (CPU) in a computer.
- Explain the function performed by the two general types of memory (ROM and RAM) used in computers.
- Explain the purpose of the low address and high address as it pertains to memory location in a computer.
- Define the terms *latch* and *tri-state IC.*
- Explain how data can be transmitted by a computer over long distances.
- Identify applications of computers in the home and in industry.

Terms to Study

arithmetic logic unit (ALU)

ASCII

binary system

bit

byte

central processing unit (CPU)

input/output (I/O)

latch

multiplexing

operating system (OS)

random-access memory (RAM)

registers

read-only memory (ROM)

tri-state IC

Do you know that each time you use a device that has a control panel, you are using a computer? When you change the channel on your television set using a remote control, you are inputting data into a computer. It might be interesting to keep a list of all the computers you use in the next 24 hours. You might be surprised just how many times you use a computer in the course of a single day.

PLUG IN TO *Social Studies*

Computer-related Careers

Some of the most rapidly growing careers in the United States are in computer-related fields. The most sought-after computer specialists include systems analysts and programmers.

Systems analysts develop methods for computerizing businesses and scientific centers. They also improve efficiency in computer systems.

Programmers write software for purposes such as education, entertainment, and increasing productivity. Applications programmers write commercial programs. Systems programmers write the complex programs that control the inner-workings of the computer.

BASIC ELEMENTS IN A COMPUTER SYSTEM

One basic need of business, industry, and government is to handle large amounts of information quickly with minimum error. Thus, they need reliable systems for gathering information, making mathematical computations, and storing information so that it can be retrieved easily.

These and many other functions are performed by computers. A *computer* is a programmable electronic device that can store, retrieve, and process data. Today, most data processing is done by digital computer systems. Digital computers range from small, portable computers to large, complicated systems. These systems are made up of many pieces of interconnected equipment. Figure 25-1 shows some of the equipment commonly found in a large computer center.

The electronic technology involved in computers of all sizes is basically the same. This chapter concentrates on microcomputers because they are becoming the standard. They are used not only in small companies, but also in large corporations that previously used much larger mainframe computers.

All computer systems require hardware, software, and an operating system. *Hardware* refers to the actual devices, such as the keyboard and the monitor. The term also includes all of the chips and other electronic devices inside the computer. *Software* refers to the programs that run on the hardware. Software programs list step-by-step instructions for the computer to follow.

The *operating system* is an important element that interprets software so that the hardware can act on it. No computer can operate without an operating system. Examples of operating systems for microcomputers include Windows NT, DOS, OS8, and Unix.

Fig. 25-1. The mass storage system at the National Center for Atmospheric Research is an example of a large-scale digital data archive. *(National Center for Atmospheric Research/University Corporation for Atmospheric Research/National Science Foundation)*

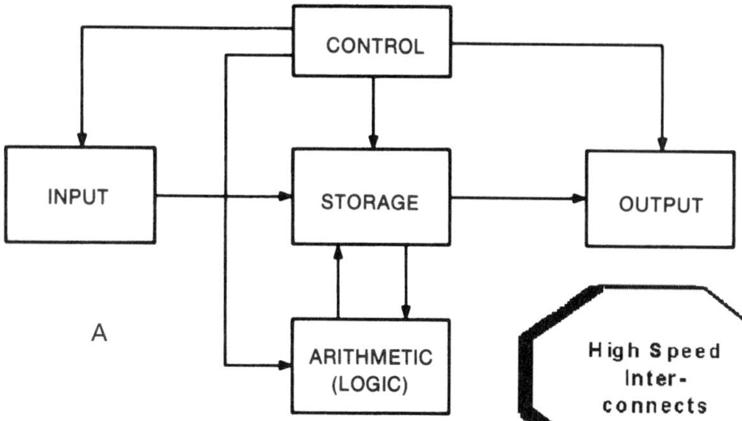

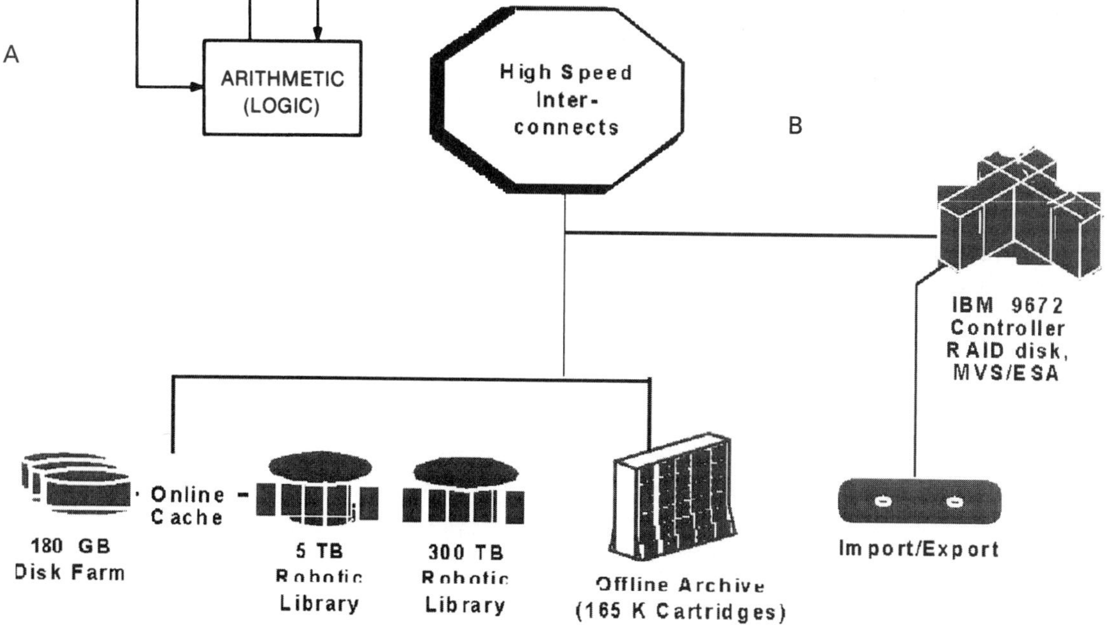

Fig. 25-2. **(A) Basic elements of a computer system; (B) a block diagram of the large-scale computer system shown in Fig. 25-1.** ***(National Center for Atmospheric Research/University Corporation for Atmospheric Research/National Science Foundation)***

Hardware

Computer systems have five basic elements: input devices, control units, arithmetic logic units, storage devices, and output devices (Fig. 25-2). Each of these elements may take more than one form and perform more than one type of function. This section discusses the basic characteristics of each element.

Input Devices

Computer input devices feed data into the computer system. Input devices can take many forms. Examples include keyboards, mice, joysticks, trackballs, scanners, magnetic tapes, CD-ROMs, various forms of disks, diskettes, removable cartridges, and even microphones (Fig. 25-3). Even an electronic home security keypad is a computer input device. The computer translates input from all of these devices into an electronic form that the computer can process. Data can also be entered into a computer by copying the data from a removable storage device.

Control Units

The heart of the computer—the part that does the actual "work"—is the control unit. In microcomputers, the control unit is generally included as part of the **central processing unit (CPU)**. The CPU controls and coordinates

LINKS

You'll find information on Storage Units later in this chapter.

Fig. 25-3. Input devices. A scanner, mouse, keyboard, and diskettes are shown at the top. A security keypad is shown at the bottom.

all the hardware in the computer. Later in this chapter, you will discover just how big a job that can be. Basically, the CPU receives input from the input devices, processes it according to instructions, and sends it to the output devices.

Arithmetic Logic Units

The control unit sends arithmetic jobs to the **arithmetic logic unit (ALU)**, which is generally included as part of the CPU in microprocessors. The ALU can do comparisons, make decisions, add, and subtract. It does these things by rapidly counting and adding electrical pulses using specially designed logic circuits.

LINKS

Logic circuits are discussed in detail in Chapter 24, "Digital Electronic Circuits."

Storage Units

Data must be stored in the computer at several stages. Internal storage is discussed later in this chapter. Long-term, external storage devices include hard disk drives and various types of removable cartridges and disks. Removable cartridges provide unlimited storage on a single drive; the only limiting factor is the number of cartridges you have. These devices include floppy disks, CD-ROMs, and magnetic tape cartridges, among others. Some removable cartridges can hold a gigabyte or more of data.

Output Devices

Computers provide output in many different forms, depending on the hardware available and the user's instructions. A computer monitor is probably the most common form of output. The next most common is "hard copy"—paper output created by a printer. Fax/modems send output over telephone connections. In addition, data can be copied to any of the external storage devices previously mentioned.

Software

Software are those application programs that tell the computer what tasks to perform. Examples include word processing, spreadsheet, financial, and database software, as well as games, computer utilities, and "front-end" software that allow people to easily use the Internet. Software programs present detailed, step-by-step logic instructions to the CPU for processing.

Operating System

The computer hardware does not automatically understand the software or the words that you enter using a keyboard. Before a computer can work with the data it receives from input devices, the data must be converted to a format the computer can recognize. The computer's **operating system (OS)** acts as an interpreter that manages translation of instructions within the computer. It ensures that the computer can understand the software instructions. After the computer performs the required tasks, the operating system translates the result into a form that the software can read. The software then presents the information to you using an output device.

COMPUTER TERMINOLOGY

Before you can understand more precisely how computers work, you need to know a few basic terms. You have probably heard the terms *byte* and *megabyte.* In most computers today, a **byte** is equal to 8 bits. The term **bit** is short for "binary digit," which is the smallest unit of information a computer can hold. Each bit can hold a single binary digit (1 or 0). Larger amounts of information are based on the byte, using powers of 2. Table 25-A shows the terminology used for larger amounts of memory. Today's microcomputers have much more internal and external memory than the original computers. Therefore, most computer storage is now given in terms of megabytes and gigabytes.

1. In general, an IBM-compatible (DOS) computer cannot read floppy diskettes that have been formatted to hold files from a Macintosh computer. Explain why.

2. Look at the Description column in Table 25-A and think about the numbers involved in using powers of 2 for larger quantities of memory or information. For example, it would seem simpler to use powers of 10, so that a kilobyte equals 1,000 bytes rather than 1,024 bytes. Why do you think powers of 2 were used in these definitions?

Table 25-A Table of computer memory size terminology

Name	Description	Abbreviation	Comments
bit	1 binary digit	(none)	Smallest unit of information in a computer
byte	8 bits	(none)	
kilobyte	1024 bytes	K or KB	2 to the 10th power (2 x 2 x 2 x 2 x 2 x 2 x 2 x 2 x 2 x 2)
megabyte	1,048,576 bytes	M or MB	2 to the 20th power
gigabyte	1,073,741,824 bytes	G or GB	2 to the 30th power; 1024 megabytes
terabyte	1,099,511,627,776 bytes	T or TB	2 to the 40th power; 1024 gigabytes
petabyte	1,125,899,906,842,624 bytes	P or PB	2 to the 50th power; 1024 terabytes

FASCINATING FACTS

The National Center for Atmospheric Research uses a computerized data archive that routinely handles more than 28 terabytes of data.

NUMBERING SYSTEMS

In addition to understanding terminology, you must have a basic idea of how computers "count." Computers use an entirely different numbering system than humans. The numbering system most humans use is called base 10 because it consists of ten digits: 0, 1, 2, 3, 4, 5, 6, 7, 8, and 9. Combinations of these digits in various positions have values that we recognize and understand how to use. For example, you know without thinking the values of numbers such as 21,450 and 37. You also know immediately that 21,450 is a higher number than 37.

Binary

Computers, however, use the **binary system**, which relies on base 2. This system is convenient for computers because it has only two digits: 0 and 1. These two digits relate easily to the two possible states of a logical circuit component. The digit 0 stands for the "off" (or "no") state. The digit 1 stands for the "on" (or "yes") state.

No matter how complex the information a computer receives, the computer processor reads and processes it as a series of 1s and 0s. Electronically, this is accomplished using two voltages that are different in value. For example, a 5-V pulse may stand for the digit 1. The absence of a pulse may stand for the digit 0. Each byte makes up one *alphanumeric character*, such as a letter, number, or other symbol.

In binary, place values do not increase by powers of 10 as you move to the right in a number. Instead, they increase by powers of 2. See Table 25-B for the place values of an 8-bit byte.

Numbers in binary are generally easy to detect—they consist entirely of 1s and 0s. For some lower numbers, such as 101, it may be hard to tell the difference. Is the number in base 10 or base 2? When the number might be misinterpreted, you should include a subscript 2 written immediately following the number to make it clear that the number is written in base 2, or binary.

For humans to "communicate" easily with computers, programmers have developed high-level programming languages. These languages use letter and number codes that have meaning to humans. The computer converts these symbols into binary codes.

However, in looking at data in memory locations, we must deal with the binary codes of the computer. The binary system is fine for the

Table 25-B Table of Equivalents for the Bits in an 8-Bit Byte

1	**1**	**1**	**1**	**1**	**1**	**1**	**1**
128	64	32	16	8	4	2	1

In binary, digit values increase from left to right in powers of 2. The leftmost digit in an 8-bit byte has a value of 1, or 2 to the 0th power; the next digit to the right has a value of 2, or 2 to the 1st power. The third digit is 2 to the 2nd power: 2 x 2 = 4. The fourth digit is 2 to the 3rd power: 2 x 2 x 2 = 8, and so on.

computer, but the human mind has a difficult time remembering hundreds of different combinations of 1s and 0s. For this reason, the octal or hexadecimal code is often used.

Octal

Octal (base 8) is a numbering system that uses eight digits: 0, 1, 2, 3, 4, 5, 6, and 7. In computer systems, octal uses three binary digits to represent an 8-bit binary number. To do this, we assign place values to the digits in the binary number. Then we divide the digits into groups of three. Starting at the right, group the first 3 bits together. These have the values 1, 2, and 4, respectively, from right to left (Fig. 25-4). The next 3 bits are grouped and again have the value of 1, 2, and 4, from right to left. The 7th bit and the 8th bit, the most significant bit (MSB), have the values of 1 and 2. Whenever a binary 1 appears in the binary number, the octal value for that place is added to the values in the group. Numbers written in octal are distinguished by the subscript 8 written immediately after the number (see Fig. 25-5). Some people use the letter O to show that a numeric value is an octal number, but O is easy to confuse with zero (0). Therefore, most people use the subscript 8.

As an example, a binary value of 10 101 001 has an octal value of 251_8 (or O251). With 8 binary bits, the octal numbers start at 000_8 and end at 377_8. More examples of octal conversions are shown in Fig. 25-5.

2 1	4 2 1	4 2 1	PLACE VALUE
1 1	1 1 1	1 1 1	BINARY NUMBER
3	7	7	OCTAL

Fig. 25-4. Octal place value.

Hexadecimal

Many computers now use the hexadecimal (hex) numbering system, which has a base of 16. In this system, the digits are 0, 1, 2, 3, 4, 5, 6, 7, 8, 9, A, B, C, D, E, and F. A combination of two hex digits can represent any of the binary numbers from 00000000 to 11111111. It is much easier to remember two digits than it is to remember eight. Numbers written in hex are identified by a $ before the leftmost digit. For example, the binary number 00011010 is written $1A in hex.

To convert from binary to hex, each group of 8 bits of data is grouped into two "nibbles," or groups of 4 bits (Fig. 25-6). Starting at the rightmost bit (the least significant bit, or LSB), the bits are given place values. The first bit on the right has a value of 1; the second bit, moving to the left, has the value of 2. The third bit is 4, and the fourth bit is 8. The values are repeated, right to left, for the next 4 bits so that they receive the values of 1, 2, 4, and 8. The leftmost bit (the MSB) has the value of 8. Each

2 1	4 2 1	4 2 1		PLACE VALUE	
1 0	1 1 1	0 1 0	0 0	0 0 0	0 1 0
	272_8			002_8	
1 1	0 0 0	1 0 1	0 0	0 0 0	0 0 0
	305_8			000_8	
1 1	1 1 0	1 1 1	1 1	1 1 1	1 1 1
	367_8			377_8	

Fig. 25-5. Examples of binary-to-octal conversions.

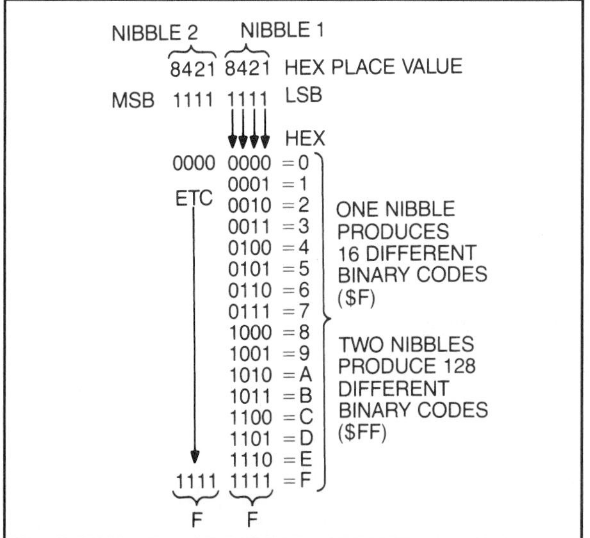

Fig. 25-6. **Hexadecimal place values.**

8 4 2 1	8 4 2 1	HEX VALUE
1 0 0 1	0 0 1 1	BINARY NUMBER
8 + 1 = 9	2 + 1 = 3	

93_{16}
or
$93

Fig. 25-7. **Binary-to-hex conversion.**

8 4 2 1 8 4 2 1	PLACE VALUE
0 0 0 1 0 0 1 1 $13	1 0 0 0 1 0 1 0 $8A
1 0 1 1 0 1 0 1 $B5	0 0 0 0 0 0 0 0 $00
1 1 0 0 1 1 1 0 $CE	1 1 1 1 1 1 1 1 $FF

Fig. 25-8. **Examples of binary-to-hex conversions.**

hex number equals the sum of place values where a binary 1 appears in that nibble. Thus, as shown in Fig. 25-7, 10010011 in binary equals 8 + 1 = $9 and 2 + 1 = $3, so the hex value is $93. If the sum of the place values is greater than 9, a single hex digit is provided using the letters A through F. These are valid numbers in base 16. Thus, 1010 is 8 + 2 = $A, and 1101 is 8 + 4 + 1 = $D. Figure 25-8 shows several more examples of this conversion.

Consider the binary number 10111101. How would this number be written in octal? In hex?

MICROCOMPUTERS

The microcomputer (often called a *personal computer* or *PC*) is made up of many ICs that together perform one or more functions (Fig. 25-9). The main IC is the device which holds the CPU. This IC is the center for all functions performed by the computer.

As mentioned previously, the computer is controlled by a list of step-by-step instructions called a *program*. Early CPUs performed each

Fig. 25-9. **A typical microcomputer workstation.**

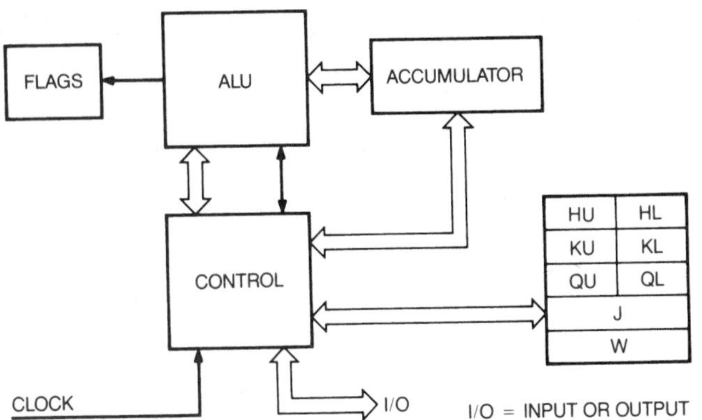

Fig. 25-10. Block diagram of a central processing unit (CPU).

function one step at a time. Modern units contain multiple processors that handle several functions simultaneously. One of them might be for mathematical operations, another for graphic generation, another for input and output of data, and a fourth for general operations. Using multiple processors, designers can increase the computer's speed by as much as four times. This is very important in office operations in which computers are connected together in a *network* to share information.

Inside the CPU

A simplified block diagram of a CPU is illustrated in Fig. 25-10. The arrows indicate the direction of data flow to and from the blocks in the diagram. The broad arrows indicate multiline data paths in parallel. One bit of binary data (0 or 1) is sent on each conductor at any given time. By grouping several of these conductors together, data can be sent in parallel: 4, 8, 16, or 32 bits at a time. This is an example of a *bus structure.* The single arrows are pathways for single bits of information, generally for control purposes.

Registers

The block labeled with letters on the right side of the diagram in Fig. 25-10 represents the **registers**. These are temporary storage spaces for data that are used later in the program.

The registers consist of a series of electronic devices called *flip-flops.* They are not connected to each other, but work together to form a "bank" of flip-flops. They store data in the form of binary 1s and 0s.

The Accumulator

The *accumulator* is a special register within the CPU. Data in the accumulator can be used to perform mathematical functions with the ALU. *Flags* are used to indicate the results of the arithmetic and logic operations. Each flag is one bit of a special 8-bit group that can be set individually to a logic 1 or 0. Flags show the result of an arithmetic operation in binary form. They report the result as either equal or unequal to zero. A flag can also indicate a carry value that is generated by adding two numbers. Flags can control the operation of a program according to the condition of the flag.

The *control circuit* gates, or transfers, information through the computer. As the name indicates, this section of the CPU controls the entire computer operation. All data leaving or entering the computer must pass through the control system.

LINKS

For more information about the operation of flip-flops, see Chapter 24, "Digital Electronic Circuits."

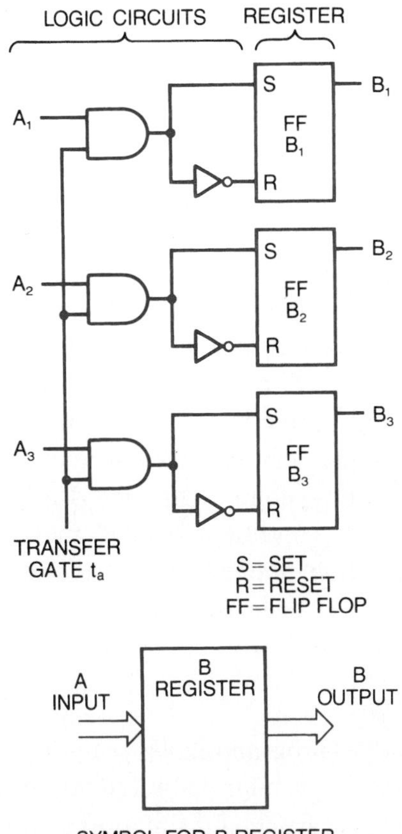

Fig. 25-11. A 3-bit register made up of three flip-flops.

The binary data is stored in groups called *register words.* Figure 25-11 shows a register capable of holding (storing) a 3-bit word. Input information A (A_1, A_2, A_3) passes to the flip-flop inputs when the transfer gate signal t_a is 1. If an input is 1, a 1 appears at the set input and the flip-flop assumes a value of 1. If the input is a 0, a 1 appears at the reset input and the flip-flop assumes a value of 0.

The concept of *register transfer* is important to data movement in a computer. Figure 25-12A illustrates the transfer of data in register A to register B. The large arrow indicates a data word of many bits. The t_1 indicates that the transfer takes place at a given time. In this case, t_1 could be set to equal 1 at some specific time known as t_1 .

Register C in Figure 25-12A contains a group of 12 binary bits. The 12-bit configuration could represent the integer number 849, the letter S, the decimal number $+0.43 \times 10^{+1}$, or an instruction to add two variables. Register D in Figure 25-12A shows one possible interpretation of register content that represents a decimal number. Figure 25-12B shows that data can be transferred to and from several different units including memory, arithmetic unit, and input-output.

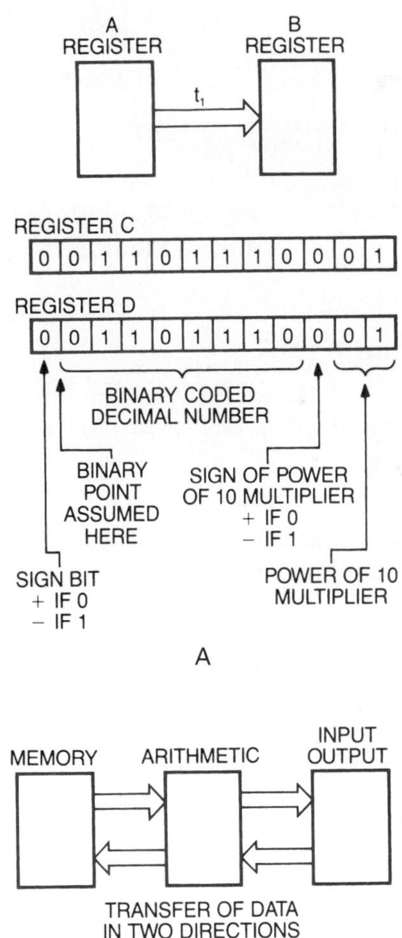

Fig. 25-12. (A) Concept of register transfer and interpretation of binary bits in a register; (B) data can be transferred in either direction among the various units.

Types of Memory

As discussed previously, data is stored in specific locations in the computer in binary format. Electronic circuits called memory perform this function. Memory is installed in the computer as individual integrated circuits or on small plug-in circuit boards that hold the memory chips. Circuit boards called SIMMs (Single Inline Memory Modules) are one popular form of memory (Fig. 25-13). In this form, many millions of bytes of data can be stored for almost instant access. SIMMs also make it easy to (add) memory in a computer.

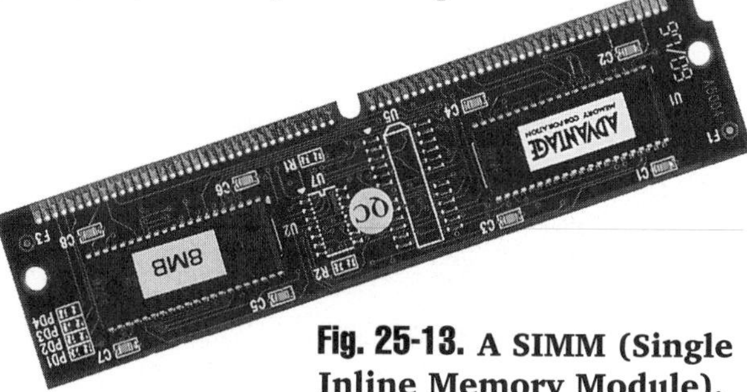

Fig. 25-13. A SIMM (Single Inline Memory Module).

ROM and RAM

The two types of memory used in computers are RAM and ROM. Each type has many design variations.

RAM stands for random access memory, and is memory that can temporarily store program information. When you turn the computer off, the information in RAM is lost. When you turn the computer on again, RAM again becomes available for temporary storage, although it does not "remember" any information it contained when you turned it off.

ROM stands for read-only memory. It is permanent memory that contains the programs that are needed to "boot up" the computer and begin operations when the computer is turned on.

Programs in ROM are called *firmware.* Each time the computer is turned on, the OS must be running for the computer to decode the instructions the operator sends into the computer. A ROM-integrated circuit has the program designed into its structure, and it cannot be changed.

PROM

Because ROM is expensive to design and manufacture, it is used when hundreds of thousands of bytes are needed. When a smaller number are needed, *programmable read-only memory* (PROM) is programmed using special equipment. The PROM-integrated circuit allows manufacturers to provide custom firmware in a less expensive format than ROM. Programming equipment electrically changes the binary condition of the memory cells, changing their states to 1 or 0 as needed. Once a PROM has been programmed, the information stored on it cannot be changed.

EPROM

Erasable programmable read-only memory (EPROM) is similar to a PROM, but can be erased. To erase the information on an EPROM, the user exposes it to ultraviolet (UV) light. EPROMs look much like PROMs. You can identify the EPROM by the window on the top of the lC. This window allows ultraviolet light to shine on the surface of the chip. Exposure to UV light changes the state of all the binary digits to 0.

A variation of the EPROM is the EEPROM, or *electrically erasable programmable read-only memory*. In EEPROMs, electrical pulses are used instead of UV light to set the bits to binary 0. The advantage of EEPROMs is that the information they contain can be altered without removing them from the circuitry.

Each type of PROM is programmed using special programmer circuitry called a *PROM programmer*. A computer checks the data as it goes through the process to make sure that the proper data is in the desired locations.

Most microcomputers have slots for printed circuit boards. Special boards can be placed in these slots to expand the function of the computer. Among these boards are sound cards, speech synthesizers, video accelerators, and special types of disk drive controllers. To operate these circuit boards, a PROM is included within the controller circuitry to run the program.

Addressing Memory

As discussed previously, the computer contains many individual bytes of memory. The operator must be able to access any one individual byte to obtain or alter the data it contains. Therefore, the computer includes a system of addresses. Each address has two parts because it requires two bytes of memory. The bytes are called the *low* and *high* addresses. As illustrated in Fig. 25-14, each part of the address contains 8 bits.

Low Address

If a single binary bit is used to mark objects, it can only identify two addresses: address 0 and address 1. It is easy to see that this is not sufficient because a computer has many more than two memory locations. If the number of bits is doubled, the combinations increase to four: 00_2, 01_2, 10_2, 11_2. This is still not enough to be usable.

If, however, the number of bits is increased to 8, or 1 byte, the combinations are increased to 256 identifiable locations. These 256 bytes, grouped together, are said to be "one page of memory." As shown in Fig. 25-15, the 8-bit binary addresses on the page begin at 00000000_2 and continue to 11111111_2. This system identifies memory locations that may contain stored data.

The code for an address is sent to a memory chip via a group of 16 conductors called the *address bus*. The low eight address-bus lines are marked on drawings with the symbols A7, A6, A5, A4, A3, A2, A1, and A0. When all eight lines are considered together (as the low address), notations are written in simpler form: A7-A0. As noted in Fig. 25-16, A0 is called the

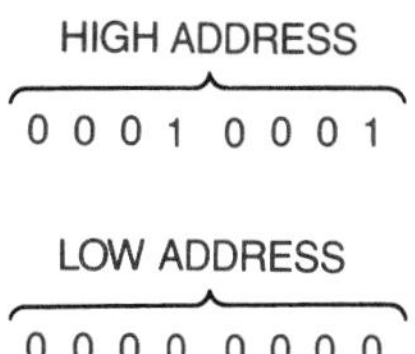

Fig. 25-14. Low and high bytes of a memory address.

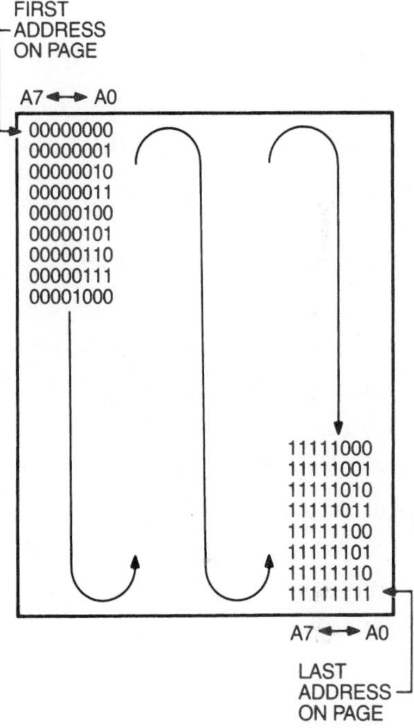

Fig. 25-15. One page of memory contains 256 bytes. The bytes are numbered in base 2, or binary. Notice the sequence of numbering in base 2. The first nine addresses correspond to the decimal (base 10) numbers 0, 1, 2, 3, 4, 5, 6, 7, and 8. What is the decimal equivalent of the last eight binary numbers on this page?

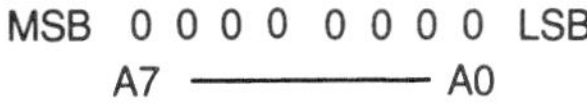

Fig. 25-16. Lower 8 bits of the address bus.

least significant bit (LSB). Just as in base 10, in base 2, the rightmost digit of any number is the least significant. A7 is the *most significant bit* (MSB). The digit in the leftmost position of any data value is the MSB, no matter how many digits are in the value. With this system, many locations can be addressed by these combinations of digits. However, many more are needed to perform even the simplest tasks. To enlarge the addressable locations, the upper eight address lines are needed.

High Address

As you have seen, the low address identifies the individual byte on a page. By adding a 1-byte high address, the computer can address 256 pages. The lines of the high address bus are marked A15, A14, A13, A12, A11, A10, A9, and A8. When the eight lines of the high address are considered as a group, they are marked A15—A8. Bit A8 is the LSB of the high address and bit A15 is the MSB (Fig 25-17).

By combining the high and low addresses (256 1-byte locations on each of 256 pages), the computer can address a total of 65,535 locations. This number is expressed as 64KB. The first page number is 0; this page contains locations 0 through 255. Zero is a valid number; therefore, there are 256 individual 8-bit memory locations. The next is page 1. It has 256 locations, as do pages 2 through 255 (Fig. 25-18).

Another way of increasing addressable memory is to use a third set of address lines. This set is used to increase memory in 64KB blocks. Two additional bits identify four 64KB blocks, 3 bits identify eight blocks, and so forth. A full set of 8 bits enable a computer to address more than 16 MB of memory. Even higher addresses are available now on some computers, enabling them to contain more than 128MB.

A program has directed a unit of data to be stored at a location identified by a low address of 01001011 and a high address of 00010110. On what page of memory is this information stored?

MSB 1 0 0 1 0 0 1 1 LSB
A15 ◄—— A8

Fig. 25-17. **High 8 bits of the address bus.**

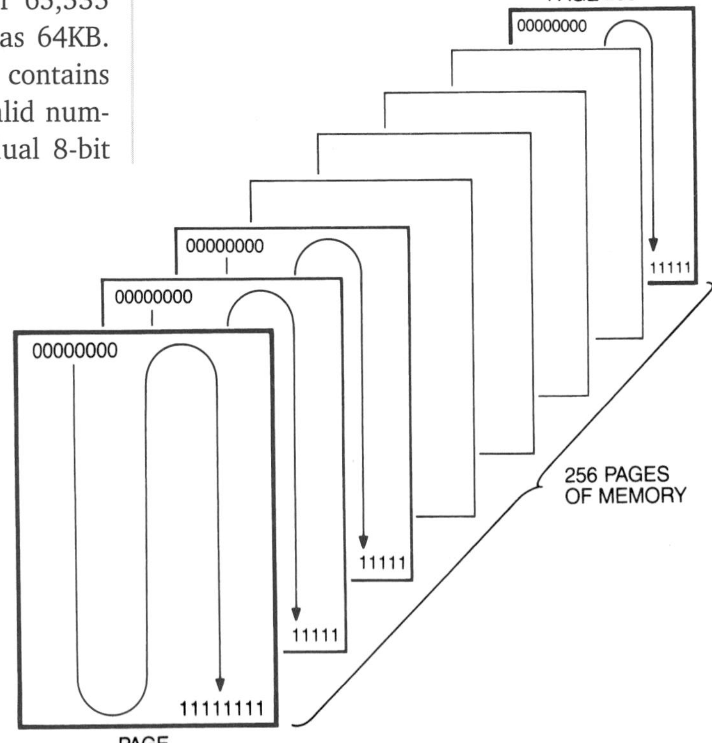

Fig. 25-18. **The high address defines the page number.**

DATA TRANSFER

A computer that uses a single 8-bit byte for data transfer is said to be an "8-bit computer." When 2 bytes are used together as data, the computer is said to be a "16-bit computer." Some computers today can transfer data in 32- and 64-bit groups. These computers have the capabilities to rapidly perform a wide variety of tasks.

Electrical Impulses

Data can be sent from point to point in a computer or from computer to computer using electrical impulses. Since digital information is limited to 1 (logical 1 or yes) or 0 (logical 0 or no), the information can be represented by voltages.

A very popular format is TTL (*t*ransistor-to-*t*ransistor *l*ogic). In this format, a 1 is represented by an energy level of 3 to 5 Vdc. Logic circuits that receive a voltage between these levels read a logical 1. They read a logical 0 when they receive 0 to 0.6 Vdc (Fig. 25-19). TTL is used primarily between the circuitry of a computer and the peripheral devices located close to the computer.

Cables

If a distance of about 2 feet or less is needed, a *ribbon cable* can be used (Fig. 25-20). This is a flat cable in which all the conductors run parallel. Over distances greater than 2 feet, the data induces voltages between the conductors. This is called *crosstalk*. If crosstalk occurs, the data received at the other end of the cable end

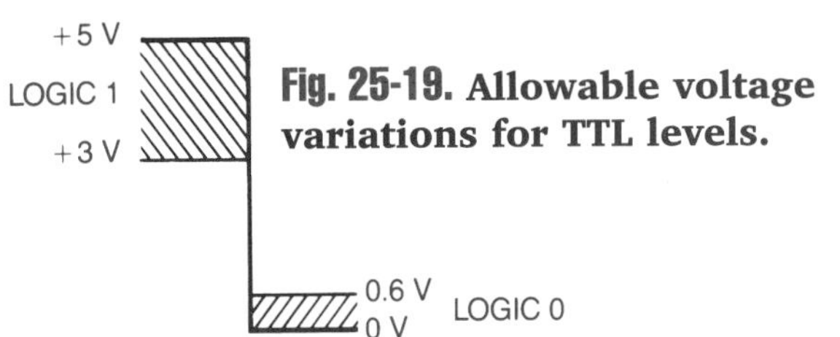

Fig. 25-19. **Allowable voltage variations for TTL levels.**

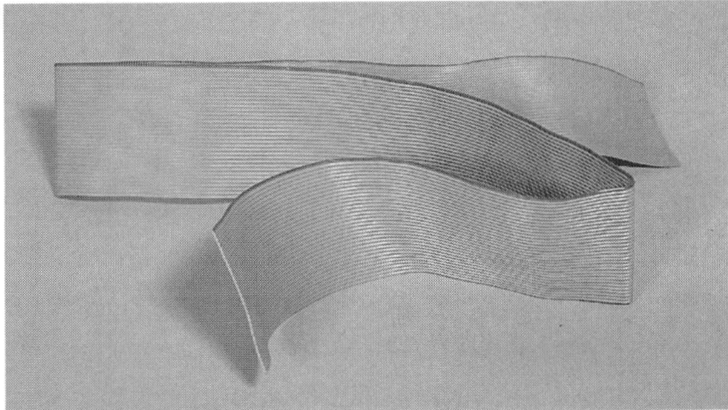

Fig. 25-20. **A typical ribbon cable.**

may not match the data that was entered. For greater distances (6 to 10 feet), a cable with twisted wire pairs can be used. In this type of cable, each data conductor is twisted with a signal return wire. This cuts crosstalk to a minimum. Both cables can also have a conductive shield to minimize interference. The ribbon cable has a flat, woven wire mesh under the insulator, and the twisted-pair cable is wrapped with a metal foil.

Transmitting Data Over Long Distances

When data is to be sent between computers that are separated by hundreds of feet or even miles, fewer conductors are desirable. TTL will not do the job. A common data format for transport over distance is called RS-232. Details on this standard version and on updated versions are published by the Electronic Industries Association.

Data from a microcomputer is sent and received over a set of three wires labeled *send*, *receive*, and *ground*. Other wires can be used for control purposes. The data is sent 1 bit at a time in rapid succession. This single-bit transmission is called *serial communication*. The number of individual bits that is sent every second is the *baud rate*. Common baud rates are 2,400, 4,800, and 9,600 bits per second. To send and receive information, the baud rate for all communications equipment connected to the cable must be set at the same rate. This can be achieved either through software or hardware switches.

Telephone networks generally provide higher-speed communications. When telephone networks are used, a modem is needed to communicate the computer signals properly to the telephone lines. Some high-speed telephone modems can now carry information at speeds in excess of 56,000 bits per second.

The 1s and 0s of the data is sent in the form of + 12 to + 15 V for a 0 and a -12 to -15 V for a 1 (Fig. 25-21). Both the sending and the receiving devices have at least three basic connections. A connector used for this format is the 25-pin D connector. A standard has been set for this connector so that data is sent on pin 2 and received on pin 3. Pin 7 is the signal ground or common return (Fig. 25-22).

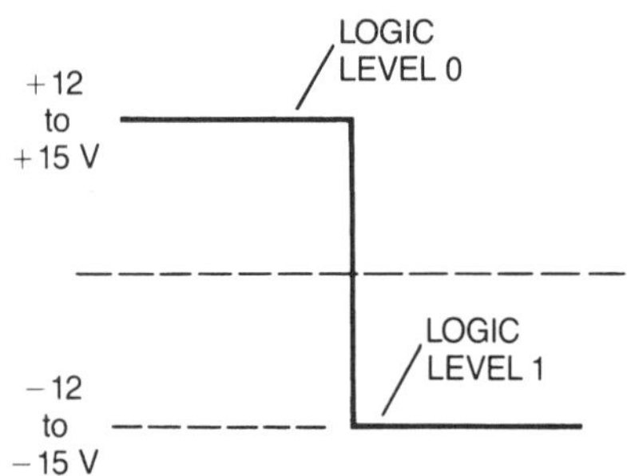

Fig. 25-21. RS-232 logic levels that represent the binary digits.

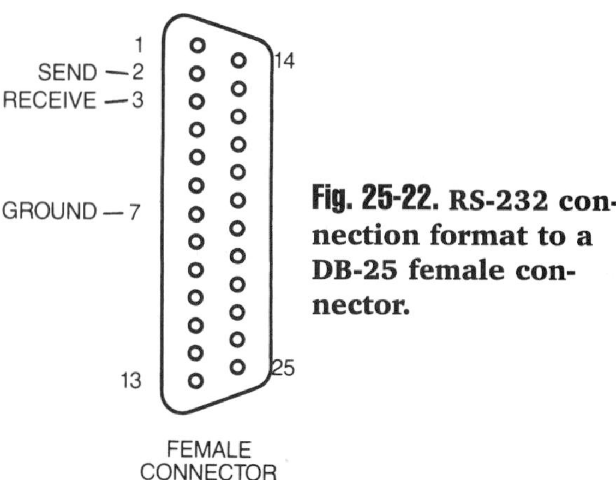

Fig. 25-22. RS-232 connection format to a DB-25 female connector.

FASCINATING FACTS

The network of networks, the Internet was formed in 1970 as part of the Department of Defense as ARPAnet (Advanced Research Projects Agency network). In the 1970s universities and companies doing defense-related research were given access. In the late 1980s most universities and many businesses around the world came on-line. In 1993 commercial providers were permitted to sell Internet connections to individuals, and millions of new users came on-line within months.

All microcomputers have at least one serial (RS-232) port. For what peripheral equipment are these ports routinely used? Why is serial communication used with this equipment? (If you don't know, look in a computer reference manual or in the online help that is supplied with many computers.)

LOGIC CIRCUITS

The pulses, or waves, of a logic circuit are rectangular in shape. Each pulse, however, has a distinct set of parts to describe the shape of the pulse or wave (Fig. 25-23). The time it takes for a change in state from 0 V to 5 V is called the *rise time* or the *leading edge*. Many circuits trigger on the rise of the pulse. The 5-V level is called the *steady high*. The steady-high level is a constant 5 V for the duration of the pulse. The time it takes to change from 5 V to 0 V is called the *fall time* or the *trailing edge*. The low level is called the *steady low*. It is near 0 V until the next rise is generated. Any one of these parts (leading edge, high, low, or trailing edge) can be used to trigger logic circuits.

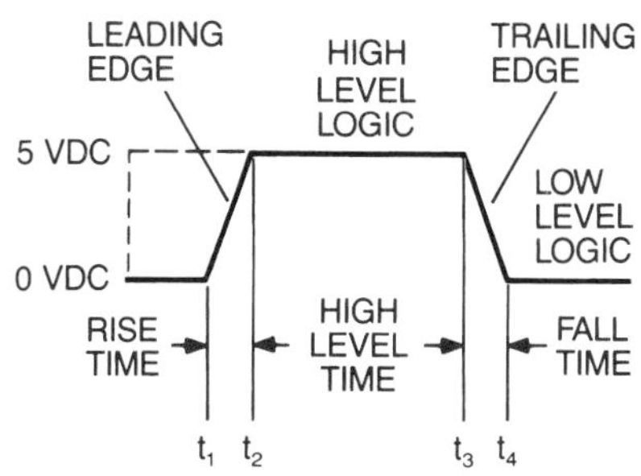

Fig. 25-23. Parts of a square wave.

Clock

The *clock* is a square-wave generator that feeds a signal to the CPU. A crystal (XTAL) is used to control the frequency so that all signals within the computer are stable and synchronized. In Fig. 25-24, the clock is outside the microprocessor IC. In other systems, it is a part of the microprocessor, but it performs the same function of timing.

The desired clock frequency (for example, 4.0 MHz), designated Ø, is sent to the microprocessor. The CPU uses these clock pulses to control all other devices. The period of each clock pulse is time state *t*. A *time state* is the amount of time it takes the computer to complete a single step in an instruction from a software program. In early microprocessors, a specific number of time states were needed for each instruction. Instructions typically took from 4 to 17 time states to be executed. To speed operation, later microprocessors handle a complete instruction in one time state.

During execution, the CPU uses various parts of the computer for different purposes. All the parts must be ready when the CPU sends a signal. The clock, therefore, is a very important part of a computer.

Timing Diagrams

When combinations of logic circuits are used, it is essential to know when the various input pulses will occur. Each time the signal enters a gate, it takes a period of time for the necessary switching to be done. This is called the *propagation delay*. If two signals are being presented to the inputs of an AND gate, they must be there at the same time. Figure 25-25A shows the pulses at input A and B arriving at the same time, t_0. The elapsed time between the rise and fall of the pulse is the same for both pulses. For this example, the time period between t_0 and t_1 is 0.5 microseconds (μs). If the pulse at A arrives late at the logic gate, a time delay is created by that circuit. In the illustration, the pulse at input A arrives later than the pulse at input B (Fig. 25-25B). If the AND gate is to function properly, the duration of pulse B must be long enough to accommodate pulse A. This can be shown by using a timing diagram as shown in Fig. 25-25C. As can be seen, if the delay of pulse A is 5μs, then pulse B must be 5μs longer than pulse A.

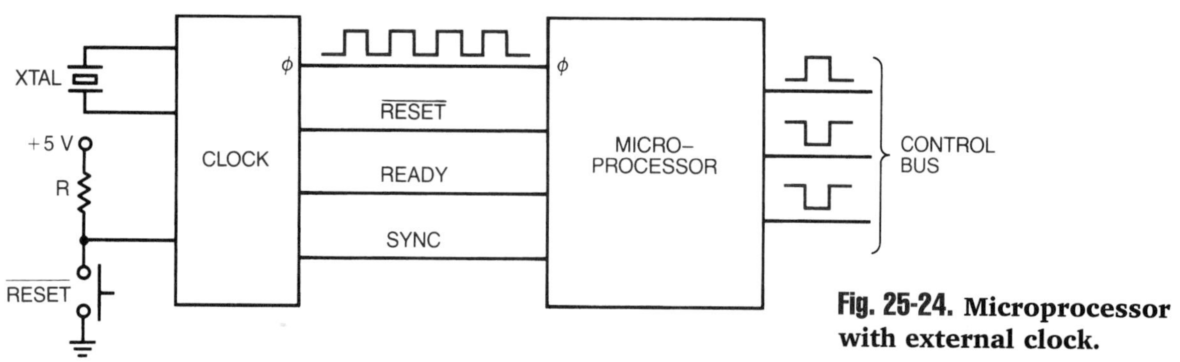

Fig. 25-24. Microprocessor with external clock.

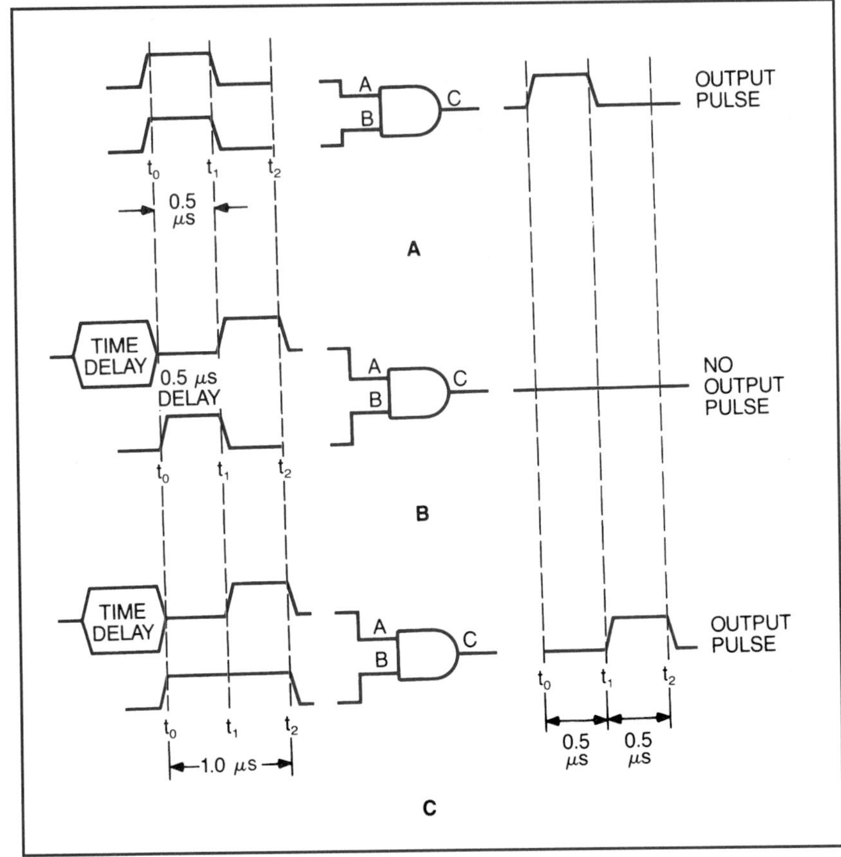

Fig. 25-25. Input timing for the correct operation of a logic gate.

The timing diagram of a machine cycle for a microcomputer is shown in Fig. 25-26. After a delay, for example, from the falling edge of the WRITE signal, the address lines (A0 to A11) become stable. Data must be valid at the output lines of memory prior to the next falling edge of the WRITE pulse. The total access time is shown at *t*. The total access time is the time when the address data is stable so that valid data can be placed on the data bus.

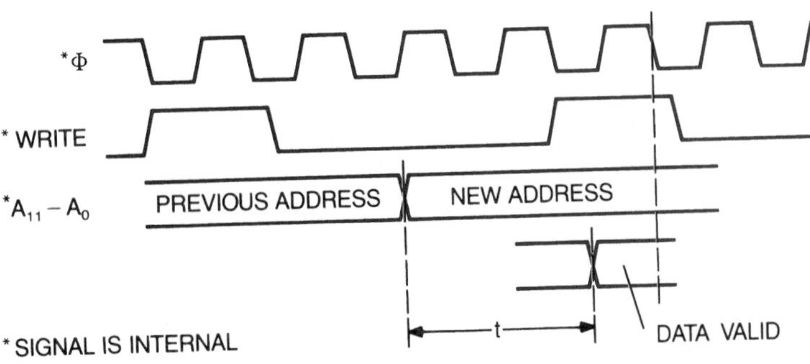

Fig. 25-26. Timing signals for an external main memory access (OP CODE FETCH).

If the clock in a computer goes bad, what effect do you think will this have on the computer?

INPUT/OUTPUT (I/O)

The microcomputer is of little use if there is no way to get information in and out of it. Thus, a microcomputer must be able to operate on data and perform some useful function. The **input/output** (I/O) subsystem of the computer monitors the flow of information in the computer.

Tri-State ICs

When it is necessary to bring data into a computer, the data must be presented to the bus only when the computer is ready to receive it. Used for this purpose the **tri-state IC** provides three states of impedance. When not enabled, it provides a very high impedance to the data bus. Thus, it does not load the data bus in this state. When the device is selected by the enabling signals, the input side accepts data and stores it. In the third state, the output side presents the data to the bus.

Latch Circuits

When a unit of data is sent to an output device, it remains there for only a short period of time. This time is typically only a few microseconds. There are few mechanical devices that can react in this time period. It is, therefore, necessary to have a **latch**, a device that can read the unit of data and hold it until the next unit of data arrives to replace it. Most microprocessors have internal latch circuitry.

If a particular unit does not have latch circuitry, then an external latching device is necessary. An external latch, such as a 74LS75 IC, is connected to four conductors that carry data. When conductors are used in groups of 4, 8, 16, or 32, each group is called a *data bus*. It is important that the data bus have only one device connected to it at any given time so that the data can be channeled to the right destination. To do this, the connections to all latches must have a very high impedance. A latch will draw little, if any, current from the bus when the impedance is high.

A latch is selected to receive data through the use of two enable pulses. Even if there is more than one latch connected to a data bus, the combination of two enable pulses can be gated to select only one device at a time. The enable pulses cause the impedance on the data line to go low. Once this occurs, the data is accepted by the device (Fig. 25-27A). The output lines send the data to the selected device and hold the logic at this level until changed. When the enable signals are removed from the latch, the input side of the latch returns to a high impedance. This sequence is illustrated in the timing diagram in Fig. 25-27B.

Device Select by Binary Codes

When the program needs to send data to an output device, it must have a method of specifying that device. The computer generates device-select (ds) codes for this purpose. Binary codes between 0 and 255 define any one of 256 different possible output devices individually. The same codes are used to define a separate set of 256 input devices.

The use of the same codes for different functions is possible because the computer also has a control bus. This bus is made up of a single conductor with a bit (0 or 1) that is set by the program instructions. The control bus is also connected to the enable pin of an IC. The enable connection allows a device to be active on the data bus when appropriate. This is important because only one device can be active on the bus during any single time state.

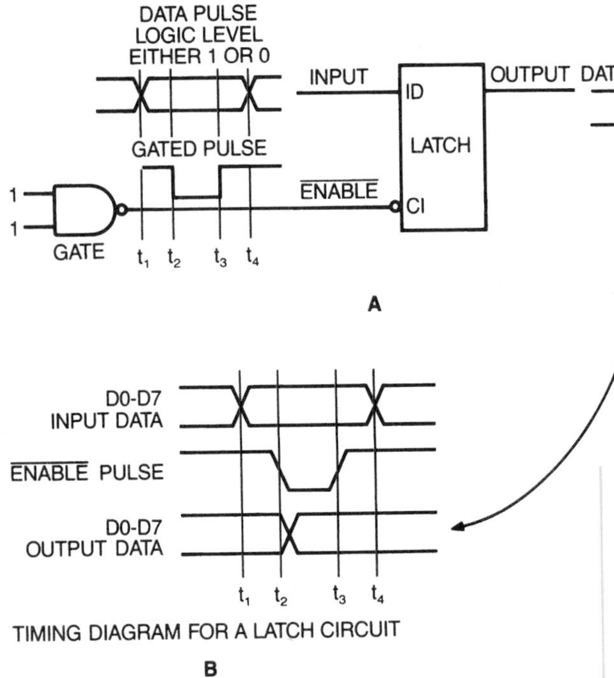

Fig. 25-27. **(A) A typical latch circuit; (B) timing diagram for the latch.**

Control lines are usually in an active state when at 0 V. This means that they are normally at a 5-V level and are changed to 0 V to activate a device. The word notation for this type of signal has a line or bar across the top of the word (i.e., $\overline{\text{RESET}}$, $\overline{\text{ENABLE}}$). One example of several control lines is read as "IN not" or "IN bar"; both terms have the same meaning. If an input instruction is given, the IN control line changes to a logic 0. In the same way, if an output is to be executed, the OUT line goes low and an output device is enabled (Fig. 25-28). This system uses any or all lower 8 bits of the address bus (A0 through A7) to identify the selected device, in this case A0. The computer now uses the address bus to supply a code for an individual device within a group of devices attached to the same bus. Using a bus for several purposes during different time states is called **multiplexing**. A group of bits, input or

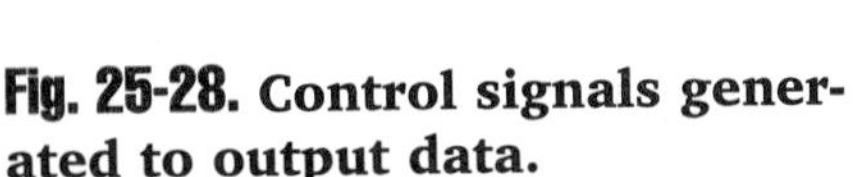
Fig. 25-28. **Control signals generated to output data.**

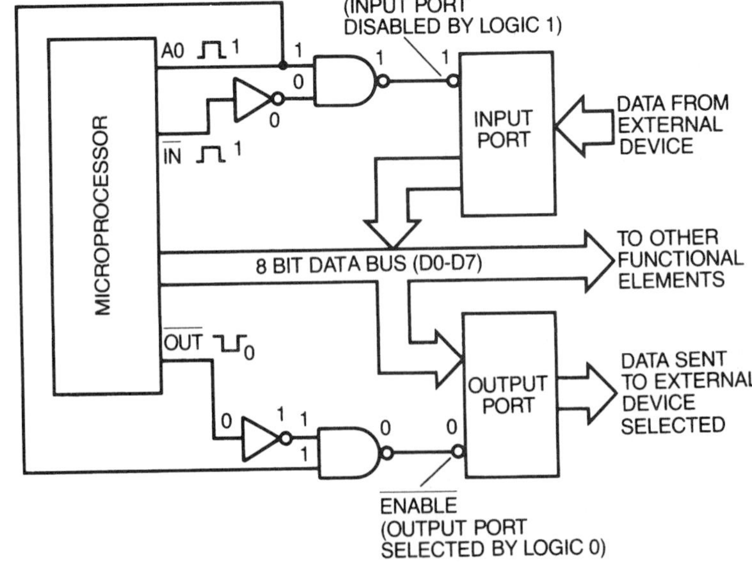

output, is selected by the control bus. Two enabling pulses are gated through a NAND circuit and enable only the input or the output port to function. Once a single device is identified, the data bus carries data to it. The data is in groups of 8, 16, or 32 bits in most microcomputers.

Device Select by Memory Map

Another system used to select I/O devices is *memory mapping*. As illustrated before, there are 65,536 addressable memory locations on a 16-bit address bus. In microcomputers that use memory mapping, some of these memory locations are set aside especially for I/O purposes. When the data from an input device is desired, the content of that particular location is read. This is done through the use of a memory READ instruction. To output data, a memory WRITE instruction is executed. The memory locations to which data is written are connected to an output device.

Output to a Video Monitor

Characters are "printed" in blocks on the TV-like video monitor. Each block is identified by a single memory address. A binary code has been assigned for each character that appears on the screen. One widely used system is the *American Standard Code for Information Interchange*, or **ASCII**. The ASCII code for each alphanumeric text character or graphic character is written to a specific address. When this address is selected, the character appears on the screen.

In addition to text printing, computers can display pictures or graphic shapes on the screen. The resolution of these images depends on the capability of the monitor. For routine work such as word processing, lower resolutions, such as 640 x 480 pixels, are sufficient. (*Pixel* is short for *picture element*, an individual point of light on a video screen.) For work that requires highly detailed graphics, higher resolutions are required. Monitor resolutions for microcomputers are now available with resolutions as high as 2048 by 768 pixels. For high-resolution monitors, thousands of memory locations are used to identify individual pixels on the video display.

Why do you suppose the number of input and output devices is limited to 256 each when they are selected by binary codes?

SINGLE-CHIP MICROCOMPUTERS

Figure 25-29A illustrates a 40-pin, single-chip microcomputer. It was designed to meet a wide range of applications that require some type of "intelligent" control. Because the number of new devices that use the same pin configuration and circuit design (referred to as the *architecture*) is constantly increased, the Mostek MK 3870 "family" of microcomputers emerged. Figure 25-29B illustrates the part numbering used to indicate the generic part type and the ROM and RAM size (the memory space available on the chip). Figures 25-29C and D show the names of the pins and the functions performed. This is only one of several families—and architectures—used for microcomputers today. Can you think of some others?

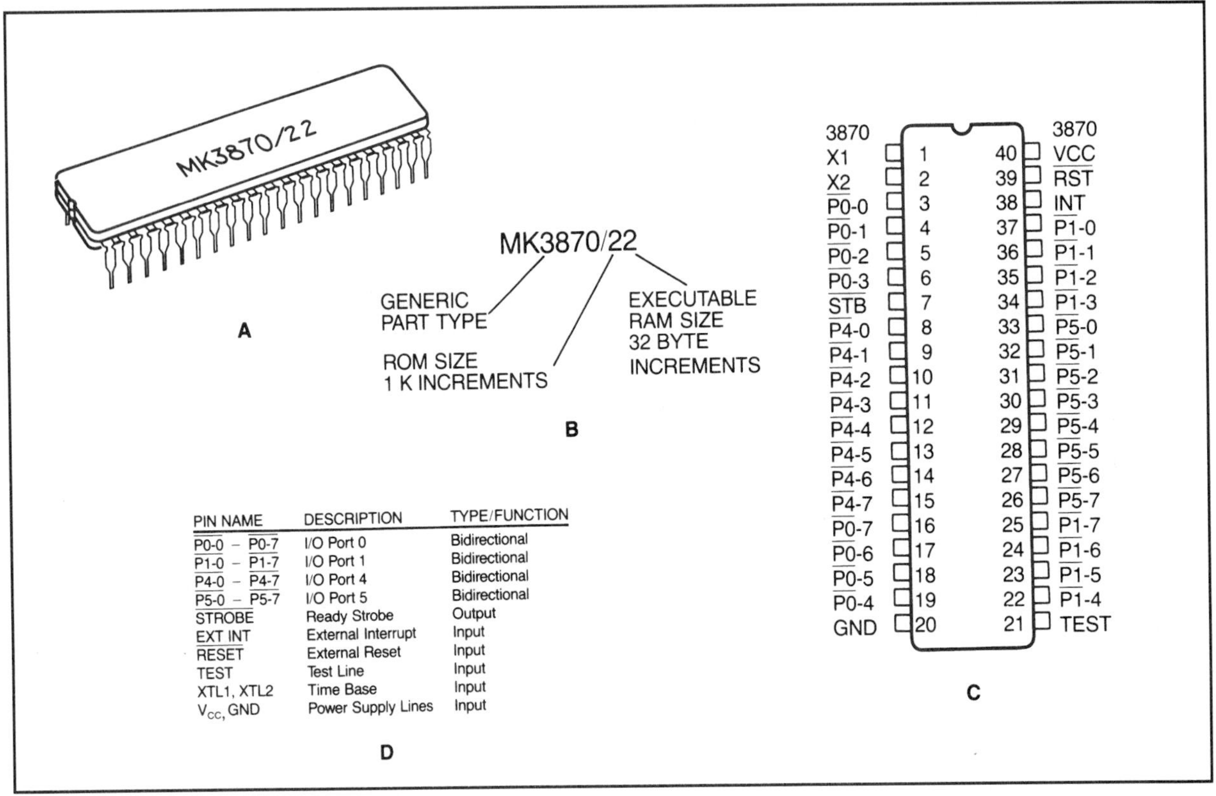

PIN NAME	DESCRIPTION	TYPE/FUNCTION
P0-0 – P0-7	I/O Port 0	Bidirectional
P1-0 – P1-7	I/O Port 1	Bidirectional
P4-0 – P4-7	I/O Port 4	Bidirectional
P5-0 – P5-7	I/O Port 5	Bidirectional
STROBE	Ready Strobe	Output
EXT INT	External Interrupt	Input
RESET	External Reset	Input
TEST	Test Line	Input
XTL1, XTL2	Time Base	Input
V_{CC}, GND	Power Supply Lines	Input

Fig. 25-29. A single-chip microcomputer: (A) the MK 3870. an 8-bit microcomputer IC; (B) part numbering example; (C) pin number and pin name; (D) summary of pin functions.

Figure 25-30 is a block diagram that shows the directions in which data is transferred in and out of the registers. Figure 25-31 illustrates the variety of registers used on this single-chip microcomputer. The *scratch pad* provides 64 x 8 bit registers that may be used as general-purpose RAM (Fig. 25-32). The MK 3870/2 has extra RAM that is addressable. Software routines can be loaded into this RAM and executed. All MK 3870 family microcomputers execute a common set of more than 70 instructions. Designers can, therefore, choose a device with the right combinations of ROM and RAM to suit their system requirements.

Information input and output is accomplished through the four parallel ports (I/O ports 0, 1, 4, and 5). These ports are shown on the block diagram (Fig. 25-30). Since information flows to and from the ports, they are called *bidirectional* or *I/O ports*. In this illustration, 8 bits are available on each port. The ports can be used individually as either TTL-compatible inputs or as latch outputs.

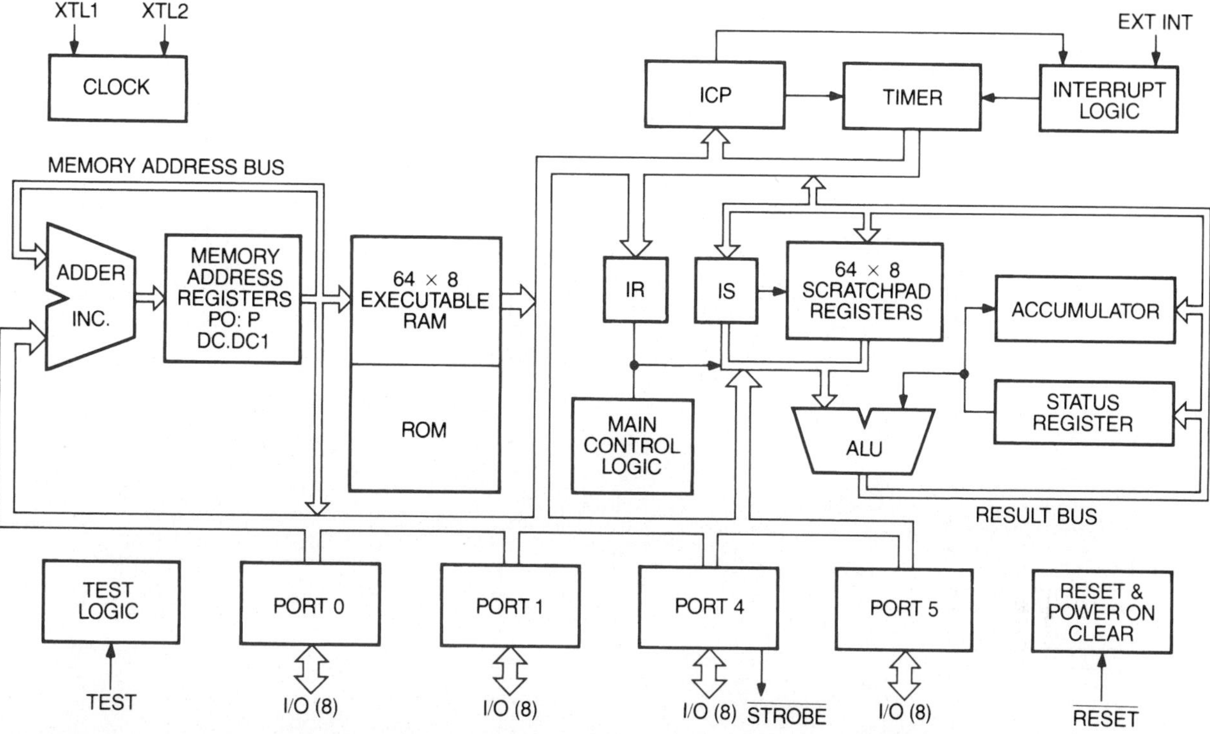

Fig. 25-30. Block diagram of the MK 3870 single-chip microcomputer.

WHAT THE FUTURE HOLDS

In the future, computers will control the production of more and more products around the world. Office work is changing rapidly because of the portability of smaller computers (laptops), telephone/modem links, local-area networks (LANs), wide-area networks (WANs), and cellular technology. These technologies allow large companies to link their offices all over the world. They also allow companies to give their workers more flexibility and freedom to work outside the office.

Computers are also used extensively in smaller applications. Electronic data processing by computer is even used in "fast-food" restaurants. No longer do the attendants write your food order on a piece of paper. As you state your order, they just press keys on a machine that contains a microprocessor. When the transaction has been completed, you receive a printed paper receipt and the order appears on a monitor where the food is prepared and then delivered to the service area. At the same time, the machine automatically stores the information electronically for future use. This information can be retrieved easily by managers for reordering or accounting purposes.

List at least five activities you perform every day that involve a computer or microprocessor in some way.

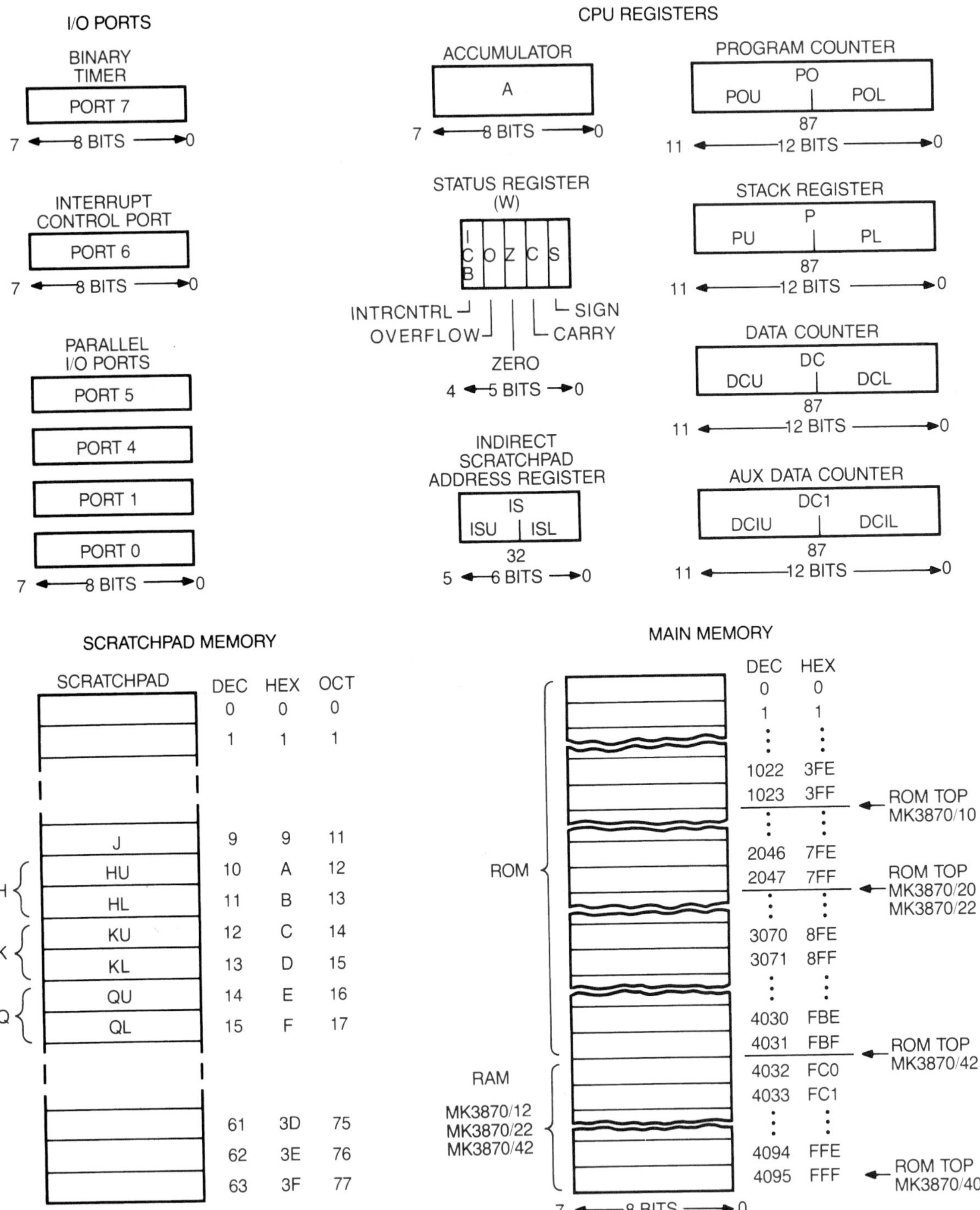

Fig. 25-31. The MK 3870 family programming model illustrating the I/O ports, CPU registers, scratch pad, and main memory.

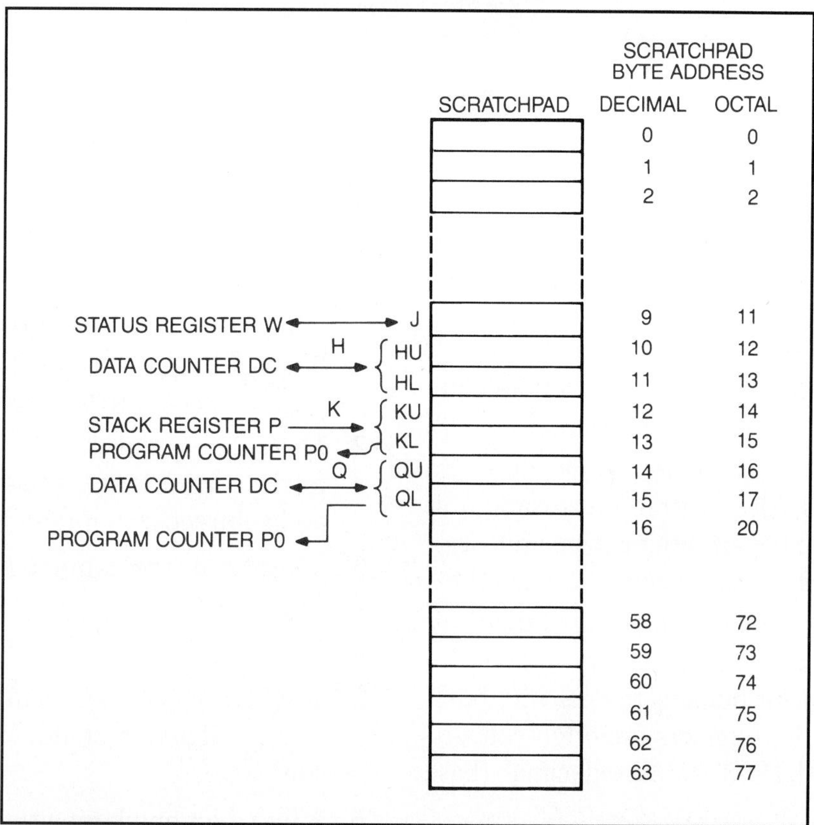

Fig. 25-32. Scratch pad register map, illustrating data transfers between registers.

Chapter *Review*

Summary

- Computer systems are made up of hardware, software, and an operating system.
- The five basic hardware components are input devices, the central processing unit (control unit), the arithmetic logic unit, storage units, and output devices.
- Unlike humans, computers do all of their calculations in base 2, the binary system. To make it easier for humans to deal with binary numbers, the numbers are often converted to octal (base 8) or hexadecimal (base 16).
- Microcomputers store data temporarily in two basic types of internal memory: random access memory (RAM) and read-only memory (ROM).
- Logic circuits in the microprocessor are responsible for its function. The clock makes sure all the signals are sent to the appropriate circuits at the appropriate time to make the most efficient use of the circuitry.
- Input and output devices allow people to enter data into a computer and receive information from the computer.

Review Main Ideas

Review this chapter's main ideas by writing, on a separate sheet of paper, the word or words that most correctly complete the following statements:

1. In computer systems, ______ are the actual devices used. ______ refers to the programs that run on the computer's hardware. The ______ interprets software so that the hardware can act on it.
2. The ______ makes it possible for the CPU to do comparisons, make decisions, add, and subtract.
3. The part of the computer that is the control unit is the ______.
4. A byte is equal to ______ bits.
5. The binary system of data uses ______ and ______ digits to stand for the "off" and "on" states.
6. A two-digit numbering system in base 16 is called ______ .
7. Hexadecimal codes use ______ written before the left-most digit.
8. ______ are temporary storage spaces for data that are used later in the program.
9. The registers consist of electronic devices called ______ .
10. Data in the ______ accumulator can be used to perform mathematical functions with the ALU.
11. A ______ is used to indicate the results of the arithmetic and logic operations in the CPU.
12. The control circuit ______ information through the computer and controls its entire operation.
13. Memory is installed in the computer as individual ______ or on small circuit boards that hold the memory chips.

14. ______ stands for random access memory, and is used for temporary storage; ______ stands for read-only memory and cannot be changed.

15. ______ is used when a smaller number of bytes is needed, and is an inexpensive alternative to ROM.

16. Memory that can be programmed and will then hold that information until erased with UV light is called ______ .

17. The code for an address is sent to a memory chip via a group of 16 conductors called the ______ .

18. The byte of memory is made up of the ______ address and the ______ address.

19. There are ______ bytes of memory in 64K computers.

20. The ______ in computer memory identifies the individual byte on a page.

21. If ______ occurs the data received at the other end of the cable end may not match the data that was entered.

22. When data is to be sent between computers separated by long distances, fewer ______ are desirable.

23. ______ is the transmission of data over wires 1 bit at a time in rapid succession.

24. The square-wave generator that sets up timing and feeds a signal to the CPU is the ______ .

25. A common input device which provides three states of a computer is the ______ .

26. An output device that reads a unit of data and holds it until the next unit of data arrives in the ______ .

27. The use of a bus structure for several purposes during different time states is called ______ .

28. Another system used to select I/O devices is called ______ .

Apply Your Knowledge

1. What is the value of $3A in binary?
2. What is the value of binary 01101111 in octal?
3. Explain how tri-state ICs are used in a computer. What is the importance of this?
4. Explain the importance of latch circuitry in a computer. How does it work?
5. What is the difference in selecting devices by binary code and selecting them by memory map?
6. Identify various applications of the computer in the home and in industry.

Make Connections

1. **Communication Skills.** Prepare a report using a word processing program. Report on the use of binary codes to represent alphanumeric codes on the display screen and on a printer.
2. **Mathematics.** Prepare a chart report on the number systems used in the world today.
3. **Science.** Construct a circuit that demonstrates why data can be sent through a conductor in only one direction during any one time period.

Chapter 20

Computerized Controls

Terms to Study

analog-to-digital converter

BASIC

digital-to-analog converter

debounce circuitry

dedicated control

EEPROM

fanout

immediate mode

interface

microcontroller

programmed mode

pull-up resistor

OBJECTIVES

After completing this chapter you will be able to:

- Define the term *dedicated control*.
- Describe one method of driving a peripheral with a microcontroller program.
- Define and describe an interface circuit for a microcontroller.
- Contrast the functions of an analog-to-digital converter with those of a digital-to-analog converter.
- Explain why a pull-up resistor is used with a switch.
- Draw a diagram of a photo-operated input and explain the operation of the circuit.

The digital computer is a very common device in our lives. How many times have you used a digital computer in the last week? Do you realize that you must include the times you used a microwave oven, played a video game, or just changed the channel on your television? This chapter focuses on how the computer is interfaced with the real world.

INTERFACING WITH THE REAL WORLD THROUGH ELECTRONICS

Computers that include a microprocessor, memory, and input/output devices are included in the category of digital computers. *Microcontrollers* are a member of this family and compute in a very special way called *dedicated control.* **Dedicated control** means the controller is programmed to perform one general function for the user. It is used in toys, automobiles, televisions, ovens, cloths irons, and many other common devices you use everyday. For larger scale control applications, such as production control in a manufacturing operation, personal computers are used in dedicated control applications. For small applications of dedicated control, inexpensive microcontrollers are used. In this chapter, you will learn about one such unit called the BASIC Stamp II®. You will learn how to connect input and output devices that convey information and perform useful functions. You will also learn how to program the BASIC Stamp II® to meet your requirements.

Data

An important concept to remember from Chapter 25 is that all data in a computer are in the form of voltage pulses. Each bit of data is either logic 1 or logic 0. Logic 1 represents a +5-V pulse that is gated by a TTL circuit. Logic circuits can supply base current to switch a transistor and the current needed to drive TTL-compatible devices. Logic 0, represents very low or zero volts and is used to turn off a transistor or other device. A computer program controls these voltage pulses and you will learn in this chapter about specific instructions for computers to carry out tasks.

Fanout

The amount of current that can be sourced, or supplied, by a TTL logic circuit is too little to operate many devices. **Fanout** is the number of TTL devices that one TTL output can drive (Fig. 26-1). In addition, the 5-V, rectangular pulse wave, used by the computer's microprocessor, is not compatible with most motors, relays, and other actuating devices. Therefore, some type of additional circuitry is needed.

Simple Interface

The **interface** is the electrical connection to a microcontroller that provides a means of communication between it, a host computer to store and process collected data, and actuating devices such as relays and motors. These electrical connections also enable the user to program the microcontroller. Figure 26-2 shows a simple interface circuit. A single data line is connected to a transistor in a common-emitter

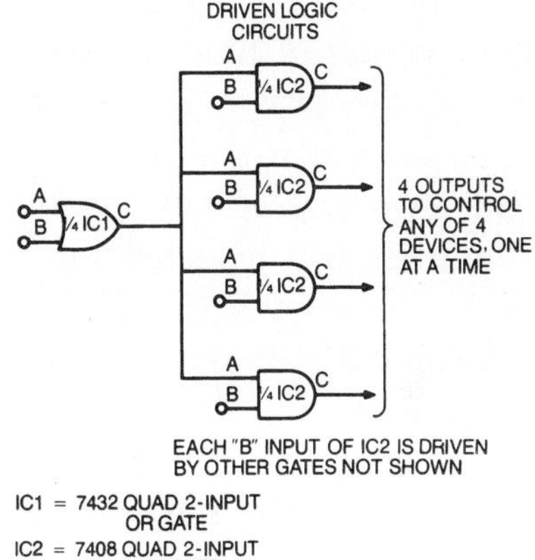

Fig. 26-1. **One TTL OR gate fanout to four TTL inputs.**

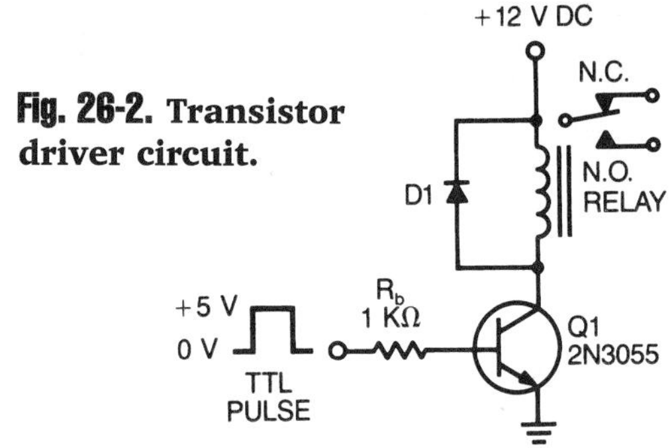

Fig. 26-2. **Transistor driver circuit.**

configuration. In this circuit, a transistor controls a relay. The base current is limited by the 1,000 resistor (R_b). This resistor can be any value between 500 and 5,000. The relay uses a 12 Vdc supply, and the relay coil provides the load for the collector circuit. A TTL voltage pulse from the computer to the emitter-base junction produces a base current that biases on the transistor. This allows collector current to energize the relay. As long as the base current continues, the relay will remain energized. If the base current stops flowing, the relay will return to the normal position. The diode D1 is connected across the relay coil to protect the circuit because the relay coil is an inductive load that can produce high transient voltages when turned on or off.

Latch

The instructions in a computer program take a very short period of time to execute. Data sent on a data line lasts only until the microprocessor begins a new scan cycle. This is, at most, only a few microseconds. For this reason, the data must be sent repeatedly or captured by a *latch*, as illustrated in Fig. 26-3. The 4-bit data bus is connected to the input side of the latch (pins 2, 3, 6, and 7). The device to be controlled is connected to the output side of the latch (pins 16, 15, 10, and 9). The action of the latch circuit is controlled by the *enable* connections that are activated by signals from the computer. Since the data are latched, the programmer must remember to send another signal that will turn off the signal to the motor controller, or other device, when its task has been completed. If logic 1 is used to turn the device on, logic 0 is sent to turn it off.

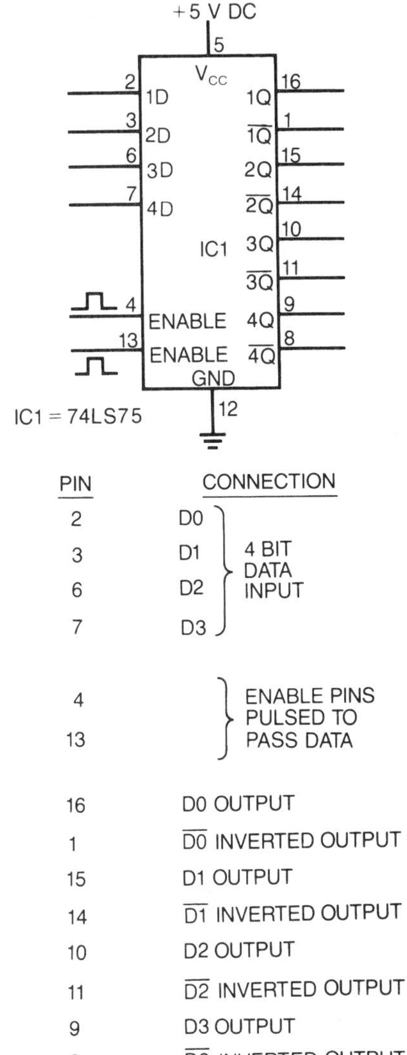

Fig. 26-3. 4-bit data latch circuit.

Explain how dedicated control differs from standard computer applications. Give several examples of each.

ANALOG-TO-DIGITAL CONVERTER (ADC)

The internal circuitry of the digital computer is limited to logic levels of 1 and 0. The outside world, on the other hand, is not limited to two logic levels. Most devices use voltages of continuously varying levels. For example, there are an unlimited number of colors, temperatures, light and dark levels, or resistances in the real

Fig. 26-4. **Typical analog-to-digital converter.**

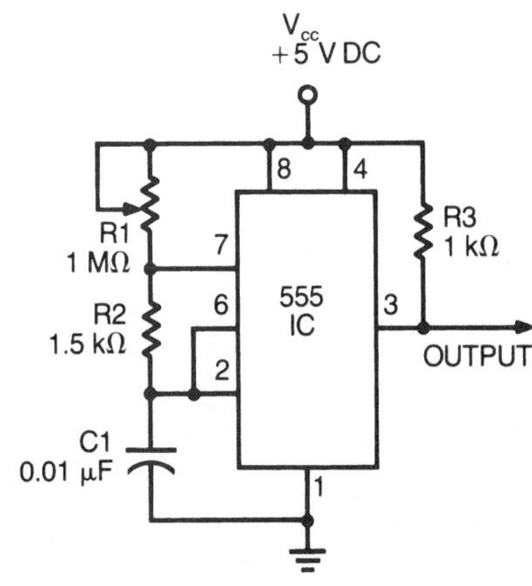

Fig. 26-5. **ADC circuit using a 555 timer IC.**

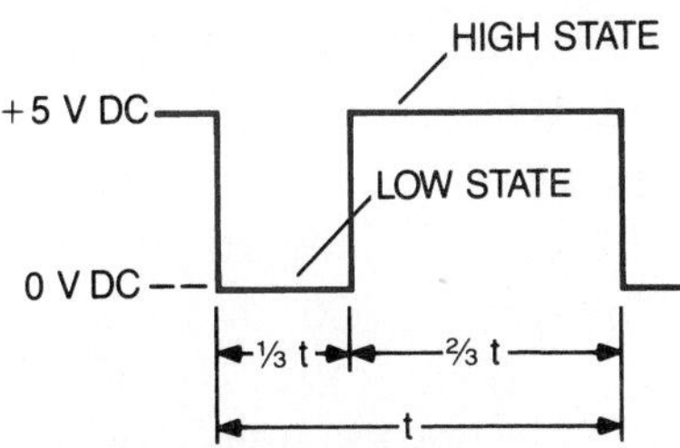

Fig. 26-6. **Graph illustrating the 66 percent duty cycle of a 555 timer circuit.**

world. Audio amplifiers and operational amplifiers are typical *analog devices* that are operated by these variable levels.

In order for a digital computer to determine one voltage or current level from another, an *analog-to-digital converter* (ADC) circuit is used. Figure 26-4 shows a typical **analog-to-digital converter** that converts analog signals into digital signals. As the ADC receives the analog signal at the input, it converts the analog signal to a digital code. The output of the ADC is digital in nature so the computer can use it. Most ADC devices can send data in either serial or parallel format, depending upon which format the receiving device needs. The digital data is formed into a data table that the program can interpret. This data can then be used in equations and graphs, on monitor screens, and in any number of applications such as printing, processing, and counting.

An inexpensive and simple ADC is in Fig. 26-5. This circuit uses a timer IC, the 555, and an *RC* network for control. Figure 26-5 illustrates a circuit that generates a free-running rectangular wave. The period of the rectangular wave at the low state is fixed. The period of the rectangular wave at the high state is determined by the value of the resistors R_1 and R_2 and the capacitor C_1. The *duty cycle* is the ratio of the length of time of a high logic level to the length of time for one full cycle (one high state plus one low state). If the time of the high pulse is twice that of the low pulse, the wave has a 66% duty cycle (Fig. 26-6). As R_1 changes, so does the duty cycle of the wave.

A computer program can determine the level of the square wave. For example, as long as the voltage pulse is low, nothing happens. When the logic level of the voltage pulse changes, an instruction in the computer program senses this and a software counter increases the count. The counter continues to count until the logic level returns to zero at the end of the duty cycle. At this point, the computer has a digital value that represents the analog numeric value of the resistance.

DIGITAL-TO-ANALOG CONVERSION (DAC)

Sometimes it is necessary for a computer to generate an analog output in order to operate a device. An analog output is generally a varied voltage level. A digital code sent by the computer to the DAC determines the voltages that vary within predetermined limits. Figure 26-7 is a typical **digital-to-analog converter** that changes digital information into a continuous variable voltage output.

For both ADCs and DACs, the accuracy of converting from one format to another is called the resolution. Each digital value represents a point on an analog scale. For example, consider an analog scale from 0.0 V to 12 V. If four bits, or one nibble, is used for the data range, then the 12 V is represented by fifteen increments. (There are sixteen combinations of four bits.) In this case, each digital numeric value between 0000 and 1111 equals decimal 0.80V (12V/15). That is, 0000 = 0.0 V, 0001 = 0.80 V, 0010 = 1.6 V, and so forth, until 1111 = 12 V (Fig. 26-8). These increments of 0.80 V are quite large, but they are usable in some applications. Note that the change from 0.0 V to 12 V is not a smooth transition, but instead has a staircase, or stepped, effect (Fig. 26-8). However, by increasing the resolution to 8-bits, the number of steps increases to 256 (00000000 to 11111111). This decreases the value of each step to 0.04706 V (12 V/255). If the computer receives data from an encoder measuring the position of a motor, this accuracy is necessary. Twelve- and 16-bit data ranges are available when even greater accuracy is necessary.

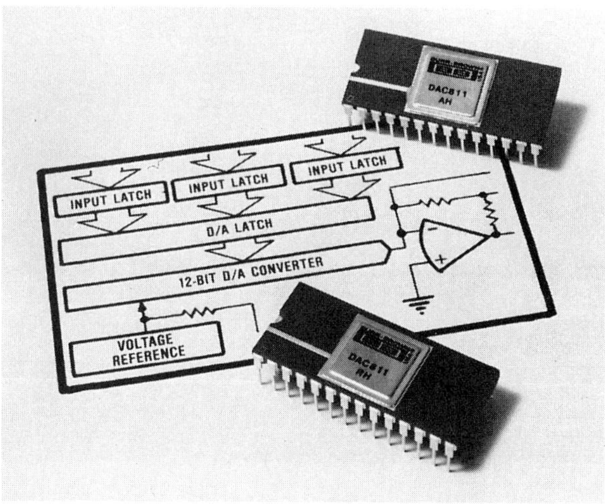

Fig. 26-7. **Example of a digital-to-analog conversion.**

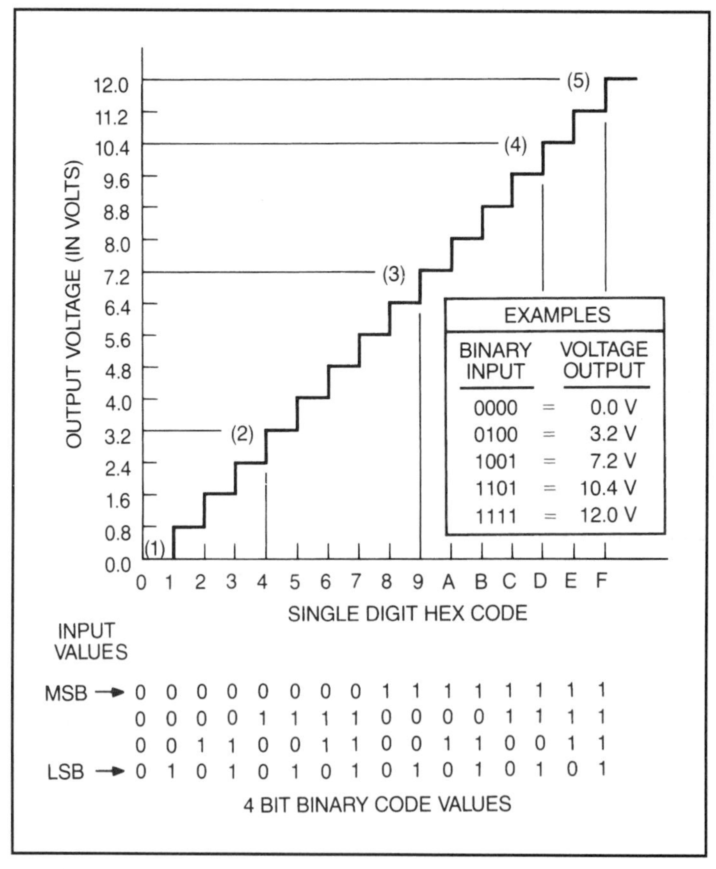

Fig. 26-8. **An example of digital-to-analog converter voltage output (analog) for each possible input code (digital), using a 4-bit input.**

1. Write a brief essay explaining the need for analog-to-digital converters when controlling an operation with a microcomputer.
2. How many steps are available in 16-bit resolution in a digital-to-analog converter?

COMPUTER PROGRAMMING

BASIC is only one of several computer languages used to control computers. A computer language is actually a set of program instruction codes in the form of words, abbreviations, and numeric characters and is written in modified forms and different modes. The instructions or commands in BASIC can be given in the *immediate mode* and in the *programmed mode*. In the **immediate mode**, or calculator mode, the computer responds immediately to instructions or commands, just like a handheld electronic calculator. The following is an example of an instruction in the immediate mode of operation:

PRINT "The cost of the transformer is 60 dollars."

When *return* or *enter* is pressed on the computer, it immediately responds by executing the instruction PRINT. This instruction directs the computer to display the statement within the quotation marks. The following would be printed on the CRT display screen:

The cost of the transformer is 60 dollars.

Review Chapter 25, "Fundamentals of Computers," for an in-depth discussion of binary and other numbering systems.

In the **programmed mode**, the execution of the computer program is deferred. The computer stores the program in memory until it is executed with a RUN command. Line numbers organize a BASIC program—a line number is a one-, two-, three-, four-, or five-digit number entered at the beginning of a program line. By convention, line numbers 10, 20, 30, and so on, are assigned to successive lines. This allows room between each for additional instructions (11,12, 21, 33, etc.) to correct and enhance the software program. The following is an example of instructions written in the programmed mode:

```
10 PRINT "The class designation letter for a capacitor is C."
20 PRINT "The class designation letter for an inductor is L."
30 PRINT "The class designation letter for a push button is S."
40 PRINT "The class designation letter for a transistor is Q."
50 END
RUN
```

After entering the RUN command (without a line number) and pressing *return*, the computer will execute the program, one line at a time, and display the following on the screen:

The class designation letter for a capacitor is C.
The class designation letter for an inductor is L.
The class designation letter for a push button is S.
The class designation letter for a transistor is Q.

Figure 26-9 shows a program that adds two numbers. The numbers, entered from the keyboard by the input statements, are added. The result is printed on the display screen, and the program ends. Each instruction code causes the computer to perform a sequence of operations that complete the instructions.

```
10  INPUT"WHAT IS THE FIRST
    NUMBER TO BE ADDED?"; X
20  INPUT"WHAT IS THE SECOND
    NUMBER TO BE ADDED?"; Y
30  Z = X + Y: PRINT"THE SUM
    OF ";X;" + ";Y;" = ";Z
40  END
```

Fig. 26-9. A sample program in BASIC in which two numbers are added.

The number of operations a computer can perform is limited by its design. Some computers can execute seventy-two different operations (for example, PRINT, +, GOTO, and IF THEN). It is the programmer's job to determine which operations the computer performs and when. Instruction codes for each computer differ slightly, much as people of different parts of a country have different dialects of a common language. It will be helpful for you to have the programming manual for your computer if you intend to program it yourself.

Another programming language is PBASIC™. This language is a custom version of BASIC, and is discussed in this chapter. PBASIC™ was developed by PARALLAX INC. for use in the BASIC Stamp II® which is used in the project section of this text and in Group 12 of the workbook.

Write a short BASIC program that will print a list of ten people's names and addresses.

The Road to BASIC

The first commercial high-level computer language was called FLOW-MATIC. It was devised in the early 1950s by Grace Hopper, a U.S. Navy computer programmer. As computers became an increasingly important scientific tool, IBM developed a language that simplified the programming of complicated mathematical formulas. Completed in 1957, FORTRAN (Formula Translating system) was the first comprehensive programming language. It was extremely effective in manipulating numbers and equations efficiently. It is still widely used in engineering and scientific applications.

COBOL (Common Business-Oriented Language) was developed in 1959 to respond to business needs such as creating, moving, and processing data files. COBOL was based on FORTRAN, but emphasized data organization and file-handling. It became the most important programming language for commercial and business-related applications. It is still used today.

A simplified version of FORTRAN, called BASIC (Beginner's All-purpose Symbolic Instruction Code), was developed in 1965 by two professors at Dartmouth College. At the time, it was considered too slow and inefficient for professional use. However, BASIC was simple to learn and easy to use. It became an important tool for teaching non-professional computer users. When microcomputer use exploded in the late 1970s and 1980s, BASIC became a universal programming language. Tens of thousands of BASIC programs are commonly used.

PULSE INPUT

In order for a computer to communicate with devices outside of itself, it must look for inputs at its pins. Those inputs can come from switches, sensors, thermocouples, or various other devices. Whatever the device, it must ultimately provide logic 0, or logic 1 to the computer. If

FASCINATING FACTS

The eight programs that run each Space Shuttle consist of about half a million separate instructions and were written by hundreds of programmers.

the device does not provide these voltage levels, it will need to be interfaced with an I/O module that does.

Switches

The BASIC Stamp II® has sixteen single pin connections that can be designated, through software instruction, as input or output pins. These are designated P0 through P15, as shown in Fig. 26-10. The location of these I/O pins on the BASIC Stamp II® board is pins 5 through 20, respectively. Any input pin, PIN 7 for example, can be used to read data or to output data. All input pins can be used to read data from a variety of devices capable of generating a logic 1 or a logic 0. All that is necessary is that the pulse generated by the input device is either a +5-Vdc or a 0-V signal when the program executes the instruction. The instruction then stores the logical value (1 or 0) of the switch as a variable that can be accessed, as needed, by a software program.

By integrating software and hardware, useful tasks can be performed. In this case, for instance, a switch can be connected to externally control the program by interrupting the computer program sequence. A 5-Vdc voltage pulse, generated by a switch, tells the program that an external device requires service of a predetermined nature. The pulse is read, and this causes the computer to run a portion of the program that performs that service. For this circuit to work, the program must check the switch periodically.

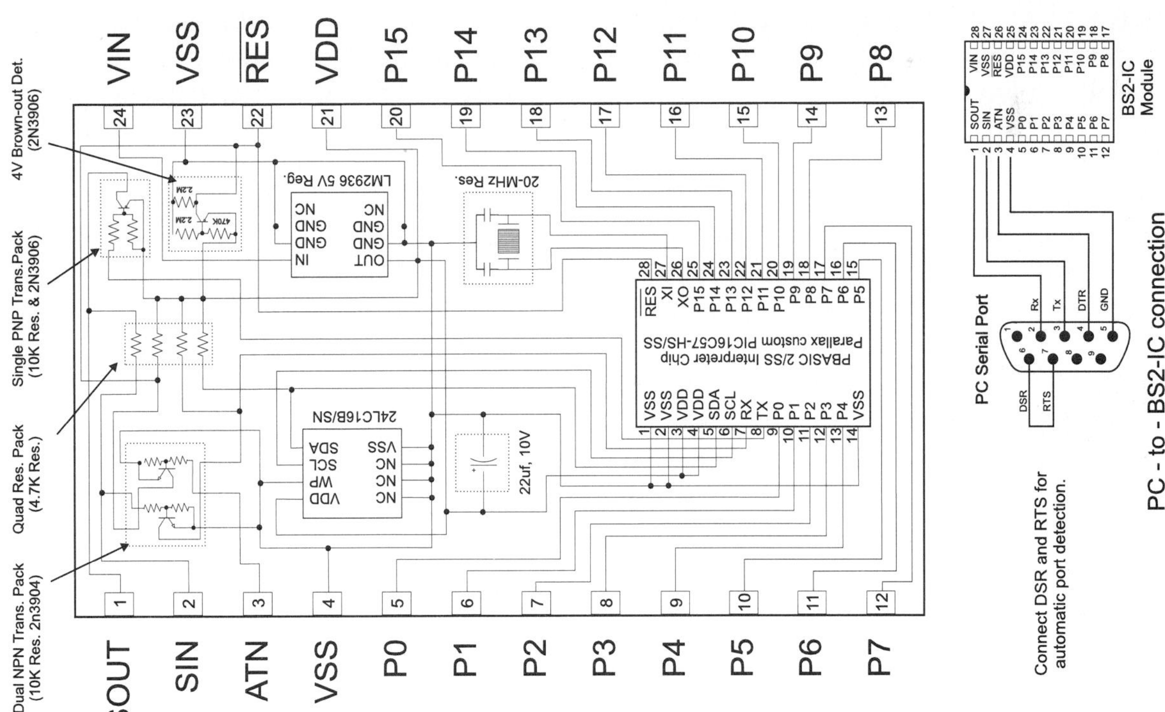

Fig. 26-10. A microcontroller: BASIC Stamp II® pin assignment.

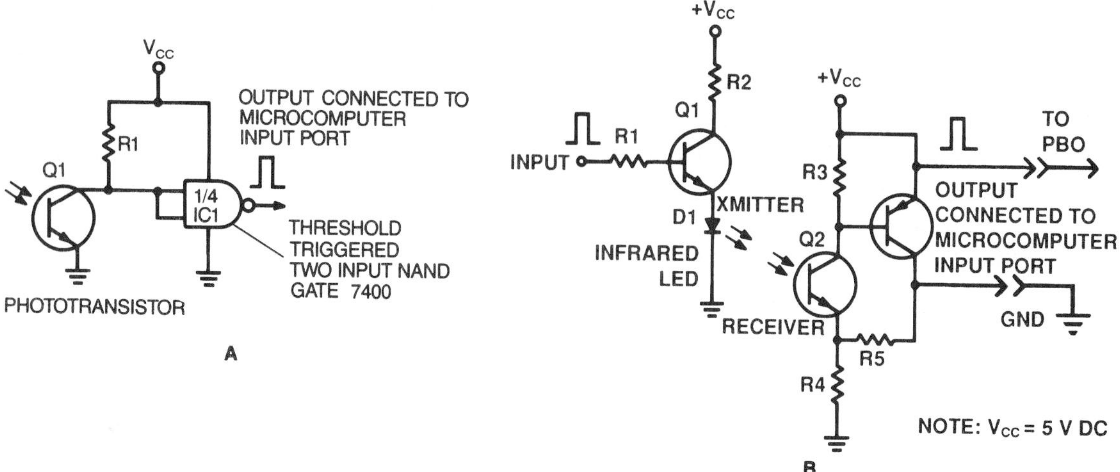

Fig. 26-11. Photo-operated devices: (A) phototransistor circuit; (B) infrared diode-transmitter receiver circuit.

Photosensitive devices that generate or switch a 5-Vdc pulse can be used as a computer input. Figure 26-11 shows two typical circuits of this type. Phototransistors (Fig. 26-11A) or an infrared, diode emitter-receiver pair (Fig. 26-11B) can generate the required logic voltage pulse. The infrared diode pair circuit can count the passage of items between the transmitter (emitter) and receiver.

Pull-up Resistor

As previously discussed, the input to TTL devices must be within the proper voltage levels. To do this, a 1-k resistor is connected as shown in Fig. 26-12. It is called a **pull-up resistor** because it makes sure that the voltage goes from 0 to +5 Vdc when needed. The value here is given at 1 k, but it can range from about 470 to several thousand ohms and still be effective. One side of the resistor, connected to 5 Vdc, provides the proper logic level. The other side of the resistor is connected to the input of the IC which is to one side of the switch. The other side of the switch is connected to the common ground. When the switch is in the normally open position, the input reads logic 1. This is because the input has very high impedance, and very little current flows through the resistor. Therefore, there is almost no voltage drop across the resistor. For all practical purposes, the input appears to be connected directly to the +5 V and is held at logic 1.

An example of this in Fig. 26-13A shows the reset function on IC1, which requires logic 0 to activate. In the normal position of S1, logic 1 is on (not) $\overline{reset}$ and will not allow the reset action to take place. When switch S1 is depressed and the contacts come together (Fig. 26-13B), current will flow through the resistor. Since it is the only resistive element in the circuit, all five volts will be dropped across the 1 k Ω. The input is then brought to ground potential, or logic 0 (Fig. 26-13C). With logic 0

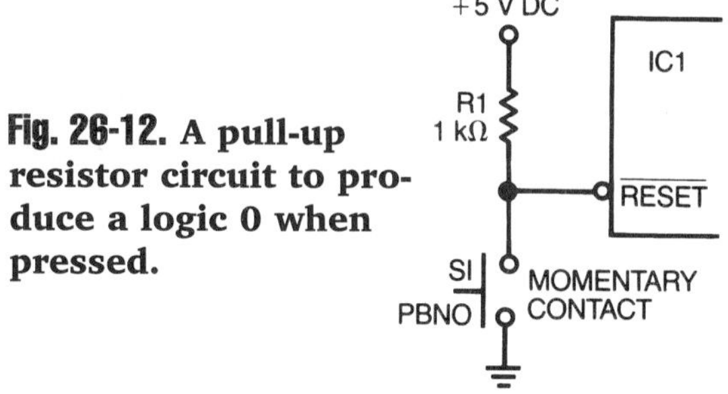

Fig. 26-12. A pull-up resistor circuit to produce a logic 0 when pressed.

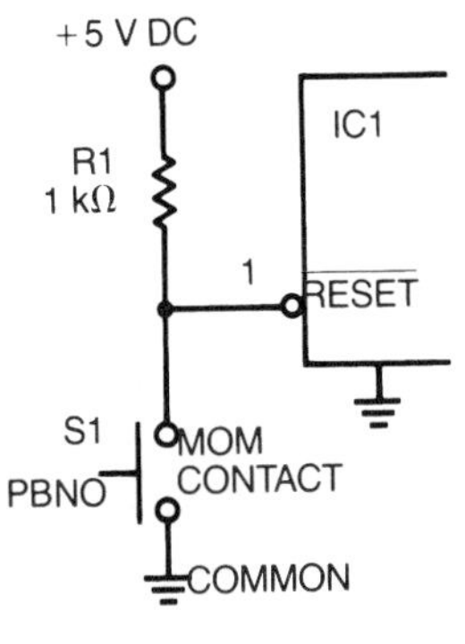

A

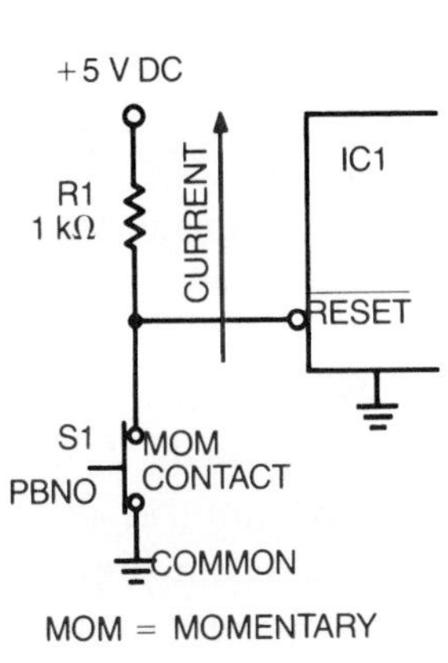

B

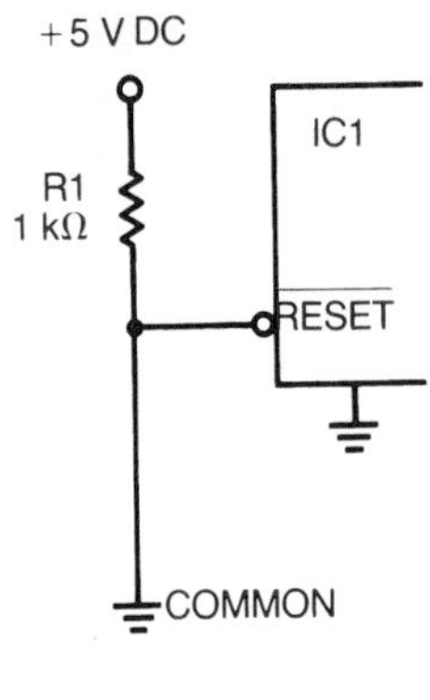

C

Fig. 26-13. Developing a logic 0 reset function with a pull-up resistor: (A) logic 1 at reset pin; (B) logic 0 at reset pin; (C) the closed push button switch effectively puts the reset pin to ground potential.

at the input of *reset*, the reset action will occur. Without this circuit, the logic level cannot be predicted when the switch is open. Therefore, the data received would be unreliable.

The preceding discussion dealt with a normally open switch that changed the logic level from 1 to 0 when it was closed. If the normally closed contacts of the pushbutton switch are used in the circuit, the logic reverses from 0 to 1. For the circuit shown in Fig. 26-14, this is important. The *set* connection on this IC requires a +5 Vdc pulse (logic 1) to function. When S1 is in the normally closed position, the *set* pin is logic 0. Pressing S1 pulls the voltage to +5 Vdc (logic 1) and activates *set*.

Contact Bounce

In the previous discussion of a pull-up resistor and push- button switch, the only pulse needed was the first one; therefore, no *debounce* circuitry was necessary. **Debounce circuitry** removes the effects of contact bounce from a circuit.

When using a switch as a computer input device, you must carefully consider the switch action. If the switch is controlling a lamp, the lamp lights when the switch is closed. However, when the switch closes, the contacts touch and then bounce for several tenths of a second. This bounce is no problem to a lamp, but a computer will have trouble. When that

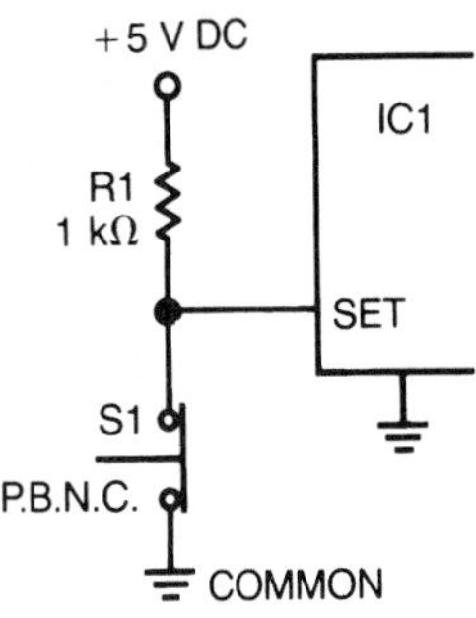

Fig. 26-14. Developing a logic 1 with a push button normally closed (PBNC) switch and pull-up resistor.

same switch is connected to a computer, the microprocessor can input data fast enough to read the bounce as a series of switch closures. A computer will see switch bounce as a voltage pulse. If it is counting the series of events, contact bounce will yield a false count.

Software Debounce

The switch must be connected between a pin that has been designated as an input and ground or the V_{SS} connection (pin 23) of the BASIC Stamp II® (Fig. 26-10). The state of the switch can then be determined by using the instruction *BUTTON*. The syntax for this instruction is *BUTTON pin,downstate,delay, rate,bytevariable,targetstate,address. Pin* designates which of the sixteen (0 through 15) I/O pins is to be connected to the push button. *Downstate* defines which logical state is examined when the push button is pressed. *Delay* specifies how long the program must delay before reading the switch again. This is the software debounce feature. The value to which *delay* may be set is between 0 and 255. The 1–254 values set the delay longer as the number is set higher. This leaves 0 and 255 as special function values. If set to 0, the depressed switch state is returned with no auto repeat and no debounce. When set to 255, debounce occurs but auto repeat is canceled. *Rate* specifies the auto repeat rate for the action to be taken as a result of the button being pressed and being held down. This feature initiates after the delay function has timed out. *Bytevariable* is the workspace, or variable, designated to store the logic level of the switch when it is read. This variable must be cleared to 0 before the instruction *BUTTON* is executed. It should also be before any program loop that results from the switch routine. *Targetstate* can be either a variable or a constant that specifies the logical state of the switch when it causes the program to shift operation. This is called a branching operation. *Address* specifies a program label location where the branching operation is to occur.

Hardware Debounce

What happens, though, if the program is designed to count a number of quickly occurring events? To make the problem more difficult, assume that the events do not happen in any predictable pattern. That is, the span of time between pulses can be from a few milliseconds to several seconds in duration. The problem, however, can be solved with the circuit found in Fig. 26-15. This is a hardware debounce solution. A single-poll, double-throw,

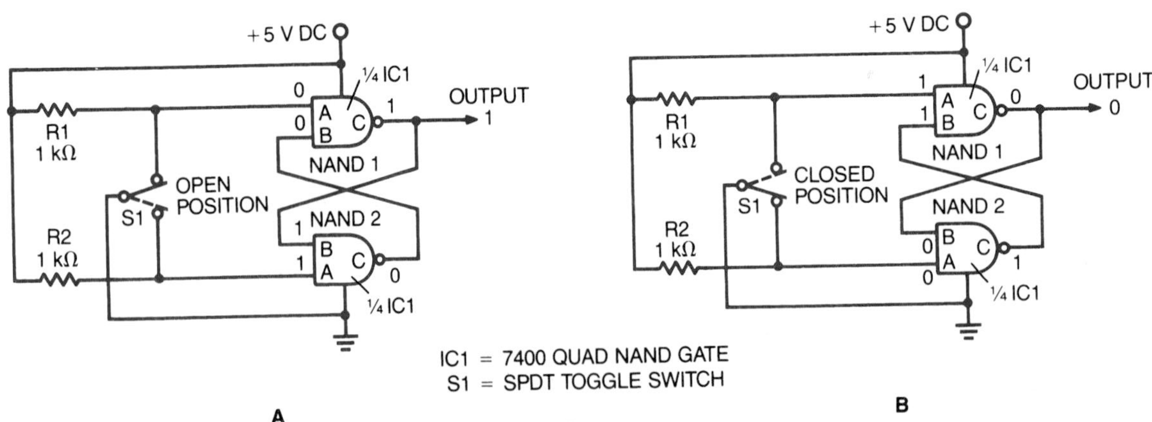

Fig. 26-15. A debounce circuit for a manually operated switch.

Fig. 26-16. **NAND gate truth table.**

INPUTS		OUTPUTS
A	B	C
0	0	1
0	1	1
1	0	1
1	1	0

momentary-contact switch is connected to the +5-Vdc source through the pull-up resistors. Each input of the NAND gates is connected to each of the two contacts of S1. The B input of each NAND gate is connected to the output of the opposite gate. The truth table in Fig. 26-16 shows that the output C will be at logic 0 if logic 1 is on both A and B inputs. When the switch is in the normally opened position (Fig. 26-15A), logic 0 is applied to input A of NAND gate #1, and logic 1 is applied to input A of NAND gate #2. Looking at the truth table again (Fig. 26-16) shows that logic 0 on either A or B input causes a logic 1 at the output. This logic 1 at the output is applied to the input B of NAND gate #2. Since input A of NAND gate #2 is logic 1 from the pull-up resistor, the output of NAND gate #2 is logic 0. This output is sent back to input B of NAND #1. Even if the switch contacts bounce, the logic 0 on input B of NAND #1 forces the output to stay at logic 1.

When the switch is changed, as in Fig. 26-15B, the input logic changes. Input A of NAND gate #2 is changed to logic 0, which changes the output to logic 1. This logic 1 is applied to the input B of NAND gate #1. The switch has already changed input A of NAND gate #1 to logic 1, and the output of NAND gate #1 becomes logic 0. This is applied to input B of NAND gate #2, thus forcing NAND gate #2's output to remain at logic 1 regardless of switch bounce.

Explain, in detail, how contact bounce can effect a software program. Use the reset switch on a computer as an example.

THE MICROCONTROLLER

A **microcontroller** is a single-chip, dedicated-control computer that has a microprocessor receiving instructions from a software program which executes those instructions one at a time. These software instructions are latched into programmable memory. This type of memory is called **EEPROM**, or *electrically erasable programmable read only memory*. This type of memory can be programmed from a host microcomputer, such as a PC. The EEPROM will retain the program in its memory even when the power is disconnected. This type of memory is *nonvolatile* memory. For the purposes of this text and Group 12 of the correlated Student Workbook, the BASIC Stamp II® microcontroller, manufactured by Parallax Inc., has been selected. The BASIC Stamp II® is contained on a single surface mounted technology (SMT) board with 24 connector pins used to access the circuitry. For user convenience, the 24 pins can be inserted into an optional carrier board that has a power connection for a 9-V battery, communications connections for a serial cable, and a reset button (Fig. 26-17A).

BASIC Stamp II® Specifications

The BASIC Stamp II® (Fig. 26-17B) has 2048 bytes of EEPROM and twenty-four bytes of RAM. The voltage required is 5–15 Vdc that can be from an external supply source. Pin 24 is the positive power terminal and pin 23 is the negative terminal. An internal 5-volt regulator maintains the correct operating voltage for the

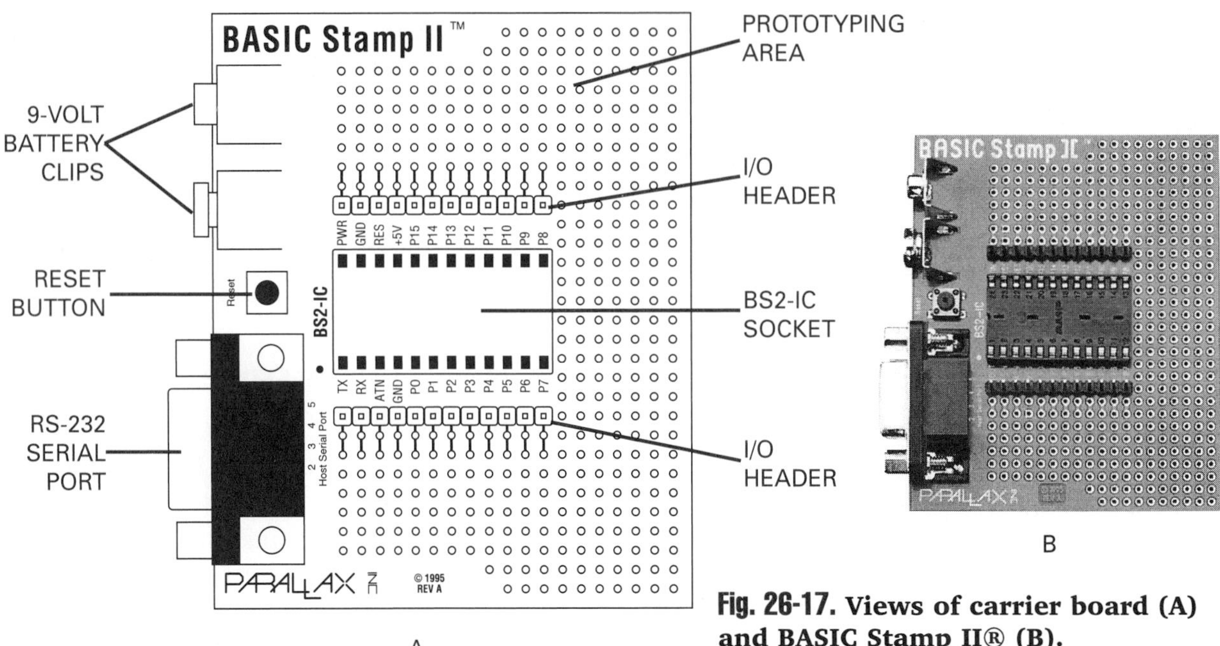

Fig. 26-17. Views of carrier board (A) and BASIC Stamp II® (B).

BASIC Stamp II® microcontroller. For independent operation, such as a robot, a single, 9-volt battery provides a very convenient power source.

When an optional carrier board is used, there is a connector compatible to standard 9-volt batteries. There are sixteen connections (pins) for inputs and outputs (I/O) on the carrier board, each of which is user definable through a software program. Programming is accomplished by first loading the BASIC Stamp II® software into a PC running MS-DOS 2.0 or higher. Next, the user's software is loaded with special instructions for controlling a device. The BASIC Stamp II® software, imbedded in the processor, interprets the user's program and changes it to numeric codes or tokens that are instructions to the BASIC Stamp II®. The user's program is automatically downloaded through a 9-pin cable connected between the computer serial port and the BASIC Stamp II® carrier board. These numeric codes (data) are saved in the 2k memory of the EEPROM. The language used is PBASIC™, a special version of BASIC, with thirty-six instructions especially designed to enable the user to control devices connected to the sixteen I/O pins of the microcontroller.

Describe the BASIC STAMP II®.

PBASIC™ PROGRAMMING

As discussed previously, a computer language is actually a set of instruction codes or program statements in the form of words, abbreviations, and alphanumeric characters. When put together in a logical sequence, they are interpreted by the microcontroller and converted into action. Figure 26-18 shows a PBASIC™ program that sends a set of binary num-

```
Start:                              'Label for beginning of program
          w1 = 0                    'clear the variable
again:    for w2 = 0 to 7           'set loop to scan LEDs
          random w1                 'get random number
          freqout w2,50,w1          'send pulses to identified (w2) LED
          next                      'pulse the next LED until all 8 are pulsed
          goto again                'start over when all have been pulsed
```

Fig. 26-18. Program to send data to LED display.

bers to a row of eight LEDs connected to the outputs P0–P7. Figure 26-19 lists some PBASIC™ instruction codes used to control devices external to the BASIC Stamp II®.

PBASIC™

The BASIC Stamp II® programming package includes software that will run in a PC with MS-DOS 2.0 or higher. After the DOS prompt appears on the screen, insert the diskette into the A drive, type *a:* and press *return*. From here the stamp directory is entered by typing *CD STAMP2* and pressing *return*. At the DOS prompt, type *STAMP2 /m* and press *return*. This runs the software and puts the display into monochrome mode for better resolution. You may now write your own program in PBASIC™ and it will appear on the PC screen. Some useful function keys supported by the editor are *Alt-L*, *Alt-S*, and *Alt-R*. Pressing *Alt-L* displays the disk program menu. Move the highlighting bar over the desired program title with the arrow keys and press *return* to load a program. Pressing *Alt-S* causes the program editor to save the program to the floppy disk in drive A. You will be asked to name the program or to accept the current name. *Alt-R* will run the current program in memory.

After writing your own program, several things will happen before the program is downloaded to the BASIC Stamp II®. First, the program is checked for errors. If any exist, the first one will be marked and the error explained. You must then correct each error as it is highlighted. After correcting the errors, save to the file disk again and press *ALT-R*. Second, the BASIC Stamp II® will be checked for proper cable and power connections. If these checks

INSTRUCTION	MICROCONTROLLER ACTION
TOGGLE	Make the specified pin an output and toggle it.
PULSIN	Measure the timing of an input pulse.
PULSOUT	Output a pulse at a specified frequency at a pin for a period of time.
BUTTON	Input the condition of a switch at a specified pin with software debounce of the switch.

Fig. 26-19. Selected PBASIC™ instructions and function.

```
Sample program

tout    var    word
start:
        freqout 11,900,450,500      'set Pin 11 to output mode and send freq for time 900
        for tout=1 to 250: next     'put a time space between tones
        goto start                  'go back and repeat
```

Fig. 26-20. Sample PBASIC™ program.

are successful, the program will be converted from PBASIC™ to tokens usable by the BASIC Stamp II®. It will then be written to the EEPROM and begin to run. This process is similar to the process of compiling a software program in other computers.

Program Structure

Programs in PBASIC™ are similar to standard BASIC but this version is specifically structured to control applications. A sample program, shown in Fig. 26-20, generates a tone that sounds like the telephone busy signal. PBASIC™ programs should always be begun by declaring the size of all variables. The syntax is the name of the variable, i.e. *tout* followed by *var* to indicate that tout is a variable and then followed by the size. In this case, the variable is a word, or two bytes, in size. The size can be one bit, a nibble (four bits), a byte, a word (two bytes), or a specified number of bits, for example, *bit(10)*.

In standard BASIC, each line needs a number. In PBASIC™, however, only key points in a program need to be identified. These key points are identified by a *label*. A label is a word with meaning and describes the program section. Labels are followed by a colon (:). *Descriptive remarks* can be included in the program at any point but they must be preceded by the single quote mark ('). This causes the interpreter to ignore descriptive remarks when generating usable code for the microcontroller. Program instructions should be written as described in the following section under selected instructions.

Discuss the differences between BASIC and PBASIC™ programming.

SELECTED INSTRUCTIONS

Dedicated control requires that pulses be exchanged between devices. The controller generates pulses to be sent to devices connected to the output pins. External devices send pulses to the controller to indicate conditions of the devices. Therefore, I/O instructions are very important in control computing. PBASIC™ has several instructions that make I/O possible. The instruction *INPUT pin* sets the designated single pin, from pins 0–15, to the input mode. When using an instruction that has a designation such as *pin*, the word *pin* is replaced by the desired number of the connection being used. For example, the instruction *INPUT 15* makes pin 15 on the BASIC Stamp II® board a single pin input. The controller can read any compatible input device connected to this pin. This device can be a switch, photo resistor, a fixed resistor or some other device of this type. In the same way, the instruction *OUTPUT pin* makes the designated connections, from 0–15 an output point. An instruction to pulse this pin will actuate a transistor, a speaker, an LED, or some other compatible output component. The instruction, *REVERSE pin*, causes the indicated pin to change from its current I/O state. To set an output pin to a known level, the instructions *LOW pin* and *HIGH pin* are used. These will make the indicated pin an output and set the voltage level to 0 V or 5 Vdc, respectively. For more I/O instructions and a more in depth discussion of this class of instructions, refer to the user's manual supplied with the BASIC Stamp II®.

Users often call upon microcontrollers to perform some unique function for which it may be necessary to write a number of lines of code. This would normally use many valuable memory locations. The PBASIC™ used in this controller has several very useful, one-line instructions that economically use memory to support these functions. Among these are *PULSIN, PULSOUT, PWM, RCT1ME.*

PULSIN

The instruction *PULSIN pin,state,variable* measures the length of time a high or low state of a pulse is present at the designated pin. When the indicated state is detected at the pin, a counter begins and continues to count until the opposite state is detected. The counter increments every two micro-seconds for a period of 65,535 counts or 131 milli-seconds. The collected data are then stored in the defined variable.

PULSOUT

The instruction *PULSOUT pin,period* makes the designated pin an output pin and changes the logic condition to the opposite of its logic condition before the instruction. The logic state remains for a time equal to *period* multiplied by 2 μsec and then returns to its original state. The period, a value between 0 and 65,535, yields a logic state duration of 0 seconds to 131 milli-seconds in 2 μsec increments. This instruction is very useful to pulse the input of a transistor or logic gate with a known timed pulse.

PWM

Using the instruction *PWM pin,duty,cycles* generates a pulse width modulated signal at the designated pin. The duty cycle of the pulse, or ratio of the logic 1 state duration to a full cycle, is determined by the value of *duty* which can be between 0 and 255. The number of pulses is controlled by *cycles*. *Cycles* designates the preset number of the 256 pulse periods that take approximately one milli-second to complete. After the instruction is completed, the designated pin changes to the high impedance input mode. This maintains voltage to any device connected to that pin. A digital-to-analog converter can be made with this instruction and an RC time constant circuit can be connected to the designated pin. The circuit (Fig. 26-21) consists of a 10-k resistor connected between the pin and the output with a 1 μF capacitor connected between the resistor output and ground. The capacitor will be charged

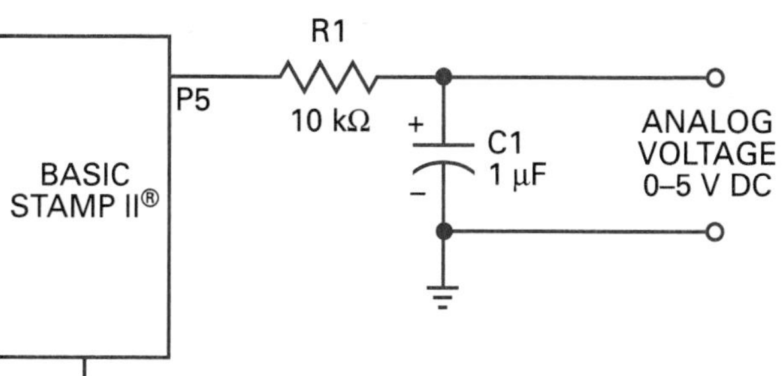

Fig. 26-21. Circuit used with PWM instruction.

between 0 V and 5 V, depending on the duty cycle established by the *PWM* instruction.

RCTIME

It is often desirable to determine the position of an object that is being controlled by a software program. This might be a robotic arm, a section of a machine, a door, or the handle of a game controller. The instruction *RCTIME pin,state,variable* and external circuitry tell the controller to get data that indicates the position of a potentiometer. If the shaft of the potentiometer is physically connected to an actuator that moves the shaft, the position of the device can be determined. Figure 26-22A is a typical external circuit. The potentiometer R_1 and capacitor C_1 build an active RC time circuit. Resistor R_2 is included for current limiting to keep pin 0 from shorting to ground if R_1 is set to zero resistance. On execution of *RCTIME*, the designated pin is moved to input mode. A counter tallies while the pin is in a logic level that conforms to the condition (1 or 0) that is specified by *state*. The counter increments every 2 μsec. The resulting count total, when the logic level does not equal *state*, is contained in the specified *variable*. Figure 26-22B shows a sample program for this. Pin 0 illustrates this and is set as an output. Logic 1 is sent to it by the first instruction *HIGH 0*. This discharges the capacitor so a known level is established. *PAUSE 1* causes pin 0 to be held high for one milli-second to discharge the capacitor. The instruction *RCTIME 0, 1, b2* clocks the discharge of the capacitor through the potentiometer. You may remember from your readings about RC circuits, that the time to charge or discharge a capacitor depends on the values of the capacitor and the resistor. Therefore, the value that is stored in variable *b2* will be a function of the circuit.

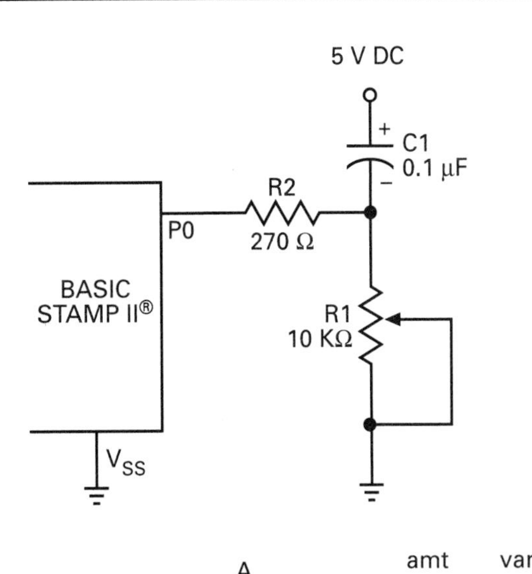


```
amt    var    word          'word size variable to be stored in amt
       HIGH 0               'discharge C1
       PAUSE 1              'delay 1 msec
       RCTIME 0,1, amt      'measure RC time constant
       DEBUG amt            'send result to host computer display
```

B

Fig. 26-22. Circuit and software.

BRAIN BOOSTER

Describe an application which uses PULSIN, PULSOUT, PWM, and RCTIME commands.

FASCINATING FACTS

The word robot *comes from Czechoslovakian writer Karel Kapek's 1921 play, R.U.R., which stood for "Rossum's Universal Robots." The play depicted mechanical beings that were manufactured as slaves. They later rose up in rebellion and killed their creators.*

CONTROLLING PERIPHERALS

One of the most common types of devices controlled by a microcontroller is a robot. Because of the distances and time involved in interplanetary travel, it is necessary to send robots to explore surfaces that, to date, are inaccessible by people. These exploration robots require many devices to make them effective when they arrive at the surface of a planet or moon. The remainder of this chapter discusses some of the circuitry needed to build a robot.

Motor Control

Motors move robot arms, wrists, and axes. Most motors require an operating current greater than the 20 ma that is produced at the output pins of the BASIC Stamp II®. It is, therefore, necessary to control a circuit that turns the motor on and off and controls speed. It may also be necessary to reverse the direction of robot axis motion. One common circuit used for this purpose is shown in Fig. 26-23. This circuit consists of four NPN transistors connected in the common emitter mode. They are controlled by the P6 and P7 outputs of the BASIC Stamp II®. The four 270-Ω resistors set the base current to a level when the 5 V logic 1 is sent to P6 or P7. The program (Fig. 26-24) is designed to send a voltage signal to either P6 or P7, but not to both at the same time. If both pins receive 5 volts, all the transistors will turn on. This can damage the external power supply, motor drives, motor—and the robot. When P6 receives a signal from the software, Q1 and Q2 both turn on. Current flows through Q2, through the motor to Q1, turning the motor in only one direction. The transistors Q3 and Q4 are turned off and will not conduct. When the voltage is removed from P6, Q1 and Q2 turn off and the motor stops. By sending logic 1 to P7, Q3 and Q4 turn on. This part of the circuit acts just like the circuit biasing on Q1 and Q2, but the current flows through the motor in the opposite polarity and the motor reverses direction. The diodes CR_1–CR_4 protect the transistors from the inductive kick (back emf) from

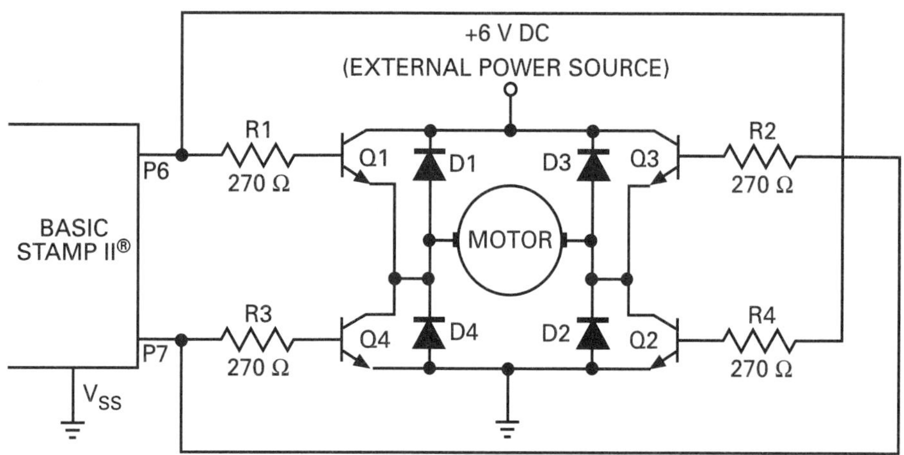

Fig. 26-23. Typical motor drive circuit with forward and reverse.

Fig. 26-24. **Motor driver software program.**

```
MOTOR:
        low 6           'set out 6 to 0 volts
        low 7           'set pin 7 to 0 volts
        high 6          'start motor
        Pause 5000      'delay
        low 6           'turn motor off
        high 7          'reverse motor direction
        pause 5000      'delay
        goto MOTOR      'start operation again
```

the motor when it is turned off. This voltage can be substantial and could damage the transistors. The diodes are forward biased by the induced voltage and they conduct current away from the transistors.

LED Control

It is often convenient to know the state of a particular pin to trace the operation of a program. This is accomplished by connecting an LED with a current limiting resistor to the pin being observed. Figure 26-25 shows the cathode of the LED connected to ground and the anode connected to a 1000-Ω resistor. The resistor is then connected to the pin on the BASIC Stamp II®. The resistor limits the current sourced by the microcontroller and helps to extend battery life as well as to protect the output latch.

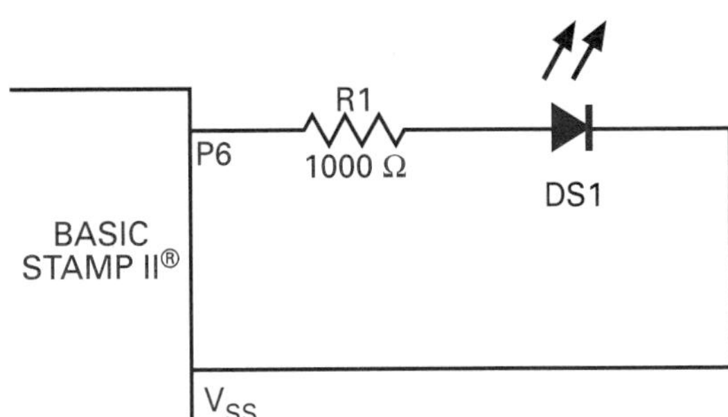

Fig. 26-25. **LED driver circuit.**

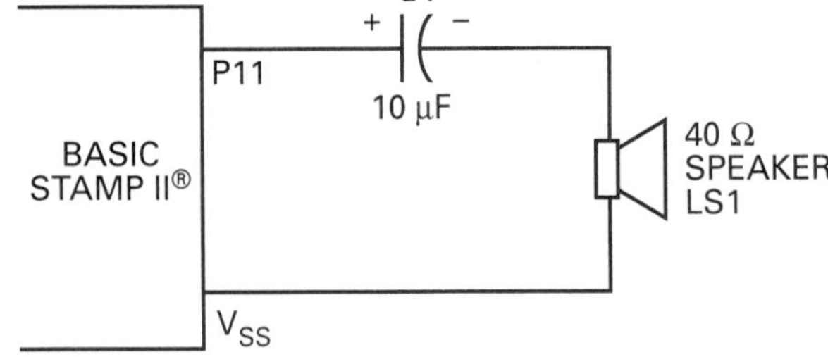

Fig. 26-26. **Speaker driver circuit.**

Speaker Control

Adding a sound feature and a speaker to a robot run by the BASIC Stamp II® is a very simple matter. One circuit used to produce sound is shown in Fig. 26-26. It consists of a 40-Ω speaker connected in series with a 10 μf, 10 Vdc, electrolytic capacitor. The capacitor is in the circuit to keep the speaker from drawing current when the speaker is not making a tone. To send a tone to the speaker, use the instruction *FREQOUT pin,milliseconds, freq1*{freq2}. The pin to which the speaker is connected is defined by *pin*. The duration of the tone is set by *milliseconds*. The tone frequency is defined by *freq1* and can be a value from 0 to 32,768—which indicates the frequency in 1 Hz increments. This instruction has an optional frequency available, *freq2*, which has the same parameters as *freq1*. A sample program that sends a tone to a speaker is shown in Fig. 26-27. The tone will last for ten seconds and then stop.

```
Start:
tone: Freqout 11,10000,933,1000
```

Fig. 26-27. Program to send a tone to a speaker for ten seconds.

Name five other peripheral devices that were not discussed here.

INPUT DEVICES

When a robot moves, it is necessary for information to be sent to the controller so that the motors can be controlled. These inputs, or sensors, include such devices as switches, photoresistors, infrared transmitter receiver pairs, ultrasonic range finders, and motion detectors. The general operation of the switch and infrared pair have been discussed previously. The BASIC Stamp II® has some special features that take advantage of the functions of these devices to provide data for sensing.

Switch Sensing

Since the instruction *RCTIME* can determine the charge or discharge rate of a resistive/capacitive circuit, this instruction can determine which individual switch in a group of switches is closed. The BASIC Stamp II® can determine which switch has been closed when used with the circuit (Fig 26- 28A) and the program shown in Fig. 26-28B. It will then transfer control of the program to a routine that acts according to the meaning of the closed switch. If switch #1 is closed, and switch #2 is open, the variable SW will contain a value near 1. If switch #2 is closed, and switch #1 is open, the variable SW will contain a value of about 28. Therefore, the software determines if the SW value is less than fifteen (<15). If the switch value is greater than fifteen (>15), the software then checks for a switch value of less than thirty (<30). It then branches to the appropriate routine. By looking at the switch circuits in

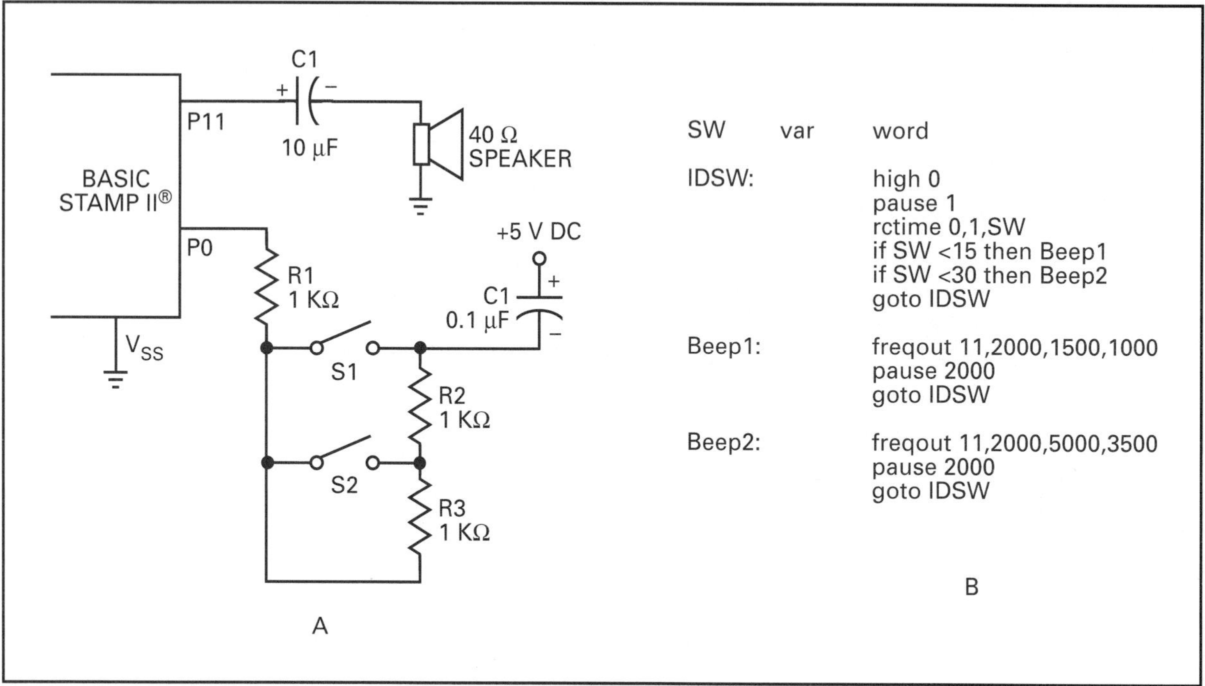

Fig. 26-28. Two-switch identification circuit and software program.

ascending order of resistance, the software will not mistake switch 2 for switch 1 because switch 1 has already been eliminated as a possibility. Each switch closure is considered by the program as an event outside the controller. This circuit can have several more switches connected to it in the same way (shown in Fig. 26-28A), thus increasing the number of external events that can determine a software's course of action. With this circuit, only one switch closure is identified at a time, so it is not a useful circuit to choose between multiple closures. If two switches should be closed at the same time, the switch nearest the capacitor will be the one chosen. When multiple closures are expected, each switch must be connected to an individual input.

Photocell Sensing

It is often desirable to determine the level of light in front of a robot because this may indicate a clear or blocked path. A photo-resistive cadmium sulfide photocell can do this. This resistive device changes its level of resistance based on the level of light that strikes it. A typical photocell will have 12 k Ω resistance in minimal light (about 0.1 ft candle) and 250 Ω at about 100 ft candles. This change in resistance can be determined by connecting the photocell between ground and an input pin, along with the software program (Fig. 26-29). The resistance value of the photocell is changed to a numeric value by the software. This value, used as a factor, determines the frequency of the tone at the speaker.

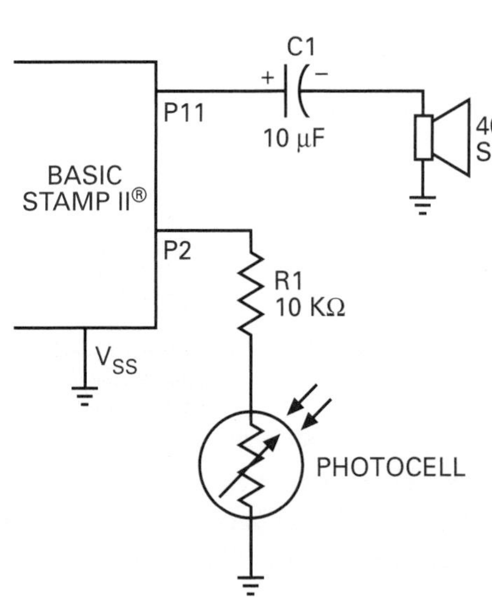

A

```
photo   var       word                  'set variable photo to word length
Value:  high 0                          'discharge capacitor
        rctime 2,500,photo              'get value of time constant
tone:   freqout 11,500,photo            'send tone to speaker with photo as frequency rate
        goto value                      'go back and do it again
```

B

Fig. 26-29. **Photocell circuit (A) with software to read the photocell (B).**

Write a short essay explaining the importance of direction, speed, and positioning inputs to a robot controller.

Chapter Review

Summary

- Microcontrollers are digital computers used for dedicated control. Dedicated control means that the controller performs only one general function. Toys, automobiles, televisions, and many other everyday devices include microcontrollers.
- All digital computers use digital data to operate. In order to control analog and higher voltage devices, microcontrollers must be interfaced with electronic devices that can translate the information coming into and leaving the controller. Analog-to-digital and digital-to-analog converters are two frequently used interface devices.
- A program is a set of instructions that tell a computer what functions to perform. Programs can be written in any one of many computer languages.

Review Main Ideas

Review this chapter's main ideas by writing, on a separate sheet of paper, the word or words that most correctly complete the following statements:

1. The term ______ describes a controller that is programmed to perform one general function for a user.
2. Each bit of data in a computer is in the form of ______.
3. The term used to describe the number of TTL devices that can be driven by a single TTL device is called ______.
4. A logic 1 represents a ______ pulse that is gated by a TTL circuit.
5. The interface connects a microcontroller with a ______ to store and process data.
6. For a digital computer to determine one voltage from another, an ______ circuit is used.
7. An ADC can send data in ______ or ______ format, depending on the format the receiving device needs.
8. An analog output is generally a ______ voltage level.
9. A ______ changes digital information into a continuous variable voltage output.
10. The accuracy of converting from either digital or analog signals is called ______.
11. The set of instructions that operates a digital computer is called a ______.

12. In ______, the computer responds immediately to commands.
13. The execution of a computer program is deferred in ______.
14. In order for a computer to communicate with peripheral devices, it must look for inputs at its ______.
15. To ensure that the input to TTL devices is within the proper voltage levels, a ______ is used.
16. Without a resistor circuit, the logic level of an open switch could not be predicted, and the data received would be ______.
17. The effects of contact bounce are removed from a circuit by ______.
18. The ______ of a switch must be determined because the contacts bounce when the switch is ______.
19. Dedicated control requires that ______ be exchanged between devices.
20. Use the instruction ______ and external ______ to indicate the position of a potentiometer.
21. To determine the state of a particular pin in order to trace the operation of a program, an ______ with a current limiting resistor is connected to the pin.

Apply Your Knowledge

1. Describe how to assign input and output pins on the BASIC Stamp II®.
2. Explain why an analog-to-digital circuit is sometimes needed to enter data into a computer.
3. List the definition for each of the PBASIC™ commands that program debounce.
4. Describe the operation of a photosensitive device.
5. Draw a schematic of a square wave with a 75% duty cycle.

Make Connections

1. **Communication Skills.** Demonstrate for the class how a microcomputer is used to handle data using a spreadsheet.
2. **Mathematics.** Write a report on the various methods of selecting pseudo-random numbers in a computer. Explain why the numbers are considered to be pseudo-randomly selected.
3. **Science.** Research the various numbering systems that have been devised. Identify the number that serves as the starting point for each system. Form a hypothesis as to why the commonalities exist.
4. **Workplace Skills.** Construct the cable needed to connect a BASIC Stamp II® carrier board to a host computer.
5. **Social Studies.** Research the first digital computer and compare it to the BASIC Stamp II®. Report how they differ and how they are alike.

Section 8

Electricity and Electronics at Work

Chapter 27

Communication Systems

Terms to Study

amplitude modulation (AM)

bandwidth

charge-coupled device (CCD)

DVD

electromagnetic waves

frequency modulation (FM)

geostationary orbit

heterodyning

high-definition television (HDTV)

intermediate frequency (IF)

Internet

modem

network

noise

pixels

radio-frequency (RF) signals

resolution

server

OBJECTIVES

After completing this chapter, you will be able to:

- describe the communication process.
- explain the importance of the feedback process in communication.
- describe the components of a stereo audio system.
- list several sources of noise and explain how they affect electronic communication.
- explain how a traditional telephone system works and contrast it with cordless and cellular technologies.
- describe the Internet and other computer networks.
- explain the differences between AM and FM.
- compare and contrast traditional television technology and digital television technology.
- describe various video systems that are now available and identify the technology that makes each system possible.

If you lived in the seventeenth century and wanted to send a letter to a friend in England, how long do you suppose it would take: a week, a month, or more? This chapter provides an overview of different types of communication systems people have developed to speed communication over long distances.

THE COMMUNICATION PROCESS

This chapter explains how electricity can change voice and music sounds, or *audio*, into analog and digital signals. A discussion of the Internet, a global network now widely used for both business and pleasure, shows how advanced communication systems have become. Selected electrical and electronic circuits provide insight on how information is transmitted and received through radio, telephone, and television. The chapter also provides ideas and facts about the use of electronics in systems such as satellite networks and fax machines.

List an example for each element of communication.

People communicate with each other using one or more of the primary senses—hearing, vision, or touch. The basic concepts in communication are shown in Fig. 27-1.

Note that communication is started by an *information sender*. This is a person who has information to give to someone else, the *information receiver*. This information can consist of various kinds of messages, including music. The information receiver is the person who receives and makes use of the information. The information goes through several processes—*encode, transmit, carry, receive,* and *decode*—before it can be received by the user.

Feedback from the information receiver lets the sender know whether the message has been received and understood. Feedback is essential to an effective communication system. Sometimes, even if the sender's message is given clearly, something happens to distort it so that the information receiver cannot understand it. Anything that distorts the message

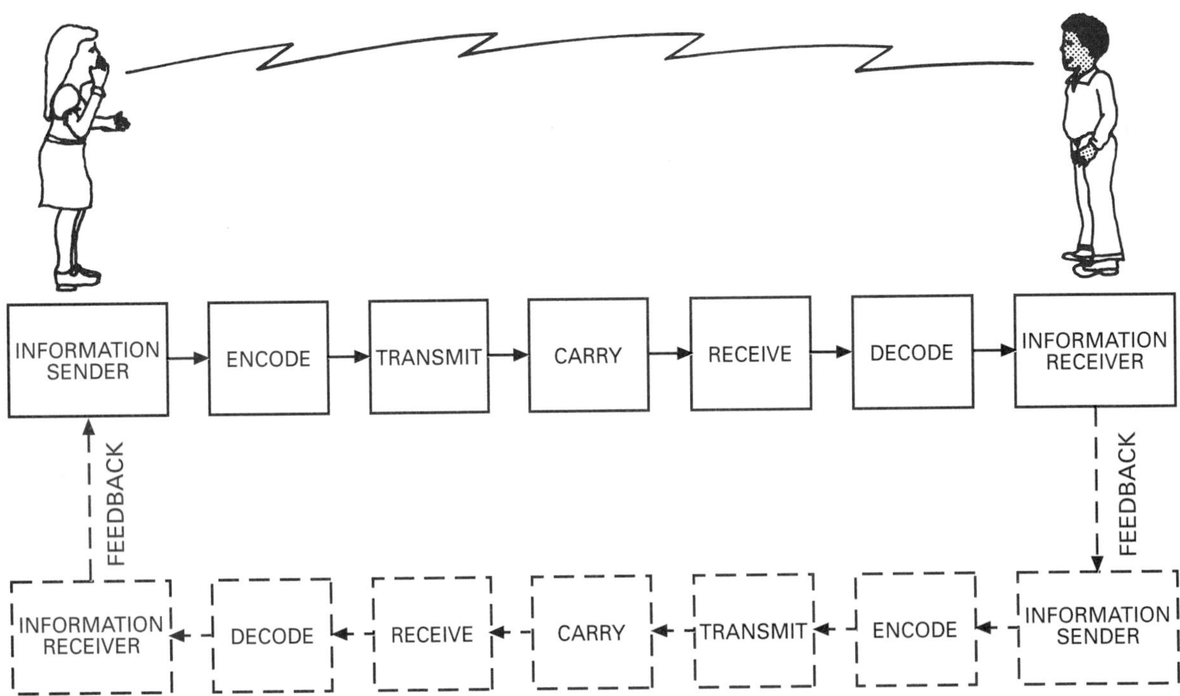

Fig. 27-1. Basic concepts in the communication process.

Information initiator makes use of human senses	Processes and elements involved					Information user makes use of human senses
	Message encoding techniques	Transmitting elements	Carrying environment	Receiving elements	Message decoding techniques	
(A) Hearing *Telephone* 	Sound waves (voice) are translated or changed to electric (audio frequency) currents.	Telephone transmitter	Wire and/or electromagnetic waves (distance between communicators: worldwide)	Telephone earphone	Reproducer (earphone) translates electric currents into sound waves.	Hearing Ears receive sound waves. Brain interprets message 
(B) Hearing *Radio* 	Microphone changes sound waves into electric currents. These currents modulate radio-frequency currents.	Radio transmitter	Electromagnetic waves (distance between communicators: worldwide and interplanetary)	Radio receiver	Detector separates audio frequencies from radio-frequency currents; receiver translates audio-frequency currents into sound waves.	Hearing Ears receive sound waves. Brain interprets message. 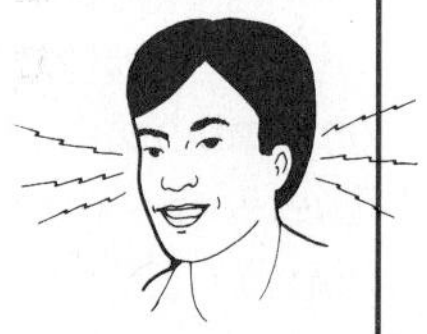
(C) Vision and touch *Computer/ Internet Facsimile* 	Keyboard or microprocessor controlled. Alphanumeric characters and images.	Transmitter or sender	Wire and electromagnetic waves (distance between communicators: worldwide)	Printer or monitor screen	Electric pulses translated into mechanical or electronic action to print alphanumeric characters or images on paper or monitor screen.	Vision Eyes receive visual image from printed paper or monitor screen. Brain interprets message. 
(D) Hearing and vision *Television* 	TV camera translates image (video) into electric currents. These currents modulate high-frequency currents. Microphone changes sound waves (audio) into electric currents, as in radio.	TV transmitter (video and audio)	Electromagnetic waves (distance between communicators: worldwide and interplanetary)	TV receiver (picture and sound)	Video detector separates video currents from audio currents; picture tube translates video currents into images; audio detector acts same as in radio above.	Hearing and Vision Ears receive sound waves from speaker. Eyes receive visual images from TV screen. Brain interprets messages. 

Fig. 27-2. Examples of communication systems.

between the sender and user is called *noise*. You will learn more about noise later in this chapter.

The use of electricity and electronics makes it possible to communicate over long distances. As shown in Fig. 27-2, telephones, radios, computers, and other devices begin and end the process of communication through the primary senses. However, electric energy is used to encode, transmit, carry, receive, and decode the messages.

AUDIO SYSTEMS

A basic audio system includes one or more microphones, an audio amplifier, one or more loudspeakers, and connecting cables. Early audio systems were *monaural*—they used only one microphone and had only one channel. Therefore, even if you used two loudspeakers, you would hear exactly the same audio from both of them.

Explain how an audio system works.

Figure 27-3 shows a stereo sound system. This system has two amplifiers, providing two channels for the sound to be amplified. This makes it possible to use two microphones and amplify the sound from each microphone separately. The microphones can be put in different places. They can be used to pick up the sounds from the left side and the right side of a stage or the sounds from a stage and the echoes from an auditorium. Therefore, this system reproduces more lifelike sounds. Other systems have four channels. These systems use four amplifiers and four speakers. They are often called *quadraphonic* or *surround-sound* systems.

Microphone

A microphone changes the energy of sound waves into electric energy. It changes the mechanical energy of the moving air striking the microphone's diaphragm into electric voltages and currents. These represent the audio frequencies and volumes of the original sound wave.

One type of microphone is the *dynamic* microphone. It has a diaphragm connected to a lightweight form with a small coil of wire wrapped around it. The form and coil are free to move over one pole of a permanent magnet (Fig. 27-4A). Sound waves striking the diaphragm cause the form and coil to move. As the coil moves, it cuts across the magnetic field of the permanent magnet,

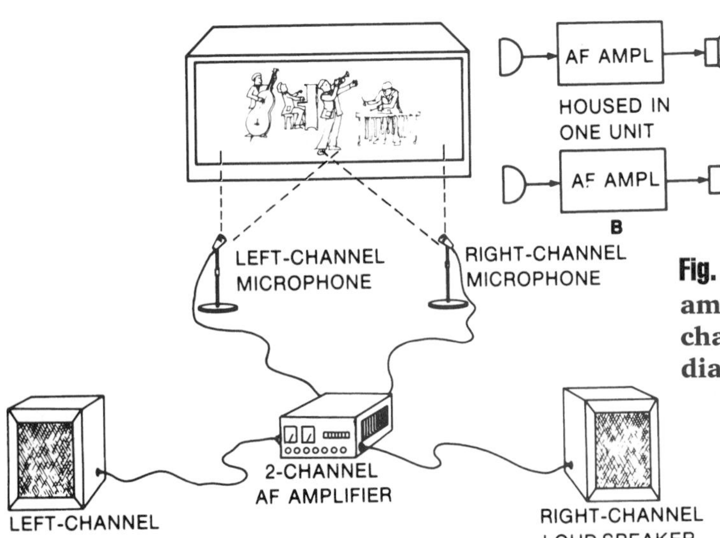

Fig. 27-3. Basic elements in a stereo amplifier system: (A) the two-channel audio amplifier; (B) block diagram.

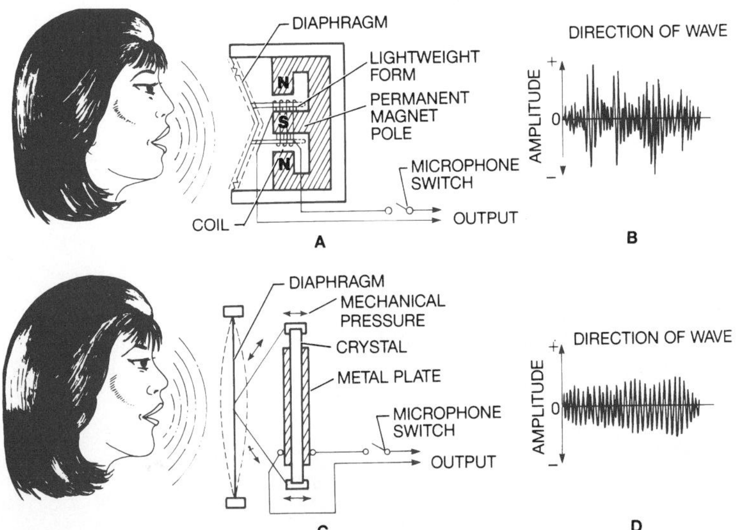

Fig. 27-4. Microphones: (A) a dynamic microphone; (B) AF voltage from a dynamic microphone; (C) a crystal microphone; (D) AF voltage from a crystal microphone.

inducing an audio-frequency (AF) voltage across it. The amount and frequency of this voltage vary according to the intensity and frequency of the sound wave. The AF voltage is an electrical representation of the sound wave (Fig. 27-4B).

The crystal microphone consists of Rochelle salt crystals or a ceramic element. The crystals have properties similar to those of quartz. When the crystal is vibrated mechanically, an alternating voltage develops between its two opposite faces (Fig. 27-4C). This is known as the *piezoelectric effect.* The AF voltages from dynamic and crystal microphones are similar. However, the output impedance of the crystal microphone is much higher.

Preamplifier

A *preamplifier* increases the output voltage of the signal source so that the main audio amplifier can reach full power. It is connected between an input device with a low output level, such as a magnetic phono cartridge or a dynamic microphone, and the main amplifier. The preamplifier may be a separate unit, or it may be part of the main amplifier.

Three-Stage Audio Amplifier

The main function of an audio amplifier is to increase the volume of the input signal (Fig. 27-5). A three-stage audio amplifier uses four transistors to provide three separate stages of amplification. Figure 27-6 shows a schematic diagram of a three-stage amplifier.

The first stage provides primary amplification of the input signal. It is made up of transistor Q1, resistors R_1, R_2, R_3, and R_4, and capacitors C_1 and C_2. The main current path is from the negative terminal of the battery (also connected to the chassis) through R_4, into the collector of Q1, out the emitter, and through resistors R_6 and R_7 and switch S1 to the positive terminal of the battery. The circuit that establishes the bias for Q1 starts at the negative terminal of the battery and goes through R_3, R_2, R_6, R_7, and S1 back to the positive terminal of the battery. The bias for the base of Q1 is developed at the junction of R_2 and R_3. Since the col-

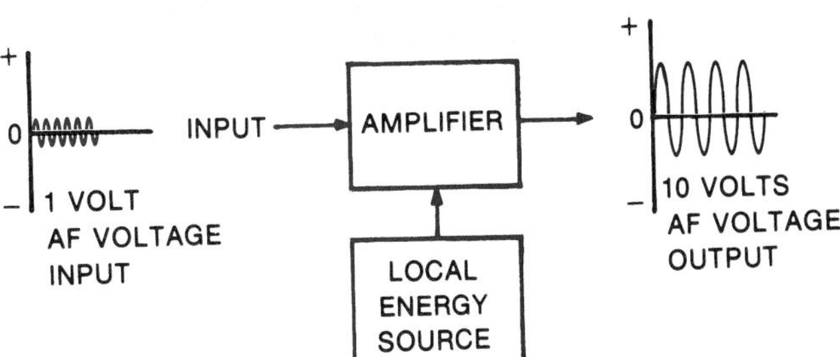

Fig. 27-5. AF input voltages are increased by the amplifier. In this case, 1 V of input voltage is increased tenfold to 10 V at the output.

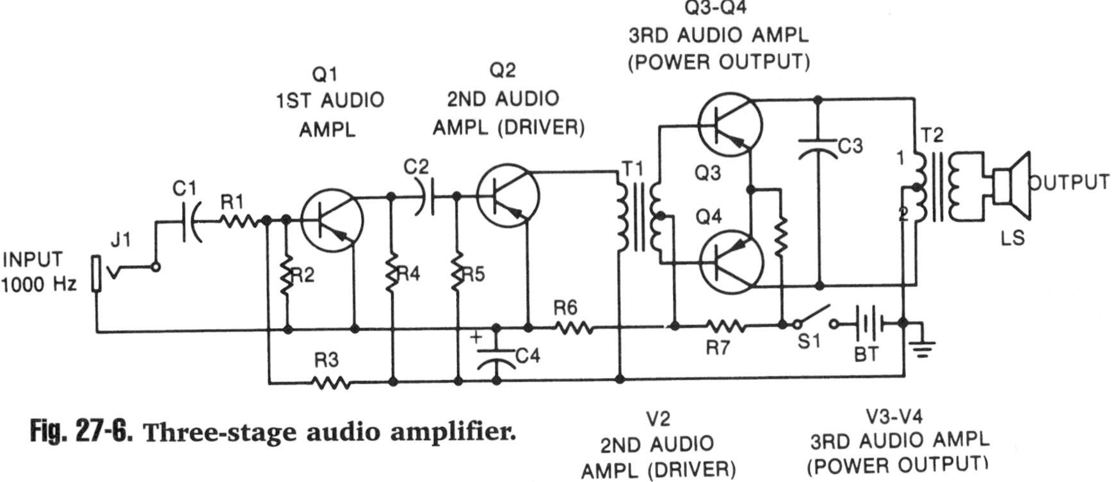

Fig. 27-6. Three-stage audio amplifier.

lector current of transistor Q1 differs according to the variations of the input signal, a varying voltage drop is produced across Q1. This process produces a voltage and current gain, which amplifies the input signal.

The second stage of the amplifier provides resistance-capacitance coupling between Q1 and Q2. The signal output of Q1 is coupled to the second-stage transistor Q2 by capacitor C_2. Resistor R_4 acts as the collector load resistor for Q1. C_2 is the dc-blocking capacitor that keeps the direct voltage at the collector of Q1 from appearing at the base of Q2. Resistor R_5 is a current-limiting resistor through which the necessary base current is applied to Q2. Transformer T1 provides interstage coupling between transistor Q2 and transistors Q3 and Q4. It acts as the circuit through which the base-emitter voltage is applied to these transistors.

Transformer coupling is also used between the output stage (transistors Q3 and Q4) and the loudspeaker voice coil. Here, transformer T2 (the output transformer) acts to match the relatively high impedance of the transistor output circuit with the low impedance of the voice coil. Since this is a step-down transformer, its coupling action produces a low voltage across the secondary winding. As a result, the value of the current in the secondary-winding circuit is great enough to energize the voice coil to the level needed to operate the loudspeaker.

The third stage provides the final output, or power amplification. This circuit is known as a push-pull-stage (Fig. 27-7). In push-pull operations, two transistors are used. The input signals to a push-pull stage are developed across the ends of the center-tapped secondary winding of interstage transformer T1. Therefore, the signal voltages applied to the input circuits of a push-pull stage from the secondary winding of T1 surge first in one direction and then in the other. When this happens, the signals are said to be 180 degrees *out of phase* with each other. The output currents pass through transformer T2.

Power Output

The power output of an amplifier is rated in a number of ways. For example, music, peak, and RMS power output are common ratings. The abbreviation *RMS* means "root mean squared." The RMS power rating is the most meaningful rating for comparing different amplifiers.

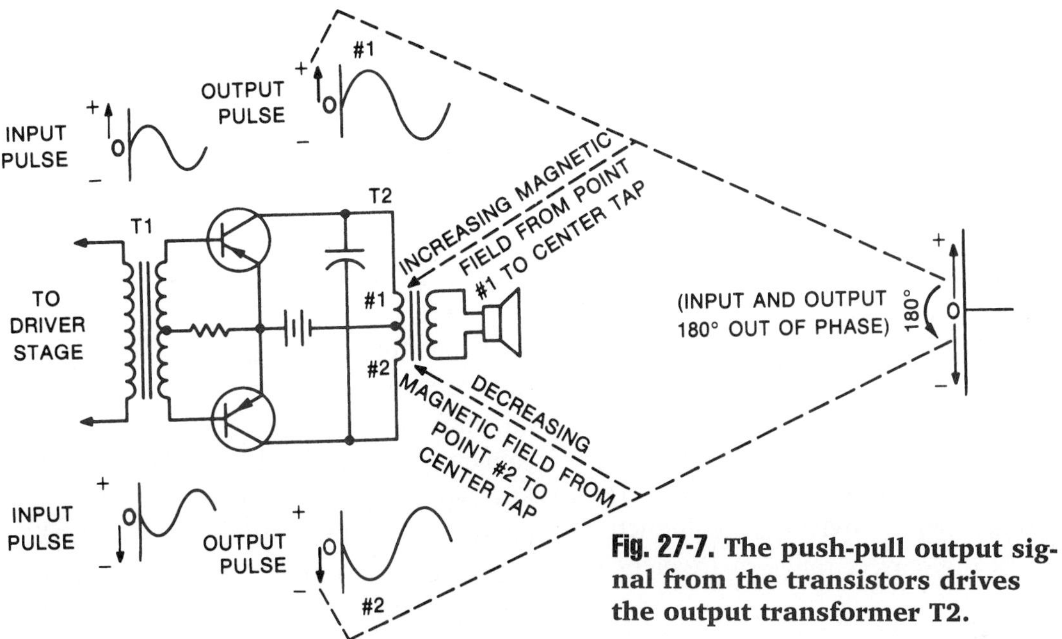

Fig. 27-7. The push-pull output signal from the transistors drives the output transformer T2.

Loudspeakers

A loudspeaker, or speaker, changes electric energy into sound. Figure 27-8 shows a cross-section of a typical speaker. In this speaker, the voice coil moves freely over, but does not touch, the permanent-magnet pole piece. The voice coil is also connected to the secondary winding of the output transformer T2 (Fig. 27-7). The amplified audio signals develop across the secondary winding of the output transformer. These signals are applied to the voice coil, energizing it. The electromagnetic field of the voice coil is then alternately attracted and repelled by the permanent magnetic field. This causes the voice coil to vibrate. The intensity of this movement varies with the audio signals originally produced by the microphone. The voice coil is connected to the speaker's cone, so it causes large amounts of air to vibrate. These vibrations produce sound waves. If the audio system is of high quality, the amplified sounds match the input signals almost exactly.

Audio systems often use three speakers: a *woofer*, *midrange speaker*, and *tweeter.* With three speakers, incoming frequencies are split. The low frequencies (20 Hz to 1,000 Hz) are handled by the woofer. The middle frequencies (800 Hz to 10 kHz) are handled by the midrange speaker. The high frequencies (3.5

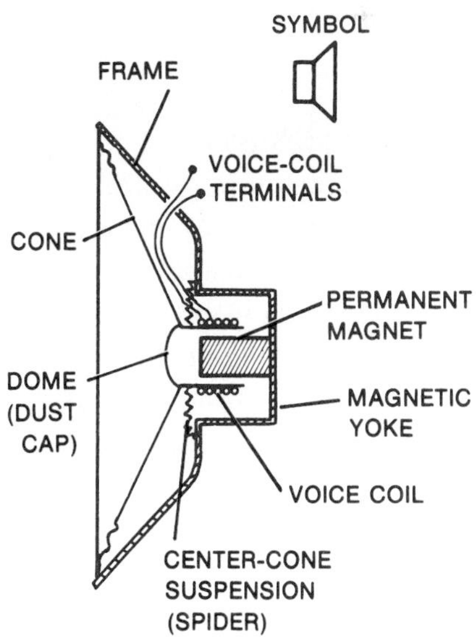

Fig. 27-8. Loudspeaker and symbol for loudspeaker.

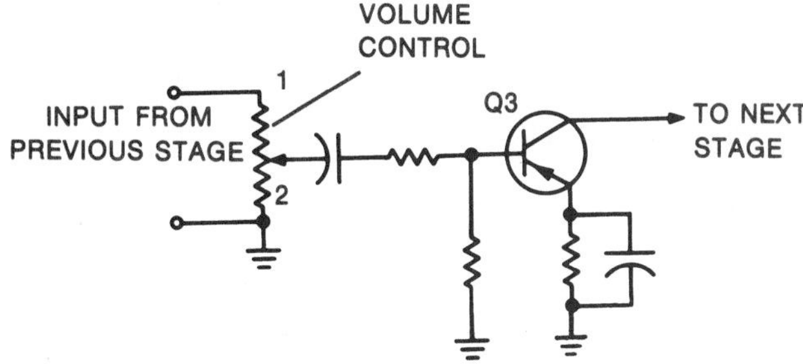

Fig. 27-9. Volume-control circuit.

kHz to 20 kHz) are handled by the tweeter. A *crossover network* is used to split the input signals into frequency bands for the speakers. This network, consisting of capacitors and inductors (coils), feeds the correct range of frequencies to each speaker.

Cables

Coaxial shielded cable is the best choice for connecting parts of an audio amplifier system. Generally, no more than 20 ft. (6.1 m) of cable can be used between a high-impedance microphone and an amplifier without distorting the sound. Connectors, such as cable plugs, phone plugs, and microphone connectors, should be fastened securely to the cable by soldering or by tightening all screws. Many problems in amplifier systems can be traced to loose connections or broken wires in the cables.

Volume and Tone Control

The *volume* is the output sound level of a circuit. The volume control of an amplifier circuit is usually a *potentiometer variable resistor*. This is usually put into the input circuit of an amplifier stage (Fig. 27-9). The volume control acts as a voltage divider. As the shaft of the potentiometer is adjusted, the sliding arm moves toward either point 1 or point 2. When it moves toward point 1, a greater part of the voltage developed between points 1 and 2 is applied to the input circuit of transistor Q3. This increases the volume of the amplifier circuit. Moving the sliding arm to point 2 decreases the volume.

Tone refers to the balance of low frequencies, or bass, and high frequencies, or treble, in the sound output. A tone control lets you adjust the frequency response of an audio amplifier to change this balance.

A basic tone-control circuit consists of a capacitor and a potentiometer variable resistor connected in series across the output of an amplifier stage. In Fig. 27-10, C_2 has a capacitance value that causes it to present a low capacitive reactance to the higher-frequency signals in the output circuit of Q2. The capacitor then tends to bypass those high-frequency signals to ground. This prevents them from reaching the speaker. With the high frequencies weakened, the amplifier produces relatively more bass response. The degree to which this happens is determined by the setting of the potentiometer.

Noise

In electric and electronic communication systems, **noise** is anything that interferes with a signal so that the quality of the output is decreased. Common sources of noise include:

- atmospheric noise—usually the result of random radio waves that induce voltages in the antenna on a receiver; often called *static*.

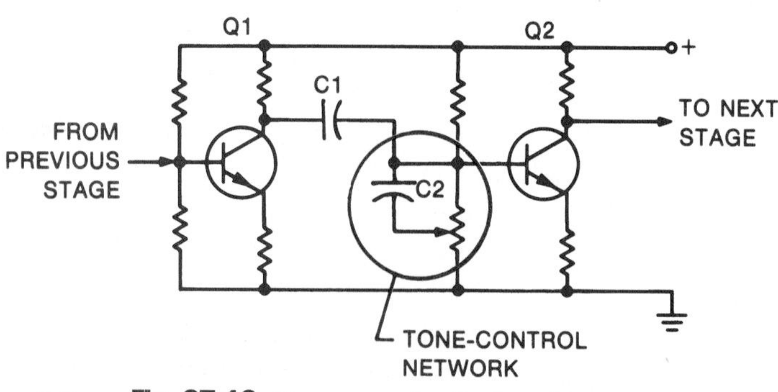

Fig. 27-10. Tone-control circuit.

- extraterrestrial noise—radio waves radiated from the sun or distant stars.
- industrial noise—noise produced by an arc discharge in products such as automobile and aircraft ignition, electric motors, and fluorescent lights.
- thermal noise—random noise generated by resistance or a resistive component in a receiver; also called *agitation* or *white* noise.
- shot noise—random variations in the arrival of electrons at the output electrode of an amplifier.
- transit-time noise—frequency distortion that occurs when the time it takes an electron to travel from an emitter to the collector of a transistor becomes relatively long.

Storing Sound

Currently, the two most common devices for storing sounds are magnetic tape and compact disc. The tape stores sound by magnetically rearranging the metal oxide particles on the tape. The compact disc stores sound using a laser beam that burns pits into a plastic disc.

Magnetic Tape

Magnetic tape has a special coating of iron oxide. Sounds are electromagnetically induced on this coating (Fig. 27-11). Two-track and

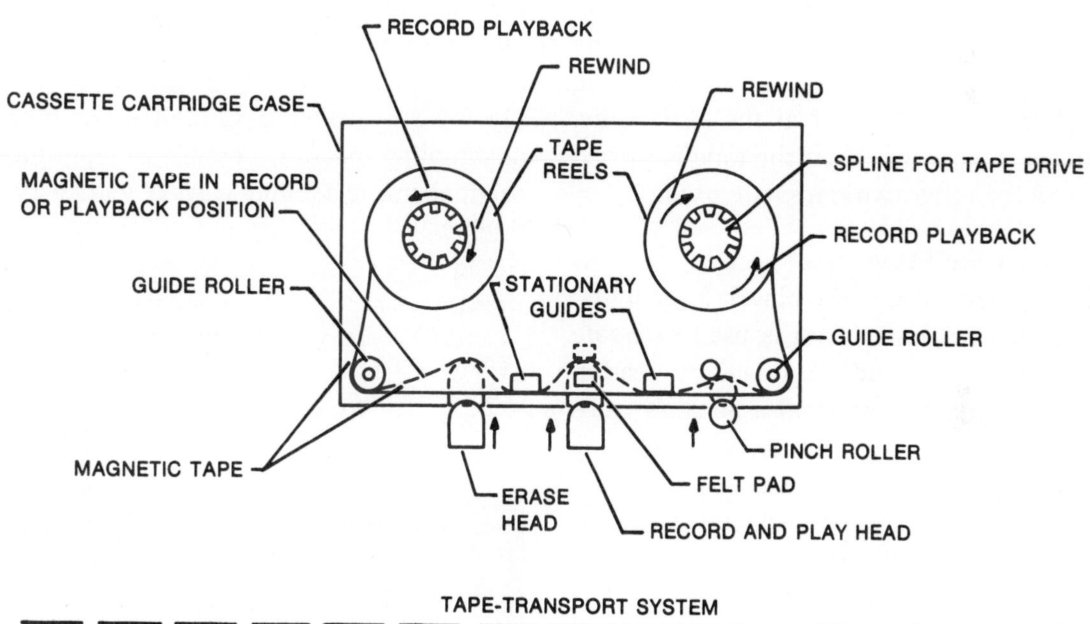

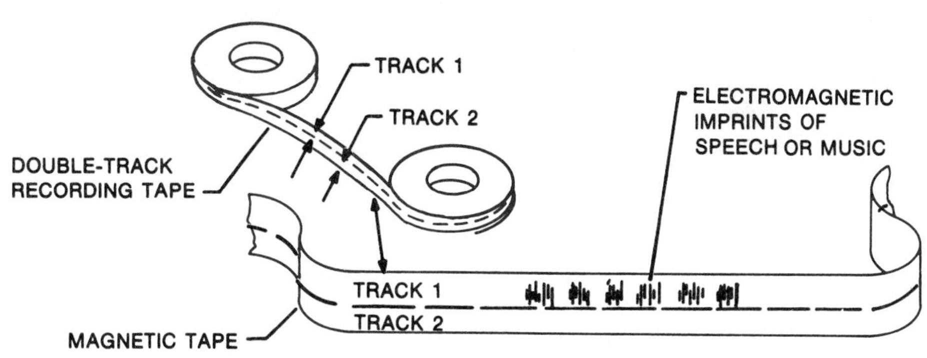

Fig. 27-11. **Cassette tape cartridge and transport system.**

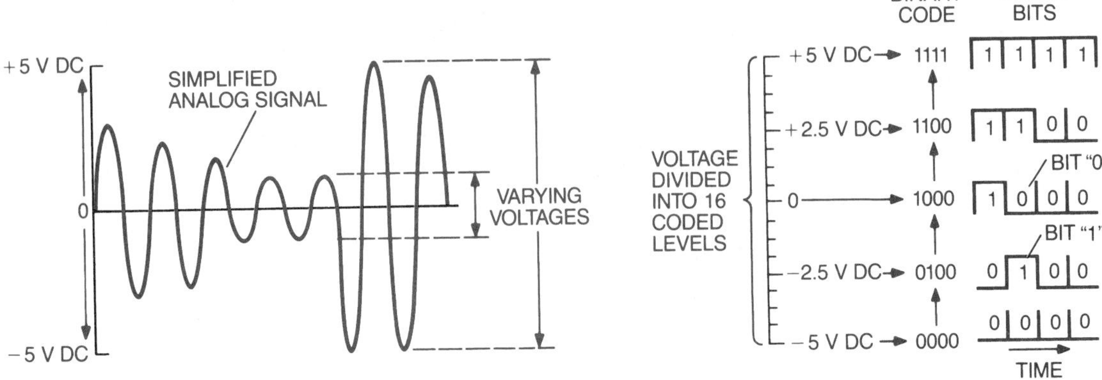

Fig. 27-12. An analog-to-digital converter in which voltage is divided into 16 levels.

four-track tapes are available today. On a two-track tape, one side is recorded and played on track 1. Then the tape is turned over, and the other track is used. For stereo, four tracks are used. Two tracks are used at the same time, one for each channel. Then the tape is turned over, and the other two tracks are used.

Compact Disc Player

The compact disc (CD) player is a digital audio system. A laser beam is used to create (record) the data on the CD and later to read it.

Audio signals derived from a person's voice or from musical instruments are very complex audio waveforms. To convert audio signals into a digital form, an analog-to-digital converter (ADC) is used (Fig. 27-12).

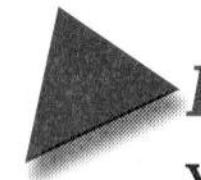

LINKS

You may wish to review the discussion of analog-to-digital converters in Chapter 26.

A simplified analog waveform is shown in Fig. 27-12. The voltage levels of this waveform are coded into 16 different levels. Each time a waveform is *sampled* (sensed), a binary code for the voltage level for that sample is established. The magnitude of the voltage is sampled 44,100 times per second. For every second, therefore, the audio signals are transformed into 1,411,200 individual binary bits (16 x 2 x 44,100 = 1,411,200). The high rate of sampling makes it possible to reproduce the original music (sound waves) with very high fidelity.

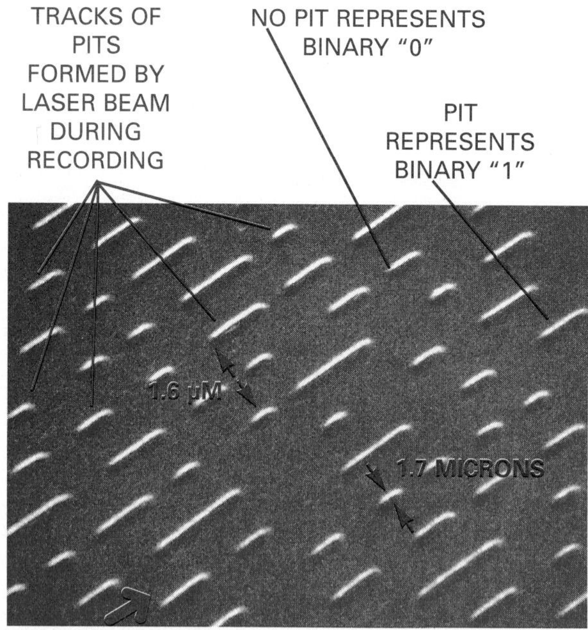

Fig. 27-13. Greatly enlarged view of the pits formed by a laser beam on a disc.

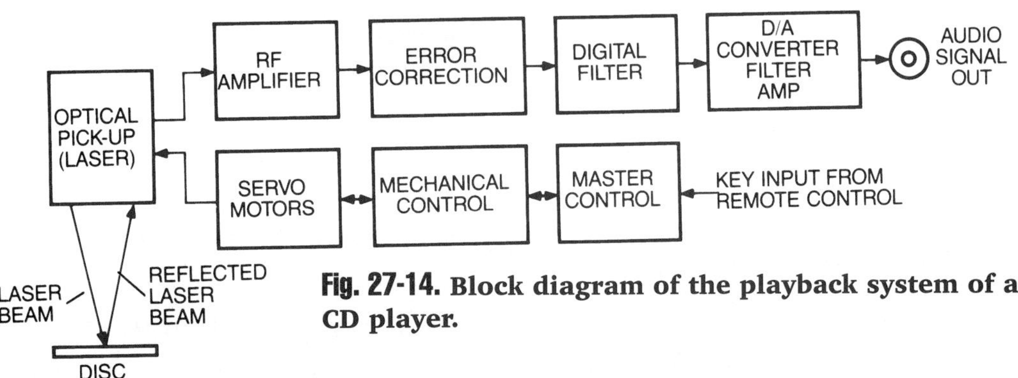

Fig. 27-14. Block diagram of the playback system of a CD player.

To record digital audio, the beam is directed at a blank plastic disc that has been coated with a thin layer of active material. The laser beam burns "pits" into the active material. The pits change the reflectivity of a tiny area on the disc. Figure 27-13 shows an enlarged view of the pits. The pits and absence of pits represent the audio recorded in digital form on a disc. The length of the pit and the distance between pits varies according to the 16-level binary code derived from the analog-to-digital converter.

Roughly 15 billion pits pass the playback head of a CD player every hour. No traditional stylus could pick up that amount of information. Therefore, the playback process on the CD player also uses a laser beam. Figure 27-14 shows a block diagram of the playback system, and Fig. 27-15 shows a cutaway view.

CD-ROM

CD-ROMs (compact-disc-read-only memory) are similar to audio CDs, except that they can store graphics, text, and video as well as audio. Installed in a computer, a CD-ROM drive offers almost instant access to a large amount of data. Most CD-ROM discs can hold up to 630 megabytes—approximately the same amount as 700 floppy disks.

LINKS

You may wish to read ahead to Chapter 32 for more detailed information about lasers.

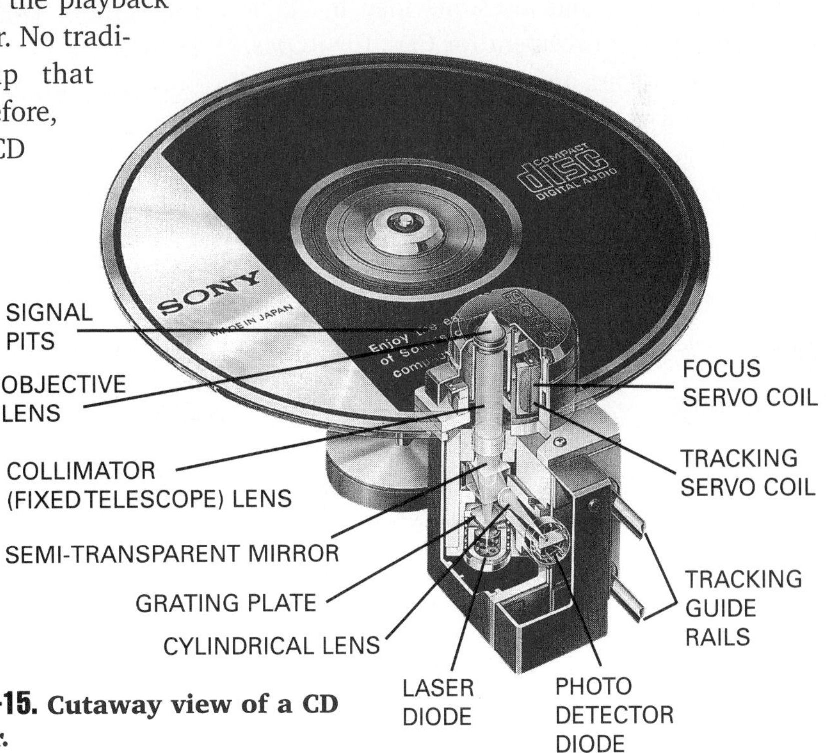

Fig. 27-15. Cutaway view of a CD player.

LINKS

You may wish to review Chapter 25 for more detailed information about computer systems and terms.

Basic CD-ROMs are read-only devices; after they have been produced, you cannot record or change information on them. However, variations of the CD-ROM, such as CD-R (compact disc-recordable) and CD-RW (compact disc-rewritable) do allow you to record data. The difference between a CD-R and a CD-RW is that you can only record, or write, information to a CD-R one time. You cannot erase the data or change it. The CD-RW discs allow you to record and erase information many times.

DVD

DVD is a newer storage technology intended for use on both computers and home entertainment systems. The term *DVD* originally stood for "digital videodisc," but it is now commonly referred to as "digital versatile disc" because it can do much more than just store video. In fact, it may become a replacement for CDs, CD-ROMs, videocassette tapes, and game cartridges.

As DVD technology matures, it is expected to have a large effect on the computer and home entertainment markets. It can hold up to eight hours of video or up to 17 gigabytes of data. DVD is currently a read-only format. However, in the next several years, its developers plan to add recordable and rewritable versions.

TELEPHONE

Telephones allow people to talk to each other even though they may be miles apart. Traditional telephone systems use telephone lines to make the necessary connections. Recently, however, fiber optics and cellular technologies have begun to change the way telephones work.

Give five examples of the ways in which telephones, when paired with computers, are powerful communication tools.

Parts of a Telephone System

A telephone system changes sound energy into electric energy. The electric energy travels through the circuit to the receiver of another telephone. There, electric energy is used to reproduce the original sounds.

Telephone Transmitter

The part of the telephone system that changes sound energy into electric energy is the *transmitter* (Fig. 27-16A). Sound waves are actually a series of varying air pressures. These exert a force on the metal diaphragm of the transmitter. An increase in pressure pushes the diaphragm against a small container of carbon granules, decreasing the electrical resistance of the carbon. This causes the current in the line to increase.

When there is no pressure on the diaphragm, the electrical resistance of the carbon granules is high. This causes the current to decrease. The diaphragm vibrates against the carbon granules in step with the sound waves. This vibrating action changes the steady direct current in the telephone circuit into pulsating, or varying, direct current. The frequencies of these pulsating currents are, when changed back to sound waves, within the hearing range of humans. They are called audio-frequency (AF) currents or analog signals (Fig. 27-16C).

Telephone Receiver

The AF currents produced in the transmitter are conducted through the circuit to the *receiver*, or earphone. In the receiver, the AF currents pass through an electromagnetic coil. The

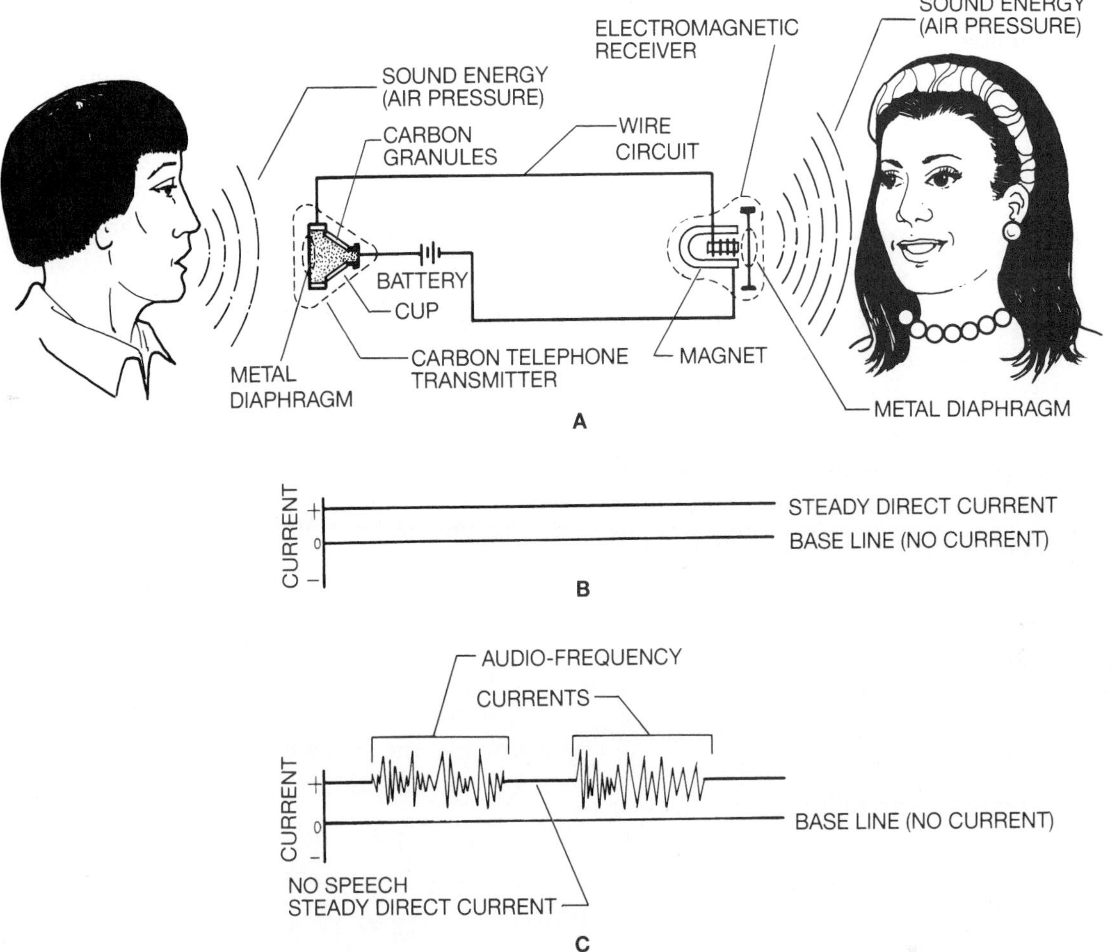

Fig. 27-16. The telephone transmitter changes sound energy into electric energy: (A) basic telephone electric circuit; (B) steady dc current flowing in wires; (C) comparison of steady dc and audio-frequency currents.

resulting electromagnetism causes a metal diaphragm in the receiver to move back and forth. It does this in step with the varying value and frequency of the pulsating current. The pushing of the diaphragm against the air produces sound waves.

Bandwidth

Humans can hear sounds that range from about 16 Hz to 20,000 Hz. However, most of the significant content of speech falls in the frequency range below 3,000 Hz. For this reason, voice-communication receivers *attenuate*, or reduce, the high audio frequencies and boost, or increase, the lower frequencies. The **bandwidth**, or frequency range, of a normal telephone channel is about 4,000 Hz.

Fiber Optics

Older telephone wires and cables are made of copper. However, as people find new ways to use the telephone, these are no longer adequate to handle the increasing number of telephone calls. The Bell Telephone Laboratories answered the need for a new type of system by developing telephone wires made from glass fibers. The popular term for this technology is fiber optics.

LINKS

You may wish to read ahead in Chapter 32. It discusses the use of lasers and fiber optics in communication systems.

Modem Communication

In addition to handling "voice calls," in which people talk to each other, telephones can transmit data in electronic form. The data can be text, graphics, or even digital music.

The term **modem** is a contraction of *modulator-demodulator*, which is an electronic device that provides an interface between digital and analog signals. Modems connect computers and other digital devices to telephone lines. The modulator in the modem at the sending end converts binary digital data into analog signals. This allows the signal to be transmitted on telephone lines. The demodulator in the modem at the receiving end converts analog signals back to digital signals for use by another digital device.

Facsimile (Fax) Transmission

Facsimile, or fax, technology allows people to transmit text and graphics around the world almost instantly. In fax transmission, a photocell performs a raster scan over a printed page or picture. The variations of print or picture density on the document cause the photocell to generate an electric video signal. The video signal modulates a carrier wave that is transmitted to a remote destination over telephone wires, radio, or cable communication links. At the remote terminal, demodulation reconstructs the video signal, which determines the density of print produced by a printing machine. The printing machine is synchronized with the raster scan at the transmitting terminal. This produces an exact copy, or facsimile, of the document.

The block diagram in Fig. 27-17 shows the data flow in a fax machine. The telephone lines provide the connection between the machine and outside information.

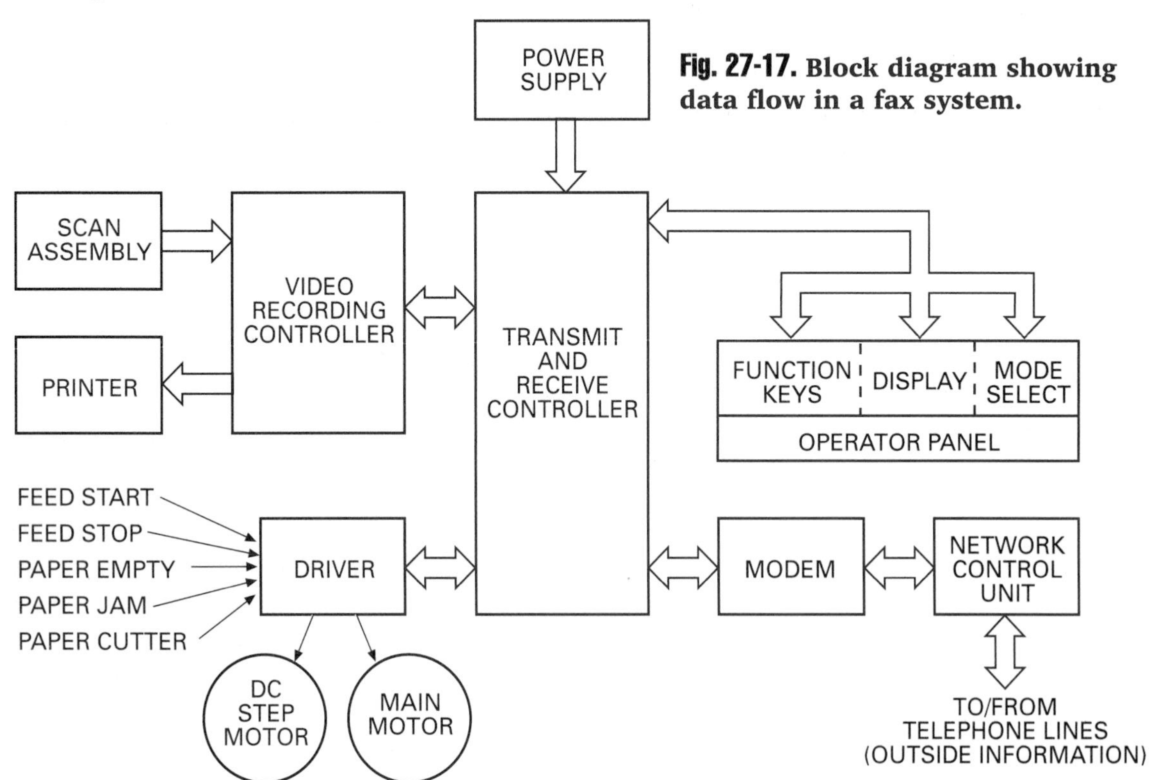

Fig. 27-17. Block diagram showing data flow in a fax system.

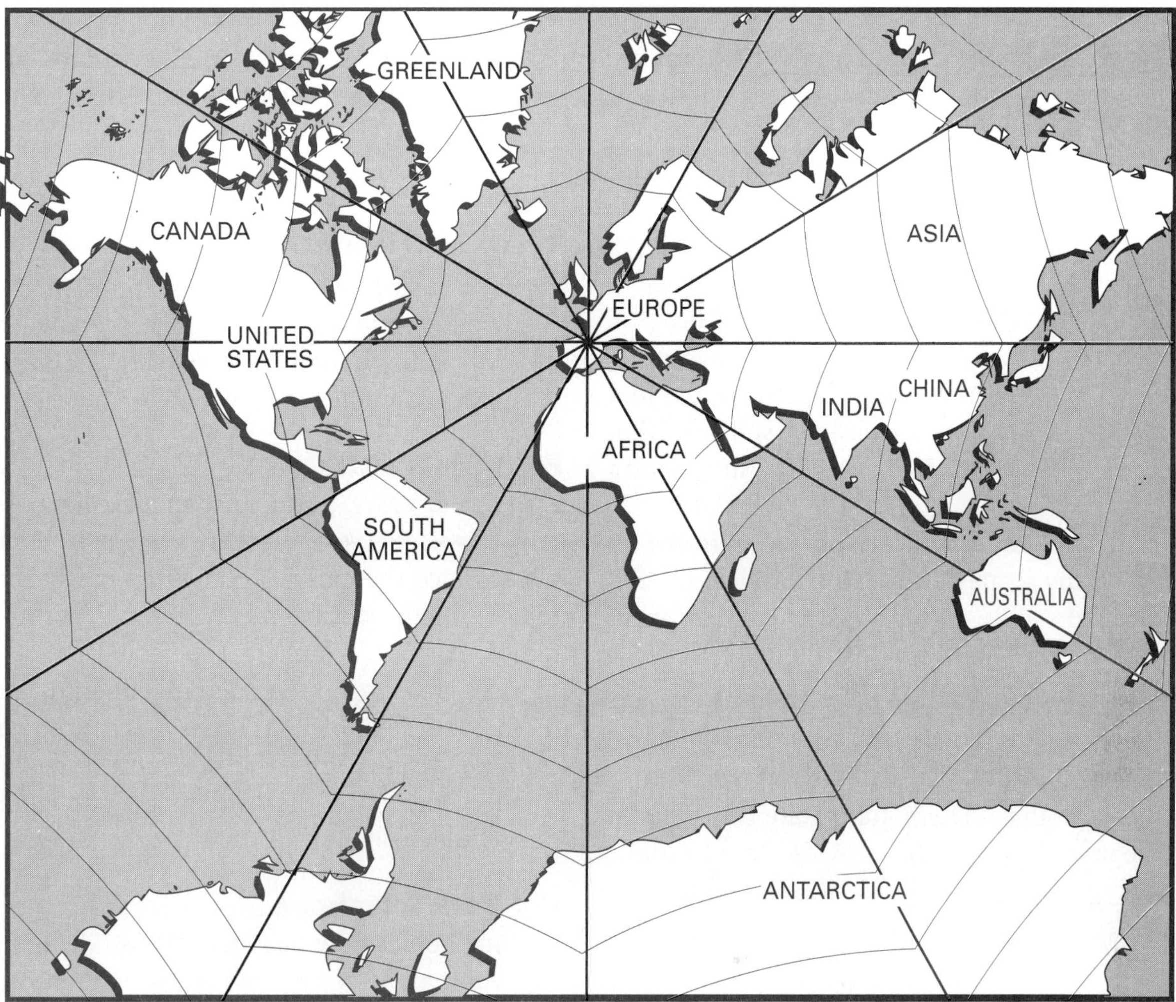

Fig. 27-18. **An artist's concept of the Internet, a global communication system.**

Computer Networks

In a sense, any computer connected to another computer for the purpose of exchanging information is a **network.** Computer networks use various electronic connection or communication devices. Most of them depend at least to some extent on telephone lines and services.

Many companies today are replacing older, more expensive mainframe computers with desktop computers connected to a network. These companies often use local area network (LAN) technology to connect all of the computers to one or more central servers. A **server** is a computer that has an enormous amount of power and memory in order to store the network software and information that can be used by all the network computers.

Companies that have offices in more than one location may use a wide area network (WAN). This type of network is similar to a LAN, except that the computers do not have to be located in the same building or area.

The Internet

The **Internet** is the world's largest computer network. It is much like a huge spider web encircling the world (Fig. 27-18). The Internet makes use of a variety of communication devices, including copper wires, fiber cables, satellites, microwaves, and underground and underwater cables.

In its original form, the Internet was difficult to use. The invention of the World Wide Web (WWW) in 1989 made use easier by creating a graphical interface. The interface provided a method to organize subject matter. This provided a way to access related subject matter by using links. The WWW developed into the first browser. A *browser* is front-end software that allows people to find specific information quickly and easily.

A complete discussion of the Internet is beyond the scope of this text. However, the following are some of the activities you can experience on the Internet using a browser.

- E-mail—You can send electronic messages (e-mail) to other people who are connected to the Internet.
- Newsgroups—These groups are like bulletin boards. Newsgroups provide an electronic forum on topics of interest to the user.
- Subject—Hypertext Markup Language (HTML) allows you to use key words and graphic images to link to related documents. In this way, you can find almost any subject matter or topic of interest anywhere in the world. It is like having a global encyclopedia only a few keystrokes away.
- Downloading—Copying files from another computer into your computer is known as *downloading.* You can also *upload* your files to another computer or server on the Internet.

Portable Communication

In recent years, portable communication systems have changed the way people live and work. Since the early 1980s, there has been a rapid growth in the use of personal portable radios, cordless telephones, personal message services and pagers, cellular telephones, and radio-controlled devices. These devices provide communication while you are "on the move." Good examples are the cordless telephone and the cellular telephone.

Cordless Telephone

Many homes now have a cordless telephone. The equipment necessary for this system consists of a base unit connected to the phone and power lines and a portable unit that can be carried about in the home or yard. The portable unit is basically a radio transmitter that can operate away from the base unit, the receiver.

Cellular Telephones

Cellular telephones, or *cell phones,* enable people to make

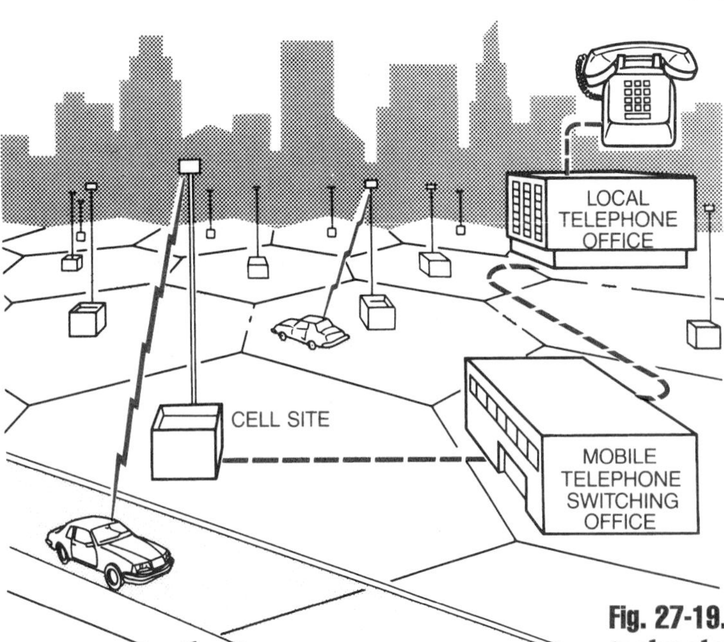

Fig. 27-19. Basic concept of cellular technology.

long-distance calls without using telephone lines. Cellular technology has made this mobile system possible. Cellular technology is based on a grid system that covers a specific geographic area (Fig. 27-19). Each cell contains a low-powered radio transmitter and control equipment located in a building called a *cell site.*

The cell site is connected by wire to a mobile telephone switching office (MTSO). The MTSO monitors the mobile units and automatically switches or "hands off" conversations in progress as the mobile unit moves from one cell to another.

RADIO COMMUNICATION

Radio communication requires a transmitter and a receiver. The transmitter is the radio station, and the receiver is the device commonly known as the radio.

Describe the difference between amplitude modulation and frequency modulation.

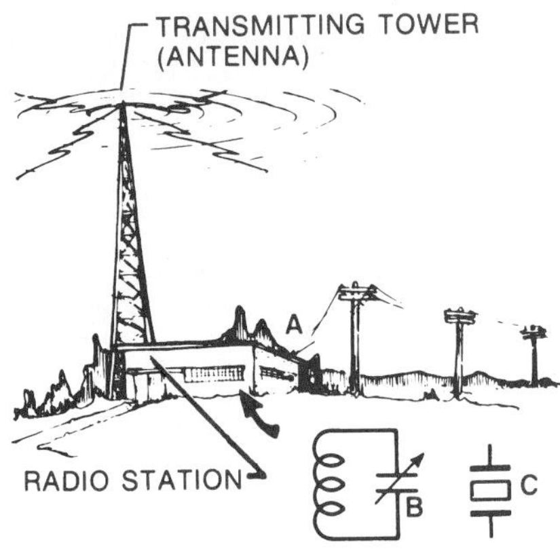

Fig. 27-20. **Transmitting station: (A) power lines (energy source); (B) oscillator using coil and capacitor; (C) oscillator using a crystal.**

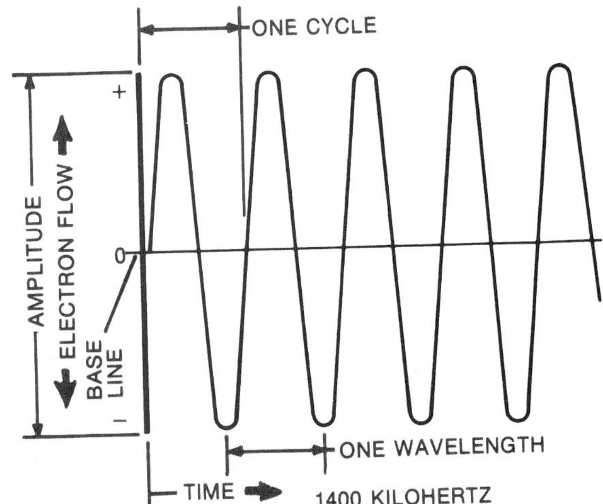

Fig. 27-21. **High-frequency electromagnetic (radio) wave.**

Radio Transmitter

The radio transmitter makes it possible to send audio signals by means of *radiated,* or emitted, energy. The radiated energy is transmitted through the air. The radio station has equipment for transmitting audio signals. It also has an energy source that produces a high voltage (Fig. 27-20A).

Radio Waves

Sound waves are transmitted through the air as varying air pressures. In a radio transmitter, a circuit is used to create waves that are somewhat like sound waves. This circuit is a high-frequency oscillator that contains a coil and a capacitor (Figs. 27-20B and C). The oscillator causes a stream of electrons to vibrate, or *oscillate.* The rapidly oscillating movement of electrons causes energy to be radiated in the form of electromagnetic waves.

Electromagnetic waves, or radio waves, are a form of radiated energy that can do work. Figure 27-21 shows the waveform of radio waves. The higher the frequency, the shorter the wavelength. *Wavelength* is the distance from a point on one wave to a similar point on the next wave.

FASCINATING FACTS

Radio waves, light, and many other forms of electromagnetic radiation do not always travel at 186,000 miles per second, the defined speed of light. The waves travel at that speed only in a perfect vacuum. When moving through the air, the waves travel much slower. When describing the maximum speed of light, we should actually call it, "the speed of light in a perfect vacuum."

Carrier Waves

Radio waves of different frequencies are assigned to individual radio stations. This type of radio wave is called a *carrier wave* because it "carries" the audio signal from the transmitter at the radio station to radio receivers.

The frequencies of the carrier waves used for different kinds of radio broadcasting are assigned by the *Federal Communications Commission* (FCC). Commercial AM broadcast stations operate within a carrier frequency band that extends from 535 to 1,605 kHz. The carrier frequency of commercial FM broadcasting extends from 88 to 108 MHz. The circuit that combines audio information with a carrier wave is known as a *modulator.*

Amplitude Modulation (AM)

When sound waves strike a microphone, they produce a varying current, which is then fed into a modulator. Modulation allows the carrier wave to carry information from one place to another by electromagnetic energy (Fig. 27-22). Note that the frequency of the carrier wave does not change after modulation. The alternating current from the microphone modulates the carrier wave by causing its amplitude, or strength, to rise and fall in a process called **amplitude modulation (AM)**. At the receiving end, this information is separated from the carrier wave by special circuits in a radio receiver.

A two-stage AM transmitter is shown in Fig. 27-23. It is designed to transmit a frequency that can be received by an ordinary AM radio. Despite its low power (400 μW), its energy can be picked up by a receiver 50 to 100 ft. (5 to 30 m) away.

Frequency Modulation

The carrier wave can also be modulated by changing its frequency in a process called **frequency modulation (FM)**. The changes in frequency are controlled by the signal from a microphone. Note that the amplitude does not change in frequency modulation (Fig. 27-24).

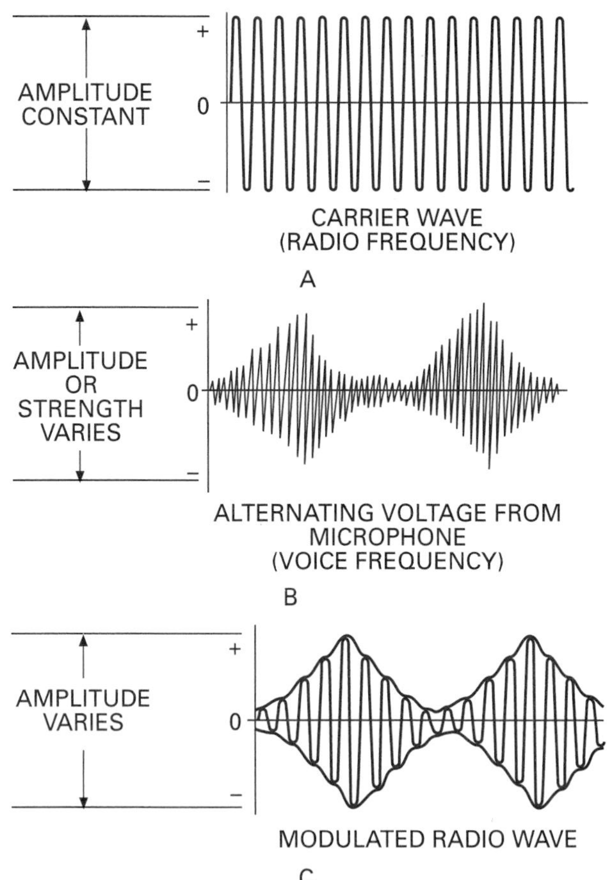

Fig. 27-22. Amplitude of carrier wave modulated by an alternating voltage: (A) carrier wave; (B) alternating voltage; (C) modulated radio wave.

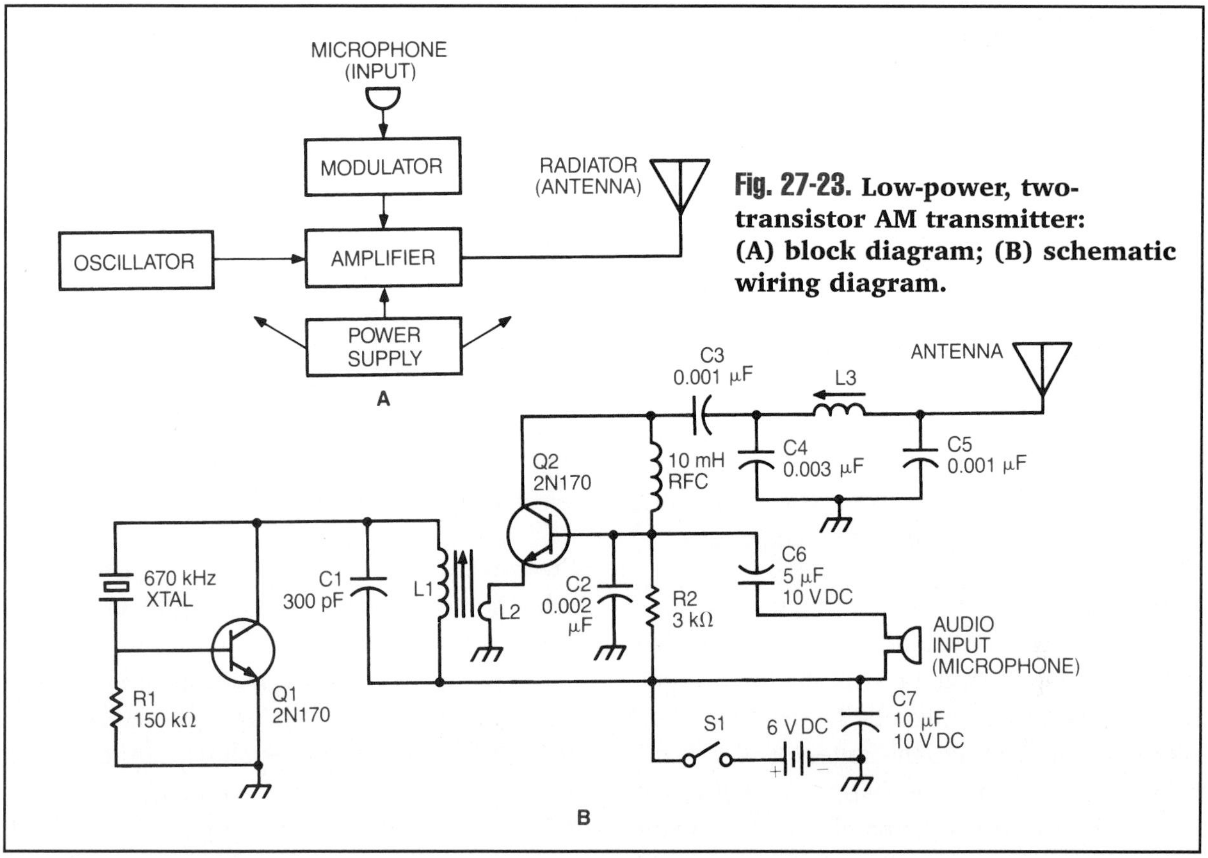

Fig. 27-23. **Low-power, two-transistor AM transmitter: (A) block diagram; (B) schematic wiring diagram.**

Amateur Radio

Anyone interested in amateur radio transmission can apply to the FCC for a license to operate a station. The FCC provides the rules and regulations for the amateur radio service. There are five classes of operator's licenses: novice, technician, general, advanced, and amateur extra. Each class of license requires a written examination that covers, for example, FCC rules and regulations, operating procedures, and radio theory. If you are interested in becoming a licensed operator, you should obtain the latest FCC rules and regulations governing the amateur radio service.

AM Radio Receiver

In an attempt to improve radio reception, Edward H. Armstrong, an American, invented a special circuit called the *superheterodyne circuit* in 1918. This circuit helped make radio receivers more selective and more sensitive. A radio receiver is *selective* when it can pick out the modulated carrier wave from the desired broadcasting station and reject all the other

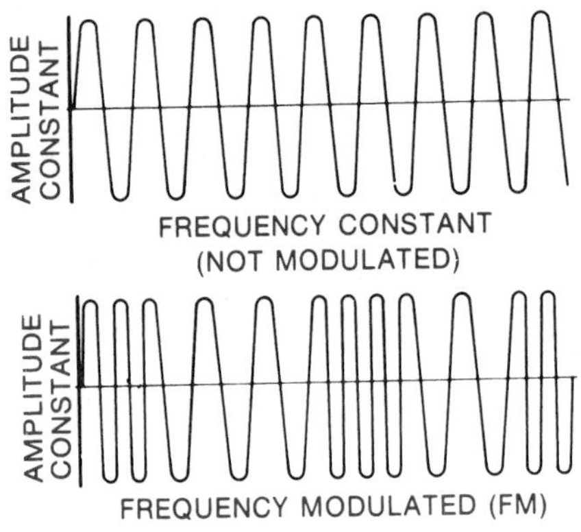

Fig. 27-24. **The effect of frequency modulation.**

carrier signals reaching its antenna. A radio receiver is *sensitive* when it can receive weak radio waves or signals.

The superheterodyne circuit has a mixer, or converter, stage and an oscillator stage called a *local oscillator.* The local oscillator generates a signal that has a higher frequency than the carrier signal. The two signals are combined in the mixer stage, resulting in an **intermediate frequency (IF)**. This process is known as **heterodyning**. In the United States, AM superheterodyne radio receivers operate at an intermediate frequency of 455 kHz. The block diagram in Fig. 27-25 shows the electronic stages in a transistorized AM superheterodyne receiver.

Antenna

The antenna of a radio receiver may be compared to the secondary winding of a transformer. The electromagnetic fields of the radio waves radiating from the antenna of a transmitter cut across the receiving antenna, inducing voltages in the antenna circuit. These voltages represent the audio information with which the carrier wave has been modulated. The voltages are applied to the tuning circuit of the radio receiver. Since these voltages "carry" information, they are called **radio frequency (RF) signals**, or radio signals.

Tuning Circuit

The tuning circuit for a superheterodyne radio receiver is shown in Fig. 27-26. The circuit consists of variable capacitor C and antenna coil L1. The variable capacitor is used to tune, or adjust, the circuit to different frequencies. When the tuning circuit is adjusted to the frequency of a station, the circuit is said to be in resonance with the frequency of that station. The circuit now offers the least impedance to signals of the station carrier frequency. However, it offers a high impedance to carrier signals of other frequencies, causing them to be rejected by the circuit.

Converter

In the converter, or oscillator-mixer, incoming RF signals are heterodyned to the intermediate frequency of 455 kHz. In most areas, several stations broadcast simultaneously. For example, three stations, each transmitting with its own frequency of 630 kHz, 980 kHz, and 1,500 kHz, could be tuned in on the receiver. However, no matter what the tuned radio frequency is, the intermediate frequency is always 455 kHz.

IF Amplifier

Because the IF amplifier amplifies only one frequency (455 kHz), it can be highly selective

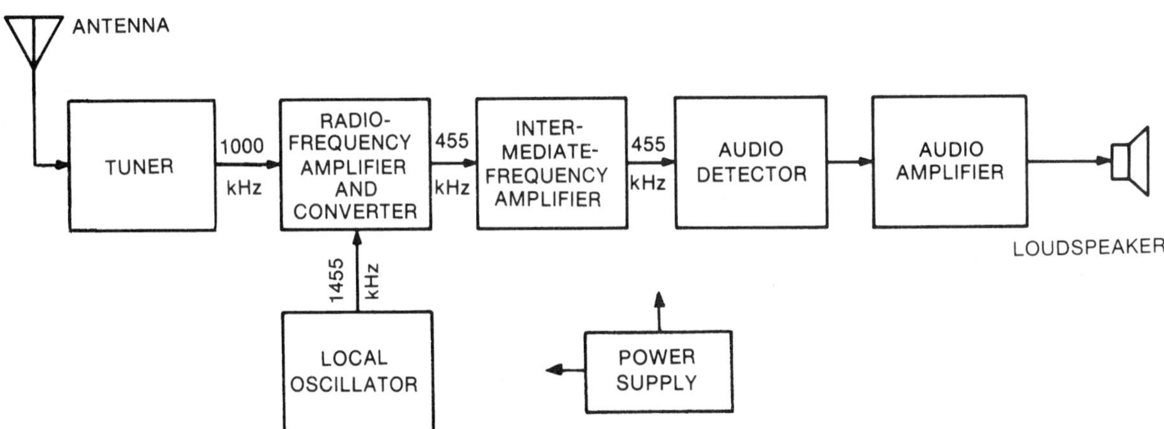

Fig. 27-25. Block diagram of an AM superheterodyne radio receiver.

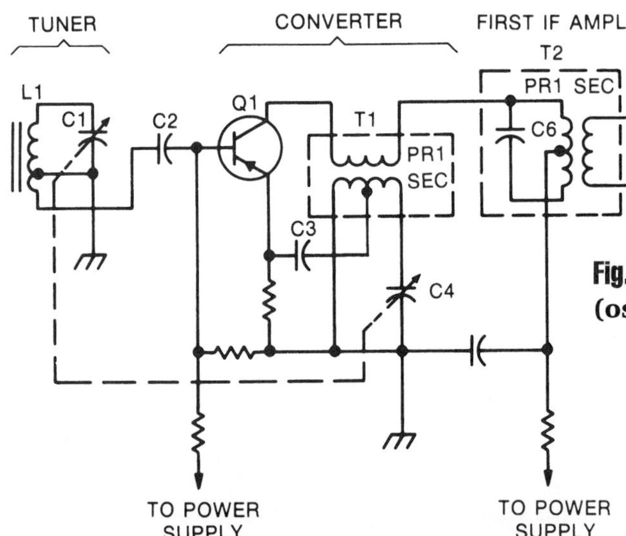

Fig. 27-26. **Tuner and converter (oscillator-mixer) circuits.**

and efficient. The IF signals are coupled to their stages by tuned transformers such as the first IF transformer T2 in Fig. 27-26. Most receivers have two IF amplifying stages before the audio signals are detected by a diode.

Audio Detector and AVC Circuit

The detector CR1, a semiconductor diode, removes the negative-going parts of the IF signals coming from the IF transformer T3 (Fig. 27-27). The audio output appears across the volume-control resistor R . It is coupled by capacitor C to the audio driver stage Q4. The amplified audio signals are next fed into a push-pull amplifier and then to the loudspeaker.

The *automatic volume control* (*AVC*) provides high gain for weak signals at the incoming IF stages. It also reduces the strength of strong signals. For example, when a strong audio signal is detected, the voltage across C becomes more positive. The positive-going AVC signal is fed to the incoming IF stages and reduces their gain. The opposite happens when weak signals are received.

Superheterodyne Receiver (FM)

The circuit stages of an FM superheterodyne receiver are shown in the block diagram of Fig. 27-28. These circuits have been discussed previously, except for the FM detector. The standard FM broadcasting band in the United States is 88 to 108 MHz. The intermediate frequency is 10.7 MHz.

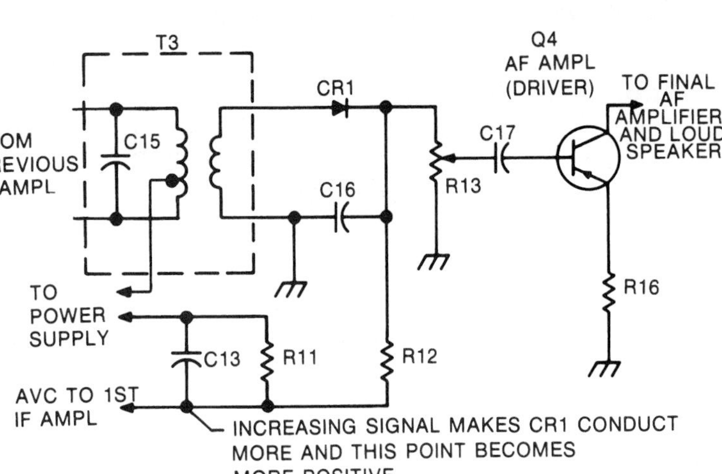

Fig. 27-27. **Detector, automatic volume control (AVC), and AF amplifier (driver).**

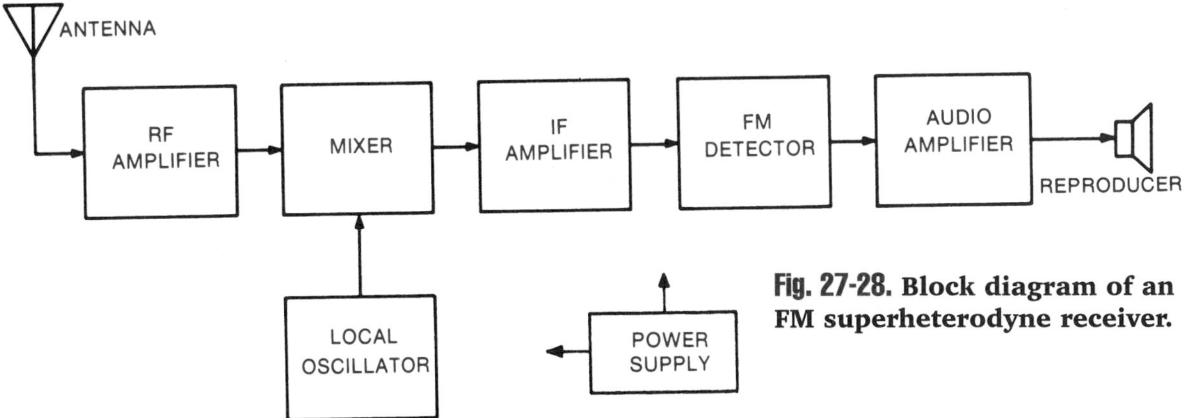

Fig. 27-28. Block diagram of an FM superheterodyne receiver.

FM Detector

Figure 27-29 is a schematic of an FM detector called a *ratio detector.* The previous IF amplifier stages provide the input to this circuit. The ratio detector recaptures the audio signals that are frequency-modulated on the carrier signal. In FM, the amplitude of the audio signal is proportional to the amount of frequency shift above and below the carrier frequency. Therefore, the ratio detector must be able to produce an output voltage that is proportional to the frequency of the carrier wave at any moment in time.

Advantages of FM over AM

Automobile ignition systems, lightning, and the switching actions involved in making and breaking circuit connections create noise in the form of radio waves. These waves cause amplitude-modulated carrier waves to be disrupted or distorted. These unwanted radio waves do not cause the frequency of a frequency-modulated carrier wave to change. Therefore, the FM radio receiver can reproduce audio that is not distorted by static.

TELEMETRY AND SATELLITES

Telemetry is the process of making measurements at one place and sending them to a distant place for recording and study. For example, weather satellites gather important information that help weather forecasters give advance warning of hurricanes and other bad weather. Figure 27-30 shows the elements of a weather satellite system.

Do communication satellites function as encoders, transmitters, receivers, carriers, decoders, or a combination of elements? Explain your answer.

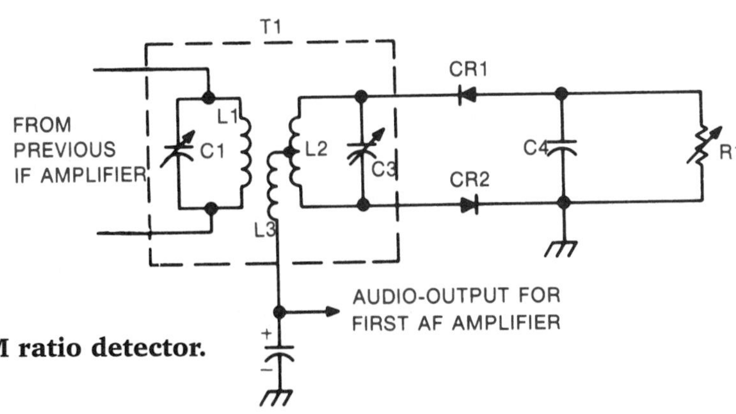

Fig. 27-29. FM ratio detector.

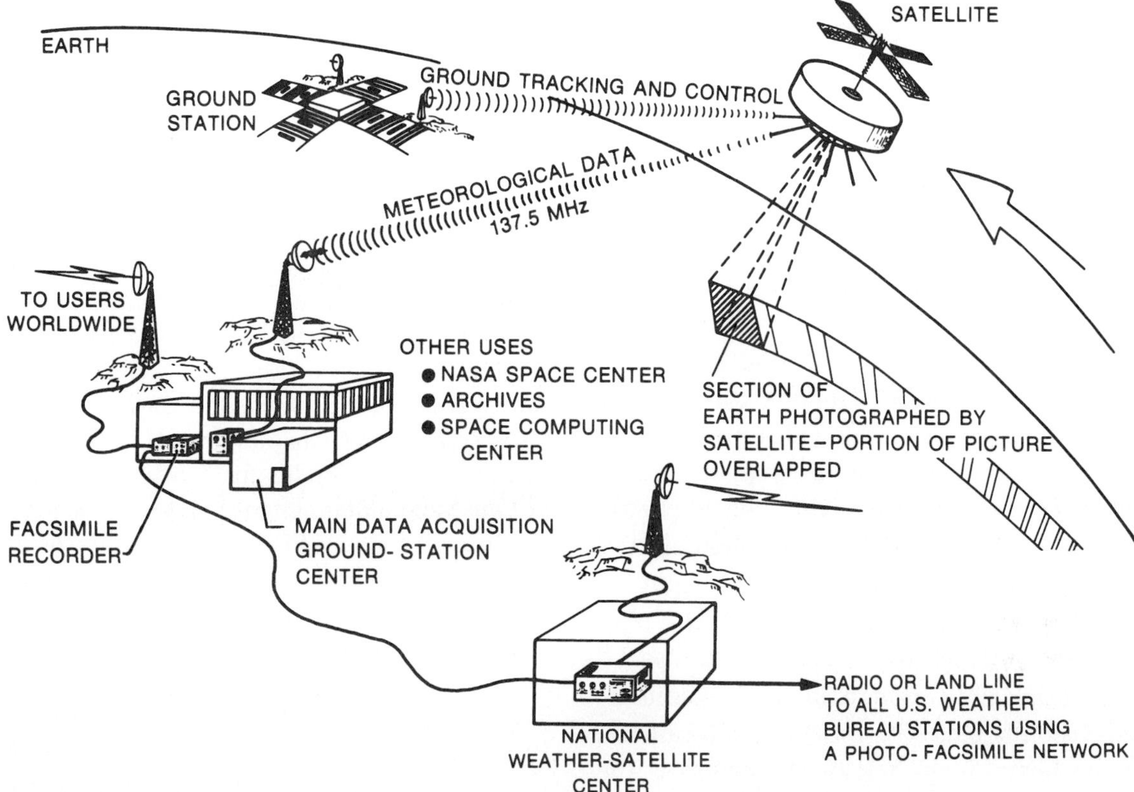

Fig. 27-30. Basic elements of a weather satellite data-acquisition system.

Because of the large amount of data transmitted by communication satellites and other spacecraft, a system of ground stations has been built to receive and pass on this information. This system is called the Space Tracking and Data Acquisition Network (STADAN). *Acquisition* means "the act of obtaining or gaining." The network sends all its data to the Communications and Computing Center at the Goddard Space Flight Center of the National Aeronautics and Space Administration (NASA) in Greenbelt, Maryland, for processing. From there, the data is sent on to the users.

TELEVISION

Of all the electronic devices available today, the television probably has more impact on the lives of people than any single electronic device. Although most current television systems use an analog video signal to reproduce images, many future systems will be digitized. One of the advantages of a digitized television signal is that it can be interfaced easily with most computers and other digital processing equipment.

How do radio and television transmission signals differ?

FASCINATING FACTS

Newton Minow, Federal Communications Commission chairman during President John F. Kennedy's administration called television a "vast wasteland" because he felt the programming had an overall dismal quality.

PLUG IN TO *Social Science*

Rating Television

Audience popularity determines how long commercially sponsored programs stay "on the air." A sponsor will pay for advertising time only if large audiences are watching. To survive, a show must hold a 30% audience share with reasonable consistency.

The most widely accepted rating of a show's popularity is conducted by the A.C. Nielsen Company. Nielsen collects and publishes two sets of statistics: the Nielsen Television Index *for network broadcasting and the* Nielsen Station Index *for local broadcasts.*

To collect data, Nielsen traditionally used an Audimeter attached to television sets in 1,200 selected households. The device registered how long a TV was on, and to which channel it was tuned. In 1987, Nielsen initiated the use of people meters. *These supposedly give a more accurate record of viewing habits. Each household member in the study punches in a personal code number on a remote-control device, which feeds viewing information to a central computer. Advertisers then have access to such information as the number of viewers, their age, sex, income level, and ethnic background.*

Generating a Television Picture

When light strikes certain substances, it produces an electric current. This photoelectric effect, discovered in 1873, eventually led to the first television system. In the photoelectric effect, electrons are emitted when light or other electromagnetic radiations strike a photosensitive surface. Two common photosensitive materials are cesium oxide and selenium.

Television begins with the television camera, and the heart of the television camera is the camera tube. The *image orthicon tube* and *vidicon* tube are two kinds in common use today.

Image Orthicon Tube

The basic parts of the image orthicon tube are shown in Fig. 27-31. The image of an object to be televised is focused by lenses onto a light-sensitive material inside the tube. As in a camera, this image is upside-down. The image causes electrons to be emitted from the light-sensitive material. The number of electrons emitted depends on the intensity of the light striking the material. A white dress on a performer, for example, causes more electrons to be emitted than a dark dress. The emitted electrons are attracted to the target, a thin glass plate. This plate is made so that the electron image is transferred to its opposite side, the side facing the scanning beam.

The areas of the target where more electrons have been emitted are less negative. In these areas, the scanning beam is repelled less. The areas where more of the electrons have been retained are more negative. In these areas, the scanning beam is repelled more. The result is a return beam of electrons that varies in density according to the details of the picture.

To reproduce an image, the scanning beam must scan, or view, the whole image. The scanning beam transfers the picture information to the multiplier section of the tube. It scans horizontally and vertically at the same time (Fig. 27-32). Therefore, each horizontal line of scanning slants downward a little. As each horizon-

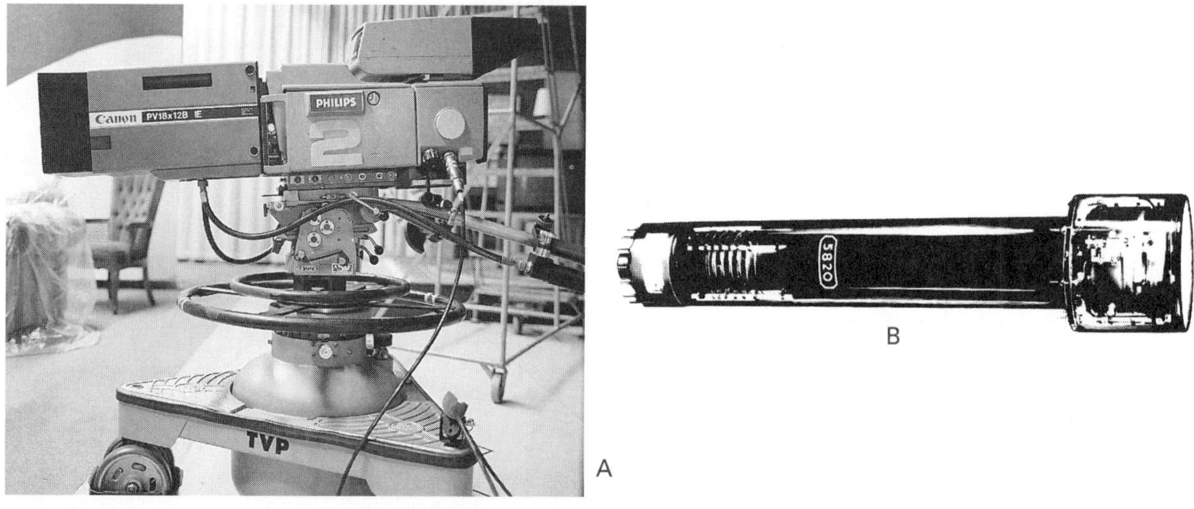

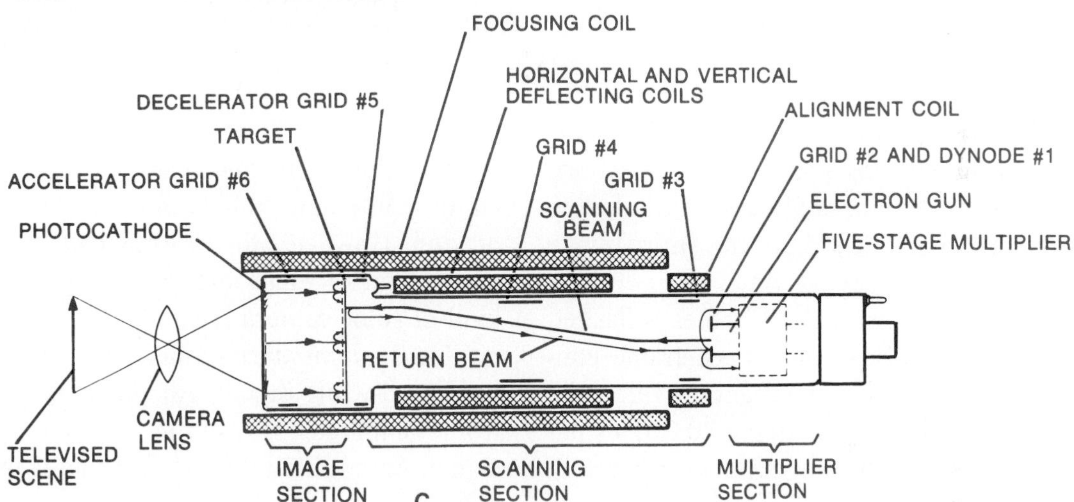

Fig. 27-31. (A) A modern television camera with a zoom lens; (B) a typical image orthicon tube; (C) schematic diagram of a typical image orthicon tube.

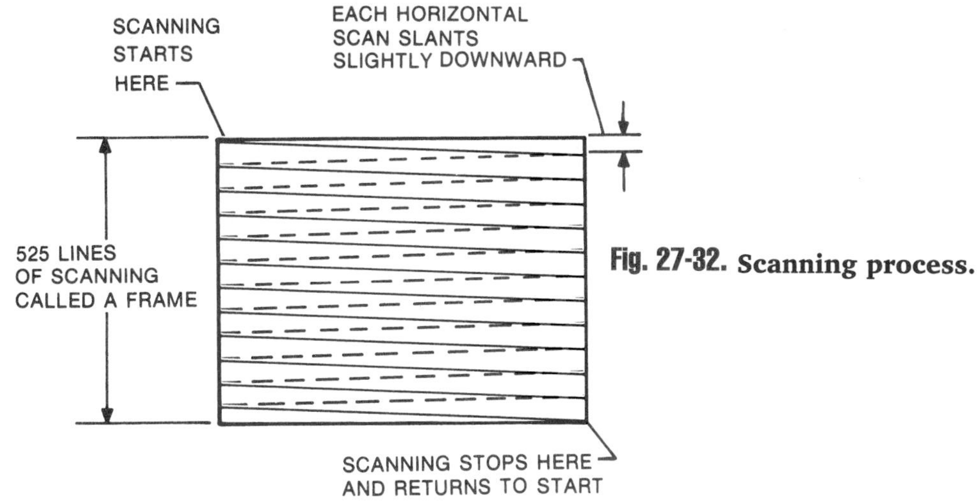

Fig. 27-32. Scanning process.

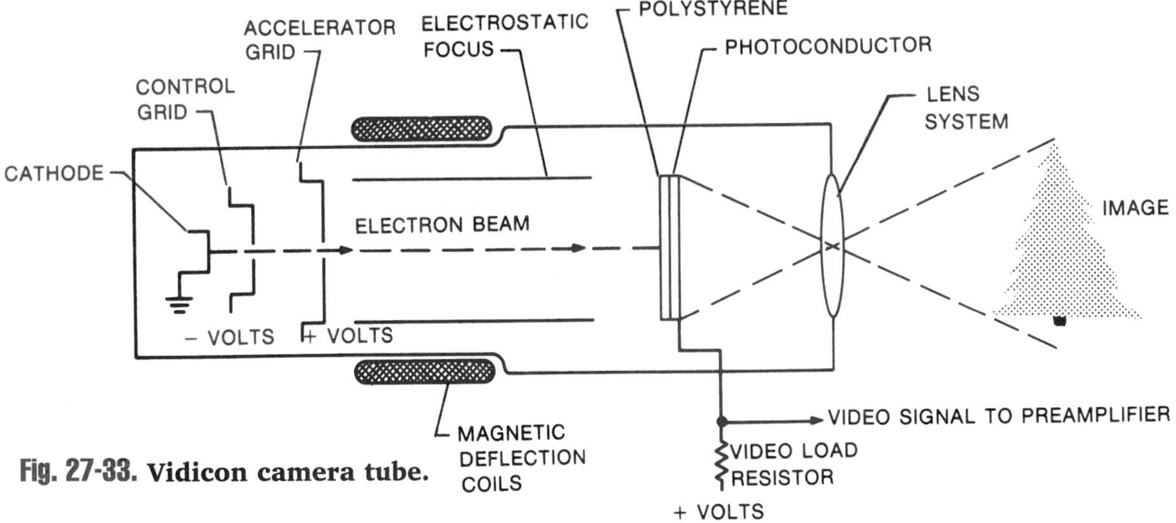

Fig. 27-33. Vidicon camera tube.

tal line is completed, the electron beam returns rapidly to the starting point of the next horizontal line. This point is always a little below the previous one. One scan of the entire image is called a *frame.*

In television, a series of 30 frames is flashed on the screen every second. Each frame consists of 525 lines. Thus, the scanning rate is 30 x 525, or 15,750 lines. Showing 30 frames per second creates the effect of continuous motion.

Vidicon Camera

Figure 27-33 shows a simplified schematic diagram of a vidicon camera tube. Unlike the image orthicon tube, the vidicon camera tube does not have a return beam.

When the camera tube is taking a picture, the electrically-sensitive material is exposed to the image. An emission of electrons converts the image into an electrically charged pattern on the surface of this material. The pattern is transferred to the polystyrene layer. There, it is scanned by an electron beam moving across and down the surface. The pattern of the electrical charge causes the electron beam, or current, to vary. This variable current forms the video signal. The video signal voltage appears across the load resistor.

Color Camera Tubes

To produce a *monochromatic,* or black-and-white, television signal, only one camera is needed. However, to transmit color, three or more camera tubes are used. One common kind of color camera has four tubes. One tube produces a *luminance* signal that is similar to a black-and-white signal. All the light reflected from the image is transmitted as the luminance signal. An additional optical system is used for color. The light is filtered so that only red colors can be seen by the red camera tube. Only green colors can be seen by the green camera tube, and only blue colors can be seen by the blue camera tube. Before transmission, the four video signals (luminance, red, green, and blue) are mixed together. Black-and-white television receivers accept the luminance signal. Color receivers separate the red, green, and blue signals and reproduce them on their screens (Fig. 27-34).

Charge-Coupled Devices

A **charge-coupled device (CCD)** is a solid-state digital imaging system that uses an array of chargeable devices much like capacitors. When the array is exposed to light, the picture elements, or **pixels,** of the array are charged.

The level of charge reached by each pixel is proportional to the level of light it receives from the image. The charge is represented by a number that ranges from 0 (no light at all) to 65,536 (very bright light). CCD arrays have as many as 500,000 pixels.

CCD imaging has an advantage over the tube systems in some situations because it is very small and rugged. These characteristics make the CCD camera ideal for "helmet cam"—a small video camera mounted on the helmets of some football quarterbacks and baseball catchers to add a different perspective to viewing these sporting events. Another use of the CCD array camera is for security. A CCD video camera can be made to look like any number of common items, such as a light switch, a smoke detector, or light fixture.

Transmission of Television Signals

The video signal of television is transmitted somewhat like the audio signal of radio. Both video and audio signals are varying direct currents. Each time the electron beam scans the image in the studio camera, the video signal is picked up and transmitted. Sound for a television program is transmitted at the same time as the picture, but on a separate frequency. In a television channel, therefore, there are two carrier frequencies: one for the picture and one for the sound.

Television and frequency-modulation broadcasting stations use high-frequency carrier waves. Television carrier waves are classified as *very-high-frequency* (vhf) waves (30 to 300 MHz) or as *ultra-high-frequency* (uhf) waves (300 to 3,000 MHz). Table 27-A lists the bandwidths and video and sound frequencies for selected television channels. There are 12 vhf channels: numbers 2 to 13. Channel 1 is not used because that frequency band has been assigned by the FCC to another kind of communication service. There are 70 uhf channels, numbers 14 to 83. These uhf channels provide for the expansion of television service, especially through the use of cable television.

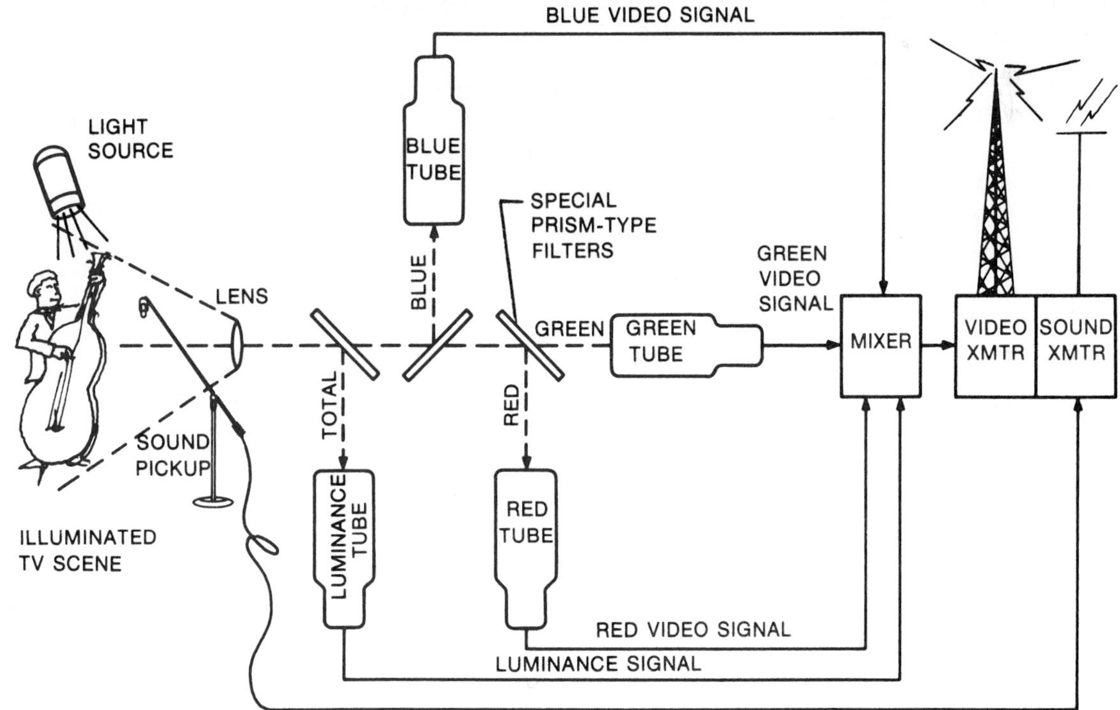

Fig. 27-34. Basic elements in television color transmission using a four-tube color camera.

Table 27-A Examples of Video and Audio Frequencies Used in Television Broadcasting

Maximum Bandwidth	Frequency (MHz)	Nature of Service	Channel Number	vhf or uhf
66-72	67.25 71.75	Video Sound	} 4	}Very-high-frequency (vhf) range
180-186	181.25 185.75	Video Sound	} 8	
204-210	205.25 209.75	Video Sound	}12	
506-512	507.25 511.75	Video Sound	}20	
542-548	543.25 547.75	Video Sound	}26	}Ultra-high-frequency (uhf) range
818-824	819.25 823.75	Video Sound	}72	

Because of the high frequencies used in television, the waves take on the characteristics of light waves. As a result, television broadcasting stations are limited to straight-line paths (Fig. 27-35). For long-distance transmission, relay towers or satellites are used. Coaxial telephone cables are also used for long-distance transmission. At the local station, the programs are transmitted over the air to local viewers.

Television Receiver

The television receiver and the radio receiver are alike in many ways. Both television and radio use the superheterodyne principle, intermediate frequencies, radio frequencies, amplification, and mixer and detector circuits. Figure 27-36 is a block diagram of a color television receiver. Black-and-white television receivers are similar, but they do not have the circuits to process the color signals.

Documenting an Invention

Lewis H. Latimer (1848-1928), a drafter and inventor, assisted Alexander Graham Bell (1847-1922) in illustrating his creation of the telephone.

Latimer, born in Boston, was the son of an escaped slave. He served in the Union navy in the American Civil War. Latimer became friends with Bell while working at a patent law firm near the school where Bell was experimenting with the telephone. Latimer drew each part of the telephone as Bell improved it, showing how the particular part worked. Bell was granted the patent for the telephone in 1876, when the device and its drawings were complete.

In 1880, Latimer went to work with the quickly developing electrical technology. In 1881, he and a coworker patented an improved method for bonding carbon filaments for the light bulb.

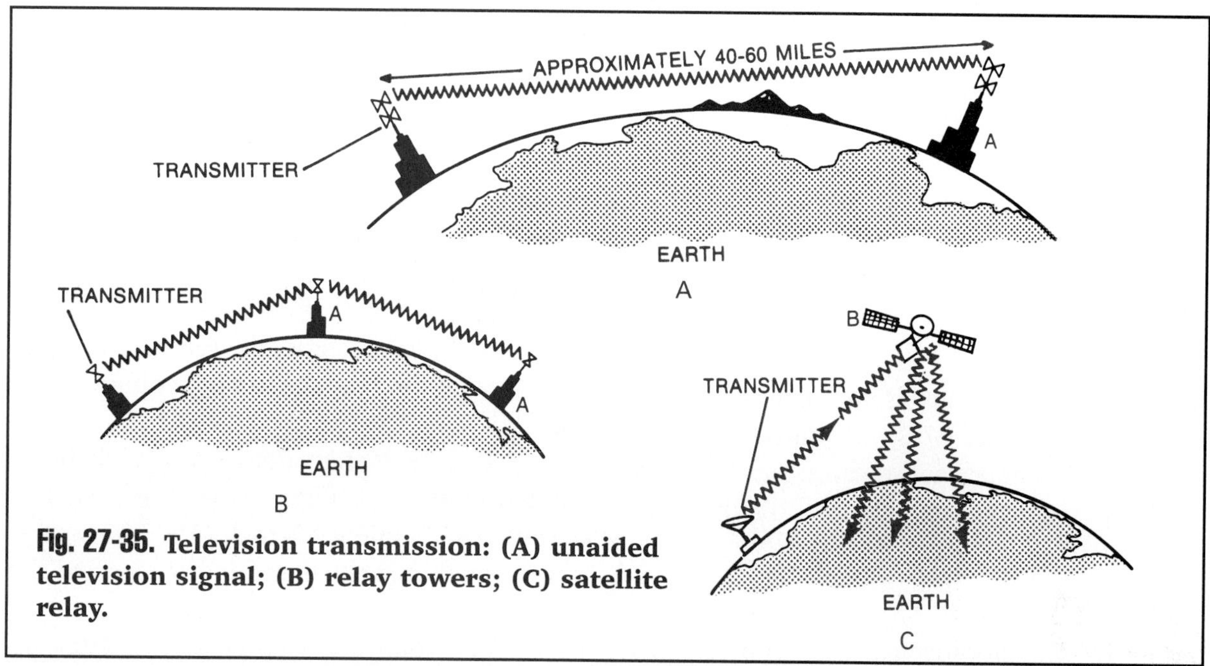

Fig. 27-35. **Television transmission: (A) unaided television signal; (B) relay towers; (C) satellite relay.**

Many televisions also provide stereo capabilities, and most have a remote-control device that allows you to scan through all the available channels. The remote control has an infrared (IR) LED that is pulsed at about 40 kHz by the internal circuitry of the controller. Each button sends a specific digital code to an IR receiver on the front panel of the television. The IR receiver sends the code to the controlling circuitry.

Television Antenna

Television antennas pick up both sound and picture signals. One of the simplest antennas

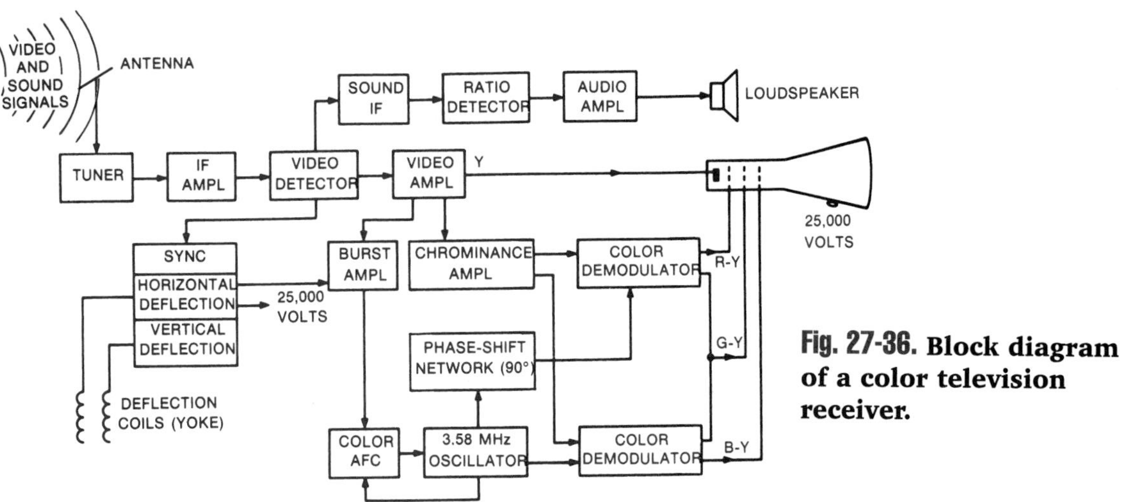

Fig. 27-36. **Block diagram of a color television receiver.**

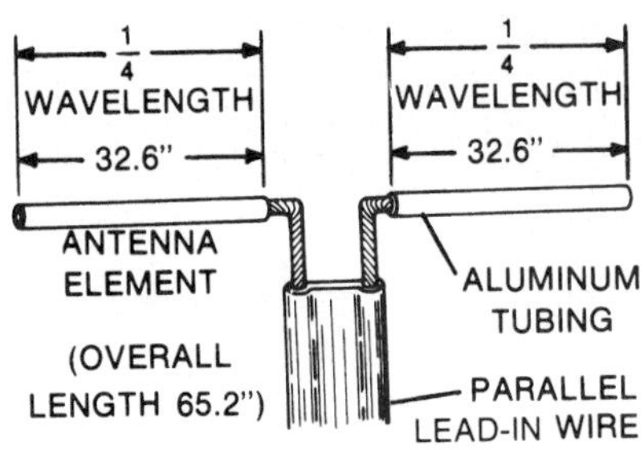

Fig. 27-37. Dipole antenna for channel 6.

for television reception is the *dipole*, or two-pole, antenna (Fig. 27-37). Connecting wires leading to the television receiver are fastened to the open ends in the middle of the dipole.

Tuner

The tuner of a television receiver is controlled by a microprocessor that also controls many other operations in the receiver. The microprocessor registers which channel you want, gets the necessary data from memory for that channel, and changes circuitry to tune to the desired channel.

Amplifiers

As in a superheterodyne radio receiver, television receiver circuits have radio frequency amplifiers, IF amplifiers, and mixer stages. The block diagram of a color television receiver shown in Fig. 27-36 has a video amplifier in addition to the audio amplifiers. The video amplifier increases the power of the video signals. Black-and-white television receivers are much the same, but do not have the circuits for the burst and chrominance amplifiers.

Sound Circuits

The FM sound circuit contains an IF amplifier, a ratio detector, and an audio amplifier. The IF amplifier amplifies the sound signals and increases the receiver's sensitivity and selectivity toward them. The ratio detector separates the audio component in the frequency-modulation system. The audio amplifier increases the audio signal enough to operate a speaker.

High-Voltage System

A very high voltage is needed to light a television screen. A system commonly used in television receivers is the *flyback high-voltage system*. This system uses part of the horizontal-deflection circuit to develop a high voltage. As the horizontal sweep flies back, or returns, to begin another pass across the television screen, its magnetic field induces a high voltage in the coil.

This flyback system does away with bulky transformers and large capacitors. This is because the frequency is high (15,750 Hz) and only a small current is needed to operate the television picture tube. Other electric elements in the high-voltage system rectify the alternating sweep circuit and smooth any variations in the circuit. A typical color television receiver applies 27,500 V to the picture tube.

Television Picture Tube

The television picture tube reproduces the image on the target of the camera tube in the television broadcasting station. Many of the elements in the television picture tube are similar to elements in the camera tube. The television picture tube produces an electron beam that scans the face of the tube. The scanning system uses either deflection plates or magnetic coils to deflect the beam.

Synchronizing and Sweep Circuits

The *synchronizing,* or *sync,* circuit receives the sync pulses from the televised picture signal. The *sweep,* or *deflection,* circuit sweeps the television picture-tube beam in synchronization with the camera-tube beam. Like the camera tube, the picture tube has vertical- and horizontal-sweep plates, or coils. These plates sweep the beam across and up and down the chemically-coated screen to make the picture.

The horizontal-sweep circuit returns the picture-tube beam to its starting point. This return sweep, called the horizontal retrace signal, is *blanked.* That is, the retrace signal cannot be seen by the viewer.

Several controls are used to adjust the operation of the sync and sweep circuits. The horizontal- and vertical-hold controls adjust oscillators within the range of the sync pulses. The height and width controls adjust the amplitude of the sweep. Other controls adjust the horizontal or vertical effect of the sweep to prevent distortion.

Electron Beam

The electron beam is actually a stream of electrons modulated according to the video signals placed on the control grid. The light intensity on the television screen depends on how many electrons strike it and the speed with which they do so. The greater the number and the higher the speed of the electrons, the greater the light intensity.

Tube Screen

On the inside of the face of the television picture tube is a coating of phosphor. This is a material that glows whenever a beam of electrons or other radiated energy strikes it. Actually, this coating continues to glow after the electron beam has passed. A material that is able to do this is said to be phosphorescent. The glow remains for only a short time, but it keeps the picture from appearing to flicker.

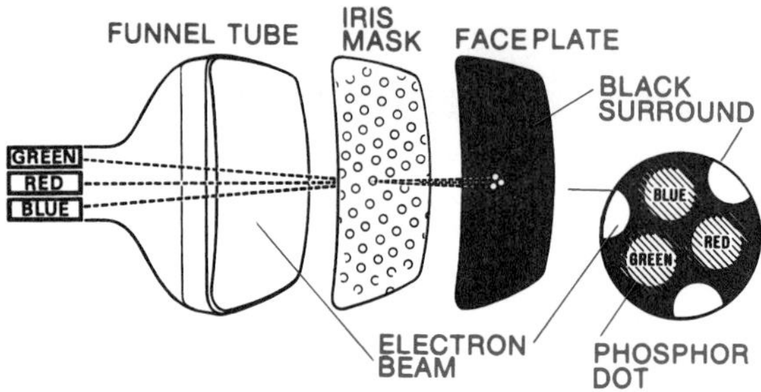

Fig. 27-38. Color television picture tube.

Table 27-B Relationship of Pixels to Screen Resolution in Fig. 27-39.

Photo No.	Resolution (Pixels)
1	512 x 480
2	256 x 240
3	128 x 120
4	64 x 60
5	32 x 30
6	16 x 15

Color Picture Tube

In one kind of color television picture tube, the video signals (green, red, and blue) are beamed to excite 1,350,000 tiny phosphor dots (Fig. 27-38). When the dots are struck by an electron beam, they glow green, red, or blue.

Digital Television and HDTV

The photos shown in Fig. 27-39 represent a new type of television picture. The photos are created using pixels. You can visualize a pixel as one square on an ordinary sheet of graph paper. As the vertical and horizontal lines on the graph paper become closer, the squares become smaller. Also, more squares fit into the same amount of space. Table 27-B shows the

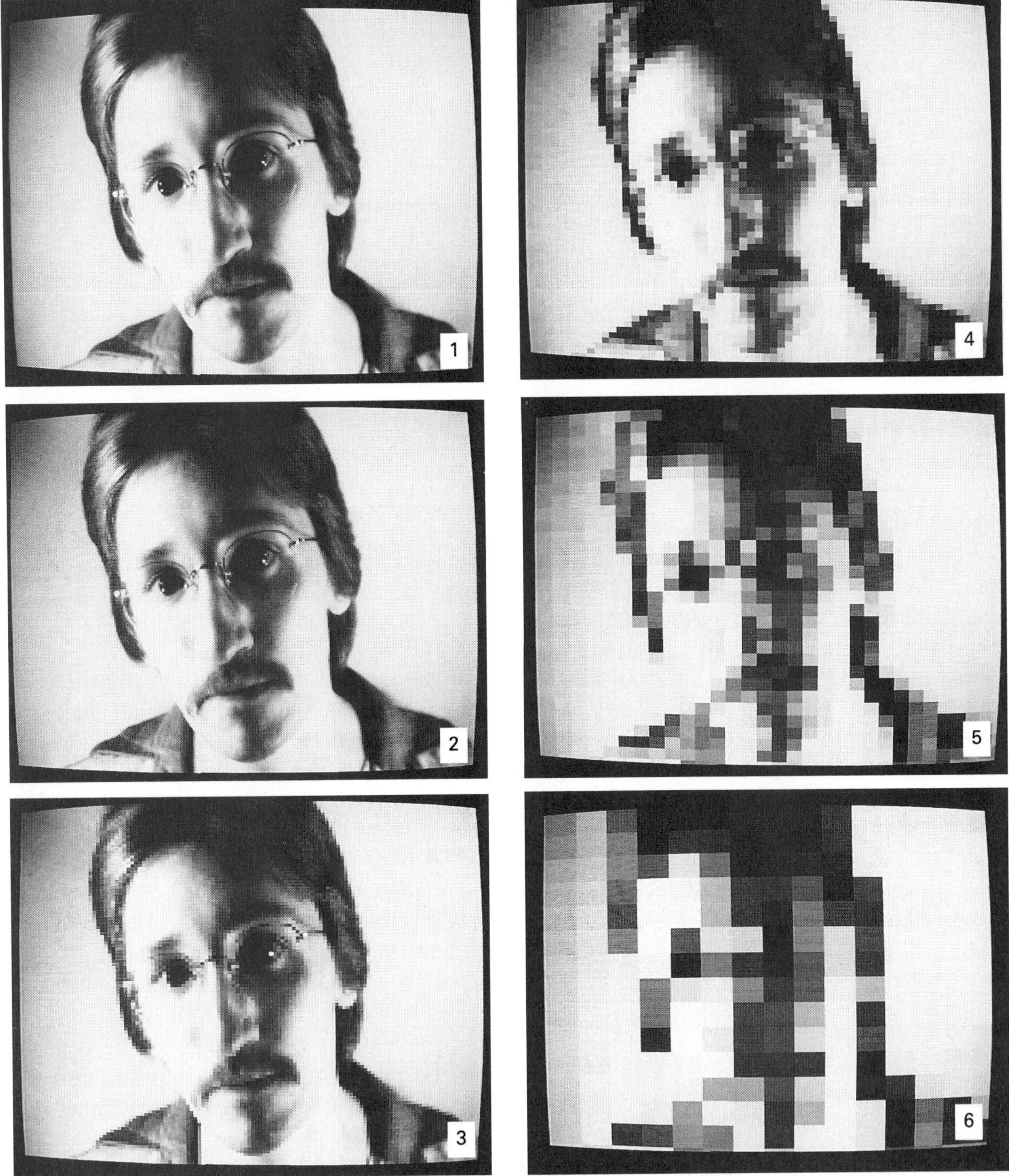

Fig. 27-39. Effect of the number of pixels on the resolution of a TV screen.

number of pixels used to create each of the photos in Fig. 27-39. As the number of pixels increases, the resolution of the picture improves. **Resolution** is a measure of an image's clarity and the amount of detail it shows. In Fig. 27-39, Photo No. 1 has the highest resolution, and Photo No. 6 has the lowest.

In a digital television picture, each pixel has its own binary code. The device that creates the pixels is the video digitizer. The digitizer also detects the intensity, or brightness, of the pixels.

High-definition television (HDTV) is one type of digital television that promises to improve the quality of television images. In traditional television, a resolution is limited to 525 horizontal lines per frame. HDTV is defined as any television system that has more than 1,000 lines per frame. Some systems can achieve even higher line resolutions. As this technology continues to evolve, it will provide viewers with images that have the clarity of a 35-mm camera. It will also have CD-quality audio and will enhance other technologies, such as on-demand movies and live conferencing.

Cable Television

Cable television was originally designed to bring television signals into communities located too far from the television station to receive signals any other way. It was also used in communities in which television signals are blocked by deep valleys or high mountains. The antennas of cable systems are located in areas with good reception. Then the signals are distributed by cable to subscribers for a fee. Today, cable television is widely used by people who could receive local channels without cable. They subscribe to cable television to increase the number and variety of channels they can receive.

Most cable companies distribute television signals through coaxial cables. These are strung over telephone poles or placed underground. Some systems have the capacity for two-way communication (Fig. 27-40). This allows subscribers to shop for merchandise, conduct banking business, reply to community surveys, and have utility meters read.

TVRO Satellite Systems

Television receive-only (*TVRO*) electronic systems allow people to view broadcast transmissions directly from satellites. Figure 27-41 illustrates the basic concepts involved in this process.

The TVRO systems use a satellite that is placed in a **geostationary orbit**, which means that the satellite, situated 22,300 miles (35,888.3 km) above the equator, revolves at the same speed as the Earth. Therefore, to an observer on Earth, the satellite appears to remain fixed in space, or stationary. Because the satellite is stationary with respect to a position on Earth, the dish-shaped antenna of the receiver can be aimed directly at it to receive its signals. You have probably seen a typical "satellite dish" in people's yards. These dishes are about 8 to 10 ft. (2.4 to 3.0 m) in diameter. However, a new set of more powerful, direct broadcast satellites (DBS) have been placed into orbit. This has enabled the industry to use receiving dishes of about 2 ft. (0.6 m) in diameter.

The satellites function as relay stations in the sky. Television programs that originate on Earth are converted into microwave signals at frequencies around 6 GHz. These signals are then transmitted in a narrow beam to the satellite. The transmission of the beam from Earth to the satellite is called an *uplink*. The satellite

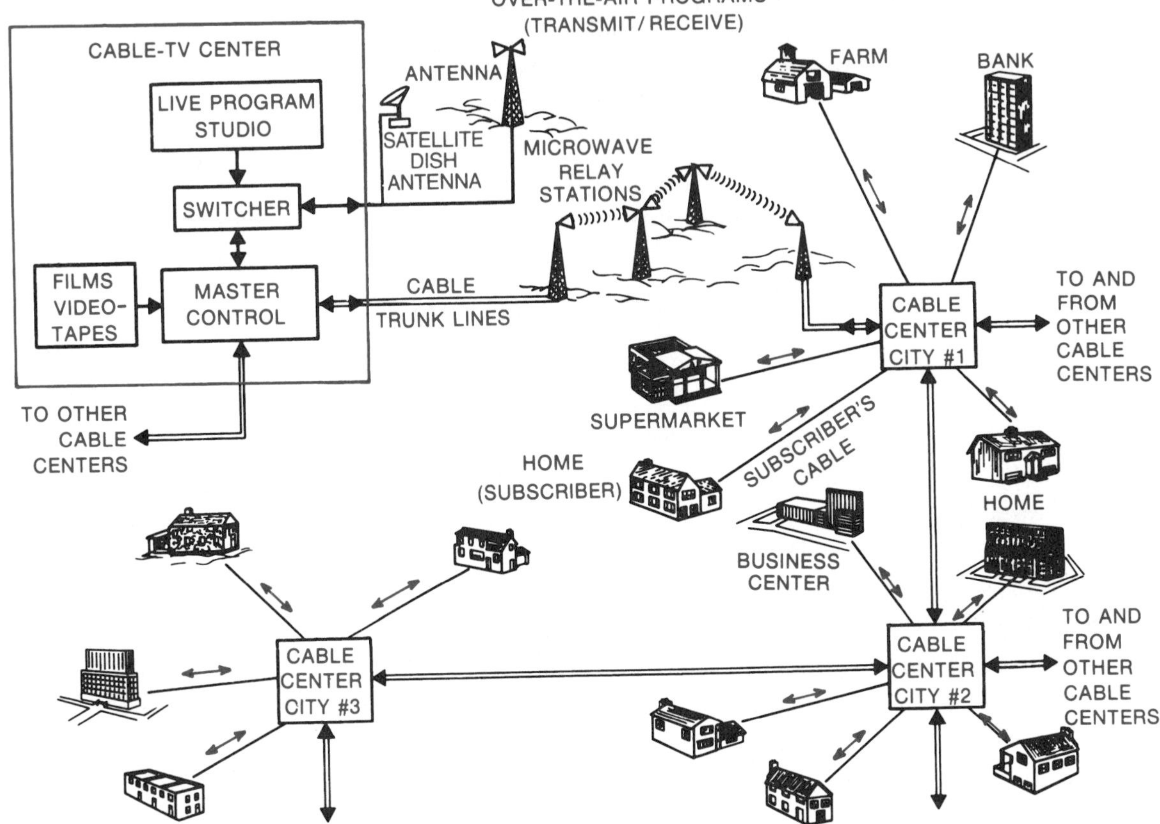

Fig. 27-40. A cable television two-way communication system.

receives the signals and passes them to a transponder. The transponder changes the signal frequencies so that they have a broader beam. The satellite amplifies the signals and transmits them back to Earth as a *downlink*. The downlink beam is transmitted at a different frequency than the uplink beam. For example, a system that has an uplink at 6 GHz may have a downlink at 4 GHz.

Television Recycling

What happens to older televisions as they become worn out or are replaced with more modern systems? People used to throw them away. However, some of the components of televisions can be dangerous to the environment. One solution is to recycle old televisions. Recycling companies have defined several different grades, or sets, of specifications for recycling televisions. In the "Whole VDT/TV Scrap" grade, for example, televisions, computer monitors, and video display terminals are accepted if they are not crushed or broken. The recyclers can reuse the picture tube or CRT. Other grades include specifications for the glass from television screens. Check with recyclers in your area for availability of television recycling.

OTHER VIDEO SYSTEMS

As video technology has become more advanced, a large number of video devices have become available. Videocassette recorders and camcorders are only two of the products made possible. Originally designed for entertainment,

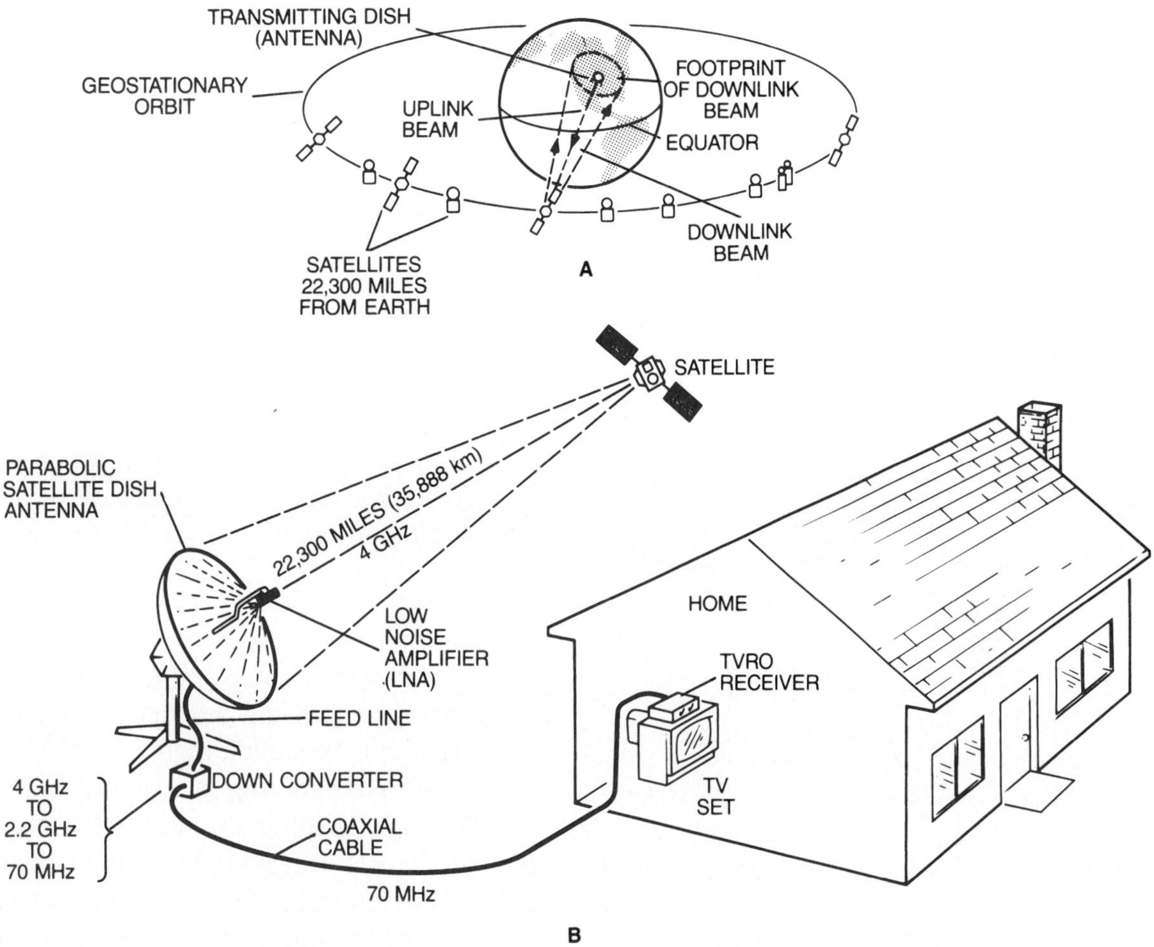

Fig. 27-41. Satellite communication: (A) satellites in geostationary orbit; (B) elements of a television receive-only system.

these and other systems are now widely used in business. Video conferencing is an example of a technology developed for business use. As it becomes less expensive, people may want to use it for everyday telephone calls.

Design a concept for a video system that combines two or more of the video systems discussed in this chapter.

Videocassette Recorders

Because videocassettes are playable on a home television set, the videocassette recorder (VCR) has moved movies into the home. Instead of waiting in line at the movie theater, you can rent a prerecorded movie on a videocassette. A videocassette is a special cartridge with two reels that contain a magnetic tape. The tape is designed to record and play back a television signal.

The VCR and video camera make it possible for people to create their own "instant replay" home video entertainment. Most home VCRs

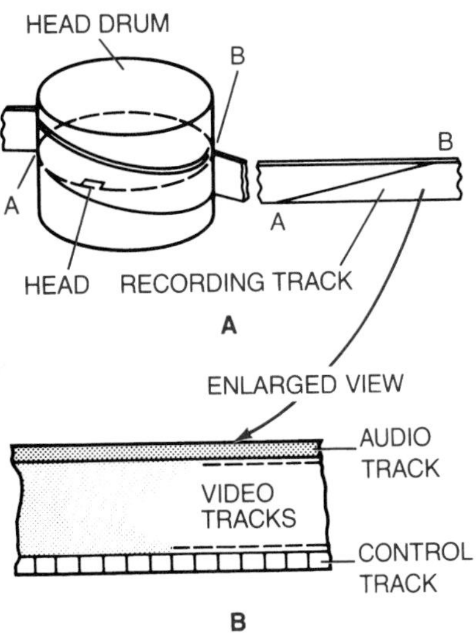

Fig. 27-42. Videocassette recorder principles: (A) helical scan recording track; (B) simplified tracks on a helical scan system.

use the helical scan method of recording videotapes (Fig. 27-42A). The tape is wrapped around a rotating drum that contains one or two heads. The heads travel at an angle to the direction of the tape travel. By selecting a cylinder head diameter, degree of tape wrap, tape speed, and cylinder head speed, it is possible to record a track long enough (across the tape at an angle—line AB in Fig. 27-42A) to hold a complete television field.

Figure 27-42B illustrates a simplified video track with sound and control signals. The control signals are needed to maintain proper timing so that a servomotor can adjust for any deviation from the desired performance.

Camcorders

The block diagram in Fig. 27-43 illustrates the various electronic circuits in a *camcorder*, or portable video camera. The lens focuses the incoming light waves onto a pickup tube. The filters correct for the proper white light balance. Horizontal and vertical deflection circuits control the scanning process as the electron beam sweeps across and down the scene being televised.

The output video signal from the preamplifier is applied to the signal processing circuits. The red, green, and blue video signals are first separated from the luminance signal. These signals are color corrected and encoded. The luminance signal passes through two low-pass filters and eventually is mixed with the color signals and the horizontal and vertical synchronizing pulses. The final video output signal must meet the standards established by the National Television System Committee (NTSC).

Video Conferencing

As computers become more powerful and smaller video cameras are developed, video conferencing is becoming more and more popular. Video conferencing is a technology that allows people in different areas of the country to "meet" without leaving their desks. A basic video conferencing system consists of a computer equipped with a video camera and video processing tools, a microphone, and a sound card.

Video conferencing has many advantages for businesses. It is much less expensive than traveling to a central conference site. It also takes less time, which is a big advantage for busy executives and employees. Therefore, it allows people in a company to meet more often, which improves communication within the company. Video conferencing also makes it easier for people to share their knowledge and ideas. Design companies, for example, can show samples to clients across the country without spending travel money. This drives down the cost of designing a product, making the end product more affordable.

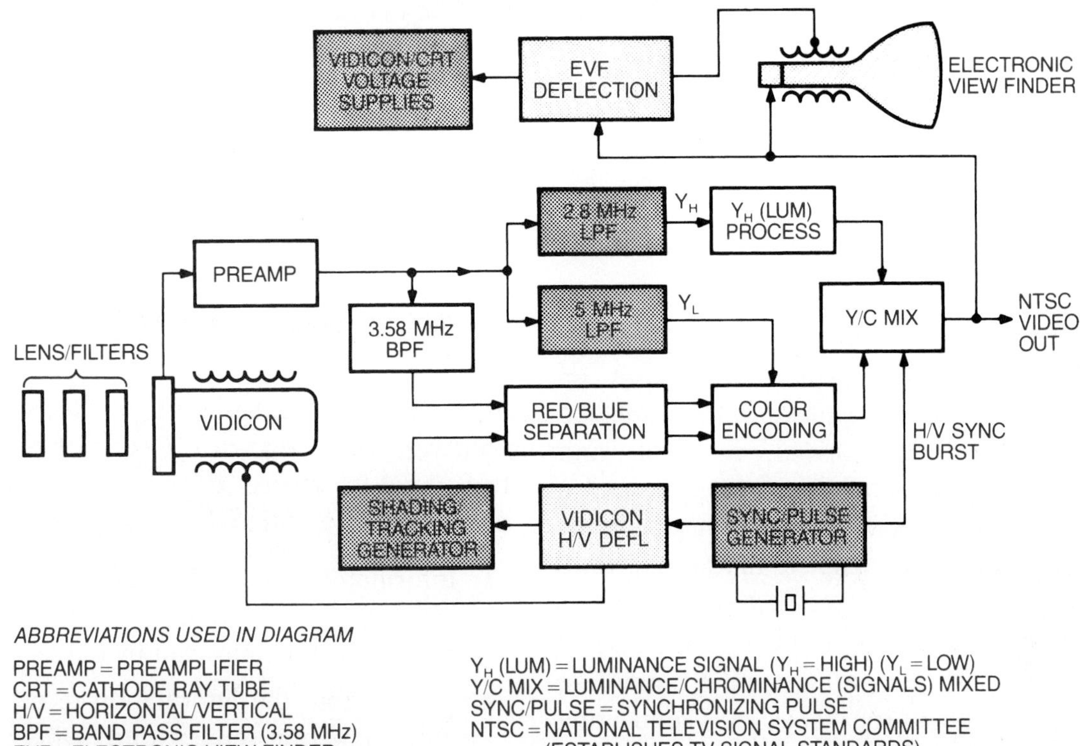

Fig. 27-43. Block diagram of a color camcorder.

Chapter *Review*

Summary

- The communication process includes an information sender; a message that is encoded, transmitted, carried, received, and decoded by an information receiver; and a feedback loop. A variety of devices employ electric energy to complete the steps in communication.
- Audio systems consist of microphones, a preamplifier, a three-stage audio amplifier, power output, speakers, and connecting cables. Volume and tone controls allow adjustment of the audio amplifier.
- Sound can be stored using many methods, including magnetic tape, compact disc, CD-ROM, and DVD. Telephone systems allow people to talk over great distances. Modems permit useful technologies such as fax machines and computer networks, including the Internet. New technology used in combination with the telephone has made portable communication such as cordless and cellular telephones possible.
- Radio transmissions can be accomplished by amplitude modulation (AM) or frequency modulation (FM).
- A television system includes a camera to generate the image, a receiver, a picture tube, and a sound system. Traditional television is an analog system. Digital televisions are being developed for greater resolution. Other video systems include videocassette recorders, camcorders, and video conferencing systems.

Review Main Ideas

Review this chapter's main ideas by writing, on a separate sheet of paper, the word or words that most correctly complete the following statements:

1. Information goes through the ______, ______, ______, and ______ processes before it can be received by the user.
2. ______ from the information receiver lets an information sender know whether the message has been received and understood.
3. The parts of a basic audio system are ______, ______, ______, and ______.
4. A ______ microphone has a diaphragm connected to a lightweight form with a small coil of wire wrapped around it.
5. Transformer coupling is used between the ______ and the loudspeaker voice coil in an audio-amplifier.
6. The most meaningful rating for comparing different amplifiers is ______ power output.
7. Three kinds of loudspeakers found in an audio system are the ______, ______, and ______.
8. A ______ network is used to split the input signals into frequency bands for speakers.

9. ______ is best for connecting the parts of an audio-amplifier system.

10. In an audio system, a basic tone-control circuit consists of a ______ and a ______ connected in series across the output of an amplifier stage.

11. The two common means of storing sound are ______ and the ______.

12. ______ has a special iron oxide coating that holds electromagnetically induced sounds.

13. In a compact disc player, it is the high rate of ______ that makes it possible to reproduce sound waves with very high fidelity.

14. A telephone transmitter changes sound energy into ______ energy.

15. In the telephone transmitter, an increase in pressure pushes the diaphragm against a small container of carbon granules, ______ the carbon's electrical resistance.

16. Pulsating electric currents generated in a telephone transmitter are called ______-frequency or ______ currents.

17. Humans can hear sounds that have a range of about ______ to ______ Hz.

18. The telephone (voice) channel or pass band (bandwidth) is about ______ Hz.

19. The ______ is an electronic device that provides interface between digital and analog signals.

20. The ______ is the world's largest computer network.

21. Radio communication requires a ______ and ______.

22. In a transmitter, the device that causes a stream of electrons to vibrate back and forth at high frequencies is called a high-frequency ______.

23. The wavelength of a radio wave depends on its frequency. The higher the frequency, the ______ the wavelength.

24. A radio wave must have a frequency of at least about ______ Hz before it can be used for communication.

25. A radio wave of a specific frequency assigned to a radio station is called a ______ wave.

26. During modulation in an AM system, the ______ of the carrier wave rises and falls according to the alternating current from the microphone.

27. When a radio receiver is able to pick out the modulated carrier wave from the desired broadcasting station, and reject all others, it is said to be ______.

28. A radio receiver is said to be ______ when it can receive especially weak radio waves or signals.

29. In an FM system, changes in ______ are controlled by the signal from a microphone.

30. The local oscillator of a superheterodyne radio circuit generates a signal that has a ______ frequency than the carrier signal.

31. The standard intermediate frequency used in a superheterodyne radio circuit is ______ kHz.

32. The automatic volume control (AVC) provides ______ gain for weak signals.

33. The intermediate frequency in FM superheterodyne receivers is ______ MHz.

34. The FM radio receiver can reproduce audio that is not distorted by ______.

35. Two common television camera tubes are the ______ and the ______.

36. In an image orthicon tube, a ______ transfers an entire image to the multiplier section of the tube.

37. Each frame on the television screen consists of ______ horizontal scanning lines.

38. In the vidicon tube, the video signal voltage appears across the ______ resistor.

39. All the light reflected from the image in a color camera tube is transmitted as the ______.

40. In a television channel there are two carrier frequencies: one for ______ and one for ______.

41. Television carrier-wave frequencies are classified as ______ and ______ waves.

42. There are ______ vhf and ______ uhf television channels.

43. A simple kind of television antenna with two poles is called a ______.

44. The television picture tube, like the camera tube, has an ______ that scans the face of the tube.

45. A charge-coupled device (CCD) is a solid-state ______ that uses an array of chargeable devices much like capacitors.

46. Both television and radio use the ______, amplification, and mixer and ______ circuits.

47. ______ is a measure of an image's clarity and the amount of detail it shows.

48. The coating on the inside surface of a television picture tube is a coating of ______.

49. Television signals are transmitted from Earth to a satellite in an ______; the satellite amplifies signals and transmits them back to Earth in a ______.

50. A tape from a videocassette recorder (VCR) has three tracks: video, audio, and ______.

Chapter *Review*

Apply Your Knowledge

1. Find out more about communication using the Internet and video conferencing and compare them. Which do you think is more practical? What are the advantages and disadvantages of each?
2. Explain the differences between CD-ROM and DVD technology. Do you think DVD will replace CD-ROM technology in the computer? Explain your reasons.
3. What is the difference between cordless and cellular telephone technology?
4. Describe three sources of noise and the affect each has on electronic communication.
5. In computer networks, what is the difference between an LAN a WAN, and a server?

Make Connections

1. **Communication Skills.** Prepare a script and production plan for a three-minute report to be placed on the local cable access channel. Your topic will be an activity in your class.
2. **Mathematics.** Using the standard speed of electromagnetic wave transmission through the atmosphere and space, calculate the time it would take for a communication signal to travel: (a) halfway around the world, (b) to the moon, (c) to Mars, (d) to a moon of Jupiter, (e) to Pluto.
3. **Science.** Research the process involved in developing color images on a charge couple device. Report your findings to the class.
4. **Workplace Skills.** Build a simple AM tuning circuit. Connect the output to an oscilloscope and demonstrate to your class what the signals look like and how many are in the air all the time.
5. **Social Studies.** In a written report, summarize the methods that people in different parts of the world developed to communicate over long distances prior to the use of electricity.

Chapter 28

Residential Wiring

OBJECTIVES

After completing this chapter, you will be able to:

- Name seven of the major elements of a basic residential wiring system.
- Identify the main elements found in a complete service entrance system for residential homes.
- Name three types of branch circuits in a house wiring system.
- Compare the difference between nonmetallic sheathed cable to armored cable used in a residential electrical system.
- Explain five advantages of remote control, low-voltage wiring.

Terms to Study

armored cable

electrical metallic tubing

general-purpose branch circuits

individual circuits

junction boxes

range circuit

remote control

service entrance

small appliance circuits

switch boxes

three-way switch

wiring devices

Do you know how safe the wiring is where you live? Do the lights dim when an electric motor or air conditioner starts? This chapter explains wiring safety and how residents receive electricity from the power company and how electricity is distributed throughout a building.

NATIONAL ELECTRICAL CODE

The electrical wiring system in a house or apartment consists of conductors, devices, and other electrical equipment such as blower motors and A/C condensers. Although this chapter discusses residential wiring, many of the same materials are found in the electrical wiring systems in other buildings.

The *National Electrical Code* (*NEC*) is a publication sponsored by the National Fire Protection Association in cooperation with the *American National Standards Institute* (ANSI). The NEC includes recommendations and guidelines on wiring materials and methods. Following these guidelines results in the safe use of electric energy for lighting, heating, communications, and other purposes. The code is reviewed and continuously revised every 3 years.

The *NEC* contains the basic minimum requirements to protect people and property from electrical hazards. Because of this, the *NEC* is very useful for developing safety standards throughout the country. These standards, as well as those of local electrical codes, are used by electricians, contractors, and inspectors. Using these standards means that all building wiring will be done in an approved way. When local governmental bodies adopt a code, its provisions become legally enforceable.

Inspect your home, office, or classroom. Can you find any wiring problems that might be code violations?

SERVICE DROP

A *service drop* of a residential wiring system is the part of the system that extends from a point outside the house to the nearest electrical company distribution line. A service drop can be installed either overhead or underground. An underground service drop is sometimes called a *service lateral*.

This line is fed from a distribution transformer. This device steps down the high voltage of a main power line to the 115 and 230 Vac used in the house in Fig. 28-1. The distribution transformer is mounted overhead on a utility pole, at ground level on a concrete pad, or underground in a vault (Fig. 28-2). The distribution transformer may supply electric energy to several houses.

Three-Wire Service

In order to supply two different voltages (115 and 230 V), the service drop is made of three conductors (Fig. 28-3). One of these, called the *neutral*, is often bare. The other two conductors are insulated.

Other Voltages

The 230-V lines brought into a house are used to operate major appliances, such as ranges, hot-water heaters, and air-conditioning units. Comparable voltages used by different electric power companies include three-phase 208 V, single-phase 220, and 240 V. The reason higher voltages are used for these purposes can best be explained by referring to power formulas.

Suppose, for example, that a 1725 W appliance operates at 115 Vac. The amount of current used by the appliance under this condition is $I = P/E$, or 1725/115 = 15 A. Now, suppose that the same appliance is operated at 230 V. The current is then 1725/230, or 7.5 A. The lower value of current (at 230 V) has two important advantages. First, the wires leading to the appliance can be smaller. This reduces the cost of installing them. Second, the lower current results in a smaller power loss in the wires. According to the formula $P = I^2R$, the loss is directly proportional to the square of the current.

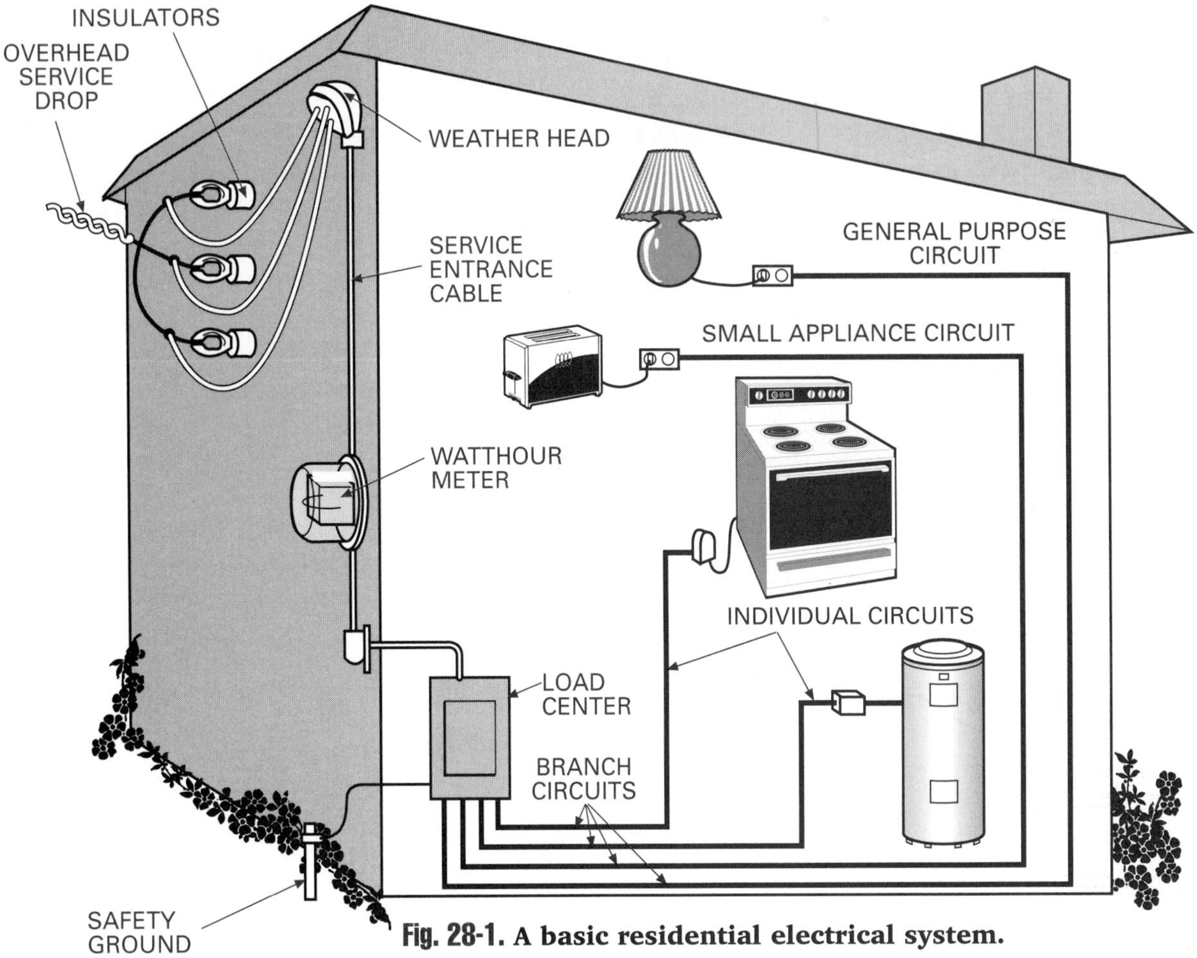

Fig. 28-1. A basic residential electrical system.

How much current is used by a 2800-W appliance that operates at 115 Vac?

SERVICE ENTRANCE

The **service entrance** includes all the parts of the assembly that bring electricity into a building. Its conductors extend from the point at which the service-drop conductors are attached at the house to the load center. The complete service entrance is generally thought to include the conductors, the watt-hour meter, and the load center. Usually the power company takes care of the wiring from the power line up to and including the meter.

Service-entrance Cables

The service-entrance cable has three conductors (Fig. 28-4). One of these, the neutral conductor, is bare. It is usually in the form of individual strands of wire. When the cable is to be connected to a terminal point, the strands of the ground conductor are twisted together to form a single-stranded conductor.

Fig. 28-2. **Distribution transformer mounted overhead on a utility pole.**

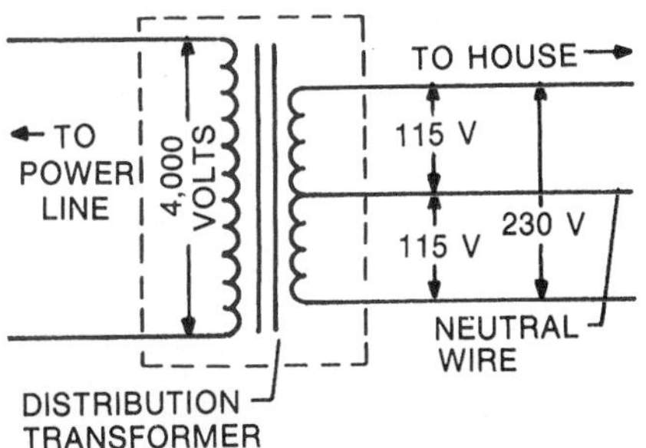

Fig. 28-3. **Voltages of a typical three-wire system used in homes.**

Watt-hour Meters

Watt-hour meters measure and record the amount of electric energy (in kilowatt-hour units) supplied to a building. The meter is usually, but not always, installed by and belongs to the power company.

LINKS

You may wish to refer to Chapter 6, which explains how to calculate a residential electric bill.

Load Center

Load centers, also called *fuse* or *circuit-breaker panels*, are the boxes from which electricity can be distributed to different locations in a house (Fig. 28-5). It contains the main breaker or fuses that can disconnect all the electrical service from a house. The load center also contains the fuses or circuit breakers that protect the branch circuits of the wiring system. The size of the load center is designated by the number of fuses or circuit breakers that can be placed into it, and by its load, or ampere, rating. The fuse panel shown in Fig. 28-5A is

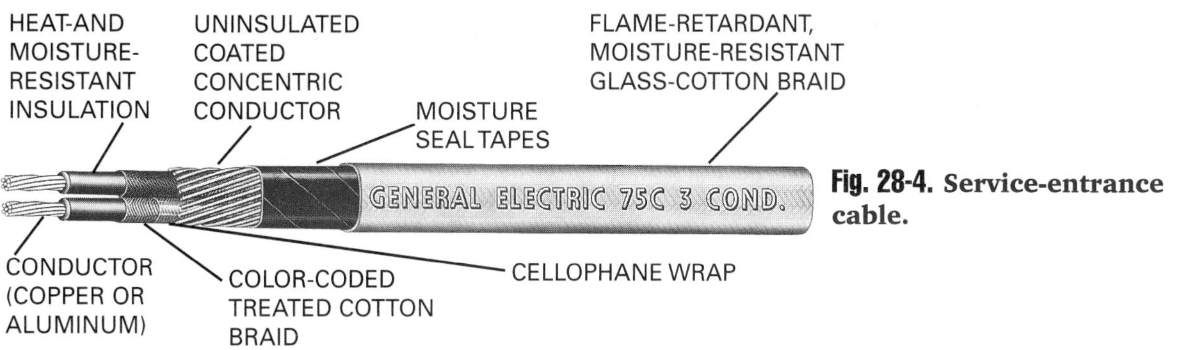

Fig. 28-4. **Service-entrance cable.**

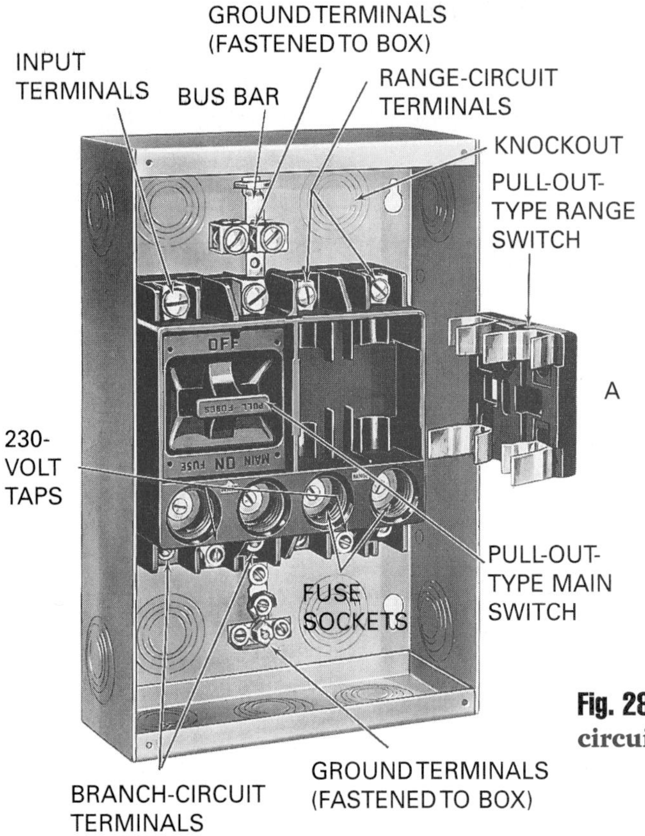

Fig. 28-5. **Load centers: (A) fuse panel; (B) circuit-breaker panel.**

rated for 60 amps. Larger load centers equipped with more fuse holders have ampere ratings of 100, 150, 200, or more amperes. The circuit-breaker panel in Fig. 28-5B is a newer style of distribution panel.

Service Grounds

To reduce the danger of shock and to protect against lightning, the bare and white conductors of the service entrance cable are grounded separately. This is done by connecting the white wires to the neutral *bus bar* in the load center. The bus bar is then grounded by being connected to a ground. The bare, or sometimes green, conductor is connected to a separate ground rod. This is the safety ground.

Explain the differences between a neutral wire and a safety ground.

BRANCH CIRCUITS

Branch circuits of a residential wiring system distribute electricity from the load center to the rooms of the house. The three common kinds of branch circuits are discussed below.

General-Purpose Circuits

General-purpose branch circuits are used for lighting and outlets. The outlets are intended to serve radios, television sets, clocks, and other small appliances used for food preparation (such as toasters, broilers, etc.). Lighting circuits are usually wired with number 12 (2.00-mm) or number 14 (1.60-mm) wire and are protected with 20-A or 15-A fuses or circuit breakers, respectively. As shown in Fig. 28-6, one wire of each general-purpose circuit is connected to the neutral bus bar of the load center. The grounded neutral wire is color coded white. The wire connected directly to the fuse or circuit breaker is generally black or red. This circuit wire is often called the *hot* wire. The grounded white wire of a two-wire, general-purpose circuit is never used as a safety grounding wire. When the cabinet or case of a product operated from such a circuit is to be grounded, a third grounding wire is used.

Small Appliance Circuits

Small appliance circuits are used for such appliances as refrigerators, toasters, broilers, coffee makers, and irons. They are wired with number 12 (2.00-mm) wire and are protected with 20A fuses or circuit breakers.

Individual Circuits

An **individual circuit** is used for only one piece of equipment. Examples of individual circuits are those used with electric ranges, dryers, water heaters, heating systems, and air conditioners.

A **range circuit** is usually a four-wire, 115/230 V circuit. It has two hot wires, a neutral wire, and a ground. There is a fuse or circuit breaker in each of the hot wires. A 230-V

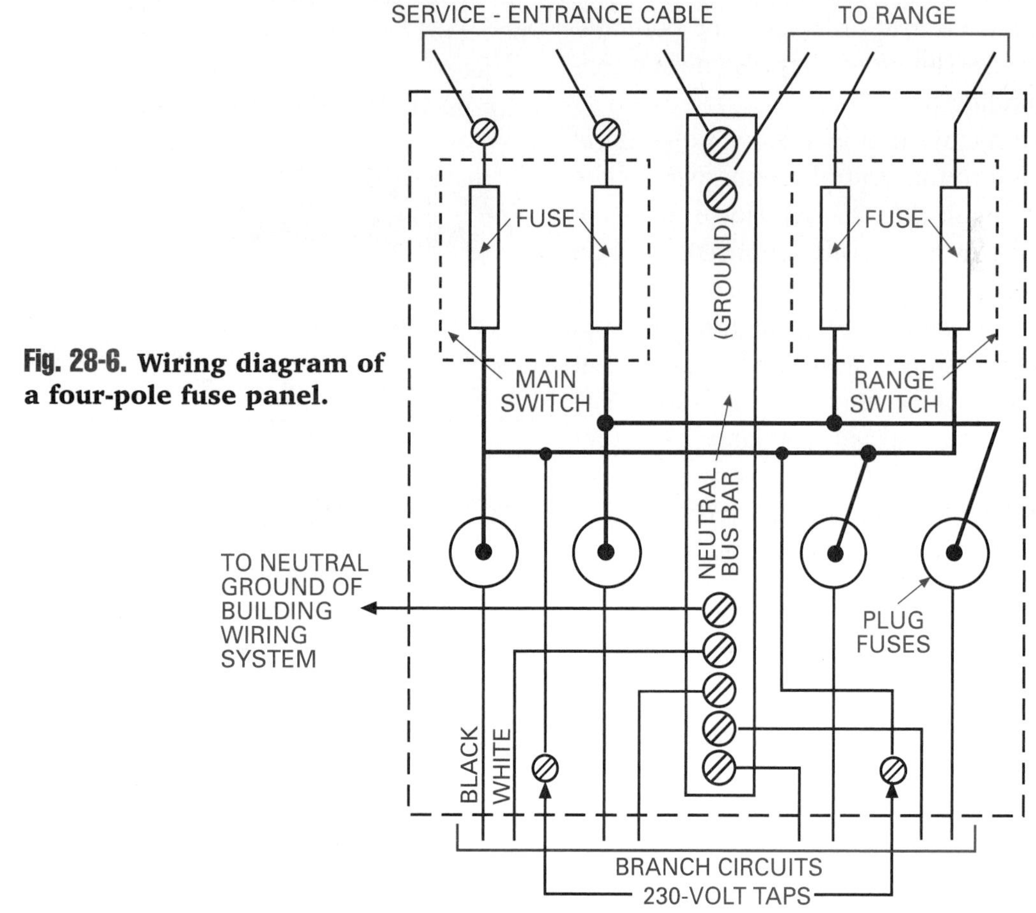

Fig. 28-6. Wiring diagram of a four-pole fuse panel.

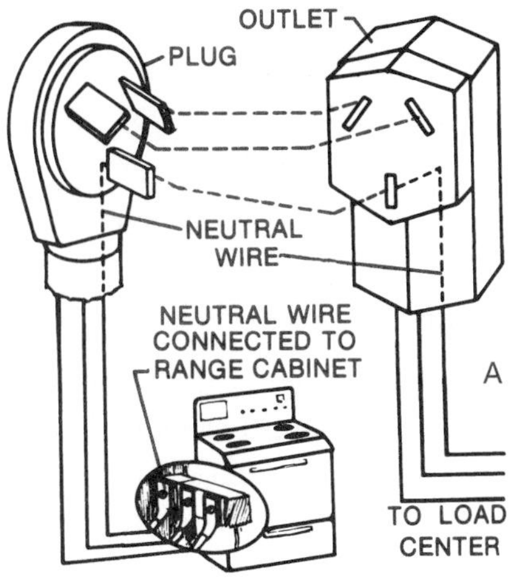

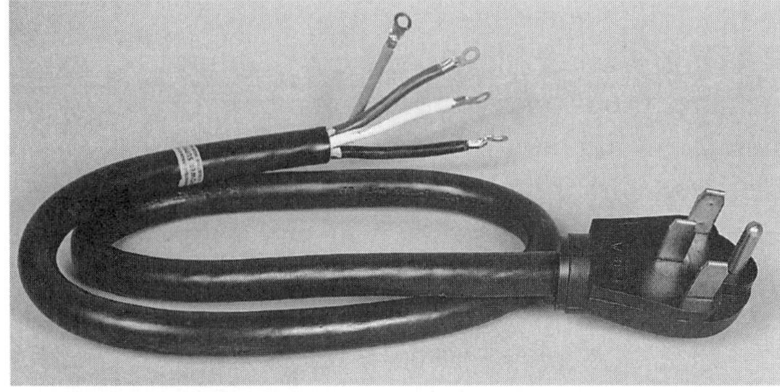

Fig. 28-7. (A) The cabinet of an electric range is grounded using a three-wire range plug and outlet; (B) four-wire range plugs contain a connection for a service ground.

individual circuit is a two-wire circuit. Both wires are hot and are protected with a fuse or circuit breaker, and there is a safety ground. Many electrical appliances and tools operated at 115 Vac such as washing machines, dishwashers, and electric drills, present potential shock hazards. Therefore, they also need a separate grounding wire.

The return current of an electric range is via a connection to the neutral ground wire of the three-wire circuit (Fig. 28-7A). However, a separate grounding wire is also connected from a terminal on the frame of the range to a grounding point of the safety grounding system. Figure 28-7B shows a four-wire plug that includes the safety ground connection. The cabinet or frame of an appliance or other equipment operated with a two-wire, 230-V circuit is grounded with an additional grounding wire. This wire is usually located in the connecting cable.

LINKS

Refer also to grounding for personal safety discussed in Chapter 2, "Safety."

Voltage Taps

The 230-V taps in a panel are for wiring additional individual branch circuits. When they are used for such a circuit, extra fuses or circuit breakers must be installed outside the panel. These are necessary because the circuit is not protected with branch-circuit fuses or circuit breakers in the panel.

How many general-purpose, small appliance, and individual circuits are in your residence, school, or office?

BRANCH-CIRCUIT WIRING

Branch circuits are wired with several kinds of approved wiring meant to protect the wire from mechanical damage. The most common wiring found in residential systems is discussed below.

Nonmetallic Sheathed Cable

Nonmetallic (NM) sheathed cable usually has two or three insulated wires and a grounding wire (Fig. 28-8). NM cable is used for all

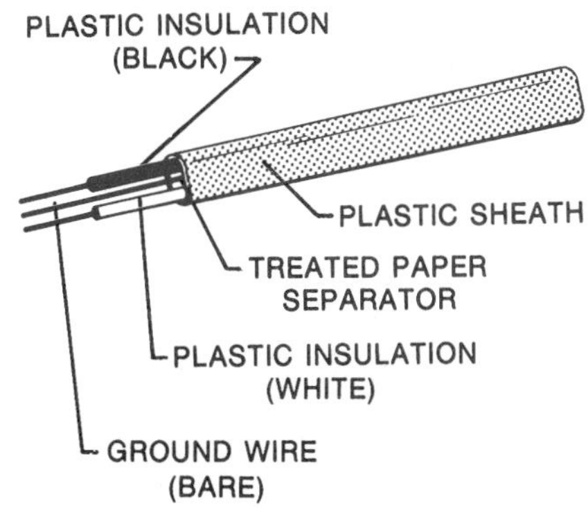

Fig. 28-8. Construction of two-conductor nonmetallic (NM) sheathed cable with a grounding wire.

kinds of indoor wiring work and is never to be buried in cement or plaster. NM cable has a moisture-resistant overall covering, which allows it to be used for both exposed and concealed work in dry, moist, and damp locations, and on outside or inside walls of masonry, block, or tile.

Nonmetallic sheathed cable is made in several sizes designated as 14/2, 12/2, 12/3, etc. The number to the left of the slash is the AWG size of the individual conductors. The other indicates the number of conductors in the cable, other than the bare grounding wire. For example, a 12/2 cable has two number 12 AWG (2.00-mm) insulated wires.

Armored Cable

Armored cable (AC) contains insulated wires in a flexible armor sheath that is a single spiral of galvanized steel tape interlocked to form a tube-like enclosure. Along its entire length (Fig. 28-9), the cable also contains a bare-copper or aluminum bonding strip in contact with the armor. The purpose of the bonding strip is to provide continuity of the circuit ground in the event that armor is broken at any point. This is necessary because the armor is part of the grounding system. AC cable is approved for use in dry locations where branch-circuit wiring must be protected against wear or be embedded in plaster. Armored/leaded cable is used in damp locations and where oil or other substances may corrode the wire insulation. In this cable, the insulated wires are covered with a lead sheath over which the armor is laid.

Electrical Metallic Tubing

Electrical metallic tubing (EMT) is thin-walled metal piping that is installed between wiring devices that are to be connected and is a code requirement in some locations. Individual lengths of *building wire* are then inserted or pulled through the tubing and connected to the devices. For long lengths of tubing, the wires are pulled through with a *fish tape* made of springy steel.

ALUMINUM CONDUCTORS

Some building electrical systems are now being wired with aluminum conductors. Aluminum has a higher resistance than copper. Because of this, conductors made of aluminum must be larger than those made of copper

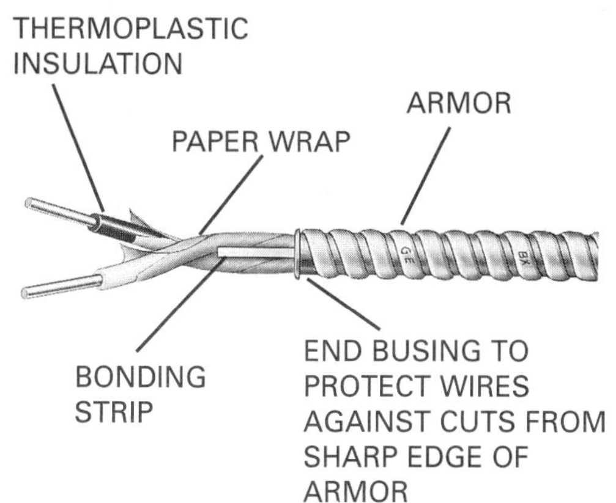

Fig. 28-9. Armored cable.

when used for the same purpose. For example, a circuit that could be wired with number 14 AWG (1.60-mm) copper conductors would be wired with number 12 AWG (2.00-mm) aluminum conductors. Aluminum wiring requires special connectors and wiring devices that must also be approved for use with aluminum conductors. Aluminum building wires and cables are identified by the word *aluminum* or by its abbreviation, *Al*. It is not considered a good practice to wire kitchens, ranges, or dryers with aluminum. In fact, it is illegal in some locations.

Write an essay that discusses aluminum as a conductor. Include why you think it has a higher resistance and why it can only be used with approved wiring devices.

WIRING DEVICES

Wiring devices are components such as switches, receptacles, and light fixtures. To complete a residential wiring system, a number of wiring devices are necessary (Fig. 28-10). These are designed and installed to provide safe, convenient, and reliable use of electricity.

Aluminum Production

Because aluminum is extremely light, corrosion resistant, and heat reflective it is often utilized in electronic components.

Aluminum is the most plentiful metal in the Earth's crust. It can be found in clay banks and most common rocks. Currently, it is not economical to extract the metal from clay. Nearly all aluminum comes from the ore bauxite. *Open-pit and shaft mining methods are used to obtain bauxite.*

To extract the aluminum, bauxite is first crushed and sometimes dried, and then shipped to treatment plants. Then impurities are removed from the ore and aluminum oxide, or alumina, is produced. Electrolysis is then used to produce pure aluminum. Alumina is dissolved in a bath, or electrolyte. Then a strong electric current is passed through the solution. This action takes out the oxygen and deposits pure aluminum on the bottom of the bath. When enough has accumulated, the molten aluminum is siphoned off and cast into pigs. About 4 to 6 pounds (2 to 3 kilograms) of bauxite yield 2 pounds (1 kilogram) of alumina, making 1 pound (0.5 kilogram) of pig aluminum.

Fig. 28-10. Selected house wiring devices.

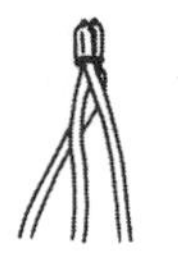

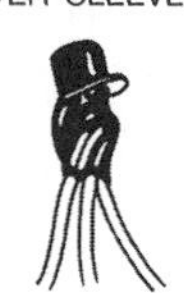

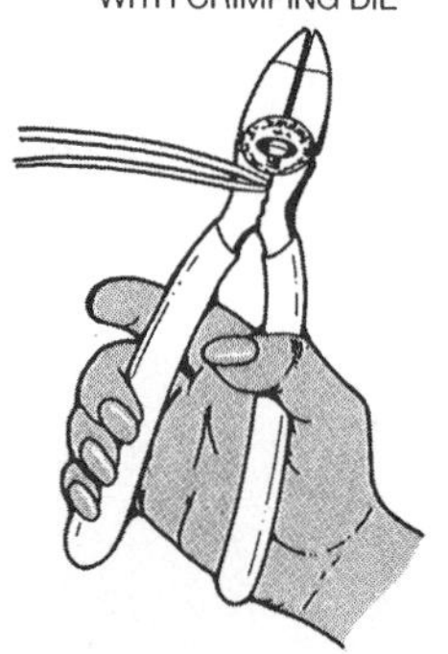

Fig. 28-11. Connecting wires with solderless, pressure connectors.

Switch Boxes

Switch boxes are used to place flush switches in code-approved metal or nonmetallic enclosures. These enclosures are also used for mounting outlets. The boxes protect the devices mechanically and prevent switch sparking from becoming a fire hazard. Nonmetallic boxes are often used in damp locations.

Junction Boxes

Junction boxes can be metal or plastic and contain connections in the branch-circuit wiring. The connections are often made with wire nuts or other solderless, pressure connectors (Fig. 28-11). Junction boxes are also used for mounting various kinds of lighting fixtures.

Fig. 28-12. Box connectors.

Box Connectors

All the cables entering a switch or a junction box must be securely fastened to the box. Some boxes have built-in cable clamps for this purpose. Otherwise, separate box connectors must be used (Fig. 28-12).

Surface-Wiring Devices

Surface-wiring devices are particularly useful to change an existing house wiring system (Fig. 28-13). They can be installed in dry indoor locations without breaking through plaster or wallboard.

Multi-Outlet Strips

A *multi-outlet strip* is a rectangular metal enclosure with preinstalled wires and outlets (Fig. 28-14A). Such strips provide a convenient way of adding to existing wiring. They are very useful for placing outlets in places that cannot be reached easily by ordinary cable or conduit wiring. They also can provide several outlets in electrical repair stations (Fig. 28-14B) and come with surge suppressing, current controlling, and switching options.

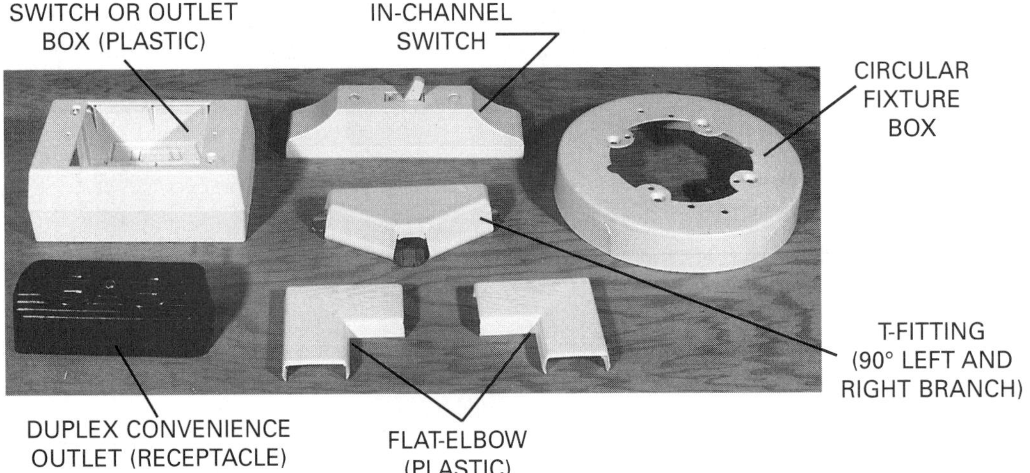

Fig. 28-13. Selected surface-wiring devices.

It is essential that the wiring devices installed on metal or nonmetallic switch boxes and junction boxes be properly grounded. The aluminum bonding strip, or wire, in an armored cable or the bare copper wire in the nonmetallic, sheathed cable provide the means to do this. The grounding wires are usually connected to the metal boxes by metal screws. When secured properly, these wires provide the grounding continuity for the entire electrical system. Before a connection is made, about 5/8 inch (16.5 mm) of the insulation is removed from each wire to expose the bare wire needed for the connection. Figure 28-15 shows a junction box with two cables joined together. Notice that the insulated hot wires (black or red) are connected together and the two insulated neutral wires (white) are connected together. They are joined tightly together by solderless twist-on connectors. The twist-on connector is insulated, but the inner surface has a metal spring that grips the wires as the connector is twisted in a clockwise direction

Fig. 28-14. (A) Multi-outlet strip equipped with grounding outlets; (B) electrical repair stations equipped with multi-outlet strips.

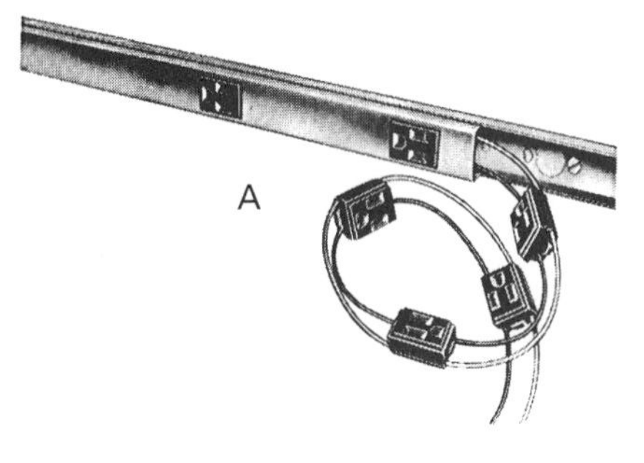

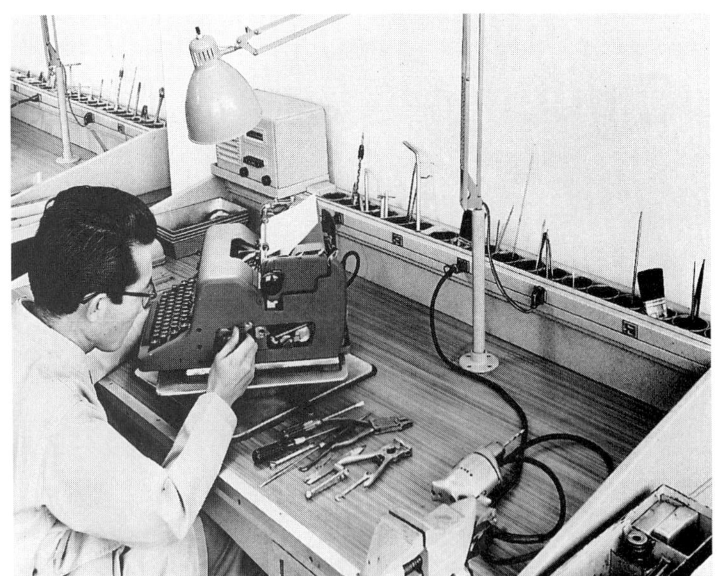

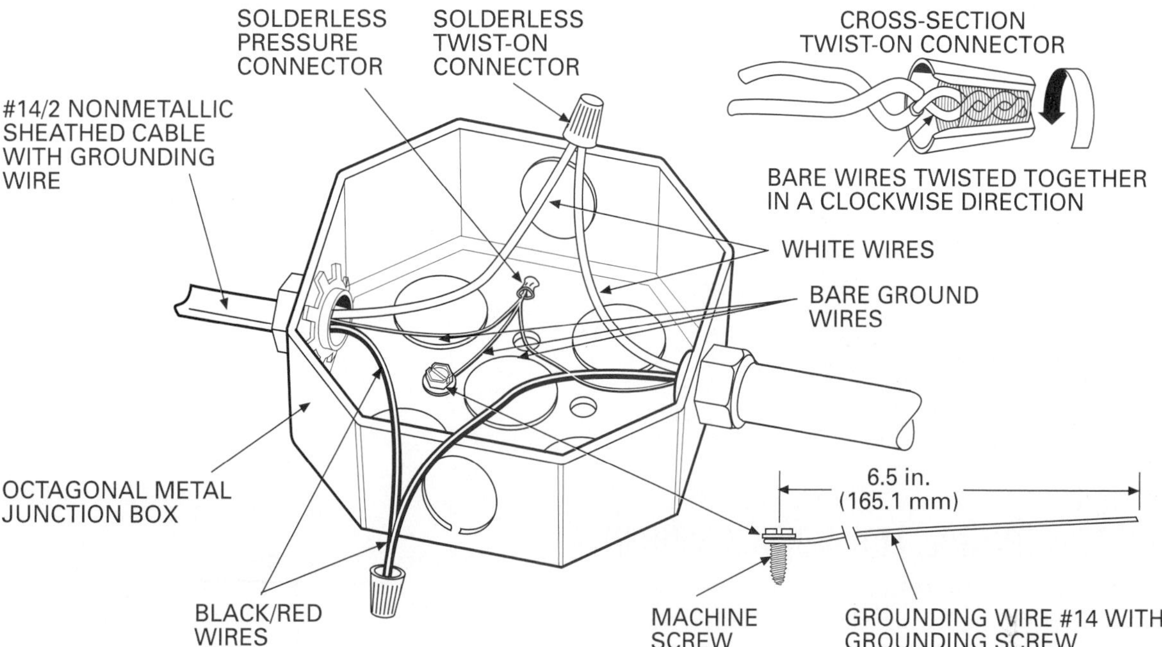

Fig. 28-15. Connecting wires in a junction box with solderless connectors and grounding wire and grounding screw.

over them. This completes the electrical circuit from one conductor to the other. The cables' bare ground wires are also connected to a special grounding wire and a green grounding screw.

Some devices, such as duplex receptacles (grounding type) used in new residential houses, have an additional green-colored hexagonal ground terminal. This terminal is used to connect a jumper grounding wire to the metal box.

BRAIN BOOSTERS

1. Discuss the methods for grounding wiring devices such as junction boxes and nonmetallic boxes.
2. Draw a circuit that controls a load from five locations. Use as many three- and four-way switches as you need.

When metal or nonmetallic boxes or enclosures are not used, a separate continuous ground wire must be installed to provide adequate grounding.

THREE- AND FOUR-WAY SWITCHES

Three-way switches are used to control loads from two points such as in lighting circuits in hallways, stairways, and doorways. The schematic diagram of a two-point control circuit using two three-way switches is shown in Fig. 28-16A. A combination of three- and four-way switches can be used to control a load from any number of places. The diagram of a three-point control circuit using two three-way switches and one four-way switch is shown in Fig. 28-16B.

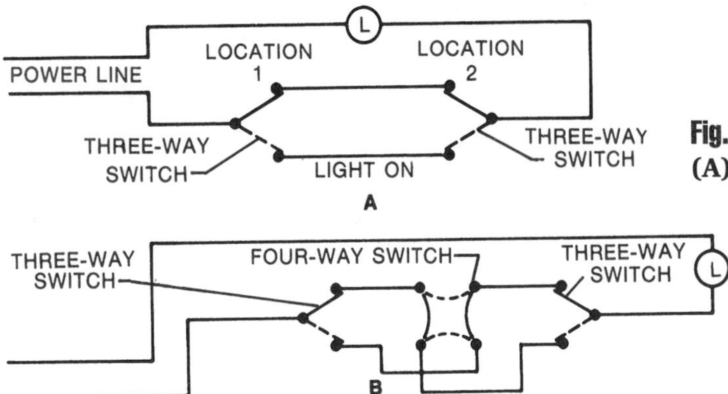

Fig. 28-16. Multipoint control circuits: (A) two-point; (B) three-point.

REMOTE CONTROL, LOW-VOLTAGE WIRING

Remote control wiring makes it possible to control lights and appliances in various rooms and areas from a central place. In such systems, local control of lights is still possible. Remote control wiring uses a low-voltage control circuit to operate line voltage circuits.

A basic remote control circuit is illustrated in Fig. 28-17A. Additional components may be added to the basic circuit as needed to accomplish desired functions, such as multipoint control or automatic control. The low voltage for the basic circuit is provided by the transformer. It steps down the line voltage to 24 Vac. The rectifier in the circuit changes the 24 Vac to about 24 Vdc. The relay is controlled by a momentary contact switch. The relay permits the switching of line voltage loads with low-voltage control circuits.

The internal wiring of one type of relay is shown in Fig. 28-17B. This relay is mechanically *latched.* That is, the relay maintains its contacts in the last position without keeping the coil energized. Latching relays are usually mounted in relay-center boxes or cabinets.

Some of the advantages of remote control wiring are:

1. Flexible switches—one relay can be controlled by many switches (multi-point switching)

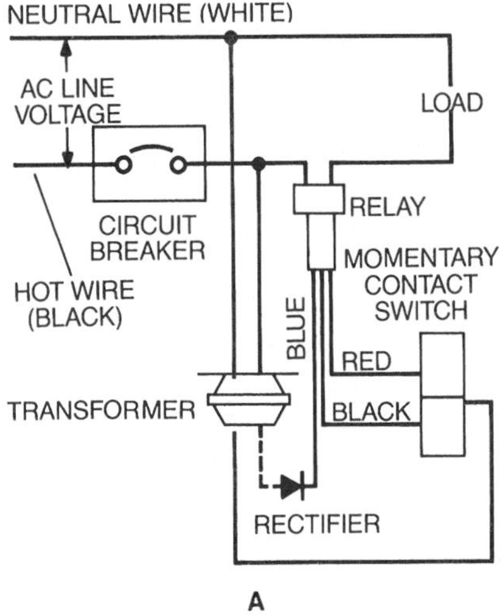

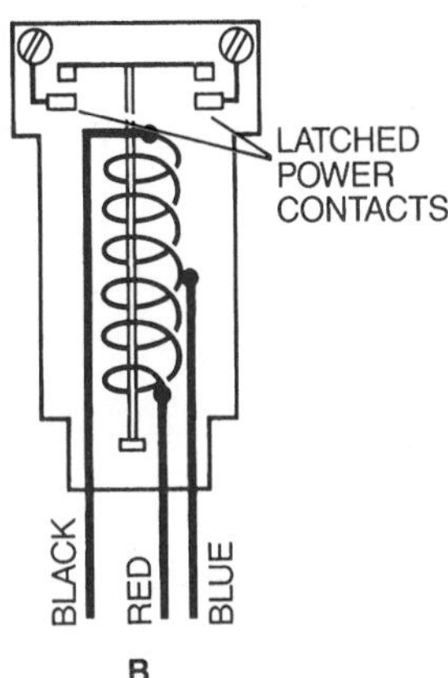

Fig. 28-17. Remote-control, low-voltage wiring: (A) basic circuit; (B) internal wiring diagram of a split-coil, mechanically latched relay.

2. Parallel relays—relays can be controlled from one or more switches
3. Master switches —many relays can be controlled from one or more locations
4. Deenergized unneeded circuits—combining a remote control system with automatic control devices, such as timers and photocells, can deenergize select circuits when not in use
5. Reduced fire danger
6. Simpler switching—remote control wiring can substitute for more complex switching. For example, smaller, low voltage and current conductors control relays for large power-handling devices
7. Less expensive—smaller, lower-voltage wire is less expensive
8. More accepted—most local electrical codes permit low-voltage wiring without conduits

Design a low voltage remote control circuit that operates an electric furnace and blower motor.

SYMPTOMS OF INADEQUATE WIRING

In recent years, the number of electrical and electronic products used in homes has greatly increased. This creates a problem of overloaded branch circuits in homes when too many appliances are connected to branch circuits. This condition wastes electricity, creates a fire hazard, and often causes appliances to work improperly.

There are several signs of inadequate wiring. The most common of those are listed below.

1. Fuses blow and circuit breakers often open because of branch-circuit overloading.
2. Heating appliances do not warm up as quickly as they should or do not heat to the right temperature.
3. Lights dim when appliances are used or when motors in appliances such as refrigerators start.
4. The television picture shrinks when appliances are used or when appliance motors start.
5. Appliance motors start having difficulty or fail to start.
6. Motors operate at a lower-than-normal speed, which causes them to overheat.

If any of these signs are present, the wiring system should be thoroughly checked by a qualified electrician. In some cases, an inadequate wiring system can be corrected by simply adding branch circuits or by using wire of a larger size in a branch circuit. In some cases, the entire wiring system may have to be replaced.

Guidelines for Safe Wiring

The following guidelines will assist you in ensuring that electrical wiring in your residence is safe.

1. Use only equipment listed by reputable testing organizations such as the Underwriters' Laboratories.
2. Follow the rules specified by the National Electrical Code for installing and wiring methods used in your residence.
3. Observe any local electrical wiring codes and ordinances.
4. Have the completed wiring installation inspected and approved by a local inspector.

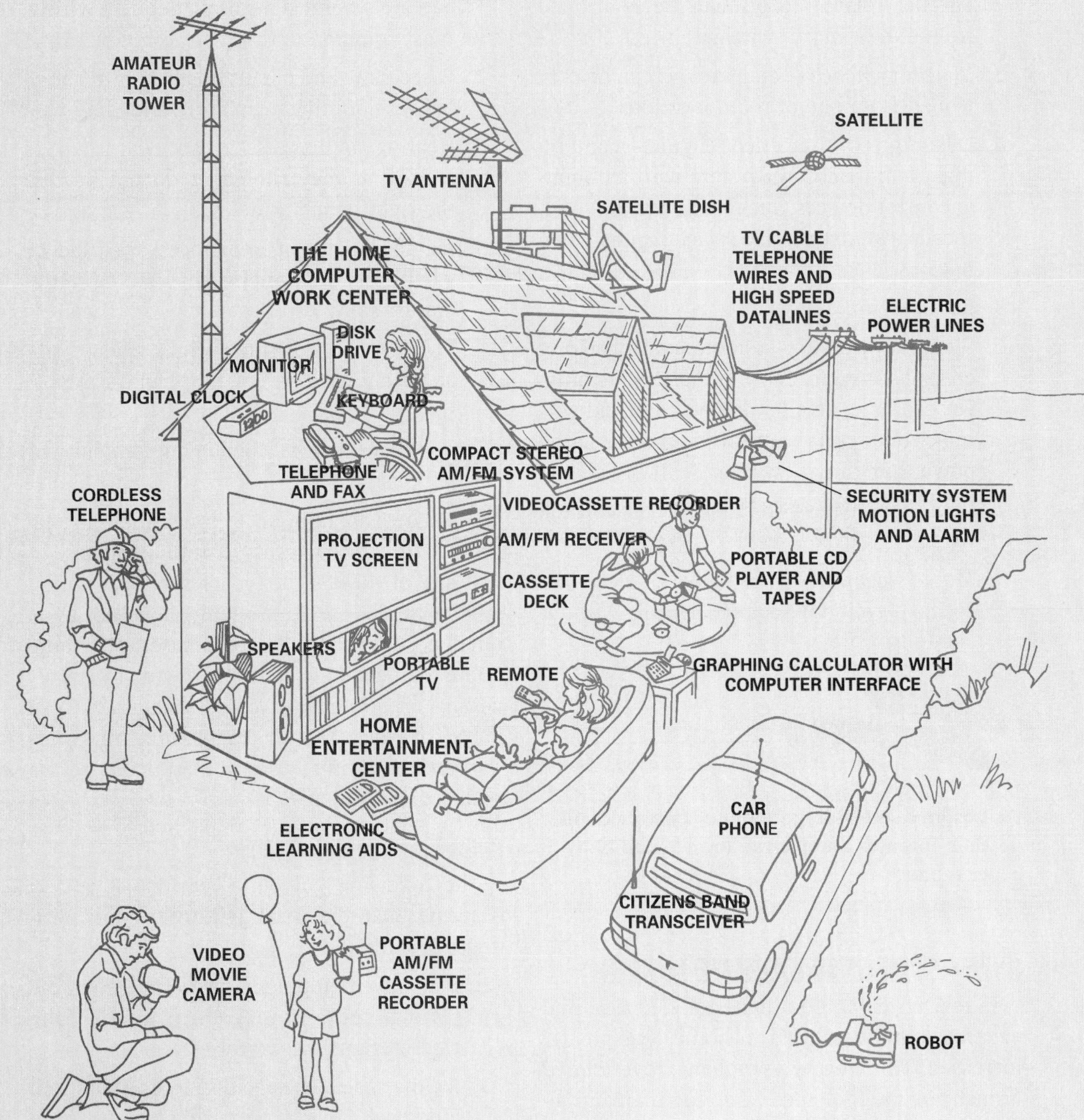

Fig. 28-18. Electronics in the information age.

PLANNING FOR THE FUTURE

The home electronic revolution has brought new electronic equipment to the homeowner. Home computers, portable television sets, compact disc players, car telephones, video games, large rear-projection television screens, and other items are being produced for the consumer market.

A variety of electronic equipment is available for consumers. Modern homes may have a space designed as a television or entertainment center, another space for a home computer or telecommunication center, and a space outside the home for portable electronic equipment (Fig. 28-18).

Modern electronic communication devices not only bring information to the individual, but also allow the individual to interact with this information. We are, no doubt, living in an information age. The home dweller, for example, may stay at home and use the computer and modem to send completed work to the office. This method of working is called telecommuting.

Each new electrical device selected for your residence may need special wiring, electrical devices, and connectors. Connecting to the Internet or telecommuting may require a special telephone line. If your plans include a security system, intercom, and sound in every room, these devices will require different types of cabling and outlets. You may wish to consider a communication wiring closet to meet your needs. A wiring closet provides a central location where various low voltage and special cables terminate, and which contains special connecting blocks that use jumper cables to help make connections. Having one location, like the load center for branch circuits, makes it easier to troubleshoot problems. It is important, therefore, to determine your needs and make written plans to meet them.

FASCINATING FACTS

The total use of electricity in residences equals about 34% of the national electric power output. With the increased use of electrical appliances, computers, and other devices in the home, this number will multiply.

Chapter *Review*

Summary

- The wiring in a house consists of conductors, devices, and other equipment such as blower motors and air conditioning condensers.
- The National Electrical Code includes recommendations and guidelines on wiring materials and methods to be used when wiring a house, apartment, office, or other building.
- The wiring in a building begins at the service drop. A service drop of a residential wiring system is the part of the system that extends from a point outside the house to the nearest electrical company distribution line.
- The service entrance includes the point where the service-drop conductors are attached at the house, the service entrance cables, the watthour meter, the service ground, and the load center. The load center serves the branch circuits.
- There are three common types of branch circuits in a building or dwelling: general purpose, small appliance, and individual.
- Remote control is one way to save money on expensive wiring and reduce the risk of fire. There are many advantages to remote control wiring, some of which are flexibility, parallel relays, master switching capability, and less expensive wiring.

Review Main Ideas

Review this chapter's main ideas by writing, on a separate sheet of paper, the word or words that most correctly complete the following statements:

1. Information about wiring materials and methods that will result in the safe use of electric energy is given in the ______.
2. A three-wire service drop is used to supply two different ______.
3. The bare wire of a service drop is called the ______ wire or conductor.
4. Electricity is distributed to different locations in a house from the ______.
5. The load center contains the ______ or ______ that protect the branch circuits of a wiring system.
6. The neutral or bare conductor of the service-entrance cable is grounded to reduce the danger of ______ and to protect against ______.
7. Three common kinds of branch circuits are ______ circuits, ______ circuits, and ______ circuits.
8. One wire of each general-purpose 115 V branch circuit is connected to the neutral ______ of the load center. The insulation of the grounded neutral wire is colored ______.
9. The grounded wire of a two-wire, general-purpose circuit is never used as a ______ wire.

10. An electric-range circuit is an example of an ______ branch circuit.
11. When voltage taps are used for wiring additional individual branch circuits, extra ______ or ______ must be installed outside the panel.
12. Nonmetallic sheathed cable is used for indoor wiring work and is never buried in ______ or ______.
13. A no. 14/3 nonmetallic sheathed cable contains ______ no. 14 AWG insulated wires.
14. Armored cable contains a ______ to provide continuity of the circuit ground in the event the armor is broken.
15. ______ protect devices and prevent switch sparking.
16. Overloaded ______ waste electricity, create a fire hazard, and often cause appliances to work improperly.

Apply Your Knowledge

1. Explain how a 230-Vac circuit saves money as compared to a 115-Vac circuit.
2. List seven elements of a basic residential wiring system.
3. Define a load center.
4. Describe at least five advantages of remote-control, low-voltage wiring.
5. Explain why it is necessary to enforce electrical codes.

Make Connections

1. **Communication Skills.** Prepare a report on how electric power is supplied and distributed in the building where you live. Begin your report with the weather or entrance head, which is usually secured to the outside of the building near where the electric power lines are attached.
2. **Mathematics.** Define the term *circular mil.* Explain to the class why this term is important in electricity and electronics.
3. **Science.** Write a short report specifying the reasons for maintaining a good ground system in a residential home.
4. **Workplace Skills.** Describe the type of load center that is installed in your home. Identify the number and type of circuits. Identify the type and size of fuses used for each circuit.
5. **Social Studies.** Explain to the class why electric codes for highly populated areas are more restrictive than electric codes for populated rural areas.

Chapter 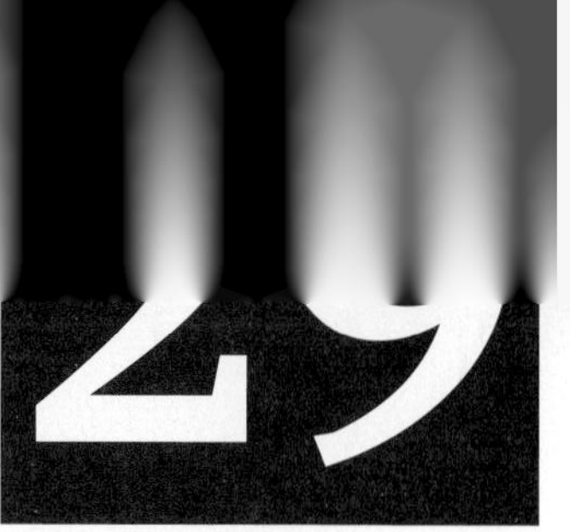 29

Electric Motors

Terms to Study

armature

brushes

commutator

field

horsepower

power factor

starter

synchronous motor

three-phase motor

universal motor

OBJECTIVES

After completing this chapter, you will be able to:

- Explain the basic electric principles of an electric motor.
- Draw and label schematic diagrams of dc series, shunt, and compound motors.
- Describe the operation of a capacitor-start motor.
- Identify three common uses for the following: shaded-pole, repulsion-start, and the synchronous motor.
- Identify four advantages of energy-efficient motors.
- Make a simple two-pole motor and show how the direction of rotation can be reversed.

What would life be like without electric motors? For one thing, it would be difficult to drill a hole in metal. The small electric hand drill is probably found in more home workshops than any other tool powered by electricity. Of all the electric energy produced in the United States, about 64 percent is used by electric motors. This chapter will help you learn how these devices work and how to maintain them.

MOTOR ACTION

A basic example of how electric energy can be changed into mechanical motion is shown in Fig. 29-1. As current passes through the large wire from left to right, a magnetic field is produced around the wire. A **field** is a space within which a definite effect (such as magnetism) exists and has a definite value at each point. The polarity of this field causes it to be repelled by the magnetic field of the permanent magnet. The wire, thus, moves away from the magnet (Fig. 29-1A).

If the direction of the current through the wire is reversed, the polarity of the magnetic field around it also reverses. Consequently, the wire moves toward the permanent magnet because of magnetic attraction (Fig. 29-1B).

An example of how simple motor action produces circular motion is shown in Fig. 29-2. A permanent magnet is mounted on a pivot so that it is free to turn between the poles of two electromagnet coils. These coils are connected in series with a battery and a reversing, or double-pole-double-throw (dpdt), switch (Fig. 29-2A). The reversing switch is used to change the polarity of the voltage applied to the coils.

When the reversing switch is closed in one direction, the current through the electromagnet coils produces magnetic poles of the polarity, shown in Fig. 29-2B. These magnetic poles attract the poles of the permanent magnet. It begins to turn in a clockwise direction. As the magnet nears the poles, the switch is opened and the electromagnet is demagnetized. Because the permanent magnet is already turning, it will continue to rotate beyond the coils.

The reversing switch is then closed to the opposite direction. The direction of current through the coils is, thereby, reversed. This, of course, changes the polarity of the electromagnets. Now the permanent magnet is repelled. It continues to turn in the clockwise direction (Fig. 29-2C).

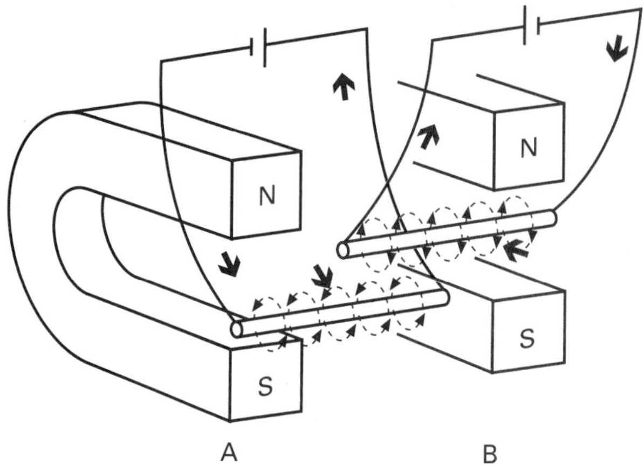

Fig. 29-1. **Basic motor action.**

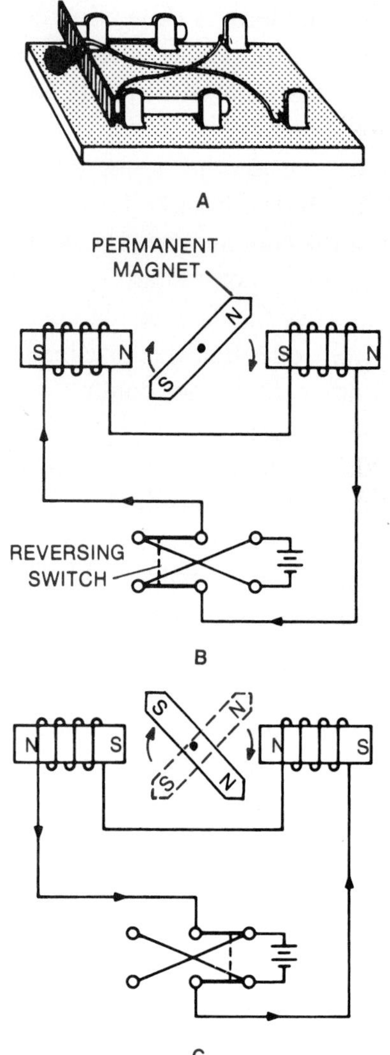

Fig. 29-2. **Producing circular motion by simple motor action.**

Write a step-by-step procedure that explains the operation of a simple motor. Be sure to number each step.

COMMUTATOR ACTION

Using a dpdt switch is not a practical way to change the magnetic polarity of a motor. **Commutators** are segmented motor cores that replace dpdt switches. The action of a commutator in a simple motor is shown in Fig. 29-3.

An **armature** is an electromagnet coil mounted on a shaft that turns inside a motor. It is placed between the poles of a permanent magnet. A *rotor* is another name for an armature. Each end of the rotor coil is connected to a segment of the commutator. The commutator is mounted on the motor shaft. The commutator has segments that are insulated from the shaft and from one another. **Brushes** are stationary contact that the commutator rubs against as it turns.

In Fig. 29-3A, commutator segment 1 is connected to the negative terminal of the battery through brush 1. Segment 2 is connected to the positive terminal of the battery through brush 2. Under this condition, the rotor electromagnet is repelled by the poles of the permanent magnet, and the rotor begins to turn in a clockwise direction. The magnetic polarity of the rotor does not change as it continues to turn to the position shown in Fig. 29-3B. Now its poles are attracted by the poles of the permanent magnet.

When the rotor reaches the position shown in Fig. 29-3C, the commutator reverses its connections to the battery. Segment 1 is connected to the positive terminal of the battery. Segment 2 is connected to the negative terminal. This changes the direction of the current through the rotor coil and reverses its magnetic polari-

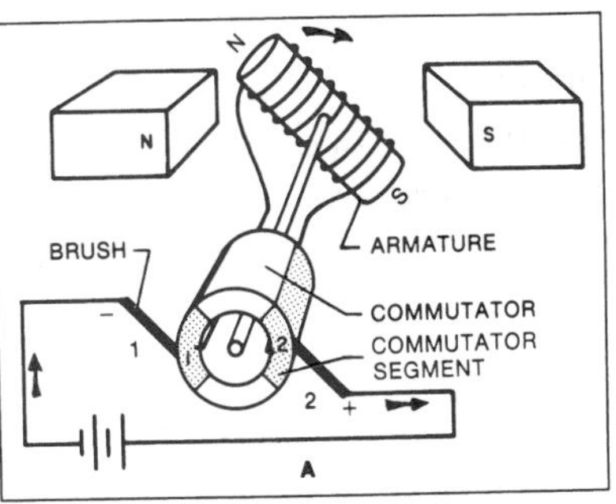

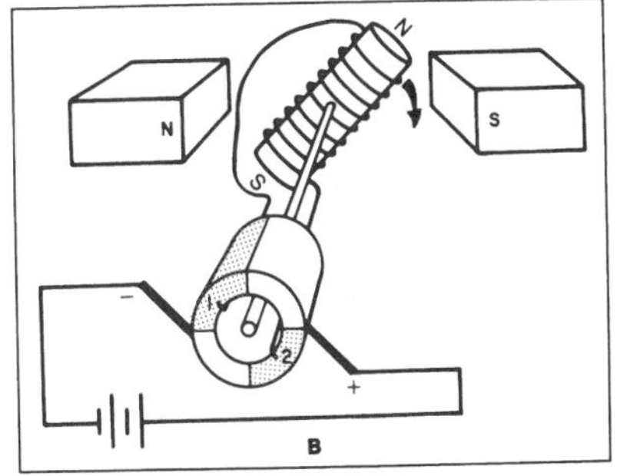

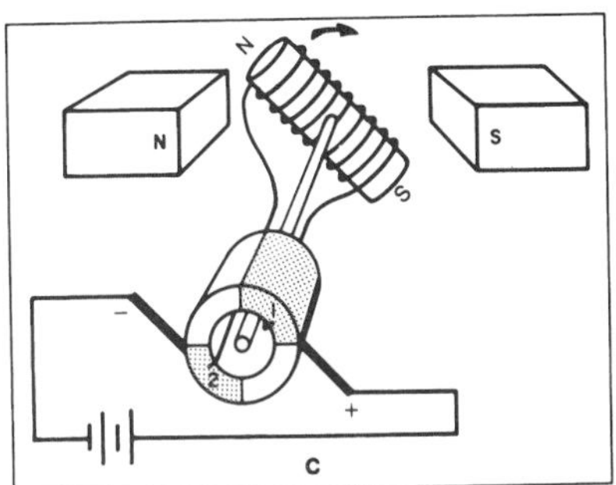

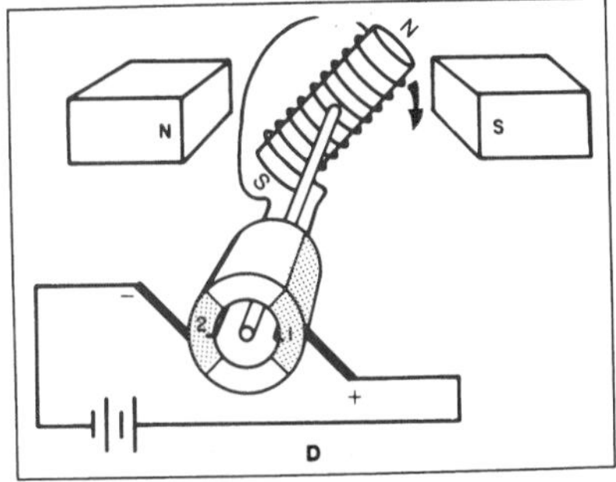

Fig. 29-3. Motor commutator action.

ty. Like magnetic poles are again near each other. Magnetic repulsion causes the rotor to continue to turn in a clockwise direction.

When the rotor reaches the position shown in Fig. 29-3D, the magnetic polarity of its poles does not change. They are attracted by the poles of the permanent magnet. This causes the rotor to continue to turn in a clockwise direction.

MOTOR SIZE

Horsepower is a mechanical unit of power that indicates the size of an electric motor. One horsepower is equal to 746 watts of electrical power and is defined as the amount of work needed to raise a weight of 550 lb (250 kg) a distance of one foot (305 mm) in one second. In the metric system, motor sizes are given in watts.

Most of the motors discussed in the following paragraphs are known as *fractional-horsepower motors.* This means that their size is less than 1 hp. Other motors used for heavy-duty work may have power ratings from 1 to 1,000 hp (0.746 to 746 kW) or more.

1. Explain the difference between brushes and commutator segments.

2. Gather data from your classroom or a local library that shows examples of the actual weight and size of three different types of motors and compare their horsepower ratings.

Horsepower

The British engineer and inventor James Watt established the value for horsepower in the eighteenth century. He concluded that horses could haul coal at an average rate of 22,000 ft-lb. per minute. He then arbitrarily raised this figure by one-half to set the standard we use today.

In modern practice, three different horsepower values are used to describe an engine's performance.

1. Indicated horsepower *is the theoretical efficiency of a reciprocating engine. This is determined from the pressure developed by the engine's cylinders.*
2. Brake or shaft horsepower *is more commonly used to indicate the practical ability of the engine, or the maximum performance. This value is determined by the indicated horsepower minus the power lost through heat, friction, and compression.*
3. Rated horsepower *is the power that an engine or motor can efficiently produce for sustained periods of time.*

Motors from various countries are often rated using different horsepower values. For example, British automobile engines are classified in rated horsepower. *However, the engine's* brake horsepower *may actually be four to six times this value. American automobile engines are quoted in* brake horsepower. *In the metric system, 1 hp is sometimes called* force de cheval or cheval-vapeur. *It is equivalent to 32,549 ft-lb. per minute, or 0.986 of the English horsepower unit.*

TYPES OF MOTORS

There are literally thousands of ways to use motors. From making tiny robotic movements to moving huge amounts of molten steel, people depend on motors to do their work. Therefore, there are also several kinds of motors. This chapter discusses direct current, universal, alternating current, split-phase, capacitor-start, and other motors.

Direct Current Motors

Direct current motors and direct current generators are very similar. The magnetic fields of dc motors are produced by stationary windings called the *field* and by rotating windings in the rotor, or armature (Fig. 29-4). The circuit through the rotor windings in the typical dc motor is completed through stationary carbon brushes. The brushes are in contact with commutator segments that are connected to the rotor windings.

Series Motors

In series dc motors, field and rotor windings are connected in series (Fig. 29-5A). Series motors have high starting *torque*, or twisting force. This makes them better than other motors at starting while connected to heavy loads. The ordinary automobile cranking motor, or starter, is a series motor. Such motors are also used in cranes, hoists, cargo winches, and train engines.

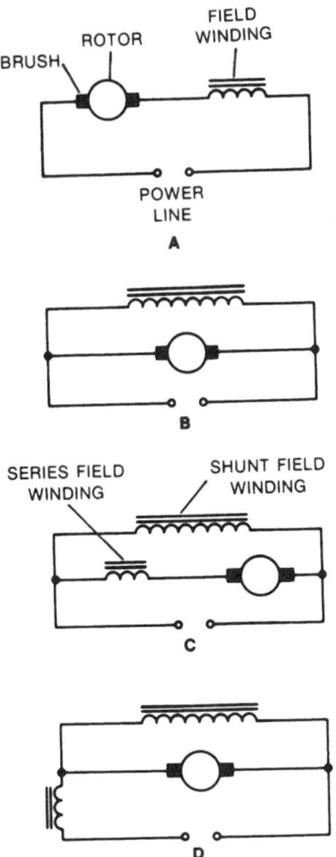

Fig. 29-5. Schematic diagrams of dc motors: (A) series motor; (B) shunt motor; (C) long-shunt compound motor; (D) short-shunt compound motor.

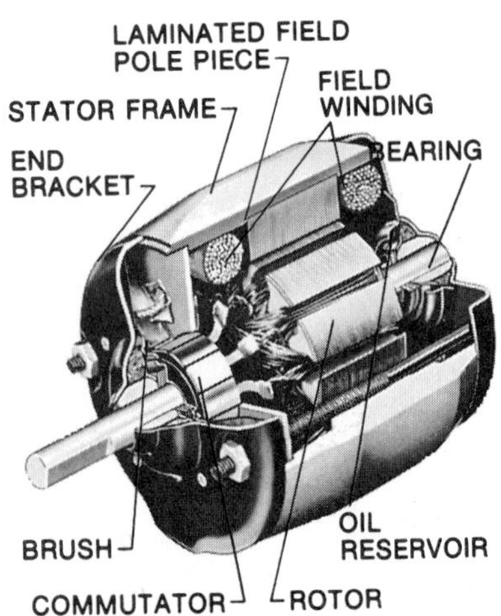

Fig. 29-4. Main parts of a dc motor.

A dc series motor should never be operated without being connected directly to a load or coupled to a load through a gear assembly. Otherwise, the motor will continue to increase its speed to a point where it may be seriously damaged or even destroyed.

Shunt Motors

In shunt motors, the field and rotor windings are connected in parallel (Fig. 29-5B). Shunt motors operate at relatively constant speeds when connected to loads that vary. They are often used in heavy-duty drill presses, lathes, conveyors, and printing presses.

Compound Motors

Compound motors are a combination of series and shunt type motors (Fig. 29-5C and D). This gives it the high torque advantage of a series motor and the constant-speed advantage of a shunt motor. Compound motors may be long-shunt or short-shunt, depending on the way the shunt field is connected.

Universal Motors

Universal motors work with either direct or alternating current, and are similar to the small, series dc motor (Fig. 29-5A). They come in sizes ranging from 1/100 to 2 hp (7.46 to 1492 W) with speeds up to 10,000 revolutions per minute (rpm).

One hundred twenty volt, universal motors are small, but provide a relatively large horsepower output. Universal motors' direction of rotation can be reversed and their speed can be easily controlled. Because of these features, universal motors are used in a variety of small appliances and portable tools. These include sewing machines, vacuum cleaners, food mixers, drills, saws, and shears (Fig. 29-6).

Alternating Current Motors

There are several different kinds of ac motors. The two most common ac motors are induction motors and synchronous motors. There are four general kinds of small, single-phase induction motors: (1) split-phase motors, (2) capacitor-start motors, (3) shaded-pole motors, and (4) repulsion-start motors. Alternating current motors that do not have windings in their rotor are known as *squirrel-cage motors.*

Split-Phase Motor

A typical *split-phase motor*, a widely used single-phase motor, is shown in Fig. 29-7. Split-phase motors come in sizes ranging from l/20 hp to 1 hp (37.3 to 746 W). Because of their relatively low torque, these motors are used with loads that are easy to start. These include fans, blowers, small pumps, washing machines, clothes dryers, drill presses, grinders, and table saws.

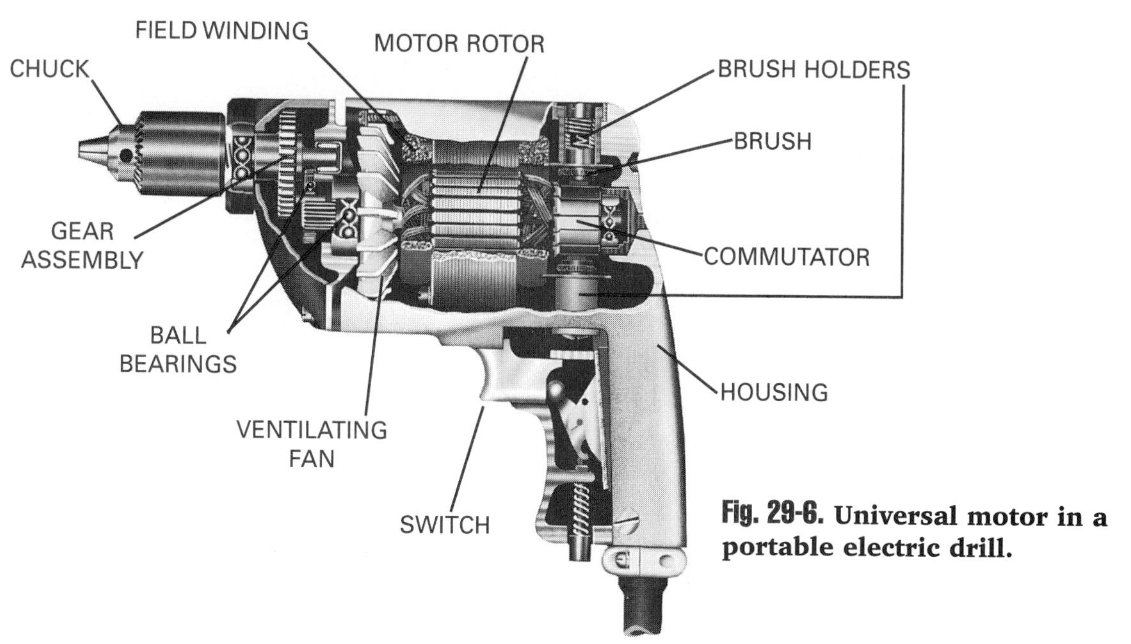

Fig. 29-6. Universal motor in a portable electric drill.

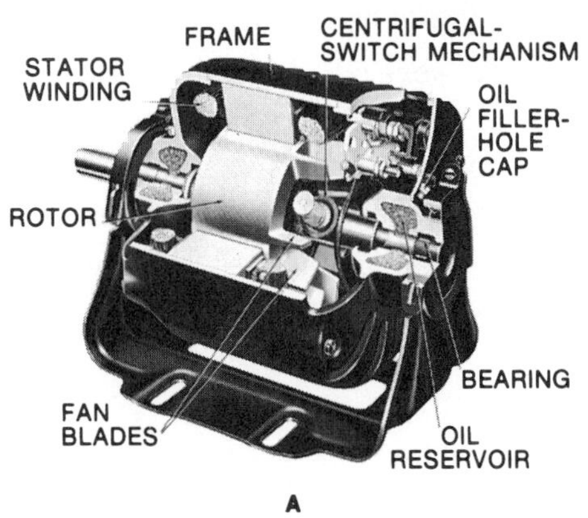

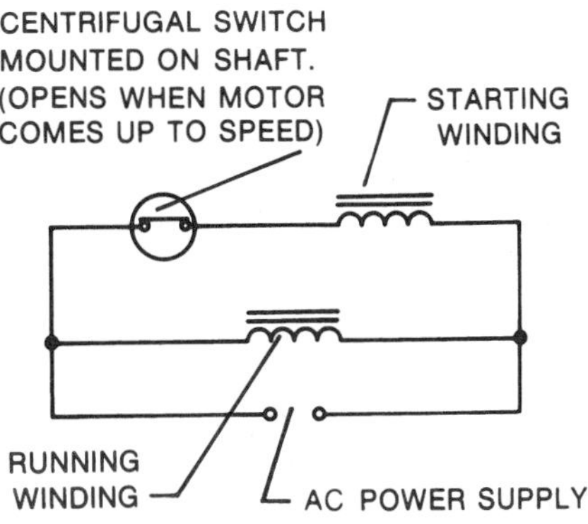

Fig. 29-7. (A) Main parts of a single-phase, split-phase induction motor; (B) schematic diagram of the split-phase field circuit.

Running Windings

In split-phase motors, electromagnetic fields are produced by current in the *running*, or *main*, winding. This winding consists of coils placed in insulated slots located around the inside of the stator, or stationary, core. These coils are connected in series. They form what are called *pole groups*, or *poles* (Fig. 29-8).

Rotors

The rotor of a split-phase motor does not have windings and is not connected to the power line in any way. The rotor is made of a laminated iron core that has copper bars connected together by end rings (Fig. 29-9). As the rotor moves through the magnetic fields of the running winding, electromagnetic induction induces current in the bars. The induced current produces a secondary (rotor) magnetic field of an opposite polarity to the running winding. The poles of the running winding always repel this field, causing the rotor to turn at a constant speed.

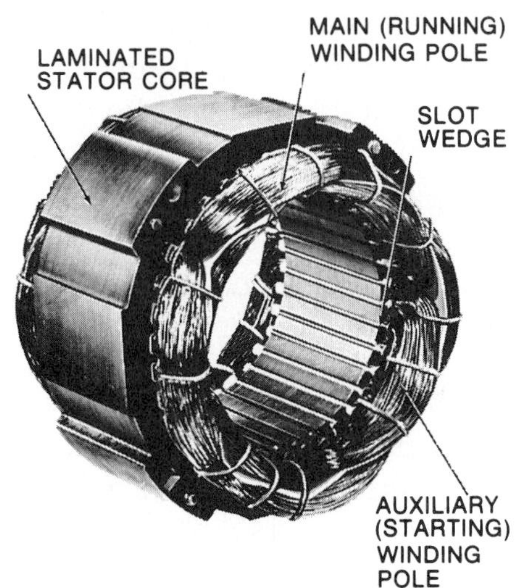

Fig. 29-8. Stator assembly (field) of a split-phase motor.

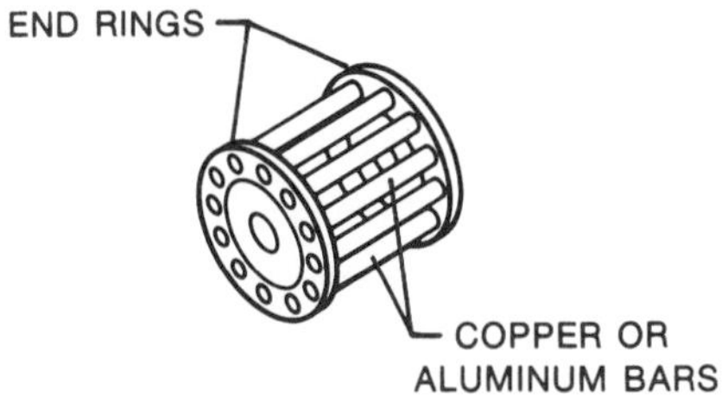

Fig. 29-9. The bars and end rings within the rotor of a split-phase motor. For clarity, the laminated rotor core through which the bars pass is not shown.

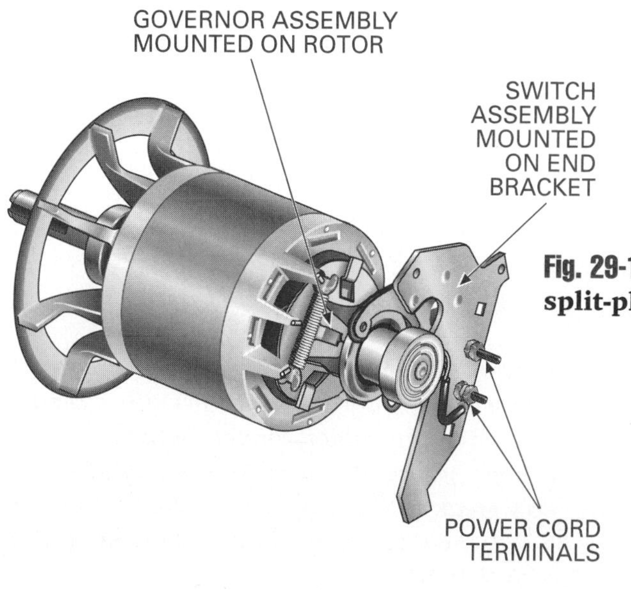

Fig. 29-10. Centrifugal switch used in a split-phase motor.

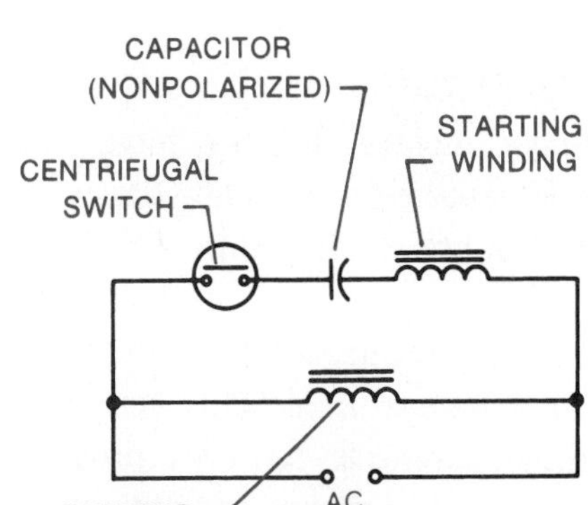

Fig. 29-11. Schematic diagram of a typical capacitor-start motor.

Starting Windings

In order to start a split-phase motor, additional magnetic energy is needed to make the rotor turn. This energy is supplied by a second set of stationary coils. These form what is called the *starting winding.* The coils are wound into the insulated slots that also contain the running winding. The starting winding is connected to the motor circuit only during the time needed for the rotor to accelerate to about 80% of its full speed. After this, the winding is disconnected from the circuit.

Centrifugal Switches

The starting winding of a split-phase motor is automatically connected into and disconnected from the motor circuit by a centrifugal switch. This switch is turned on and off by a governor assembly mounted on the motor shaft (Fig. 29-10). When the motor is turned off or is running at a slow speed, the switch is turned on. The starting winding is now connected to the motor circuit. As the rotor speeds up, centrifugal force causes the governor assembly to move toward the rotor. This turns off the switch, causing the starting winding to be disconnected from the motor circuit. The click-like sound heard very soon after a split-phase motor is turned on or off is made by the movement of the governor assembly.

Capacitor-Start Motors

A typical *capacitor-start induction motor* is similar to a split-phase motor. A capacitor-start motor contains an unpolarized electrolytic capacitor connected in series with the starting winding (Fig. 29-11) to increase starting torque. Because of this high torque, capacitor-start motors are used with hard-to-start loads. These include stokers, refrigerators, air conditioners, air compressors, and heavy-duty pumps.

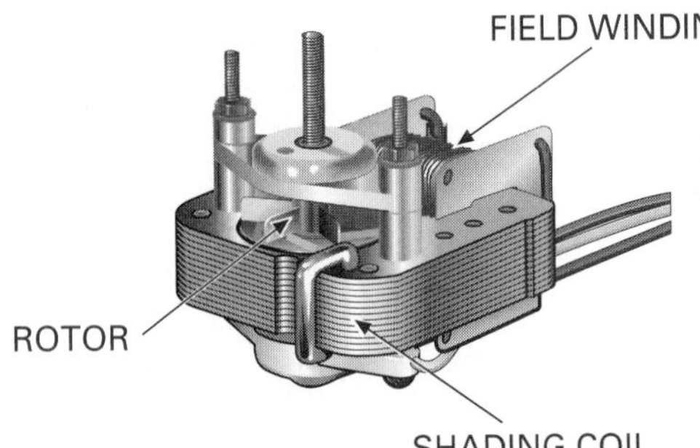

Fig. 29-12. **A two-pole, shaded-pole induction motor used in a record player.**

Shaded-Pole Motor

Shaded-pole motors do not have starting windings or capacitors. Instead, these motors have heavy copper *shading coils*. Shading coils are wound around each pole piece or part of it (Fig. 29-12). The induced current in these coils produces a magnetic field. This acts with the field-winding magnetic field to provide the rotor with a rotating magnetic field.

Shaded-pole motors are usually less than 1/4 hp (187 W) in size. Because of their low torque, they are used with light loads, including small fans, VCRs, and tape cassette players.

Synchronous Motors

A **synchronous motor** keeps in step with the frequency of the ac power source. For this reason, the motor runs at a more constant speed than other motors. A shading coil makes these motors self-starting (Fig. 29-13). A common use of synchronous motors is in electric clocks. Such motors are also used to operate time switches, business machines, and several kinds of control mechanisms.

FASCINATING FACTS

Although a synchronous motor's constant speed benefits certain machines, this motor cannot be used in applications where the mechanical load becomes very great. If the load causes the motor to slow down, the motor will actually "fall out of step" with the current's frequency and come to a stop.

Three-Phase Induction Motors

Three-phase motors are induction motors and have field windings that consist of three separate sets of coils. Each of these is energized by one phase line, or *leg* of a three-phase power system (Fig. 29-14A). When the motor is connected to a three-phase line, the alternating currents in the coils produce what is called a *revolving,* or *rotating magnetic field* (Fig. 29-

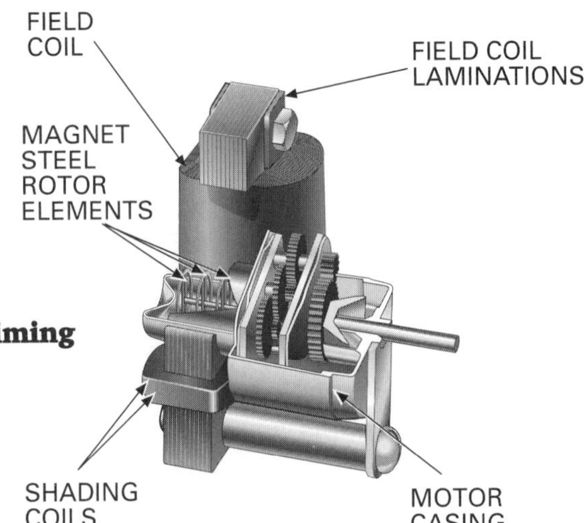

Fig. 29-13. **A small synchronous timing motor.**

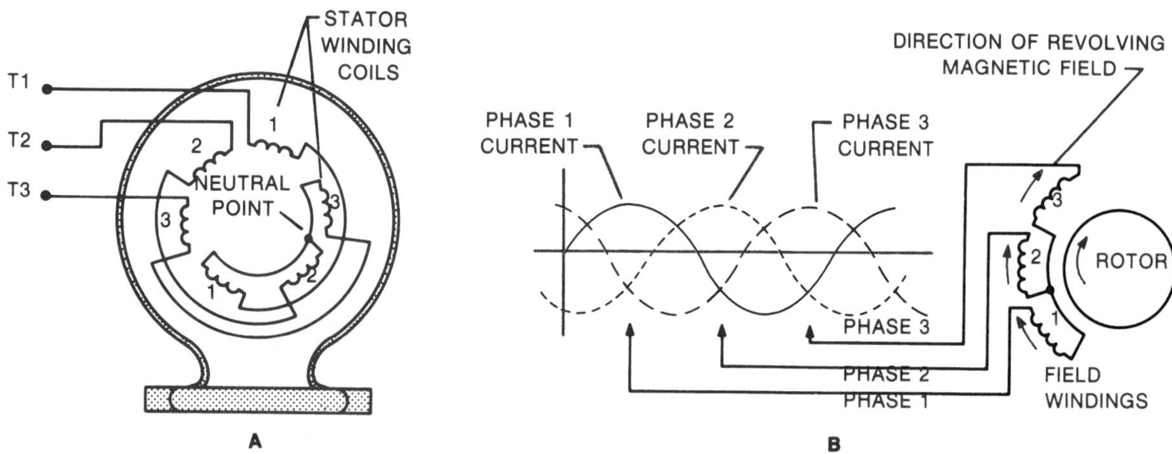

Fig. 29-14. The three-phase induction motor: (A) coil connections; (B) basic operation.

14B). This magnetic field acts on the secondary magnetic field produced by the induced rotor current to keep the rotor moving.

A three-phase motor can be reversed by simply interchanging any two of the three power-line conductors. A **starter** is a type of electromechanical relay used to reverse the power-line conductors to a three-phase motor.

Because of their simple construction, high torque, and high efficiency, three-phase induction motors are found in all kinds of heavy-duty industrial equipment. Most three-phase motors operate at a voltage of 208 V or higher.

Repulsion-start Motors

Repulsion-start motors have field windings and wire-wound rotors similar to the rotor of a dc or universal motor (Fig. 29-15). The rotor windings, however, are not connected to a power line. These motors also have commutators and brushes.

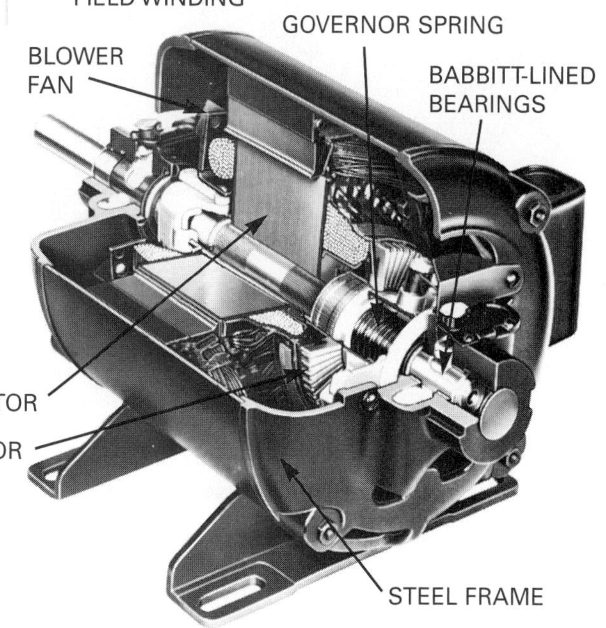

Fig. 29-15. Cutaway view of a 1/2-hp (373-watt) repulsion-start induction motor.

When the motor is turned on, the magnetic fields of the field winding induce a voltage in the rotor winding. The secondary magnetic fields produced by the rotor current are then repelled by the field winding's electromagnetism. This action provides the motor with its starting torque. After the motor reaches about 75% of its full speed, a centrifugal mechanism short-circuits the rotor commutator segments. The motor then continues to work as an ordinary induction motor.

The most common repulsion-start motors range from 1/2 to 8 hp (373 to 5970 W) in size. They are operated at 220 V. Because of their extremely high torque, these motors are used with heavy-duty loads, such as large air-conditioning units, compressors, and pumps.

Stepper Motors

Another common electric motor is the stepper motor. Most electric motors rotate in continuous directions and often at synchronous speeds. A stepper motor, however, is designed to move in short, equal steps, stop, and then start again in either direction. An example of a stepper motor is one that moves a robot arm a short distance, stops, then returns the robot arm to its previous position. Another example is the stepper motor that moves the data-sensing lever back and forth in a computer's hard drive. A stepper motor, as the term implies, rotates in small steps. Some stepper motors are quite small and require only 5 to 24 Vdc to operate and rotate each 2 to 7.5 degree step. The more steps taken in a rotation of 360 degrees the more smoothly the motor turns.

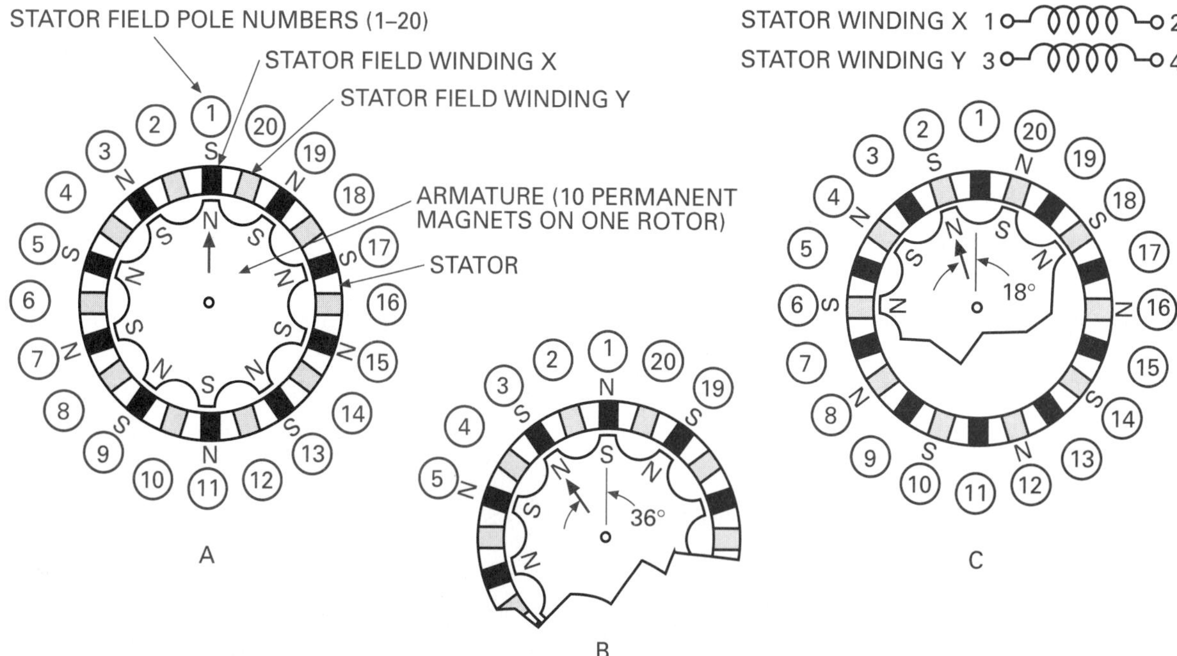

Fig. 29-16. Sketch of the poles in a small stepper motor during a partial rotation: (A) starting position; (B) first step 18°; (C) second step 18° + 18° = 36° counterclockwise rotation.

Fig. 29-17. Making the connections to test a motor before certifying it as an energy-efficient motor.

The basic principle of reversing switches is also used in stepper motors. A stepper motor, for example, has magnetic field poles in its stator that can be energized with direct current and permanent magnets that can rotate on a shaft as an armature. A small stepper motor often uses a computer software program and transistors or a driver-integrated circuit to control the switching action rather than a mechanical reversing switch.

For a stepper motor to be useful, it must be able to move in small steps in either a clockwise or a counterclockwise direction. This requires a number of stator poles surrounding an armature made up of permanent magnets. Figure 29-16 shows a simplified drawing of a stepper motor. The stepper motor has twenty stator poles (numbered in small circles 29-16A) and ten poles on the armature. Also shown are two stator windings X and Y. The X winding has the end of the coil wires labeled #1 and #2; the Y winding has the coil wires labeled #3 and #4. When energized by a dc current, stator winding X produces the polarity shown in ten poles on the stator that are cross-hatched diagonally. Winding Y, when energized, produces the polarity in the other 10 poles that are not cross-hatched. It is important to remember that the windings can be switched on and off and the current can be reversed in each winding, causing the polarity to change accordingly. The polarities of the ten armature magnets are permanent and do not change during rotation.

Energy-Efficient Motors

Electric motors use about 64% of all the electricity produced in the United States. Three-fourths of this amount is used by motors that operate pumps, blowers, fans, and machine tools. These devices are often used in the chemical, metal, paper, food, and petroleum industries (Fig. 29-17).

Because electric motors use such large amounts of electric energy, they also have great potential for energy conservation. The U.S. Department of Energy considers the development and use of energy-efficient motors by industry an important factor in the nation's energy conservation efforts.

Motors rated from 1 through 20 hp (.746 to 14.92 kW) have the greatest conservation potential. Motor efficiency is a measure of mechanical work output compared to the electrical power input. **Power factor** is a measure of how well a motor uses the current it draws.

Energy-efficient motors have several important features. For one, they operate at cooler temperatures. Thus, less electric energy is lost to heat energy and motors can be sized more closely to load requirements. For example, if the equipment takes a 5 hp (3.7 kW) motor, the motor closest to that requirement should be used. Otherwise, the motor would be overloaded or under-loaded and waste energy. Using the right-sized motor for the load is important in maintaining an efficient power factor.

The rotors of energy-efficient motors are made with more aluminum. This reduces losses resulting from current flowing in the rotor bars. More copper is used in the stators, reducing motor losses. More steel and thinner laminations are used to reduce the stator and rotor losses. The air gap between the rotor and stator is less in an energy-efficient motor. A 1-hp (746-W), energy-efficient motor will use 70 W less under continuous operation than a standard motor of the same wattage.

Choose six of the motors discussed here and write an essay that discusses the advantages of each.

CONTROLLING MOTORS

In order to use motors efficiently, they must be controlled. They need to be reversed, sped up and down, and stepped in order to do the work they are designed to do.

Reversing Direction

The direction of rotation of a dc motor or a universal motor can be reversed by interchanging the connections to the rotor windings or to the field windings. This is sometimes done by using a reversing switch connected to the brushes. An example of such a circuit arrangement using a dpdt switch with a series motor is shown in Fig. 29-18.

Split-phase motors are usually reversed by interchanging the connections to the starting winding. A number of different kinds of reversing switches can be used for this purpose. The motor can also be reversed by interchanging the connections to the running winding. If both winding connections are interchanged, the motor will continue to run in its original direction. To reverse a three-phase induction motor, simply interchange two phases of the incoming three-phase power.

Controlling Speed

The speed of a dc motor can be varied by connecting a *rheostat* in series with its field or in parallel with its rotor (Fig. 29-19A). Three-phase rheostats or inverters can be used to vary the speed of three-phase induction motors.

The speed of a universal motor connected to an alternating current can be controlled with a silicon-controlled rectifier (Fig. 29-19B). The main speed controlling components of the circuit are *potentiometer* R1 and *capacitor* C1. The

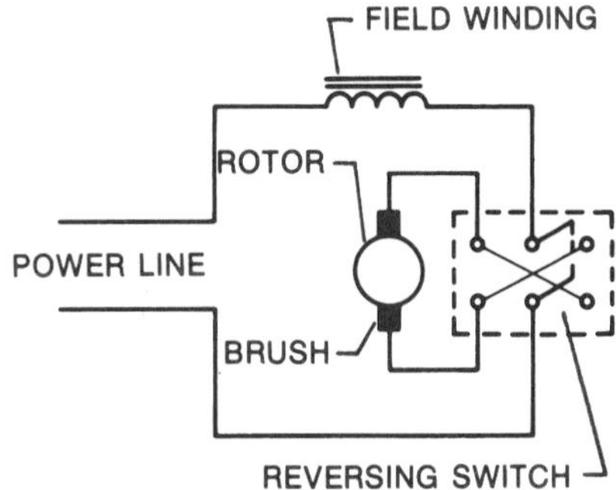

Fig. 29-18. **A method of reversing a series motor.**

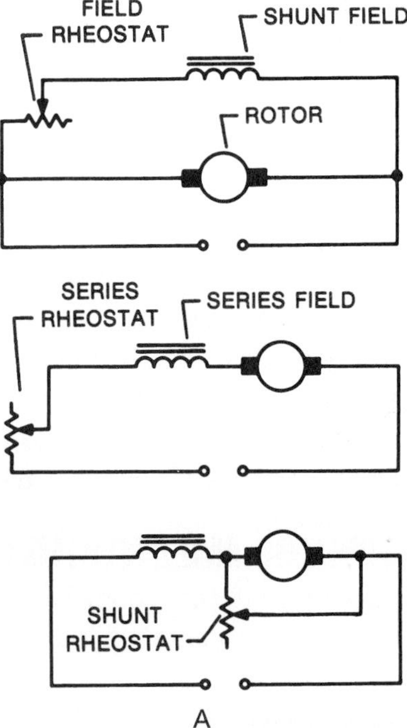

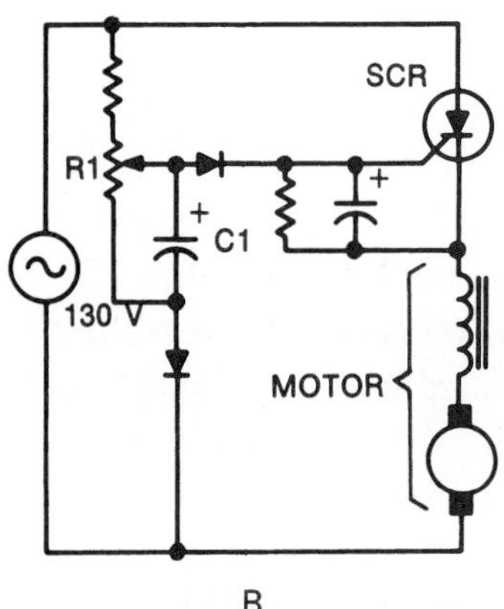

Fig. 29-19. **(A) Speed control of dc motors with a rheostat; (B) speed-control circuit using a silicon-controlled rectifier.**

potentiometer is used to control the charge and the discharge rates of the capacitor. This, in turn, controls the voltage developed across the capacitor and the voltage applied to the gate of the silicon-controlled rectifier. By adjusting R1, the amount of current conducted by the silicon-controlled rectifier can be varied. As a result, the speed of the motor also changes.

The speed of a split-phase motor, unlike that of a universal motor, cannot be varied gradually by reducing the voltage across the field without losing too much torque. For this reason, the speed may be step-controlled by changing the number of running winding poles. In a three-speed motor, for example, the running winding is arranged so that it can be switched to form

LINKS

Chapter 17 explains silicon-controlled rectifiers in detail.

two, four, or six poles. The approximate speed that each of these connections produces can be found by using the formula

$$\text{rpm} = \frac{120f}{N}$$

where rpm = revolutions per minute of the shaft

f = frequency of the operating current (expressed in hertz)

N = number of running-winding poles

This formula gives what is called the *synchronous speed* of a motor. In a nonsynchronous motor, the actual speed is always less than the synchronous speed. This is due to *slip*, which results from losses of electrical and mechanical energy in the motor.

Rotating Stepper Motors

Refer again to Fig. 29-16. Assume that winding X is energized and winding Y is turned off. Notice that the north pole of the vertical arrow on the armature is pointing to the south pole of the stator (Fig. 29-16A). The rotation begins the instant that winding X is turned off and winding Y is turned on. Winding Y produces a south pole in the stator winding at position #2 (Fig. 29-16B). The magnetic forces of the newly created south pole and the armature's north pole cause the armature to move counterclockwise 18 degrees to position #2. This magnetic action also occurs between the other stator and rotor poles in the motor. To continue rotating, winding Y is turned off, and winding X is turned on. The current, however, in winding X is reversed, causing the stator pole in position #3 to become a south pole. This causes the armature to continue its counterclockwise rotation another 18 degrees for a total of 36 degrees (Fig. 29-16C). Thus, the armature rotates in step with the stator poles as they are switched in polarity. This step-by-step rotation continues as long as the windings are switched continuously. The rate at which the stator windings are turned on and off controls the speed of the rotation. To reverse the direction of a stepper motor, you only need to reverse the direction of current in one of the windings.

To switch the polarity of the two windings in a stepper motor (with leads 1, 2, 3, and 4), four transistors controlled by a computer software program are often used. More complex stepper motors are controlled by computer software and an integrated circuit that can easily adjust the speed, direction, and operating mode. Many types of stepper motors are available to meet the needs of industry. Figure 29-20 shows a sketch of a stepper motor.

CONVERTERS AND INVERTERS

A device that can change electrical energy from one form to another is called a *converter*.

1. Discuss the different ways to control speed, direction, and range of motion in motors.

2. Can you identify other electrical devices discussed in the text that could be classified as a converter or inverter? Name the devices and explain your rationale for the choice.

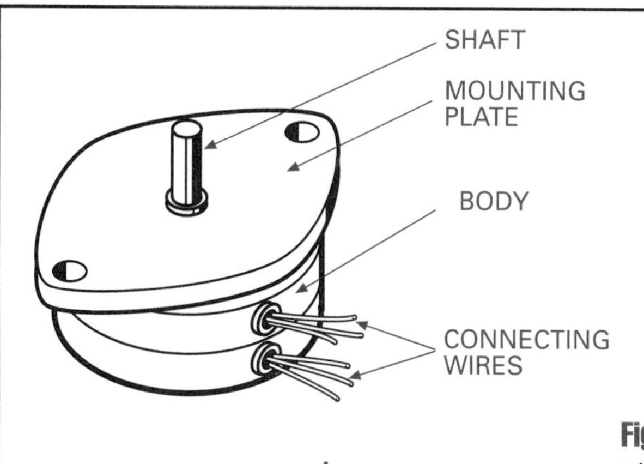

SPECIFICATIONS

12 VDC, 36.0 Ω COIL, 15 DEGREES PER STEP

BODY: 2.25″ (57.1 mm) DIAMETER × 0.95″ (24.1 mm) HIGH

SHAFT: 0.25″ (63 mm) DIAMETER × 0.65″ (69.9 mm) LONG

WIRES: 5/X

B

Fig. 29-20. Sketch of one type of small stepper motor.

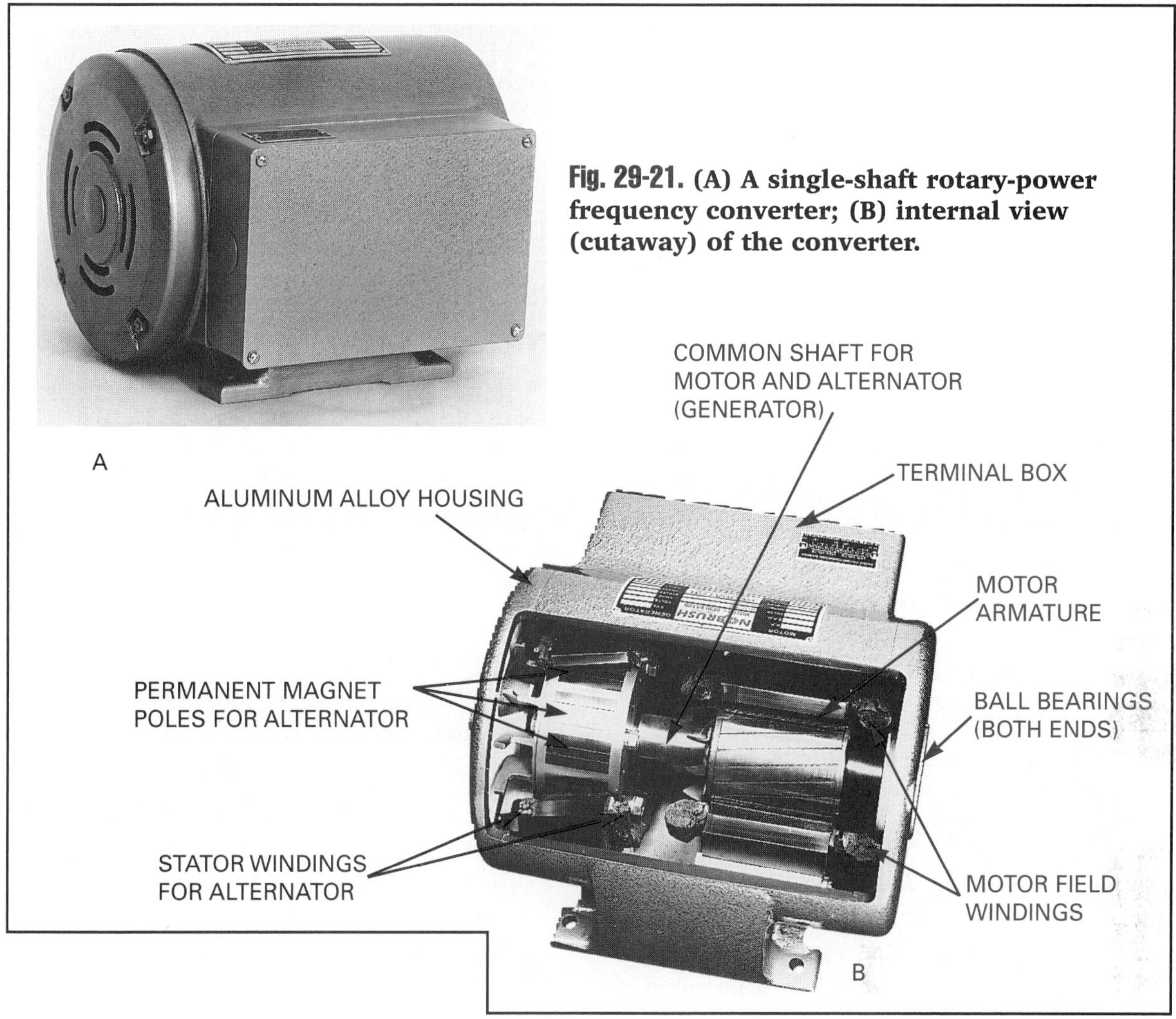

Fig. 29-21. (A) A single-shaft rotary-power frequency converter; (B) internal view (cutaway) of the converter.

An example is an ac-motor and a dc-generator that are connected together to change alternating current to direct current. A device, producing the opposite effect, that changes direct current to alternating current is called an *inverter*. Electrical devices are often used to change direct current to a direct current at a different voltage or to alternating current of a desired voltage and frequency. These devices can be a rotary (motor-generator) or solid state (electronic).

Figure 29-21A shows a power frequency converter. This rotary converter takes electrical input at one frequency and voltage and provides electrical output at a different frequency and usually at a different voltage. Single-shaft rotary-power converters have both the motor and generator rotors on the same shaft (Fig. 29-21B). These devices are manufactured in a single assembly. A motor generator, for example, that converts frequency with a single common shaft speed must have a magnetic pole count ratio according to the frequency shift requirement. A motor with 12 poles must be matched to a generator of 10 poles to achieve a 60 to 50 Hz conversion.

Modern rotary converters can be designed to provide non-standard high or low voltages. These devices are used, for example, in avionics, flight simulation, shipboard power, and high frequency tools.

MOTOR CARE AND MAINTENANCE

Electric motors are very dependable machines. However, to provide the safest and most efficient service they must be properly maintained. The following suggestions will help:

1. Before installing a new or different motor, read the information on its nameplate. Make sure that the right voltage, system (ac or dc), and wiring configuration are being used.
2. Always ground the metal frame of a motor. This is especially true if a motor is used in a damp location or near any grounded metal object.
3. If water enters the frame of a motor, the motor should be powered down and completely dried and checked for shorts before being put back into operation.
4. Never try to stop a motor by grabbing its shaft.
5. Never over oil motors. Apply a little oil to the bearings if necessary. Too much oil may damage insulation and cause dirt and dust to collect. Some motors have sealed bearings that do not need to be oiled or greased.
6. To prevent ordinary motors from overheating, keep the air openings in its frame open and any cooling fans operable.
7. Replace the brushes in universal motors when they become shorter than 1/4 in. (6.35 mm). Badly worn brushes will cause too much arcing at the commutator and contaminate it. Too much arcing is also caused by loose brush holders, loose commutator segments, and shorts in the rotor winding.
8. Unplug any motor that seems to be working at a speed slower than normal. Otherwise, the motor could become overheated. The motor may be overloaded or it may have a shorted winding. Check the motor speed with a tachometer.
9. Turn a split-phase motor off at once if it hums but fails to start after being turned on. This is very often due to a defective centrifugal switch mechanism. If the motor is left on, the running winding may burn out quickly.
10. An induction motor draws much more than the normal working current as it comes up to full speed. For this reason, such a motor should be protected with a dual-element time-delay fuse. The correct sizes of the fuses to be used with several single-phase, 115-V induction motors are given in Table 29-A. Small portable motors are adequately protected by the fuse or circuit breaker in the panel supplying their power.

Write a preventive maintenance procedure for two of the motors discussed in this chapter.

Table 29-A Size of Fuses used with Single-Phase, 115-V Induction Motors

Size of Motor (horsepower)	Equivalent (watts)	Fuse Size (amperes)
⅙	124	*
¼	187	*
⅓	249	*
½	373	*
¾	560	*
1	746	*
1½	1119	25
2	1492	30

*Motors rated at 1 hp (746W) or less are considered protected by the branch-circuit fuse in the panel supplying their power. In most cases this will be a 15- or 20-A fuse.

Chapter *Review*

Summary

- Motors change electrical energy into mechanical energy. This is done by using magnetic fields that alternately repel and attract each other to produce a circular motion of a rotor, or armature. These magnetic fields can be either permanent or electrically induced, but one of the two needed in a motor must alternate.
- To alternate a magnetic field, the polarity of the field must change. There are several different methods used to accomplish this. One method uses a reversing switch. Others use commutators and alternating currents.
- In order for motors to be used practically, they must be reversed, sped up and down, and stepped. Most motors can be controlled in these ways, but the exact method of doing so depends on the type of motor.
- Motor care and maintenance will extend the life of a motor and improve its efficiency. Always determine power and application requirements before installing motors. Motors that stall periodically should be removed for servicing.

Review Main Ideas

Review this chapter's main ideas by writing, on a separate sheet of paper, the word or words that most correctly complete the following statements:

1. Electric motors change ______ energy into ______ energy.
2. The size of an electric motor is given in terms of a ______ unit of power called the horsepower. One horsepower is equal to ______ watts of electrical power.
3. In series dc motors, ______ and ______ windings are connected in series.
4. The series dc motor has a high starting torque, or ______ force.
5. In shunt motors, the field and the rotor windings are connected in ______.
6. Compound motors are a combination of ______ and ______ type motors.
7. Universal motors work with either ______ or ______ current.
8. The two most common motors are ______ motors and ______ motors.
9. The rotor of a split-phase motor does not have ______ and is not connected to the ______.
10. A capacitor-start motor contains an unpolarized electrolytic capacitor connected in series with the ______ winding to increase starting torque.

11. ______ have field windings that consist of three separate sets of coils.
12. In repulsion-start motors, the ______ are not connected to a power line.
13. A ______ is designed to move in short, equal stops, and then start again in either direction.
14. The direction of rotation of a dc motor or a universal motor can be reversed by interchanging the connections to the ______ windings or to the ______ windings.

Apply Your Knowledge

1. Explain how a simple motor works.
2. Compare the advantages of a synchronous motor, a repulsion-start motor, and a shaded-pole motor.
3. How can the speed of a universal motor be controlled?
4. Give six procedures that are helpful in using and maintaining electric motors.
5. Describe four advantages of energy efficient motors.

Make Connections

1. **Communication Skills.** Write a report on the contribution of the English scientist Michael Faraday (1791-1867) to the development of the electric motor.
2. **Mathematics.** Define the term *horsepower*. Specify the size of the following motors in watts: 0.25 horsepower, 0.50 horsepower, and 2.5 horsepower.
3. **Science.** Write a brief report on how a rotating magnetic field is created in a three-phase alternating current motor.
4. **Technology Skills.** Research the type of motor used in an ac-operated clock system. Explain why this type of motor is used.
5. **Social Studies.** Electric motors use about 64% of the electric energy produced in the United States. In your own words, describe the impact the electric motor has had on society.

Chapter

Heating, Refrigeration , and Air Conditioning

Terms to Study

bimetallic thermostat

British thermal unit (Btu)

dielectric heating

electric arc furnace

electric arc welding

heat pump

induction heating

joule

power formula

radiant heating cable

refrigerant

resistance alloys

resistance heating

resistance wire

OBJECTIVES

After completing this chapter, you will be able to:

- Define *resistance heating.*
- Describe common uses for electric heat in the home and in industry.
- Explain the principles of induction, infrared, and dielectric heating.
- Compare and contrast resistance welding and electric arc welding.
- Explain how an electric-arc furnace is used to melt metals during the steelmaking process.
- Explain the fundamental principle upon which refrigerators and air conditioners operate.
- Name the main parts of an electric refrigerator system and explain their function.
- Explain how a typical air conditioner works.

This chapter discusses the variety of ways electricity is used to produce a heating or cooling effect. You will learn that some significant changes are occurring in the way refrigerants are being used to provide the cooling effect in refrigeration and other types of air conditioning.

HEATING

In the last few years, great strides have been made in manufacturing energy-efficient heat-producing equipment and appliances. Many devices are more reliable, last longer, and are more easily controlled with the use of solid-state electronics. Sensing devices, for example, are being installed in modern clothes dryers to detect humidity levels. When the proper humidity levels are reached, the dryer automatically shuts off—thus saving time and energy.

Most people think of heating as *space heating*, or warming the air temperature in a room, home, or other enclosed area. However, heat is used for more than just space heating. Electric energy can be converted into useful heat energy in a number of ways. In the home, electricity is used to produce heat for small appliances such as irons and toasters. It may also provide heat for clothes dryers and ranges. Electricity is also used to produce the heat needed for several industrial processes. These include welding, heat treatment, and steelmaking. In addition, special electronic generators produce the heat needed for certain medical treatments.

Typical range tops use coils which heat by resistance. Investigate the difference in range tops with a radiant element, halogen element or induction element.

Resistance Heating

Resistance heating uses the energy of moving electrons as they collide with particles in a conductor. If enough of these collisions occur, the friction that results heats the conductor (Fig. 30-1).

The amount of heat produced in any conductor by resistance heating is related to the **power formula**, $P = I R$. The heat produced is directly proportional to the resistance of the conductor and to the square of the current. This means, for example, that if the current is doubled, the amount of heat will be four times the original heat. However, if the current remains constant and the resistance is doubled, only twice as much heat will be produced.

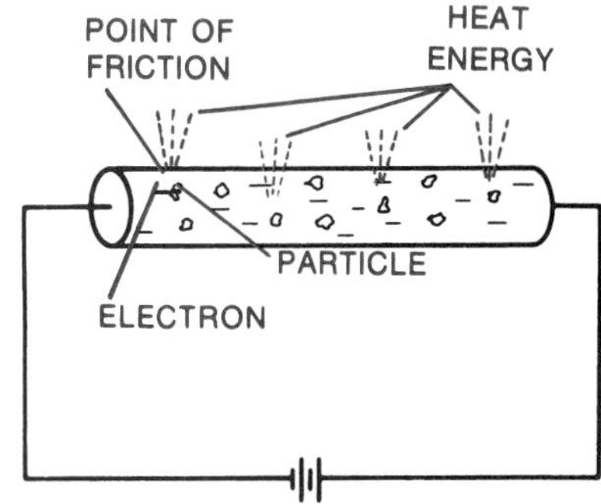

Fig. 30-1. Resistance heating.

FASCINATING FACTS

Relative humidity is the amount of moisture in the air at a specific temperature compared with the amount the air could hold at that temperature. Temperatures from 71° F with 70% relative humidity to 83.5° F with 30% relative humidity seem to be the most comfortable.

Resistance Alloys

The conductors used for resistance heating are made of two or more materials. These combinations of materials, called **resistance alloys,** have the amount of resistance needed to produce heat. The most common resistance alloys are made of nickel and chromium; copper and nickel; and nickel, chromium, and iron.

LINKS

Power formulas are discussed in more detail in Chapter 6, "Ohm's Law and Power Formulas."

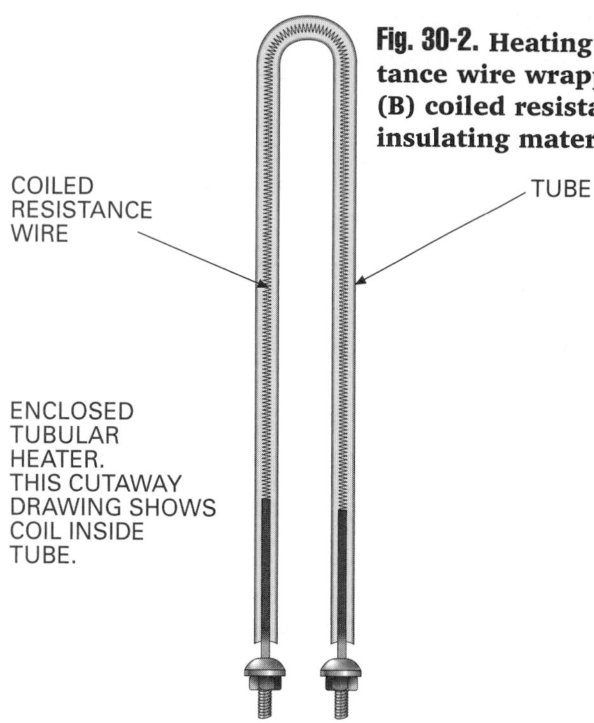

Fig. 30-2. Heating elements: (A) coiled resistance wire wrapped around insulators; (B) coiled resistance wire embedded in insulating material within a metal tube.

Heating Elements

In many heat-producing appliances and products, the resistance alloy is in the form of a round or flat wire called **resistance wire**. In some products, this wire is wrapped around an insulator such as porcelain or mica to form a heating element (Fig. 30-2). In other products, the resistance wire is shaped like a coil or a rod and is enclosed in a protective metal tube. In this kind of heating element, the resistance wire is usually embedded in a material such as magnesium oxide. This material is both a good electrical insulator and a good conductor of heat. Table 30-A gives the wattage, resistance, and wire sizes for wire heating elements made of a nickel/chrome alloy called Nichrome.

Radiant Heating

Radiant heating cable consists of resistance wire insulated with a heat-resistant rubber or thermoplastic compound. This cable is commonly used for space heating. The cable is placed in a concrete floor or in a plaster ceiling. In ceiling installation, the cable is first stapled to plasterboard and then covered with plaster

Radiant heating cable for space heating comes in several lengths. Each length is controlled by a thermostat. Each has a certain wattage rating, usually in the 400- to 5,000-W range. The heating elements are color-coded to indicate their wattage rating. They operate at either 115 V or 230 V.

Panels are also available for radiant space heating. Some panels are available in sizes of 2 x 4 ft (60.9 x 121.9 cm). They are designed to fit into a ceiling grid, replacing ceiling tiles. Two 220-Vac radiant-heating panels, each rated at 750 W, can provide sufficient heat to keep an average-size bedroom warm in winter.

Electric Water Heater

A typical electric water heater is shown in Fig. 30-3. The heating elements consist of resistance wire insulated from, and enclosed within, copper tubes. As the water in the tank heats, it cir-

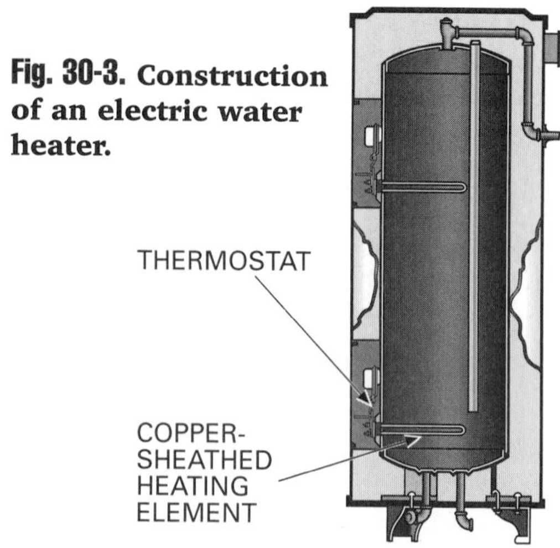

Fig. 30-3. Construction of an electric water heater.

Table 30-A Values for 115-V Operation

Wattage Required (W)	Approximate Resistance at 75°F (24°C) (Ω)	Recommended Sizes of Nichrome Wire [AWG]	
		Minimum	Maximum
100	118.100	30	26
150	78.732	30	26
200	59.050	29	25
250	47.240	28	24
300	39.366	28	24
350	33.742	27	23
400	29.525	26	22
450	26.244	24	20
500	23.620	24	20
550	21.472	23	19
600	19.683	23	19
650	18.170	23	19
700	16.871	22	18
750	15.745	22	18
800	14.762	22	18
850	13.894	21	17
900	13.122	21	17
950	12.431	21	17
1,000	11.810	20	16

culates around the elements. It slowly rises to the top of the tank, where the hot-water outlet pipe is located. The water is automatically kept at the right temperature by bimetallic thermostats. You will learn more about bimetallic thermostats later in this chapter.

Resistance Welding

Resistance heating is used for spot welding and seam welding. In spot welding, the metals to be welded are placed between two welder electrodes. Current passes through the metals from one electrode to the other. The resistance of the metals to the current produces a very high tem-

Fig. 30-4. **Resistance welding in industry. Sparks from this process show as long white streaks in the photograph.**

perature. This welds the metals together at the point of electrode contact (Fig. 30-4).

Seam welding is a special kind of spot welding. On a seam welder, the electrodes are wheels between which the metals to be welded are passed. As with spot welding, current passes through the metals, welding them together along a seam.

Electric Arc

A simple way of producing an electric arc is shown in Fig. 30-5. To start the arc, the ends of the carbon rods are brought into contact with each other. The heat produced by the current through the point of contact causes a small amount of the carbon to vaporize, or turn to gas. If the ends of the rods are then separated, current will *arc*, or pass, through the carbon vapor. This produces a very bright light and a very high temperature.

Electric Arc Welding

In **electric arc welding**, heat is produced by an arc formed between a welding rod and the metal object being welded. The rod is connected to one terminal of the welder. The metal object is connected to the other terminal (Fig. 30-6).

To start the arc, the welding rod is touched to the metal. The heat produced as current passes through the point of contact and vaporizes some of the rod, forming an arc. The heat of the arc is great enough to melt the welding rod and the workpiece. The extra metal needed for the weld is supplied by the rod.

The Electric Arc Furnace

An **electric arc furnace** is used to melt metals during the steelmaking process. In such a furnace, carbon electrodes produce heat using electric arcs. The electrodes are lowered through the top of the furnace. The furnace

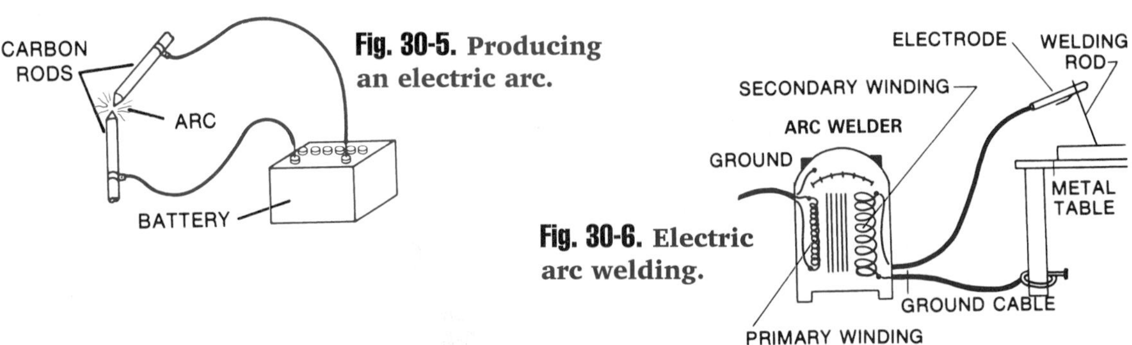

Fig. 30-5. **Producing an electric arc.**

Fig. 30-6. **Electric arc welding.**

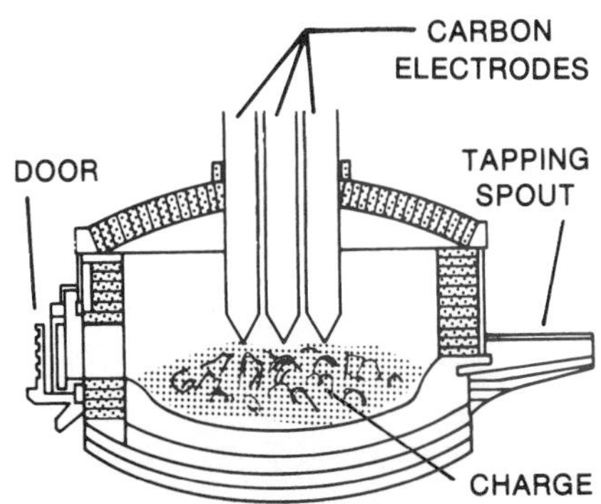

Fig. 30-7. The electric arc furnace.

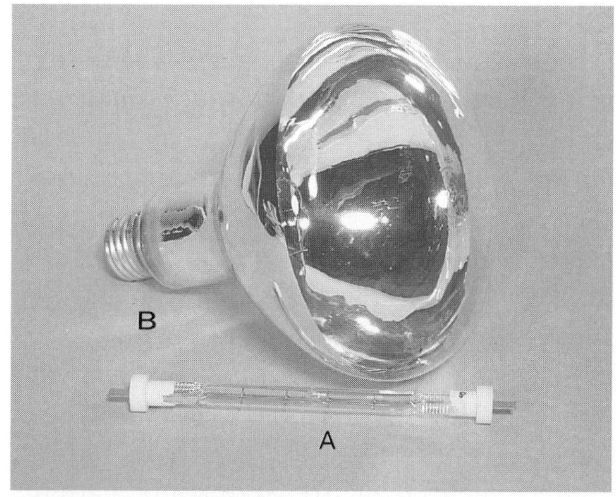

Fig. 30-8. Infrared lamps: (A) tubular quartz; (B) reflector.

contains a *charge*, or mixture, of scrap metals to be melted (Fig. 30-7). Arcs are then produced as current moves from the electrodes to the charge, melting the metal.

Infrared Heating

The prefix *infra* means "below." *Infrared rays* are a form of electromagnetic radiation that people cannot see. They have a frequency just below that of the lowest-frequency visible light, which is red.

The infrared rays used for heating are usually produced by filament lamps (Fig. 30-8). The filaments of these lamps operate at a much lower temperature than the filaments of ordinary incandescent lamps. Under this condition, the lamps produce little visible light but large quantities of infrared rays. When these rays strike a material that absorbs them, the energy of the rays is changed into heat with a high degree of efficiency (Fig. 30-9).

Fig. 30-9. The enamel finish applied to an automobile body is baked in an infrared oven. Banks of infrared lamps provide the necessary amount of heat.

Induction Heating

Induction heating takes place when a current is induced in a material by electromagnetic flux, just as in a transformer. In an induction heater, a high-frequency alternating current from 5,000 to 500,000 Hz is passed through a heater, or work, coil. This coil can be compared to the primary winding of a transformer.

As the magnetic field around the heater coil cuts across a metallic object to be heated, voltages are induced in the object because of electromagnetic induction. These voltages, in turn, produce high-frequency alternating currents that move through the object. The resistance of the metal to these induced currents causes it to become heated very rapidly.

The amount of heat produced and the depth to which this heat will penetrate an object are easily controlled by induction heating. For these reasons, this type of heating is used in a variety of industrial processes. These include forging, heat treatment, and soldering (Fig. 30-10).

LINKS

You may wish to review the principals of induction in Chapter 12, "Electromagnetic Induction and Inductance."

Fig. 30-10. **An induction heater.** ***(Jenzano)***

Dielectric Heating

Dielectric heating is used to produce heat in nonmetallic materials. In their simplest form, the working elements of a dielectric heater consist of two metal plates. The plates are connected to a high-frequency generator. The generator usually has an output with a range of 10 to 30 MHz. This produces a high-frequency electrostatic field between the plates and through the material to be heated.

The electrostatic field causes electrons within atoms of the material to be very rapidly drawn or distorted out of their normal paths. The electrons are forced to move first in one direction and then in the other (Fig. 30-11).

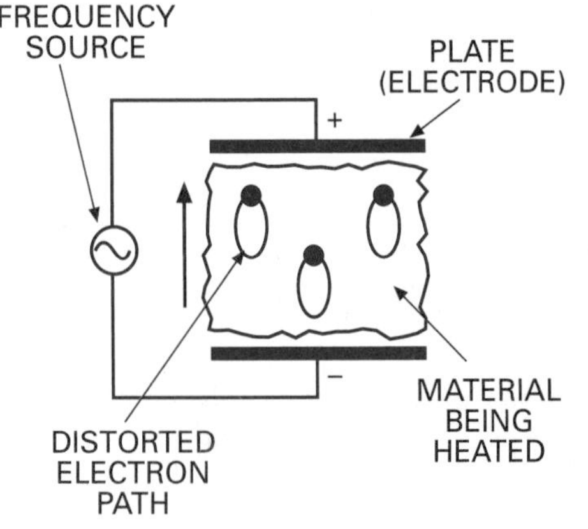

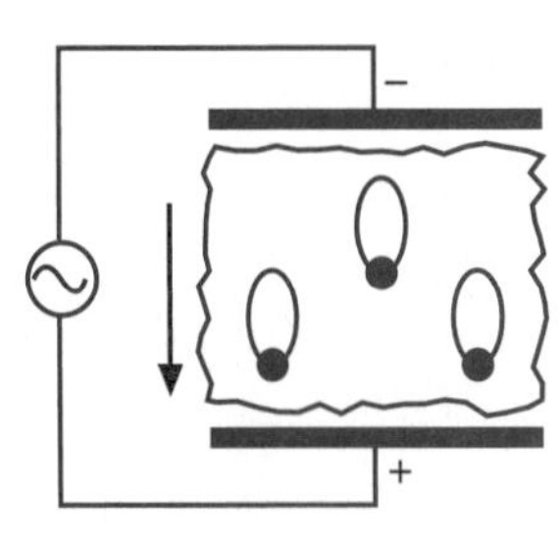

Fig. 30-11. **Principle of dielectric heating.**

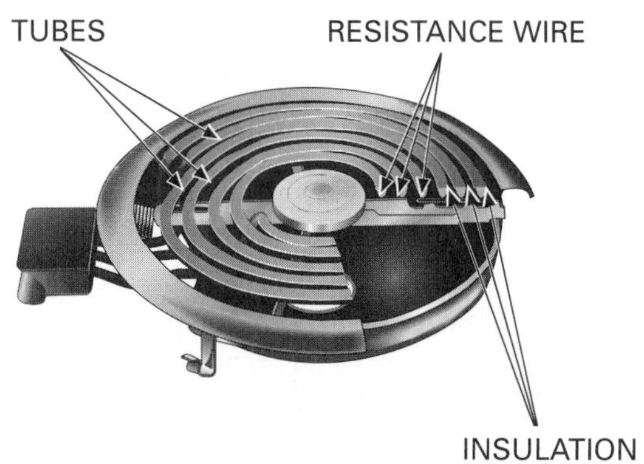

Fig. 30-12. Construction of a heating element for an electric range.

The friction that results from this back-and-forth movement of the electrons causes the material to be heated very quickly.

Dielectric heaters produce heat that can be controlled easily. In addition, the heat is evenly distributed through all parts of a nonmetallic material. For these reasons, these heaters are widely used for drying glue and heating wood, plastic, clay, and liquids.

Electric Range

A typical surface heating element for an electric range is shown in Fig. 30-12. The resistance wire of this unit is placed in a metal tube. It is insulated from the tube by an electrical insulator that is also an excellent conductor of heat.

Temperature Control

Current to each heating element is controlled by a heat-selector switch. This switch connects the sections of the heating element to the power line in different ways. For example, three-heat control of a simple two-ring heating element is shown in Fig. 30-13. Some ranges may have six or more heating-element rings that provide up to seven different levels of heat.

Oven Thermostat

A *thermostat* is a device used to control heating appliances automatically to maintain a desired temperature. It acts as a switch that can turn a circuit on or off. The temperature of the range oven is usually controlled by a bellows thermostat. The main parts of such a thermostat are the bulb, the capillary tube, the bellows, a switch control arm, and the switch contacts (Fig. 30-14). The switch contacts are connected in series with the oven power supply line. The bulb is filled with a liquid or a gas that expands when heated.

To set the oven temperature, the heat-selector dial is turned to the correct position. When the oven temperature rises above this setting, the pressure in the bulb forces the bellows to expand. The movement of the bellows is coupled to an arm that opens the circuit between the switch contacts (Fig. 30-14A). As the oven

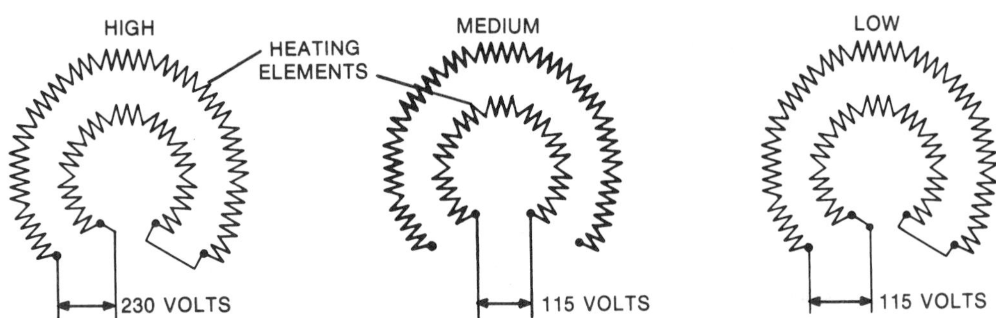

Fig. 30-13. Switching action for a two-ring, three-heat heating element for a range surface unit.

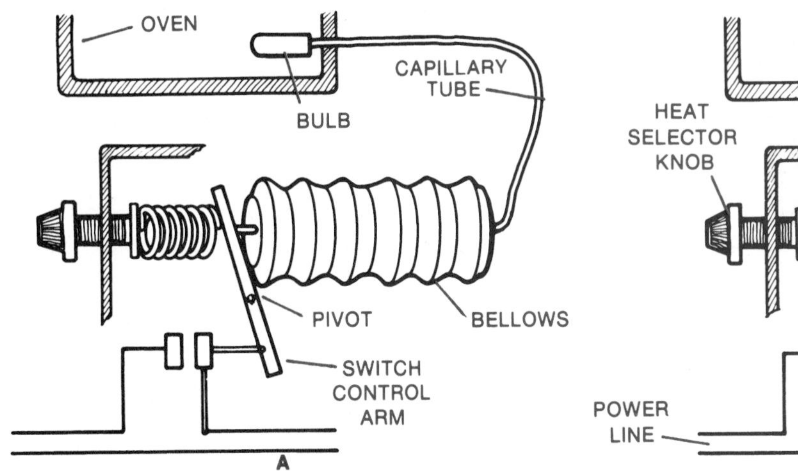

Fig. 30-14. Basic operation of a bellows oven thermostat.

cools, the bellows contract and the switch is turned on (Fig. 30-14B). This action of the thermostat is repeated as often as needed to keep the oven at the temperature shown on the heat-selector dial.

Bimetallic Thermostat

The moving part of the **bimetallic thermostat** used in many appliances is a strip of two different kinds of metal. These metals, which are rigidly joined together, expand at different rates when heated. As a result, the strip bends as it is heated. This bending is used to make or break a switching contact. The bimetallic strip may be straight or spiral.

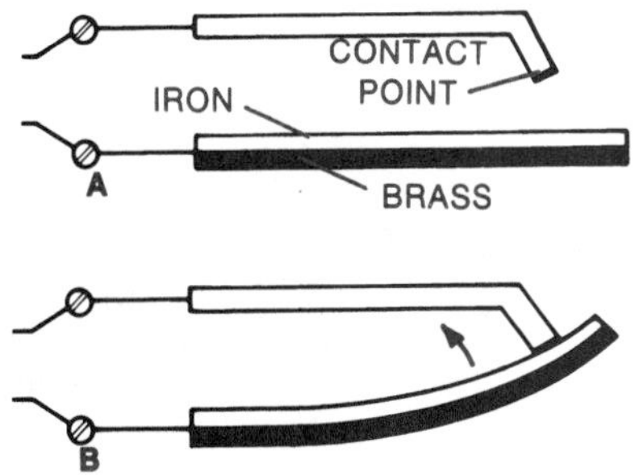

Fig. 30-15. Operation of a bimetallic thermostat: (A) open; (B) closed.

Figure 30-15 shows how a simple bimetallic thermostat works. In this thermostat, the bimetallic strip is placed near a fixed contact point (Fig. 30-15A). The thermostat is now in an "open" position.

When the thermostat is heated, the brass part of the bimetallic strip expands more than the iron. Because of this, the strip bends upward. It closes that part of the circuit connected to its terminals (Fig. 30-15B). When the strip cools, it bends back to its original position. This movement opens the circuit.

REFRIGERATORS

The operation of refrigerators depends on the fact that liquids absorb heat when they *evaporate*, or turn into gas (Fig. 30-16). In cooling appliances, a liquid called the **refrigerant** is circulated through tubes made of thin metal. These tubes form the *evaporator coils.*

The refrigerant has a very low boiling point and evaporates in the evaporator coils. As a result, the coils are cooled. They absorb heat from the air around them. After each evaporation, or cooling, cycle, the refrigerant *vapor*, or gas, is compressed and cooled. This changes the vapor back into a liquid. The refrigerant is then ready to be used again.

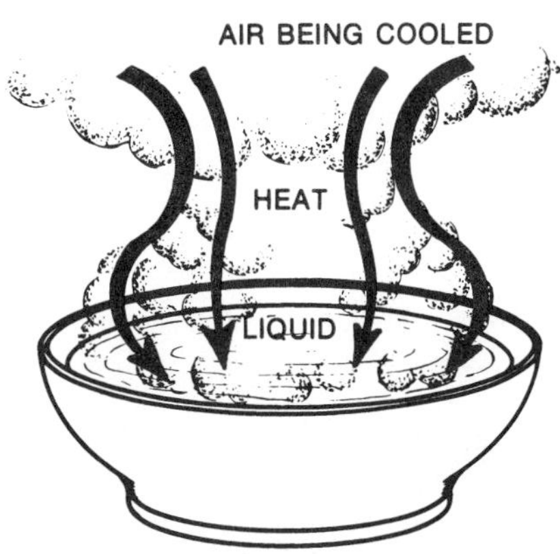

Fig. 30-16. Liquids absorb heat when they evaporate.

Types of Refrigerants

In the past, a refrigerant called CFC-12, known by the trade name *freon*, was widely used in various refrigerators, air conditioners, and other cooling devices. CFC-12 is a compound of chlorine, fluorine, and carbon. Some scientists, however, have concluded that CFC-12 and other chlorofluorocarbons may contribute to the depletion of the ozone layer that surrounds the earth. Some data indicate that these types of substances contribute to possible global warming and have other toxic characteristics.

Over 150 countries are concerned about protecting the earth's ozone layer. Therefore, new substitutions for refrigerants, such as hydrofluorocarbon (HFC134A), are now replacing CFC-12 in many devices. The new refrigerants are safer for human health and the environment. Other harmful refrigerants are being phased out.

Parts of a Refrigerator

Figure 30-17 shows the main parts of an electric refrigerator system. When the temperature in the refrigerator cabinet rises above a specific setting, the motor-driven compressor begins to operate. The compressor pumps the refrigerant vapor from the evaporator coils. This reduces the pressure in these coils, allowing the liquid refrigerant to move into them through the capillary tube. A bellows thermostat is used to control the temperature in the cabinet.

Cooling Cycle

As the liquid refrigerant evaporates, it absorbs heat from the freezer compartment and from the refrigerator cabinet. This causes an increase in vapor pressure, which forces the heated vapor through the compressor and into the condenser. In the condenser, the vapor cools. As the pressure reduces, the vapor changes back into a liquid. The liquid refrigerant is then forced back into the evaporator coils. Here it evaporates once again to complete the cooling cycle. This action is continuous while the refrigerator is in operation.

As more and more heat is absorbed from the cabinet, the temperature within the cabinet drops. This causes the temperature-control-thermostat contacts to open. This, in turn,

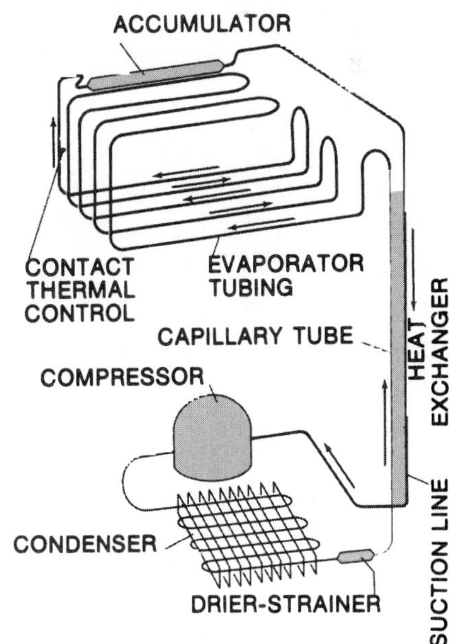

Fig. 30-17. Main parts of an electric refrigerator system.

turns the compressor motor off. The refrigerator remains idle until the cabinet temperature once again rises enough to cause the thermostat contacts to close and start the compressor motor.

Motor-Starting Relay

The refrigerator compressor and the motor that drives it are contained in a sealed unit. The motor-starting winding is connected and disconnected from the motor circuit by means of a relay mounted on the compressor housing.

When the cooling cycle starts, the line voltage is applied to terminals L1 and L2 of the relay (Fig. 30-18). Current passes through the running winding of the motor and the relay coil. The high starting current in the circuit causes the relay contacts to close. This completes the circuit through the starting winding of the motor.

The current through the running winding decreases in value after the motor has accelerated to full speed. This causes the starting-relay electromagnet to lose strength. The starting contacts open. The motor then continues to operate with its starting winding disconnected from the circuit.

Overload Protection

The refrigerator motor-starting relay often has a protective device. This device opens the motor circuit when the motor becomes overloaded. An overloaded motor runs more slowly than usual. If it is allowed to continue running, it may be seriously damaged by the increased amount of current passing through its running winding.

To prevent damage, a bimetallic thermostat is connected in series with the motor circuit. When an abnormally high current passes through the element, its temperature increases until the thermostat contacts open. This automatically turns the motor off.

AIR CONDITIONERS

Air conditioners operate on the same principle as refrigerators. Like refrigerators, they use a refrigerant, and they use the same basic cooling cycle. The three main types are room air conditioners, central air conditioning systems, and heat pumps.

Although air conditioners are primarily used for cooling air, what are other potential applications?

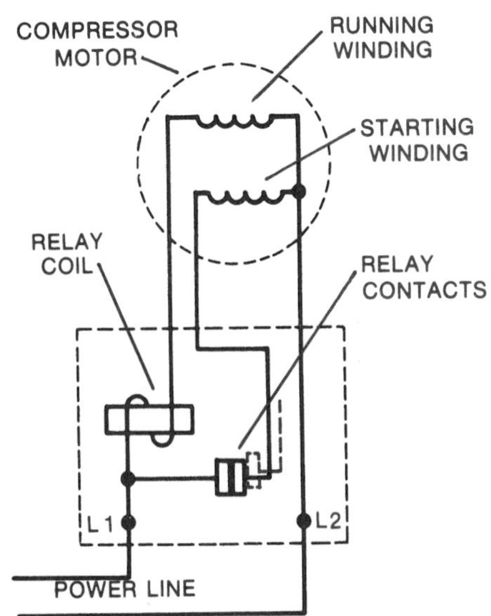

Fig. 30-18. Circuit of a refrigerator motor-starting relay.

James Joule

Physicist James Prescott Joule made important discoveries in electricity. Joule proved that heat was a form of energy. Investigating heat emitted from an electrical circuit, he formulated Joule's law of electric heating. This states that the amount of heat produced each second in a conductor by electrical current is proportional to the conductor's resistance and to the square of the current.

Joule also proved the equivalence of mechanical energy and heat. Therefore, in an isolated system, work can be converted into heat at a ratio of one to one. Joule expressed the numerical relationship between heat and mechanical energy in the joule. The joule equals one watt-second, or about 0.000948 Btu.

Joule and physicist William Thomson found that the temperature of a gas falls when it expands without doing any work. This principle, known as the Joule-Thomson effect, is a basic operating principle for refrigeration and air conditioning systems.

Room Air Conditioner

A room air conditioner is a refrigeration unit that cools, dehumidifies, cleans, and circulates the air in an enclosed space. In general, the main parts of a typical room air-conditioning unit are similar to those of a refrigerator (Fig. 30-19).

Room air is drawn through the filter to remove dirt, dust, and other particles. The air then passes through an evaporator-coil assembly, where it is cooled. The cool air is then recirculated through the room by blowers.

As the air moves past the cooled evaporator coils, some of the moisture in it *condenses*, or collects, on the coils. As a result, the air leaves the coils in a *dehumidified*, or drier, condition, making the room more comfortable. The moisture that collects at the coils is discharged from the air conditioner as water or water vapor.

The cooling capacity of a room air conditioner is given in terms of the **British thermal unit (Btu)**. The metric unit is the **joule**. One British thermal unit is the quantity of heat needed to

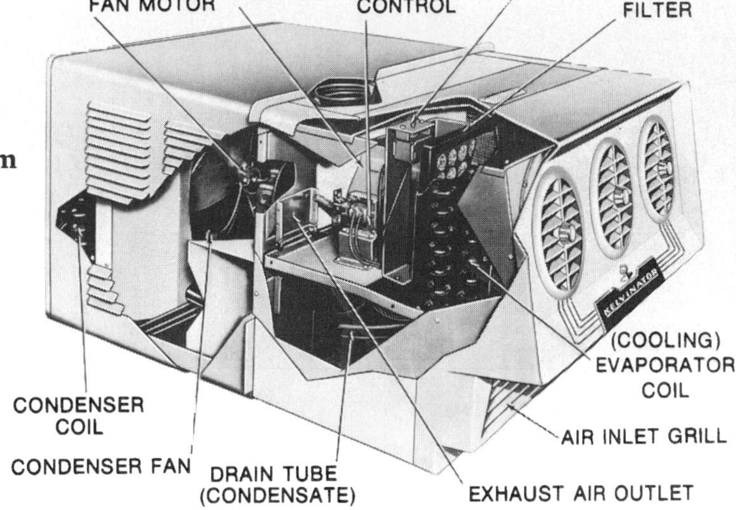

Fig. 30-19. **A typical room air conditioner.**

raise the temperature of one pound of water one degree Fahrenheit. The Btu rating of an air conditioner is a measure of how much heat it will remove from a room during a certain period of time, usually 1 hour. The effectiveness of an air conditioner is also indicated by the amount of air it can move in a given period of time. The unit usually used is cubic feet per minute (cfm).

Central Air Conditioner

A central air-conditioning system cools all areas of a house. In such a system, the condensing unit generally is located outside the house. When used in a house with a forced-air furnace, the evaporator-coil assembly is usually located above the furnace in the *plenum*, or main hot/cool output air duct (Fig. 30-20).

Heat Pump

The electrically driven **heat pump** is a combination heating and cooling unit. It transfers heat from one location to another in much the same way as a refrigerator.

Winter Operation

When the heat pump is used as a heater, a liquid refrigerant is passed through an outdoor coil (Fig. 30-21A). As the refrigerant evaporates, it absorbs heat from the outside air. The heat-carrying refrigerant is then compressed. This raises its temperature to 100°F (37.8°C) or higher. The refrigerant is then passed through the indoor coil. This acts as a heating unit. Cool air in the rooms is circulated over the warm coils. The warm air is then recirculated through the rooms.

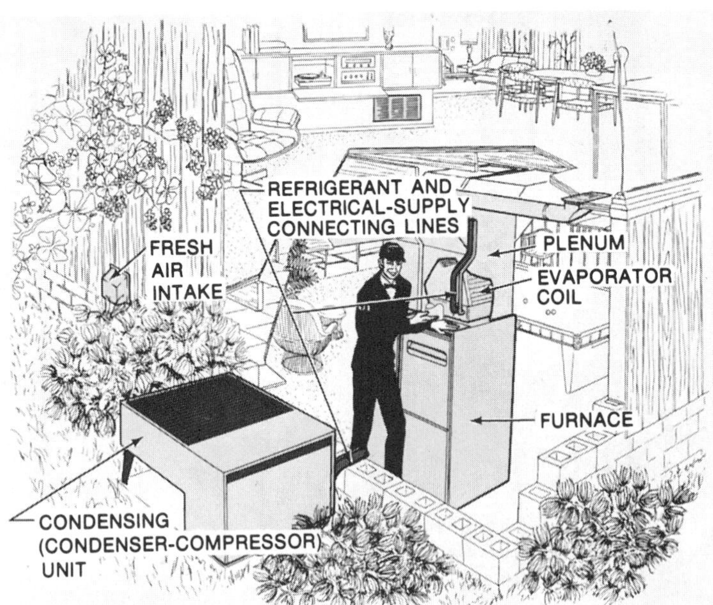

Fig. 30-20. Main parts of a central air-conditioning system for a home.

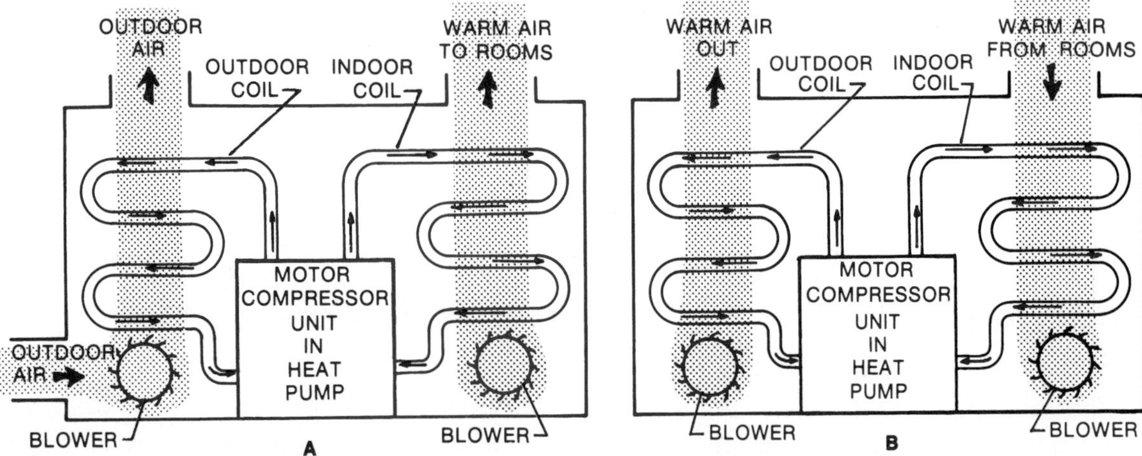

Fig. 30-21. Basic operation of a heat pump: (A) winter operation; (B) summer operation.

During very cold weather, a heat pump may not be able to absorb enough heat from the outside air to provide proper space heating. In that case, auxiliary resistance-heating units are used. These units are automatically turned on by a thermostat when the temperature drops below a certain point. The units operate until the heat pump alone is once again able to provide enough heat.

Summer Operation

During the summer, the operation of the heat pump is automatically reversed by a temperature-control thermostat. Then warm air in the rooms is circulated over the indoor coil, which is cooled by the evaporation of the refrigerant (Fig. 30-21B). The refrigerant then carries the heat it has absorbed from the rooms to the outdoor coil. There the heat is released. The refrigerant condenses to begin another cooling cycle.

Chapter *Review*

Summary

- Resistance heating occurs when electrons collide with particles in a conductor, producing friction that results in heat. The amount of heat produced is directly proportional to the resistance of the conductor and to the square of the current.
- Resistance heating is used for space heating and for heat-producing products such as clothes dryers, irons, and toasters. It is also used for resistance-welding processes such as spot welding and seam welding.
- Other types of heating used in the home and in industry include infrared, induction, and dielectric heating.
- Heat can also be used to control thermostats, which keep the temperature of an appliance at a steady level. Two common types of thermostats are bellows and bimetallic thermostats.
- Refrigerators and air conditioners operate on the principle that liquids absorb heat when they evaporate. The cooling cycle begins by circulating refrigerants through evaporator coils, where they vaporize into gas. Then the refrigerant is compressed and cooled, which changes it back into a liquid state.

Review Main Ideas

Review this chapter's main ideas by writing, on a separate sheet of paper, the word or words that most correctly complete the following statements:

1. The amount of heat produced in a conductor by resistance heating is directly proportional to the ______ of the conductor and to the ______ of the current.
2. The most common materials used in making resistance-heating wires are alloys consisting of various combinations of ______, ______, ______, and ______.
3. Resistance heating is used for ______ welding and ______ welding.
4. In an electric arc furnace, arcs are produced as ______ moves from the electrode to a charge, melting the metal.
5. Infrared rays are ______ radiation having a frequency just below that of the lowest-frequency ______ light.
6. Induction heating is produced by high-frequency ______ passing through a heater coil.
7. Dielectric heating is used to produce heat in ______ materials.
8. A thermostat acts as a ______, that can turn on or off.
9. The heat-sensing bulb of a bellows thermostat is filled with a liquid or a gas that ______ when heated.

10. The operation of a bimetallic thermostat depends on the fact that some metals ______ at different rates when heated.
11. The operation of refrigerators depends on the fact that all liquids absorb ______ when they ______.
12. In cooling appliances, a liquid known as a ______ is circulated through thin metal tubes called ______ coils.
13. In the past, the most commonly used refrigerant was ______ or CFC-12.
14. The ______ pumps the refrigerant vapor from the evaporator coils.
15. As refrigerant evaporates, it absorbs heat, causing an ______ in vapor pressure, forcing vapor into the ______.
16. The purpose of the condenser in a refrigerator is to ______ the evaporated refrigerant and change it back into a ______.
17. When a refrigerator's motor overloads, a protective device in the ______ opens the motor circuit.
18. When air in an air conditioner moves past cooled evaporator coils, the moisture ______, leaving the coils ______.
19. The cooling capacity of an air conditioner is given in terms of the ______ unit.
20. The electrically driven heat pump is a combination ______ and ______ unit.

Apply Your Knowledge

1. Besides space heating, describe three uses of heating in the home and industry.
2. Describe the similarities and differences between bellows and bimetallic thermostats.
3. Compare and contrast resistance welding and electric arc welding.
4. In what way are resistance heating and dielectric heating similar?
5. Name at least one difference between a regular room air conditioner and a heat pump.

Make Connections

1. **Communication Skills.** In a brief written report, describe the cooling-cycle process in a refrigerator.
2. **Mathematics.** Demonstrate mathematically for the class that when the resistance heating effect of electric current is doubled, the amount of heat will be four times the original heat.
3. **Science.** In a brief written report, describe the types of electric currents that occur in a metallic object that is heated by induction.
4. **Workplace Skills.** Draw a schematic diagram of a two-element heating unit for an electric range. Describe how three different temperatures for cooking can be obtained.
5. **Social Studies.** Search the Internet for the topic "Global Warming." Research the effects of the refrigerant CFC-12 on global warming.

Chapter 31

Producing Chemical Reactions

Terms to Study

chemical reaction

electrolysis

electrolyte

electrolytic furnace

electroplating

Hall process

plating pen

OBJECTIVES

After completing this chapter, you will be able to:

- Define *electrolysis.*
- Explain the process of electroplating a metal object.
- Describe the Hall process of extracting aluminum from bauxite.
- Name five metals that are extracted from their ores by electrolysis.

This chapter discusses the relationship between chemistry and electricity. You will learn that electric energy can be used to increase the value of some metals through a process called *electroplating.* An ordinary bracelet made from brass, for example, increases in value when it is electroplated with gold or silver. You will also learn how aluminum is extracted from its ore.

CHEMICAL REACTIONS AND ELECTRICITY

A **chemical reaction** occurs when two or more chemicals act on each other. This reaction usually results in changes to the materials. For example, when oxygen reacts with iron, a new substance called *iron oxide* (rust) is formed. When paper is burned, it changes into ashes and smoke. In each case, the composition of the molecules in the substance changes. The new substances have properties that are different from the those of the original materials.

The chemical reactions that take place when electric energy is applied to a substance are caused by the movement of charges through the substance. A liquid that ionizes to form positive and negative ions and that can conduct these charges is called an **electrolyte**.

LINKS

See Chapter 16, "The Electric Power Industry," for a review of electrolytes.

ELECTROLYSIS

Electrolytes cause chemical changes by a process known as **electrolysis.** In this process, electrolytes react with other solutions or with materials such as metals to change them. Electrolysis is widely used in industry to perform the following tasks:

- producing specific gases and chemicals.
- electroplating.
- refining copper.
- extracting metals such as aluminum and magnesium from their ores.

Electrolysis of Water

Pure water is a very poor conductor of electricity because it does not easily ionize. However, when certain chemical compounds are added to pure water, the solution becomes a good conductor. When a voltage is applied to such a solution, electrolysis occurs. The water breaks down into its two chemical elements, the gases hydrogen and oxygen.

In one industrial process for making hydrogen and oxygen gases, sulfuric acid is added to

1. **Examine the role a *catalyst* plays in producing a chemical reaction.**
2. **Chlorine is a gas obtained from salt water through electrolysis. What other elements are simultaneously produced in this process?**

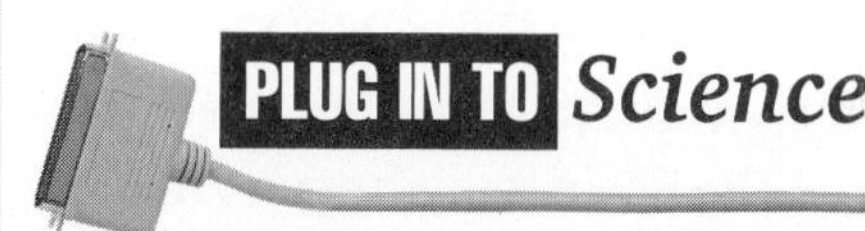

PLUG IN TO *Science*

Sulfuric Acid

One chemical highly valued for producing chemical reactions is sulfuric acid (H_2SO_4). Used in making thousands of everyday products, it has been called the most important industrial chemical.

Also called hydrogen sulfate, sulfuric acid is a highly corrosive, dense, colorless liquid with an oily consistency. In its concentrated form, it looks like a clear, rather heavy syrup. Because of its oily appearance, people years ago called it oil of vitriol. This name is now used only for the commercial grade of concentrated sulfuric acid.

Two general properties largely account for its corrosive action. First, it has a strong affinity for water. If a solid substance contains water, sulfuric acid can pull the water molecules into itself, drying the substance.

Secondly, sulfuric acid in dilute solutions has a tendency to ionize the hydrogen and sulfate (SO_4) groups, leaving the negatively charged sulfate ion free to attach itself to other atoms. Sulfuric acid reacts readily with many metals and with carbon, sulfur, and other substances. For these reasons, it is widely used as an electrolyte in storage batteries and in electroplating baths, and to clean oil and grease off metals.

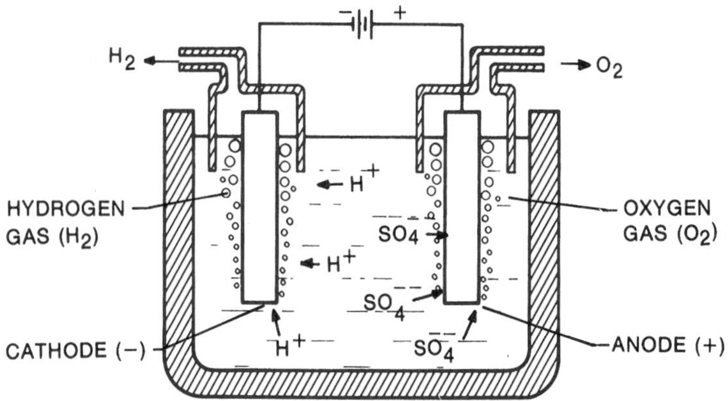

Fig. 31-1. Producing hydrogen and oxygen gases by electrolysis.

pure water. Each molecule of the acid ionizes into two positive hydrogen ions (H^+) and one negative sulfate ion having two negative charges (SO_4^{--}). The positive hydrogen ions are attracted to the cathode (negative electrode). At the same time, the negative sulfate ions are attracted to the anode (positive electrode) (Fig. 31-1). At the cathode, each hydrogen ion obtains an electron. It thereby becomes a neutral atom of hydrogen. Hydrogen, being a gas, rises from the solution. It collects at the cathode. At the anode, the negative sulfate ions deposit their excess electrons and become neutral. However, the sulfate reacts at once with the hydrogen from the water, forming sulfuric acid again. The oxygen in the water, which remains after the hydrogen has been removed, rises through the solution. It collects at the anode. This process continues as the sulfuric acid is recycled through the ionization and sulfate stages.

Electroplating

Electroplating is a form of electrolysis used to coat a base material (usually a metal) with a thin layer of another metal. The base metal can be iron, steel, brass, or a similar metal. It can be electroplated with one or more metals, including cadmium, chromium, copper, gold, silver, and tin (Fig. 31-2).

Cadmium plating protects the base metal from corrosion. Chromium is used to improve the appearance of base metals and to provide a much harder surface. Copper plating is generally used to provide an undercoating over which other metals can be plated. Gold plating is used for decorative purposes and for improving the conductivity of contact points in some electrical and electronic equipment. Tin plating allows easy soldering of surfaces. It is also used frequently on surfaces likely to come in contact with food.

Fig. 31-2. Electroplating unit in an industrial plant.

FASCINATING FACTS

A very rare element, rhodium, is used as an electroplate to protect silver from tarnishing. Rhodium has a high reflective quality. Along with gold, platinum, and palladium, rhodium is considered a noble metal since it very rarely forms compounds with other elements.

The Plating Solution

Electroplating requires the use of a plating solution. This solution is an electrolyte that contains a compound of the plating metal. In copper plating, for example, the electrolyte consists mainly of a solution of water and copper sulfate ($CuSO_4$).

Basic Operation

The construction and operation of a basic copper-plating assembly is shown in Fig. 31-3. Here the object to be plated (the cathode) and a piece of pure copper (the anode) are suspended in the electrolyte and connected to a battery.

The copper sulfate in the solution ionizes into positive copper ions (Cu^{++}) and negative sulfate ions (SO_4^{--}). The copper ions are attracted to the cathode. At the cathode, they obtain the needed electrons and become metallic copper. This copper metal is deposited on the object being plated. The sulfate ions are attracted to the anode. There they combine with copper atoms to form copper sulfate. This goes into solution and ionizes. Thus, the copper sulfate replaces the copper removed from, or plated out of, the electrolyte. During this time, the current in the external circuit consists of electrons that move into the cathode from the negative terminal of the battery and electrons that move into the positive terminal of the battery from the anode.

As the electroplating process continues, the pure copper that forms the anode is gradually used up and must be replaced. At the same time, the coating of copper deposited on the object being plated gets thicker.

The time needed to electroplate a base metal to a given thickness depends on the kind of plating metal used and the density of the ions in the electrolyte. Ion density is indicated by the value of the current in the external circuit. For this reason, a practical electroplating assembly has meters that show the current in the external circuit and the voltage at which the circuit is operating. The circuit also has controls for increasing and decreasing the plating current as needed to electroplate different metals and adjust the plating rate.

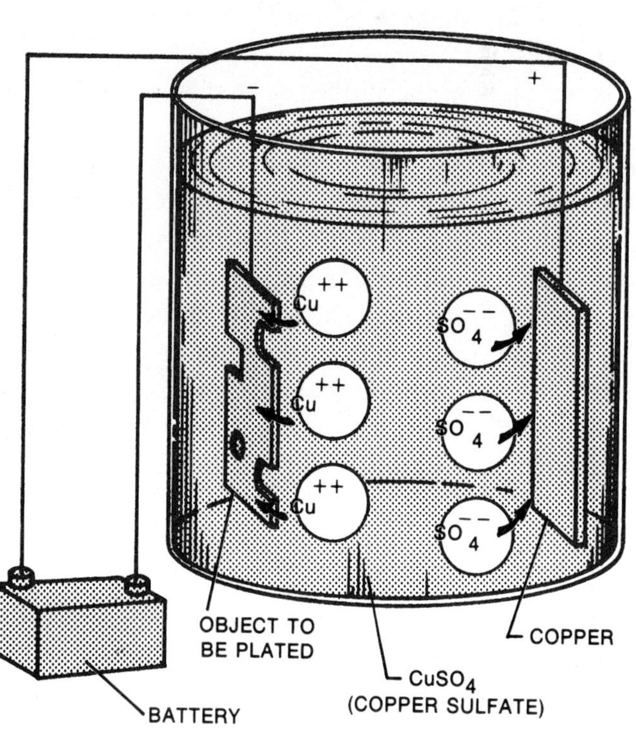

Fig. 31-3. The copper-plating circuit.

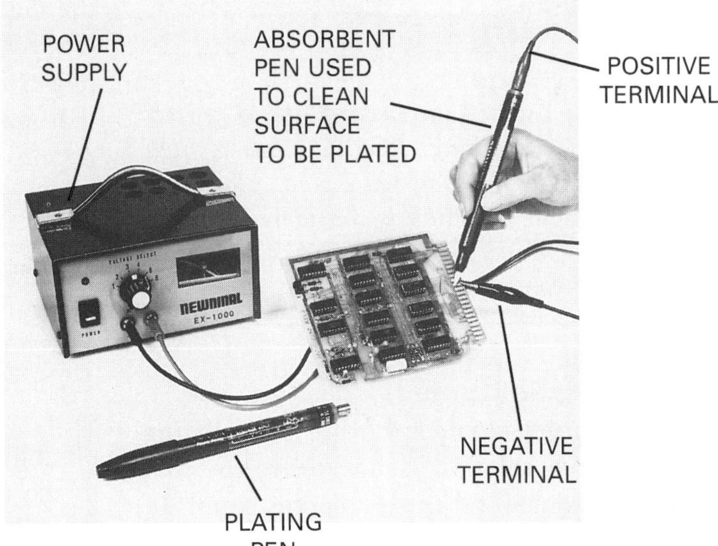

Fig. 31-4. Repairing the surface of terminals by plating with disposable plating pens.

Partial Plating

Partial plating is useful in applications such as plated terminals on printed circuit boards. Plated portions of these boards can be repaired using a compact electroplating system and a **plating pen** (Fig. 31-4). This system uses disposable cartridge pens and a power supply capable of supplying 12 Vdc at 5.0 A. Each plating pen contains its own nontoxic solution. Several types of pens are available for plating; examples include pens for plating with copper, gold, and silver.

Before plating, the surface to be plated must be polished and then wiped clean. The part to be plated is connected to the negative (-) terminal of the power supply. An absorbent pen is connected to the positive (+) terminal. The absorbent pen operates on 12 Vdc and chemically cleans (degreases) the surface to be plated. This is done by rubbing the absorbent pen point back and forth on the surface to be plated. After degreasing, the surface is rinsed clean with water. The absorbent pen is then disconnected from the positive terminal of the power supply and is replaced by the plating pen. The operating voltage is adjusted for the plating to be applied. For example, for a silver plating pen, the voltage is adjusted to a value between 5.0 and 6.0 Vdc. The plating process is accomplished by moving the pen lightly back and forth on the surface to be plated. In about 2 or 3 minutes, a silver plate is deposited on the surface. After the plating process is complete, the part is rinsed with water and wiped with a soft, clean cloth.

SAFETY TIP

Before rinsing surfaces, always turn off the power supply and disconnect the object to be plated.

EXTRACTING METALS FROM ORES

The production of aluminum is the most common process in which electrolysis is used to extract metals from ore. The first step in producing aluminum is to change the ore (*bauxite*) into a white powder called *alumina* (aluminum oxide). Then, in a method known as the **Hall process,** alumina is dissolved in molten *cryolite* (sodium aluminum fluoride) to form the electrolyte. The heat needed to keep the electrolyte in liquid form is produced by the current needed for electrolysis. Carbon rods and a carbon lining make up the cathode and the anode. This system is referred to as an **electrolytic furnace** (Fig. 31-5).

As a result of the ionization of aluminum oxide, both negative oxygen ions and positive aluminum ions are present in the electrolyte. The aluminum ions are attracted to the cathode. There they obtain electrons and are discharged to form aluminum atoms. The molten aluminum then collects at the bottom of the furnace and is withdrawn.

Several other metals are extracted from their ores by electrolysis. These include magnesium, sodium, potassium, and cesium.

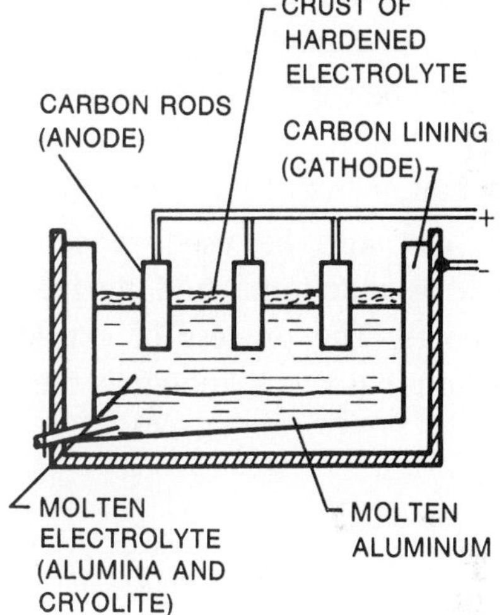

Fig. 31-5. **Basic construction of an electrolytic furnace used for producing aluminum.**

Chapter *Review*

Summary

- An electrolyte is a liquid that ionizes to form positive and negative ions and that can conduct electrical charges.
- Electrolysis is a process in which electrolytes react with other solutions or with materials such as metals to change them.
- Electrolysis can be used to plate base materials with a thin coating of another substance, usually a metal.
- Electrolysis can also be used to separate metals such as aluminum and magnesium from their ores. The process of extracting aluminum from bauxite is known as the Hall process.

Review Main Ideas

Review this chapter's main ideas by writing, on a separate sheet of paper, the word or words that most correctly complete the following statements:

1. A ______ changes the composition of the molecules in a substance.

2. A solution that will produce positive and negative ions and that will conduct these charges is called an ______.

3. Electrolytes react with other solutions or materials to change them in the process known as ______.

4. Electrolysis is used for producing specific ______ and chemicals, electroplating, refining copper, and ______ metals from their ores.

5. Pure water is a very poor conductor because it does not ______ easily.

6. As a result of electrolysis, water is broken down into the gases ______ and ______.

7. Cadmium plating is done to protect a base metal from ______.

8. In using a plating pen, the surface to be plated must first be ______ and ______.

9. The plating pen is connected to a ______ terminal, and the part to be plated is connected to a ______ terminal.

10. Alumina is dissolved in molten cryolite to form an electrolyte in a method known as the ______.
11. Metals extracted from their ores by electrolysis include ______, ______, ______, ______, and ______.

Apply Your Knowledge

1. Explain the purpose of electrolytes in the electroplating process.
2. Explain why the time needed to electroplate a base metal to a given thickness depends partly on the voltage at which the circuit operates.
3. Why is it important that the electrolyte solution for a given plating process contain a compound of the plating metal?
4. Explain how electrolysis could be used to refine copper.

Make Connections

1. **Communication Skills.** Identify various objects in and around your home that have been electroplated. Prepare a chart that classifies the objects by the metal used for electroplating.
2. **Mathematics.** Refer to the Social Studies Connection below. Compute the percentage of metal in each category that comes from the United States, compared to the metals imported from other countries.
3. **Science.** Research the process of plating an automobile bumper, or another part of an automobile, with the metal chromium. Write a report on the process.
4. **Technology Skills.** Identify a practical application for a saltwater rheostat. Demonstrate the use of a saltwater rheostat.
5. **Social Studies.** The electronic industries in the United States use a variety of metals, such as aluminum, magnesium, iron, copper, tin, lead, gold, silver, and chromium. Prepare a wall chart identifying the countries that are major producers of these metals.

Chapter 32

Lamps, Lasers, and Fiber Optics

OBJECTIVES

After completing this chapter, you will be able to:

- Specify the frequency range and wavelength of visible light.
- Explain the process of generating a laser beam in a ruby laser.
- Describe the way in which a light wave is transmitted in a fiber optic cable.
- Identify five elements in a fiber optic communication system.

Terms to Study

ballast

clad

coherent light

incandescent

incoherent light

laser

lumens

reflection

refraction

multiplexing

Have you ever watched a sunset? It's quite a sight. On a clear day the sun appears overhead as a bright light in the sky. As it sets in the evening, colored rays appear, especially red. This chapter provides an answer for the changing colors seen during a sunset. It also discusses a variety of devices that use light energy, including lasers and fiber optics.

LAMPS

Electric energy can be changed into light energy in a number of ways. The first practical device for producing light by electricity was the carbon-filament lamp invented by Thomas Edison in 1879. Since then, many kinds of lamps and other lighting devices have been developed.

Lamps come in all shapes, sizes, and brightnesses. From the torches of long ago to today's halogen headlights, mankind has used lamps to light the way for centuries. This chapter discusses several of the most commonly used electric lamps.

Incandescent Lamps

The word **incandescent** means "glowing from intense heat." In an *incandescent lamp*, current passes through a filament of a metal, tungsten. This causes the filament to become incandescent, or white hot (Fig. 32-1).

When an incandescent lamp is made, most of the air is removed from the glass bulb before it is sealed. This is done to keep oxygen from coming into contact with the filament because oxygen would cause it to burn out quickly. A mixture of the gases nitrogen and argon is placed in the bulb to lessen the filament's evaporation caused by its high temperature. This lengthens the life of the lamp.

The life of general-service incandescent lamps for home use ranges from 750 to 1,500 hours. The average life of a lamp is printed on the carton.

Incandescent lamps come in a variety of sizes and shapes. The electrical size of a general-service incandescent lamp is given in watts and in *lumens* (lm). The wattage rating indicates the amount of energy the lamp uses. The **lumen** rating indicates the amount of light a lamp produces. A 100 W light bulb gives off about 1,000 lm.

It is interesting to note that the white light given off by the light bulb represents a very small range of the electromagnetic spectrum shown in Figure 32-2. Two important characteristics of these waves are the wavelength and frequency. Wavelength is the distance from one crest of the wave to the next crest. Frequency is the number of crests of a wave that pass a single point in one second. As the frequency increases, the wavelength gets shorter.

LINKS

Chapter 27, "Communication Systems", discusses the speed of electromagnetic waves (about 186,000 miles per second).

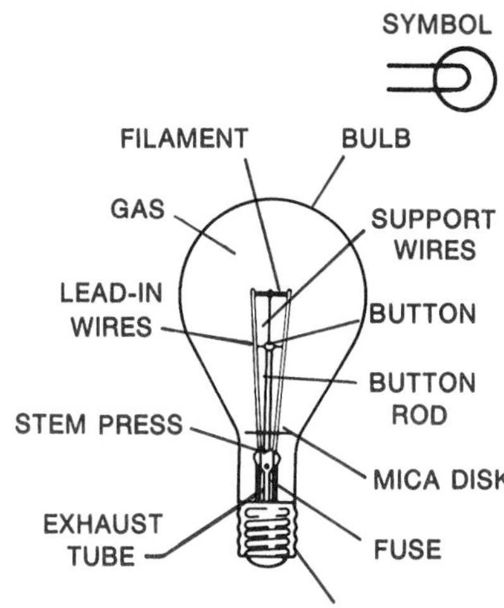

Fig. 32-1. Construction of a typical general-purpose, incandescent lamp.

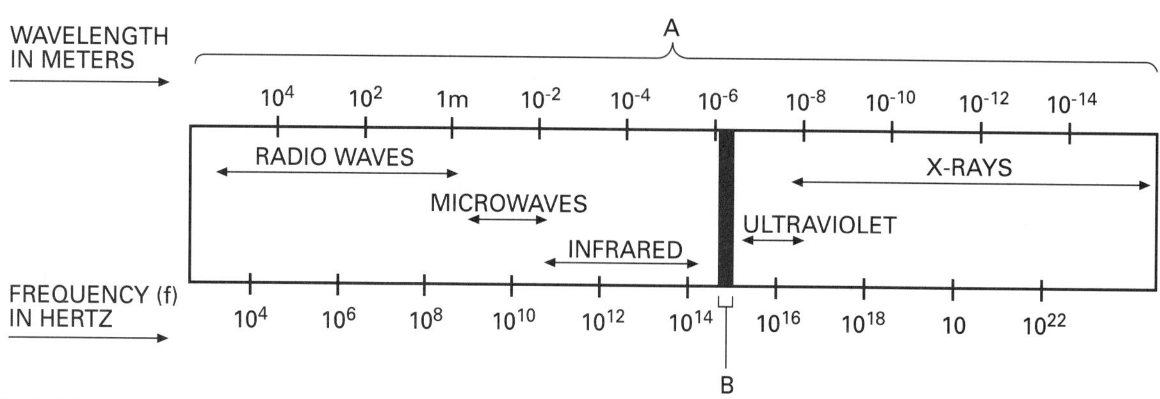

Fig. 32-2. (A) Electromagnetic spectrum; (B) range of visible light.

A close look at Figure 32-2 shows that radio waves have longer wavelengths than microwaves. The shortest wavelengths and highest frequencies in the electromagnetic spectrum are associated with x-rays. The range of visible white light waves falls between infrared and ultraviolet light, which are not visible to the human eye.

Fluorescent Lamps

Fluorescent lamps belong to a group of light sources known as *electric-discharge lamps*. Such lamps produce light by passing current through a gas. A typical fluorescent lamp is a glass tube with a tungsten filament sealed into each of its ends. The inner surface of the tube is coated with a chemical called *phosphor* (Fig. 32-3). When the lamp is manufactured, most of the air is removed from the tube and small amounts of argon gas and mercury remain sealed inside it.

FASCINATING FACTS

When Thomas Edison (1847-1931) was working on the invention of the incandescent lamp, he studied the entire history of lighting. He filled 200 notebooks with more than 40,000 pages of notes on gas illumination alone. He confidently told the public he would develop a device in six weeks. It took him slightly more than a year.

Perhaps the most difficult step was finding the proper material for the filament. At first, carbonized bamboo seemed most successful. For nine years, millions of Edison lamp bulbs were made with bamboo filaments.

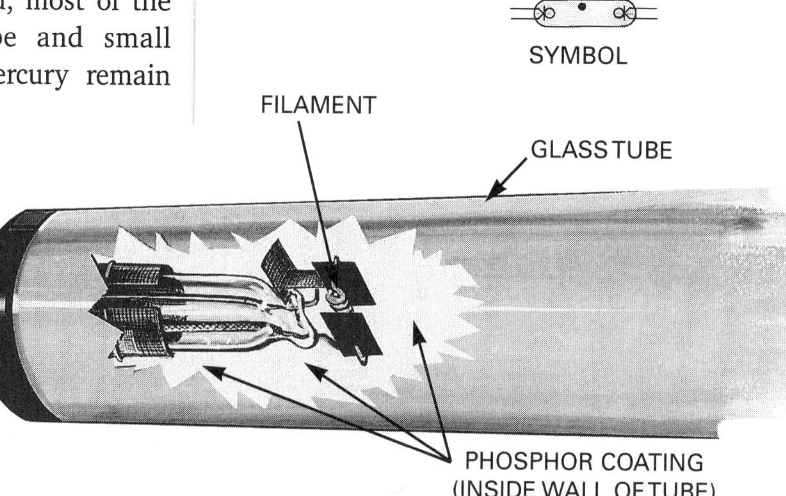

Fig. 32-3. View of one end of a typical preheat fluorescent lamp.

Lamp Circuit

To operate what is called a *preheat* fluorescent lamp, a ballast and a starter must be used. The **ballast** is a coil of insulated magnet wire wound around an iron core. The *starter* acts as a switch that is automatically turned off when its moving element, a bimetal strip, reaches a certain temperature.

In a lamp circuit, the filaments, the ballast, and the starter are connected in series (Fig. 32-4). When the main lamp switch is turned on, current passes through the circuit and heats the filaments. As a result, the temperature in the lamp increases. This causes the mercury to turn to gas. The current is also heating the bimetal strip in the starter. After a brief preheat period, the starter opens the circuit.

Suddenly stopping current produces several thousand volts across the terminals of the ballast by self-inductance. This high voltage also appears across the ends of the fluorescent tube. In an *instant-start* fluorescent lamp, a step-up transformer produces the high voltage needed to ionize the gas in the lamp. The high voltage ionizes the gas within the tube. The ionization makes the gas a good conductor. Because of its high inductance, the ballast also limits the current through the lamp circuit to a safe value.

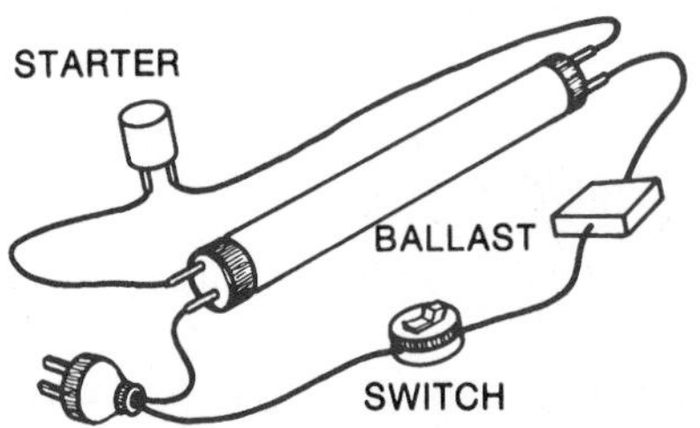

Fig. 32-4. The preheat fluorescent-lamp circuit.

LINKS

Bimetal-strip switches (thermostats) are discussed in Chapter 30.

How Light Is Produced

Current flowing through the ionized gas mixture in the lamp excites the mercury atoms (Fig. 32-5). Their energy is released in the form of *ultraviolet light*, which people cannot see. When this light strikes the inside surface of the lamp tube, the phosphor *fluoresces*, or glows. This glow is the visible light produced by the fluorescent lamp.

Efficiency

Fluorescent lamps have been available since 1938. They are common in homes, stores, offices, factories, and schools. One of the main reasons for their popularity is their efficiency. A fluorescent lamp wastes much less energy in the form of heat than does an incandescent lamp. The modern fluorescent lamp system (which includes a ballast or transformer) produces approximately three times more light than an incandescent lamp of the same wattage rating. Fluorescent lamps are now available in the shape of common incandescent lamps with screw-in bases. These lamps are used in ordinary lamp sockets. They have self-contained ballasts built into the lamp.

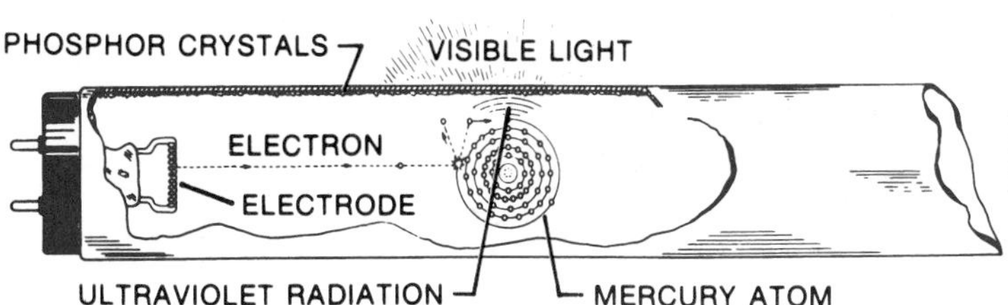

Fig. 32-5. How light is produced by a fluorescent lamp.

Mercury Lamp

The mercury, or *mercury vapor* lamp, is widely used to produce large amounts of light (Fig. 32-6). Mercury lamps are used for street and bridge lighting, parking-area lighting, and lighting for other places where the lamp fixture must be mounted high above the ground. The *arc tube*, or inner bulb, of this lamp is made of quartz, or rock crystal. The arc tube contains argon gas and a small amount of mercury. The outer bulb regulates and maintains the temperature of the inner bulb while the lamp is working.

To start the lamp, a high voltage from a transformer called a ballast, is applied to the lower main electrode and to the starting electrode next to it. This produces a glow discharge between these electrodes that heats the mercury, causing it to turn to gas. The gas mercury presents a high-resistance path for current through the arc tube from one main electrode to the other. The current, which is seen as a brilliant arc, is the source of light.

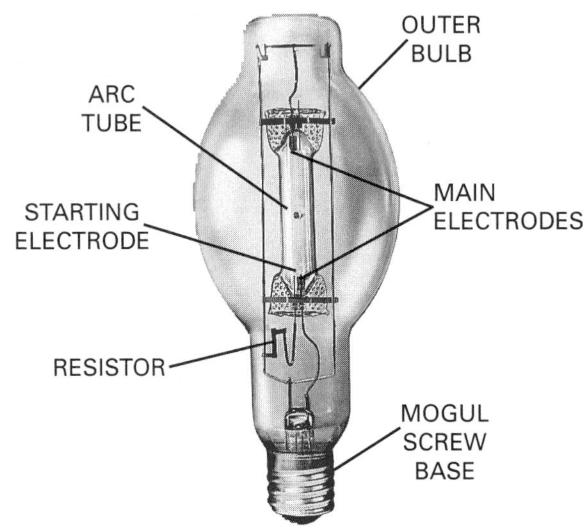

Fig. 32-6. The mercury vapor lamp.

Glow Lamps

Glow lamps have two electrodes placed very close to each other in a sealed glass bulb containing argon or neon gas (Fig. 32-7A). When the right dc voltage is applied to the electrodes, valence electrons within the atoms of gas gain enough energy to escape their parent atoms. Because of this ionization, positive gas ions are produced. As these ions move toward the negative electrode, some of them collide and regain electrons. This causes energy in the form of visible light to be released near the negative electrode. When a glow lamp operates with alternating voltage, the electrodes become alternately negative and positive. This happens so quickly that both electrodes seem to glow

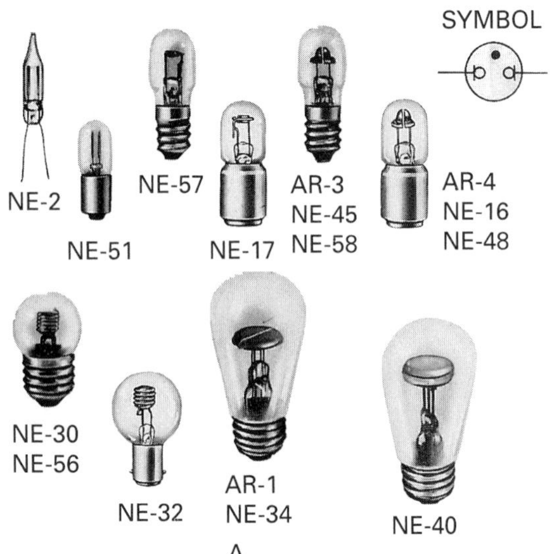

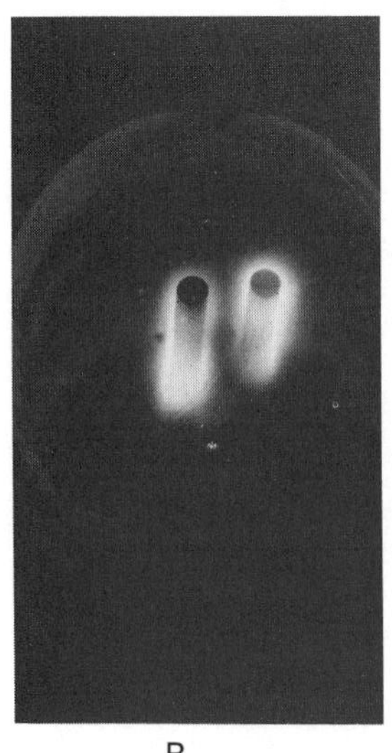

Fig. 32-7. (A) Neon and argon glow lamps; (B) both electrodes seem to glow when the lamp is operated on alternating voltage. The numbers and letters beneath the lamps refer to manufacturers' standard types.

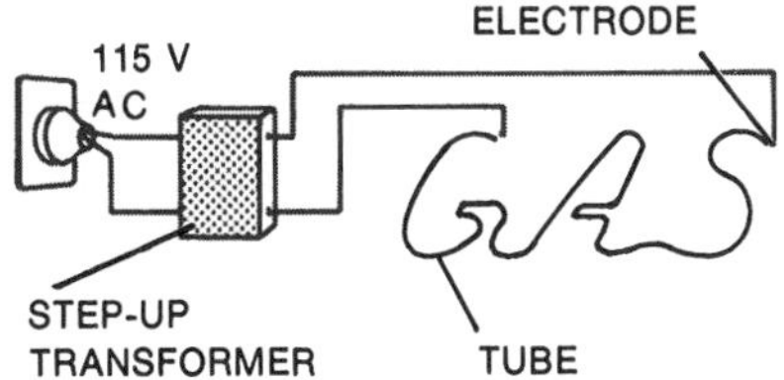

Fig. 32-8. **A step-up transformer is used with a neon sign.**

(Fig. 32-7B). The voltage needed to start a glow lamp is called the *starting*, the *striking*, or the *ionization voltage*. Typical glow lamps have a starting voltage of about 60 V.

Neon glow lamps of 1/25 to 3 W in size are used as test and pilot lights. One advantage of these lamps is that they start instantly when the right voltage is applied. For this reason, glow lamps are very often used as high-speed indicators. When operated at 120 V, small glow lamps such as the neon NE-2 require a current-limiting resistor. This resistor has a resistance of about 100,000. It is connected in series with one of the lamp leads or terminals. The resistor may be connected externally or may be contained in an indicator-light assembly. Larger glow lamps often have high-resistance, current-limiting coils in their bases.

Neon Signs

Neon signs have electrodes sealed into each end of glass tubes filled with neon gas. The longer the tube, the higher the voltages that must be applied to the electrodes to sufficiently ionize the gas. In common neon signs, a voltage of 10,000 V or more may be needed. This high voltage is supplied by a step-up transformer (Fig. 32-8).

By using argon, helium, or a mixture of these and other gases, lights of different colors are produced. Different colors are also created with tinted glass tubes or coatings on the inside surfaces of the tubes.

Explain how you think lumens, wattage, and a lamp's distance from the lighted surface are related.

LASERS

The word **laser** is actually an acronym for **L**ight **A**mplification by **S**timulated **E**mission of **R**adiation. Thus, a laser system is a way to produce and amplify light.

The color of light seen depends on the frequency of the electromagnetic radiation of visible light that reaches our eyes. Each color and shade of color has its own frequency. The frequency range of visible light extends from about 400 million MHz (red) to about 700 million MHz (violet). The wavelengths vary from about 700 nm (red) to about 400 nm (violet) (Fig. 32-9). The abbreviation nm stands for nanometer, or one billionth of a meter.

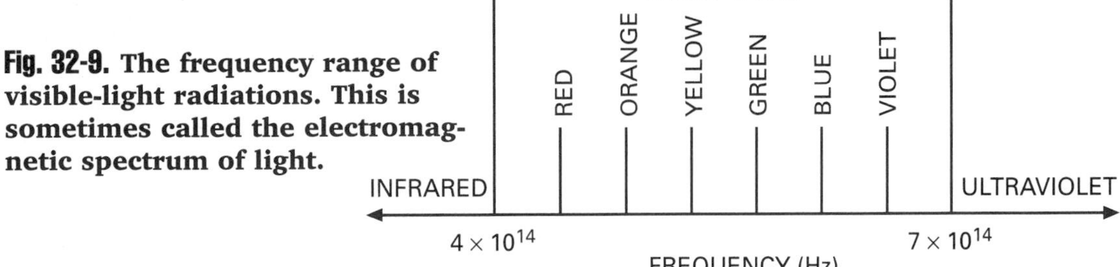

Fig. 32-9. **The frequency range of visible-light radiations. This is sometimes called the electromagnetic spectrum of light.**

Visible Light

The **incoherent light** produced by incandescent lamps consists of electromagnetic radiation over a wide range of frequencies. The *visible* (white) *light* from the lamp is actually made of a combination of colors. Our eyes see this as white light. Such light is called incoherent, or *polychromatic*, light. An important characteristic of incoherent light is that it spreads out very quickly after leaving the source (Fig. 32-10A). Consequently, this light loses much of its energy as it travels over distances.

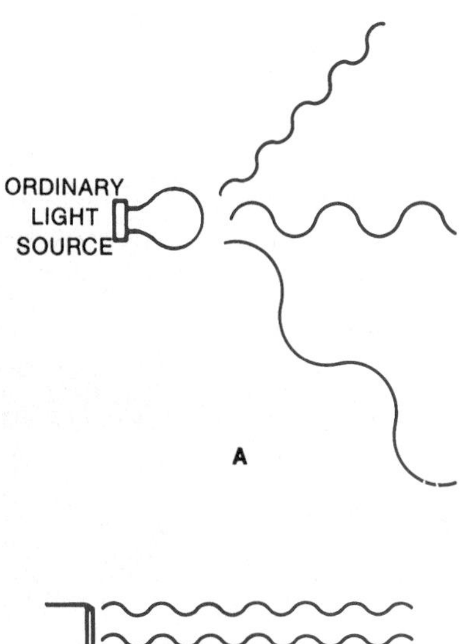

Fig. 32-10. Patterns of light radiation: (A) incoherent light; (B) coherent-light radiation of a laser beam.

Coherent Light

Coherent light is the light produced by a laser that has light waves very nearly the same frequency and are, therefore, monochromatic, or one color. They are also in *phase*, or in step with each other (Fig. 32-10B). A very important characteristic of coherent light is that it does not spread out, although it may be far from its source. The light, therefore, stays highly concentrated. Thus, it has much energy as it strikes a target area.

Laser Beam

The light output from a laser is in the form of a narrow beam of light (Fig. 32-11). The color of the beam depends on the kind of laser used. A laser beam can be modulated by many different electrical signals over long distances. Thus, it can be used in communications systems. Laser beams are also used to locate objects and to measure their distances from fixed points. More powerful laser beams are used for welding, metal cutting, and surgical cutting.

Fig. 32-11. Laser beam used to measure air velocity inside a combustion chamber.

Types of Lasers

There are several types of lasers being used today. Common types are designated as solid-state, gas, liquid, and semiconductor, or injection, lasers. The active material in the original solid-state laser was a ruby rod.

Ruby Lasers

The *ruby laser* is a good example of how a solid-state laser operates. The principal parts of a simple ruby laser assembly are shown in Fig. 32-12. The active material of the laser is a polished ruby rod. The ends of the ruby rod are coated with a silver compound that acts as a light-reflecting surface or mirror. One end is heavily coated. The other end is less heavily coated to form a semitransparent surface. The ends' surfaces and the rod form what is referred to as an *optical resonant cavity*.

The ruby rod is made of crystalline aluminum oxide in which a small number of aluminum ions are replaced by chromium ions. The chromium ions produce the laser effect.

An intense, incoherent light produced by the xenon flash lamp, excites the chromium ions into higher, unstable energy levels. This process is known as *pumping*. As a result of pumping, the chromium ions change their physical state by transferring electrons from one shell to another at a higher energy level.

When the flash lamp is turned off, incoherent light energy is no longer directed toward the ruby rod, and some of the electrons within the chromium ions return to their original, lower energy levels. Because of this action, the electrons release energy in the form of incoherent light, which resonates (surges back and forth through the rod) as it strikes the reflecting surfaces of the rod ends. As a result, more and more electrons are stimulated into releasing energy in the form of light that is in phase with the stimulating light. The light eventually becomes intense enough to escape the semitransparent end of the rod as a powerful beam of visible, monochromatic, red coherent light. The entire cycle of events takes place within a few milliseconds.

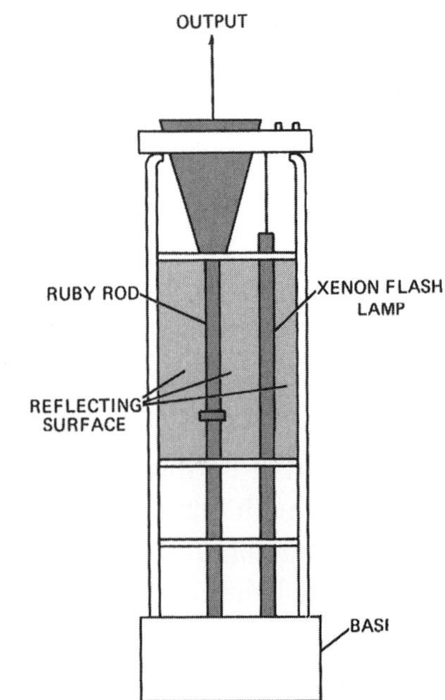

Fig. 32-12. Principal parts of a simple ruby laser.

In all solid-state lasers, coherent light is generated from this change in the energy levels of electrons within the atoms or ions of the active material when the material is exposed to an intense, incoherent light. The ruby laser and most other solid-state lasers are designed to produce an intermittent, or pulsed output, of coherent light. Some solid-state lasers, however, such as the ones that use calcium tungstate doped with neodymium as the active material, produce a continuous output. These are named *continuous wave* (CW) *lasers*.

Other Types of Lasers

Gas lasers use glass tubes instead of ruby rods. The glass tube can be filled with various gases, such as argon, helium, neon, and carbon dioxide (CO_2). Its operating principle, however, is the same as that of the ruby laser—the optical pumping is done with a flash lamp. The liquid laser uses a solvent-type liquid, such as ethanol, containing an organic dye. Therefore, liquid lasers are similar to ruby lasers in operation, except that the active medium is a liquid.

Solid State Lasers

One type of solid state laser is similar to a *light emitting diode* (LED). The junction of the diode includes a layer of gallium arsenide (GaAs) between two layers of aluminum gallium arsenide (ALGaAs). Both ends of the diode are highly polished with one silver coated end that reflects the radiation out the other end as a beam of coherent light. The light amplifica-

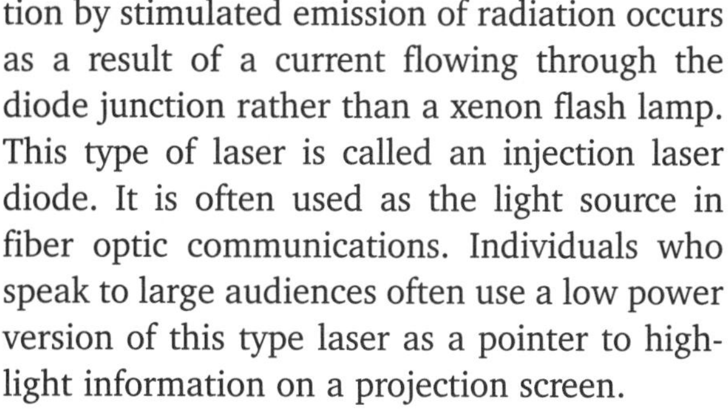

tion by stimulated emission of radiation occurs as a result of a current flowing through the diode junction rather than a xenon flash lamp. This type of laser is called an injection laser diode. It is often used as the light source in fiber optic communications. Individuals who speak to large audiences often use a low power version of this type laser as a pointer to highlight information on a projection screen.

Laser Applications

The rapid development of lasers has made possible a variety of applications in civilian, space, and military activities. Some of these applications are now realities, while others are still the subject of intense scientific and technological investigation.

Medical Applications

In medicine, lasers are used to cut and attach internal body tissues that cannot otherwise be reached without first cutting through surface tissue. Lasers are also used in conjunction with microscopes to observe previously unknown processes of body cell behavior.

Industrial Applications

As a machine tool, the laser beam is useful in cutting, shaping, drilling, and welding metal, minerals, plastics, and fabrics. The beam is also being used in industry to make precise measurements. In scientific research, the laser opens an entirely new method of examining various structures and processes (Fig. 32-13). These and other applications are examples of how electronic and optical research and technology have been combined and used.

Communication Applications

The application of lasers to the field of communications is promising. Some experts feel that lasers will eventually replace most ordinary microwave systems used in multi-channel information transmission. In this connection,

PLUG IN TO *Social Studies*

Laser Scanners in Supermarkets

A combination of optical laser scanner and high-speed digital computer has changed the checkout and inventory system used in many supermarkets.

The laser scanner is designed to sense the Universal Product Code (UPC) that is printed on an item. The code consists of a series of dark and light vertical bars that identify both the product type (e.g., chocolate ice cream) and the brand name. Each individual bar has a corresponding number printed alongside or under the bar. These numbers allow a clerk to keyboard the code if the scanner malfunctions.

As the clerk moves an item across a small, glass-covered opening in the counter, a helium-neon laser sends a light beam up to the code. The dark bars absorb the light, and the lighter colored bars reflect the light to a detector that reads the bar patterns by measuring these reflections. This data is converted into electrical signals. These are relayed to a central computer where the specific product is identified from the coded information. With this information, the computer can ring up the item's price, print it on a sales slip, and update the store's inventory.

Fig. 32-13. Four powerful laser beams are focused through chamber ports on a tiny deuterium pellet to trigger a fusion reaction.

the extremely high frequencies of laser carrier beams 470 million megahertz or higher, have a distinct advantage over typical microwave carriers. This is because a laser beam, with its higher frequency, can be modulated to provide a much greater bandwidth. Hence, the beam can be modulated to provide greater numbers of single-sideband frequencies, each of which can be used as a carrier for audio, video, and data communications that are coded light pulses called *bits*.

Explain how a laser operates.

FIBER OPTICS

Fiber optic cables are rapidly taking the place of copper wires and cables as the medium for transmitting information in the United States and in other countries around the globe. Fiber optic systems use light waves to transmit signals rather than electrons moving through copper wires.

Optical Fibers

Manufacturers use glass and plastic to make optical fibers. The material primarily used to produce glass fibers is Silicon Dioxide (SiO_2). It has a glass-like appearance, but is stronger and better able to transmit light over long distances. Flexible and solid fibers are both available.

The glass used in modern fiber optic systems is very pure. Impurities within the glass reduce its effectiveness in transmitting the light waves.

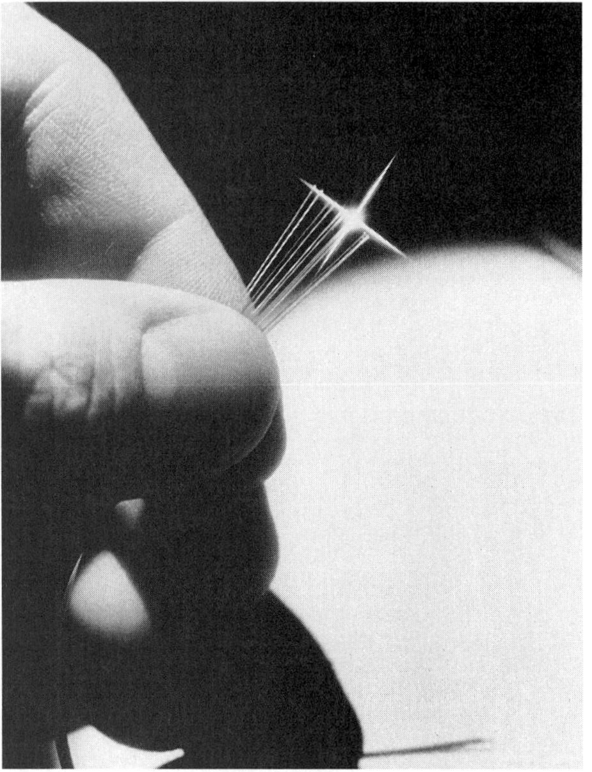

Fig. 32-14. **Hair-thin glass fiber (light guide).**

Some fibers are often the size of a human hair (Fig. 32-14). Optical fibers or light guides prevent light from escaping the surface by making the characteristics of the fiber's outer surface different from the surface of the inner core. An understanding of light wave behavior as it passes through various materials will help to explain these characteristics.

Reflection and Refraction

Light waves travel in straight paths until they encounter another medium such as a smooth surface. A smooth surface, like a mirror, will reflect the light, producing an image. In other words, a **reflection** is the return of the light wave motion. A **refraction** of a light wave occurs if it passes diagonally from one medium to another of different density, when its velocity changes, causing it to bend. As the velocity changes, the wavelength changes. This occurs as a light wave passes through layers of differing densities within the same medium or from air into water.

Figure 32-15 shows what happens when a white light wave passes through a prism. Each color in the visible spectrum is associated with a specific wavelength. As the white light wave passes through the prism, it is dispersed or refracted into various colors. The extent that light is dispersed as it passes through a medium is called the *index of refraction*. The index of refraction is expressed as the ratio of the speed of light to the speed at which a ray of light passes through a medium. The index is a characteristic of the medium that light passes through. Notice in Fig. 32-15 that the index of refraction is greater for shorter wavelengths.

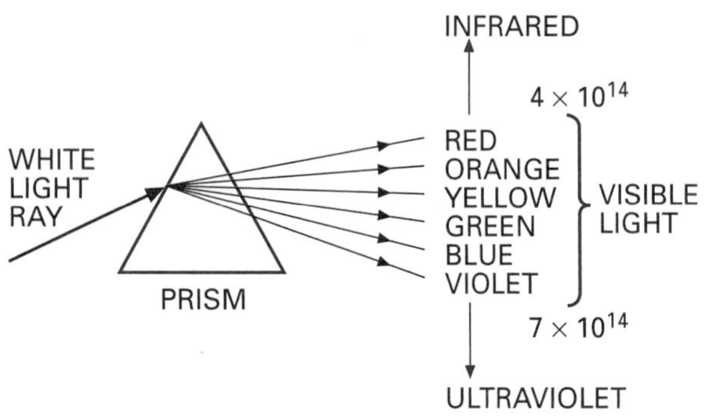

Fig. 32-15. **White light is separated into many colors when passed through a prism.**

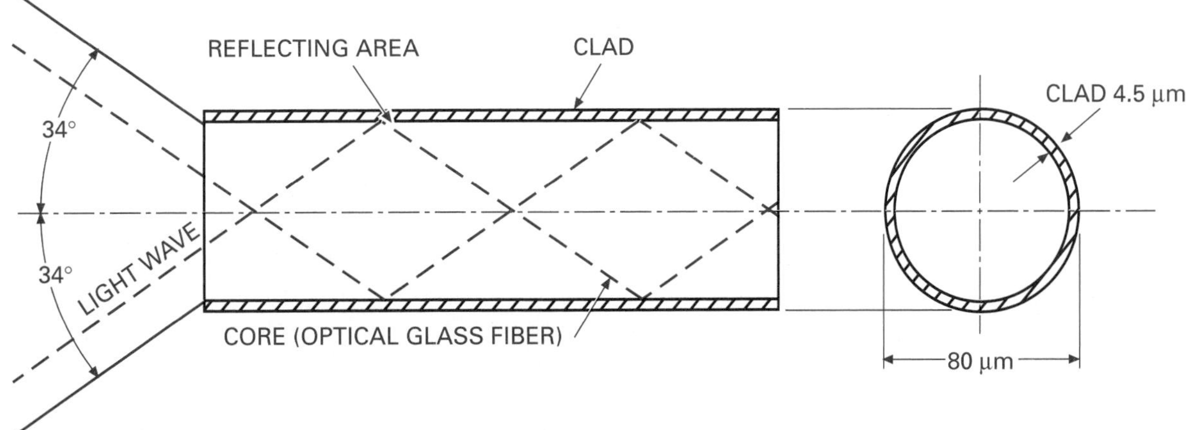

Fig. 32-16. Cross section of a glass fiber showing light rays bouncing off the reflecting area where the clad and core meet.

Violet light, for example, is dispersed the most, and red the least. White light can be produced by a correct mixture of red, blue, and green light waves. These colors are called the *primary colors of light*. A color television tube, for example, uses these three colors to produce a screen that appears white. White light, therefore, is comprised of many colors. When the light from an evening sunset passes through air of differing densities, a reddish glow is often seen. A rainbow is another illustration of sunlight that has been refracted by rain to produce many colors. The fact that light, or electromagnetic radiation, is composed of many wavelengths makes it an ideal transmission medium for communications systems.

Clad

To prevent light waves from escaping the inner core of the glass fiber, the outer surface is manufactured with a different index of refraction than the inner core through which light passes. The **clad** is the outer surface of an optic fiber. The clad, which surrounds the inner core, is fused to it. This provides a reflective surface where the inner core and the clad surfaces meet (Fig. 32-16). Experience indicates that, if the light waves enter a fiber optic material at an angle of 34 degrees or less from the center of the core, they will bounce off the reflective surface of the clad. The angle at which the light ray is reflected from the clad is called the *angle of incidence*. This angle remains constant as the light ray bounces off the reflective surface on its way through the core. If the light rays entering the core are at an angle greater than 34 degrees from the centerline of the core, the rays will pass through the clad and be lost.

Fiber Optic Cables

Fiber optic cables come in several different sizes and shapes. The material surrounding the glass fibers varies according to the intended function of the cable. Some cables are designed as *multimode cables*. Multimode cables can accommodate several light waves traveling down the core. Networks often use these cables because of their increased bandwidth capacity. *Singlemode* cables are designed to use one light wave with a wavelength of 1300 nanometers (nm) or one billionth of a meter. This high frequency provides high modulation capability.

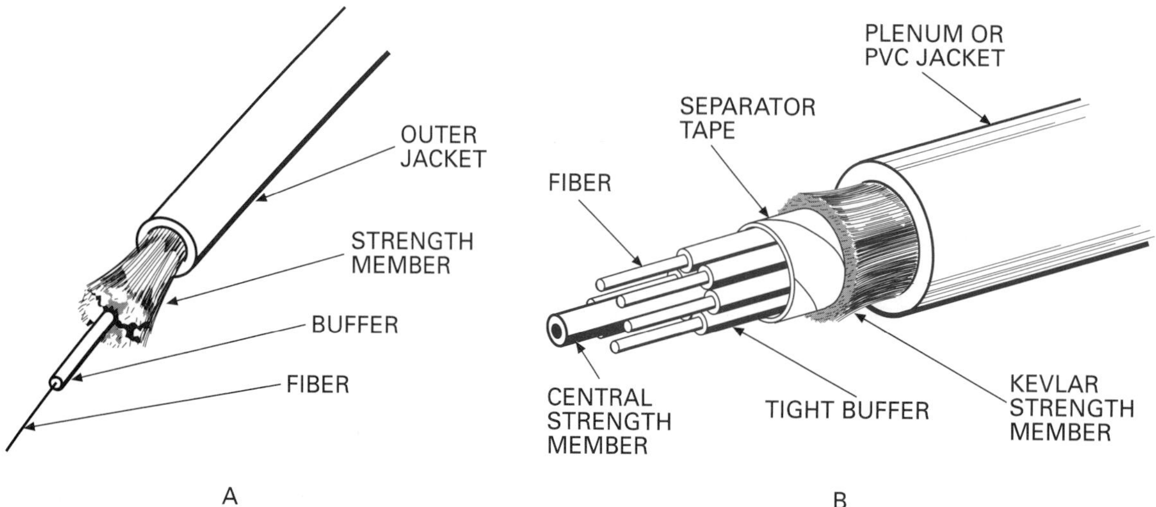

Fig. 32-17. Examples of the construction of optical cables designed for indoor environments: (A) simplex cable; (B) multifiber cable. (*Kevlar is a registered trademark of E. I. Dupont.*)

Therefore, these cables are used in long distant communication systems.

The inner core and clad of a typical fiber cable (light guide) are coated with a resin to prevent water and other impurities from penetrating the glass fiber. Over the protective coating are several layers of plastic. These layers, called a buffer coating, are applied to preserve strength. Next, a layer of DuPont Kevlar® strands provide extra strength. Finally, the entire cable is covered with a PVC (polyvinyl chloride) jacket. Figure 32-17 shows the construction of two optical cables meant for indoor plant environments where water is not present. Figure 32-18 shows a cross section of a fiber optic cable designed for carrying microwave signals. It provides details on the size of the core and related construction. Fiber cables also function as submarine cables. These cables, of course, have special coverings to meet the adverse underwater conditions.

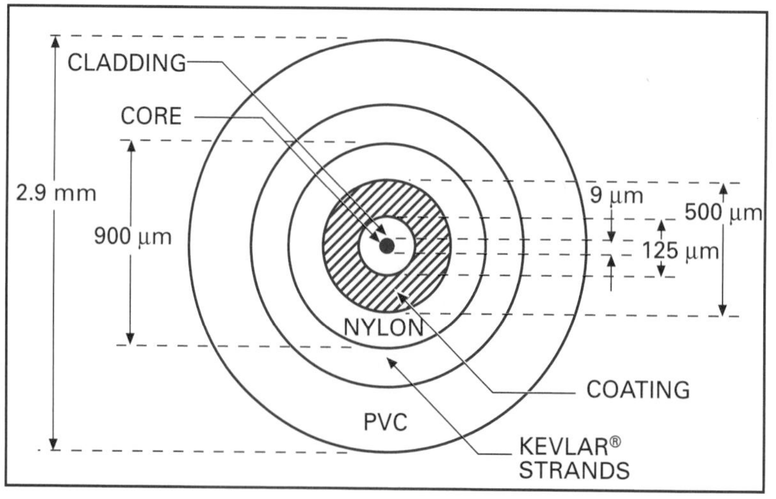

Fig. 32-18. Cross section of a single-mode fiber simplex cable.

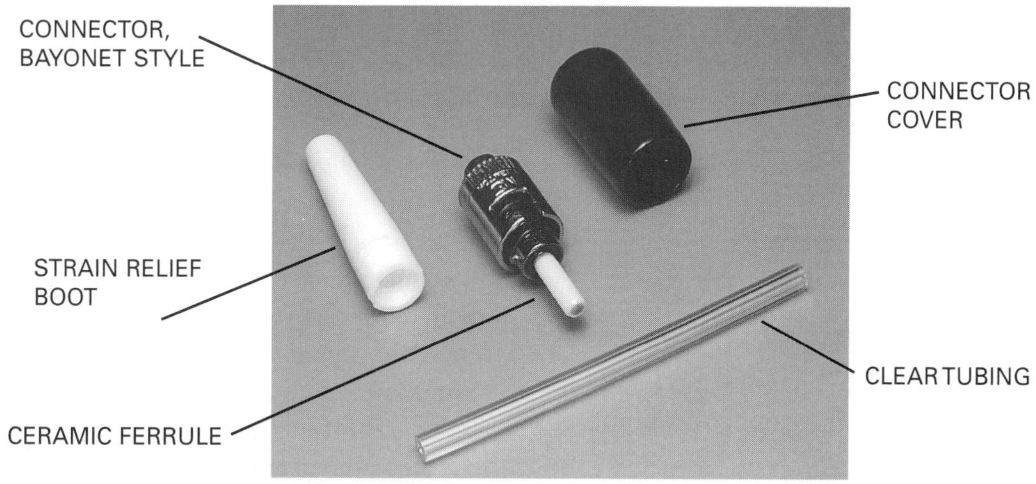

Fig. 32-19. Parts of an ST® hot melt fiber optic connector. (*ST is a registered trademark of AT&T.*)

Fiber Optic Losses

Fiber optic cables are subject to several types of losses. These include impurities in the glass, discrepancies in the material, bend losses (occurring whenever the cable is bent smaller than a 1-inch radius), dispersions of various light rays traveling in the same core, and types of connectors.

Fiber Optic Connectors

As in copper wires and cable, it is necessary to make a physical connection that can be disconnected. This can be done between wires or cables to another or to a device or component. The situation is similar with fiber optic cables. Bare fiber, however, cannot be connected directly to equipment. It is, therefore, necessary to use special optical fiber connectors on the ends of all fiber optic cables. Making a connection between two cables so that light rays can pass through without signal loss is difficult. Any misalignment of the fiber end causes loss of signal strength, or attenuation.

Several methods are used to make connections to minimize signal loss. These methods include splicing, fusing with heat, and joining mechanically. There are also reusable optical connectors. Before any method is used, the optical fibers must be cut at a precise angle to the core center line and the end surfaces must be highly polished.

Connecting Optical Fibers

One type of splice is made by joining the ends together with special tools and heating the joint so that the two ends fuse together. Another method uses a coupling gel that has the same characteristics as the optical fiber. The two ends are joined with the gel placed in between. A tube, previously placed over the cable is moved over the splice to hold it in place until the gel sets and becomes hard. Figure 32-19 shows the parts of an ST® hot melt connector. This connector is a keyed bayonet style and used with multimode cable. It uses a special preloaded hot melt adhesive and the ceramic ferrule provides stability through temperature changes. There are also many other types of connector styles.

Materials and Tools

There are a number of tools needed to successfully create a fiber optic connection. Some of the materials and tools needed to do this are a cable stripper, crimp tool, eye piece or light scope (to view imperfections), lint free cloths, a polishing device and material, scissors, scribe tools, and a manual.

Fiber Optic Communication System

Because the use of telephones and other data communications is increasing dramatically, telephone wires use modems to transmit voice signals in digital form. An analog-to-digital (ADC) converter changes the analog signals at a rate of 56,000 bits/s. Other circuits combine twenty-four digital phone lines into a single-bit stream of 1.5 Mbit/s. Faster rates include 274 Mbit/s. With fiber optics, speeds of 2.5 Gbit/s are possible.

LINKS

See the explanation of modems and ADC in Chapter 27.

This technology uses laser light beams or other light sources, optical transmitters and receivers, and optical glass fibers to communicate. These systems use light waves to transmit signals as coded pulses (bits) rather than electrons moving through copper wires. A laser, for example, transmits pulses of electric energy representing audio and video signals in the form of a light beam. The light beam is carried to its destination through optical glass fibers called light guides (Fig. 32-20). Some optical glass fibers can transmit pulses of light waves about 30 miles (50 kilometers). After this distance, the signals need to be restructured, synchronized, and amplified before sending them farther. The function of the optical receiver is to detect the transmitted light wave signals, decode, and then amplify them for conversion to sound, video, data, or other output.

Modulation

There are several different methods to code information into a laser beam. **Multiplexing**, or rapidly changing the time when various signals are transmitted, is often used. Wavelength division multiplexing can also carry several signals at different wavelengths and in the same direction through an optical fiber. For example, one wavelength at 1300 nm and another at 850 nm can travel in the same fiber. This doubles the capacity (or bandwidth) of the fiber.

Advantages of Fiber Optics

There are several advantages of fiber optic cables over copper wire and cable. Light waves

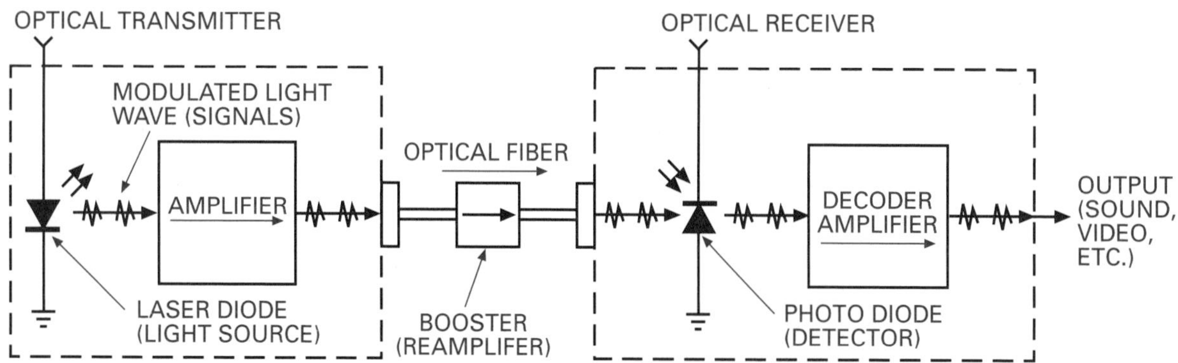

Fig. 32-20. Simplified fiber optical communication system.

can transmit many more signals than copper wires. This is because light waves carry high frequencies. There are fewer losses in transmitting signals and fiber optic cables are not affected by other electrical fields. Smaller cables can be used because of this increased capacity. Glass fibers are much lighter than copper wires and fiber optic cables are a secure medium. This means that if a signal is tapped into, light loss is unavoidable and the connection is shut down.

Fiber Optic Transmission Standards

Fiber optic technology has developed in a manner very similar to other technologies. Each manufacturer had its own format by which to transmit and receive signals. It was, therefore, difficult to communicate with different manufacturers. To communicate effectively, it became necessary to speak the same language.

Early standards were formulated for specific applications, such as telephone transmission rates. Telephone wires and cables once designed to carry analog signals now transmit a variety of signals such as computer data, facsimile signals, and other high-speed data.

Unfortunately, there is a mixture of formats that transmit electronic signals to represent information. Therefore, it is necessary to use adapters to convert signals to a standardized format for interchanging data. The devices to do this are called routers and gateways. The routers and gateways change various unique formats into a standard interchange mode, such as *Asynchronous Transfer Mode* (ATM). Once this is done, data is converted for transmission through long distance transporting systems. Two common standard systems are the Synchronous Optical Network (SONET), and Fiber Distributed Data Interface (FDDI). When signals are transmitted and received, they must be converted back to a standard interchange and later to the original format so the message can be understood.

Consider the use of fiber optics as an information highway. Describe how data is transmitted across a fiber optic cable.

Chapter *Review*

Summary

- There are many types of electric lamps. Some common ones are incandescent, fluorescent, mercury vapor, glow, and neon lamps.
- Laser is an acronym for Light Amplification by Stimulated Emission of Radiation. Lasers produce coherent light, or light that consists of waves that are very nearly the same frequency. Therefore the light stays highly concentrated and is useful in fields from communication systems to welding.
- Four types of lasers are solid-state, gas, liquid, and semiconductor.
- Fiber optics systems transmit signals with light waves rather than electrons. The fibers in these systems are made from very pure glass.
- Fiber optics systems are subject to losses from glass impurities, material inconsistencies, bends, and ray dispersion. Nonetheless, there are advantages of fiber optics over traditional copper transmission lines: (1) light waves can transmit more signals, (2) fiber optics are not affected by other electrical fields, (3) glass fibers are much lighter than copper wires, and (4) fiber optics are a secure medium.

Review Main Ideas

Review this chapter's main ideas by writing, on a separate sheet of paper, the word or words that most correctly complete the following statements:

1. The word *incandescent* means "______ from intense heat."
2. In an incandescent lamp, current passes through a filament of a metal, ______.
3. A mixture of the gases nitrogen and argon is placed in the bulb to lessen the ______ caused by its high temperature.
4. The lumen rating indicates the ______ of light a ______ produces.
5. The range of visible white light falls between ______ and ______ light, which are not visible to the human eye.
6. Fluorescent lamps belong to a group of light sources known as ______.
7. In an *instant-start* fluorescent lamp, a ______ produces the high voltage needed to ionize the gas in the lamp.
8. Typical glow lamps have a starting voltage of about ______ V.
9. The word *laser* is actually an acronym for ______.
10. An important characteristic of ______ is that it spreads out very quickly after leaving the source.
11. ______ is light produced by a laser that has light waves of very nearly the same frequency and are, therefore, monochromatic, or one color.

12. The ends' surfaces and the rod form what is referred to as an ______.
13. The material primarily used to produce glass fibers is ______.
14. ______ is the return of light wave motion.
15. A ______ of a light wave occurs if it passes diagonally from one medium into another of different density when its velocity changes, causing it to ______.
16. The ______ is the outer surface of an optic fiber.
17. A fiber optic communication system uses light sources, optical ______ and ______, and optical glass ______.
18. ______ is rapidly changing the time when various signals are transmitted.
19. ______ and ______ change various fiber optic formats into a standard interchange mode.

Apply Your Knowledge

1. Explain how an incandescent lamp works.
2. Describe the construction of a common fluorescent lamp tube.
3. Explain the difference between coherent and incoherent light.
4. Describe the *pumping* process in a ruby laser.
5. List five types of signal losses in fiber optic cables.

Make Connections

1. **Communication Skills.** Research the development of the incandescent lamp produced by the American inventor Thomas Edison. Comment on the human characteristics he exhibited in his effort to reach his goal.
2. **Mathematics.** Write the speed of light using the scientific notation method. Explain why this method is preferred by engineers and scientists.
3. **Science.** Devise a simple experiment to demonstrate the various colored rays that can be obtained from sunlight.
4. **Workplace Skills.** Draw a schematic diagram of a common fluorescent lamp. Write an explanation on how the lamp produces light.
5. **Social Studies.** Research the impact of fiber optics on communications in the United States. Summarize your findings in a brief written report.

Hands-On Experiences

DESIGNING AND MANUFACTURING PRODUCTS

RESEARCH AND DEVELOPMENT

The term *research* means a thorough study or investigation of some topic or topics. In industry, one main purpose of research is the improvement of existing products and processes. The other is the *development* of new ones that are useful. New products and processes are most often made possible because of discoveries made during research. Thus, research and development, or R&D, are closely related.

In electricity and electronics, many areas are constantly being researched in an effort to improve and expand technology. These include such areas as molecular electronics, solid-state physics, bio (life) functions, space exploration, data processing, and magnetics.

Research and engineering personnel work closely together. The engineers develop the ideas provided by research scientists into usable products and systems. An example of a product system that is the result of close cooperation among scientists, engineers, and technicians is shown in Fig. 1.

Research scientists and engineers are also responsible for tests and evaluations done on each piece of equipment or material developed. Environmental, electrical, mechanical, and physical-chemical tests are made. These provide data on the quality, performance, and dependability of a product.

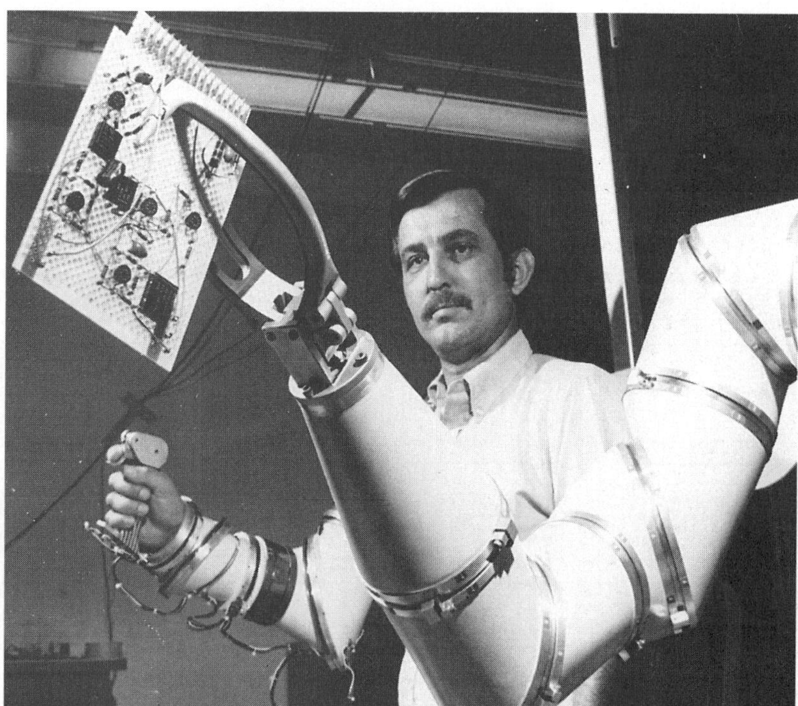

Fig. 1. **As a result of space suit research, the exoskeletal arm, as it is called, uses a complex system of electronic devices to duplicate the movement of a human operator's arm.**

The main function of a research and development company is to design and develop new products and/or materials. However, such a company needs a *manufacturing division* to support its other activities. The manufacturing division develops experimental models, prototype equipment, and fabrication and assembly procedures.

An *engineering services division* provides test equipment, finds out whether spare parts are available, and trains personnel. Engineering services are directed largely to marketing and sales departments. In places where many of a company's products are being used, engineering *field services* are provided for customers.

Developing an Electronic Device

Figure 2 shows six main steps in developing an idea into a practical electronic device.

Step 1: The Idea

Thinking of the idea is the first step. Ideas for developing new products or materials come from two main sources: (1) from the company's employees and (2) from a customer who has a need.

For example, a customer may have a problem and ask for help to solve it. This idea or need is usually expressed in a written report or proposal. It gives specifications of the new product or material and an estimate of the time needed to produce it. The research and development company and the customer come to an agreement about the factors and costs involved. Then the work begins.

Step 2: Exploring Ideas

Research and development generally involves solving new problems. Thus, much exploratory work is done first to find out if an idea is practical.

In the electronics industry, the exploratory process is called *experimental circuit development*. This is a design and development process in which scientists and engineers test theories. They want to find out whether their ideas work electronically. Rough pencil sketch-

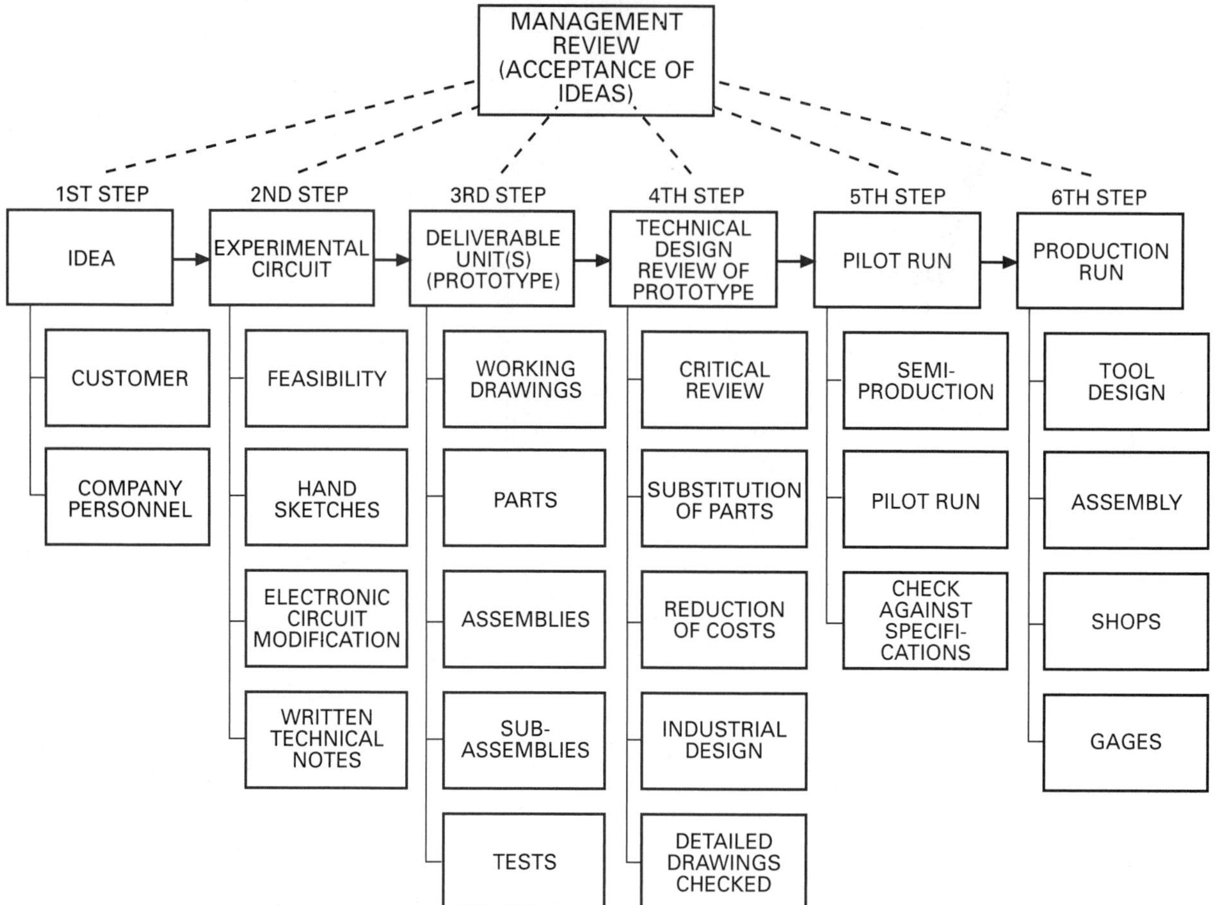

Fig. 2. Steps in the development of an idea into a practical electronic device.

es are used to record ideas about new circuits. Technical notes are written down to record the results. It may take six to eight months before a satisfactory circuit is found and accepted. At this point, a customer may want to order two or three *prototypes,* or working models.

Computer-aided design (CAD) and computer-aided manufacturing (CAM) are very common today. In industry, CAD/CAM reduces time and effort during the exploratory phase of developing a new product.

Step 3: Deliverable Prototypes

The ideas found to be practical during Step 2 are developed into a prototype. Working drawings of the parts are made. These include the specifications and the designs for supporting parts, assemblies, and subassemblies. Tests are done during the development of the prototype to make sure that it continues to work and to improve the way it works. The prototype is then delivered to the customer.

Step 4: Reviews

The *technical design review* involves taking a critical look at the whole process up to this point. Representatives from various divisions—such as manufacturing, engineering, purchasing, quality control, technical writing, and others—ask questions about the units already developed. This review is often called *value engineering*. This is a team approach to solving technological problems. It is the goal of value engineering to eliminate unneeded parts and to substitute new, lower-cost materials where possible. At this point, detailed drawings are reviewed. Any remaining mechanical and electrical design problems are solved.

Step 5: The Pilot Run

During a pilot run, a small number of devices are manufactured, just as they would be in mass production. The pilot run is done to check the devices against the original specifications. The pilot run also provides a means of checking all tools and fixtures to be used in the manufacturing process.

During the pilot run, the *process sheets* are brought up to date (Fig. 3). These explain in detail the operations that must be performed to complete the work. They include details on the time of manufacture, number of parts, sequence of operations, and assembly operations. When all parts of this step have been completed, the device is released to production.

Step 6: Production

Producing the device in the amounts needed is the sixth and last step in the development process. Manufacturing activities may include sheet-metal work, machining, welding, brazing, painting, and finishing. The model and tool shop has facilities for making models and building breadboards and experimental equipment, including printed circuits and modular and chassis units.

Some precise work is done in *clean rooms*. These have a controlled environment in which dust and other foreign particles are kept to a minimum and a constant temperature is maintained.

MANUFACTURING

The manufacture of electronic products is one of the major industries in the United States. In contrast to research and development, most electronic products are mass produced.

MANUFACTURING PROCESS SHEET
(Sheet ___ of ___ Sheets)

Employee # ________

Raw Material Code ________

Part Number ______, Part Name ____________

Job # ____ Sub-assembly ____ Quantity ________

Location		Tool	Routing	Standard Time			Parts Per Hour	Materials	Labor	Overtime Hours	Total
Operation Number	Group			Standard Unit	Hours/Unit	Total					
				Standard Cost Total→							

Processed by ________ Date ______

Fig. 3. A manufacturing process sheet.

Product manufacturing generally follows these seven basic steps: (1) developing finished drawings, (2) ordering raw materials and parts, (3) planning and controlling the manufacturing process, (4) fabricating the parts, (5) assembling the parts, (6) testing the parts (assemblies or subassemblies) against specifications, and (7) inspecting and shipping the final product.

Two important forms have been developed to control product manufacturing. These are the manufacturing process sheets and the manufacturing process specifications.

Manufacturing Process Sheets

Figure 3 shows a manufacturing process sheet. Note that the sheet has much information about the part, such as name, number, standard time of manufacture, and routing.

The *standard time* is the actual time it takes to do an operation. The times for operations are estimated before any parts are manufactured. Later these estimates are checked against the pilot run and corrected as necessary.

Manufacturing Process Specifications

These specifications describe briefly the steps to be followed during the manufacture of certain assemblies (Fig. 4). They tell the workers exactly what equipment is used, the specific operations in the process, and the maintenance needed to ensure quality. Manufacturing process specifications are kept up to date as changes and improvements are made.

Quality Control

The purpose of *quality control* is to make sure that manufactured products meet specifications. Carrying out the sequence of operations listed on the manufacturing process and specifications sheets is a major factor in quality control. This ensures that each part will be treated in the same way.

The inspection department is important in quality control. It is responsible for checking tolerances, specifications, and the like on parts as they move through the plant. The final inspection often includes touch-up and testing.

Quality control is the responsibility of all persons in the company rather than of one person or department. Whatever the job, it should be done well. This is the foundation of any effective quality-control program.

Component Manufacturing

The component manufacturer plays a key role in the electronics industry. These companies employ anywhere from a couple dozen to several thousand people. The range of components they make is as varied as the parts of electronic equipment. Some companies specialize in one item, such as a filter. Others make a variety of items, such as terminal boards, interlock receptacles, anode connections, test prods, and many others.

Several elements are common to the organization of all companies, whatever their size. The product manufacturing process always includes five areas: (1) sales, (2) engineering, (3) prototype development, (4) production, and (5) shipping. These areas are quite similar in both small component manufacturers and large product manufacturers.

DESIGNING AND BUILDING YOUR OWN PROJECT

Building a project can be one of the more interesting and useful activities in an electricity and electronics course. Such a project will let you use some of the theoretical and practical ideas you have learned in this course. It will also give you an opportunity to do planning—the very important steps involved in "thinking through" a job. You will work with devices and processes used in various electrical and electronics activities. Finally, successfully finishing a project can be very satisfying to you. It shows you were both able and willing to do a job in the responsible way of an efficient worker.

I. Purpose

To outline a process for dip soldering printed-circuit boards.

NOTE: Dip solder is used primarily for boards over 3 1/2" in width and for boards with standoffs or terminal lugs on the circuit side. Boards with heavy land areas are better suited for wave soldering and, whenever possible, should be wave soldered.

II. Equipment

Dip-soldering machine

Soldering flux formula no. 22

Holding fixtures (4 available — 2 for boards of 1 3/4" to 2 3/4" and 2 for boards 3" to 4 1/2"

63-37 solder

Flux remover type 6x

Cleaner

Thermometer

Scraper (stainless steel with plastic handle for removing dross)

III. Preparation for dip soldering

A. Turn thermostat knob on soldering pot to no. 5 (setting of 500°). This setting will heat the solder from 440 to 470°F. Caution: This is the correct temperature for dip soldering and should be checked with a thermometer to verify prior to soldering.

B. After checking the solder temperature, jog the sweep arm of the dip-soldering machine into a perpendicular position. Attach a holding fixture set to the correct width of the board to be soldered (or a dummy board of the exact width). Adjust the holding fixture to the correct height, without the board. The top of the solder should be even with the bottom of the notch in the holding fixture. With a dummy board, the board should rest on the top of the solder. Caution: Do not allow the board to submerge. Turn the machine on and return the board to its original position. The holding fixture should be to the right side of the machine.

IV. Operation

A. Adjust two fixtures to the correct width.

B. Insert boards and dip in soldering flux no. 22.

C. Clean dross from top of soldering pot with the scraper.

D. Hang the fixture and press the start button. Allow to complete one cycle.

E. Remove the board from the fixture and clean with cleaner. Reclean with flux remover type 6x.

V. Touch-up and inspection

A. Return the board to bench to be checked and touched up as required.

B. After check and touch-up, return it to the ventilation booth to be spray-cleaned with cleaner.

C. Submit boards to Quality Control.

After acceptance, boards must be placed in plastic bags.

VI. Maintenance of dip-solder machine

When the solder on the boards assumes a dull, dingy, or stringy (solder skips on the land pattern) appearance, change the solder in the pot.

Approvals:

Originator — Manufacturing Engineering

Production — Quality Control

Fig. 4. Manufacturing process specifications for printed-circuit boards: dip-soldering operation.

Projects in This Book

This book includes basic, intermediate, and advanced projects. The Basic Projects, which begin on page 553, provide experiences that relate to specific chapters in this text. The Intermediate Projects, starting on page 577, provide opportunities for you to test your ability to solve more technical problems. Advanced Projects, beginning on page 603, give you the situation and allow you to design the solution on your own.

Original Projects

Planning and designing an original project gives you a good opportunity to use theoretical information, knowledge of devices and materials, and skills related to problem solving. Figure 5 shows the steps in solving a problem. These steps are similar to the steps in the scientific method of research. The scientific method is an organized procedure for solving science problems.

Even the simplest original project can test your ability to transform an idea into a practical product. Engineers must have this ability. It is also important for anyone else who does technical work. Have a questioning attitude about the project.

- What is it to be?
- How can it be done?
- How can it be done better?
- Can it be done in a simpler way?

These are questions that challenge the mind. In design work, as in many other kinds of technical jobs, imagination is valuable. See Fig. 6.

Two examples of project-idea sketches are shown in Figs. 7 and 8. Here you see the way an idea is expanded and refined until an overall final plan is developed. This process is really one of choosing. You test several ideas on paper until you find what seems to be the most practical solution to the problem.

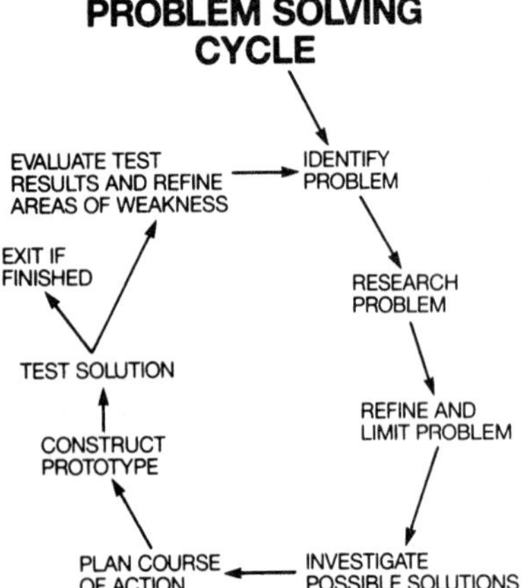

Fig. 5. An example of the process of problem solving.

An example of a project plan sheet is shown in Fig. 9 (page 551). When this is filled out, it will have all the information needed for building the project according to the ideas developed on the project-idea sketch sheet. Similar planning sheets can be used with any project.

SAFETY TIP

Always consider safety when designing or building any project. A project should be made in such a way that it is free of shock and fire hazards. If instructions are given, they should be clear so that the project can be used easily.

Fig. 6. Things to Consider When Choosing a Project

- **Function.** The project should satisfy a need.
- **Interest and ability.** Being interested in a job will encourage you to gain a new skill. However, choosing a project that is too difficult can be discouraging. If you're not sure you can make a project, choose a simpler one. Build on this experience and then go on to more challenging projects.
- **Plans and materials.** Unless a project is of your own design, its plans should include a schematic or pictorial diagram. There should also be a list of materials and any special hints for doing the job. Make sure that all the components and materials you will need are available.
- **Tools and facilities.** Make sure the project can be done with the tools, equipment, and space available to you. The project should not require special tools, hazardous operations or processes, or equipment not likely to be found in your school lab.
- **Cost.** The total cost of a project should be included in any plan. Otherwise, you may discover that the project will cost more than you are willing or able to pay.

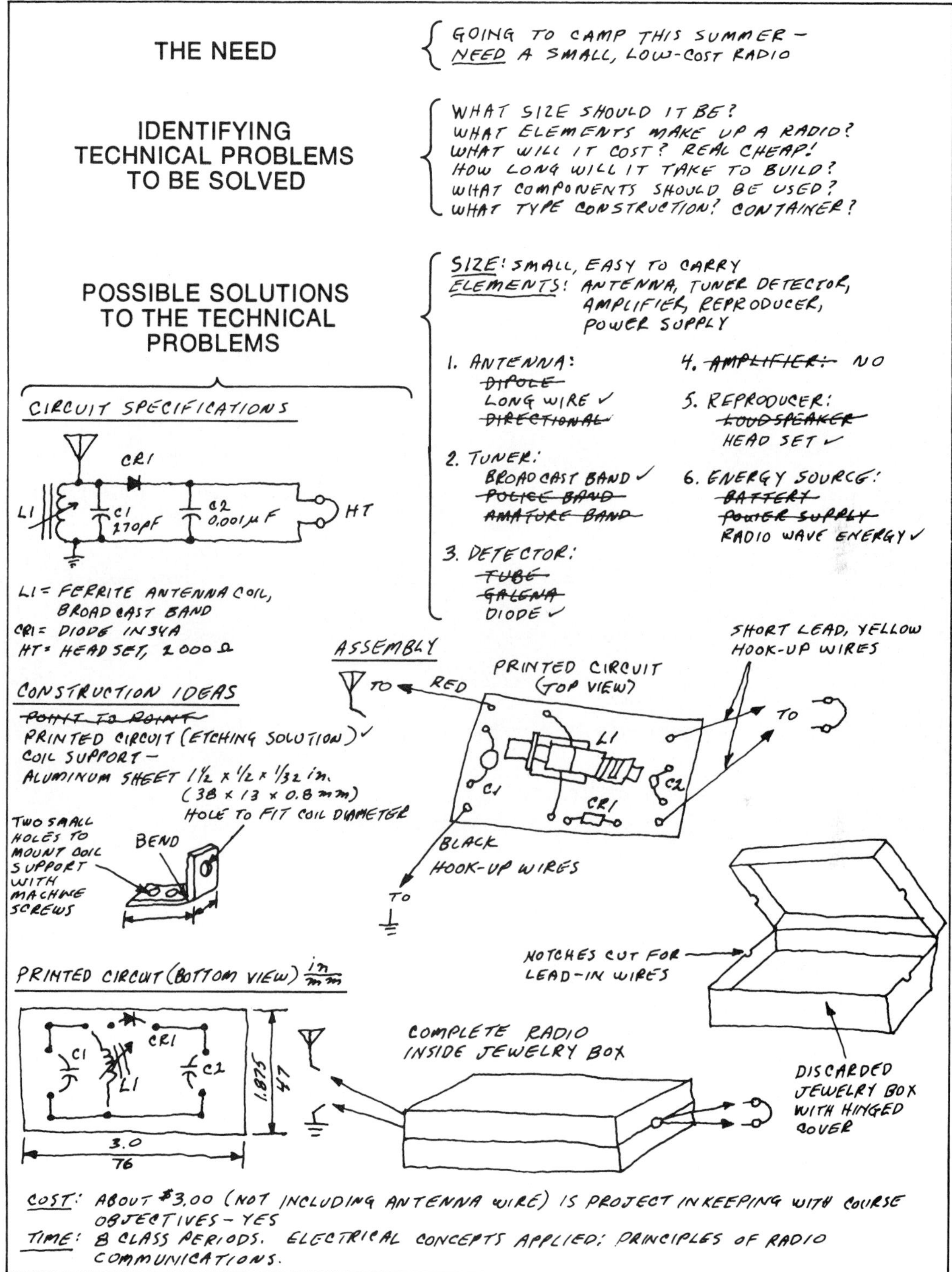

Fig. 7. An example of a project-idea sketch.

OBJECTIVE; TO LEARN TO PROGRAM A ROBOT THROUGH A MAZE.

IDENTIFYING TECHNICAL PROBLEMS TO SOLVE
- SIZE OF ROBOT
- LOCATION AND NUMBER OF WHEELS
- HOW TO POWER
- WHAT ARE MAJOR COMPONENTS?

POSSIBLE SOLUTION
2 MOTOR GEAR DRIVE
CONTROLLER – BASIC STAMP II
BATTERIES
BASE - PLASTIC, FIBERGLASS BOARD, OR EQUIVALENT
SIZE OF MAZE

- CONSTRUCTION STEPS
  - IDENTIFY ALL PARTS
  - FIND SOURCE AND COST FOR EACH
  - LAY OUT ALL PARTS TO IDENTIFY LOCATION AND BASE SIZE
  - DESIGN AND FABRICATE BASE
  - ASSEMBLE AND TEST

- ESTIMATED COST

BASIC STAMP PACKAGE	\$170.00
MOTOR DRIVER KIT	23.00
DUAL MOTOR W/GEAR	40.00
BATTERIES WITH HOLDERS	10.00
WHEELS	4.00
MISC. PARTS & SUPPLIES	10.00
	\$257.00

- PROGRAM DESIGN
  - LIST STEPS NEEDED TO CONTROL ROBOT
  - IDENTIFY STRATEGY FOR RUNNING A MAZE
  - LIST CODES FOR RUNNING MOTORS AND READING SWITCHES
  - DESIGN FLOW CHART
  - WRITE AND TEST SOFTWARE PROGRAM

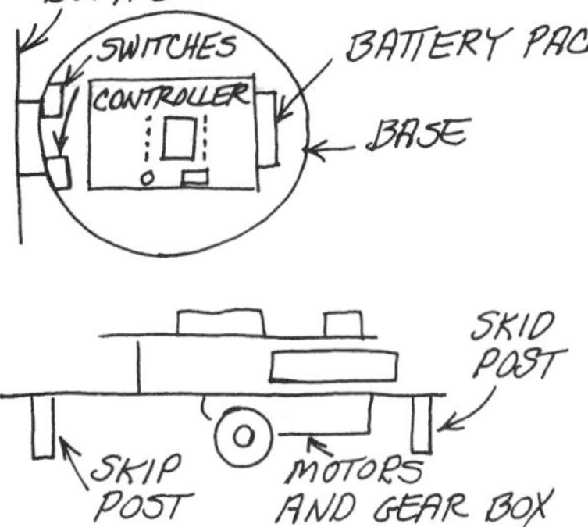

- CIRCUIT & PARTS IDEAS SEE INTERMEDIATE PROJECT 17
- ESTIMATED TIME: ABOUT 1 MONTH GROUP OF 3-4 STUDENTS WORKING TOGETHER
- CONTACT LOCAL COMPANY FOR SUPPORT
- ENTER ROBOTICS CONTEST DESIGN MAZE TO CONTEST SPECIFICATIONS

Fig. 8. **A project-idea sketch for a robot.**

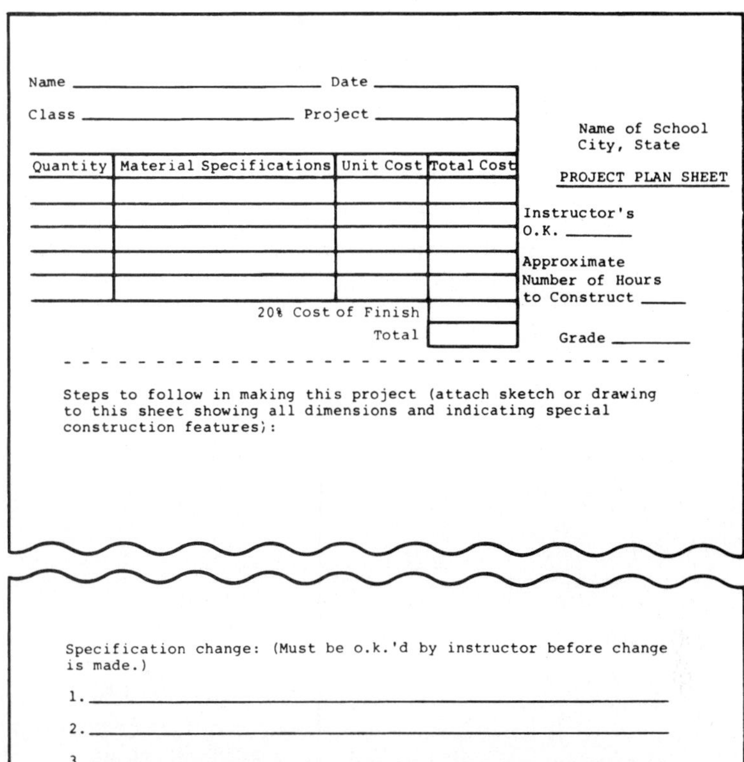

Name ________ Date ________

Class ________ Project ________

Name of School
City, State

PROJECT PLAN SHEET

Quantity	Material Specifications	Unit Cost	Total Cost
		20% Cost of Finish	
		Total	

Instructor's O.K. ________

Approximate Number of Hours to Construct ________

Grade ________

Steps to follow in making this project (attach sketch or drawing to this sheet showing all dimensions and indicating special construction features):

Specification change: (Must be o.k.'d by instructor before change is made.)

1. ________

2. ________

3. ________

Electrical Theory Involved: (Clip other sheets to plan if needed.)

Fig. 9. A project plan sheet.

Project Kits

A kit usually has all the materials needed for a project. The kit generally has instructions that show and tell how the project is to be assembled and wired.

Experimental Activities

The main purpose of an experimental activity is to show how a circuit or device works. This kind of activity can be very useful for demonstrating, in a practical way, how theory and practice are related.

A number of ideas for experiments are given in the projects included in this section. You will find that building these projects and showing them to your class will be interesting and informative. More experiments are given in the *Student Workbook for Electricity and Electronics Technology.*

Repair Projects

Repairing an existing circuit or device can give you much useful experience. In doing repair jobs, you will learn about troubleshooting (finding problems in components or circuits). This is similar to what professionals do. Thus, your experience can be put to use in a future career.

Most repair jobs require a knowledge of theory, components and materials, tools, test instruments, and testing methods. Procedures for using the basic test instrument, the volt-ohm-milliammeter, and various digital meters are given in various chapters of this book. Other test instruments and troubleshooting methods are also discussed throughout the book.

Repair projects should include a report (Fig. 10). In addition to being a record of your work, such a report gives you an opportunity to describe a job in writing. This can be a very useful experience. Written reports must be prepared as part of many technical jobs.

REPAIR-JOB REPORT

Name ______________________ Date __________

Product or Circuit Worked On ______________________

____

For Whom This Job Was Done ______________________

____

Description of Defect or Defects ______________________

____

____

____

General Procedures Used in Doing the Job ______________________

____

____

____

____

____

____

____

____

____

Instruments Used __________ __________

__________ __________

Time Spent in Doing the Job ______________________

Approved Yes __________ No __________

Fig. 10. Repair job report.

BASIC PROJECTS

The following projects provide practical hands-on experiences in the use of circuit diagrams, tools, and test equipment. Your teacher may assign them as individual exercises or as group projects.

BASIC PROJECT 1: NEON LAMP HANDITESTER

This device can be used to conveniently check for the presence of voltage at receptacle outlets and terminal s and between ungrounded appliance and tool housings and ground. A voltage of at least 60 V is needed to light the lamp. A neon direct-current (dc) voltage will cause only the electrode of the neon lamp connected to the negative terminal to glow. Thus, the handitester can be used to test dc-voltage polarity.

Materials Needed

1 clear, hard plastic (Lucite or vinyl) tube, ½ in. (12.7 mm) OD, ¼ in. (6.35 mm) ID, 4 in. (101.6 mm) long
1 neon lamp, type NE-2
1 carbon-composition or film resistor; 220,000 Ω; ½ watt
12 in . (305 mm) test-prod wire, black, no . 18 AWG (1.00 mm)
12 in . (305 mm) test-prod wire, red, no. 18 AWG (1.00 mm)
1 test solderless prod, black, 4 in. (102 mm) handle
1 test solderless prod, red, 3 in. (76 mm) handle
plastic insulating tubing (spaghetti)
epoxy cement
clear liquid casting plastic (optional)

Procedure

1. Make the neon lamp, resistor, and test-prod wire connections shown in Fig. BP-1. Make sure that all solder joints and bare wires of the assembly are completely insulated.
2. Carefully put the assembly into the plastic tube.
3. Seal the ends of the tube with epoxy cement or with some other suitable sealing compound. Put the cement or sealing compound on the lamp end of the tube a little at a time to keep it from covering a part of the lamp.

NOTE: A more attractive handitester can be made by sealing the testprod wire end of the tube with clear liquid casting plastic after the parts have been put into it. You then place the tube upright and fill it with the casting plastic from the lamp end.

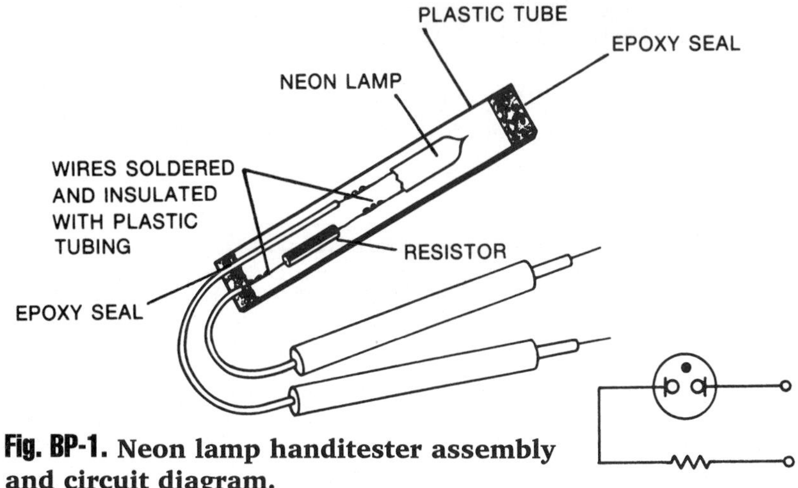

Fig. BP-1. **Neon lamp handitester assembly and circuit diagram.**

BASIC PROJECT 2: LOUDSPEAKER "FADER OR BALANCE" CONTROL

The following circuit shows a very practical use of a potentiometer. It can be used to adjust the volume of two loudspeakers, such as the front and rear loudspeakers in an automobile.

Materials Needed

potentiometer, 350 Ω, 2 W

loudspeaker, if necessary. If this loudspeaker is to be added to an existing system, it should have the same electrical characteristics as the loudspeaker already in use.

hook-up wire, no. 20 AWG (0.8 mm), as needed

Procedure

1. Wire the circuit shown in Fig. BP-2. If the chassis of an automobile is used as one conductor of the loudspeaker circuit, the common wire is not necessary. In this case, one terminal of each loudspeaker is connected to a nearby part of the chassis as shown by the dotted lines.

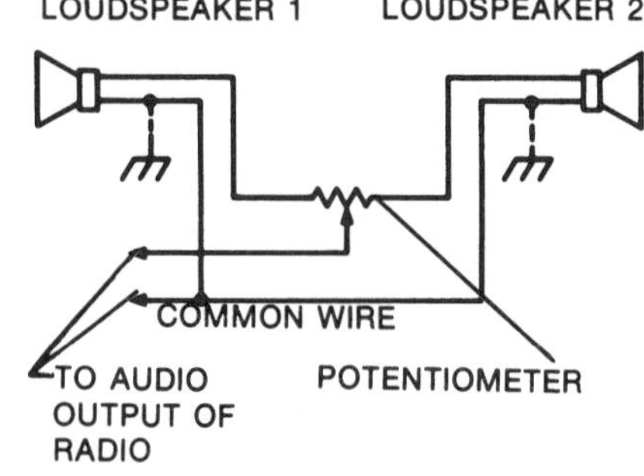

Fig. BP-2. **Circuit used for a loudspeaker or balance control.**

2. Test the circuit by turning the shaft of the potentiometer first in one direction and then the other. As this is done, one of the loudspeakers should approach full volume while the volume of the other loudspeaker decreases. At the center position, the loudspeakers' volume should be equal.

SAFETY TIP

Many devices using loudspeakers—such as radios, televisions, and phonographs—should not be expanded into multiple-speaker systems. Often these devices are directly connected to the power lines. This means that the metal chassis and the speaker wires can be at a dangerous potential, or voltage, with respect to ground. Manufacturers of such equipment almost always disapprove of modifications. Check with a qualified technician before making any connections to these circuits.

Some equipment can be damaged by connecting the wrong speaker load. Again, check with a qualified technician or service center before attaching extra speakers to such equipment.

BASIC PROJECT 3: RELAXATION OSCILLATOR

In this interesting circuit, the charge-discharge action of a capacitor causes a neon lamp to flash on and off. When the resistance or the capacitance of the circuit is varied, the time constant is changed. This causes the lamp to flash at a different rate.

Materials Needed

circuit board, prepunched, 6 x 8 in. (152 x 203 mm)
1 potentiometer, 1 megohm, ½ W
1 carbon-composition or film resistor, 1 megohm, ½ W
2 capacitors, 0.5 μF, 200 WVDC
1 neon lamp, type NE-2
90-V battery or a dc power supply

Procedure

1. Wire the circuit shown in Fig. BP-3.
2. Connect the battery or the power supply to the circuit. If a power supply is used, adjust it to produce an output voltage of 90 V.

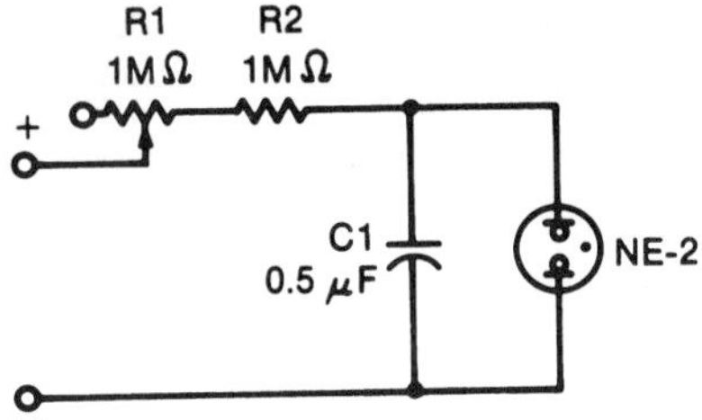

Fig. BP-3. **Schematic diagram for a relaxation oscillator project.**

3. Adjust the potentiometer so that the lamp flashes about once per second. The time between flashes can be calculated using the formula $T = RC$. T is time in seconds, R is resistance in ohms, and C is capacitance in farads. This is because 63.2 percent of 90 V is the approximate ionization voltage of an NE-2 lamp.
4. By using the formula $T = RC$, figure the total resistance of R1 and R2 when the lamp is flashing at the rate of one flash per second.
5. Adjust the potentiometer so that the lamp flashes at a faster speed. Has the total resistance of R1 and R2 been increased or decreased? Explain your answer.
6. Adjust the potentiometer to again get one flash per second. After disconnecting the battery or the power supply from the circuit, connect the other capacitor in parallel with C1. Does this cause the lamp to flash more or fewer times per second? Explain the reason for this.

BASIC PROJECT 4: ELECTROMAGNETISM EXPERIMENTER

This assembly can be used as an electromagnet, a magnetizer, a demagnetizer, and for doing several very interesting experiments with electromagnetism. It can be operated at 115 V ac or with dc voltages of up to 6 V.

Materials Needed

1 piece of wood, ¾ x 7½ x 10½ in. (19 x 191 x 267 mm)
2 round plastic or Masonite disks, ⅛ x 4 in. (3.18 x 102 mm) in diameter
enough sheet metal (transformer iron) to form a laminated core 1½ x 1½ x 4 in. (38 x 38 x 102 mm). The laminations from a large, discarded transformer are excellent for this purpose.
1 piece of band iron, ⅛ x 1½ x 7 in. (3.18 x 38 x 178 mm)
1 round soft-iron rod, 1 x 4 in. (25.4 x 102 mm)
1 round soft-iron rod, 1 x 10 in. (25.4 x 254 mm)
1 aluminum ring or washer, 1½ or 2 in. (38 or 50 mm) in diameter
1 round sheet-copper or sheet-aluminum disk, 2½ in. (63.5 mm) in diameter
350 ft (107 m) PE magnet wire, no. 14 (1.60 mm)

18 ft (5.5 m) PE magnet wire, no. 18 (1.00 mm)
10 in. (254 mm) bare wire or PE magnet wire, no. 14 (1.60 mm)
5 ft (1.5 m) two-conductor service cord, no. 16 (1.25 mm)
2 ft (0.6 m) parallel lamp cord, no. 18 (1.00 mm)
1 pilot lamp, no. 40 or no. 46
1 miniature lamp socket, screw base
1 heavy-duty attachment plug
1 15-A surface toggle switch, spst, with mounting screws
2 wire nuts, medium size
5 insulated staples
3 flat-head machine screws, no. ¼—20 x 1 in. (M6 x 1)
3 round-head wood screws, no. 8 x ¾ in.
cambric cloth, medium thick
insulating varnish
plastic electrical tape
low-voltage, battery-eliminator dc power supply

Procedure A (Basic Construction)

1. Assemble the electromagnet coil core and the coil form as shown in Fig. BP-4, A.
2. Tightly wind the coil in layers with the no. 14 (1.6 mm) PE magnet wire (Fig. BP-4, B). Insulate every third layer of the wire with a single layer of cambric cloth. After the coil is wound, apply a heavy coating of insulating varnish to its outside surfaces. Let it dry thoroughly.
3. Assemble materials and connect the circuit shown in Fig. BP-4, C.

Procedure B (The Electromagnet)

Plug the experimenter into a 115-V ac receptacle outlet and turn the switch on. By holding an object made of a magnetic material near the coil-core pole piece, you will find that a strong electromagnetic filed exists in the area near this pole piece.

SAFETY TIP

The coil will overheat if operated continuously for an extended time. When this occurs, the circuit should be turned off and the coil allowed to cool.

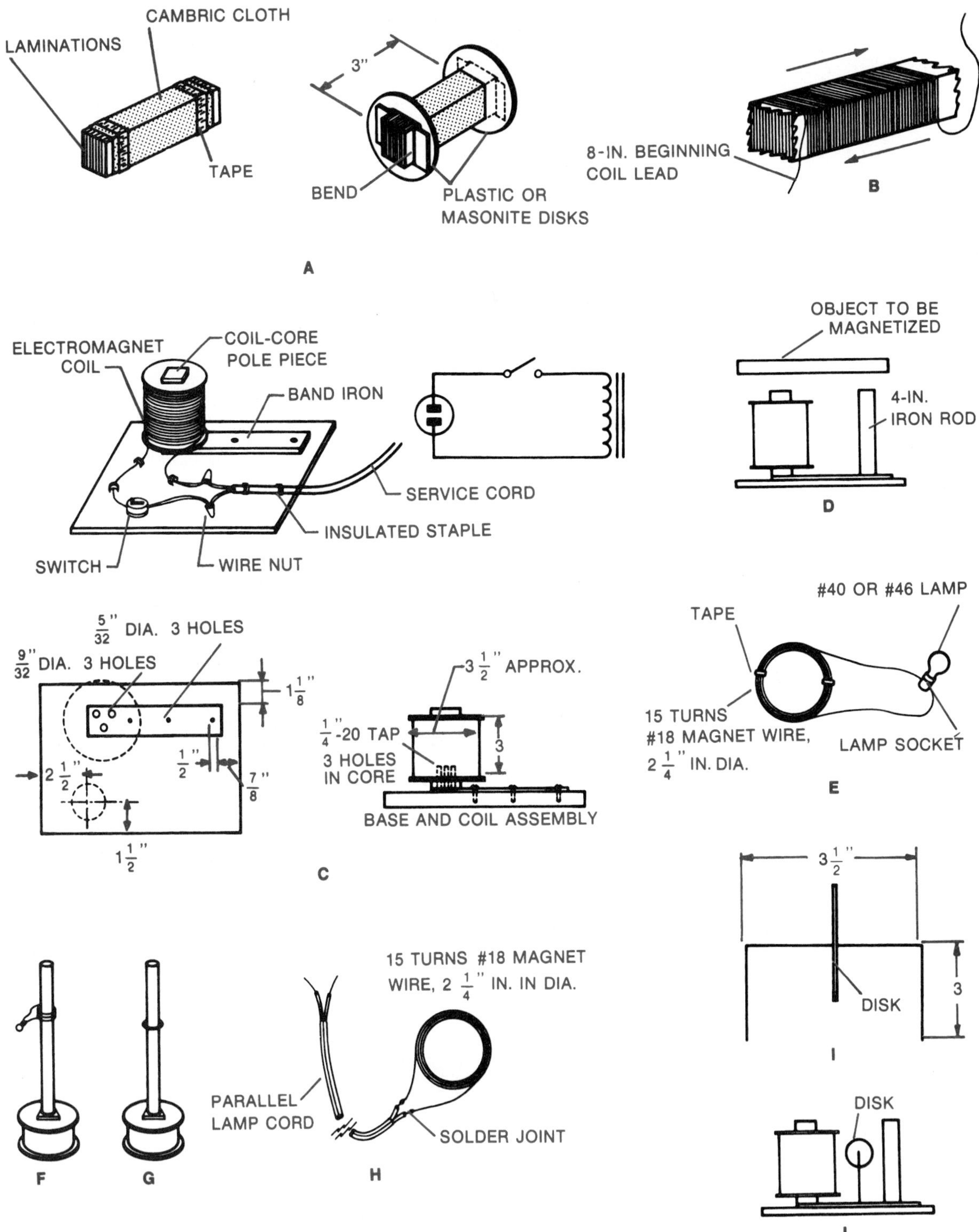

Fig. BP-4. Assemblies used for the electro-magnetism experimenter.

Procedure C (Magnetizing and Demagnetizing)

1. To use the experimenter as a magnetizer, connect it to the power supply adjusted to a dc voltage of 6 V. Put the object to be magnetized across its pole pieces. One of the pole pieces if formed by placing the 4 in. (102 mm) iron rod over the band iron in an upright position (Fig. BP-4, D). The rod can be moved as needed for the right spacing. Turn the switch on and, after a few seconds, turn it off. The object should now be magnetized. Explain this process of magnetization.
2. To use the experimenter as a demagnetizer, plug it into a 115-V ac receptacle outlet and turn the switch on. Hold the object to be demagnetized a little above the coil-core pole piece. Then slowly move it away from the pole piece for a distance of about 18 in. (457 mm). Repeat if necessary. Explain this process of demagnetization.

Procedure D (Transformer Action)

1. Assemble the coil and lamp shown in Fig. BP-4, E.
2. Place the 10 in. (254 mm) iron rod on the coil-core pole piece in an upright position. Plug the experimenter into a 115-V ac receptacle outlet. Turn the switch on.
3. Slowly move the coil down over the iron rod. Do not let it touch the rod (Fig. BP-4, F). As you do this, the lamp will begin to burn more and more brightly. Explain how energy is transferred from the electromagnet coil to the lamp circuit.
4. You will notice that the iron rod becomes quite warm after the experimenter is operated in this way for a short time. This is an example of induction heating. Induction heating is described in Chapter 30, "Heating, Refrigeration, and Air Conditioning."

Procedure E (The Jumping Ring and Coil)

1. Place the 10 in. (254 mm) iron rod on the coil-core pole piece in an upright position. Plug the experimenter into a 115-V ac receptacle outlet and turn the switch on.
2. Put the aluminum ring down over the end of the rod (Fig. BP-4, G). Explain why the ring remains suspended above the coil-core pole piece. Turn the switch off.
3. Drop the ring over the iron rod and onto the coil-core pole piece. Put one hand over the end of the rod and turn the switch on. What happens? What causes this?
4. Form the coil shown in Fig. BP-4, H.

5. Put the coil over the iron rod. With the switch on, slowly make and break contact between the ends of the lamp cord connected to it. Why does the coil jump upward when the ends of the lamp cord are in contact? Why does the coil remain stationary when the ends are not touching?

Procedure F (Motor Action)

1. Drill a 1⁄16 in. (1.6 mm) hole through the center of the copper or aluminum disk. Pass the piece of no. 14 (1.6 mm) wire through the hole. Bend the wire as shown in Fig. BP-4, I.
2. Put the legs of the disk-support wire into small holes drilled into the wood base of the experimenter. The disk should be midway between the coil and the 4 in. (102 mm) iron rod (Fig. BP-4, J).
3. Plug the experimenter into a 115-V ac receptacle outlet. Turn the switch on.
4. Spin the disk with your fingers. It will continue to rotate. This is a simple induction motor. Induction motors are described in Chapter 29, "Electric Motors."

BASIC PROJECT 5: PHOTO-CONTROLLED RELAY

By following the procedure outlined below, you will be able to see the effect an increase in light intensity has on the resistance of a typical photoconductive cell. You can then use the cell to see how it controls a circuit. The relay used in this circuit is a magnetically controlled switch. For more information about relays, see Chapter 11, "Magnetism."

Materials Needed

circuit board, prepunched, 4 x 6 in. (102 x 152 mm)
1 photoconductive cell, Clairex no. CL704M or equivalent
1 resistor, carbon-composition or film, 10,000 Ω, ½ W
1 potentiometer, 100,000 Ω, ½ or 1 W
2 transistors, p-n-p, SK-3004, GE-2, or equivalent
1 miniature sensitive relay, coil resistance 500 Ω, sensitivity about 40 mw, spdt, Radio Shack no. 275 B 004 or equivalent
1 pilot lamp, no. 40
1 miniature lamp socket, screw-base
a small desk lamp with a 60-W incandescent light bulb
9-V transistor radio battery or a dc power supply
6-V battery or a dc power supply with a current rating of at least 200 mA
volt-ohm-milliammeter

Procedure

1. On a sheet of paper, prepare a table like the one in BP-5A.

Dark Resistance	Normal Resistance	High-Intensity Resistance

Fig. BP-5A. Make a table like this one for recording your results.

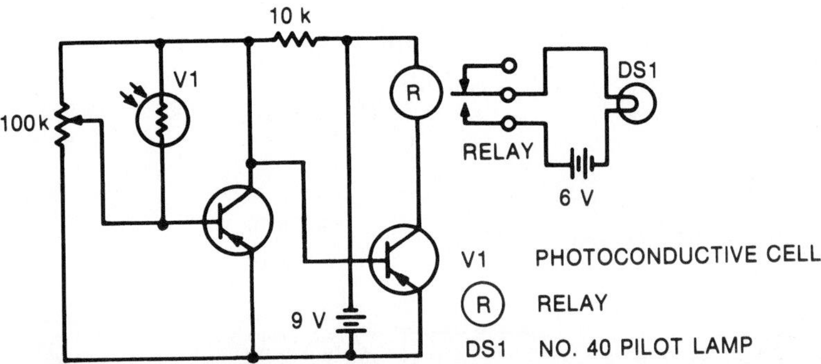

Fig. BP-5B. Schematic diagram for photo-controlled relay circuit.

2. Connect the photoconductive cell across the VOM. Using the instrument as an ohmmeter, cover the cell with your hand. Measure its resistance. Record this resistance in the dark-resistance column of the table.
3. Remove your hand from the cell. Measure its resistance. Record this resistance in the normal-resistance column of the table.
4. Turn the desk lamp on. Move it toward the cell. As you do this, you will note that the resistance of the cell gradually decreases to a minimum value. Record this resistance in the high-intensity column of the table.
5. Connect the circuit shown in Fig. BP-5B.
6. Put the 60-W lamp about 1 ft. (300 mm) from the cell. Adjust the potentiometer until the relay contacts in the pilot lamp circuit are closed. When the relay contacts close, the pilot lamp will light.
7. Turn off the 60-W lamp and all other nearby lamps. This will cause the relay contacts to open. The pilot lamp will go out.
8. By moving the lighted 60-W lamp closer to and farther away from the cell, you will note that the pilot lamp can be made to go on and off. Explain the reason for this.

BASIC PROJECT 6: THERMOCOUPLE

This activity shows how a voltage can be generated by heating the junction of two different kinds of metals.

Materials Needed

1 piece each of sheet copper, tinplate, galvanized steel, brass, and zinc, 20 to 26 gage, 1 x 4 in. (25 x 102 mm)
1 wax candle or electric heating element
galvanometer with center-zero scale

Procedure

1. Lightly tin both surfaces to be joined. Then solder the copper and tinplate together as shown in Fig. BP-6A.

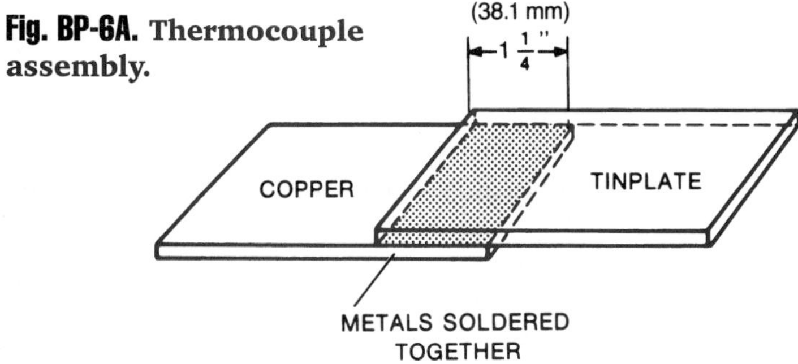

Fig. BP-6A. Thermocouple assembly.

2. Connect the unsoldered ends of the thermocouple to the galvanometer.
3. Heat the junction of the metals with the candle or electric heating element. Observe the reading of the galvanometer. Do not apply heat any longer than is needed for the pointer of the galvanometer to reach a steady, maximum position. Otherwise, the solder may melt.
4. Prepare on a piece of paper a table like the one in Fig. BP-6B.
5. Record the maximum reading of the galvanometer on the table. Also indicate on the table the voltage polarity of the thermocouple, as, for example, copper (+) and tinplate (-).
6. Repeat steps 1 through 5 with the other combinations of metals given in the table.
7. Refer to the data given in the table. Which of the metal combinations produced the most current, as indicated by the reading of the galvanometer?

Metals	Maximum Reading of Galvanometer	Polarity	
Copper-tinplate Copper-galv. steel Copper-brass Copper-zinc Tinplate-galv. steel		copper ( ) copper ( ) copper ( ) copper ( ) tinplate ( )	tinplate ( ) galv. steel ( ) brass ( ) zinc ( ) galv. steel ( )
Tinplate-brass Tinplate-zinc Galv. steel-brass Galv. steel-zinc Brass-zinc		tinplate ( ) tinplate ( ) galv. steel ( ) galv. steel ( ) brass ( )	brass ( ) zinc ( ) brass ( ) zinc ( ) zinc ( )

Fig. BP-6B. Make a table like this to record the galvanometer readings and thermocouple polarities.

BASIC PROJECT 7: THE SAFE CURRENT-CARRYING CAPACITY OF WIRES

The amount of current that any wire can conduct safely depends mainly on the wire's cross-sectional area. If too much current flows for that size of wire, the wire becomes overheated. This can damage the insulation and perhaps cause a fire. A simple experiment can show the fire hazard created when undersized wires are used to carry the load current of a circuit.

Materials Needed

3½ in. (89 mm) PE magnet wire, no. 18 (1.00 mm)
4 in. (102 mm) PE magnet wire, no. 21 (0.71 mm)
4 in. (102 mm) PE magnet wire, no 25 (0.45 mm)
4 in. (102 mm) PE magnet wire, no. 30 (0.25 mm)
2 in. (51 mm) PE magnet wire, no. 36 (0.125 mm)
1 no. 6 dry cell, 1½ V

Procedure

1. Remove ½ in. (12.7 mm) of insulation from each end of the wires. Join them as shown in Fig. BP-7, A.
2. Connect the joined wires to the dry cell as shown in Fig. BP-7, B. This connection short-circuits the cell. Therefore, do not leave the wires connected any longer than needed to finish this procedure. Otherwise, the cell will be quickly ruined.

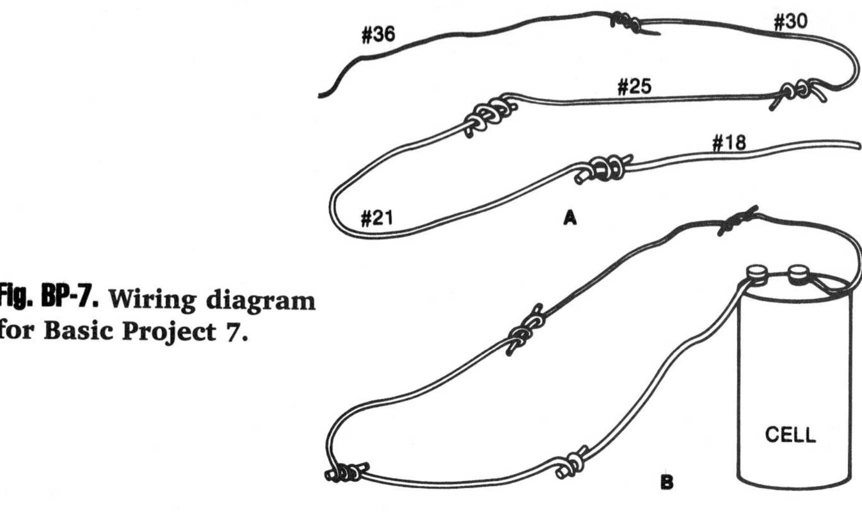

Fig. BP-7. Wiring diagram for Basic Project 7.

SAFETY TIP

Touch the no. 36 (0.125 mm) wire quickly and carefully. It will be hot enough to cause painful burns.

3. After a few seconds, touch the no. 18 (1.00 mm), the no. 21 (0.71 mm), the no. 25 (0.45 mm), the no. 30 (0.25 mm), and the no. 36 (0.125 mm) wires. Which of these wires is the hottest? Why?
4. Using the power formula $P = I^2R$, explain why each of the wires is heated to a different temperature although the current is the same.

BASIC PROJECT 8: SOLDERING EXPERIENCES

Soldering is done to ensure good electrical connections. The following soldering jobs will give you experience in this important wiring activity.

Materials Needed

1 three- or four-lug terminal strip
1 phono plug
6 in. (150 mm) untinned insulated wire, any size, no. 18 through no. 22 AWG (1.00 through 0.63 mm)
18 in. (460 mm) tinned or untinned insulated wire, any size, no. 18 through no. 22 AWG (1.00 through 0.63 mm)
6 in. (150 mm) microphone cable
soldering iron, 40 or 60 W
solder wire. rosin core

Procedure

Do one or more of the following soldering jobs. NOTE: You will find these jobs illustrated in Chapter 21, "Soldering."

1. tinning a wire (Fig. 21-8)
2. soldering two wires together (Fig. 21-13)
3. soldering wires to the lugs of a terminal strip (Fig. 21-10)
4. soldering a microphone cable to a phone plug (Fig. 21-12)

After you have finished these, show them to your instructor for approval. The terminal strip can be conveniently mounted on a baseboard.

BASIC PROJECT 9: SALTWATER RHEOSTAT

This kind of rheostat is not used in practical circuits. It does, however, clearly show the effect of ionization in a liquid solution.

Materials Needed

medium-size glass or plastic tumbler filled with water, preferably distilled
2 pieces of sheet copper or aluminum, ½ x 2¼ in. (13 x 57 mm)
1 pilot lamp, no. 46
1 miniature lamp socket, screw-base
common table salt in shaker
glass or plastic stirring rod
2, 6-V batteries, or a low-voltage power supply with a current rating of at least 0.4 A
volt-ohm-milliammeter

Procedure

1. Bend the pieces of sheet metal. Put them in the tumbler containing the distilled water, as shown in Fig. BP-9.
2. With the volt-ohm-milliammeter adjusted to a range of about 0 to 250 mA, connect the circuit as shown in Fig. BP-9. Explain why there is now no current in the circuit.
3. Slowly sprinkle salt into the water and stir. The current should gradually increase to a value that causes the lamp to light dimly.

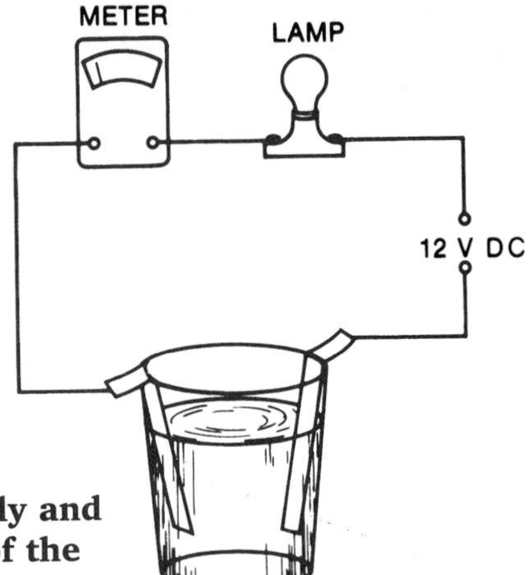

Fig. BP-9. Tumbler assembly and circuit for construction of the saltwater rheostat.

4. Keep adding salt to the solution until the current has increased to 250 mA. The lamp should now be working normally.
5. Explain the ionization process that changed the nonconducting solution into an electrolyte able to conduct ionic charges.

BASIC PROJECT 10: WIRING WITH NONMETALLIC SHEATHED CABLE

These wiring jobs will help you learn how several common wiring devices are used in the home. They will also give you experience in connecting a load to three- and four-way switches.

Materials Needed

baseboard, ¾ x 18 x 30 in. (19 x 457 x 762 mm)
1 junction box, octagonal, 4 x 1½ in. trade size
3 switch boxes, utility type, 4 x 2⅛ x 1½ in. trade size
1 junction-box cover
3 switch-box covers
1 switch, spst, flush mount with grounding screw
2 three-way switches, flush mount with grounding screw
1 four-way switch, flush mount with grounding screw
2 duplex receptacles with grounding screw
1 lampholder, keyless, medium-base, to fit 4 in. junction box
8 ft. (2.4 m) nonmetallic sheathed cable, no. 14/3 with grounding wire
5 ft (1.5 m) nonmetallic sheathed cable, no. 12/2 with grounding wire
4 ft (1.2 m) nonmetallic sheathed cable, no. 14/2 with grounding wire
1 rubber or plastic handle attachment plug with cable connector, three-prong
1 incandescent lamp, medium-base, 40-, 60-, or 100-W
assortment of solderless twist-on connectors (wire nuts); solderless pressure connectors (crimp type); metal or plastic straps (staples); box connectors for nonmetallic sheathed cables; grounding wires and grounding screws. A typical grounding screw with a slotted washer head has the following specifications: ⅜ in. length, 10-32 (9.52 mm, 5D mm-0.8P mm).

Procedure

1. Mount the three switch boxes (a, b, c) and the junction box (d) on the baseboard as shown in Fig. BP-10, Part A.
2. Connect the 4 ft. cable to the junction box as shown in Fig. BP-10, B. The junction box is also used for mounting the lampholder.
3. *Job 1.* Wire and install the duplex receptacles in switch boxes a and b. Check the wiring by tracing the circuit.

4. *Job 2.* Wire a circuit to control the lamp using the spst flush switch mounted in switch box b.

SAFETY TIP

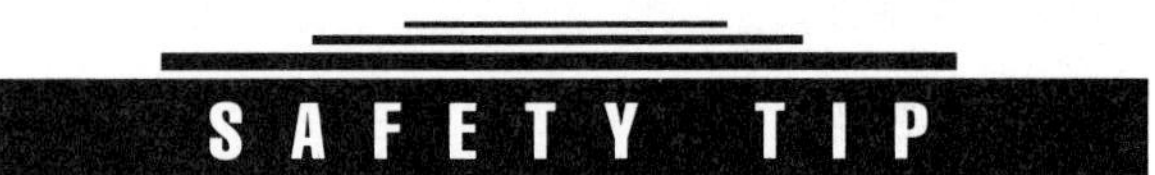

Never connect switches to the grounded (white) conductor of a building wiring system.

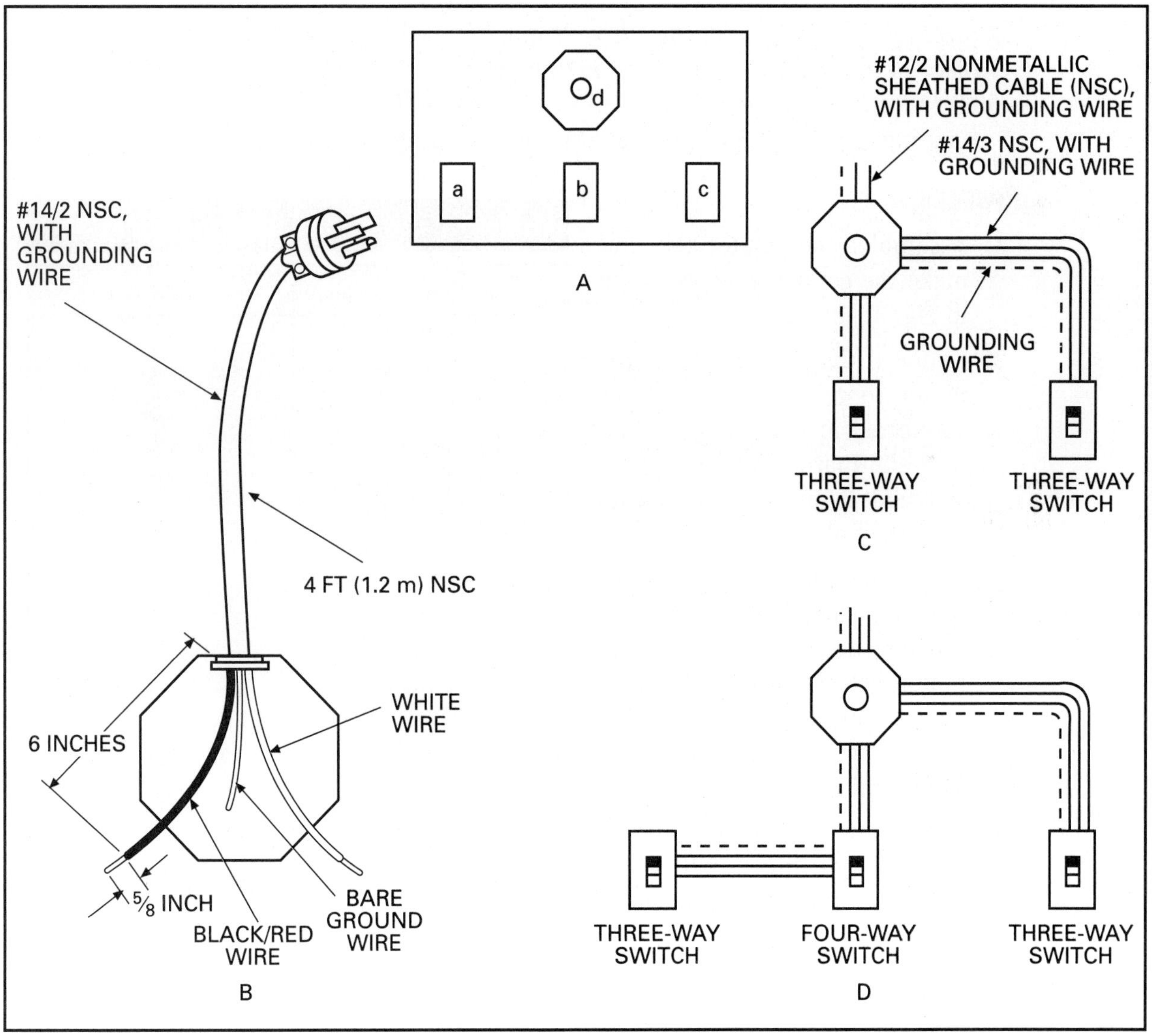

Fig. BP-10. **Assemblies for Basic Project 10.**

5. *Job 3*. Wire a two-point control circuit (Fig. BP-10, C).
6. *Job 4*. Wire a three-point control circuit (Fig. BP-10, D).

BASIC PROJECT 11: UNIVERSAL 2-POLE MOTOR

As you have learned, electromagnetic energy is used to produce mechanical motion in an electric motor. By making this motor, you will learn more about the basic parts of motors and how they work.

Materials Needed

1 piece of wood, ¾ x 4½ x 6 in. (19 x 114 x 152 mm)
1 piece of band iron, ⅛ x ½ x 11⅝ in. (3.18 x 13 x 295 mm)
1 piece of band iron, ⅛ x ½ x 6¾ in. (3.18 x 13 x 170 mm)
2 pieces of sheet brass, ⅜ x 3⅛ in. (9.5 x 79 mm)
2 pieces of tin plate, ¼ x 1¼ in. (6 x 32 mm)
1 piece of tin plate, 1½ x 3 in. (38 x 76 mm)
60 ft (18 m) PE magnet wire, no. 25 (0.45 mm); 40 ft (12 m) for the field winding and 20 ft (6 m) for the armature winding

Procedure

1. Assemble the materials and make the connections as shown in Fig. BP-11.
2. Connect the terminals of the motor to two D-size dry cells connected in series or to a battery-eliminator power supply adjusted to produce an output voltage of about 3 V.
3. Adjust the brushes so that maximum speed is obtained. The brushes should contact the commutator segments with the same amount of pressure. As brush pressure is increased, the motor speed will decrease.
4. Show how the motor can be reversed by using a dpdt switch wired as a reversing switch.

Metric Conversion Table for Fig. BP-11

in.	mm
⅛	3
3/16	5
¼	6.5
5/16	8
⅜	10
½	13
⅝	16
¾	19
1	25
1 ¼	31
1 9/16	39.5
2	51
2 ¾	70
3	76
3 ⅛	79
3 ½	89
4 ½	115
6	152
11 ⅝	297

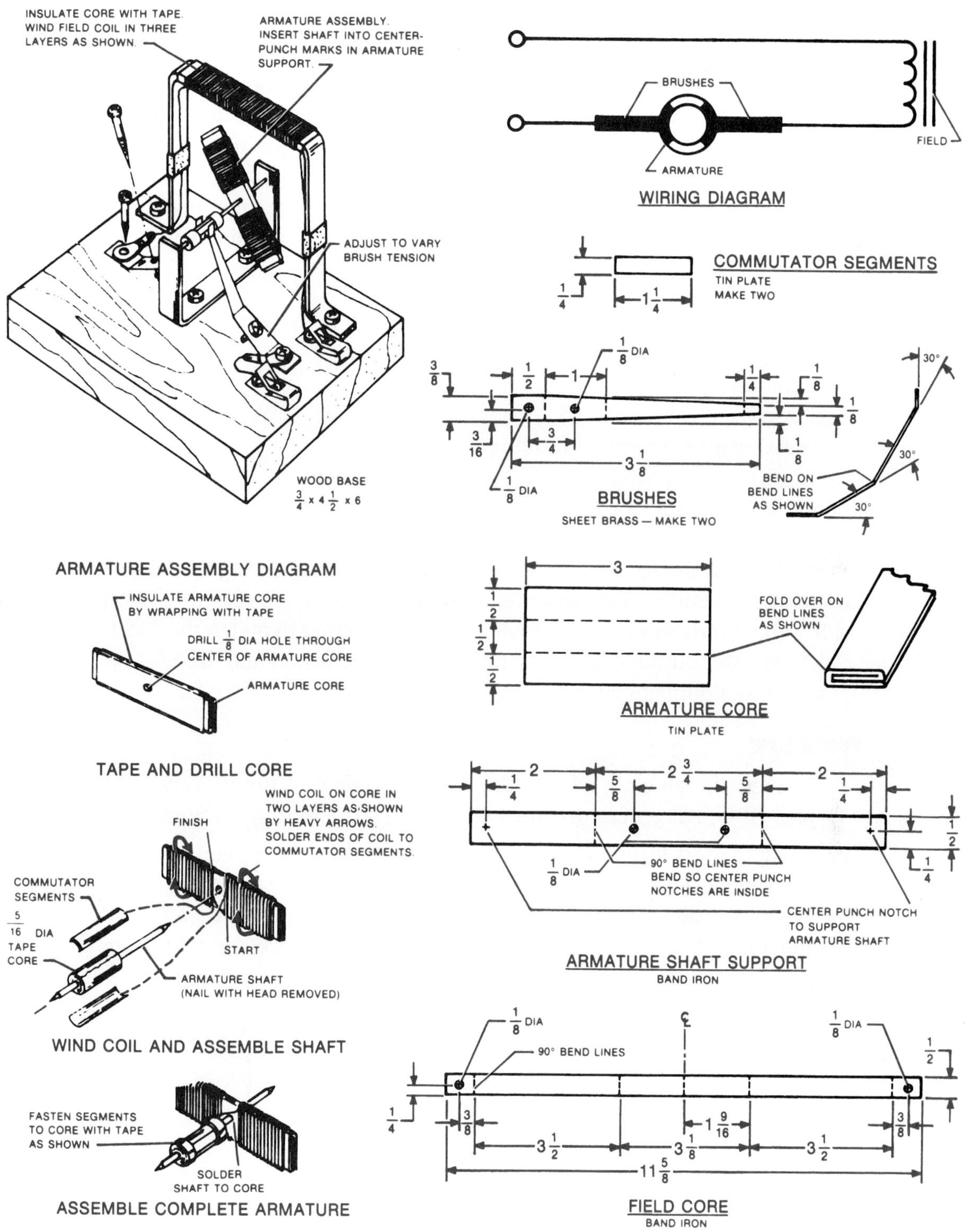

Fig. BP-11. Assembly and working drawings for Basic Project 11.

BASIC PROJECT 12: AM RADIO TRANSMISSION

This activity will show you how radio communication is accomplished (Fig. BP-12). You will send a tone over a very short distance. Radio communication of all types is controlled and licensed by the Federal Communications Commission (FCC). However, the transmitter described here is of such low power that no license is required to operate it. Standard connections and functions will be specified, but it may become necessary for you to refer to the operator's manual for your specific equipment.

Materials Needed

small, portable AM radio
RF signal generator
AF generator
oscilloscope
cassette recorder/player with output jack
10 ft. (304.8 mm) no. 18-no. 22 copper antenna wire
connector and cable for RF generator output
2 ft. (60.9 mm), two-conductor cable with connectors to match the output jacks of the AF generator on one end and the audio-input jacks of the RF generator on the other end
2 ft. (60.9 mm), two-conductor cable with connectors to match the auxiliary speaker output of the cassette recorder on one end and the RF generator audio-input connection on the other end

Procedure

1. Connect the antenna wire to the center terminal of the RF output of your RF signal generator. This will act as a transmitting antenna.
2. Using the test leads for the oscilloscope, connect the antenna wire to the vertical input terminal of the oscilloscope. The common ground lead of the oscilloscope is connected to the common ground on the RF generator. The oscilloscope will allow you to observe the waveform being transmitted to the radio.

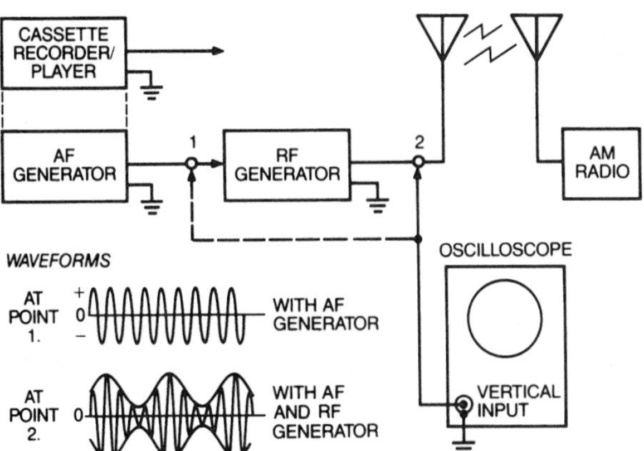

Fig. BP-12. Equipment used to construct a low-power AM radio transmitter.

3. Connect the output of the AF generator to the audio input of the RF generator and set the selector switch of the RF generator to "external modulation."
4. Coil the antenna wire in a loose circle about 1 ft. (30.48 mm) in diameter, and place it on the work surface with the AM radio near the coil.
5. Turn on the oscilloscope, RF generator, audio generator, and the AM radio.
6. Set the audio generator for 500 Hz and observe the waveform at point 1.
7. Set the tuning dial of the radio to a frequency that is not used in your area.
8. Set the frequency range selector of the RF generator to the AM band position (535 to 1605 kHz). Sweep through this band by adjusting the tuning control knob on the RF generator until a tone is heard in the radio.
9. Observe the oscilloscope and adjust the output level of the RF generator and the AF generator to get a pattern that closely approximates that shown at point 2 (Fig. BP-12).

SAFETY TIP

The equipment in this experiment may cause interference to radio and television reception in the area. If so, turn the equipment off. Set the power level of the RF generator to the minimum necessary to observe the signal on the oscilloscope. Try to eliminate interference by reorienting the antenna or relocating the test equipment.

10. Once you have an acceptable tone in the radio, sweep through the range of the AF generator. Observe the changes in the waveform on the oscilloscope and listen to the various tones in the radio.
11. Replace the AF generator with the cassette recorder/player. Do this with the cable compatible to the cassette player and the RF generator. Put in a previously recorded cassette tape and press the play button. Adjust the volume-level control until the pattern on the oscilloscope matches what you observed when there was an acceptable tone from the audio generator. Notice how the audio pattern on the carrier wave matches what you hear in the radio.
12. Draw on a separate sheet of paper the waveform observed when the cassette recorder is connected in step 11. Discuss the drawing with your instructor.

BASIC PROJECT 13: LOGIC PROBE

A very simple but useful digital instrument is the logic probe. It is used to test the logic level at important points in 5-Vdc digital circuits. The logic probe uses an AND gate and a pair of light-emitting diodes (LEDs). LED DS1 will light when the test logic is 0, and LED DS2 will light when the test logic is 1.

Materials Needed

IC1 1-7408 Quad-two input AND gate
DS1 and DS2 two-light-emitting diodes, 1.5 Vdc, 5 mA or more
R1 and R2 21,000-Ω carbon-composition ¼-W resistors
1 piece of prototyping circuit board or breadboarding unit, 1 x 3 in. (25 x 75 mm)
1 test probe
2 miniature alligator clips
2 pieces of red test lead wire, 20-24 gage, 14 in. (355.6 mm)
1 piece of black test lead wire, 20-24 gage, 14 in. (355.6 mm)
assorted lengths of solid-strand insulated hookup wire, gage to fit the selected circuit board
signal generator capable of producing a square wave at 5 Vdc

NOTE: The logic probe circuit obtains its power from the circuit being tested.

Procedure

1. Assemble the circuit on a breadboard, as shown in Fig. BP-13.
2. Connect the alligator clip, pin 14, to a source of +5 Vdc.
3. Connect pin 7 to the common ground of the circuit being tested, the square-wave generator.
4. Set the output of the generator to 5 V and the frequency from 1 to 5 Hz. The LEDs should blink.
5. Increase the generator frequency so that both LEDs appear to be on all the time. (Note the frequency at which this occurs.) Actually, the two LEDs are alternating on and off. This indicates a rapid train of logic pulses, a series of 1s and 0s. This can be verified by using an oscilloscope.

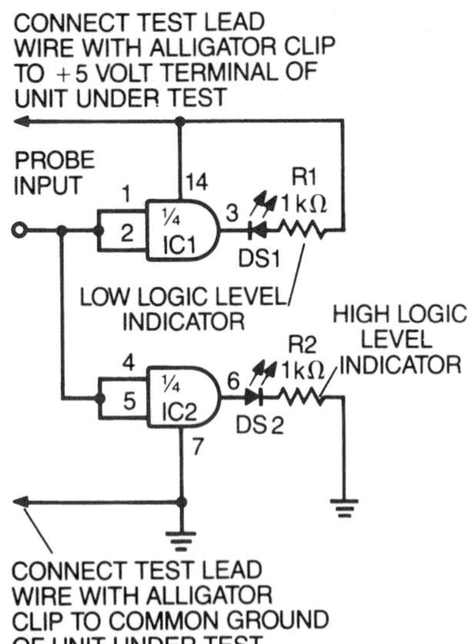

Fig. BP-13. **Schematic diagram of a logic probe.**

BASIC PROJECT 14: SURFACE ROUGHNESS GAUGE

Often data must be converted from an analog to a digital format in order for the digital computer to handle it. This activity illustrates one method of doing this data acquisition and the use of the data to rank a value on an appropriate linear scale (Fig. BP-14). The crystal phonograph stylus will produce a voltage when it is moved over a rough surface. The op-amp increases the signal strength and feeds it to the diode for rectification. This half-wave signal is fed to the always active AND gate. The AND gate will have an output pulse when the input level exceeds about 3 Vdc. Lesser voltages will be rejected. The output pulse will be sent to one of the pins of the BASIC Stamp® II. The pulse duration and frequency can be detected by the microcontroller and sent to the host computer for display using the instruction DEBUG. This is done through the interaction of the hardware and the software, as discussed in Chapter 26, "Computerized Controls."

Materials Needed

Y1 crystal conical stylus, 1.5-2.5 g tracking or equivalent
R1 25 kΩ linear taper potentiometer, 5 W
IC1 741 operational amplifier
D1 germanium diode, 1N34A or equivalent
IC2 74LS08 quad 2 input AND gate
BT1 9-V battery
1 two-conductor 9-V battery clip connector
BT2 5-V power supply
1 prototype IC plug board for assembling the circuit hookup wire as needed

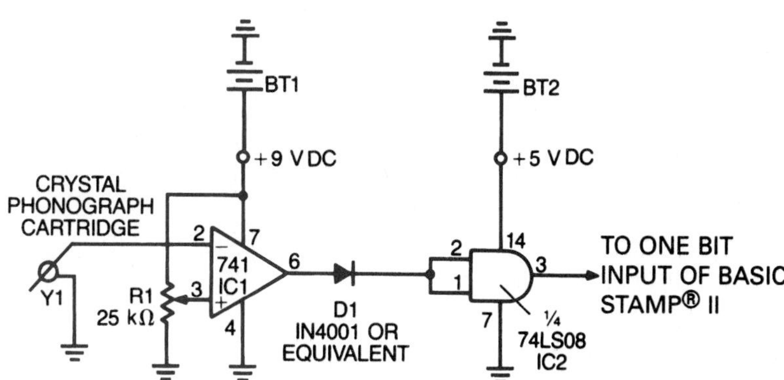

Fig. BP-14. Schematic diagram for constructing the surface roughness gauge circuitry.

Procedure

1. Construct the circuit as shown in Fig. BP-14 and connect the output of the two-input AND gate to a l-bit PB input port of a computer.
2. Move the crystal stylus over a moderately rough surface (such as a piece of wood) and observe the output of the AND gate with an oscilloscope.
3. Adjust the potentiometer, R1, to obtain enough gain from the op amp so that a rectangular wave is present at the output of the AND gate.
4. Test some other surfaces to establish a comparison scale.
5. Experiment with a computer program in order to collect and tabulate data on various types of surfaces.

BASIC PROJECT 15: REACTION TIMER

This circuit is a simple electronic game (Fig. BP-15A). When the program is run and after a short delay, the computer will turn the LED, DS1, on. When the LED lights up, the push button S1 (PBNO) is pressed. A numerical value will appear on the CRT display screen, which corresponds to the time it takes to press the button. The object of the game is to get this number as low as possible.

In this activity, you will learn how to connect an interface circuit, which has a flip-flop and a bilateral switch, to the BASIC Stamp® II pins (Fig. BP-15B). When this is done, you will experiment with writing a computer program to operate the electronic game.

Materials Needed

IC1 DUAL JK flip-flop, 74LS73
IC2 QUAD bilateral switch, 4066
2 14-pin DIP sockets
DS1 light-emitting diode (LED), maximum forward current 50 mA or higher
R1, R2 1,000 Ω resistors, ¼ watt, 5 percent
S1 PBNO momentary contact switch
prototyping IC plug board
No. 22 solid hookup wire, cut to length as needed
5 Vdc power supply, 150 mA minimum output (NOTE: Do not use the power supply from the microcomputer)
game port socket (16 pin) located on microcomputer, or equivalent input/output connector

Procedure

1. Construct the interface and timer circuit on the prototype board as shown in Fig. BP-15B. NOTE: Be sure to provide a common ground.
2. Connect pin 5 of the flip-flop to an output pin of the BASIC Stamp® II.

3. Connect the push button PBNO (S1) timing switch to the bilateral switch control (pin 13) on IC2, 4066. (See Appendix G of the *Electricity and Electronics Technology Student Workbook.)* The corresponding switch, pin 1, is connected to an input pin of the BASIC Stamp® II.
4. Connect pin 8 of the JK flip-flop $\bar{Q}$ to a second bilateral switch control (pin 5) on IC2, 4066, and the corresponding switch on IC2, 4066, (pin 3) to an input pin on BASIC Stamp® II. This will enable the BSII to determine the level of $\bar{Q}$ when the program starts.
5. Use the flowchart in Fig. BP-15C to guide you in writing, on a separate sheet of paper, the computer program to operate the Reaction Timer .

Fig. BP-15A. **A student testing a circuit for an electronic game**

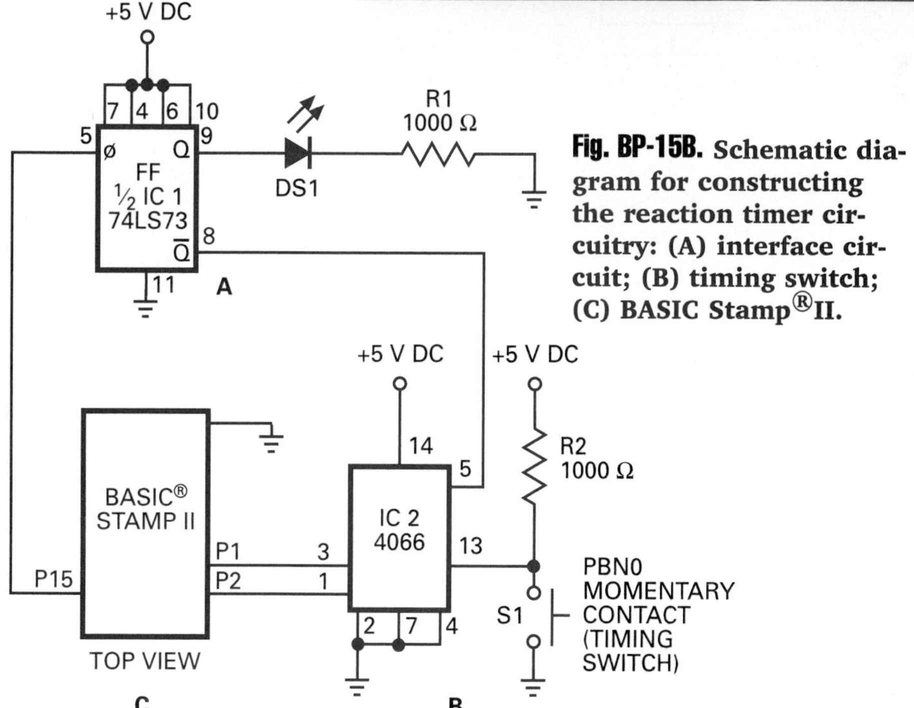

Fig. BP-15B. **Schematic diagram for constructing the reaction timer circuitry: (A) interface circuit; (B) timing switch; (C) BASIC Stamp®II.**

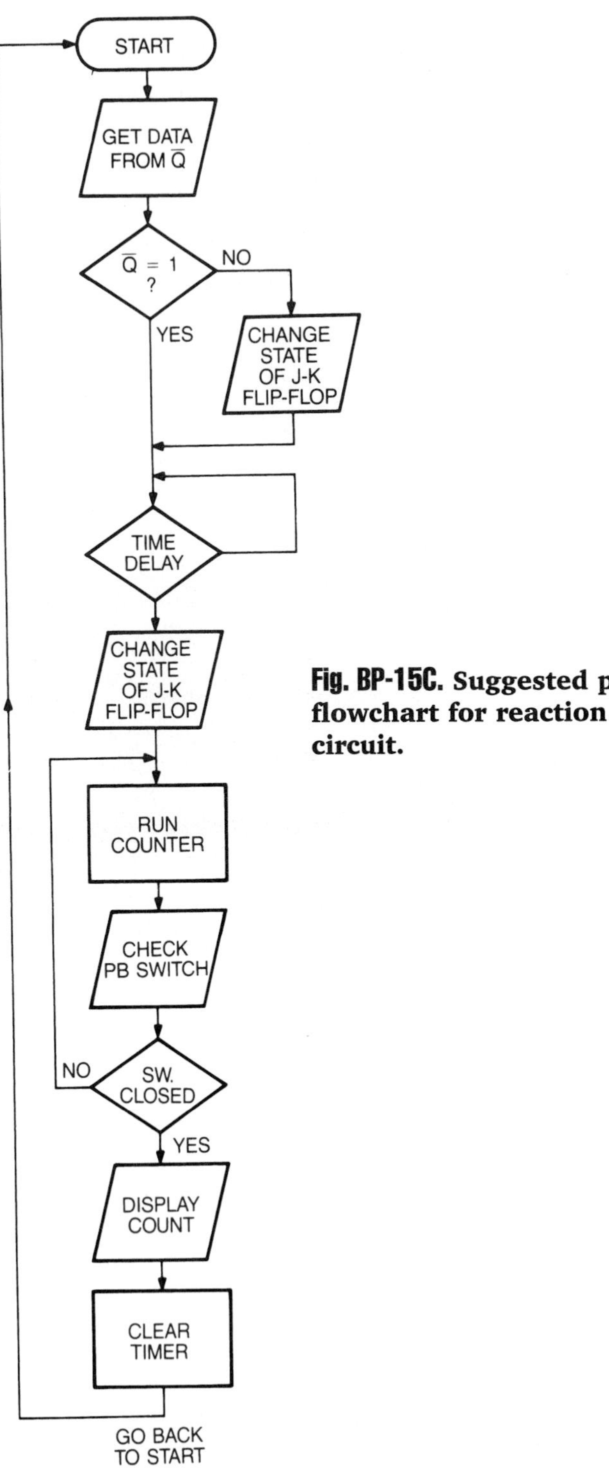

Fig. BP-15C. **Suggested program flowchart for reaction timer circuit.**

INTERMEDIATE PROJECTS

The Intermediate Projects provide opportunities for you to test your ability to solve more technical problems and to use your ingenuity in designing attractive, functional, and safe products.

INTERMEDIATE PROJECT 1: TRANSISTOR TESTER

Transistors, both n-p-n and p-n-p, can be tested using the circuit in Fig. IP-1. It will test the devices for short circuits, excessive leakage, and gain.

Parts List

BT1, battery, 6 V or 4 size-C dry cells connected in series
Ml, milliammeter, dc, 0 to 1 mA
R1 and R3, resistors, carbon-composition or film, 220 kΩ, ½ W
R2, potentiometer, 10 kΩ, ½ W
S1, switch, slide-type, TPDT (p-n-p or n-p-n selector control)
S2 and S3, switches, SPST, push button, normally open (N.O.) momentary contact type (p-n-p and n-p-n gain test controls)

Using the Tester

1. To calibrate the tester before testing any transistor, first adjust switch S1 to the p-n-p position. Then bring the alligator clips of test leads C and E into contact with each other. Next adjust R2 until the meter reads 1.

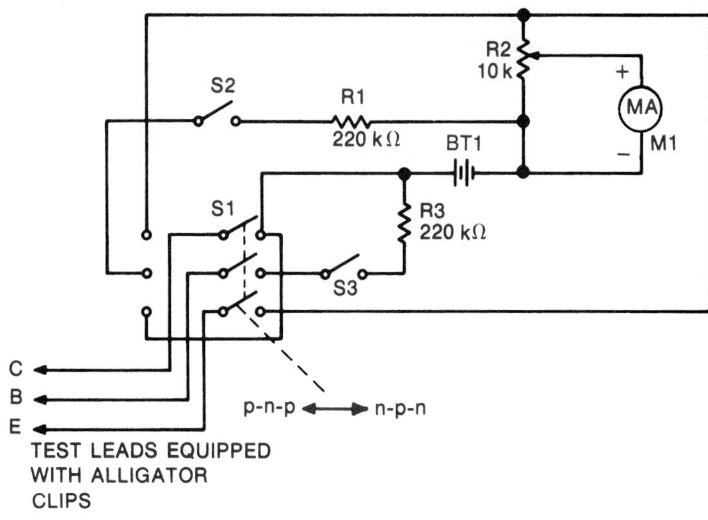

Fig. IP-1. Transistor tester.

2. To test a p-n-p transistor, adjust S1 to the p-n-p position. Then connect the test leads to the proper leads of the transistor:
   a. If the meter shows a reading of 1, the transistor is short circuited.
   b. If the meter shows a reading of more than 0.2 but less than 1, the leakage current of the transistor is excessive.
   c. To test for gain, depress S2. If the meter reading is significantly higher than the reading before S2 was depressed, the transistor is good.

3. To test an n-p-n transistor, adjust S1 to the n-p-n position. Then follow the procedures given for testing a p-n-p transistor, using S3 to provide an indication of gain.

INTERMEDIATE PROJECT 2: CONTINUITY TESTER

To determine broken or open connections as well as circuits shorted to ground, the continuity tester in Fig. IP-2 can be used. The presence of a completed circuit is indicated by either a buzzer signal or an indicator lamp.

Parts List

BT1. battery, 6 V or 4 size-C dry cells connected in series
DS1, pilot lamp, no. 47
LS1. buzzer, 6 V
S1, switch, toggle, SPDT with center off position

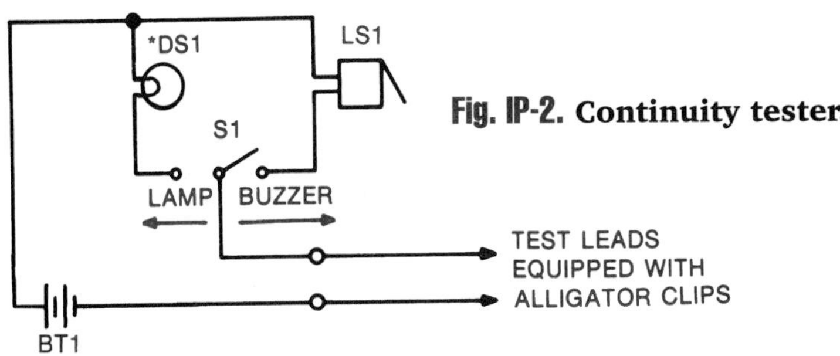

Fig. IP-2. Continuity tester.

INTERMEDIATE PROJECT 3: LAMP-DIMMER CONTROL

The circuit in Fig. IP-3 allows you to vary the light output of a I00-W incandescent lamp from zero to full brilliance. It is very convenient for controlling the operation of a study lamp or a lamp that is used as a nightlight. This circuit can be used with any 60- to 200-W incandescent lamp. However, if you expect to use a lamp larger than 100 W, fuse Fl should be increased to 2 A.

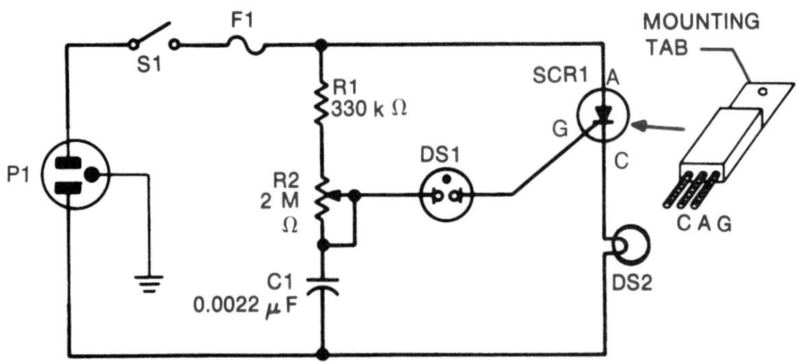

Note 1: The center terminal (A) of the silicon controlled rectifier is internally connected to the mounting tab of the device. Therefore, as a safety precaution, the mounting tab must be completely insulated from any metallic enclosure within which the circuit is placed.

Note 2: For the safest operation, the shaft of the potentiometer, R2, should be equipped with a plastic control knob.

Fig. IP-3. Lamp-dimmer control.

Parts List

P1, 3-conductor polarized connector with male contacts, rated: 15 A, 125 Vac

C1, capacitor, 0.0022 μF, 200 WVDC

DS1, neon lamp, NE-2

DS2, incandescent lamp, medium-base, 100 W, with socket

Fl, fuse, AGC, 1 A

R1, resistor, carbon-composition or film, 330 kΩ, ½ W, 5 or 10 percent tolerance

R2, potentiometer, 2 MΩ, ½ W

S1, switch, SPST (individual unit or mounted on R2)

SCR1, silicon-controlled rectifier, International Rectifier IR106B1, NTE5455 or equivalent

INTERMEDIATE PROJECT 4: LOUDSPEAKER MICROPHONE

The amplifier circuit illustrated in Fig. IP-4 uses a loudspeaker as a microphone. Many common types of audio amplifiers with which the microphone can be used require high impedance. The transistor amplifier circuit shown here provides this impedance. This dynamic-type microphone can be used for recording amateur radio-telephone work and for public address systems.

Parts List

C1, electrolytic capacitor, 10 μF, 10 WVDC
C2, capacitor, 0.1 μF, 200 WVDC
R1, composition resistor, 470 kΩ, ½ W
R2, composition resistor, 27 kΩ, ½ W
S1, switch, toggle, spst
Q1, transistor, Type 2N107 or equivalent
LS1, PM loudspeaker, 2½ inch (6.35 cm)
P1, 2-conductor plug
BT1, battery, 9 Vdc
cable, 3 ft., shielded microphone, single conductor

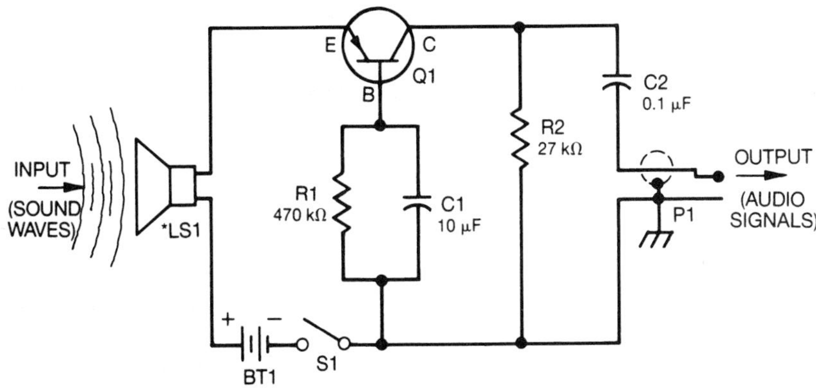

Fig. IP-4. Loudspeaker microphone.

INTERMEDIATE PROJECT 5: MINI-ORGAN

This project (Fig. IP-5) is a multivibrator circuit similar to that used in several models of electronic organs. The fundamental output frequency is determined by the RC time constant established by C1 and R1. Changes in this frequency are produced when other capacitors (C2 through C8) are connected into the circuit by switches S2 through S8. A different tone is produced with each unique combination of resistance.

Parts List

BT1, battery, 6 V or 4 size-C dry cells connected in series
C1, capacitor, 0.005 μF, 200 WVDC
C2 to C8, capacitors, 0.02 μF, 200 WVDC
C9, capacitor, 0.001 μF, 200 WVDC
C10, capacitor, 0.2 μF, 200 WVDC
CR1, germanium diode, 1N54 or SK3087
LS1, loudspeaker, 2½ in. (63.5 mm), 8 Ω
Q1, transistor, n-p-n, 2N388 or SK3011

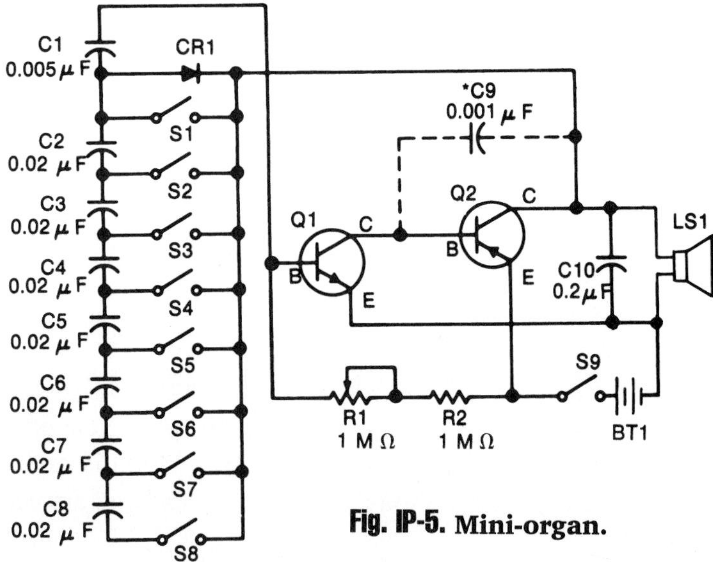

Fig. IP-5. Mini-organ.

Q2, transistor, p-n-p, 2N408 or SK3003
R1, potentiometer, 1 MΩ, ½ W (tone control)
R2, resistor, carbon-composition or film, 1 MΩ, ½ W
S1 to S8, switches, SPST, push button, normally open (N.O.) momentary contact type
S9, switch, SPST (individual unit or part of R1)

INTERMEDIATE PROJECT 6: LIGHT-CONTROLLED SWITCH

The relay-control circuit (Fig. IP-6)is operated by changes in the intensity of the light that strikes the active surface of a photoconductive cell. The relay contacts are used as a switch to control the operation of a load such as an incandescent lamp or a small motor.

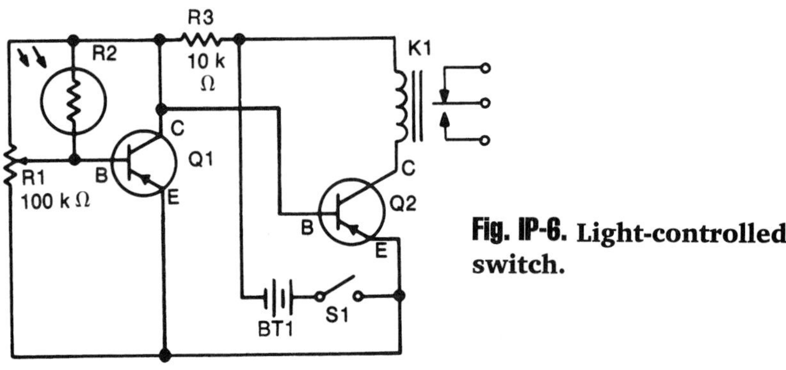

Fig. IP-6. Light-controlled switch.

Parts List

BT1, battery, 9 V
K1, relay, miniature sensitive, SPDT, 500-Ω coil, with contact rating 600 mA at 50 Vac/dc or 2 A at 117 Vac
Q1 and Q2, transistors, p-n-p, GE-2 or SK3004 or equivalent
R1, potentiometer, 100 kΩ, ½ W (sensitivity control)
R2, photoconductive cell, G.C. Electronics J4-805 or equivalent
R3, resistor, carbon-composition or film, 10 kΩ, ½ W
S1, switch, SPST (individual unit or part of R1)

INTERMEDIATE PROJECT 7: ORNAMENTAL LAMP FLASHER

The lamp flasher circuit in Fig. IP-7 produces the attractive effect of neon lamps turning on and off in random patterns. This circuit is useful for lighting jack-o'-lantern eyes, for holiday decorations, and for attracting attention to displays.

Parts List

P1, three-conductor polarized connector with male contacts, rated: 15 A, 125 Vac
C1 and C2, capacitors, 0.1 μF, 200 WVDC
C3, capacitor, 0.12 μF, 200 WVDC
CR1, rectifier diode, 1 A, 200 PIV
DS1 and DS2, neon lamps, NE-2
F1, fuse, AGC, 1/2 A
R1 and R2, resistors, carbon-composition or film, 10 MΩ, 1/2 W
S1, switch, SPST

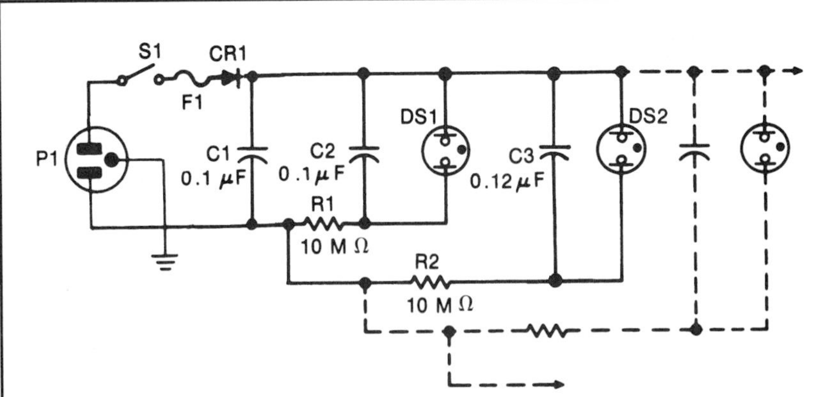

Dotted lines show circuit connections for additional lamp. More lamps can be connected into the circuit in a similar way.

All resistors are 10 MΩ and all lamps are type NE-2. Additional capacitors suggested are 0.15 μF, 0.18 μF, 0.22 μF, 0.27 μF, 0.33 μF, and so on.

Fig. IP-7. Ornamental lamp flasher.

INTERMEDIATE PROJECT 8: LIGHT-INTENSITY TRACKER

This circuit (Fig. IP-8) will allow your computer to detect the presence of light within the environment or to find the highest intensity of light. A computer program must be written to run the hardware, and this will determine the way in which this circuit can be used. The circuit can be applied to robotic devices or scientific experiments, or it can be used to experiment with rudimentary computer "vision."

NOTE: The cadmium disulfide (CdS) cell is mounted on the shaft of a slow-moving motor and is rotated to gather data to determine the relative light levels in the area. Although the motor turns slowly (2 RPM), the wires connecting the cell will wrap around the motor shaft. Therefore reverse the motor after every revolution.

Parts List

R1, resistor, carbon-composition or film, 1,000 Ω, ¼ W
R2, resistor, carbon-composition or film, 100 KΩ, ¼ W
R3, cadmium disulfide cell, resistance range 30-100 Ω
S1, switch, SPST
Dl., D2, diodes, 1N34A or equivalent
Q1, transistor, n-p-n, 2N3055 or equivalent
K1, relay, 12 Vdc coil, DPDT, SIGMA R2-12DC-SCO or equivalent
motor, 6 Vdc, 4 r/min, Hankscraft motors 184, 6 Vdc or equivalent
IC1, dual JK flip-flop, 4027 or equivalent. This IC should be mounted on a 16-pin DIP socket.
BT1, battery, 12 Vdc
BT2, battery, 6-Vdc or 4 1.5-Vdc dry cells connected in series

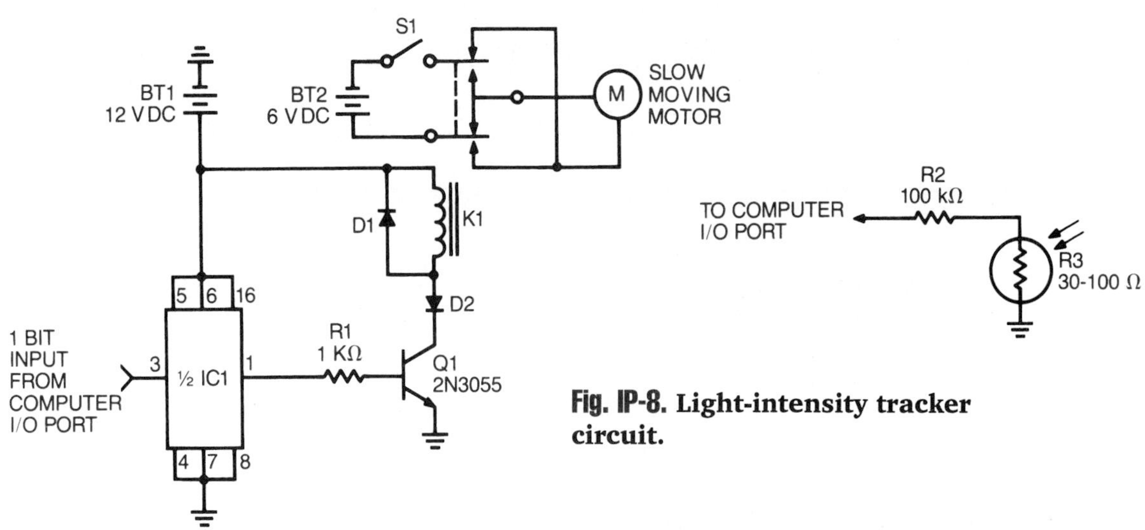

Fig. IP-8. Light-intensity tracker circuit.

INTERMEDIATE PROJECT 9: AUDIO AMPLIFIER

The audio amplifier, as in Fig. IP-9, takes a low-level audio-frequency signal as an input and boosts that signal so that a speaker can be driven to produce sound. The operational amplifier in integrated-circuit form is ideal for this function. It is small, requires few additional parts, can be operated on a simple power source, and the gain can be easily set and controlled. This circuit is useful since it can be used to amplify the signals from other projects you might make.

Parts List

R1, resistor, carbon-composition or film, 220 Ω, ½ W
R2, resistor, carbon-composition or film, 3 Ω, 1W
C1, electrolytic capacitor, 10 μF, 25 WVDC
C2, electrolytic capacitor, 500 μF, 25 WVDC
C3, C4, disk capacitor, 0.22 μF, 50 WV DC
C5, electrolytic capacitor, 1,500 μF, 25 WVDC
IC1, operational amplifier, LM 383 or equivalent
BT1, dry cell, 9 Vdc, with connector clip
LS1, speaker, 4Ω, 10 W

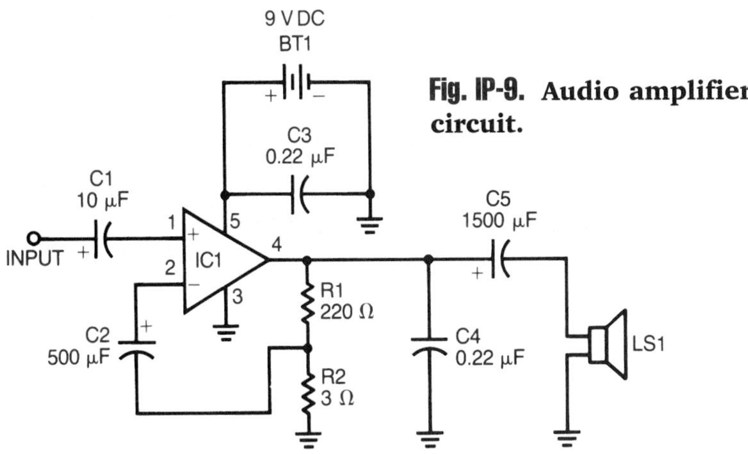

Fig. IP-9. Audio amplifier circuit.

INTERMEDIATE PROJECT 10: ONE-TRANSISTOR RADIO

Figure IP-10 illustrates a reliable low-cost radio. For best results use an outside antenna at least 50 ft. (15.25 m) long and a ground connection. Select radio stations by adjusting the antenna coil, L1.

NOTE 1: A transistor can be wired directly into a circuit without the use of a socket. This procedure, however, is not recommended.

NOTE 2: Try substituting a variable capacitor with a maximum capacity of 250 to 500 pF for the 270 pF fixed capacitor in the schematic diagram. This may improve the selectivity of the set.

Parts List

L1, broadcast-band ferrite antenna coil
BT1, battery, 3 to 9 Vdc
C1, capacitor, 270 pF, 200 WVDC
C2, capacitor, 0.02 μF, 200 WVDC
C3, capacitor, 0.001 μF, 200 WVDC
R1, resistor, carbon composition, 220 kΩ, 0.5 W
J1 and J2, insulated tip jacks
S1, toggle switch, spst
Q1, transistor, p-n-p, 2N107 or equivalent
Dl, crystal diode, type 1N34A or equivalent
HT1, magnetic headset, 2 to 5 kΩ

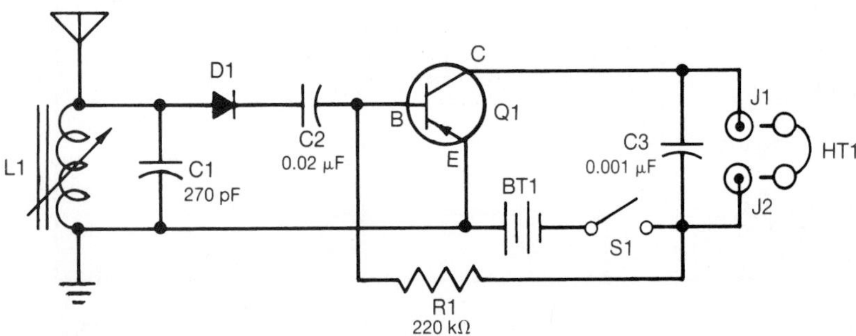

Fig. IP-10. **One-transistor radio.**

INTERMEDIATE PROJECT 11: CODE-PRACTICE OSCILLATOR/SIREN

This oscillator circuit (Fig. IP-11) provides adequate volume for either individual or group code practice. It is useful in preparing for the Amateur Radio Operator licenses given by the Federal Communications Commission (FCC). The siren feature of the circuit is an interesting example of the charge-discharge function of a capacitor.

NOTE 1: To operate as a code-practice oscillator, close switch S1 to position 1 and key with S3.

NOTE 2: To operate as a siren, close switch S1 to position 2 and close S3 momentarily.

Parts List

BT1, battery, 9 Vdc
C1, electrolytic capacitor, 50 μF, 15 WVDC
C2, capacitor, 0.02 μF, 200 WVDC
LS1, loudspeaker, 8 Ω
Q1, transistor, n-p-n, 2N497 or equivalent
Q2, transistor, p-n-p, TR-01 or equivalent

R1 and R6, resistors, 4700 Ω, 0.5 W
R2, resistor, 68 kΩ, 0.5 W
R3, resistor, 56 kΩ, 0.5 W
R4, resistor, 470 Ω, 1.0 W
R5, audio potentiometer, 1 kΩ, volume control, equipped with spst off-on switch S2
S1, switch, spdt, function control
S2, switch, spst, individual unit or mounted upon R5
S3, telegraph key

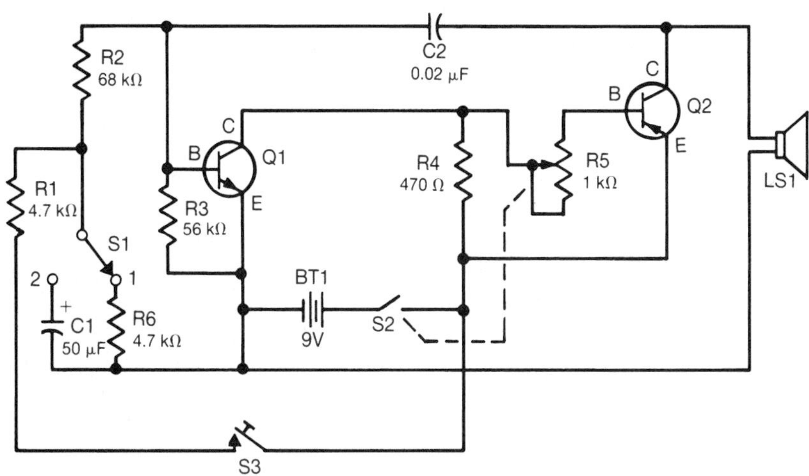

Fig. IP-11. Code-practice oscillator/siren.

INTERMEDIATE PROJECT 12: SIGNAL TRACER

Fig. IP-12 shows a test instrument that is very useful in troubleshooting either audio or radio-frequency circuits. It may also be used as a general-purpose audio amplifier.

NOTE 1: The test lead used with the signal tracer should be made of single-conductor, shielded cable.

NOTE 2: To operate as an audio-signal tracer, close switch S1 to position 1.

NOTE 3: To operate as a radio-frequency signal tracer, close switch S1 to position 2.

Parts List

BT1, battery, 9 Vdc
C1, capacitor, 0.01 μF, 200 WVDC
C2, capacitor, 0.05 μF, 200 WVDC
CR1, diode, 1N34A or equivalent
J1, jack, open-circuit type, 0.25 in. phone plug

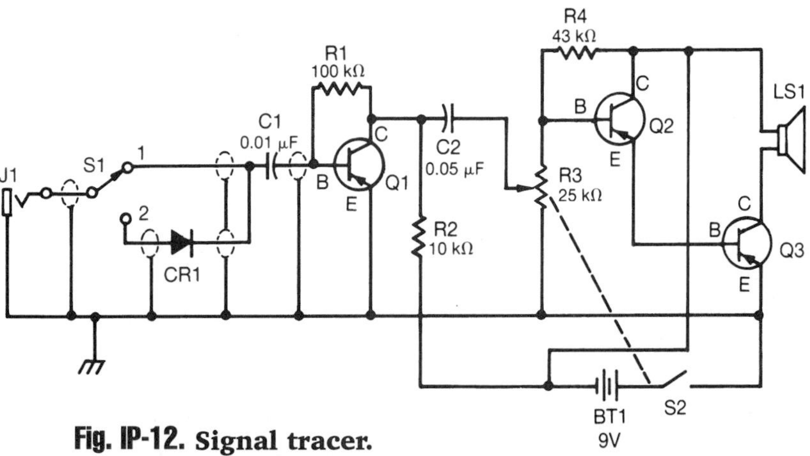

Fig. IP-12. Signal tracer.

LS1, loudspeaker, 2 or 3 in., 8 Ω
Q1, transistor, p-n-p, 2N1414 or equivalent
Q2, transistor, p-n-p, 2N109 or equivalent
Q3, transistor, p-n-p, TR01, NTE 104 or equivalent
R1, resistor, 100 kΩ, 0.5 W
R2, resistor, 10 kΩ, 0.5 W
R3, audio potentiometer, 25 kΩ, volume control, equipped with spst off-on switch S2
R4, resistor, 43 kΩ, 0.5 W
S1, switch, spdt, function control
S2, switch, spst, individual unit or mounted upon R3

INTERMEDIATE PROJECT 13: MOTOR-SPEED CONTROLLER

This 110 Vac device (Fig. IP-13) will provide reliable speed control for power tools and appliances. Electric drills, blenders, and other devices offer much more versatility when speed control is available.

SAFETY TIP

This speed controller can be used only with the universal-type motors. Be sure the motor has brushes and a commutator.

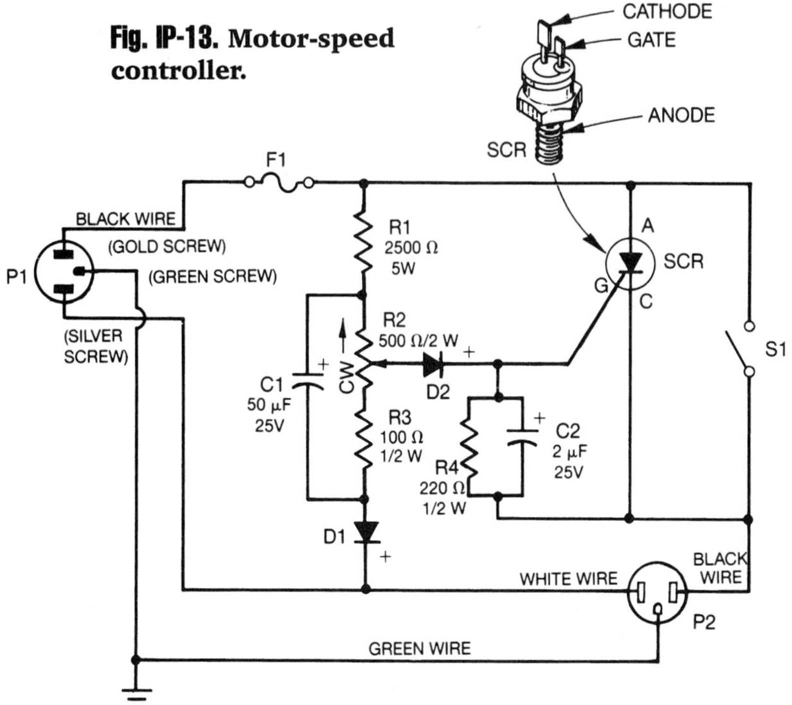

Fig. IP-13. Motor-speed controller.

The heart of the speed control is the silicon controlled rectifier (SCR). It will conduct from cathode to anode when the proper signal is applied to the gate terminal. Like any rectifier, it will not conduct from anode to cathode.

Diode Dl and the associated resistors form a variable, direct-current power supply. As the slider on the 500-ohm potentiometer is moved in the direction of the arrow, more direct-current voltage is applied to the gate of the silicon-controlled rectifier. The effect of this increased gate voltage is to enable the rectifier to "turn-on" much earlier during the positive alternation. The earlier the moment of turn-on, the more energy will be delivered to the motor, which will run faster.

Because the silicon-controlled rectifier can never turn on during the negative alternation, maximum speed will be approximately half the normal speed without the controller in the circuit. Switch S1 is therefore provided to bypass the rectifier for full-speed operation.

The motor-speed controller also provides speed regulation. When the work load on the motor increases, this will tend to slow the motor down. As the motor speed drops, the induced electromotive force across the motor windings will decrease. The induced electromotive force is fed back to the gate circuit. The polarities of applied gate voltage and induced electromotive force are arranged to oppose one another. The result is that when the feedback drops, the applied gate voltage is able to turn on the silicon-controlled rectifier sooner, which tends to make the motor run faster. The overall effect is that the motor speed tends to remain constant.

When using the speed controller, be very careful not to allow the motor to overheat. Most universal-type motors generate a lot of heat, and they are provided with built-in cooling fans to overcome this problem. When the motor is operating at reduced speed, the fan does not work efficiently. Occasionally running the tool at full speed with no load is recommended to allow the cooling fan to quickly lower the temperature. NOTE: The SCR must be mounted to a heat-sink, using a mica washer for electrical isolation between anode (case) and the heat-sink.

Parts List

C1, electrolytic capacitor, 50 μF, 25 WVDC
C2, electrolytic capacitor, 2 μF, 25 WVDC
R1, resistor, 2500 Ω, 5 W
R2, potentiometer, 500 Ω, 2 W
R3, resistor, 220 Ω, 0.5
D1 and D2, silicon rectifiers, International rectifier 8D4, NTE 116 or equivalent
F1, fuse or circuit breaker, 5 A
S1, switch, spst, 110 Vac, 5 A (close for high speed)
SCR, silicon-controlled rectifier, General Electric C20B, NTE 5492 or equivalent
P1, three-conductor polarized plug with male contacts, 125 Vac, 15 A
P2, three-conductor polarized receptacle with female contacts, 125 Vac, 15 A

INTERMEDIATE PROJECT 14: DESIGN AND CONSTRUCTION OF SMALL POWER TRANSFORMERS

The procedure outlined in this activity is a general guide to the design and construction of step-down transformers with an output from 10 to 100 W and operating at a primary voltage of 115 Vac, 60 Hz. Designing and constructing a small power transformer can be an interesting and worthwhile experience. The project is very practical since many devices can be operated from this type of transformer (Fig. IP-14).

The procedure for this project is followed by a sample design and construction problem. The problem illustrates how to use the procedure to determine the electrical and physical characteristics of a typical small power transformer.

Procedure

1. Determine the output (secondary-winding) voltage and current. These values of voltage and current will depend upon the specific purpose for which the transformer is to be used. The approximate power output rating of the transformer (in watts) is then found by using the $P = EI$ formula.

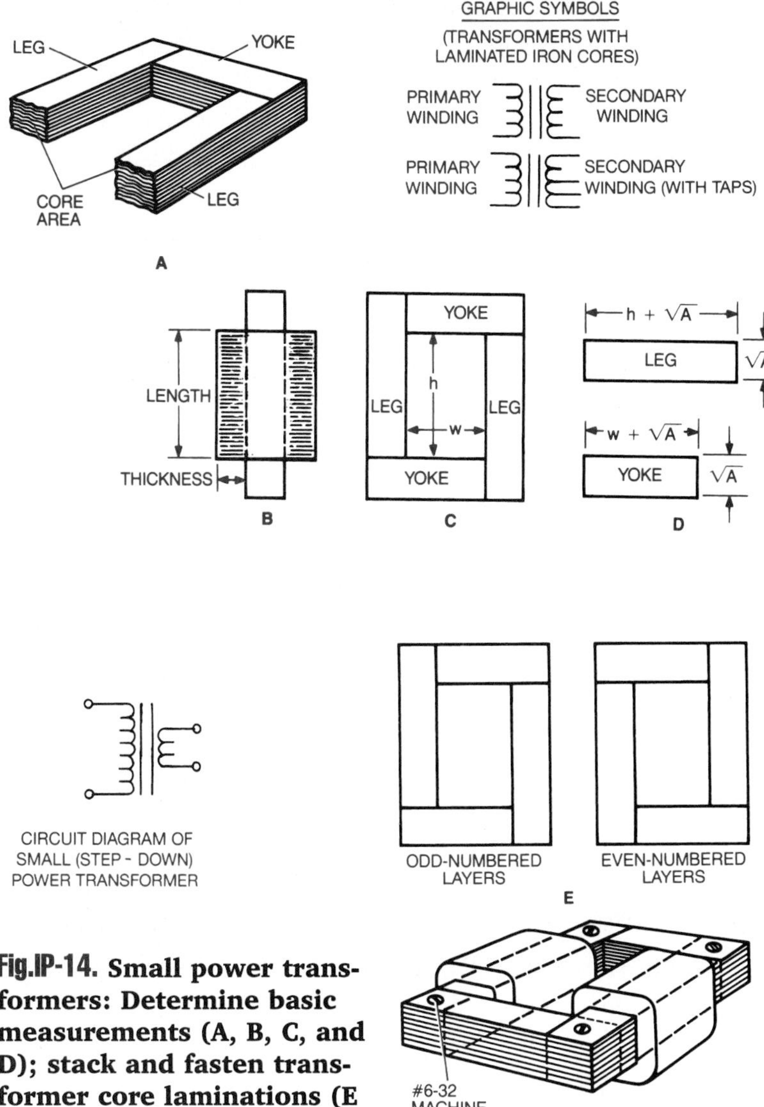

Fig.IP-14. Small power transformers: Determine basic measurements (A, B, C, and D); stack and fasten transformer core laminations (E and F).

2. Determine the cross sectional area (IP-14, Part A) of the transformer core from Table 1, "Selecting Core Area for Transformers."
3. Find the number of turns of wire necessary for the primary winding by using the following formula:

$$T_p = \frac{6.25 \times E_p}{A}$$

T_p = number of primary turns

E_p = the primary voltage

A = the cross-sectional area of the core in square inches

Table 1. Selecting Core Area for Transformers

Power output of transformer, watts	Core area, sq. in. (cm^2)
10	0.25 (1.6 cm^2)
20	0.49 (3.2 cm^2)
25	0.60 (3.9 cm^2)
30	0.70 (4.5 cm^2)
50	1.0 (6.5 cm^2)
75	1.77 (11.4 cm^2)
100	2.25 (14.5 cm^2)

4. Find the number of turns of wire necessary for the secondary winding by using the following formula:

$$T_S = \frac{T_P \times E_S \times 1.1}{E_P}$$

T_S = number of secondary-winding turns
T_P = number of primary-winding turns
E_S = secondary-winding voltage
E_P = the primary voltage

NOTE: The factor of 1.1 compensates for losses in the transformer, which are estimated to be about 10 percent.

5. Determine the primary and secondary winding currents by using the following formula:

$$I_P = \frac{W \times 1.1}{E_P}$$

$$I_S = \frac{W}{E_S}$$

I_P = primary-winding current in amperes
W = power-output rating of the transformer in watts
E_P = the primary voltage
E_S = secondary-winding voltage
I_S = secondary-winding current in amperes

6. Find the size of plain enameled magnet wire to be used for the primary and secondary winding from Table 2. The size of the wire used for each winding will depend upon the value of the primary and secondary winding currents as determined in step 5.

Table 2. Determining Turns per Inch in a Transformer Coil

Wire size (AWG)	Current-carrying capacity (at 750 cir mils per amp), amp	Approximate turns per inch (25.4 mm) per layer (enameled wire)
12	8.70	11.5
14	5.40	15.0
15	4.30	16.5
16	3.40	18.5
17	2.70	20.5
18	2.10	23.0
19	1.70	26.0
20	1.30	29.0
21	1.10	32.5
22	0.85	36.0
23	0.75	40.0
24	0.54	45.0
25	0.43	50.0
26	0.34	57.0
27	0.27	64.0
28	0.21	72.0
29	0.19	80.0
30	0.13	90.0

7. Determine the length of the primary and secondary winding coils using Fig. IP-14, Part B, and Table 3. Coils should be of the same length.

8. Determine the number of layers of wire in each coil by using the following formula:

$$N_1 = \frac{T_t}{T_1}$$

N_1 = the number of layers of wire in a coil

T_t = the total number of turns in a coil as found in steps 3 and 4.

T_1 = the number of turns of wire in each layer. Multiply the turns per inch of a particular size of wire (Table 2, column 3) by the length of a coil (step 7). Count a fractional part of a layer as a whole layer.

9. Estimate the thickness of each coil (Fig. IP-14, B) by multiplying the number of layers of wire in a coil by the diameter of the wire used.

Table 3. Determining Length of Transformer Coils

Power output of transformer, watts	Length of each coil, in. (mm)
10	1 ¼ (31.5 mm)
20	1 ⅝ (41.3 mm)
25	1 ⅞ (47.6 mm)
30	2.0 (50.8 mm)
50	2 ¼ (57.2 mm)
75	2 ½ (63.5 mm)
100	2 ¾ (69.9 mm)

10. Determine the size of the transformer opening or window (Fig. IP-14, C). The height (h) of the window should be ½ inch greater than the length of the coils as found in step 7. The width (w) of the window should be 1 inch greater than the sum of the thicknesses of each coil as found in step 9.

11. Determine the size of the core laminations in Fig. IP-14, D. The use of a core having a square cross section is recommended. The width of the laminations will be equal to the square root of the cross-sectional area of the core (step 2). The length of one-half of the laminations required will be equal to the height of the window plus the width of the laminations. The length of the remaining half of the laminations required will be equal to the width of the window plus the width of the laminations. The total number of laminations will depend upon the thickness of the sheet metal used.

12. Prepare the laminations from No. 22 to 26 sheet metal. Black sheet steel or transformer sheet steel is best, but ordinary galvanized sheet steel can also be used. The cores of discarded power transformers are an excellent source of lamination material.

13. Insulate each lamination by applying a thin coat of insulating varnish to its surfaces.

14. Form the transformer core by stacking the laminations as shown in Fig. IP-14, E. After the core has been formed, secure the laminations in place by wrapping each yoke and leg of the core with a double layer of plastic electrical tape or friction tape.

15. Wind the primary and the secondary winding coils. See Appendix B of the *Student Workbook* for more information. Each layer of each coil should be insulated (covered) with a single layer of thin cambric cloth or insulating paper.

Table 4. Design and Construction of a Power Transformer

Information needed	Step of procedure used	Result
Output voltage Output current Power output	 1. Power = *EI* or 10 x 4	10 volts 4 amp 40 watts (Design and construct a 50-watt transformer.)
Cross-sectional area of core	2. 1 sq. in. (from Table 1)	1 sq. in. (6.5 cm^2)
Number of primary turns	3. $T_p = \frac{6.25 \text{ x } E_p}{A} = \frac{6.25 \text{ x } 115}{1}$ or	719 turns
Number of secondary turns	4. $T_s = \frac{T_p \text{ x } E_s \text{ x } 1.1}{E_p} = \frac{719 \text{ x } 10 \text{ x } 1.1}{115}$ or	69 turns
Primary current	5. $I_p = \frac{W \text{ x } 1.1}{E_p} = \frac{50 \text{ x } 1.1}{115}$ or	0.48 amp
Secondary current	5. $I_s = \frac{W}{E_s} = 50/10$ or	5 amp
Size of primary-winding wire	6. See Table 2	No. 24 AWG
Size of secondary-winding wire	6. See Table 2	No. 14 AWG
Length of winding coils	7. See Table 3	2 ¼ in. (57.2 mm)
Number of layers of wire in primary coil	8. N_l (primary) $= \frac{T_t}{T_l} = \frac{719}{45 \text{ x } 2.25}$ or	8 layers
Number of layers of wire in secondary coil	8. N_l (secondary) $= \frac{T_t}{T_l} = \frac{69}{15 \text{ x } 2.25}$ or	3 layers
Approximate thickness of primary coil	9. 8 x 0.02 or	0.16 in. (4.1 mm)
Approximate thickness of secondary coil	9. 3 x 0.064 or	0.192 in. (4.8 mm)
Height of transformer core window	10. 2.25 in. + 0.5 in. or	2.75 in. (69.9 mm)
Width of transformer core window	10. 0.16 in. + 0.192 in. + 1 in. or	1.352 in. (34.3 mm)
Width of core laminations	11. √ 1 or	1 in. (25.4 mm)
Size of one-half of the laminations	11. 1 in. x (2.75 in. + 1 in.) or	1 x 3.75 in. (25.4 x 95.5 mm)
Size of remaining half of laminations	11. 1 in. x (1.352 in.+ 1 in.)	1 x 2.352 in. (25.4 x 60.3 mm) (approx. 2 ⅜ in.)

16. Remove the coils from their forms, or mandrels, and dip them into insulating varnish. Allow the varnish to dry thoroughly.
17. Remove the yokes from the legs of the core stack (step 14) and slip the coils over the legs.
18. Reassemble the transformer yokes to the legs to complete the core.
19. Reinforce the core, as shown in Fig. IP-14, Part F, with machine screws.
20. Mount or enclose the transformer as desired.

Sample Problem

Design and construct a power transformer to operate a 10 V, 4 A ac motor. The transformer is to be operated from a 115 Vac source. Table 4 contains the information needed, procedure, and results.

INTERMEDIATE PROJECT 15: POWER SUPPLY AND FILTER DESIGNS

Electricity is delivered to homes and businesses in the form of alternating current (ac). However, the operation of most electronic devices, such as radios, TV receivers, and tape recorders, requires direct current (dc). Batteries could be used, but they have limited operational life and power output capabilities. A circuit, therefore, must be used to change the ac to dc at a constant value. Figure IP-15 shows a typical rectifier circuit (A) to change ac to dc and three filter circuits (B, C, and D).

Part 1: Power Supply (Circuits A and B)

The power supply rectifier circuit shown in Fig. IP-15, A, when connected to the filter circuit in Fig. IP-15, B, changes 120 Vac to a specific value of dc. This power supply is designed to power a specific electronic device. The rectifier circuit can be found usually as a subassembly in a power supply chassis or enclosure. The transformer T1 changes the voltage from 120 Vac to the rated secondary Vac for the transformer used. In choosing a transformer, select one that has a secondary voltage slightly above the required dc voltage. You may wish to design your own transformer (see Intermediate Project 14). The bridge rectifier CR1 changes the ac to a pulsating, full-wave dc. This rectifier circuit is designed to be used with the filter circuits shown in Fig. IP-15, Parts B, C, and D.

A basic filter circuit is shown in Fig. IP-15, B. It smoothes out the pulsation from the rectified voltage to produce a dc of constant voltage level, for example, 12 Vdc. This voltage is then applied to the load circuit RL. Capacitors C1 and C2 and resistor R1 make up the filter network.

This type of power supply works well in applications where the load is fairly constant. See also the discussion on filter action and bridge rectifiers in Chapter 17, "Semiconductors and Diodes."

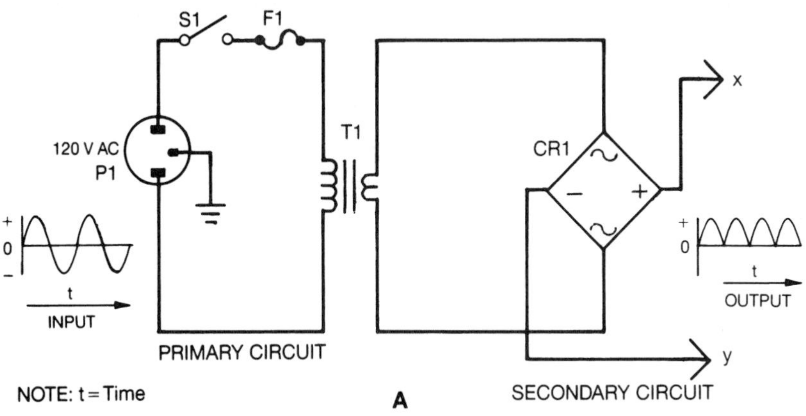

Fig. IP-15. **Filter designs for dc power supplies: (A) rectifier circuit, (B) basic filter network, (C) filter with fixed voltage regulator, (D) filter with variable voltage output.**

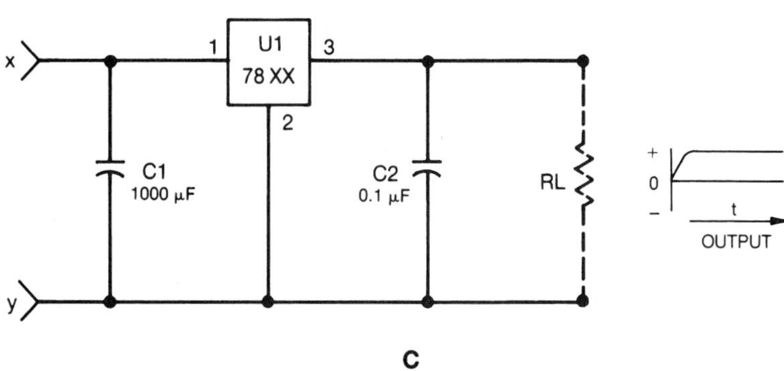

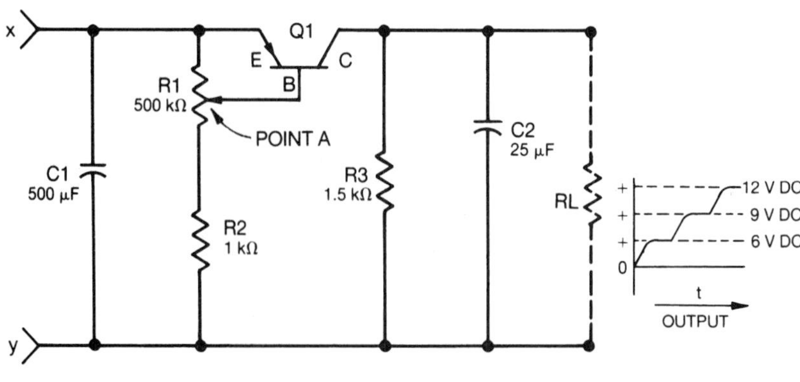

Parts List for Part 1 (Circuits A and B)

- C1 and C2, electrolytic capacitors, 500 μF, 50 WVDC
- CR1, bridge rectifier, 2-3 A, 50 PIV, NTE 166, Archer 2761146 or equivalent
- Fl, fuse, 1 A or matched to the transformer rating
- P1, 3-conductor polarized plug with male contacts, rated 15 A, 120 Vac
- R1, resistor, 22 Ω, 5 W
- S1, switch, SPST
- T1, power transformer, 120 Vac primary, secondary to match design output voltage
- Fuse holder, assorted fasteners, chassis for mounting and enclosing components, miscellaneous

Part 2: Power Supply (Circuits A and C)

The power supply rectifier circuit shown in Fig. IP-15, A, when connected to the filter circuit in Fig. IP-15, C, changes 120 Vac to a regulated value of dc.

A filter with a voltage regulator is shown in Fig. IP-15, C. This filter circuit will deliver a regulated dc voltage, which is set by the selection of the integrated circuit, U1. The IC is a voltage regulator which has one specific dc output voltage. This is a good power supply for digital work since the regulator holds or "clamps" the voltage at a fixed level. Capacitor C1 filters the rectified dc voltage from the rectifier circuit, Fig. IP-15, A. Capacitor C2 removes voltage spikes from the output of the regulator. This is very important since digital circuits might read voltage spikes as digital pulses and the data would be incorrect. Integrated circuit U1 is selected for a desired voltage, such as 5 Vdc for the 7805 IC. Other common voltages are available, such as 6.2 V, 8 V, 9 V, 12 V, 15 V, and 24 V. The type number designation for the IC is given as 78XX, with the voltage value replacing the "XX." For example, a 7812 IC is a 12 Vdc regulator and a 7806 is a 6.2 Vdc regulator.

Parts List for Part 2 (Circuits A and C)

- C1, electrolytic capacitor, 1000 μF, 35 WVDC
- C2, tantalum capacitor, 0.1 μF, 35 WVDC
- CR1, bridge rectifier, 2-3 A, 50 PIV
- Fl, fuse, 1 A, or matched to the transformer rating
- P1, 3-conductor polarized plug with male contacts, rated 15 A, 120 Vac
- S1, switch, SPST
- T1, power transformer, 120 Vac primary, secondary to match design output voltage
- U1, voltage regulator, integrated circuit 78XX (type depends on required voltage)
- Line cord, fuse holder, assorted fasteners, chassis for mounting and enclosing components, miscellaneous

Part 3: Power Supply (Circuits A and D)

The power supply rectifier circuit shown in Fig. IP-15, A, when connected to the filter circuit in Fig. IP-15, D, changes 120 Vac to a variable voltage dc output.

The filter with a variable voltage output is shown in Fig. IP-15, D. This circuit allows the user to set the output to a desired level within its designed output range. The voltage is set by adjusting potentiometer R1. The voltage drop at point A sets the operating point of Q1 and therefore the output voltage level for the load. A meter connected to the output will enable the user to set and monitor the output voltage. Transistor Q1 is used as a regulator so that as the load current changes, the voltage will remain at its set value, within limitations. This circuit can be used to power experiments, circuits being repaired, or in hobby applications.

Parts List for Part 3 (Circuits A and D)

C1, electrolytic capacitor, 500 μF, 50 WVDC
C2, electrolytic capacitor, 25 μF, 50 WVDC
CR1, bridge rectifier, 2-3 A, 50 PIV
F1, fuse, 1 A or matched to the transformer rating
P1, 3-conductor polarized plug with male contacts, rated 15 A, 120 Vac
Q1, transistor, p-n-p, 2N230, NTE 104 or equivalent
R1, potentiometer, 500 kΩ, 0.5 W, linear taper
R2, resistor, 1 kΩ, 0.5 W
R3, resistor, 1.5 kΩ, 0.5 W
S1, switch, SPST
T1, power transformer, 120 Vac primary, secondary to match design output voltage
Fuse holder, assorted hardware, chassis for mounting, enclosure hardware, miscellaneous

INTERMEDIATE PROJECT 16: PLANT SOIL MOISTURE PROBE

This project, Fig. IP-16, is used to show the relative water content of plant soil. It works due to the conductivity of the moisture within the soil. The more moisture present, the greater the conductivity. If the moisture content is too low, the red LED will glow. If the level is either correct or above the desired level, the green LED will glow. The circuit diagram is shown in Fig. IP-16A. The desired level is set by adjusting the potentiometer through experimentation with a soil sample. Do not use a potted plant for adjustment, as too much water could be added before the proper level is determined, and the plant might drown.

The foil pattern in Fig. IP-16B is full scale. It can be used to make the printed circuit (PC) board. Although U1 (IC4011) can be soldered directly to the board, it is recommended that a 14-pin DIP (dual in-line package) socket be used. The two large foil areas make up the probe. The two small

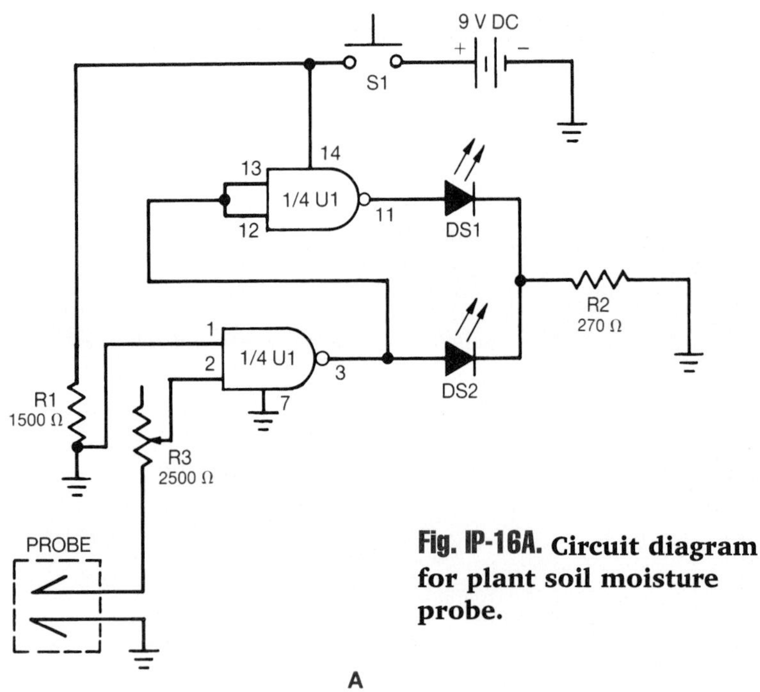

Fig. IP-16A. Circuit diagram for plant soil moisture probe.

holes drilled in the PC board, near the foil area, are used for a plastic tie wrap to secure the battery. An insulator strip should be put between the battery case and the PC board to prevent an electrical short. Mount the push-button switch on the PC board with a silicone adhesive so that the button extends beyond the edge of the PC board at the opposite end from the probe.

Once the project is complete, insert the probe into a sample of plant soil that has the proper moisture content. Adjust the potentiometer until the red LED glows. Next, adjust the potentiometer in the opposite direction until the green LED glows. Stop adjusting the potentiometer when the LEDs just change condition. This is the desired water level.

Parts List

DS1, LED, red, general-purpose type
DS2, LED, green, general-purpose type
R1, resistor, 1.5 kΩ, 0.25 W
R2, resistor, 270 Ω, 0.25 W
R3, potentiometer, 2500 Ω, miniature trimmer, 0.5 W
S1, switch, PBNO, miniature momentary contact
U1, integrated circuit 4011, CMOS quad, two input NAND gate
BT1, battery, 9 Vdc, with two lead connector
IC DIP socket, 14-pin, soldering type for PC use (optional)
Tie wrap, insulating paper, silicon adhesive, copper circuit PC board 5.5 in. x 1 in. (14 cm x 2.54 cm), miscellaneous

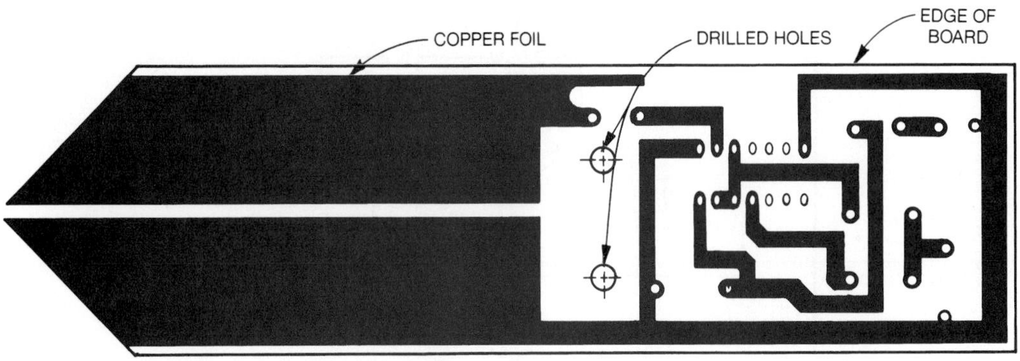

Fig. IP-16B. **Copper foil pattern on pc board (foil side up).**

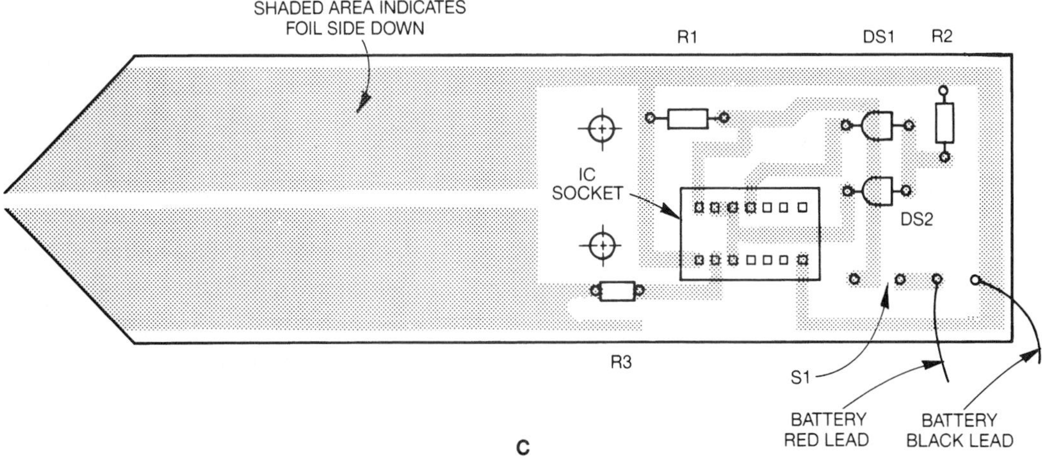

Fig. IP-16C. **Part layouts on the pc board (foil side down).**

INTERMEDIATE PROJECT 17: EXPERIMENTAL ROBOT

The experimental robot in Fig. IP-17 is designed as a minimal robot base. It is powered by a 9 Vdc battery through a 5 Vdc regulator for the motors and sensor circuitry. The controller is the BASIC Stamp® II with the optional carrier board. It will require you to write a software program in the PBASIC™ language to run this project. A start switch (S1, PBNO) is used. This is done so that when the power is applied to the drive and the controller, the robot will not start moving before you are ready. It will be necessary to include a subroutine at the beginning of the software program to service this switch, until it has been pressed indicating you want it to start. DS1-4 are included to help in the observation and analysis of motor operation.

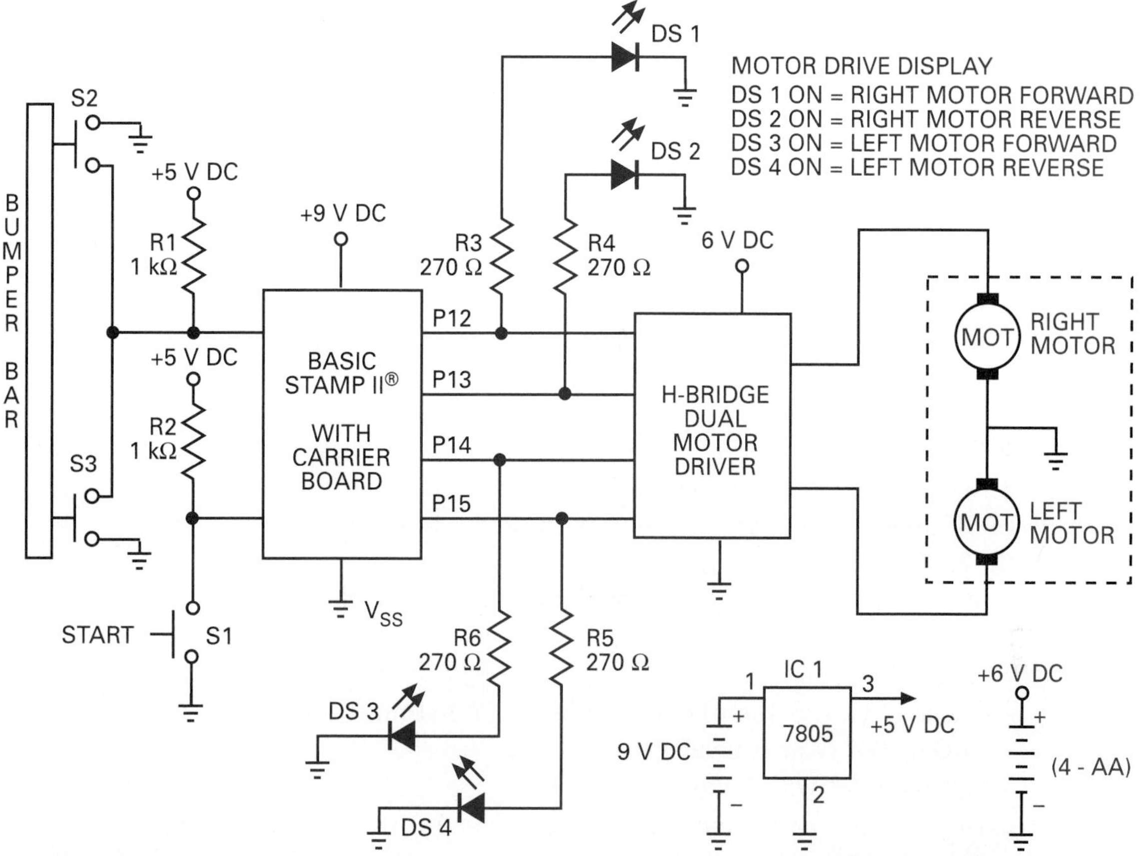

Fig. IP-17. Schematic diagram of the experimental robot. "Pin" connections are suggested only.

SAFETY TIP

Never allow the software program to pulse both the forward and reverse connections for the same motor simultaneously. This will cause the H-bridge circuit (motor driver circuit) to become short circuited and damage the transistors.

This project, as shown, is intended to be a basic platform . You may add your own ideas to make a unique robot project. There is ample capacity for you to add features and functions that enhance the operation. As you get an idea for a feature, refer to the various chapters of this text, and to other reference materials, such as the BASIC Stamp® Manual V1.8, to help you design and build your personal small robot.

Consider adding sensors to have the robot seek out and "hide" in dark spaces, or seek out and stop on bright lights. Another idea is to make your robot a mobile smoke or carbon monoxide detector. With the addition of devices available from local electronics stores or mail order catalogs, the robot might deliver voice messages. These are only a few ideas. The authors leave the rest to your imagination.

Parts List

dual motor gearbox, Tamiya #70097 or equivalent
dual H-bridge motor driver, Mondo-tronics #3-301 or equivalent
4 resistors, 270 Ω, ¼ W
2 resistors, 1 kΩ, ¼ W
4 LEDs
3 switches, PBNO mom contact
BASIC Stamp® II microcontroller with carrier board
IC, 7805
2 9 Vdc batteries with connector clip and mounting clamp
base, ⅛ in. (3.175 mm) thick fiberglass project board, 5-6 in. (127 - 152.4 mm) in diameter
2 wheels, 1 in. (25.4 mm) diameter, 3/32 in. (2.381 mm) axle diameter, DU-BRO #100TW or equivalent
4 "AA" batteries with holder
plastic for front switch bumper: ¾ in. (19.05 mm) wide x ¼ in. (6.35 mm) thick. Length to fit your own design.
miscellaneous machine screws, aluminum angle for mounts, aluminum rod for skid posts

NOTE: This parts list includes the basic parts and materials to build the project. You are encouraged to expand the robot platform with ideas of your own, and of course you'll then need additional parts.

ADVANCED PROJECTS

This section will give you some technical problems to solve. These problems are designed so that you will use what you have learned in your electricity and electronics course. There is no one best solution to any of the problems. The result depends on several factors. Some of these factors are your experience, the materials available (commercial and scrap), and the tools available to you. Keep in mind that the basic purpose of these engineering and technical problems is to provide you with situations where you will apply your knowledge, skills, and creativity.

ADVANCED PROJECT 1: POTENTIOMETER PLACEMENT

Given This Situation: An amateur radio operator wishes to drill two holes, on the same horizontal axis, to accommodate two potentiometers that have ¾-in. diameter knobs. The potentiometer will be turned frequently by young adults. Each potentiometer operates independently of the other, and each adjustment is critical.

Solve This Problem: How close should the holes be located from each other on the horizontal axis to allow easy access to either potentiometer?

ADVANCED PROJECT 2: ELECTROMAGNET

Given This Situation: A coil of No. 20 PE magnet wire is to be wound around a stove bolt core ¼ in. diameter by 2 in. long. The length of the coil is to be 1¼ in. The coil is to operate at 1.5 Vdc at 1.3 A.

Solve This Problem: Determine the number of turns of wire needed to lift 1 pound of soft iron. The range of accuracy for this problem is ±25 turns.

ADVANCED PROJECT 3: TEMPORARY RESISTOR

Given This Situation: An electronic device is inoperative because of the failure of a l-W, I5-Ω resistor. You are in a location where there are no manufactured spare parts available. However, on your worktable there are a blotter, two pieces of writing paper, one ordinary wooden pencil, a jackknife, a small 3-in. piece of No. 25 bare copper wire, and an ohmmeter.

Solve This Problem: Manufacture a temporary replacement part for the resistor from any of the materials mentioned that are on your worktable.

ADVANCED PROJECT 4: RADIO BATTERY TESTER

Given This Situation: An electronics technician wishes to test an A battery in a portable receiver while it is operating under normal conditions. However, the location of the A battery in the receiver makes it impossible to touch the poles of the battery with the regular voltmeter test prods .

Solve This Problem: Design a method to easily measure the voltage of the A battery when it is normally operating in a portable radio. Assume a standard flashlight battery is used and pressed into place against two metal terminals within a plastic cavity made especially for the battery.

ADVANCED PROJECT 5: AGITATOR

Given This Situation: When performing the etching process in making a printed-circuit board, the board is placed in the etchant in a plastic container. The acid etchant must be agitated for the etching process to be efficient. Depending upon the temperature and strength of the solution, this process could take several hours. For this reason, some method of mechanical agitation is necessary. The speed of the action should be controllable so that the agitation does not become violent and spill the solution, thereby making a mess. Use materials that will not be affected by the corrosive nature of the solution and that can be found in the shops.

Solve This Problem: Design and construct a suitable agitator that will reliably perform this task with only occasional observation by the operator .

ADVANCED PROJECT 6: FIXED-VALUE RESISTORS

Given This Problem: An electronic instrument operates on a balanced circuit pair, which must have exactly the same amount of current in each of two fixed-value resistors. One resistor has been damaged and broken in two pieces, and it cannot be repaired. The only replacement parts available are several resistors of a value higher than the broken part. After attempting many of the possible combinations of series and parallel circuits, it is found that no combination will give the exact value .

Solve This Problem: Change the value of a fixed-value resistor so as to make a duplicate of the broken resistor.

ADVANCED PROJECT 7: ELECTRONIC DETECTOR

Given This Problem: It is necessary to break a doorway in a masonry wall to make a passage between two rooms. It is known that the wall contains several large cast-iron pipes, and it is necessary to avoid the area in which these pipes are located.

Solve This Problem: Design and build an electronic circuit that will detect and react to the presence of cast iron in the wall.

ADVANCED PROJECT 8: TESTING AN ELECTRICAL SYSTEM

Given This Situation: In an industrial situation, it is desirable to monitor the condition of each leg of a high-voltage, three-phase electrical system. This is especially important in the delta configuration. Monitoring should be easily accomplished, and only normal operation or failure conditions need to be indicated.

Solve This Problem: Design a method of visually testing a 240-V, three-phase electrical system.

ADVANCED PROJECT 9: CIRCUIT CONSTRUCTION

Given This Situation: Your science teacher wishes to use a microcontroller to collect data related to the swing of a pendulum. The BASIC Stamp® II is the available system.

Solve This Problem: Construct a circuit that will provide data to the microcontroller and, through it, to the display of the host computer.

ADVANCED PROJECT 10: ILLUMINATED NOTE PAD

Given This Situation: An amateur astronomer wishes to take notes on her night observations made through her telescope. Once her eyes have adjusted to the dark, she needs to avoid white light.

Solve This Problem: Design a method of holding an illuminated note pad to enable note taking without interfering with night vision.

ADVANCED PROJECT 11: WATER LEVELS IN FISH TANKS

Given This Situation: A man who raises tropical fish as a hobby uses individual tanks for each species to protect the young. The water evaporates from the tanks and must be replenished periodically.

Solve This Problem: Design a method of automatically maintaining the water level of the fish tanks, drawing from a reservoir of properly conditioned water.

ADVANCED PROJECT 12: ELECTRONIC LEVELING DEVICE

Given This Situation: The foreman of a fence installation crew notices the difficulty involved in plumbing the steel posts used for stockade-type fencing. He wishes to increase the efficiency of the post setters by reducing the time taken to plumb each post in two directions. If this can be done, labor cost can be reduced and profit increased.

Solve This Problem: Design an electronic leveling device. The device must operate in two directions while giving separate indicators for each direction where the post is out of plumb.

ADVANCED PROJECT 13: LIGHTING CONTROL

Given This Situation: By local ordinance, the residents of a particular community must illuminate the front walk of each house during the hours between dusk and dawn. Timers are used generally but prove unsatisfactory, as the timers must be adjusted with each change of daylight duration. To avoid this problem, many residents leave their light on constantly.

Solve This Problem: Design an electronic system that will control an exterior light to turn off during daylight hours and turn on during dusk.

ADVANCED PROJECT 14: WARNING SIGNAL FOR MOPEDS

Given This Situation: Several young moped operators in a school injured themselves recently while riding into turns at too sharp a tilt angle, causing loss of control. The electronics technology teacher wants to make a tilt angle warning device available to student moped riders.

Solve This Problem: Design a circuit that can be installed on the moped so that the rider will be warned as the critical tilt angle is approached. In this way the rider may avoid a dangerous fall.

ADVANCED PROJECT 15: HEARING AID BATTERY TESTER

Given This Situation: A hearing impaired person finds that the battery that supplies power to his "in-the-ear" hearing aid fails without warning. This occurs even though the battery tests "good" using a standard voltmeter.

Solve The Problem: Design a device that will test the hearing aid battery under actual operating load conditions. The battery, however, must be removed from the hearing aid to do this test.

ADVANCED PROJECT 16: WARNING SIGNAL FOR FREEZERS

Given This Situation: A family finds that, for one reason or another, the door of the upright freezer has been left ajar on several occasions. This has resulted in the thawing of the food and considerable spoilage.

Solve This Problem: Design an electronic device that will attach to the door of the freezer to indicate when the door has been left ajar for a set amount of time. The device must not interfere with the normal access to the freezer.

ADVANCED PROJECT 17: TEMPERATURE MONITOR

Given This Situation: A photographer wants to monitor the temperature of the chemical solutions used to develop prints in her darkroom. By the very nature of the process she finds it very difficult to read a standard thermometer by the dim safelight.

Solve This Problem: Design a circuit that will enable the photographer to monitor the temperature of the chemical solutions used during print development in the darkroom.

APPENDIX A: METRIC CONVERSION FACTORS

When you know	You can find	If you multiply by
Length		
inches	millimeters	25.4
feet	centimeters	30.48
yards	meters	0.9144
miles	kilometers	1.609
millimeters	inches	0.039
centimeters	inches	0.39
meters	yards	1.094
kilometers	miles	0.621
Area		
square inches	square centimeters	6.45
square feet	square meters	0.0929
square yards	square meters	0.836
square miles	square kilometers	2.59
acres	hectares	0.405
square centimeters	square inches	0.155
square meters	square yards	1.196
square kilometers	square miles	0.386
hectares	acres	2.47
Weight (mass)		
ounces	grams	28.3
pounds	kilograms	0.454
short tons	metric tons	0.907
grams	ounces	0.0353
kilograms	pounds	2.2
metric tons	short tons	1.1
Liquid volume		
ounces	milliliters	29.6
pints	liters	0.473
quarts	liters	0.946
gallons	liters	3.785
milliliters	ounces	0.0338
liters	pints	2.113
liters	quarts	1.057
liters	gallons	0.264
Temperature		
degrees Fahrenheit	degrees Celsius	0.6 (after subtracting 32)
degrees Celsius	degrees Fahrenheit	1.8 (then add 32)

APPENDIX B: STANDARD LETTER SYMBOLS FOR UNITS OF MEASUREMENT

Unit	Symbol	Notes*
ampere	A	SI unit of electric current.
ampere (turn)	A	SI unit of magnetomotive force.
ampere-hour	Ah	Also A·h.
ampere per meter	A/m	SI unit of magnetic field strength.
atomic mass unit (unified)	u	The (unified) atomic mass unit is defined as one twelfth of the mass of an atom of the carbon-12 nuclide. Use of the old atomic mass unit (amu), defined by reference to oxygen, is deprecated.
baud	Bd	In telecommunications, a unit of signaling speed equal to one element per second. The signaling speed in bauds is equal to the reciprocal of the signal element length in seconds.
bel	B	
becquerel	Bq	SI unit of activity of a radionuclide.
billion electronvolts	GeV	The name *gigaelectronvolt* is preferred for this unit.
bit	b	In information theory, the bit is a unit of information content equal to the information content of a message, the *a priori* probability of which is one-half. In computer science, the name bit is used as a short form of *binary digit*.
bit per second	b/s	
byte	B	A byte is a string of bits, usually eight bits long, operated on as a unit. A byte is capable of holding one character in the local character set.
candela	cd	SI unit of luminous intensity.
candela per square inch	cd/in^2	Use of the SI unit, cd/m^2, is preferred.
candela per square meter	cd/m^2	SI unit of luminance.
candle	cd	The unit of luminous intensity has been given the name candela; use of the name candle for this unit is deprecated.
centi	c	SI prefix for 10^{-2}.
centimeter	cm	

* The notes give exact definitions (indicated by the symbol $\underline{\Delta}$) for many of the units and give conversion factors in other cases. The conversion factors indicated with the equals sign are accurate to the number of figures shown. For more accurate conversion factors and other general information about units, see ANSI/IEEE Std 268-1992.

centistokes	cSt	cSt ≜ mm^2/s. The name centistokes is deprecated (see ANSI/IEEE Std 268-1992).
circular mil	cmil	cmil ≜ $(\pi/4) \cdot 10^{-6}$ in^2
coulomb	C	SI unit of electric charge.
curie	Ci	Ci ≜ 3.7 X 10^{10} Bq. A unit of activity of a radionuclide. Use of the SI unit, the becquerel, is preferred.
cycle per second	Hz	See hertz.
deci	d	SI prefix for 10^{-1}.
decibel	dB	
degree (temperature):		
degree Celsius	°C	SI unit of Celsius temperature. The degree Celsius is a special name for the kelvin, used in expressing Celsius temperatures or temperature intervals.
degree Fahrenheit	°F	Note that the symbols for °C, °F, and °R are comprised of two elements, written with no space between the ° and the letter that follows. The two elements that make the complete symbol are not to be separated.
degree Kelvin		See kelvin.
degree Rankine	°R	
deka	da	SI prefix for 10.
electronvolt	eV	
erg	erg	erg ≜ 10^{-7} J. Deprecated (see ANSI/IEEE 268-1992).
exa	E	SI prefix for 10^{18}.
farad	F	SI unit of capacitance.
femto	f	SI prefix for 10^{-15}.
footcandle	fc	fc ≜ lm/ft^2. The name lumen per square foot is also used for this unit. Use of the SI unit of illuminance, the lux (lumen) per square meter, is preferred.

footlambert	fL	fL ∆ $(1/\pi)$cd/ft^2. A unit of luminance. One lumen per square foot leaves a surface whose luminance is one footlambert in all directions within a hemisphere. Use of the SI unit, the candela per square meter, is preferred.
gauss	G	The gauss is the electromagnetic CGS unit of magnetic flux density. Deprecated (see ANSI/IEEE Std 268-1992).
giga	G	SI prefix for 10^9.
gigaelectronvolt	GeV	
gigahertz	GHz	
gilbert	Gb	The gilbert is the electromagnetic CGS unit of magnetomotive force. Deprecated (see ANSI/IEEE Std 268-1992).
gray	Gy	SI unit of absorbed dose in the field of radiation dosimetry.
hecto	h	SI prefix for 10^2.
henry	H	SI unit of inductance.
hertz	Hz	SI unit of frequency.
horsepower	hp	hp ∆ 550 ft·lbf/s = 746 W. The horsepower is an anachronism in science and technology. Use of the SI unit of power, the watt, is preferred.
hour	h	
joule	J	SI unit of energy, work, and quantity of heat.
joule per kelvin	J/K	SI unit of heat capacity and of entropy.
kelvin	K	In 1967, the CGPM gave the name *kelvin* to the SI unit of temperature, which had formerly been called *degree kelvin*, and assigned it the symbol K (without the symbol °).
kilo	k	SI prefix for 10^3. The symbol K shall not be used for kilo. The prefix kilo shall not be used to mean 2^{10} (that is, 1024).
kilobit per second	kb/s	
kilobyte	kB	kB ∆ 1000 bytes.
kilogauss	kG	Deprecated (see ANSI/IEEE Std 268-1992).
kilohertz	kHz	
kilohm	kΩ	
kilovar	kvar	

kilovolt	kV	
kilovoltampere	kVA	
kilowatt	kW	
kilowatthour	kWh	Also kW·h.
lambert	L	L $\triangleq$ $(1/\pi)$cd/cm^2. A CGS unit of luminance. One lumen per square centimeter leaves a surface whose luminance is one lambert in all directions within a hemisphere. Deprecated (see ANSI/IEEE Std 268-1992).
liter	L	L $\triangleq$ 10^{-3} m^3. In 1979, the CGPM approved L and l as alternative symbols for the liter. Because of frequent confusion with the numeral 1, the letter symbol l is not recommended for US use (see Federal Register notice of December 20, 1990, vol. 55, no. 245, p. 52242). The script *l* shall not be used as a symbol for liter.
lumen	lm	SI unit of luminous flux.
lumen per square foot	lm/ft^2	A unit of illuminance and also a unit of luminous exitance. Use of the SI unit, lumen per square meter, is preferred.
lumen per square meter	lm/m^2	SI unit of luminous exitance.
lumen per watt	lm/W	SI unit of luminous efficacy.
lumen second	lm.s	SI unit of quantity of light.
lux	lx	lx $\triangleq$ lm/m^2. SI unit of illuminance.
maxwell	Mx	The maxwell is the electromagnetic CGS unit of magnetic flux. Deprecated (see ANSI/IEEE Std 268-1992).
mega	M	SI prefix for 10^6. The prefix mega shall not be used to mean 2^{20} (that is, 1 048 576).
megabit per second	Mb/s	
megabyte	MB	MB $\triangleq$ 1 000 000 bytes.
megaelectronvolt	MeV	
megahertz	MHz	
megohm	MΩ	
meter	m	SI unit of length.
mho	S	The name *mho* was formerly given to the reciprocal ohm. Deprecated; see siemens (S).
micro	μ	SI prefix for 10^{-6}.

microampere	μA	
microfarad	μF	
microgram	μg	
microhenry	μH	
microinch	μin	
microliter	μL	See note for liter.
micrometer	μm	
micron	μm	The name micron is deprecated. Use micrometer.
microsecond	μs	
microwatt	μW	
mil	mil	mil $\triangleq$ 0.001 in.
milli	m	SI prefix for 10^{-3}.
milliampere	mA	
millihenry	mH	
millipascal second	mPa·s	SI unit-multiple of dynamic viscosity.
millisecond	ms	
millivolt	mV	
milliwatt	mW	
minute (time)	min	Time may also be designated by means of superscripts as in the following example: $9^h\ 46^m\ 30^s$.
nano	n	SI prefix for 10^{-9}.
nanoampere	nA	
nanofarad	nF	
nanosecond	ns	
newton	N	SI unit of force.
newton meter	N·m	
newton per square meter	N/m^2	SI unit of pressure or stress; see pascal.
oersted	Oe	The oersted is the electromagnetic CGS unit of magnetic field strength. Deprecated (see ANSI/IEEE Std 268-1992).
ohm	Ω	SI unit of resistance
pascal	Pa	Pa $\triangleq$ N/m^2. SI unit of pressure or stress.
peta	P	SI prefix for 10^{15}.
phot	ph	ph $\triangleq$ lm/cm^2. CGS unit of illuminance. Deprecated (see ANSI/IEEE Std 268-1992).
pico	p	SI prefix for 10^{-12}.

APPENDIX B: STANDARD LETTER SYMBOLS FOR UNITS OF MEASUREMENT (Continued)

picofarad	pF	
picowatt	pW	
rad	rd	rd $\triangleq$ 0.01 Gy. A unit of absorbed dose in the field of radiation dosimetry. Use of the SI unit, the gray, is preferred.
radian	rad	SI unit of plane angle.
rem	rem	rem $\triangleq$ 0.01 Sv. A unit of dose equivalent in the field of radiation dosimetry. Use of the SI unit, the sievert, is preferred.
revolution per minute	r/min	Although use of rpm as an abbreviation is common, it should not be used as a symbol.
revolution per second	r/s	
roentgen	R	A unit of exposure in the field of radiation dosimetry.
second (time)	s	SI unit of time.
siemens	S	S $\triangleq$ Ω^{-1}. SI unit of conductance.
slug	slug	slug $\triangleq$ lbf· s^2/ft = 14.594 kg.
stilb	sb	sb $\triangleq$ cd/cm^2. A CGS unit of luminance. Deprecated (see ANSI/IEEE Std 268-1992).
stokes	St	Deprecated (see ANSI/IEEE Std 268-1992).
tera	T	SI prefix for 10^{12}.
terabyte	TB	TB $\triangleq$ 10^{12} B.
tesla	T	T $\triangleq$ N/(A·m) $\triangleq$ Wb/m^2. SI unit of magnetic flux density (magnetic induction).
therm	thm	thm $\triangleq$ 100 000 Btu.
(unified) atomic mass unit	u	The (unified) atomic mass unit is defined as one-twelfth of the mass of an atom of the carbon-12 nuclide. Use of the old atomic mass unit (amu), defined by reference to oxygen, is deprecated.
var	var	IEC name and symbol for the SI unit of reactive power.
volt	V	SI unit of voltage.
volt per meter	V/m	SI unit of electric field strength.
voltampere	VA	IEC name and symbol for the SI unit of apparent power.
watt	W	SI unit of power.
watt per meter kelvin	W/(m·K)	SI unit of thermal conductivity.
watt per steradian	W/sr	SI unit of radiant intensity.

watt per steradian square meter	$W/(sr \cdot m^2)$	SI unit of radiance.
watthour	Wh	
weber	Wb	Wb $\triangleq$ V·s. SI unit of magnetic flux.

CAPACITANCE

Capacitance is the ability to store electric energy. The basic unit of capacitance is the farad (F). The capacitance of capacitors is most often given in microfarads (μF) or picofarads (pF).

For capacitors connected in parallel, the total capacitance is:

$$C_t = C_1 + C_2 + C_3 + \ldots + C_n$$

For capacitors connected in series, the total capacitance is:

$$C_t = \cfrac{1}{\cfrac{1}{C_1} + \cfrac{1}{C_2} + \cfrac{1}{C_3} + \cdots \cfrac{1}{C_n}}$$

CAPACITIVE REACTANCE

The opposition of a capacitor to the flow of alternating current is called capacitive reactance. The formula for figuring capacitive reactance is:

$$X_c = \frac{1}{2\pi f C}$$

where X_c = capacitive reactance in ohms
f = frequency in hertz
C = capacitance in farads
π = 3.14 (pi), a constant

IMPEDANCE

The total opposition to current in a circuit containing a combination of resistance and capacitive reactance is called impedance. It is measured in ohms. The letter symbol for impedance is Z.

For any circuit, impedance can be calculated by using the formula:

$$Z = \frac{V_t}{I_t}$$

INDUCTIVE REACTANCE

Inductive reactance is the opposition to a changing current in a coil that is caused by the inductance of the coil. The unit of inductive reactance is the ohm. Its letter symbol is X_L. The value of inductive reactance depends on the inductance of the coil and the frequency of the current that passes through it. The formula for inductive reactance is:

$$X_L = 6.28 fL$$

where f = the frequency of the current in hertz
L = the inductance of a coil in henrys
6.28 = 2π , or 2 times 3.14

KILOWATT-HOUR FORMULA

The kilowatt-hour (kWh) is the unit of electric energy on which electric companies base their bills. Kilowatts multiplied by the number of hours used is equal to kilowatt-hours. The amount of energy (in kilowatt-hours) used by equipment or some other product can be computed by the following formula:

$$kWh = \frac{P \times h}{1000}$$

where P = power in watts
h = time in hours

OHM'S LAW

By using the letter symbols for voltage (E), current (I), and resistance (R), Ohm's law can be expressed by the formulas given here.

To solve for E (voltage), use the formula:

$$E = I \times R$$

To solve for I (current), this same formula can be written as:

$$I = \frac{E}{R}$$

To solve for R (resistance), arrange it as

$$R = \frac{E}{I}$$

POWER FORMULAS

The power formula shows the relationships between electric power (P), voltage (E), and current (I) in a dc circuit. The basic power formula is:

$$P = E \times I$$

The relationship between power, current, and resistance is expressed as:

$$P = I^2R$$

The relationship between power, voltage, and resistance is expressed as:

$$P = \frac{E^2}{R}$$

RESISTANCE IN A PARALLEL CIRCUIT

All parallel resistance combinations can be calculated to find an equivalent total resistance value by using the following general formula.

$$R_t = \frac{1}{\frac{1}{R_1} + \frac{1}{R_2} + \frac{1}{R_3} + \dots \frac{1}{R_n}}$$

where R_t = total resistance
R_1 and R_2 = parallel resistances
R_n = any number of resistors

If there are two resistances of unequal value:

$$R_t = \frac{R_1 \times R_2}{R_1 + R_2}$$

If all the resistances are equal in value:

$$R_t = \frac{\text{value of one resistance}}{\text{number of resistances}}$$

RESISTANCE IN A SERIES CIRCUIT

Since a series circuit has only one path for current to flow, the current must pass through all the resistors in the circuit. Therefore, the total resistance is the sum of all the resistors in the series circuit:

$$R_t = R_1 + R_2 + R_3 + \dots + R_n$$

NOTE: The small "n" in R_n stands for any number of resistors.

RESISTANCE IN A SERIES-PARALLEL CIRCUIT

See Chapter 9 for examples of how to figure resistance in a series-parallel circuit.

SYNCHRONOUS SPEED OF A MOTOR

The speed of a split-phase motor, unlike that of a universal motor, cannot be varied gradually by reducing the voltage across the field without losing too much torque. For this reason, the speed may be step-controlled by changing the number of running winding poles. In a three-speed motor, for example, the running winding is arranged so that it can be switched to form two, four, or six poles. The approximate speed that each of these connections produces can be found by using the formula

$$rpm = \frac{120f}{N}$$

where rpm = revolutions per minute of the shaft
f = frequency of the operating current (expressed in hertz)
N = number of running-winding poles

TIME CONSTANT

The time required for the current in a series *RL* circuit to increase to 63.2 percent of its maximum (steady state) value is known as the -*RL* time constant of the circuit. It is expressed as the formula:

$$t = \frac{L}{R}$$

where t = time constant, in seconds
L = inductance, in henrys
R = resistance, in ohms

APPENDIX D American National Standard Code for Information Interchange (ASCII)

(This material is used with permission from the American National Standards Institute, Inc., publication ANSI X 3.4–1977.)

<table>
<tr><td colspan="5">b_7 →</td><td>0</td><td>0</td><td>0</td><td>0</td><td>1</td><td>1</td><td>1</td><td>1</td></tr>
<tr><td colspan="5">b_6 →</td><td>0</td><td>0</td><td>1</td><td>1</td><td>0</td><td>0</td><td>1</td><td>1</td></tr>
<tr><td colspan="5">b_5 →</td><td>0</td><td>1</td><td>0</td><td>1</td><td>0</td><td>1</td><td>0</td><td>1</td></tr>
<tr><td>Bits b_4 ↓</td><td>b_3 ↓</td><td>b_2 ↓</td><td>b_1 ↓</td><td>COLUMN → ROW ↓</td><td>0</td><td>1</td><td>2</td><td>3</td><td>4</td><td>5</td><td>6</td><td>7</td></tr>
<tr><td>0</td><td>0</td><td>0</td><td>0</td><td>0</td><td>NUL</td><td>DLE</td><td>SP</td><td>0</td><td>@</td><td>P</td><td>`</td><td>p</td></tr>
<tr><td>0</td><td>0</td><td>0</td><td>1</td><td>1</td><td>SOH</td><td>DC1</td><td>!</td><td>1</td><td>A</td><td>Q</td><td>a</td><td>q</td></tr>
<tr><td>0</td><td>0</td><td>1</td><td>0</td><td>2</td><td>STX</td><td>DC2</td><td>"</td><td>2</td><td>B</td><td>R</td><td>b</td><td>r</td></tr>
<tr><td>0</td><td>0</td><td>1</td><td>1</td><td>3</td><td>ETX</td><td>DC3</td><td>#</td><td>3</td><td>C</td><td>S</td><td>c</td><td>s</td></tr>
<tr><td>0</td><td>1</td><td>0</td><td>0</td><td>4</td><td>EOT</td><td>DC4</td><td>$</td><td>4</td><td>D</td><td>T</td><td>d</td><td>t</td></tr>
<tr><td>0</td><td>1</td><td>0</td><td>1</td><td>5</td><td>ENQ</td><td>NAK</td><td>%</td><td>5</td><td>E</td><td>U</td><td>e</td><td>u</td></tr>
<tr><td>0</td><td>1</td><td>1</td><td>0</td><td>6</td><td>ACK</td><td>SYN</td><td>&</td><td>6</td><td>F</td><td>V</td><td>f</td><td>v</td></tr>
<tr><td>0</td><td>1</td><td>1</td><td>1</td><td>7</td><td>BEL</td><td>ETB</td><td>'</td><td>7</td><td>G</td><td>W</td><td>g</td><td>w</td></tr>
<tr><td>1</td><td>0</td><td>0</td><td>0</td><td>8</td><td>BS</td><td>CAN</td><td>(</td><td>8</td><td>H</td><td>X</td><td>h</td><td>x</td></tr>
<tr><td>1</td><td>0</td><td>0</td><td>1</td><td>9</td><td>HT</td><td>EM</td><td>)</td><td>9</td><td>I</td><td>Y</td><td>i</td><td>y</td></tr>
<tr><td>1</td><td>0</td><td>1</td><td>0</td><td>10</td><td>LF</td><td>SUB</td><td>*</td><td>:</td><td>J</td><td>Z</td><td>j</td><td>z</td></tr>
<tr><td>1</td><td>0</td><td>1</td><td>1</td><td>11</td><td>VT</td><td>ESC</td><td>+</td><td>;</td><td>K</td><td>[</td><td>k</td><td>{</td></tr>
<tr><td>1</td><td>1</td><td>0</td><td>0</td><td>12</td><td>FF</td><td>FS</td><td>,</td><td><</td><td>L</td><td>\</td><td>l</td><td>|</td></tr>
<tr><td>1</td><td>1</td><td>0</td><td>1</td><td>13</td><td>CR</td><td>GS</td><td>-</td><td>=</td><td>M</td><td>]</td><td>m</td><td>}</td></tr>
<tr><td>1</td><td>1</td><td>1</td><td>0</td><td>14</td><td>SO</td><td>RS</td><td>.</td><td>></td><td>N</td><td>^</td><td>n</td><td>~</td></tr>
<tr><td>1</td><td>1</td><td>1</td><td>1</td><td>15</td><td>SI</td><td>US</td><td>/</td><td>?</td><td>O</td><td>_</td><td>o</td><td>DEL</td></tr>
</table>

(A) Standard Code.

Col/ Row		Mnemonic and Meaning[1]	Col/ Row		Mnemonic and Meaning[1]
0/0	NUL	Null	1/0	DLE	Data Link Escape (CC)
0/1	SOH	Start of Heading (CC)	1/1	DC1	Device Control 1
0/2	STX	Start of Text (CC)	1/2	DC2	Device Control 2
0/3	ETX	End of Text (CC)	1/3	DC3	Device Control 3
0/4	EOT	End of Transmission (CC)	1/4	DC4	Device Control 4
0/5	ENQ	Enquiry (CC)	1/5	NAK	Negative Acknowledge (CC)
0/6	ACK	Acknowledge (CC)	1/6	SYN	Synchronous Idle (CC)
0/7	BEL	Bell	1/7	ETB	End of Transmission Block (CC)
0/8	BS	Backspace (FE)	1/8	CAN	Cancel
0/9	HT	Horizontal Tabulation (FE)	1/9	EM	End of Medium
0/10	LF	Line Feed (FE)	1/10	SUB	Substitute
0/11	VT	Vertical Tabulation (FE)	1/11	ESC	Escape
0/12	FF	Form Feed (FE)	1/12	FS	File Separator (IS)
0/13	CR	Carriage Return (FE)	1/13	GS	Group Separator (IS)
0/14	SO	Shift Out	1/14	RS	Record Separator (IS)
0/15	SI	Shift In	1/15	US	Unit Separator (IS)
			7/15	DEL	Delete

[1](CC) Communication Control; (FE) Format Effector; (IS) Information Separator.

(B) Control characters used in the ASCII code for information interchange affect the recording, processing, transmission, and interpretation of data.

APPENDIX D (continued)

Column/Row	Symbol	Name
2/0	SP	Space (Normally Nonprinting)
2/1	!	Exclamation Point
2/2	"	Quotation Marks (Diaeresis)[2]
2/3	#	Number Sign[3]
2/4	$	Dollar Sign
2/5	%	Percent Sign
2/6	&	Ampersand
2/7	'	Apostrophe (Closing Single Quotation Mark; Acute Accent)[2]
2/8	(	Opening Parenthesis
2/9	)	Closing Parenthesis
2/10	*	Asterisk
2/11	+	Plus
2/12	,	Comma (Cedilla)[2]
2/13	-	Hyphen (Minus)
2/14	.	Period (Decimal Point)
2/15	/	Slant
3/0 to 3/9	0 . . . 9	Digits 0 through 9
3/10	:	Colon
3/11	;	Semicolon
3/12	<	Less Than
3/13	=	Equals
3/14	>	Greater Than
3/15	?	Question Mark
4/0	@	Commercial At[3]
4/1 to 5/10	A . . . Z	Uppercase Latin Letters A through Z
5/11	[	Opening Bracket[3]
5/12	\	Reverse Slant[3]
5/13	]	Closing Bracket[3]
5/14	^	Circumflex[3]
5/15	_	Underline
6/0	`	Opening Single Quotation Mark (Grave Accent)[2,3]
6/1 to 7/10	a . . . z	Lowercase Latin letters a through z
7/11	{	Opening Brace[3]
7/12	|	Vertical Line[3]
7/13	}	Closing Brace[3]
7/14	~	Tilde[2,3]

[2]The use of the symbols in 2/2, 2/7, 2/12, 5/14, 6/0, and 7/14 as diacritical marks is described in A5.2 of Appendix A.

[3]These characters should not be used in international interchange without determining that there is agreement between sender and recipient. (See Appendix B5.)

(C) Graphic characters used in the ASCII code for information interchange provide a visual representation normally handwritten, printed, or displayed.

Glossary

A

adapter. A two-to-three wire device which makes it possible to use a grounding plug in a two-socket outlet.

aiding voltage sources. In this type of cell connection, the total voltage is equal to the sum of the individual voltages. Although voltage is increased, the current capacity of the cells in series is equal only to the cell with the lowest capacity.

alligator clips. Clips that have spring-loaded jaws with teeth to hold a wire securely. They allow a firm wire connection to be made to a terminal point. They are used with instrument test leads and other temporary wiring systems.

alternating current (ac). Current that is produced by a voltage source that changes polarity, or alternates, with time.

alternator. Another name for an alternating current (ac) generator.

ammeter. A testing device used to measure current.

ampacity. The safe current-carrying capacity of a wire.

ampere. The unit of measure for electrical current. Abbreviated as *amp*.

amplifier. A device or circuit that increases the output value of voltage or current.

amplitude modulation (AM). A process in which the alternating current from the microphone modulates the carrier wave by causing its amplitude, or strength, to rise and fall.

analog-to-digital converter. A device that converts analog signals to a digital code. It can send data in series or parallel format.

anode. A negatively charged electrode.

ANSI. An acronym for American National Standards Institute. An organization responsible for standardizing symbols such as the ones used in electronics and electricity.

apprenticeship program. A way to learn the electrical trade. A person must complete a four- or five-year program, which may include classroom instruction and on-the-job training.

arithmetic logic unit (ALU). Often part of the CPU in microprocessors, it does comparisons, makes decisions, adds, and subtracts.

armature. An electromagnet coil mounted on a shaft that turns inside a motor. It is placed between the poles of a permanent magnet. Also called a rotor.

armored cable (AC). Contains insulated wires in a flexible armor sheath that is a single spiral of galvanized steel tape interlocked to form a tube-like enclosure. The cable also contains a bare-copper or aluminum bonding strip in contact with the armor.

ASCII. An acronym for American Standard Code for Information Interchange. A widely used code system in which the ASCII code for each alphanumeric text character or graphic character is written to a specific address. When this address is selected, the character appears on the screen.

atom. The basic unit of matter which is made of tiny particles called electrons, protons, and neutrons.

atomic fission. A process in which an atomic particle, the neutron, strikes the nucleus of a uranium atom and splits.

audio cable. Cable commonly used in telephone circuits, in intercommunications circuits, and in circuits connecting the loudspeakers of public-address systems to amplifiers. It is made up of two or more solid or stranded wires that are color-coded, insulated, and placed in an outer jacket.

B

ballast. A coil of insulated magnet wire wound around an iron core.

bandwidth. The frequency range of a telephone channel.

base. The input junction to the emitter.

BASIC. One of several computer languages used to control computers. The instructions or commands in this language can be given in the immediate mode and in the programmed mode.

battery. An electrical device formed when two or more cells are connected in series or in parallel.

beta. The letter symbol ß. The letter designation for the amplification factor of a transistor.

bimetallic thermostat. A thermostat consisting of two different kinds of metal. The strip bends as it is heated. This bending is used to make or break a switching contact.

binary system. The system computers use. It relies on base 2.

bit. The smallest unit of information a computer can hold. Each bit can hold a single binary digit (1 or 0).

block diagram. A convenient way to quickly give information about complex electrical systems in which each block represents electrical circuits that perform specific system functions.

Boolean algebra. The method of mathematics used to describe logic circuits.

Boolean logic. A system of algebraic equations that represents the function of logic circuits.

booster battery. An additional battery that is temporarily connected in parallel to an automobile battery that won't deliver enough current to start a car.

branch circuit. A circuit that distributes electricity from the load center to other parts of the system.

breadboard. A flat wooden board used as a base for mounting components.

bridge. A full-wave rectifier that uses four diodes in specific configuration.

British thermal unit (Btu). One Btu is the quantity of heat needed to raise the temperature of one pound of water one degree Fahrenheit. It is also the unit used to rate the cooling capacity of a room air conditioner.

brush. Stationary contact that the commutator rubs against as it turns.

byte. A unit of measurement equal to 8 bits.

C

cable. A bundle of insulated wires through which electric current can be sent. Cables are most often used for connecting components or products that are some distance apart. They may be unshielded or shielded.

capacitance. The ability of a circuit or a device to store electric energy.

capacitive reactance. The opposition of a capacitor to the flow of alternating current.

capacitor. A device used where a specific value of capacitance is needed. The basic form is two conductors separated by an insulating material.

capacity. A cell's ability to deliver a given amount of current to a circuit.

career. A series of related jobs in which you train and pursue a life's work.

cathode. A positively charged electrode.

cell. A single unit that changes chemical energy into electric energy in the form of voltage.

central processing unit (CPU). The unit that controls and coordinates all the hardware in the computer.

charge-coupled device (CCD). A solid-state digital imaging system that uses an array of chargeable devices much like capacitors.

chassis ground. A common circuit return to a structure, such as the frame of an air, space, or land vehicle that is not electrically connected to earth.

chemical reaction. A reaction that occurs when two or more chemicals act on each other. It usually results in changes to the materials.

circular mil. A unit of measurement equal to the area of a circle with a diameter of 0.001 in. (0.0254 mm).

clad. The reflective outer surface of an optic fiber. The clad and the optic fiber are fused together, allowing light to reflect off the optic fiber.

class designation letters. Letters used to identify components on a schematic diagram.

closed loop. A circuit through which current can flow.

coaxial cable. Cable with a center conductor of Copperweld wire covered by a polyethylene insulation of a very specific diameter.

cogeneration. A system in which the unused steam and hot water discharged by the turbine are used for air conditioning, heating, cooking, and other commercial and industrial applications before being recycled to the boiler.

coherent light. Light that has light waves of very nearly the same frequency. The waves are in phase with each other.

cold solder joint. A dull-colored solder joint. Such a joint does not provide a good electrical contact.

collector. The output junction to the emitter.

combination circuit. A series-parallel circuit.

combinatorial circuit. A circuit arranged in such a way that two circuits can be fed into the OR circuit so that either circuit gives an output when the right inputs are present.

commutator. A device needed to produce a direct current. It is connected to the ends of the armature coil or coils. In its basic form, a commutator is a ring-like device made up of metal pieces called segments.

comparators. Circuits that use AND gates to compare quantities, conditions, or states to determine equality.

compound. A substance formed when different elements are combined.

computer-generated diagrams. Horizontal and vertical lines plotted by a computer that represent wires connected to graphic symbols.

conductor. A material or object in a circuit that provides an easy path through which electrons can move through the circuit.

cords. Wires covered with insulation used to connect electrical and electronic equipment and appliances to outlets.

counter electromotive force (cemf). The opposing capacitor voltage. This voltage opposes the battery voltage that is applied to the circuit.

covalent bond. A bond in which an atom shares each of its valence electrons with those of nearby atoms.

current. The flow of electrons.

cycle. A complete wave alternating current.

D

debounce circuitry. Circuitry that removes the effects of contact bounce from a circuit.

dedicated control. A type of control in which the controller is programmed to perform one general function for the user.

degaussing. The process of removing magnetism from an object; sometimes called demagnetizing.

desoldering. The melting and removal of molten solder.

diagnostic software. A set of computer programs that will troubleshoot computers. It will electronically test the circuitry to determine the quality of operation.

diagram. A picture that shows how different parts are connected to form a circuit.

dielectric. An electrical insulator that does not easily conduct an electric current.

dielectric heating. Heat produced by a device used to produce heat in nonmetallic materials. In its simplest form, the working elements of a dielectric heater consist of two metal plates.

digital electronics. A technology that employs low voltage pulses to operate electronic equipment such as computers, stereos, and televisions. It has replaced many electronic functions previously performed by analog circuitry.

digital multimeter. An instrument that measures electrical properties and displays the results in numeric format.

digital-to-analog converter. A device that changes digital information into a continuous variable voltage output.

diode. A semiconductor that is made as a solid unit.

direct current (dc). Current that is produced in a circuit by a steady voltage source.

discrete circuits. Circuits connected to the solderless breadboard and made up of separate components connected by leads.

domains. The areas in magnetic materials where atoms align.

doping. A process that adds small amounts of other substances such as arsenic. It creates free electrons in the semiconductor through covalent bonding.

dry cell. A cell that contains a paste electrolyte.

dual-trace oscilloscopes. A type of oscilloscope that can display two wave patterns at the same time. This system allows technicians to compare related patterns for amplitude and timing.

DVD (digital video disc). A newer storage technology intended for use on both computers and home entertainment systems.

E

earth ground. The type of connection made by fastening a wire to a metal rod that is driven into the earth.

eddy currents. Currents induced in the core of a transformer.

EEPROM. An acronym for Electrically Erasable Programmable Read Only Memory. It latches software instructions into programmable memory.

electric arc furnace. A furnace used to melt metals during the steelmaking process. In such a furnace, carbon electrodes produce heat using electric arcs. The electrodes are lowered through the top of the furnace.

electric arc welding. A welding process in which heat is produced by an arc formed between a welding rod and the metal object being welded. The rod is connected to one terminal of the welder. The metal object is connected to the other terminal.

electric circuit. A combination of parts connected to form a complete path through which electrons can move.

electric generator. A machine that produces a voltage by means of electromagnetic induction.

electric shock. A physical reaction of the nerves to electric current.

electrical engineer. A person trained in the planning and building of electrical systems.

electrical metallic tubing (EMT). Thin-walled metal piping that is installed between wiring devices that are to be connected. This is a code requirement in some locations.

electrical system. A network of several interconnected series-parallel circuits.

electrician. A person who installs and maintains electrical systems for a variety of purposes, including climate control, security, and communications.

electrode. A terminal that conducts electric current.

electrolysis. A process in which electrolytes react with other solutions or with materials such as metals to change them.

electrolyte. A liquid that ionizes to form positive and negative ions and that can conduct these charges. A solution made with acids, bases, or salts that conducts electricity well.

electrolytic furnace. A system in which the heat needed to keep the electrolyte in liquid form is produced by the current needed for electrolysis. Carbon rods and a carbon lining make up the cathode and the anode.

electromagnetic waves. A form of radiated energy that can do work. Also called radio waves.

electromagnetism. The type of magnetism produced by the current.

electron. A negatively charged particle that revolves around the nucleus of an atom.

electroplating. A form of electrolysis used to coat a base material (usually a metal) with a thin layer of another metal.

electrostatic field. An area which stores energy. It exists between the plates of a charged capacitor.

element. The substance in which all of the atoms are alike.

emitter. The common conductive region to both the base and the collector.

equivalent resistance. R_{eq}. The electrical equivalent of the circuit components it represents. It is convenient to use in analyzing parallel and more complex circuits.

eutectic solder. A solder that is 63% tin and 37% lead. It has the lowest melting point of any tin-lead combination.

excitation current. The direct current used to energize the field windings and start a generator.

F

Fahnestock clips. Clips with a spring lever that are often used for making temporary connections to experimental models.

fanout. The number of transistor-transistor logic (TTL) devices that one TTL output can drive.

farad (F). The basic unit of capacitance.

Federal Communications Commission license. A license granted by the Federal Communications Commission (FCC), which regulates radio, television, wire, and cable communications.

ferrites. Hard, ceramic, magnetic materials that are very efficient and have a high electrical resistance.

field. A space within which a definite effect (such as magnetism) exists and in which there is a definite value at each point.

flip-flops (FF). Also known as bistable multivibrators. They are the basic circuits used in most binary counters. These circuits have two stable output states. They can be switched from one output state to the other.

flux. The core of the wire-shaped solder used for electrical and electronic solder. When the solder melts, the flux flows over the surfaces to be soldered. It acts as a cleaner to remove any oxides coating the surfaces. Also, the invisible lines of force that make up the magnetic field.

fossil fuel. Natural gas.

frequency counter. An instrument that can be used with oscilloscopes to study circuit frequencies. These instruments count the number of cycles present at the input leads for a short period of time and then display that count. This process is repeated periodically, and the display is updated.

frequency modulation (FM). A process in which the carrier wave can also be modulated by changing its frequency.

full-adders. Circuits that can add two bits and carry a bit from a prior adding circuit. A basic full-adder circuit is a combination of two half-adders plus another logic gate.

function generators. Generators that provide signal voltages in a variety of waveforms and shapes at very precise, selectable frequencies.

fuse. A safety device that works as a switch to turn a circuit off when the current goes over its specified value.

G

gate. A region of p-type semiconductor material that forms a peninsula within a block of n-type semiconductor material.

general-purpose branch circuits. Circuits used for lighting and outlets.

geostationary orbit. An orbit in which a satellite revolves at the same speed as the Earth.

geothermal energy. Energy supplied by the earth's heat.

graphic symbols. Letters, drawings, or figures that represent components in diagrams.

grid. A transmission network.

grommets. Devices made of rubber or plastic used to protect wires and cords that must pass through holes in metal.

ground. An electrical connection between the surface and the earth, or between a circuit and some metal object that takes the place of the earth.

ground wire. The green wire connected to the third prong of an attachment plug. It provides for a safe grounding.

ground-fault circuit interrupter (GFCI). A device that gives extra personal protection against electric shock.

grounding wire. Usually a wire that connects metal cabinets and certain parts of circuits, such as those in a house wiring system, to ground rods. It is either solid or stranded, and either green or bare.

H

half-adders. Digital circuits capable of adding two one-bit binary numbers.

Hall process. A method in which alumina is dissolved in molten cryolite (sodium aluminum fluoride) to form an electrolyte.

hand wiring. A way of wiring that involves using wires to connect the terminals of sockets and other devices.

heat pump. An electrically driven pump that is a combination heating and cooling unit. It transfers heat from one location to another in much the same way as a refrigerator.

heat sink. A device that absorbs or dissipates excess heat that could otherwise damage the semiconductor device.

heater cords. Cords that are insulated with materials that can withstand high temperatures. Such cords are used with heat-producing appliances and with soldering irons and guns.

henry. The unit of inductance.

heterodyning. A process in which two signals are combined in the mixer stage, resulting in an intermediate frequency (IF).

hexadecimal. The base-16 number system having the characters 0 through 9 and A, B, C, D, E, and F as numbers.

high-definition television (HDTV). A television system that has more than 1,000 lines per frame.

hole. The void in the crystal-lattice structure caused by doping with indium or another acceptor impurity.

horsepower. A mechanical unit of power that indicates the size of an electric motor. One horsepower is equal to 746 watts of electrical power. It is defined as the amount of work needed to raise a weight of 550 lb. (250 kg) a distance of one foot (305 mm) in one second.

hydrometer. A common lead-acid battery-testing device used to measure the specific gravity of the electrolyte in a lead-acid cell.

I

immediate mode. A mode in which the computer responds immediately to instructions or commands, just like a hand-held electronic calculator.

impedance. The total opposition to current in a circuit containing a combination of resistance and capacitive reactance.

incandescent. That state in which an item glows from intense heat.

incoherent light. Light that has electromagnetic radiation over a wide range of frequencies. This is the type of light produced by incandescent lamps.

individual circuit. A circuit used for only one piece of equipment.

inductance. The voltage induced by self-inductance. Also called a counter voltage or a counter electromotive force (cemf).

induction heating. Heating that takes place when a current is induced in a material by electromagnetic flux. In an induction heater, a high-frequency alternating current from 5,000 to 500,000 Hz is passed through a heater or work coil.

inductor. A coil that is wound to have a certain amount of inductance.

input/output (I/O). A subsystem of the computer that monitors the flow of information in the computer.

insulator. A material that does not easily conduct an electric current. This material contains valence electrons that are tightly bound to the nuclei of their atoms.

integrated circuit (IC). A circuit that contains components such as diodes, transistors, and resistors formed into a common block of base material called the substrate.

interface. The electrical connection to a microcontroller that provides a means of communication between it, a host computer, and actuating devices such as relays and motors.

intermediate frequency (IF). A frequency that consists of two signals combined in the mixer stage.

Internet. The world's largest computer network. It makes use of a variety of communication devices, including copper wires, fiber cables, satellites, microwaves, and underground and underwater cables.

inverters. Also called complementary gates. These gates have only one input.

ion. A charged atom in which the negative and positive changes are not equal.

J

job. The work you have been hired to do at a given time in your life.

joule. The metric unit for measuring cooling capacity.

junction. A point formed when three or more circuit elements such as resistors are connected.

junction boxes. Metal or plastic boxes that contain connections in the branch-circuit wiring. Also used for mounting various kinds of lighting fixtures.

K

kilowatt-hour (kWh). The unit of electric energy on which electric companies base their bills. A unit of energy equal to that expended in one hour at a rate of one kilowatt.

Kirchhoff's current law. The law that states that the total current flowing out of a junction point equals the current flowing into that junction point.

Kirchhoff's voltage law. The law that states that, in a series circuit, the sum of the voltage drops across each load is equal to the total voltage applied to the circuit.

L

lamp cord. A cord made with two parallel, stranded wires. Each wire is separately covered with rubber or plastic, and the two are joined into what looks like a single unit.

laser. An acronym for Light Amplification by Stimulated Emission of Radiation. A laser system is a way to produce and amplify light.

latch. A device that can read the unit of data and hold it until the next unit of data arrives to replace it.

Lenz's law. The law that states that when a current is induced by moving a conductor across lines of magnetic force, the induced current in the conductor creates a magnetic field that opposes the lines of magnetic force that produced it. Additionally, the induced current and its own magnetic field change direction depending upon which way the conductor is moved in the magnetic field.

load. The device in a circuit that changes the energy of moving electrons into some other useful form of energy.

logic gates. Electronic circuits that process data by comparing signal inputs logically, or in a predetermined manner.

logic levels. Pulses of voltage, or threshold levels, that are detected by logic probes.

logic probes. Test instruments used mostly to troubleshoot different families of integrated circuits, or ICs. Logic probes detect pulses of voltage, or logic levels, and store and display them.

lug. A metal connector that provides a convenient terminal for connecting wires and components.

lumen. The rating that indicates the amount of light a lamp produces.

M

magnetic field. The field that surrounds the magnet.

magnetism. The attraction between certain metals.

magnetizer. A device used to make permanent magnets commercially.

magnetohydrodynamic conversion. A system of energy production that uses magnetic induction and high temperature ionized gas.

matter. Anything that occupies space and has mass or weight.

microcontroller. A single-chip, dedicated-control computer in which a microprocessor receives instructions from a software program, which executes those instructions one at a time.

modem. A device that connects computers and other digital devices to telephone lines. The term is a contraction of modulator-demodulator, which is an electronic device that provides an interface between digital and analog signals.

modulated signals. Signals that vary according to voices or music.

molecules. Tiny particles that make up matter.

multimeter. A meter that can measure current, voltage, and resistance.

multiple connection. Parallel connection.

multiplexing. Rapidly changing the time when various signals are transmitted. Also, using a bus for several purposes during different time states.

N

network. Any computer connected to another computer for the purpose of exchanging information. Also, several interconnected series-parallel circuits.

node. A terminal, or connection point, between two or more parts of a circuit.

noise. In electric and electronic communication systems, anything that interferes with a signal so that the quality of the output is decreased.

O

occupation. A lifetime profession.

ohm. The basic unit of resistance.

ohmmeter. A meter that measures electrical resistance. The digital display is the same as that in voltmeters and ammeters. It can be either analog or digital.

Ohm's law. The law that presents the relationship between voltage, current, and resistance. Georg Simon Ohm discovered this relationship in 1827.

open circuit. A situation that occurs when a wire is broken or disconnected at some point, causing the current flow to stop.

open loop. An open circuit in which the circuit is not complete.

operating system (OS). A system that acts as an interpreter in a computer and that manages translation of instructions within the computer.

operational amplifiers (op amps). Versatile, high-gain integrated-circuit devices that can perform mathematical operations, such as addition, subtraction, and integration. They also function as comparators, oscillators, and linear amplifiers.

opposing voltage source. A voltage source in which the voltage to the load becomes the difference between the two source voltages.

optoelectronics. A field of study that combines optics and electricity.

oscilloscope. A CRT device that will show deviations in wave patterns and help find problems. Its main purpose is to show a waveform, or picture, of a voltage signal.

OSHA. An acronym for the Occupational Safety and Health Act. This act was passed by the U.S. Congress in 1970. Its purpose is to ensure that every worker in the nation has safe and healthful working conditions.

overload. A situation that occurs when the current increases until the total current is greater than the ampere rating of the fuse or circuit breaker. At this point the fuse blows or the circuit breaker trips.

P

parallel circuit. A circuit in which the loads are connected between the two conductors that lead to the energy source. There may be two or more paths, or loops, through which electrons can flow.

Peltier Effect. The observation that the polarity of the current flowing through the junctions will determine which junction will cool and which will heat.

permeability. The ability to conduct magnetic lines of force easily.

phase. A term used to describe a stage in a waveform. For example, when the waveform produced by voltage and current rises and falls at the same instant, voltage and current are said to be in phase with each other.

photodiodes. Semiconductor devices that detect the presence of light.

photoresistor. A device that changes resistance in relationship to changes in the light level exposed to the photo-sensitive surface.

photovoltaic cell. A cell that generates electricity directly from sunlight and uses no fuel.

pictorial diagram. A sketch of circuit parts showing how the parts are connected and where they will be located within the assembly.

pixels. The picture elements of an array. The level of charge reached by each pixel is proportional to the level of light it receives from the image.

plates. The two conductors which are the basic form of a capacitor.

plating pen. A disposable cartridge pen with its own nontoxic solution. It is used with a power supply to repair plated portions of boards.

polarity. The electrical charge of an energy source. It is indicated on diagrams by common electrical symbols of positive (+) and negative (-) signs.

polarized plug. An attachment plug in which one prong is wider than the other. The difference in prong size ensures that the connection will be made in only one way. Also, a three-pronged grounded plug.

potentiometer. A resistor that is used to vary the value of the voltage applied to a circuit.

power. The time rate of doing work. In an electric circuit, power may also be defined in two other ways. First, it is the rate at which electric energy is delivered to a circuit. Second, it is the rate at which an electric circuit does the work of converting the energy of moving electrons into some other form of energy. The basic unit of power is the watt (W).

power factor. A measure of how well a motor uses the current it draws.

power formula. The formula $P = I^2R$. This formula states that the heat produced is directly proportional to the resistance of the conductor and to the square of the current. Also, a formula that shows the relationships between electric power (P), voltage (E), and current (I) in a dc circuit. The basic power formula is: $P = E \times I$.

preventive maintenance. The maintenance performed by electricians who inspect equipment and locate problems before breakdowns occur.

primary cell. A cell that cannot be recharged.

programmed mode. A mode in which the execution of the computer program is deferred. The computer stores the program in memory until it is executed with a RUN command.

protons. Particles that are positively charged and held tightly together in the nucleus of the atom.

pull-up resistor. A resistor that makes sure that the voltage goes from 0 to +5 Vdc when needed.

pyrometers. Meters designed to measure thermocouple voltages.

R

radiant heating cable. Cable commonly used for space heating that consists of resistance wire insulated with a heat-resistant rubber or thermoplastic compound.

radio frequency (RF) signals. Voltages in the antenna circuit that represent the audio information with which the carrier wave has been modulated.

RAM. An acronym for Random Access Memory. It can temporarily store program information.

range circuit. A circuit that is usually a four-wire, 115/230-V circuit with two hot wires, a neutral wire, and a ground. There is a fuse or circuit breaker in each of the hot wires.

reactor. The place in a nuclear power plant where a controlled amount of heat is produced by nuclear chain reactions.

reflection. The return of light wave motion.

refraction. The passing of a light wave diagonally from one medium to another of different density. The light wave bends. As the velocity changes, the wavelength changes.

refrigerant. A liquid that is circulated through the evaporator coils in cooling appliances.

registers. Memory devices that store bits of data as words, or groups of bits. Also, temporary storage spaces for data that are used later in the computer program.

relay. A magnetically operated switch that opens or closes one or more of the contacts between its terminals.

remote control. A type of wiring that makes it possible to control lights and appliances in various rooms and areas from a central place. It uses a low-voltage control circuit to operate line voltage circuits. In such systems, local control of lights is still possible.

resistance. An opposing force between one thing and another. In an electric circuit it is the property of a conductor (wire) opposing the passage of current.

resistance alloys. Combinations of materials in which the conductors used for resistance heating are made of two or more materials. These combinations have the amount of resistance needed to produce heat.

resistance heating. A way of heating that uses the energy of moving electrons as they collide with particles in a conductor.

resistance wire. A round or flat wire made of the resistance alloy that is found in many heat-producing appliances and products.

resistor. A device with a known value of resistance. It is a common part of many electric and electronic circuits. Used to control voltage and current, it operates at fixed or variable values.

resolution. A measure of an image's clarity and the amount of detail it shows.

retentivity. The ability to hold a magnetic-domain alignment for a long time.

rheostat. A variable resistor that is generally made of resistance wire.

ROM. An acronym for Read-Only Memory. It is permanent memory that contains the programs needed to "boot up" the computer and begin operation.

S

saturation. A condition that occurs with magnets when all the domains are aligned and the magnetic field becomes as strong as possible.

scale. The relationship that accurately shows the sizes and locations of parts in relation to each other, whether they are shown full size, smaller, or larger than actual size.

schematic diagram. A diagram that communicates information about electric and electronic circuits.

secondary cell. A cell that can be recharged.

series circuit. A circuit that provides only one path, or one loop, through which electrons can move from one terminal of the energy source to the other.

series-parallel circuit. A system in which series and parallel circuits are combined to form one circuit with several paths, or loops, that allow electrons to flow through more than one load.

server. A computer with an enormous amount of power and memory used to store the network software and information that can be used by all the network computers.

service cords. Cords that have two or three stranded, bare-copper wires. These wires are each insulated and put together in a round outer jacket made of rubber or plastic.

service entrance. Includes all the parts of the assembly that bring electricity into a building. Its conductors extend from the point at which the service-drop conductors are attached at the house to the load center. The complete service entrance is generally thought to include the conductors, the watt-hour meter, and the load center.

shelf life. The period of time during which a cell can be stored without losing more than about 10% of its original capacity.

short circuit. A situation that occurs if two uninsulated live wires contact each other in such a way as to bypass the load, causing an excessive amount of current to flow in the circuit.

shorts. Wires that touch when they be insulated from each other.

shunt connection. A parallel connection.

signal generator. An electronic device that produces a variety of electronic control waveforms.

signal generators. Test instruments that supply different voltage frequencies. These voltages, or signals, can then be applied to certain circuits such as radio receivers. This is done to test and adjust the circuits for efficient operation.

silicon-controlled rectifiers (SCR). Four-layer semiconductor devices equipped with three external connections.

sine wave. The shape of the waveform that represents the rotation of the coil.

slip rings. Rings that are connected to the ends of the armature coil.

small appliance circuits. A circuit wired with number-12 (2.00-mm) wire and protected with 20-A fuses or circuit breakers.

soft soldering. Electrical soldering used to make a good electrical contact between the soldered surfaces.

solar array. A device consisting of several solar panels connected together.

solder. Metal with a low melting point that is used in the process of joining metals.

solder wick. A solder-absorbing, braided copper ribbon. It can also be used to desolder a joint.

soldering. The process of joining metals using another metal that has a low melting point.

soldering gun. A solder-melting tool that has a step-down transformer.

soldering iron. A tool used to melt solder.

solenoid. A wire wound into a coil that concentrates the magnetic field produced by current flowing in a wire. It has magnetic poles and a magnetic field with the same properties as those of a permanent magnet.

starter. A type of electro-mechanical relay used to reverse the power-line conductors to a three-phase motor.

superconductivity. A condition in which the temperature of the conductor is reduced to absolute zero (-273.16^{o} C, or -459.69^{o} F), making the resistance of the conductor zero.

surface mount technology. A type of technology in which components can be mounted directly on the surface of printed circuit boards without drilling holes, allowing components to be mounted very close to each other.

switch boxes. Enclosures used for mounting outlets. The boxes protect the devices mechanically and prevent switch sparking from becoming a fire hazard. They are also used to place flush switches in code-approved metal or nonmetallic enclosures.

synchronous motor. A motor that keeps in step with the frequency of the ac power source. It runs at a more constant speed than other motors.

T

technician. Someone who has technical knowledge about a specific field.

terminal strips. Strips of thin phenolic. They consist of solder lugs and a mounting lug or lugs.

thermistor. Also called a thermal (heat) resistor. A device designed so the resistance decreases when temperature increases.

thermocouples. Solid-state devices used to change heat energy into voltage. They consist of two different metals joined at a junction.

three-phase motors. Induction motors that have field windings and consist of three separate sets of coils. Each of these is energized by one phase line, or leg, of a three-phase power system.

three-way switches. Switches used to control loads from two points, such as in lighting circuits in hallways, stairways, and doorways.

thru-hole soldering. Soldering in which the leads of some components are inserted through drilled holes and soldered to the printed circuit on the other side of the baseboard.

thyristor. A device used in switching or current-control circuits in which a trigger voltage is applied to their control electrode.

tinning. The process of melting a thin coat of solder over the surfaces of the tip of a soldering iron. This allows most of the heat to be conducted from the tip to the surface being soldered. This process is also used to make the pin plate for a tin can.

tolerance. The variation by which the actual resistance of a resistor may be greater or less than its rated value.

total current. The sum of the branch circuit currents.

transformer. A device used to change the value of voltage or current in an electrical system. If it reduces a voltage, it is called a step-down transformer. If it increases a voltage, it is called a step-up transformer. If it does not change the value of the voltage, it is called an isolation transformer.

transistor. The semiconductor device used to control current and amplify input voltages or currents.

transmission lines. Overhead wires supported by high towers that transmit electric energy from power plants.

tri-state IC. A device used when it is necessary to bring data into a computer. Data must be presented to the bus only when the computer is ready to receive it. The device provides three states of impedance. When not enabled, it provides a very high impedance to the data bus. When the device is selected by the enabling signals, the input side accepts data and stores it. In the third state, the output side presents the data to the bus.

troubleshooting. The process of determining why a device or a circuit is not working.

truth tables. Tables that describe the operation of a circuit in terms of the relationship between its input and its output.

U

Underwriters' knot. A type of knot used to tie wires when an attachment plug does not have a cord clamp. The knot keeps the wires from pulling loose from the screw terminals if the cord is pulled.

universal motors. Motors that work with either direct or alternating current. They are similar to the small, series dc motor. They come in sizes ranging from 1/100 to 2 hp (7.46 to 1492 W) with speeds up to 10,000 revolutions per minute.

V

valence electrons. Electrons that are in the outermost shell of an atom.

virtual instrumentation. A system in which computers perform the function of electronic instruments.

voltage. An electromotive force that causes electrons to move.

voltage divider. A network used to supply different values of voltage from one energy source.

voltage drop. The automatic division of the total voltage applied to a series circuit among the loads and devices in the circuit. The voltage across any load is the amount needed to force the circuit current through the resistance of that load.

voltage rating. The voltage that the cell produces when it is not connected to a circuit.

voltaic chemical cell. A combination of materials used to change chemical energy into electric energy in the form of voltage.

voltmeter. A testing device used to measure voltage. The meter test leads or wire connections are attached to the tested circuit in parallel with the device that is expected to have a voltage drop.

W

watt (W). The basic unit of power.

watt-hour meter. A device that measures the electric energy in kilowatt-hours supplied by electric power companies to users.

wet cell. A cell that contains a liquid electrolyte.

wire nuts. Devices used to make solderless connections in house wiring systems, motors, and various appliances.

wire strippers. Tools used to remove insulation from wires.

wire wrapping. The process of making an electrical connection by tightly coiling a wire around a metal terminal.

wiring devices. Components such as switches, receptacles, and light fixtures.

word. A unit of data consisting of 16 bits or 2 bytes.

working voltage. The largest value of direct voltage that can safely be applied to a capacitor.

Index

C

D

E

T

Photo Credits

Special Acknowledgement:
Ron Rockwell/NIDUS Corp.
Electronic Update and New Technical Illustration

3M Telecom Systems Div., 537
Allen-Bradley Co., 75(b), 77(b), 144(lc)
Alpha Wire Div. of Loral Corp., 278
American Electric Heater Co., 308, 309(b)
American Iron and Steel Institute, 505(lc)
Analog Devices, 266
R.B. Annis Co., 151(trc)
AT&T, 534(rc)
Atlantic Finishing Inc., 518(rc)
Avco Everett Research Laboratory, 226(b)
Roger B. Bean, 40, 575(t)
Bell Atlantic Mobile Systems, 436
Keith M. Berry, 16, 18(lc), 20, 238(lc), 240(lc), 279, 296(brc), 298, 310(trc), 345(brc), 373, 377(brc), 383(trc), 445(tl), 468(rc), 505(trc)
Black Box Corp., 280, 342
W.H. Brady Co., 300(lc)
Brook Motor Co., 299(brc)
Bud Radio, Inc., 151(tlc)
Burr-Brown Corp., 399(tlc), 400(trc)
Bussmann Manufacturing Div./McGraw-Edison Co., 35, 36, 37
Chrysler Corporation, 313(brc), 505b(Jeep)
Circle Design/Carol Spengle, 476, 550
Controls Co. of America/Motor Div., 484(lc)
Corbis-Bettmann, 62
Colorado Video, Inc., 452
Cutler-Hammer, 38(t)
Decker Manufacturing Co., 485
Delco Products & App. Div./General Motors Corp., 276(tlc), 486(lc)
Edsyn, Inc., 345(brc)
Essex international Corp., 168(lc)
Exxon, 533
Fairbanks Morse and Co, 204(rc), 206(rc)
FauceHot Heater Co., 502(rc)
Fayette Manufacturing Corp., 222, 223(lc)
Fermilab/Batavia, IL, 49(t)
Ford Motor Co., 530(rc)
David R. Frazier Photolibrary, 21
Generac Corp., 205(lc)
General Electric Corp., 218(lc), 277(b), 465(b), 469(brc), 474(rc), 488(b), 528(b)
General Industries Co., 488(t)
Georator Corp., 495
Global Specialties Corp., 296(t)
Gould Inc., Electric Motor Div., 491
HC Protek, 318(rc), 319, 320, 321(lc), 332
Harvey Hubbel, Inc., 38(b)
Hexacon Electric Co., 309(lc)
Hewlett-Packard, 377
Hunter Assoc., 520
Jenzano, 506(rc)
Ideal Industries, Inc., 286(brc), 471(t)
Ingersoll Rand, 14
International Rectifier, 235(trc), 246, 254
Interstate Voice Products, 328(lc)
Kelvinator, Inc., 511
Keystone Carbon Co., 256(brc)
Kurz & Root Co, 275
Lead Magazine, Lead Industries Assoc., Inc., 216(rc)
Lewis Research Center/NASA, Cleveland, 221(b)
Lockheed Missiles & Space Co., Inc., 223(rc)
Magnaflux Corp., 153

Master Magnetics, Inc., 141(lc)
Midwest Electric Manufacturing Co., 471(b)
Mishima Productions, Inc., 189, 190(rc)
Minneapolis-Honeywell Regulator Co., 257
Mostek Corp., 386(b), 390, 391, 392, 393,
NASA, 225(t), 543
National Center for Atmospheric Research/University Corporation for Atmospheric Research/National Science Foundation, 371, 372(b)
National Instruments Corp., 325, 326, 327, 328(t)
Ohmite Manufacturing Co., 74(b), 76(t), 253
OK Machine and Tool Corp. 301(trc), 302(b)
Ortel Corp., 536(b)
Pace, Inc., 343, 344, 346(b), 347, 348, 349
Parallax, Inc., 403, 408
Brent Phelps, 18(rc)
Potomac Electric Power Co., 217, 219
Elizabeth Purcell, 168(brc), 211(lc), 311(lc), 311(trc), 422(t), 423, 425, 433, 471(t)
Radiant Communications Corp., 536(t)
Radio Shack, a Div. of Tandy Corporation, 39, 285
RCA Corp., 445(tr,b), 456
Cloyd Richardson, 580, 585, 586, 587, 588, 590, 596, 600
Richwoods High School, 575(t)
RCA Corp., 259, 261(tr), 456
Robbins & Myers, Inc., 487(tlc)
Siemens Opto, 262, 263, 264(lc), 265
Sencore, Inc., 137
Sony Corp. of America, 430(br), 431
Sprague Electric Co., 129(rc), 131(tl)
Square D Co., 466
The L.S. Starrett Co., 272(brc)
Tappan Stove Co., 507(lc)
Tetronix, 331(rc)
Texas Instruments, Inc., 232
3M Telecom Systems Div., 537
Triad-Ultrad Distributors Div./Litton Precision Products, Inc., 160(rc), 163(tlc)
Triplett Corp., 321(rc)
Underwriters' Laboratories, 42
Unimation, Inc., 504(tlc)
U.S. Bureau of Reclamation, 213(t)
Vector Electronic Co., Inc., 297(tlc), 303
Wagner Electric Corp., 486(lc), 489(b)
Westinghouse Electric Corp., 203, 526(b), 528(trc)
Dana White/Dana White Productions, 22, 380
Wiremold Co., 472
Zenith Radio Corp., 451(b)